DICTIONARY OF

GEOPHYSICS, ASTROPHYSICS, AND ASTRONOMY

COMPREHENSIVE DICTIONARY OF PHYSICS

Dipak Basu
Editor-in-Chief

PUBLISHED VOLUMES

Dictionary of Pure and Applied Physics
Dipak Basu

**Dictionary of Material Science
and High Energy Physics**
Dipak Basu

**Dictionary of Geophysics, Astrophysics,
and Astronomy**
Richard A. Matzner

A VOLUME IN THE
COMPREHENSIVE DICTIONARY
OF PHYSICS

DICTIONARY OF

GEOPHYSICS, ASTROPHYSICS, AND ASTRONOMY

Edited by

Richard A. Matzner

CRC Press
Boca Raton London New York Washington, D.C.

Library of Congress Cataloging-in-Publication Data

Dictionary of geophysics, astrophysics, and astronomy / edited by Richard A. Matzner.
     p. cm. — (Comprehensive dictionary of physics)
     ISBN 0-8493-2891-8 (alk. paper)
      1. Astronomy—Dictionaries. 2. Geophysics—Dictionaries. I. Matzner, Richard A.
(Richard Alfred), 1942- II. Series.

QB14 .D53 2001
520′.3—dc21   2001025764

Visit the CRC Press Web site at www.crcpress.com

PREFACE

This work is the result of contributions from 52 active researchers in geophysics, astrophysics and astronomy. We have followed a philosophy of directness and simplicity, while still allowing contributors flexibility to expand in their own areas of expertise. They are cited in the contributors' list, but I take this opportunity to thank the contributors for their efforts and their patience.

The subject areas of this dictionary at the time of this writing are among the most active of the physical sciences. Astrophysics and astronomy are enjoying a new golden era, with remarkable observations in new wave bands (γ-rays, X-rays, infrared, radio) and in new fields: neutrino and (soon) gravitational wave astronomy. High resolution mapping of planets continuously yields new discoveries in the history and the environment of the solar system. Theoretical developments are matching these observational results, with new understandings from the largest cosmological scale to the interior of the planets. Geophysics mirrors and drives this research in its study of our own planet, and the analogies it finds in other solar system bodies. Climate change (atmospheric and oceanic long-timescale dynamics) is a transcendingly important societal, as well as scientific, issue. This dictionary provides the background and context for development for decades to come in these and related fields. It is our hope that this dictionary will be of use to students and established researchers alike.

It is a pleasure to acknowledge the assistance of Dr. Helen Nelson, and later, Ms. Colleen McMillon, in the construction of this work. Finally, I acknowledge the debt I owe to Dr. C.F. Keller, and to the late Prof. Dennis Sciama, who so broadened my horizons in the subjects of this dictionary.

Richard Matzner
Austin, Texas

CONTRIBUTORS

Tokuhide Akabane
Kyoto University
Japan

David Alexander
Lockheed Martin Solar & Astrophysics Laboratory
Palo Alto, California

Suguru Araki
Tohoku Fukushi University
Sendai, Japan

Fernando Atrio Barandela
Universidad de Salamanca
Salamanca, Spain

Nadine Barlow
University of Central Florida
Orlando, Florida

Cecilia Barnbaum
Valdosta State University
Valdosta, Georgia

David Batchelor
NASA
Greenbelt, Maryland

Max Bernstein
NASA Ames Research Center
Moffett Field

Vin Bhatnagar
York University
North York, Ontario, Canada

Lee Breakiron
U.S. Naval Observatory
Washington, D.C.

Roberto Casadio
Universita di Bologna
Bologna, Italy

Thomas I. Cline
Goddard Space Flight Center
Greenbelt, Maryland

Vladimir Escalante
Instituto de Astronomia
Morelia, Mexico

Chris L. Fryer
Los Alamos National Laboratories
Los Alamos, New Mexico

Alejandro Gangui
Observatoire de Paris
Meudon, France

Higgins, Chuck
NRC-NASA
Greenbelt, Maryland

May-Britt Kallenrode
University of Luneburg
Luneburg, Germany

Jeff Knight
United Kingdom Meteorological
Berkshire, England

Andrzej Krasinski
Polish Academy of Sciences
Bartycka, Warsaw, Poland

Richard Link
Southwest Research Institute
San Antonio, Texas

Paolo Marziani
Osservatorio Astronomico di Padova
Padova, Italy

Richard A. Matzner
University of Texas
Austin, Texas

Norman McCormick
University of Washington
Seattle, Washington

Nikolai Mitskievich
Guadalajara, Jalisco, Mexico

Curtis Mobley

Sequoia Scientific, Inc.

Mercer Island, Washington

Robert Nemiroff

Michigan Technological University

Houghton, Michigan

Peter Noerdlinger

Goddard Space Flight Center

Greenbelt, Maryland

Gourgen Oganessyan

University of North Carolina

Charlotte, North Carolina

Joel Parker

Boulder, Colorado

Nicolas Pereyra

Universidad de Los Andes

Merida, Venezuela

Zoltan Perjes

KFKI Research Institute

Budapest, Hungary

Patrick Peter

Institut d' Astrophysique de Paris

Paris, France

Morris Podolak

Tel Aviv University

Tel Aviv, Israel

Casadio Roberto

Università di Bologna

Bologna, Italy

Eric Rubenstein

Yale University

New Haven, Connecticut

Ilya Shapiro

Universidade Federal de Juiz de Fora

MG, Brazil

T. Singh

I.T., B.H.U.

Varanasi, India

David P. Stern

Goddard Space Flight

Houston, Texas

Virginia Trimble

University of California

Irvine, California

Donald L. Turcotte

Cornell University

Ithaca, New York

Kelin Wang

Geological Survey of Canada

Sidney, Canada

Zichao Wang

University of Montreal

Montreal, Quebec, Canada

Phil Wilkinson

IPS

Haymarket, Australia

Mark Williams

University of Colorado

Boulder, Colorado

Fabian Wolk

European Commission Joint Research Institute

Marine Environment Unit-TP 690

Ispra, Italy

Paul Work

Clemson University

Clemson, South Carolina

Alfred Wuest

IOS

Sidney, British Columbia, Canada

Shang-Ping Xie

Hokkaido University

Sapporo, Japan

Huijun Yang

University of South Florida

St. Petersburg, Florida

Shoichi Yoshioka
Kyushu University
Fukuoka, Japan

Stephen Zatman
University of California
Berkeley, California

Editorial Advisor

Stan Gibilisco

A

Abbott, David C. Astrophysicist. In 1976, in collaboration with John I. Castor and Richard I. Klein, developed the theory of winds in early type stars (CAK theory). Through hydrodynamic models and atomic data, they showed that the total line-radiation pressure is the probable mechanism that drives the wind in these systems, being able to account for the observed wind speeds, wind mass-loss rates, and general form of the ultraviolet P-Cygni line profiles through which the wind was originally detected.

Abelian Higgs model Perhaps the simplest example of a gauge theory, first proposed by P.W. Higgs in 1964. The Lagrangian is similar to the one in the Goldstone model where the partial derivatives are now replaced by gauge covariants, $\partial_\mu \to \partial_\mu - ieA_\mu$, where e is the gauge coupling constant between the Higgs field ϕ and A_μ. There is also the square of the antisymmetric tensor $F_{\mu\nu} = \partial_\mu A_\nu - \partial_\nu A_\mu$ which yields a kinetic term for the massless gauge field A_μ. Now the invariance of the Lagrangian is with respect to the gauge $U(1)$ symmetry transformation $\phi \to e^{i\Lambda(x)}\phi$ and, in turn, the gauge field transforms as $A_\mu(x) \to A_\mu(x) + e^{-1}\partial_\mu\Lambda(x)$, with $\Lambda(x)$ being an arbitrary function of space and time. It is possible to write down the Lagrangian of this model in the vicinity of the true vacuum of the theory as that of two fields, one of spin 1 and another of spin 0, both of them being massive (plus other higher order interaction terms), in complete agreement with the Higgs mechanism.

Interestingly enough, a similar theory serves to model superconductors (where ϕ would now be identified with the wave function for the Cooper pair) in the Ginzburg–Landau theory. See Goldstone model, Higgs mechanism, spontaneous symmetry breaking.

Abelian string Abelian strings form when, in the framework of a symmetry breaking scheme

$G \to H$, the generators of the group G commute. One example is the complete breakdown of the Abelian $U(1) \to \{1\}$. The vacuum manifold of the phase transition is the quotient space, and in this case, it is given by $\mathcal{M} \sim U(1)$. The first homotopy group is then $\pi_1(\mathcal{M}) \sim \mathbb{Z}$, the (Abelian) group of integers.

All strings formed correspond to elements of π_1 (except the identity element). Regarding the string network evolution, exchange of partners (through intercommutation) is only possible between strings corresponding to the same element of π_1 (or its inverse). Strings from different elements (which always commute for Abelian π_1) pass through each other without intercommutation taking place. See Abelian Higgs model, homotopy group, intercommutation (cosmic string), Kibble mechanism, non-Abelian string, spontaneous symmetry breaking.

aberration of stellar light Apparent displacement of the geometric direction of stellar light arising because of the terrestrial motion, discovered by J. Bradley in 1725. Classically, the angular position discrepancy can be explained by the law of vector composition: the apparent direction of light is the direction of the difference between the earth velocity vector and the velocity vector of light. A presently accepted explanation is provided by the special theory of relativity. Three components contribute to the *aberration of stellar light* with terms called diurnal, annual, and secular aberration, as the motion of the earth is due to diurnal rotation, to the orbital motion around the center of mass of the solar system, and to the motion of the solar system. Because of annual aberration, the apparent position of a star moves cyclically throughout the year in an elliptical pattern on the sky. The semi-major axis of the ellipse, which is equal to the ratio between the mean orbital velocity of earth and the speed of light, is called the aberration constant. Its adopted value is 20.49552 sec of arc.

Abney's law of additivity The luminous power of a source is the sum of the powers of the components of any spectral decomposition of the light.

A-boundary   (or atlas boundary) In relativity, a notion of boundary points of the space-time manifold, constructed by the closure of the open sets of an atlas A of coordinate maps. The transition functions of the coordinate maps are extended to the boundary points.

absolute humidity   One of the definitions for the moisture content of the atmosphere — the total mass of water vapor present per unit volume of air, i.e., the density of water vapor. Unit is g/cm^3.

absolute magnitude   *See* magnitude.

absolute space and time   In Newtonian Mechanics, it is implicitly assumed that the measurement of time and the measurement of lengths of physical bodies are independent of the reference system.

absolute viscosity   The ratio of shear to the rate of strain of a fluid. Also referred to as molecular viscosity or dynamic viscosity. For a Newtonian fluid, the shear stress within the fluid, τ, is related to the rate of strain (velocity gradient), $\frac{du}{dz}$, by the relation $\tau = \mu \frac{du}{dz}$. The coefficient of proportionality, μ, is the *absolute viscosity*.

absolute zero   The volume of an ideal gas at constant pressure is proportional to the absolute temperature of the gas (Charles' Law). The temperature so defined corresponds to the thermodynamic definition of temperature. Thus, as an ideal gas is cooled, the volume of the gas tends to zero. The temperature at which this occurs, which can be observed by extrapolation, is *absolute zero*. Real gases liquefy at temperatures near absolute zero and occupy a finite volume. However, starting with a dilute real gas, and extrapolating from temperatures at which it behaves in an almost ideal fashion, absolute zero can be determined.

absorbance   The (base 10) logarithm of the ratio of the radiant power at a given wavelength incident on a volume to the sum of the scattered and directly transmitted radiant powers emerging from the volume; also called optical density.

absorptance   The fraction of the incident power at a given wavelength that is absorbed within a volume.

absorption coefficient   The absorptance per unit distance of photon travel in a medium, i.e., the limit of the ratio of the spectral absorptance to the distance of photon travel as that distance becomes vanishingly small. Units: $[m^{-1}]$.

absorption cross-section   The cross-sectional area of a beam containing power equal to the power absorbed by a particle in the beam $[m^2]$.

absorption efficiency factor   The ratio of the absorption cross-section to the geometrical cross-section of the particle.

absorption fading   In radio communication, fading is caused by changes in absorption that cause changes in the received signal strength. A short-wave fadeout is an obvious example, and the fade, in this case, may last for an hour or more. *See* ionospheric absorption, short wave fadeout.

absorption line   A dark line at a particular wavelength in the spectrum of electromagnetic radiation that has traversed an absorbing medium (typically a cool, tenuous gas between a hot radiating source and the observer). *Absorption lines* are produced by a quantum transition in matter that absorbs radiation at certain wavelengths and produces a decrease in the intensity around those wavelengths. *See* spectrum. *Compare with* emission line.

abstract index notation   A notation of tensors in terms of their component index structure (introduced by R. Penrose). For example, the tensor $T(\theta, \theta) = T_a^b \theta^a \otimes \theta_b$ is written in the abstract index notation as T_a^b, where the indices signify the valence and should not be assigned a numerical value. When components need to be referred to, these may be enclosed in matrix brackets: $(v^a) = (v^1, v^2)$.

abyssal circulation   Currents in the ocean that reach the vicinity of the sea floor. While the general circulation of the oceans is primarily driven by winds, *abyssal circulation* is mainly driven by density differences caused by temperature and salinity variations, i.e., the thermohaline circulation, and consequently is much more sluggish.

abyssal plain   Deep old ocean floor covered with sediments so that it is smooth.

acceleration   The rate of change of the velocity of an object per unit of time (in Newtonian physics) and per unit of proper time of the object (in relativity theory). In relativity, *acceleration* also has a geometric interpretation. An object that experiences only gravitational forces moves along a geodesic in a spacetime, and its acceleration is zero. If non-gravitational forces act as well (e.g., electromagnetic forces or pressure gradient in a gas or fluid), then acceleration at point p in the spacetime measures the rate with which the trajectory C of the object curves off the geodesic that passes through p and is tangent to C at p. In metric units, acceleration has units cm/sec^2 ; m/sec^2.

acceleration due to gravity (g)   The standard value (9.80665m/s^2) of the acceleration experienced by a body in the Earth's gravitational field.

accreted terrain   A terrain that has been accreted to a continent. The margins of many continents, including the western U.S., are made up of *accreted terrains*. If, due to continental drift, New Zealand collides with Australia, it would be an accreted terrain.

accretion   The infall of matter onto a body, such as a planet, a forming star, or a black hole, occurring because of their mutual gravitational attraction. *Accretion* is essential in the formation of stars and planetary systems. It is thought to be an important factor in the evolution of stars belonging to binary systems, since matter can be transferred from one star to another, and in active galactic nuclei, where the extraction of gravitational potential energy from material which accretes onto a massive black hole is reputed to be the source of energy. The efficiency at which gravitational potential energy can be extracted decreases with the radius of the accreting body and increases with its mass. Accretion as an energy source is therefore most efficient for very compact bodies like neutron stars (R $\sim$ 10 km) or black holes; in these cases, the efficiency can be higher than that of thermonuclear reactions. Maximum efficiency can be achieved in the case of a rotating black hole; up to 30% of the rest energy of the infalling matter can be converted into radiating energy. If the infalling matter has substantial angular momentum, then the process of accretion progresses via the formation of an accretion disk, where viscosity forces cause loss of angular momentum, and lets matter drift toward the attracting body.

In planetary systems, the formation of large bodies by the accumulation of smaller bodies. Most of the planets (and probably many of the larger moons) in our solar system are believed to have formed by accretion (Jupiter and Saturn are exceptions). As small objects solidified from the solar nebula, they collided and occasionally stuck together, forming a more massive object with a larger amount of gravitational attraction. This stronger gravity allowed the object to pull in smaller objects, gradually building the body up to a planetismal (a few kilometers to a few tens of kilometers in diameter), then a protoplanet (a few tens of kilometers up to 2000 kilometers in diameter), and finally a planet (over 2000 kilometers in diameter). *See* accretion disk, active galactic nuclei, black hole, quasi stellar object, solar system formation, star formation, X-ray source.

accretionary prism (accretionary wedge)
The wedge-shaped geological complex at the frontal portion of the upper plate of a subduction zone formed by sediments scraped off the top of the subducting oceanic plate. The sediments undergo a process of deformation, consolidation, diagenesis, and sometimes metamorphism. The wedge partially or completely fills the trench. The most frontal point is called the toe or deformation front. *See* trench.

accretion disk   A disk of gas orbiting a celestial body, formed by inflowing or accreting matter. In binary systems, if the stars are sufficiently close to each other so that one of the stars is filling its Roche Lobe, mass will be transferred to the companion star creating an *accretion disk*.

In active galactic nuclei, hot accretion disks surround a supermassive black hole, whose presence is part of the "standard model" of active galactic nuclei, and whose observational status is becoming secure. Active galactic nuclei are thought to be powered by the release of potential gravitational energy by accretion of matter onto a supermassive black hole. The accretion disk dissipates part of the gravitational potential energy, and removes the angular momentum of the infalling gas. The gas drifts slowly toward the central black hole. During this process, the innermost annuli of the disk are heated to high temperature by viscous forces, and emit a "stretched thermal continuum", i.e., the sum of thermal continua emitted by annuli at different temperatures. This view is probably valid only in active galactic nuclei radiating below the Eddington luminosity, i.e., low luminosity active galactic nuclei like Seyfert galaxies. If the accretion rate exceeds the Eddington limit, the disk may puff up and become a thick torus supported by radiation pressure. The observational proof of the presence of accretion disks in active galactic nuclei rests mainly on the detection of a thermal feature in the continuum spectrum (the big blue bump), roughly in agreement with the predictions of accretion disk models. Since the disk size is probably less than 1 pc, the disk emitting region cannot be resolved with present-day instruments. *See* accretion, active galactic nuclei, big blue bump, black hole, Eddington limit.

accretion, Eddington   As material accretes onto a compact object (neutron star, black hole, etc.), potential energy is released. The Eddington rate is the critical accretion rate where the rate of energy released is equal to the Eddington luminosity: $G \dot{M}_{Eddington} M_{accretor} / R_{accretor} = L_{Eddington} \Rightarrow \dot{M}_{accretion} = \frac{4\pi c R_{accreting\ object}}{\kappa}$ where κ is the opacity of the material in units of area per unit mass. For spherically symmetric accretion where all of the potential energy is converted into photons, this rate is the maximum accretion rate allowed onto the compact object (*see* Eddington luminosity). For ionized hydrogen accreting onto a neutron star ($R_{NS} = 10$ km $M_{NS} = 1.4 M_{\odot}$), this rate is: $1.5 \times 10^{-8} M_{\odot}$ yr^{-1}. *See also* accretion, Super-Eddington.

accretion, hypercritical   *See* accretion, Super-Eddington.

accretion, Super-Eddington   Mass accretion at a rate above the Eddington accretion limit. These rates can occur in a variety of accretion conditions such as: (a) in black hole accretion where the accretion energy is carried into the black hole, (b) in disk accretion where luminosity along the disk axis does not affect the accretion, and (c) for high accretion rates that create sufficiently high densities and temperatures that the potential energy is converted into neutrinos rather than photons. In this latter case, due to the low neutrino cross-section, the neutrinos radiate the energy without imparting momentum onto the accreting material. (Syn. hypercritical accretion).

Achilles   A Trojan asteroid orbiting at the L4 point in Jupiter's orbit (60° ahead of Jupiter).

achondrite   A form of igneous stony meteorite characterized by thermal processing and the absence of chondrules. *Achondrites* are generally of basaltic composition and are further classified on the basis of abundance variations. Diogenites contain mostly pyroxene, while eucrites are composed of plagioclase-pyroxene basalts. Ureilites have small diamond inclusions. Howardites appear to be a mixture of eucrites and diogenites. Evidence from micrometeorite craters, high energy particle tracks, and gas content indicates that they were formed on the surface of a meteorite parent body.

achromatic objective   The compound objective lens (front lens) of a telescope or other optical instrument which is specially designed to minimize chromatic aberation. This objective consists of two lenses, one converging and the other diverging; either glued together with transparent glue (cemented doublet), or air-spaced. The two lenses have different indices of refraction, one high (Flint glass), and the other low (Crown glass). The chromatic aberrations of the two lenses act in opposite senses, and tend to cancel each other out in the final image.

achronal set   (semispacelike set) A set of points S of a causal space such that there are no two points in S with timelike separation.

acoustic tomography   An inverse method which infers the state of an ocean region from measurements of the properties of sound waves passing through it. The properties of sound in the ocean are functions of temperature, water velocity, and salinity, and thus each can be exploited for acoustic tomography. The ocean is nearly transparent to low-frequency sound waves, which allows signals to be transmitted over hundreds to thousands of kilometers.

actinides   The elements of atomic number 89 through 103, i.e., Ac, Th, Pa, U, Np, Pu, Am, Cm, Bk, Cf, Es, Fm, Md, No, Lr.

action   In mechanics the integral of the Lagrangian along a path through endpoint events with given endpoint conditions:

$$I = \int_{t_a, x_a^j, \mathcal{C}}^{t_b, x_b^j} L\left(x^i, dx^i/dt, t\right) dt$$

(or, if appropriate, the Lagrangian may contain higher time derivatives of the point-coordinates). Extremization of the action over paths with the same endpoint conditions leads to a differential equation. If the Lagrangian is a simple $L = T - V$, where T is quadratic in the velocity and V is a function of coordinates of the point particle, then this variation leads to Newton's second law:

$$\frac{d^2 x^i}{dt^2} = -\frac{\partial V}{\partial x^i}, i = 1, 2, 3.$$

By extension, the word *action* is also applied to field theories, where it is defined:

$$I = \int_{t_a, x_a^j}^{t_b, x_b^j} \mathcal{L}\sqrt{|g|} d^n x,$$

where $\mathcal{L}$ is a function of the fields (which depend on the spacetime coordinates), and of the gradients of these fields. Here n is the dimension of spacetime. *See* Lagrangian, variational principle.

activation energy (ΔH_a)   That energy required before a given reaction or process can proceed. It is usually defined as the difference between the internal energy (or enthalpy) of the transition state and the initial state.

activation entropy (ΔS_a)   The *activation entropy* is defined as the difference between the entropy of the activated state and initial state, or the entropy change. From the statistical definition of entropy, it can be expressed as

$$\Delta S_a = R \ln \frac{\omega_a}{\omega_I}$$

where ω_a is the number of "complexions" associated with the activated state, and ω_I is the number of "complexions" associated with the initial state. R is gas constant. The *activation entropy* therefore includes changes in the configuration, electronic, and vibration entropy.

activation volume (ΔV)   The *activation volume* is defined as the volume difference between initial and final state in an activation process, which is expressed as

$$\Delta V = \frac{\partial \Delta G}{\partial P}$$

where ΔG is the Gibbs energy of the activation process and P is the pressure. The *activation volume* reflects the dependence of process on pressure between the volume of the activated state and initial state, or entropy change.

active continental margin   A continental margin where an oceanic plate is subducting beneath the continent.

active fault   A fault that has repeated displacements in Quaternary or late Quaternary period. Its fault trace appears on the Earth's surface, and the fault has a potential to reactivate in the future. Hence, naturally, a fault which had displacements associated with a large earthquake in recent years is an *active fault*. The degree of activity of an active fault is represented by average displacement rate, which is deduced from geology, topography, and trench excavation. The higher the activity, the shorter the recurrence time of large earthquakes. There are some cases where large earthquakes take place on an active fault with low activity.

active front   An active anafront or an active katafront. An active anafront is a warm front at which there is upward movement of the warm sector air. This is due to the velocity component crossing the frontal line of the warm air being larger than the velocity component of the cold air. This upward movement of the warm air usually produces clouds and precipitation. In general, most warm fronts and stationary fronts are active anafronts. An active katafront is a weak cold frontal condition, in which the warm sector air sinks relative to the colder air. The upper trough of active katafront locates the frontal line or prefrontal line. An active katafront moves faster than a general cold front.

active galactic nuclei (AGN)   Luminous nuclei of galaxies in which emission of radiation ranges from radio frequencies to hard-X or, in the case of blazars, to γ rays and is most likely due to non-stellar processes related to accretion of matter onto a supermassive black hole. *Active galactic nuclei* cover a large range in luminosity ($\sim 10^{42} - 10^{47}$ ergs s^{-1}) and include, at the low luminosity end, LINERs and Seyfert-2 galaxies, and at the high luminosity end, the most energetic sources known in the universe, like quasi-stellar objects and the most powerful radio galaxies. Nearby AGN can be distinguished from normal galaxies because of their bright nucleus; their identification, however, requires the detection of strong emission lines in the optical and UV spectrum. Radio-loud AGN, a minority (10 to 15%) of all AGN, have comparable optical and radio luminosity; radio quiet AGN are not radio silent, but the power they emit in the radio is a tiny fraction of the optical luminosity. The reason for the existence of such dichotomy is as yet unclear. Currently debated explanations involve the spin of the supermassive black hole (i.e., a rapidly spinning black hole could help form a relativistic jet) or the morphology of the active nucleus host galaxy, since in spiral galaxies the interstellar medium would quench a relativistic jet. *See* black hole, QSO, Seyfert galaxies.

active margins   The boundaries between the oceans and the continents are of two types, active and passive. *Active margins* are also plate boundaries, usually subduction zones. Active margins have major earthquakes and volcanism; examples include the "ring of fire" around the Pacific.

active region   A localized volume of the solar atmosphere in which the magnetic fields are extremely strong. *Active regions* are characterized as bright complexes of loops at ultraviolet and X-ray wavelengths. The solar gas is confined by the strong magnetic fields forming loop-like structures and is heated to millions of degrees Kelvin, and are typically the locations of several solar phenomena such as plages, sunspots, faculae, and flares. The structures evolve and change during the lifetime of the active region. Active regions may last for more than one solar rotation and there is some evidence of them recurring in common locations on the sun. Active regions, like sunspots, vary in frequency during the solar cycle, there being more near solar maximum and none visible at solar minimum. The photospheric component of active regions are more familiar as sunspots, which form at the center of active regions.

adiabat   Temperature vs. pressure in a system isolated from addition or removal of thermal energy. The temperature may change, however, because of compression. The temperature in the convecting mantle of the Earth is closely approximated by an *adiabat*.

adiabatic atmosphere   A simplified atmosphere model with no radiation process, water phase changing process, or turbulent heat transfer. All processes in *adiabatic atmosphere* are isentropic processes. It is a good approximation for short-term, large scale atmospheric motions. In an adiabatic atmosphere, the relation between temperature and pressure is

$$\frac{T}{T_0} = \left(\frac{p}{p_0}\right)^{\frac{AR}{C_p}}$$

where T is temperature, p is pressure, T_0 and p_0 are the original states of T and p before adiabatic processes, A is the mechanical equivalent of heat, R is the gas constant, and C_p is the specific heat at constant pressure.

adiabatic condensation point   The height point at which air becomes saturated when it is lifted adiabatically. It can be determined by the adiabatic chart.

adiabatic cooling　In an adiabatic atmosphere, when an air parcel ascends to upper lower pressure height level, it undergoes expansion and requires the expenditure of energy and consequently leading to a depletion of internal heat.

adiabatic deceleration　Deceleration of energetic particles during the solar wind expansion: energetic particles are scattered at magnetic field fluctuations frozen into the solar wind plasma. During the expansion of the solar wind, this "cosmic ray gas" also expands, resulting in a cooling of the gas which is equivalent to a deceleration of the energetic particles. In a transport equation, *adiabatic deceleration* is described by a term

$$\frac{\nabla \cdot \mathbf{v}_{\text{sowi}}}{3} \frac{\partial}{\partial T} (\alpha T U)$$

with T being the particle's energy, T_0 its rest energy, U the phase space density, $\mathbf{v}_{\text{sowi}}$ the solar wind speed, and $\alpha = (T + 2T_0)/(T + T_0)$.

Adiabatic deceleration formally is also equivalent to a betatron effect due to the reduction of the interplanetary magnetic field strength with increasing radial distance.

adiabatic dislocation　Displacement of a virtual fluid parcel without exchange of heat with the ambient fluid. *See* potential temperature.

adiabatic equilibrium　An equilibrium status when a system has no heat flux across its boundary, or the incoming heat equals the outgoing heat. That is, $dU = -dW$, from the first law of thermodynamics without the heat term, in which dU is variation of the internal energy, dW is work. *Adiabatic equilibrium* can be found, for instance, in dry adiabatic ascending movements of air parcels; and in the closed systems in which two or three phases of water exist together and reach an equilibrium state.

adiabatic index　Ratio of specific heats: C_p/C_V where C_p is the specific heat at constant pressure, and C_V is the specific heat at constant volume. For ideal gases, equal to (2+*degrees of freedom*)/(*degrees of freedom*).

adiabatic invariant　A quantity in a mechanical or field system that changes arbitrarily little even when the system parameter changes substantially but arbitrarily slowly. Examples include the magnetic flux included in a cyclotron orbit of a plasma particle. Thus, in a variable magnetic field, the size of the orbit changes as the particle dufts along a guiding flux line. Another example is the angular momentum of an orbit in a spherical system, which is changed if the spherical force law is slowly changed. *Adiabatic invariants* can be expressed as the surface area of a closed orbit in phase space. They are the objects that are quantized (=mh) in the Bohr model of the atom.

adiabatic lapse rate　Temperature vertical change rate when an air parcel moves vertically with no exchange of heat with surroundings. In the special case of an ideal atmosphere, the *adiabatic lapse rate* is 10° per km.

ADM form of the Einstein–Hilbert action
In general relativity, by introducing the ADM (Arnowitt, Deser, Misner) decomposition of the metric, the Einstein–Hilbert action for pure gravity takes the general form

$$S_{EH} = \frac{1}{16\pi G}$$
$$\int d^4x \, \alpha \, \gamma^{1/2} \left[K_{ij} K^{ij} - K^2 + \, ^{(3)}R \right]$$
$$- \frac{1}{8\pi G} \sum_a \int_{t_a} d^3x \, \gamma^{1/2} \, K + \frac{1}{8\pi G}$$
$$\sum_b \int dt \int_{x_b^i} d^2x \, \gamma^{1/2} \left(K\,\beta^i - \gamma^{ij}\alpha_{,j} \right) ,$$

where the first term on the r.h.s. is the volume contribution, the second comes from possible space-like boundaries Σ_{t_a} of the space-time manifold parametrized by $t = t_a$, and the third contains contributions from time-like boundaries $x^i = x_b^i$. The surface terms must be included in order to obtain the correct equations of motion upon variation of the variables γ_{ij} which vanish on the borders but have non-vanishing normal derivatives therein.

In the above,

$$K_{ij} = \frac{1}{2\alpha} \left(\beta_{i|j} + \beta_{j|i} - \gamma_{ij,0} \right)$$

is the extrinsic curvature tensor of the surfaces of constant time Σ_t, | denotes covariant differentiation with respect to the three-dimensional metric γ, $K = K_{ij}\,\gamma^{ij}$, and $^{(3)}R$ is the intrinsic scalar curvature of Σ_t. From the above form of the action, it is apparent that α and β^i are not dynamical variables (no time derivatives of the lapse and shifts functions appear). Further, the extrinsic curvature of Σ_t enters in the action to build a sort of kinematical term, while the intrinsic curvature plays the role of a potential. *See* Arnowitt–Deser–Misner (ADM) decomposition of the metric.

ADM mass   According to general relativity, the motion of a particle of mass m located in a region of *weak* gravitational field, that is far away from any gravitational source, is well approximated by Newton's law with a force

$$F = G\,\frac{m\,M_{ADM}}{r^2}\,,$$

where r is a radial coordinate such that the metric tensor g approaches the usual flat Minkowski metric for large values of r. The effective ADM mass M_{ADM} is obtained by expanding the time-time component of g in powers of $1/r$,

$$g_{tt} = -1 + \frac{2\,M_{ADM}}{r} + \mathcal{O}\left(\frac{1}{r^2}\right)\,.$$

Intuitively, one can think of the ADM mass as the total (matter plus gravity) energy contained in the interior of space. As such it generally differs from the volume integral of the energy-momentum density of matter. It is conserved if no radial energy flow is present at large r.

More formally, M can be obtained by integrating a surface term at large r in the ADM form of the Einstein–Hilbert action, which then adds to the canonical Hamiltonian. This derivation justifies the terminology. In the same way one can define other (conserved or not) asymptotical physical quantities like total electric charge and gauge charges. *See* ADM form of the Einstein–Hilbert action, asymptotic flatness.

Adrastea   Moon of Jupiter, also designated JXV. Discovered by Jewitt, Danielson, and Synnott in 1979, its orbit lies very close to that of Metis, with an eccentricity and inclination that are very nearly 0 and a semimajor axis of 1.29×10^5 km. Its size is $12.5 \times 10 \times 7.5$ km, its mass, 1.90×10^{16} kg, and its density roughly $4\ \mathrm{gcm}^{-3}$. It has a geometric albedo of 0.05 and orbits Jupiter once every 0.298 Earth days.

ADV (Acoustic Doppler Velocimeter)   A device that measures fluid velocity by making use of the Doppler Effect. Sound is emitted at a specific frequency, is reflected off of particles in the fluid, and returns to the instrument with a frequency shift if the fluid is moving. Speed of the fluid (along the sound travel path) may be determined from the frequency shift. Multiple sender-receiver pairs are used to allow 3-D flow measurements.

advance of the perihelion   In unperturbed Newtonian dynamics, planetary orbits around a spherical sun are ellipses fixed in space. Many perturbations in more realistic situations, for instance perturbations from other planets, contribute to a secular shift in orbits, including a rotation of the orbit in its plane, a precession of the perihelion. General relativity predicts a specific advance of the perihelion of planets, equal to 43 sec of arc per century for Mercury, and this is observationally verified. Other planets have substantially smaller advance of their perihelion: for Venus the general relativity prediction is 8.6 sec of arc per century, and for Earth the prediction is 3.8 sec of arc per century. These are currently unmeasurable.

The binary pulsar (PSR 1913+16) has an observable periastron advance of 4.227°/year, consistent with the general relativity prediction. *See* binary pulsar.

advection   The transport of a physical property by entrainment in a moving medium. Wind advects water vapor entrained in the air, for instance.

advection dominated accretion disks   Accretion disks in which the radial transport of heat becomes relevant to the disk structure. The advection-dominated disk differs from the standard geometrically thin accretion disk model because the energy released by viscous dissipation is not radiated locally, but rather advected toward the central star or black hole. As a consequence, luminosity of the advection dominated disk can be much lower than that of a standard thin accretion disk. Advection dominated disks are expected to form if the accretion rate is above the Eddington limit, or on the other end, if the accretion rate is very low. Low accretion rate, advection dominated disks have been used to model the lowest luminosity active galactic nuclei, the galactic center, and quiescent binary systems with a black hole candidate. *See* active galactic nuclei, black hole, Eddington limit.

advective heat transfer (or advective heat transport)   Transfer of heat by mass movement. Use of the term does not imply a particular driving mechanism for the mass movement such as thermal buoyancy. Relative to a reference temperature T_0, the heat flux due to material of temperature T moving at speed v is $q = v \, \rho c (T - T_0)$, where ρ and c are density and specific heat, respectively.

aeolian   *See* eolian.

aerosol   Small size (0.01 to 10 μm), relatively stable suspended, colloidal material, either natural or man-made, formed of solid particles or liquid droplets, organic and inorganic, and the gases of the atmosphere in which these particles float and disperse. Haze, most smokes, and some types of fog and clouds are *aerosols*. Aerosols in the troposphere are usually removed by precipitation. Their residence time order is from days to weeks. Tropospheric aerosols can affect radiation processes by absorbing, reflecting, and scattering effects, and may act as Aitken nuclei. About 30% of tropospheric aerosols are created by human activities. In the stratosphere, aerosols are mainly sulfate particles resulting from volcanic eruptions and usually remain there much longer. Aerosols in the stratosphere may reduce insolation significantly, which is the main physics factor involved in climatic cooling associated with volcanic eruptions.

aesthenosphere   Partially melted layer of the Earth lying below the lithosphere at a depth of 80 to 100 km, and extending to approximately 200 km depth.

affine connection   A non-tensor object which has to be introduced in order to construct the covariant derivatives of a tensor. Symbol: $\Gamma^{\alpha}_{\beta\gamma}$. Under the general coordinate transformation $x^{\mu} \longrightarrow x'^{\mu} = x^{\mu} + \xi^{\mu}(x)$ the *affine connection* possesses the following transformation rule:

$$\Gamma^{\alpha'}_{\beta'\gamma'} = \frac{\partial x^{\alpha'}}{\partial x^{\mu}} \frac{\partial x^{\nu}}{\partial x^{\beta'}} \frac{\partial x^{\lambda}}{\partial x^{\gamma'}} \Gamma^{\mu}_{\nu\lambda} + \frac{\partial x^{\alpha'}}{\partial x^{\tau}} \frac{\partial^2 x^{\tau}}{\partial x^{\beta'} \partial x^{\gamma'}}$$

while for an arbitrary tensor $T^A = T^{\mu_1 \ldots \mu_k}_{\nu_1 \ldots \nu_k}$ one has

$$T^{\alpha_{1'} \ldots \alpha_l}_{\beta_{1'} \ldots \beta_{k'}} = \frac{\partial x^{\alpha_{1'}}}{\partial x^{\mu_1}} \cdots \frac{\partial x^{\alpha_{l'}}}{\partial x^{\mu_l}} \frac{\partial x^{\nu_1}}{\partial x^{\beta_{1'}}} \\ \cdots \frac{\partial x^{\nu_k}}{\partial x^{\beta_{k'}}} T^{\mu_1 \ldots \mu_l}_{\nu_1 \ldots \nu_k}$$

The non-tensor form of the transformation of affine connection guarantees that for an arbitrary tensor $T^{\alpha\beta\ldots\gamma}_{\rho\nu\ldots\alpha}$ its covariant derivative

$$\nabla_{\mu} T^{\alpha\beta\ldots\gamma}_{\rho\nu\ldots\alpha} = T^{\alpha\beta\ldots\gamma}_{\rho\nu\ldots\alpha,\mu} + \Gamma^{\alpha}_{\sigma\mu} T^{\sigma\beta\ldots\gamma}_{\rho\nu\ldots\alpha} + \cdots \\ - \Gamma^{\sigma}_{\rho\mu} T^{\alpha\beta\ldots\gamma}_{\sigma\nu\ldots\alpha} - \cdots$$

is also a tensor. (Here the subscript "μ" means $\partial / \partial X^{\mu}$.) Geometrically the affine connection and the covariant derivative define the parallel displacement of the tensor along the given smooth path. The above transformation rule leaves a great freedom in the definition of affine connection because one can safely add to $\Gamma^{\alpha}_{\beta\gamma}$ any tensor. In particular, one can provide the symmetry of the affine connection $\Gamma^{\alpha}_{\beta\gamma} = \Gamma^{\alpha}_{\gamma\beta}$ (which requires torsion tensor = 0) and also metricity of the covariant derivative $\nabla_{\mu} g_{\alpha\beta} = 0$. In this case, the affine connection is called the Cristoffel symbol and can be expressed in terms of the sole metric of the manifold as

$$\Gamma^{\alpha}_{\beta\gamma} = \frac{1}{2} g^{\alpha\lambda} \left(\partial_{\beta} g_{\gamma\lambda} + \partial_{\gamma} g_{\beta\lambda} - \partial_{\lambda} g_{\beta\gamma} \right)$$

See covariant derivative, metricity of covariant derivative, torsion.

African waves   During the northern hemisphere summer intense surface heating over the Sahara generates a strong positive temperature gradient in the lower troposphere between the equator and about 25° N. The resulting easterly thermal wind creates a strong easterly jet core near 650 mb centered near 16° N. *African waves* are the synoptic scale disturbances that are observed to form and propagate westward in the cyclonic shear zone to the south of this jet core. Occasionally African waves are progenitors of tropical storms and hurricanes in the western Atlantic. The average wavelength of observed African wave disturbance is about 2500 km and the westward propagation speed is about 8 m/s.

afternoon cloud (Mars)   *Afternoon clouds* appear at huge volcanos such as Elysium Mons, Olympus Mons, and Tharsis Montes in spring to summer of the northern hemisphere. Afternoon clouds are bright, but their dimension is small compared to morning and evening clouds. In their most active period from late spring to early summer of the northern hemisphere, they appear around 10h of Martian local time (MLT), and their normal optical depths reach maximum in 14h to 15h MLT. Their brightness seen from Earth increases as they approach the evening limb. Afternoon clouds show a diurnal variation. Sometimes afternoon clouds at Olympus Mons and Tharsis Montes form a W-shaped cloud together with evening clouds, in which the afternoon clouds are identified as bright spots. The altitude of afternoon clouds is higher than the volcanos on which they appear. *See* evening cloud, morning cloud.

aftershocks   Essentially all earthquakes are followed by a sequence of *"aftershocks"*. In some cases aftershocks can approach the main shock in strength. The decay in the number of aftershocks with time has a power-law dependence; this is known as Omori's law.

ageostrophic flow   The flow that is not geostrophic. *See* geostrophic approximation.

agonic line   A line of zero declination. *See* declination.

air   The mixture of gases near the Earth's surface, composed of approximately 78% nitrogen, 21% oxygen, 1% argon, 0.035% carbon dioxide, variable amounts of water vapor, and traces of other noble gases, and of hydrogen, methane, nitrous oxide, ozone, and other compounds.

airfoil probe   A sensor to measure oceanic turbulence in the dissipation range. The probe is an axi-symmetric airfoil of revolution that senses cross-stream velocity fluctuations $u = |\mathbf{u}|$ of the free stream velocity vector $\mathbf{W}$ (see figure). *Airfoil probes* are often mounted on vertically moving dissipation profilers. The probe's output is differentiated by analog electronic circuits to produce voltage fluctuations that are proportional to the time rate of change of u, namely $\partial u(z)/\partial t$, where z is the vertical position. If the profiler descends steadily, then by the Tayler transformation this time derivative equals velocity shear $\partial u/\partial z = V^{-1} \, \partial u(z)/\partial t$. This microstructure velocity shear is used to estimate the dissipation rate of turbulent kinetic energy.

airglow   Widely distributed flux predominately from OH, oxygen, and neon at an altitude of 85 to 95 km. Airglow has a brightness of order 14 magnitudes per square arcsec.

air gun   An artificial vibration source used for submarine seismic exploration and sonic prospecting. The device emits high-pressured air in the oceanic water under electric control from an exploratory ship. The compressed air is conveyed from a compressor on the ship to a chamber which is dragged from the stern. A shock produced by expansion and contraction of the air in the water becomes a seismic source. The source with its large capacity and low-frequency signals is appropriate for investigation of the deeper submarine structure. An *air gun* is most widely used as an acoustic source for multi-channel sonic wave prospecting.

Airy compensation   The mass of an elevated mountain range is "compensated" by a low density crustal root. *See* Airy isostasy.

Airy isostasy   An idealized mechanism of isostatic equilibrium proposed by G.B. Airy in 1855, in which the crust consists of vertical rigid rock columns of identical uniform density ρ_c independently floating on a fluid mantle of a higher density ρ_m. If the reference crustal thickness is H, represented by a column of height H, the extra mass of a "mountain" of height h must be compensated by a low-density "mountain root" of length b. The total height of the

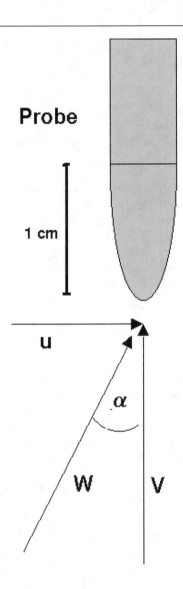

Geometry of the airfoil probe, α is the angle of attack of the oncoming flow.

record of surface waves. An Airy phase appears at a transition between normal dispersion and reverse dispersion. For continental paths an Airy phase with about a 20-sec period often occurs, while for oceanic paths an Airy phase with 10- to 15-sec period occurs, reflecting the thickness of the crust.

Airy wave theory   First-order wave theory for water waves. Also known as linear or first-order theory. Assumes gravity is the dominant restoring force (as opposed to surface tension). Named after Sir George Biddell Airy (1801–1892).

Aitken, John   (1839–1919) Scottish physicist. In addition to his pioneering work on atmospheric aerosol, he investigated cyclones, color, and color sensations.

Aitken nucleus count   One of the oldest and most convenient techniques for measuring the concentrations of atmospheric aerosol. Saturated air is expanded rapidly so that it becomes supersaturated by several hundred percent with respect to water. At these high supersaturations water condenses onto virtually all of the aerosol to form a cloud of small water droplets. The concentration of droplets in the cloud can be determined by allowing the droplets to settle out onto a substrate, where they can be counted either under a microscope, or automatically by optical techniques. The aerosol measured with an Aitken nucleus counter is often referred to as the *Aitken nucleus count*. Generally, Aitken nucleus counts near the Earth's surface range from average values on the order of 10^3 cm^{-3} over the oceans, to 10^4 cm^{-3} over rural land areas, to 10^5 cm^{-3} or higher in polluted air over cities.

rock column representing the mountain area is then $h + H + b$. Hydrostatic equilibrium below the mountain root requires $(\rho_m - \rho_c)b = \rho_c h$.

Airy phase   When a dispersive seismic wave propagates, the decrease of amplitude with increasing propagation distance for a period whose group velocity has a local minimum is smaller than that for other periods. The wave corresponding to the local minimum is referred to as an *Airy phase* and has large amplitude on a

Alba Patera   A unique volcanic landform on Mars that exists north of the Tharsis Province. It is less than 3 km high above the surrounding plains, the slopes of its flanks are less than a quarter of a degree, it has a diameter of ≈ 1600 km, and it is surrounded by an additional 500 km diameter annulus of grabens. Its size makes it questionable that it can properly be called a volcano, a name that conjures up an image of a distinct conical structure. Indeed from the ground on Mars it would not be discernible because the horizontal dimensions are so large. Nevertheless, it is interpreted as a volcanic structure on the basis that it possesses two very large summit craters from which huge volumes of lava have erupted from the late Noachian until the early Amazonian epoch; hence, it might be the largest volcanic feature on the entire planet. The exact origin is unclear. Possible explanations include deep seated crustal fractures produced at the antipodes of the Hellas Basin might have subsequently provided a conduit for magma to reach the surface; or it formed in multiple stages of volcanic activity, beginning with the emplacement of a volatile rich ash layer, followed by more basaltic lava flows, related to hotspot volcanism.

albedo   Reflectivity of a surface, given by I/F, where I is the reflected intensity, and πF is the incident flux. The Bond albedo is the fraction of light reflected by a body in all directions. The bolometric Bond albedo is the reflectivity integrated over all wavelengths. The geometric albedo is the ratio of the light reflected by a body (at a particular wavelength) at zero phase angle to that reflected by a perfectly diffusing disk with the same radius as the body. *Albedo* ranges between 0 (for a completely black body which absorbs all the radiation falling on it) to 1 (for a perfectly reflecting body).

The Earth's albedo varies widely based on the status and colors of earth surface, plant covers, soil types, and the angle and wavelength of the incident radiation. Albedo of the earth atmosphere system, averaging about 30%, is the combination of reflectivity of earth surface, cloud, and each component of atmosphere. The value for green grass and forest is 8 to 27%; over 30% for yellowing deciduous forest in autumn; 12 to 18% for cities and rock surfaces; over 40% for light colored rock and buildings; 40% for sand; up to 90% for fresh flat snow surface; for calm ocean, only 2% in the case of vertically incident radiation but can be up to 78% for lower incident angle radiation; 55% average for cloud layers except for thick stratocumulus, which can be up to 80%.

albedo neutrons   Secondary neutrons ejected (along with other particles) in the collision of cosmic ray ions with particles of the upper atmosphere. *See* neutron albedo.

albedo of a surface   For a body of water, the ratio of the plane irradiance leaving a water body to the plane irradiance incident on it; it is the ratio of upward irradiance to the downward irradiance just above the surface.

albedo of single scattering   The probability of a photon surviving an interaction equals the ratio of the scattering coefficient to the beam attenuation coefficient.

Alcyone   Magnitude 3 type B7 star at RA 03h47m, dec $+24°06'$; one of the "seven sisters" of the Pleiades.

Aldebaran   Magnitude 1.1 star at RA 04h25m, dec $+16°31'$.

Alfvénic fluctuation   Large amplitude fluctuations in the solar wind are termed *Alfvénic fluctuations* if their properties resemble those of Alfvén waves (constant density and pressure, alignment of velocity fluctuations with the magnetic-field fluctuations; *see* Alfvén wave). In particular, the fluctuations $\delta\mathbf{v}_{\text{sowi}}$ in the solar wind velocity and $\delta\mathbf{B}$ in magnetic field obey the relation

$$\delta\mathbf{v}_{\text{sowi}} = \pm\frac{\delta\mathbf{B}}{\sqrt{4\pi\varrho}}$$

with ϱ being the solar wind density. Note that in the definition of Alfvénic fluctuations or Alfvénicity, the changes in magnetic field and solar wind speeds are vector quantities and not the scalar quantities used in the definition of the Alfvén speed.

Obviously, in a real measurement it will be impossible to find fluctuations that exactly fulfill the above relation. Thus fluctuations are classified as Alfvénic if the correlation coefficient between $\delta\mathbf{v}_{\text{sowi}}$ and $\delta\mathbf{B}$ is larger than 0.6. The magnetic field and velocity are nearly always observed to be aligned in a sense corresponding to outward propagation from the sun.

Alfvénicity   *See* Alfvénic fluctuation.

Alfvén layer   Term introduced in 1969 by Schield, Dessler, and Freeman to describe the region in the nightside magnetosphere where region 2 Birkeland currents apparently originate. Magnetospheric plasma must be (to a high degree of approximation) charge neutral, with equal densities of positive ion charge and negative electron charge. If such plasma convects earthward under the influence of an electric field, as long as the magnetic field stays constant (a fair approximation in the distant tail) charge neutrality is preserved.

Near Earth, however, the magnetic field begins to be dominated by the dipole-like form of the main field generated in the Earth's core, and the combined drift due to both electric and magnetic fields tends to separate ions from electrons, steering the former to the dusk side of Earth and the latter to the dawn side. This creates *Alfvén layers,* regions where those motions fail to satisfy charge neutrality. Charge neutrality is then restored by electrons drawn upwards as the downward region 2 current, and electrons dumped into the ionosphere (plus some ions drawn up) to create the corresponding upward currents.

Alfvén shock   *See* intermediate shock.

Alfvén speed   In magnetohydrodynamics, the speed of propogation of transverse waves in a direction parallel to the magnetic field B. In SI units, $v_A = B/\sqrt{(\mu\rho)}$ where B is the magnitude of the magnetic field [tesla], ρ is the fluid density [kg/meter^3], and μ is the magnetic permeability [Hz/meter].

Alfvén's theorem   *See* "frozen-in" magnetic field.

Alfvén wave   A hydromagnetic wave mode in which the direction (but not the magnitude) of the magnetic field varies, the density and pressure are constant, and the velocity fluctuations are perfectly aligned with the magnetic-field fluctuations. In the rest frame of the plasma, energy transport by an Alfvén wave is directed along the mean magnetic field, regardless of the direction of phase propagation. Large-amplitude *Alfvén waves* are predicted both by the equations of magnetohydrodynamics and the Vlasov–Maxwell collisionless kinetic the-

ory, without requiring linearization of the theory.

In magnetohydrodynamics, the characteristic propagation speed is the Alfvén speed $C_A = B/\sqrt{4\pi\rho}$ (cgs units), where B is the mean magnetic field and ρ is the gas density. The velocity and magnetic fluctuations are related by $\delta\mathbf{V} = \mp\delta\mathbf{B}/\sqrt{4\pi\rho}$; the upper (lower) sign applies to energy propagation parallel (antiparallel) to the mean magnetic field. In collisionless kinetic theory, the equation for the characteristic propagation speed is generalized to

$$V_A^2 = C_A^2 \left[1 + \frac{4\pi}{B^2} \left(P_\perp - P_\parallel - \Pi \right) \right],$$

where $P_\perp$ and $P_\parallel$ are, respectively, the pressures transverse and parallel to the mean magnetic field,

$$\Pi = \frac{1}{\rho} \sum_\alpha \rho_\alpha \left(\Delta\mathbf{V}_\alpha \right)^2 .$$

ρ_α is the mass density of charge species α, and $\Delta\mathbf{V}_\alpha$ is its relative velocity of streaming relative to the plasma. Alfvén waves propagating through a plasma exert a force on it, analogous to radiation pressure. In magnetohydrodynamics the force per unit volume is $-\nabla \langle\delta\mathbf{B}\rangle^2 /8\pi$, where $\langle\delta\mathbf{B}\rangle^2$ is the mean-square magnetic fluctuation amplitude. It has been suggested that Alfvén wave radiation pressure may be important in the acceleration of the solar wind, as well as in processes related to star formation, and in other astrophysical situations.

In the literature, one occasionally finds the term "Alfvén wave" used in a looser sense, referring to any mode of hydromagnetic wave. *See* hydromagnetic wave, magnetoacoustic wave.

Algol system   A binary star in which mass transfer has turned the originally more massive component into one less massive than its accreting companion. Because the time scale of stellar evolution scales as M^{-2}, these systems, where the less massive star is the more evolved, were originally seen as a challenge to the theory. Mass transfer resolves the discrepancy. Many *Algol systems* are also eclipsing binaries, including Algol itself, which is, however, complicated by the presence of a third star in orbit around the eclipsing pair. Mass transfer is proceeding on the slow or nuclear time scale.

Allan Hills meteorite   A meteorite found in Antarctica in 1984. In August of 1996, McKay et al. published an article in the journal *Science,* purporting to have found evidence of ancient biota within the Martian meteorite ALH 84001. These arguments are based upon chemically zoned carbonate blebs found on fracture surfaces within a central brecciated zone. It has been suggested that abundant magnetite grains in the carbonate phase of ALH 84001 resemble those produced by magnetotactic bacteria, in both size and shape.

allowed orbits   *See* Störmer orbits.

all sky camera   A camera (photographic, or more recently, TV) viewing the reflection of the night sky in a convex mirror. The image is severely distorted, but encompasses the entire sky and is thus very useful for recording the distribution of auroral arcs in the sky.

alluvial   Related to or composed of sediment deposited by flowing water (alluvium).

alluvial fan   When a river emerges from a mountain range it carries sediments that cover the adjacent plain. These sediments are deposited on the plain, creating an *alluvial fan.*

alongshore sediment transport   Transport of sediment in a direction parallel to a coast. Generally refers to sediment transported by waves breaking in a surf zone but could include other processes such as tidal currents.

Alpha Centauri   A double star (α-Centauri A, B), at *RA* $6^h 45^m 9^s$, declination $-16°42'58''$, with visual magnitude -0.27. Both stars are of type G2. The distance to α-Centauri is approximately 1.326 pc. In addition there is a third, M type, star (Proxima Centauri) of magnitude 11.7, which is apparently bound to the system (period approximately 1.5 million years), which at present is slightly closer to Earth than the other two (distance = 1.307 pc).

α effect   A theoretical concept to describe a mechanism by which fluid flow in a dynamo such as that in the Earth's core maintains a magnetic field. In mean-field dynamo theory, the magnetic field and fluid velocities are divided into mean parts which vary slowly if at all and fluctuating parts which represent rapid variations due to turbulence or similar effects. The fluctuating velocities and magnetic fields interact in a way that may, on average, contribute to the mean magnetic field, offsetting dissipation of the mean field by effects such as diffusion. This is parameterized as a relationship between a mean electromotive force ϵ due to this effect and an expansion of the spatial derivatives of the mean magnetic field $\mathbf{B}_0$:

$$\epsilon_i = \alpha_{ij} B_{0j} + \beta_{ijk} \frac{\partial B_0 j}{\partial x_k} + \cdots$$

with the first term on the right-hand side, usually assumed to predominate, termed the "alpha effect", and the second term sometimes neglected. $\nabla \times \boldsymbol{\epsilon}$ is then inserted into the induction equation for the mean field. For simplicity, α is often assumed to be a scalar rather than a tensor in mean-field dynamo simulations (i.e., $\boldsymbol{\epsilon} = \alpha \mathbf{B}_0$). For α to be non-zero, the fluctuating velocity field must, when averaged over time, lack certain symmetries, in particular implying that the time-averaged helicity ($\mathbf{u} \cdot \nabla \times \mathbf{u}$) is non-zero. Physically, helical fluid motion can twist loops into the magnetic field, which in the geodynamo is thought to allow a poloidal magnetic field to be created from a toroidal magnetic field (the opposite primarily occurring through the ω effect). *See* magnetohydrodynamics.

alpha particle   The nucleus of a ^{4}He atom, composed of two neutrons and two protons.

Altair   Magnitude 0.76 class A7 star at RA 19h50.7m, dec $+8°51'$.

alternate depths   Two water depths, one subcritical and one supercritical, that have the same specific energy for a given flow rate per unit width.

altitude   The *altitude* of a point (such as a star) is the angle from a horizontal plane to that point, measured positive upwards. Altitude 90° is called the *zenith* (q.v.), 0° the *horizontal,* and $-90°$ the *nadir.* The word "altitude" can also be used to refer to a height, or distance above or below the Earth's surface. For this usage, *see* elevation. Altitude is normally one coordinate of the three in the topocentric system of coordinates. *See also* azimuth and zenith angle.

Amalthea   Moon of Jupiter, also designated JV. Discovered by E. Barnard in 1892, its orbit has an eccentricity of 0.003, an inclination of 0.4°, a precession of 914.6° yr^{-1}, and a semimajor axis of 1.81×10^5 km. Its size is $135 \times 83 \times 75$ km, its mass, 7.18×10^{18} kg, and its density 1.8 g cm^{-3}. It has a geometric albedo of 0.06 and orbits Jupiter once every 0.498 Earth days. Its surface seems to be composed of rock and sulfur.

Amazonian   Geophysical epoch on the planet Mars, 0 to 1.8 Gy BP. Channels on Mars give evidence of large volumes of water flow at the end of the Hesperian and the beginning of the *Amazonian* epoch.

Ambartsumian, Viktor Amazaspovich (1908–1996) Soviet and Armenian astrophysicist, founder and director of Byurakan Astrophysical Observatory. *Ambartsumian* was born in Tbilisi, Georgia, and educated at the Leningrad State University. His early work was in theoretical physics, in collaboration with D.D. Ivanenko. Together they showed that atomic nuclei cannot consist of protons and electrons, which became an early indication of the existence of neutrons. The two physicists also constructed an early model of discrete space-time.
   Ambartsumian's achievements in astrophysics include the discovery and development of invariance principles in the theory of radiative transfer, and advancement of the empirical approach in astrophysics, based on analysis and interpretation of observational data. Ambartsumian was the first to argue that T Tauri stars are very young, and in 1947, he discovered stellar associations, large groups of hot young stars. He showed that the stars in associations were born together, and that the associations themselves were gravitationally unstable and were expanding. This established that stars are still forming in the present epoch.

ambipolar field   An electric field amounting to several volts/meter, maintaining charge neutrality in the ionosphere, in the region above the E-layer where collisions are rare. If that field did not exist, ions and electrons would each set their own scale height — small for the ions (mostly O^+), large for the fast electrons — and densities of positive and negative charge would not match. The *ambipolar field* pulls electrons down and ions up, assuring charge neutrality by forcing both scale heights to be equal.

Amor asteroid   One of a family of minor planets with Mars-crossing orbits, in contrast to most asteroids which orbit between Mars and Jupiter. There are 231 known members of the Amor class.

ampere   Unit of electric current which, if maintained in two straight parallel conductors of infinite length, of negligible circular cross-section, and placed 1 m apart in vacuum, produces between these conductors a force equal to 2×10^{-7} N/m of length.

Ampere's law   If the electromagnetic fields are time independent within a given region, then within the region it holds that the integral of the magnetic field over a closed path is proportional to the total current passing through the surface limited by the closed path. In CGS units the constant of proportionality is equal to $4\,\pi$ divided by the speed of light. Named after A.M. Ampere (1775–1836).

amphidrome (amphidromic point)   A stationary point around which tides rotate in a counterclockwise (clockwise) sense in the northern (southern) hemisphere. The amplitude of a tide increases with distance away from the *amphidrome,* with the amphidrome itself the point where the tide vanishes nearly to zero.

Am star   A star of spectral type A as determined by its color but with strong heavy metal lines (copper, zinc, strontium, yttrium, barium, rare earths [atomic number = 57 to 71]) in its spectrum. These stars appear to be slow rotators. Many or most occur in close binaries which could cause slow rotaton by tidal locking. This slow rotation suppresses convection and allows chemical diffusion to be effective, producing stratification and differentiation in the outer layers of the star, the currently accepted explanation for their strange appearance.

anabatic wind   A wind that is created by air flowing uphill, caused by the day heating of the mountain tops or of a valley slope. The opposite of a katabatic wind.

analemma   The pattern traced out by the position of the sun on successive days at the same local time each day. Because the sun is more northerly in the Northern summer than in Northern winter, the pattern is elongated North-South. It is also elongated East-West by the fact that civil time is based on the mean solar day. However, because the Earth's orbit is elliptical, the true position of the sun advances or lags behind the expected (mean) position. Hence, the pattern made in the sky resembles a figure "8", with the crossing point of the "8" occurring near, but not at, the equinoxes. The sun's position is "early" in November and May, "late" in January and August. The relation of the true to mean motion of the sun is called the equation of time. *See* equation of time, mean solar day.

Ananke   Moon of Jupiter, also designated JXII. Discovered by S. Nicholson in 1951, its orbit has an eccentricity of 0.169, an inclination of 147°, and a semimajor axis of 2.12×10^7 km. Its radius is approximately 15 km, its mass, 3.8×10^{16} kg, and its density 2.7 g cm^{-3}. Its geometric albedo is not well determined, and it orbits Jupiter (retrograde) once every 631 Earth days.

Andromeda galaxy   Spiral galaxy (Messier object M31), the nearest large neighbor galaxy, approximately 750 kpc distant, centered at RA $00^h 42.7^m$, dec $+41°16'$, Visual magnitude 3.4 , angular size approximately 3° by 1°.

anelastic deformation   Solids creep when a sufficiently high stress is applied, and the strain is a function of time. Generally, the response of a solid to a stress can be split into two parts: elastic part or instantaneous part, and anelastic part or time-dependent part. The strain contributed by the anelastic part is called *anelastic deformation*. Part of the anelastic deformation can be recovered with time after the stress is removed

(retardation strain), and part of it becomes permanent strain (inelastic strain). Anelastic deformation is usually controlled by stress, pressure, temperature, and the defect nature of solids. Two examples of anelastic deformation are the attenuation of seismic waves with distance and the post-glacial rebound.

anemometer   An instrument that measures windspeed and direction. Rotation *anemometers* use rotating cups, or occasionally propellers, and indicate wind speed by measuring rotation rate. Pressure-type anemometers include devices in which the angle to the vertical made by a suspended plane in the windstream is an indication of the velocity. Hot wire anemometers use the efficiency of convective cooling to measure wind speed by detecting temperature differences between wires placed in the wind and shielded from the wind. Ultrasonic anemometers detect the phase shifting of sound reflected from moving air molecules, and a similar principle applies to laser anemometers which measure infrared light reemitted from moving air molecules.

angle of repose   The maximum angle at which a pile of a given sediment can rest. Typically denoted by ϕ in geotechnical and sediment transport studies.

angle-redshift test   A procedure to determine the curvature of the universe by measuring the angle subtended by galaxies of approximately equal size as a function of redshift. A galaxy of size D, placed at redshift z will subtend an angle

$$\theta = \frac{D\Omega_o^2(1+z)^2}{2cH_o^{-1}}$$
$$\left(\Omega_o z + (\Omega_o - 2)\left[(\Omega_o z + 1)^{1/2} - 1\right]\right)^{-1},$$

in a universe with mean density Ω_o and no cosmological constant. In models with cosmological constant, the angle also varies in a defined manner but cannot be expressed in a closed form. However, since galaxies are not "standard rods" and evolve with redshift, this test has not been successful in determining cosmological parameters.

Ångström (Å)   A unit of length used in spectroscopy, crystallography, and molecular structure, equal to 10^{-10} m.

angular diameter distance   Distance of a galaxy or any extended astronomical object estimated by comparing its physical size to the angle subtended in the sky: if D is the diameter of the galaxy and δ the angle measured in the sky, then $d_A = D/\tan\delta \simeq D/\delta$. For a Friedmann model with density Ω_o in units of the critical density, and zero cosmological constant, the angular diameter d_A of an object at redshift z can be given in closed form:

$$d_A = \frac{2cH_o^{-1}}{\Omega_o^2(1+z)^2} \left(\Omega_o z + (\Omega_o - 2)\left[(\Omega_o z + 1)^{1/2} - 1\right] \right) .$$

Other operational definitions of distance can be made (*see* luminosity distance) depending on the intrinsic (assumed to be known) and the observed properties to be compared.

angular momentum $\mathbf{L} = \mathbf{r} \times \mathbf{p}$, where $\times$ indicates the vector cross product, $\mathbf{r}$ is the radius vector from an origin to the particle, and $\mathbf{p}$ is the momentum of the particle. $\mathbf{L}$ is a pseudovector whose direction is given by $\mathbf{r}, \mathbf{p}$ via the right-hand rule, and whose magnitude is

$$|L| = |r||p|\sin\theta ,$$

where θ is the angle between $\mathbf{r}$ and $\mathbf{p}$. For a body or system of particles, the total angular momentum is the vectorial sum of all its particles. In this case the position is generally measured from the center of mass of the given body. *See* pseudovector, right-hand rule, vector cross product.

angular velocity (ω)   The angle through which a body rotates per unit time; a pseudovector with direction along the axis given by the right-hand rule from the rotation.

anisotropic   A material whose properties (such as intrinsic permeability) vary according to the direction of flow.

anisotropic scattering   Scattering that is not spherically symmetric.

anisotropic turbulence   *See* isotropic turbulence.

anisotropic universe   A universe that expands at different rates in different directions. The simplest example is Kasner's model (1921) which describes a space that has an ellipsoidal rate of expansion at any moment in time. Moreover, the degree of ellipticity changes with time. The generic Kasner universe expands only along two perpendicular axes and contracts along the third axis.

anisotropy   The opposite of isotropy (invariance under rotation), i.e., variation of properties under rotation. For example, if a rock has a fabric such as layering with a particular orientation, then phases of seismic waves may travel at different speeds in different directions through the rock, according to their alignment with the fabric. The wave speed along an axis varies when the axis is rotated through the rock with respect to the fabric, i.e., it is anisotropic. In terms of the material properties of the rock, this would be associated with an elasticity tensor that varies under rotation. This occurs in the real Earth: for example, wave speeds are observed to be faster in the upper mantle under the ocean in the direction perpendicular to the mid-ocean ridges. The Earth's inner core has been determined to be anisotropic, with (to a first approximation) faster wave speeds parallel to the Earth's rotation axis than in directions perpendicular to it. Many other physical properties may also be anisotropic, such as magnetic susceptibility, diffusivity, and turbulence.

annual flood   The maximum discharge peak flow during a given water year (October 1 through September 30) or annual year.

annular eclipse   A solar eclipse in which the angular size of the moon is slightly too small to obscure the entire solar photosphere. As a result, a ring ("annulus") of visible photosphere surrounds the dark central shadow of the moon. *Annular eclipse* occurs when the moon is near apogee, giving it a smaller angular size.

anomalistic month   *See* month.

anomalistic year   *See* year.

anomalous resistivity   For a fully ionized collision-dominated plasma, such as the solar corona, the extremely low value of the classical resistivity ensures that the rate of energy release is negligible since the field lines are prevented from diffusing through the plasma. In a turbulent plasma, the resistivity can be enhanced via the correlation of particles over length scales much larger than the usual plasma length scale, the Debye length. This increases the collision frequency and, consequently, the resistivity. This turbulently enhanced resistivity is known as *anomalous resistivity.*

anomaly   *See* mean anomaly, true anomaly.

anomaly, South Atlantic   The region above the southern Atlantic Ocean, in which the radiation belt descends to heights lower than elsewhere, so that near-earth satellites, nominally below the radiation belt, are likely to encounter peak radiation levels there.

The "anomaly" is caused by the non-dipole components of the main magnetic field of the earth, which create a region of abnormally weak magnetic field there (in the eccentric dipole model of the Earth's field, the dipole is furthest away from that region).

Each ion or electron trapped along a field line in the Earth's field has a mirroring field intensity B_m at which its motion along the line is turned around. Such particles also drift, moving from one field line to the next, all the way around the Earth. If in this drift motion the mirror point where the particle is turned back (and where the field intensity equals B_m) passes above the *South Atlantic anomaly,* it probably reaches an altitude lower there than anywhere else. The radiation belt thus extends lower in this region than elsewhere, and the loss of belt particles by collisions with atmospheric molecules is likely to occur there.

anorthosite   Mafic igneous rock type that consists predominantly of the mineral plagioclase (silicates of feldspar group) that seems to have differentiated at high temperature at the crust-mantle boundary, where plagioclase crystallized before separating from the main magma body and rose through the crust in a semi-molten state. *Anorthosites* are rare on Earth, but appear to be more common on the moon. *See* igneous.

anoxia   The condition arising from insufficient ambient oxygen to support biological respiration, or the effect of such lack.

Antarctic circle   The latitude $66°32'S$. South of this line the sun does not rise on the southern winter solstice and does not set on the day of the southern summer solstice.

antarctic circumpolar current   South Ocean current circling the Antarctic continent eastward. The largest oceanic current in terms of volume. Also called the West Wind Drift. Spans $40°$ to $60°$ South. Very close to the Antarctic continent is the East Wind Drift, driven by prevailing easterly winds near the continent.

Antarctic ozone depletion   A rapid and accelerating decrease in the ozone over Antarctica each September and October, as the so-called "ozone hole", which is due to the chemical activity of the chlorine atoms contained in the chlorofluorocarbons (CFCs or "Freons"). It was first reported on May 16, 1985, by J.C. Farman et al. from the British Antarctic Survey published in the British journal *Nature.* Field campaigns incorporating remote sensing, in situ and satellite observations, have now clearly demonstrated that man-made CFCs and some other halogenated industrial compounds are responsible for this dramatic loss of ozone. These chemicals are released into the atmosphere where their long lifetimes (50 to 100 years) allow them to be transported to the middle and upper stratosphere, where they can be decomposed by shortwave solar radiation to release their chlorine and bromine atoms. These free radicals are extremely reactive and can destroy ozone readily, but in most parts of the atmosphere they react to form harmless "reservoir" compounds. In the Antarctic, however, very low temperatures in the late winter and early spring stratosphere permit the formation of natural Polar Stratospheric Cloud (PSC) particles, which provide sites for surface reactions in which the reservoir halogens revert to ozone-destroying radicals with the help of sunlight. The severity of the ozone loss is also due, in part, to the special meteorology of the Antarctic winter stratosphere, which isolates the ozone hole, preventing the replenishment of ozone and the dilution of ozone destroying compounds. Thus, the ozone hole results from the combination of a range of special local and seasonal conditions with man-made pollution; its appearance in recent years simply corresponds to the build-up of anthropogenic halogenated gases in the atmosphere.

The production of CFCs and some other compounds potentially damaging to ozone is now limited by the Montreal Protocol and its amendments. However, the lifetimes of these gases are long, and although it is thought that stratospheric chlorine levels will peak in the next few years, recovery of the ozone hole may not be detectable for a number of years, and full recovery, to pre-ozone hole conditions, may not occur until the middle of the twenty-first century.

Antarctic ozone hole   A large annual decrease in the ozone content of the ozone layer over the Antarctic region during the southern hemisphere spring. Discovered in 1985, the ozone hole presumably appeared in the early 1980s and continued to increase in severity, size, and duration through the 1990s. In recent years, up to two-thirds of the total amount of ozone has been lost by mid-October, largely as a result of losses of over 90% in the layer between 14 and 22 km where a large fraction of the ozone is normally found. The onset of the ozone losses occurs in September, and the ozone hole usually recovers by the end of November.

Antarctic Zone   In oceanography, the region in the Southern Ocean northward of the Continental Zone (which lies near the continent). It is separated from the Continental Zone by a distinct oceanographic front called the Southern Antarctic Circumpolar Current front.

Antares   0.96 magnitude star, of spectral type M1, at RA 16h 29m 24.3, dec $-26°25'55''$.

anthropic principle   The observation that humankind (or other sentient beings) can observe the universe only if certain conditions hold to allow human (or other sentient) existence. Whenever one wishes to draw general conclusions from observations restricted to a small sample, it is essential to know whether the sample should be considered to be biased and, if so, how. The *anthropic principle* provides guidelines for taking account of the kind of bias that arises from the observer's own particular situation in the world. For instance, the *Weak* Anthropic Principle states that as we exist, we occupy a special place of the universe. Since life as we know it requires the existence of heavy elements such as C and O, which are synthesized by stars, we could not have evolved in a time less than or of the order of the main sequence lifetime of a star. This principle can be invoked to explain why the age of astronomical objects is similar to the Hubble time. This time scale would represent the lapse of time necessary for life to have evolved since the Big Bang. On the other hand, in the Steady State Cosmology, where the universe has no origin in time, the coincidence mentioned above has no "natural" explanation.

In the more controversial *strong* version, the relevant anthropic probability distribution is supposed to be extended over an ensemble of cosmological models that are set up with a range of different values of what, in a particular model, are usually postulated to be fundamental constants (such as the well-known example of the fine structure constant). The observed values of such constants might be thereby explicable if it could be shown that other values were unfavorable to the existence of anthropic observers.

Thus the Strong Anthropic Principle states that the physical properties of the universe are as they are because they permit the emergence of life. This teleological argument tries to explain why some physical properties of matter seem so fine tuned as to permit the existence of life. Slight variations in nuclear cross-sections could have inhibited the formation of heavy elements in stars. A different fine-structure constant would lead to a different chemistry and presumably life would not exist.

anticyclone   A wind that blows around a high pressure area, in the opposite sense as the Earth's rotation. This results in a clockwise rotation in the Northern Hemisphere and counterclockwise in the Southern Hemisphere.

anticyclonic   Any rotation that is opposite the sense of the locally measured Earth rotation: clockwise in the Northern Hemisphere, counterclockwise in the Southern Hemisphere.

antidune   Dunes that form in rivers or canals at relative high flow speeds. Dunes and *antidunes* are similar in shape, but the water surface above a dune is out of phase with the bed, whereas the water surface above antidunes is in phase with the bed. Antidunes and the corresponding surface waves often march gradually upstream.

antinode   A point on a standing wave where the field has maximum amplitude. For a standing water wave, this corresponds to a point with maximum vertical motion. For a standing wave transverse on a string, the *antinode* corresponds to a point which has maximum motion in a direction normal to that axis defined by the string.

antiparticle   A particle having the same mass as a given elementary particle and a charge equal in magnitude but opposite in sign.

apastron   In planetary motion, the farthest distance achieved from the gravitating central star. Generically one says *apapse*. Specific applications are *aphelion,* when referring to the motion of planets in our solar system; *apogee,* when referring to orbits around the Earth. Similar constructions are sometimes invented for orbits about the moon or other planets.

aperture correction   The difference between the photometric magnitude of an object as measured with two different-sized apertures.

When making photometric measurements of stars on an image, the resulting magnitudes are often referenced to the light measured in a fixed-size aperture (perhaps a few arcseconds in diameter). However, this aperture is usually smaller than the full profile of the star (which can be as large as an arcminute or more for light that is still detectable above the background). The *aperture correction* is the difference between the small, measurement aperture and a larger, reference aperture that is large enough to include any frame-to-frame variations that may be due to seeing or other variable effects. The aper-

ture correction is calculated for each frame and added to the magnitude of the objects in the frame to get a total magnitude. The aperture correction can be calculated by modeling the stellar profile, then integrating it out to infinity (or some large radius), or it may be calculated by simply measuring a number of isolated bright stars in an image using the small and large apertures and taking the average difference.

aperture synthesis   Method whereby the information-gathering capability of a large aperture is achieved by two or more smaller apertures operating together as interferometers.

aphelion   The point in an elliptical orbit around the sun that is farthest from the sun. (The perihelion is the point closest to the sun.) The time of *aphelion* passage for the Earth is around July 4.

aphotic zone   That portion of the ocean where light is insufficient for plants to carry on photosynthesis.

Ap index   The planetary index for measuring the strength of a disturbance in the Earth's magnetic field defined over a period of one day from a set of standard stations around the world. *See* geomagnetic activity.

apoapsis   The point in an elliptical orbit where the orbiting body is the farthest distance from the body being orbited. (The periapsis is the point of the shortest distance.) When the sun is the central body, the point of *apoapsis* is called the aphelion.

Apollo asteroid   One of a family of minor planets with Earth-crossing orbits. The majority of asteroids orbit between Jupiter and Mars, but the Apollos cross Earth's orbit and thus pose at least the potential for collision with Earth. It is estimated that there are at least 2000 Earth-crossing *Apollo asteroids* with diameters of 1 km or larger, and at least 10^6 larger than 50 meters.

Impact with an asteroid 1 km in size would deposit about 10^{21} J of energy if it impacted the Earth. This is about 10^5 Mtons of equivalent nuclear weapons, equivalent to exploding a good fraction of all the nuclear weapons on Earth at one instant. The crater produced would be about 10 km across. This event would raise matter into the atmosphere that would cause dramatic surface cooling by blocking sunlight for at least several years.

There are 240 known Apollos. *See* Amor asteroid, Aten asteroid.

apparent horizon A spacelike topological 2-sphere from which the outgoing null rays all have zero expansion. In gravitational theories, especially in general relativity, a horizon is a boundary between events visible from infinity and those that are not. The surface of a black hole, for instance, consists of those marginally trapped rays (which just fail escape to infinity); these constitute the event horizon. Generators of the event horizon are not truly identified until the evolution of the spacetime is complete into the future. A more local definition is the *apparent horizon,* the outermost surface defined by the null rays which instantaneously are not expanding. *See* event horizon, trapped surface.

apparent magnitude *See* magnitude.

apparent optical property (AOP) A ratio of radiometric quantities that depends both on the inherent optical properties and on the directional nature of the ambient light field and which is spatially and temporally stable. Applied in oceanography to describe a water body; examples include the average cosine of the light field, the irradiance reflectance, the remote sensing reflectance, and the diffuse attenuation coefficients.

apparent solar time Time based on the diurnal motion of the true (observed) sun, as opposed to mean solar time, to which it is related by the equation of time. The rate of diurnal motion undergoes seasonal variations because of the obliquity of the ecliptic, the eccentricity of the Earth's orbit, and irregularities in the Earth's orbit.

apse Line connecting the pericenter to the apocenter of an orbit, the longest axis of the orbit.

Ap star A chemically peculiar star of temperature classification *A*, which is a slow rotator *and* has a strong gravitational field. *Ap stars* have a pattern of overabundance including silicon, chromium, strontium, and europium and other rare earths. Their magentic fields are measured by the polarization induced in their spectral lines by the Zeeman effect; the fields have been measured up to 34000 Gauss (compared to ≈ 1G for the sun). Present understanding is that the slow rotation and the magnetic field together suppress convection to allow chemical segregation and enhancement in the surface layers of the stars.

aquifer A highly pervious geological formation, empirically defined as a geologic formation saturated with water and sufficiently permeable to transmit "significant" quantities of water under normal field conditions. On land, water enters an *aquifer* through precipitation or influent streams and leaves an aquifer through springs or effluent streams. An unconfined aquifer is a geologic formation in which the upper boundary of the saturated zone is the water table. A confined aquifer is an aquifer that is overlain by a confining bed with significantly lower hydraulic conductivity (an aquitard); water in a well or piezometer within a confined aquifer will rise above the top of the confined aquifer to the potentiometric surface. A perched aquifer is a region in the unsaturated zone that may be temporarily saturated because it overlies an area with lower hydraulic conductivity such as an aquitard or aquiclude.

aquitard A semipervious geological formation that transmits water very slowly as compared to an aquifer.

Arago point One of three points on the sky in a vertical line through the sun at which the polarization of skylight vanishes. Usually located at about 20° above the antisolar point (the point opposite the sun on the sky). *See* Babinet point, Brewster point.

arcade A configuration of coronal loops spanning a magnetic neutral line. The loops are often perpendicular to the neutral line but can be sheared due to the forces of differential rotation.

A coronal arcade is frequently associated with a filament channel.

archaeoastronomy   The study of the astronomical knowledge and techniques of prehistorical societies by studies of archaeological structures.

archaeomagnetism   The study of the Earth's magnetic field using archaeological artifacts. Historical magnetism uses explicit historical measurements of the Earth's magnetic field, which (despite claims that the compass was invented as far back as the second century BC) are only useful back to around 1600 AD. Paleomagnetism relies on measurements of the magnetization of geological materials, such as lava flows and lake bed sediments, and tends to have coarser resolution in time. *Archaeomagnetism* attempts to bridge the gap between the two by providing measurements of field older than historical but with better resolution than paleomagnetism. A magnetic measurement may be obtained from an excavation from, for example, a kiln whose last firing may be determined using radiocarbon dating. The kiln may record the magnetic field of that time through thermoremanent magnetization.

Archean   The period in the Earth's evolution prior to 2.5 billion years ago.

Archimedes' principle   An object partially or totally submerged in a liquid is buoyed up by a force equal to the weight of the displaced liquid.

Archimedian spiral   Shape of the interplanetary magnetic field line. Physically, the solar magnetic field is frozen into the radially streaming solar wind (*see* frozen-in flux theorem). Because the footpoint of the field line is fixed on the sun, the sun's rotation winds up the field to a spiral with constant distances between neighboring windings.

Mathematically, such a spiral is called an *Archimedian spiral*. In polar coordinates (r, φ) it is described as

$$r = v_{\text{sowi}} \cdot \frac{\varphi - \varphi_0}{\omega_\odot} + r_0$$

with v_{sowi} being the solar wind speed, $\omega_\odot$ the angular speed of the sun in the equatorial plane, r_0 the source height of the plasma parcel, and φ_0 its source longitude. With $\psi = \omega_\odot r / v_{\text{sowi}}$ the path length s along the spiral is

$$s = \frac{1}{2} \cdot \frac{v_{\text{sowi}}}{\omega_\odot} \left(\psi \cdot \sqrt{\psi^2 + 1} + \left(\ln \left\{ \psi + \sqrt{\psi^2 + 1} \right\} \right) \right).$$

For sufficiently large distances, the garden hose angle δ, that is the angle between the magnetic field line and a radius vector from the sun, can be written as

$$\tan \delta = \frac{\omega_\odot r}{v_{\text{sowi}}}.$$

At Earth's orbit, the garden hose angle is about 45° for the average solar wind speed of 400 km/s, and the distance s to the sun along the Archimedian magnetic field spiral is about 1.15 AU.

arc minute   A measure of angular size, abbreviated arcmin or $'$. There are 60 *arc minutes* in 1 arc degree. On the surface of the Earth 1 arc minute of latitude corresponds very closely to a north-south distance of 1 nautical mile (1852 m).

arc second   A measure of angular size in the plane of the sky, abbreviated arcsec or $''$. There are 60 *arc seconds* in 1 arc minute and, therefore, 3600 arc seconds in 1 arc degree. One arc second corresponds to about 725 km on the surface of the sun, as viewed from the Earth.

Arctic circle   The latitude 66°32'N. North of this line the sun does not rise on the northern winter solstice and does not set on the day of the northern summer solstice.

arctic oscillation (AO)   Dominant mode of atmospheric sea level pressure (SLP) variability in the Northern Hemisphere, most pronounced in winter. At its positive phase, the AO features a deepened Icelandic low and Azores high in the North Atlantic but a weakened Aleutian low in the North Pacific. Surface air temperature rises over northern Eurasia but falls over high-latitude North America. The AO involves changes in the latitude and strength of westerly jet in the troposphere and in the intensity of polar vortex in the lower stratosphere.

Arcturus −0.2 magnitude star, of spectral type K2, at RA 14^h 15^m 39.6^s, dec $+19°10'57''$.

argon Inert (noble) gas which is a minor (0.94%) constituent of the Earth's atmosphere. Atomic number 18, naturally occurring atomic mass 39.95, composed of three naturally occurring isotopes A^{36} (0.34%), A^{38} (0.06%), and A^{40} (99.60%). A^{40} is produced by decay of K^{40}, and potassium-argon dating is used to date the solidification of rocks, since the gas escapes from the melt, but is then regenerated by the decaying potassium.

argument of periapse The angle from the ascending node of an orbit to the periapse.

Ariel Moon of Uranus, also designated UI. It was discovered by Lassell in 1851. Its orbit has an eccentricity of 0.0034, an inclination of 0.3°, a semimajor axis of 1.91×10^5 km, and a precession of 6.8° yr^{-1}. Its radius is 576 km, its mass is 1.27×10^{21} kg, and its density is 1.59 g cm^{-3}. Its geometric albedo is 0.34, and it orbits Uranus once every 2.520 Earth days.

Arnowitt–Deser–Misner (ADM) decomposition of the metric In a four-dimensional space-time Ω, with Lorentzian metric tensor g, consider any one-parameter (t) family of space-like hypersurfaces Σ_t with internal coordinates $\vec{x} = (x^i, i = 1, 2, 3)$ and such that, by continuously varying t, Σ_t covers a domain $D \subseteq \Omega$ of non-zero four-dimensional volume. Inside D, on using $(t, \vec{x})$ as space-time coordinates, the proper distance between a point $A_{\vec{x}}$ on Σ_t and a point $B_{\vec{x}+d\vec{x}}$ on Σ_{t+dt} can be written according to the Pythagorean theorem

$$ds^2 = \gamma_{ij} \left(dx^i + \beta^i \, dt \right) \left(dx^j + \beta^j \, dt \right) - (\alpha \, dt)^2,$$

where γ is the metric tensor (pull back of g) on Σ_t, α (lapse function) gives the lapse of proper time between the two hypersurfaces Σ_t and Σ_{t+dt}, β^i (shift vector) gives the proper displacement tangential to Σ_t between $A_{\vec{x}}$ and the

point $A'_{\vec{x}}$, which has the same spatial coordinates as $A_{\vec{x}}$ but lies on Σ_{t+dt}.

In matrix notation one has for the covariant metric tensor

$$g = \begin{bmatrix} -\alpha^2 + \beta_k\,\beta^k & \beta_i \\ \beta_j & \gamma_{ij} \end{bmatrix},$$

where latin indices are lowered by contraction with γ_{ij}. The contravariant metric (the matrix inverse of g) is given by

$$g^{-1} = \begin{bmatrix} -\frac{1}{\alpha^2} & \frac{\beta^i}{\alpha^2} \\ \frac{\beta^j}{\alpha^2} & \gamma^{ij} - \frac{\beta^i\,\beta^j}{\alpha^2} \end{bmatrix},$$

where $\gamma^{ik}\,\gamma_{kj} = \delta^i_j$. Any geometrical quantity can then be decomposed in an analogous way. For instance, the determinant of g becomes $g = \alpha^2\gamma$, where γ is the determinant of the 3-dimensional metric γ.

The above forms (in four dimensions) are also called 3+1 splitting of space-time and can be generalized easily to any dimension greater than one. There is a large amount of freedom in the choice of this splitting which reflects the absence of a unique time in general relativity (multifingered time). *See* ADM form of the Einstein–Hilbert action.

array seismic observation A seismic observation system improving S/N (signal to noise) ratio of seismic waves by deploying many seismometers in an area and stacking their records, giving appropriate time differences. It is also possible to identify a location of a hypocenter of an earthquake by obtaining direction of arrived seismic waves and apparent velocity. As a large-scale array system, there is the LASA (Large Aperture Seismic Array) in Montana, where more than 500 seismometers were deployed in an area about 200 km in diameter.

arrow of time A physical process that distinguishes between the two possible directions of flow of time. Most of the equations that describe physical processes do not change their form when the direction of flow of time is reversed (i.e., if time t is replaced by the parameter $\tau = -t$, then the equations with respect to τ are identical to those with respect to t). Hence, for every solution $f(t)$ of such equations ($f(t)$ represents here a function or a set of functions), $f(-t)$ is also a solution and describes a process that is, in principle, also possible. Example: For a planet orbiting a star, the time-reversed motion is a planet tracing the same orbit in the opposite sense. However, for most complex macroscopic processes this symmetry is absent; nature exhibits histories of directed events in only one direction of time, never the reverse. This is known as the *arrow of time.* The arrow of time is provided by the expansion of the universe, the thermodynamics of the physical system, or the psychological process. The most famous example is the entropy in thermodynamics: All physical objects evolve so that their entropy either increases or remains constant. The question of whether an arrow of time exists in cosmology is a theoretical problem that has not been solved thus far. Observations show that the universe is expanding at present, but the Einstein equations allow a time-reversed solution (a contracting universe) as well. Note also that at a microscopic level certain quantum particle interactions and decays are not time reversal invariant, and thus define a direction of time. However, no completely convincing connection has yet been made to the large-scale or cosmological arrow of time.

ascending node   For solar system objects, the right ascension of the point where the orbit crosses the ecliptic travelling to the North; in other systems, the equivalent definition.

aseismic front   An ocean-side front line of an aseismic wedge-shaped region located between a continental plate and an oceanic plate subducting beneath an island arc such as the Japanese islands. An *aseismic front* is almost parallel to a trench axis and a volcanic front. Very few earthquakes whose hypocentral depths range from 40 to 60 km between the oceanic and the continental plates occur on the continental side of the aseismic front. This is thought to be because temperature is high and interplate coupling is weak on the continental side of the aseismic front. These are closely related to slow velocity structure of the uppermost mantle beneath the island arc, detected from an analysis of observed *Pn* waves.

aseismic region   A region with very few earthquakes.

asperity   Earthquakes occur on faults. Faults are approximately rough planar surfaces. This roughness results in *asperities* that impede displacements (earthquakes) on the fault. An extreme example of an asperity would be a bend in a fault.

association   An obvious collection of stars on the sky that are part of, or contained within, a constellation.

A star   Star of spectral type A. Vega and Sirius are examples of A stars. A0 stars have color index = 0.

asterism   A small collection of stars (part of a constellation) that appear to be connected in the sky but form an association too small to be called a constellation.

asteroid   Small solid body in orbit around the sun, sometimes called *minor planet.* Asteroids are divided into a number of groups depending on their reflection spectrum. The major classes are *C-type,* characterized by low albedo (0.02 to 0.06) and a chemical composition similar to carbonaceous chondrites; *S-type,* which are brighter (albedo between 0.07 and 0.23) and show metallic nickel-iron mixed with iron and magnesium silicates; and *M-type* with albedos of 0.07 to 0.2 which are nearly pure nickel-iron. C-type asteroids comprise about 75% of all main belt asteroids, while S-type comprise about 17%. Additional rare classes are E (enstatite), R (iron oxide?), P (metal?), D (organic?), and U (unclassifiable). Asteroids are also classified according to location. *Main belt* asteroids lie in roughly circular orbits between Mars and Jupiter (2 to 4 AU from the sun). The *Aten* family has semimajor axes less than 1.0 AU and *aphelion* distances larger than 0.983 AU. These form a potential hazard of collision with Earth. The *Apollo* family has semimajor axes greater than 1.0 AU and *perihelion* distances less than 1.017 AU. *Amor* asteroids have perihelia between 1.017 and 1.3 AU. *Trojan* asteroids lie at the L4 and L5 Lagrange points of Jupiter's orbit around the sun. *Centaurs* have orbits that bring them into the outer solar system. Obervationally, the distinction between asteroids and comets is that comets display a coma and tail.

Some asteroids are probably dead *comets* which have lost most of their icy material due to their many passages around the sun. Some asteroids have been found to show comet-like characteristics, and the asteroid Chiron (for which the Centaur asteroids were named) has now been reclassified as a comet on this basis. The largest asteroid is Ceres, which has a diameter of about 950 km. The asteroids within the asteroid belt, however, are believed to be left-over debris from the formation of the solar system, which was never allowed to accrete into a planet due to the gravitational influence of nearby Jupiter. Images taken by spacecraft show that asteroids are generally irregular, heavily cratered objects. Some may be solid rock, although many are likely collections of small debris ("rubble piles") held together only by their mutual gravity.

asteroid classification   A classification of asteroids according to their spectra and albedo: C-type, apparently similar to carbonaceous chondrite meteorites; extremely dark (albedo approximately 0.03). More than 75% of known asteroids fall into this class. S-type, albedo .10-.22; spectra indicating metallic nickel-iron mixed with iron- and magnesium-silicates; approximately 17% of the total. M-type, albedo .10-.18; pure nickel-iron.

asteroid orbital classification   Main Belt: asteroids orbiting between Mars and Jupiter roughly 2 to 4 AU from the sun; Near-Earth Asteroids (NEAs): asteroids that closely approach the Earth; Aten asteroids: asteroids with semimajor axes less than 1.0 AU and aphelion distances greater than 0.983 AU; Apollo asteroids: asteroids with semimajor axes greater than 1.0 AU and perihelion distances less than 1.017 AU; Amor asteroids: asteroids with perihelion distances between 1.017 and 1.3 AU; Trojans asteroids: asteroids located near Jupiter's Lagrange points (60° ahead and behind Jupiter in its orbit).

Asterope   Magnitude 5.8 type B9 star at RA 03^h45^m, dec $+24°33'$; one of the "seven sisters" of the Pleiades.

asthenosphere   The inner region of a terrestrial planet which undergoes ductile flow (also

called solid state convection). In the Earth, the *asthenosphere* is composed of the lower part of the mantle and is the region between 100 and 640 km depth. It is marked by low seismic velocities and high seismic-wave attenuation. The ability of the asthenosphere to flow over long time periods (thousands to millions of years) helps to transport heat from the deep interior of a body and leads to plate tectonic activity on Earth as the rigid outer lithosphere rides atop the asthenosphere.

Astraea   Fifth asteroid to be discovered, in 1845. Orbit: semimajor axis 2.574 AU, eccentricity 0.1923, inclination to the ecliptic 5°.36772, period 4.13 years.

astrochemistry   Chemistry occurring under extraterrestrial conditions including: reactions of atoms, ions, radicals, and neutral molecules in the gas phase, and reactions of such species in ices on metal or mineral surfaces and in/on ices on grains, comets, and satellites, especially induced by impinging atoms, ions, and photons.

astrometric binary   A binary star system that reveals itself as a single point of light whose position or centroid shifts with the orbit period. A famous example is Sirius, recognized by Bessell in 1844 as having a very faint companion of roughly its own mass, accounting for the shift of its position with a 50-year period. Improved angular resolution or sensitivity can turn an *astronometric binary* into a visual binary. *See* binary star system, visual binary system.

astrometry   The measurement of positions and motions of celestial objects.

astronomical latitude   Defined as the angle between the local vertical, as defined by gravity, and the Earth's equatorial plane, counted positive northward and negative southward. (*See also* latitude.) *Astronomical latitude* is generally within $10''$ arc of geodetic latitude in value. The local vertical, in this sense, is the normal to the geoid; in simple terms, it is the upwards line defined by the plumb bob. The difference between astronomical latitude and geodetic latitude is due to small, local gravity variations. These are caused by mass concentrations, such as mountains, lakes, and large ore deposits, which cause the plumb line to deviate slightly from the normal to the ellipsoid.

astronomical refraction   The apparent angular displacement toward the zenith in the position of a celestial body, due to the fact that the atmosphere over any observer is apparently a planar slab with density decreasing upward. The effect vanishes overhead and is largest near the horizon, where it becomes as much as $30'$. The fact that the sun is refracted to appear above its true angular position contributes measurably to the length of the apparent day. Also called atmospheric scintillation.

astronomical scintillation   Any irregular scintillation such as motion, time dependent chromatic refraction, defocusing, etc. of an image of a celestial body, produced by irregularities in the Earth's atmosphere. The effects have periods of 0.1 to 10 sec and are apparently caused by atmospheric irregularities in the centimeter to decimeter and meter ranges, within the first 100 m of the telescope aperture.

astronomical tide   Fluctuations in mean water level (averaged over a time scale of minutes) that arise due to the gravitational interaction of (primarily) the earth, moon, and sun. May also be used to refer to the resulting currents.

astronomical twilight   *See* twilight.

astronomical unit (AU)   The mean distance between the sun and the Earth (1.4959787×10^8 km). This is the baseline used for trigonometric parallax observations of distances to other stars.

astronomy, infrared   The observation of astronomical objects at infrared (IR) wavelengths, approximately in the range from 1 to 200 μm, that provide information on atomic motions that cause changes in charge distribution. The mid-infrared spans approximately the range from 2.5 to 25 μm and includes fundamental transitions for bond stretching and bending of most interstellar molecules. Longer and shorter wavelengths, known as the far and near IR, respectively, correspond to low frequency motions of

groups of atoms and overtones of far and mid-IR features.

astronomy, infrared: interstellar grains, comets, satellites, and asteroids   Absorption, reflection, and emission at infrared (IR) wavelengths provide astronomers with unique molecular information for molecules not visible at other wavelengths, such as radio, because they lack a permanent dipole moment, or are solids, such as ices on interstellar grains or solar system bodies. IR spectroscopy of these solid materials, measured in absorption and reflection, respectively, have supplied most remotely measured information about the mineralogy and chemical composition of interstellar grains and solar system surfaces. Most spectra of outer solar system bodies have been measured in reflected sunlight in the near IR because solar radiation diminishes with increasing wavelength so they are dark in the mid-IR.

astronomy, ultraviolet: interstellar   The observation of astronomical objects and phenomena at ultraviolet (UV) wavelengths, approximately in the range from 100 to 4000 Å, provide information on the electronic transitions of materials, molecules, and reactive species. UV absorption of interstellar materials have helped to put constraints on the form and distribution of most carbon bearing species in the galaxy. *See* diffuse interstellar bands (DIBs).

asymmetry factor   In scattering, the mean cosine of the scattering angle.

asymmetry parameter   Asymmetry factor.

asymptotic   The (normalized) angular shape of the radiance distribution at depths far from the boundary of a homogeneous medium; the directional and depth dependencies of the asymptotic radiance distribution decouple and all radiometric variables (e.g., irradiances) vary spatially at the same rate as the radiance, as governed by the inherent optical properties only. *See* diffuse attenuation coefficient.

asymptotically simple space-time   A space-time ($\mathcal{M}$, g) is said to be asymptotically simple if there exists a space-time $(\tilde{\mathcal{M}}, \tilde{g})$, such that $\mathcal{M}$ is a submanifold of $\tilde{\mathcal{M}}$ with boundary $\mathcal{I}$ and

- $\tilde{g}_{ab} = \Omega^2 g_{ab}$, $\Omega > 0 \in \mathcal{M}$

- On $\mathcal{I}$, $\Omega = 0$ and $\nabla_a \Omega \neq 0$

- Any null geodetic curve in $\mathcal{M}$ has two endpoints in $\mathcal{I}$

- In a neighborhood of $\mathcal{I}$, the space-time is empty (or has only electromagnetic fields)

asymptotic diffuse attenuation coefficient
The value of the diffuse attenuation coefficient in the asymptotic regime; it depends on the inherent optical properties only.

asymptotic flatness   The assumption in theoretical/analytical descriptions of gravitational fields, that the gravitational potential goes to zero at spatial infinity, i.e., far away from its sources. In general relativity, the gravitational field is reflected in curvature of spacetime, so requiring flatness has a direct connection to requiring vanishing gravitational effects. In situations with a nonvanishing central mass m, *asymptotic flatness* requires the metric approach flat $+ O(Gm/c^2 r)$. Thus, a space-time Ω with Lorentzian metric g is said to be asymptotically flat (at spatial infinity) if a set of spherical coordinates (t, r, θ, ϕ) can be introduced, such that g approaches the Minkowski tensor for large r:

$$\lim_{r \to +\infty} g = \text{diag}\left(-1, 1, r^2, r^2 \sin^2 \theta\right).$$

asymptotic giant branch (AGB) star   Star of low or intermediate mass ($\sim$ 0.8 to 5 solar masses) in the advanced evolutionary phase where the primary energy sources are fusion of hydrogen (by the CNO cycle) to helium and of helium (by the triple-alpha process) to carbon in thin shells surrounding an inert carbon-oxygen core. The phase is important for two reasons. First, the star develops several zones of convection which cross back and forth so as to mix to the surface products of the interior nuclear reactions, including nitrogen from the CNO cycle, carbon from the triple-alpha process, and the products of the s process, including barium and,

sometimes, technitium, thus confirming the occurrence of these reactions. The longest-lived isotope of Tc has a half life less than a million years, showing that the reactions must be occurring recently. Second, the star expels a wind of up to 10^{-6} to 10^{-4} solar masses per year, and this mass loss both terminates the interior nuclear reactions and determines that the core will become a *white dwarf* rather than igniting carbon fusion. The phase lasts only about 0.01% of the longest, main-sequence, phase. The name derives from the location of these stars on the HR diagram in a diagonal strip that approaches tangentially at high luminosity to the main red giant branch. AGB stars are much brighter and more extended, but cooler on the surface, than the same stars were on the main-sequence. *See* CNO cycle, convection, HR diagram, main sequence star, red giant, s process, triple-alpha process, white dwarf.

asymptotic regime   In oceanography, depths at which the rate of decay with depth of all radiometric variables, given by the asymptotic diffuse attenuation coefficient, depends only on the inherent optical properties.

Aten asteroid   A member of a class of asteroids with Venus-crossing orbits, in contrast to the majority of asteroids that orbit between Mars and Jupiter. There are 30 known members of the Aten class.

Atlas   A moon of Saturn, also designated SXV. It was discovered by R. Terrile in 1980 in Voyager photos. Its orbit has an eccentricity of 0, an inclination of 0.3°, and a semimajor axis of 1.38×10^5 km. Its size is roughly 20×10 km, and its mass has not yet been determined. It appears to be a shepherd satellite of Saturn's A ring and orbits Saturn once every 0.602 Earth days. Also, magnitude 3.8 type B9 star at RA 03h49m, dec $+24°03'$; "Father" of the "seven sisters" of the Pleiades.

atmosphere   The gaseous envelop surrounding the Earth and retained in the Earth's gravitational field, which contains the troposphere (up to about 10 to 17 km), stratosphere (up to about 55 km), mesosphere (up to about 80 km), and ionosphere (up to over 150 km). The total mass of the *atmosphere* is about 5.3×10^{18} kg, which is about one-millionth of the total mass of Earth. At sea level, average pressure is 1013.25 hPa, temperature 288.15 K, and density is 1.225 kg/m^3. The density of the atmosphere decreases rapidly with height, and about three-quarters of the mass of the atmosphere is contained within the troposphere. The atmosphere has no precise upper limit. Formally one defines the top of the atmosphere at 1000 km altitude, which is also the highest observed altitude of aurora.

atmosphere effect Whenever a gas that is a weak absorber in the visible and a strong absorber in the infrared is a constituent of a planetary atmosphere, it contributes toward raising the surface temperature of the planet. The warming results from the fact that incoming radiation can penetrate to the ground with relatively little absorption, while much of the outgoing longwave radiation is "trapped" by the atmosphere and emitted back to the ground. This is called the *atmosphere effect*. This warming is commonly referred to as the "greenhouse effect".

atmospheric angular momentum As wind flows in the atmosphere, an air parcel rotates about the Earth's axis, so the atmosphere contains angular momentum. In tropical easterlies, friction with the Earth's surface transfers angular momentum to the atmosphere; in the mid-latitiude westerlies in both hemispheres, angular momentum is transferred from the atmosphere to the surface. Over long periods of time, the angular momentum of the atmosphere is in a steady state. Thus, there must be angular momentum transport from the tropics to mid-latitude in the two hemispheres. In the tropics, the mean meridional circulation plays an important role in the meridional transport of atmospheric angular momentum; and at mid-latitudes transient eddies and stationary eddies play a major role. Short term variations in the total *atmospheric angular momentum* can be observed in the rotation rate of the soled Earth.

atmospheric conductivity Conductivity of the atmosphere, determined by ion concentration and ion mobility. The conductivity in-

creases roughly exponentially with height because ion mobility depends on the number of collisions between air particles and thus increases with increasing height. Since the mobility of small ions is much larger than that of large ones, aerosol particles form a sink for small ions, reducing the atmospheric conductivity.

atmospheric electric field The atmospheric electric field on the ground is about -100 V/m with strong variations depending on weather conditions and the availability of dust particles. With increasing height, the *atmospheric electric field* decreases because the conductivity increases. The atmospheric electric field is part of the global electric circuit which can be conceptualized as a spherical capacitor formed by the terrestrial surface and the bottom of the ionosphere filled with a slightly conductive medium, the atmosphere. Thunderstorms work as generators, driving a current from the surface to the bottom of the ionosphere. The circuit is closed through the fair weather atmosphere which acts as a resistor.

atmospheric noise Radio noise produced by natural electrical discharges below the ionosphere and reaching the receiving point, where it is observed, along normal propagation paths between the Earth's surface and the ionosphere. Distant lightning has usually been thought to be the main source for this noise. *See* galactic noise.

atmospheric pressure The ambient air pressure at a particular time and location. Expressed as an absolute pressure (i.e., relative to a vacuum). *See also* gauge pressure. "Standard" *atmospheric pressure* is taken as 14.7 lb/in^2 or 101.3 kPa.

atmospherics A lightning stroke transmits a wide range of electromagnetic radiation, the most familiar being visible light. The electromagnetic emissions are short-lived, like the optical emissions. Those that can be reflected by the Earth's ionosphere can propagate to remote locations in the earth-ionosphere waveguide where they can be observed. At frequencies used for early high frequency radio communications ($\sim$ 1 to 30 MHz) the propagated lightning signal was heard as a sharp, short duration crackle on a radio receiver. This bursty crackle of interference was called an *atmospheric,* to distinguish it from the internal and local site interference. The sum of many atmospherics from remote lightning strokes all over the world produces a steady background noise limit at these radio frequencies called atmospheric noise. Atmospherics were observed at lower frequencies and used as a measure of thunderstorm activity. Early receivers for this application were sometimes caller spheric receivers.

atmospheric tide   Oscillations in any atmospheric field with periods that are simple integer fractions of either a lunar or a solar day. In addition to being somewhat excited by the gravitational potential of the sun and moon, atmospheric tides are strongly forced by daily variations in solar heating. The response of these forcings is by internal gravity waves. Unlike ocean tides, *atmospheric tides* are not bound by coastlines but are oscillations of a spherical shell.

atomic mass   The mass of an isotope of an element measured in *atomic mass* units. The atomic mass unit was defined in 1961, by the International Union of Pure and Applied Physics and the International Union of Pure and Applied Chemistry, as $1/12$ of the mass of the carbon isotope counting 6 neutrons (and 6 protons) in its nucleus.

atomic number   The number of protons in the nucleus of a given element.

atomic structure calculations — one-electron models   The calculation of possible states of an electron in the presence of an atomic nucleus. The calculations consist in obtaining the electron distribution or wave function about the nucleus for each state. This is achieved by solving the Schrödinger equation for the electron wave function in a fixed Coulomb potential generated by the nucleus of the atom. The quantified nature of the possible solutions or states appear naturally when the conditions of continuity and integrability are applied to the wave functions. An important characteristic of the one-electron models is that they can be solved

exactly; the wave functions may be expressed in terms of spherical harmonics and associated Laguerre polynomials. Relativistic treatment is done through Dirac's equation. Dirac's equation leads to the fine structure as a relativistic correction to Schrödinger's solution. Another important result of Dirac's equations is that even for non-relativistic cases one finds that the electron has two possible states, generally interpreted as two possible states of intrinsic angular momentum or spin.

atomic time   Time as measured by one or more atomic clocks, usually a cesium-beam atomic clock or a hydrogen maser. Measured since January 1, 1958, it is the most uniform measure of time available and has, therefore, replaced Universal Time as the standard.

attenuation coefficient   In propogation of a signal, beam, or wave through a medium, with absorption of energy and scattering out of the path to the detector, the *attenuation coefficient* α is

$$\alpha = d^{-1} \ln (S/S_0) ,$$

where this is the natural logarithm, and S and S_0 are the current intensity and the initial intensity. Since α is an inverse length, it is often expressed in terms of decibel per meter, or per kilometer. *See* beam attenuation coefficient, diffuse attenuation coefficient.

attenuation efficiency factor   The sum of the absorption plus scattering efficiency factors.

aulacogen   Mantle plumes create regions of elevated topography which typically have three rift valleys at about 120° apart; these are *aulacogen.* These are also known as triple junctions, and they participate in the formation of new ocean basins. An example is the southern end of the Red Sea. Typically two arms participate in the opening of an ocean, and the third is known as a failed arm. The St. Lawrence river valley is a failed arm associated with the opening of the Atlantic Ocean.

aurora   Polar lights. The aurora borealis (northern lights) and aurora australis (southern lights). Energetic electrons are trapped from the solar wind and spiral around the field lines of the

Earth's magnetic field. They enter the Earth's upper atmosphere where the field lines intersect the atmosphere, i.e., in the polar regions. There they excite atoms in the high thin atmosphere at altitudes of 95 to 300 km. The red and green colors are predominantly produced by excitations of oxygen and nitrogen. The polar lights are typically seen within 5000 km of the poles, but during times of intense solar activity (which increases the electron population), they can become visible at midlatitudes as well. Any body that possesses both a magnetic field and an atmosphere can produce *aurorae*. Aurorae are commonly seen not only on Earth but also the Jovian planets of Jupiter and Saturn.

aurora australis   Southern light, aurora in the southern hemisphere. *See* aurora.

aurora borealis   Northern light, aurora in the northern hemisphere. *See* aurora.

auroral cavity   A region on magnetic field lines which guides the aurora, typically within 10,000 km or so of Earth, where abnormally low ion densities are observed at times of strong aurora, presumably caused by it.

auroral electrojet   A powerful electric current, flowing in the auroral oval in the ionospheric E-layer, along two branches that meet near midnight. The branches are known as the eastward and westward *auroral electrojets,* respectively, and the region in which they meet, around 2200 magnetic local time, is the Harang discontinuity.

The electrojets are believed to be Hall currents in the ionospheric E-layer and to be a secondary effect of the currents linking Birkeland currents of region 1 with those of region 2. Because of Fukushima's theorem, the magnetic disturbance due to the Birkeland current sheets on the ground is very weak, and the main signature of their circuit — which can be quite strong — comes from the electrojets. The usual way of estimating the current flowing in that circuit — which is a major signature of substorms — is therefore by means of the AE, AL, and AU indices which gauge the strength of the electrojets.

auroral oval   Circular region several degrees wide around the geomagnetic pole at a geomagnetic latitude of about $\pm 70°$, its center shifted by about 200 km towards the nightside; the region in which aurora is observed at any instant, covering the region of the diffuse aurora, which is also where the discrete aurora can be seen. The *auroral oval* can be seen in satellite images in UV as a closed circle. From Earth, in visible light, in the auroral oval aurora can be seen nearly each night, during polar night for a full 24 hours. Shapes and structure of the aurora vary with local time: with a rather diffuse auroral brightening between local noon and midnight, quiet arcs during the evening hours up to around 21 local time, followed by homogeneous or rayed bands or draperies, which after about 3 local time, are complemented by patches at the southern rim of the auroral oval. These patches, together with short arcs, dominate the appearance of the aurora during the morning hours. The size of the auroral oval varies greatly; it grows during magnetic storms and may sometimes extend well beyond the region where aurora is ordinarily seen (auroral zone). At magnetically quiet times the oval shrinks and may assume a non-circular "horsecollar" shape, narrower near noon. Physically, the auroral oval is related to upward flowing Birkeland currents coupling the ionosphere and magnetosphere. *See* Birkeland current.

auroral zone   The region where auroras are ordinarily seen, centered at the magnetic pole and extending between magnetic latitudes 66° and 71°. The *auroral zone* is generally derived from ground observations of discrete aurora, but it also approximates the statistical average of the auroral oval, averaged over many nights.

autumnal equinox   The epoch at the end of Northern hemisphere summer on which the sun is located at the intersection of the celestial equator and the ecliptic; on this day, about September 21, the night and day are of equal length throughout the Earth. The date of *autumnal equinox* is the beginning of the Southern hemisphere spring. Autumnal equinox also refers to a direction of the celestial sphere: 12^h RA, 0° declination, antipodal to the direction of the vernal equinox. *See* vernal equinox. After autumnal equinox, in the Northern hemisphere, the period of daylight becomes shorter and the nights longer, until the winter solstice.

available potential energy   (Lorenz, 1955) The energy that could be obtained by some well-defined process. Such process is usually an adiabatic (or isentropic) redistribution of mass without phase changes to a statically stable state of rest. The estimate of mean available potential energy is about 11.1×10^5 J m^{-2} in the Earth atmosphere and is of order of 10^5 J m^{-2} in a typical mid-latitude ocean gyre.

avalanche   In Earth science, the sudden slumping of earth or snow down a steep slope.

average cosine   Mean cosine of radiance or scattering.

average matter-density   The mean amount of mass in a unit of volume of space. The relativity theory taken to the extreme would require that the distributions of matter density and of velocities of matter are specified down to the size of single stars, and then a cosmological model is obtained by solving Einstein's equations with such a detailed description of matter. This approach would be mathematically intractable; moreover, sufficiently precise observational data are not available except for a small neighborhood of the solar system in our galaxy. Hence, for the purposes of cosmology, average values of physical quantities over large volumes of space must be given. *Average matter density* $\bar{\rho}$ must also include the rest mass equivalent to radiation. In cosmology, the averaging volume is taken to be of the size of several galaxies at least, possibly of several clusters of galaxies. If the universe, represented in this way, is spatially homogeneous (*see* homogeneity), then $\bar{\rho}$ does not depend on which volume is used to evaluate it and so it is well defined at least in the mathematical sense. If the universe is inhomogeneous, then the value of $\bar{\rho}$ depends on the averaging volume, and choosing the right volume becomes a problem that has not yet been solved in a general way.

averaging   The mathematical procedure of calculating an average value of a given quantity.

In cosmology, average values of various quantities with respect to the volume of space are used in order to avoid introducing too detailed mathematical models of the real universe — they would be too difficult to handle. *Averaging* is straightforward only for scalars (such as matter-density, pressure, or rate of volume expansion; *see* average matter-density). For vectors (such as the velocity of matter-flow) and tensors (*see* tidal forces for an example of a tensor) this simple procedure does not work; for example, the sum of two vectors attached to different points of a curved space does not transform like a vector under a change of the coordinate system. In particular cases, a suitable concept of averaging of such objects can be found by careful consideration of the physical processes being described.

Avogadro's number   The number of atoms or molecules in an amount of substance whose total mass, when expressed in grams, equals its atomic mass: $N_A = N/n = 6.02214199(47) \times 10^{23}$ molecules/gm-mole, a fundamental constant of nature. N is the total number of molecules and n is the number of gram-moles. Named after Amadeo Avogadro (1776–1856).

away polarity   One of two possible polarities of the interplanetary magnetic field, corresponding to magnetic field lines which, at the points where they are anchored in the sun, point away from it. In interplanetary magnetic sectors with *away polarity,* magnetic field lines linked to the northern polar cap of the Earth come from the sun and contain polar rain, whereas those linked to the southern polar cap extend into the outer solar system and contain none.

AXAF   Acronym of Advanced X-ray Astrophysics facility, a space-borne astronomical observatory launched in July 1999, devoted to the observation of soft and medium energy X-rays, and renamed "Chandra" to honor Subrahmanyan Chandrasekhar. Imaging resolution is 0.5 to 1 sec of arc (comparable to that of ground-based telescopes without adaptive optics), over the photon energy range of 0.2 to 10 keV. The field of view is 31 x 31 square arcmin. Two grating spectrometers yield a maximum spectral resolving power (E/ΔE) $\sim$ 1000 over the energy range from 0.09 to 10 KeV. Chandra provides an order of magnitude improvement in resolution and two orders of magnitude improvement in sensitivity over the imaging performances of the Einstein observatory (HEAO-2). The improvement in spectral resolving power is also very significant: for comparison, the spectrometers on board the Japanese X-ray observatory ASCA, operating since 1993, had maximum energy resolving power $E/\Delta E \approx 50$ between 0.5 and 12 KeV. Chandra is expected to detect supernova remnants in M31, to resolve single galaxies in the Virgo Cluster, and distant quasars that may contribute to the diffuse X-ray background. The Chandra spectrometers are, in principle, able to resolve emission lines and absorption edges from hot plasmas, such as the intra-cluster medium in clusters of galaxies, making feasible a study of their physical properties and of their chemical composition, and to resolve the profile of the prominent iron K lines, which, in active galactic nuclei, are thought to be produced in the innermost regions of an accretion disk.

axial dipole principle   A fundamental principle established in paleomagnetic studies, which states that the axis of the geomagnetic dipole nearly coincides with Earth's rotational axis at all geological times. The principle makes it possible to use paleomagnetic data to constrain the position of continents in the geological past relative to Earth's rotational axis.

axionic string   Axions are scalar-like fields which have been proposed to solve the strong CP-problem in QCD. They are present in many grand unified models and also in superstring theories. Axions behave essentially as the phase of a scalar field and would be expected to have very small values in vacuum in order to solve the CP-problem. However, in some instances, such as phase transitions, they play the role of the phase of a Higgs field (although there is no Higgs field in most models), and thus could be forced to undergo a variation by an amount of 2π, just like the phase of an ordinary Higgs field would, thereby being responsible for the appearance of *axionic* cosmic strings. During the evolution of a network of these strings, they would radiate some energy in the form of axion particles, whose remnant density is calculable given a specific model. This is one means of constraining the actual mass of the axion particle. *See* cosmic string, CP problem, global topological defect.

azimuth   The *azimuth* of a line is the angle from a vertical plane passing through North to that line, measured positive eastwards. Thus, the points North, East, South, and West on the compass are, in turn, at $0°$, $90°$, $180°$, and $270°$ azimuth. Alternatively, it is possible to use the range from $-180°$ to $+180°$, in which case West is $-90°$. Azimuth is part of the topocentric system of coordinates. *See also* altitude. In critical applications, it is necessary to distinguish *true North* (as defined by the Earth's rotation axis) from *magnetic North*. Azimuth can also be defined relative to another marker besides North, such as the direction of motion of an aircraft. In that case, for example, $0°$ is "dead ahead" and "to the right" is $90°$. In this extended usage, one does *not* refer to the topocentric system.

B

Baade–Wesselink method   A method of determining the distance to pulsating stars, which can also be observed spectroscopically. One estimates the surface temperature and hence the surface brightness from the color index ($B - V$ color), based on simultaneous measurements of the blue and visual magnitude at bright and dim epochs in the star's pulsation. One computes the square root of the ratio of the observed flux to the surface brightness for each of these epochs. This is an estimate of the angular diameter of the star. Spectroscopy yields the surface velocity of the star (via blue and red shifting of spectral lines), which leads to a determination of the total difference in radius of the star between the observations. The combination of these observations allows a determination of the physical size of the star, and thus of its absolute magnitude and distance. The method has also been applied to the expanding envelope of type II supernovae.

Babinet point   One of three points on the sky in a vertical line through the sun at which the polarization of skylight vanishes. Usually located at about 20° above the sun. *See* Arago point, Brewster point.

baby universe   A theory regarding matter that falls into a black hole that subsequently evaporates. It states that this matter may go into a separate space-time, which could detach from the universe at one location and reattach elsewhere. This may not be particularly useful for space travel, but the existence of *baby universes* introduces a randomness so that even a complete unified theory would be able to predict much less than expected. However, averaging over ensembles of universes containing baby universes may lead to predictions via expected values of certain measured quantities, such as the cosmological constant.

backarc spreading   In some subduction zones (ocean trenches) an area of sea-floor spreading occurs behind the subduction zone, thus creating new oceanic crust. The Sea of Japan is an example. The foundering oceanic plate pulls away from the adjacent continental margin, and *backarc spreading* fills the gap.

back scattering   Scattering through angles greater than 90°.

backscattering coefficient   The integral over the hemisphere of backward directions of the volume scattering function.

backscattering fraction   The ratio of the backscattering coefficient to the scattering coefficient.

backshore   The relatively flat portion of a beach profile which lies between the steeper beachface and the dunes, cliffs, or structures behind the beach. Subaerial during non-storm conditions.

Backus effect   A particular type of nonuniqueness that can occur in the inversion of geomagnetic data on a spherical surface, where instead of knowing the full vector magnetic field only the magnitude of the field B is known and not its orientation. Historically, early satellite measurements from platforms such as the POGOs are of this form, as it was difficult to obtain high quality measurements of satellite orientation. The effect occurs even if there is perfect knowledge of B on a spherical surface but can be alleviated by knowledge of B in a shell or by knowledge of the position of the magnetic equator. The source of the error is the existence of magnetic fields that are perpendicular to Earth-like fields everywhere on a spherical surface, and which, therefore, can be added or subtracted from the Earth's field without changing B. As the Earth's field is predominantly axial dipolar, the error terms associated by the Backus effect have strong sectoral variation in a band around the equator. *See* nonuniqueness.

Baily's beads   A phenomenon appearing at the onset and at the conclusion of a solar eclipse, in which the photosphere is almost totally eclipsed, except for a few locations on the limb of the moon, which allow view of the photosphere through lunar valleys. If only one bead is apparent, a "diamond ring effect" is produced.

Ballerina model   Shape of the heliospheric current sheet as proposed by H. Alfvén in the early 1970s. The wavy neutral line on the source surface is carried outwards by the solar wind, resulting in a wavy heliospheric current sheet that resembles the skirt of a dancing ballerina. The waviness of this current sheet is described by the tilt angle. *See* heliospheric current sheet, source surface, tilt angle.

Balmer series   The series of lines in the spectrum of the hydrogen atom which corresponds to transitions between the state with principal quantum number $n = 2$ and successive higher states. The wavelengths are given by $1/\lambda = R_H(1/4 - 1/n^2)$, where $n = 3, 4, \ldots$ and R_H is the Rydberg constant for hydrogen. The first member of the series ($n = 2 \leftrightarrow 3$), which is often called the H_α line, falls at a wavelength of 6563 Å. *See* Rydberg constant.

banner cloud   An altocumulus lenticularis cloud, (lenticular cloud) which forms on the lee side of the top of a mountain and remains a stationary feature with one edge attached to the ridge of the mountain.

bar   A unit of pressure, defined as 10^6 dyne/cm^2 (10^5 N/m^2). The bar is commonly divided into 1000 mb. The pressure of atmosphere at sea level is about 1013 hPa $= 1.013$ bar.

bar detectors   Solid bars, made of metallic alloys, for detecting gravitational waves. The technology was pioneered by J. Weber. The sensitivity of $h = 10^{-18}$ (relative strain) has been reached in second-generation cryogenic detectors. At this level of sensitivity, no gravitational wave event has yet been reliably detected. *See* LIGO.

barium release   An experimental procedure near-Earth space physics, in which barium is evaporated by a thermite charge, usually in a sunlit region above the (denser) atmosphere, creating a greenish cloud. Barium atoms are rapidly ionized (within 10 sec or so), and the ions form a purple cloud, which responds to

magnetic and electric fields and therefore often drifts away from the neutral one.

The technique was pioneered by Gerhard Haerendel in Germany, and many releases have been conducted from rockets above the atmosphere, often with the barium squirted out by a shaped explosive charge. Typically releases are made after sunset or before sunrise, so that the rocket rises into sunlight but the clouds are seen against a dark sky. Effects of electric fields have been observed, striations like those of the aurora, even abrupt accelerations along field lines. Some releases have taken place in more distant space, notably an artificial comet in the solar wind, produced in 1984 by the AMPTE mission.

Barnard's star   Star of spectral type M3.8, 5.9ly distant with magnitude m $= 9.5$ and absolute magnitude M $= 13.2$; located at Right Ascension 17^{h}58^m, declination +04°41′. Discovered in 1916 by E.E. Bernard, it has the largest known proper motion: 10.29 arcsec/year. Measurements by van de Kamp had suggested that there were perturbations of the motion corresponding to an associated planet, but recent observations using the Hubble space telescope have not confirmed this claim.

baroclinic atmosphere or ocean   An atmosphere or ocean in which the density depends on both the temperature and the pressure. In a baroclinic atmosphere or ocean, the geostrophic wind or current generally has vertical shear, and this shear is related to the horizontal temperature or density gradient by the thermal wind relation.

baroclinic instability   A wave instability that is associated with vertical shear of the mean flow and that grows by converting potential energy associated with the mean horizontal temperature or density gradient.

baroclinic wave   Wave in the baroclinic flow.

barotropic atmosphere or ocean   An atmosphere (or ocean) in which the density depends only on the pressure. In the *barotropic atmosphere or ocean,* the geostrophic wind or current is independent of height.

barotropic instability A wave instability associated with the horizontal shear in a jet-like current and that grows by extracting kinetic energy from the mean flow field.

barotropy In fluid mechanics, the situation in which there is no vertical motion, and the gradients of the density and pressure field are proportional, and the vorticity (as measured in an inertial frame) is conserved.

barred galaxies Disk galaxies showing a prominent, elongated feature, often streaked by absorption lanes due to interstellar dust. Prominent bars are observed in about $\frac{1}{3}$ of disk galaxies; approximately $\frac{2}{3}$ of galaxies do, however, show some bar-like feature. A bar can contribute to a substantial part, up to $\frac{1}{3}$, of the total luminosity of a galaxy. The bar photometric profile is quite different from the photometric profile of galaxies: the surface brightness along the bar major axis is nearly constant but decreases rapidly along the minor axis. The bar occupies the inner part of the galaxy rotation curve where the angular speed is constant; bars are therefore supposed to be rotating end over end, like rigid bodies.

barrier island An elongated island separated from a coast by a shallow bay or lagoon. Generally much longer in the longshore direction than cross-shore direction. The Outer Banks of North Carolina and much of the east coast of Florida provide good examples.

barriers When an earthquake is caused by a rupture on a fault, inversions of seismic waves indicate some portions of the fault do not rupture; these are barriers.

Barycentric Coordinate Time (TCB) Barycentric Dynamical Time has been deemed by the International Astronomical Union (IAU) to be an inferior measure of time in one sense: Its progress depends on the mass of the sun and the mean radius and speed of the Earth's motion around the sun, and, to a smaller extent, on the mean gravitational perturbations of the planets. Therefore, in 1991 the IAU established a time standard representing what an SI clock would measure in a coordinate system, such that the barycenter of the solar system was stationary in this nearly inertial system, as was the clock, but the clock was so far removed from the sun and planets that it suffered no gravitational effect. That time is $TCB = TDB + L_B \cdot (JD - 2443144.5) \cdot 86400$ sec, where $L_B = 1.550505 \cdot 10^{-8}$ by definition as of mid-1999, and JD stands for the Julian Date in TDT. Presumably, the "constant", b, is subject to revision when and if the mass content of the solar system or the properties of the Earth's orbit are redetermined. *See* Barycentric Dynamical Time.

Barycentric Dynamical Time (TDB) In 1977, *Dynamical Time* was introduced as two forms, *Terrestrial Dynamical Time* (TDT) (q.v.) and TDB, on the basis of a 1976 IAU resolution. The difference between these two consists of periodic terms due to general relativity. TDB is commonly used for the determination of the orbits of the planets and their satellites, except those of the Earth. It is particularly suited to this purpose because it is adjusted from TDT in such a way as to represent what a clock on the geoid would measure if the Earth orbited the sun in a circular orbit of radius 1 *astronomical unit* (q.v.), while TDT contains relativistic effects of the eccentricity of the Earth's orbit. *See also* Barycentric Coordinate Time (TCB). Approximately, $TDB = TDT + 0.001652825 \cos(g)s$ where g is the mean anomaly (q.v.) of the Earth in its orbit. *See* Ephemeris Time, dynamical time.

baryogenesis Period of the early evolution of the universe when baryons were generated from quarks. Observationally, the universe is made of normal matter, containing baryons and leptons with no observational evidence of significant amounts of antimatter anywhere. Similarly to the successful predictions of *nucleosynthesis,* the unfulfilled goal is to build a scenario where, starting from a baryon symmetric state, quark and lepton interactions lead to an excess of matter over antimatter as the photon temperature drops. It will suffice to produce an excess of 1 baryon every 10^9 antibaryons to give rise to a universe made of normal matter and a baryon-to-photon ratio $\eta \sim 10^{-9}$, as observed.

In 1967 Sakharov identified the three necessary ingredients to dynamically evolve a baryon asymmetry: (i) Baryon Number Violation. If a baryon number is conserved in all interactions, the absence of antimatter indicates asymmetric initial conditions. (ii) C and CP Violation. If not, B-violating interactions will produce excesses of baryons and antibaryons at the same rate, thereby maintaining zero baryon number. (iii) Non-thermal equilibrium, otherwise the phase space density of baryons and antibaryons are identical. So far, a successful model has not been made because it requires physics beyond the standard model of particle interactions.

basalt   Volcanic mafic igneous rocks containing minerals such as pyroxene and olivine. The most common volcanic rock. Produced by about a 20% melting of the mantle.

basaltic lava   A form of molten rock that emerges in volcanic outflows at a temperature of 1000°C to 1250°C. Examples include eruptions in the Hawaiian chain.

basement   In many parts of the continents the surface rocks are sediments. Sediments are products of erosion. These sediments lie on top of either metamorphic or igneous rocks. The boundary is termed *basement*. The depth to basement can range from a few meters to a maximum of about 20 km.

basic MUF   Defined as the highest frequency by which a radiowave can propagate between given terminals on a specified occasion, by ionospheric refraction alone, and may apply to a particular mode, e.g., the E-layer basic MUF. The basic MUF depends on the critical frequency (fc) of the ionosphere at the mid-point of the path and the angle of incidence (I) of the radiowave on the ionosphere, and to a good approximation MUF = fc sec(I). The factor, sec(I), is called the obliquity factor for the circuit because it relates the vertical incidence ionospheric information to the oblique incidence path. A further correction is required to allow for a curved Earth and ionosphere. For a given radiowave takeoff angle, the obliquity factor reduces as the reflecting layer height increases; thus it is greatest for the E and Es layers and least for the F2 layer. It is conventional to use a standard 3000 km obliquity factors, M(3000)F2, for the F region and convert this to other pathlengths. The M(3000)F2 can be measured directly from ionograms. *See* ionospheric radio propagation path, operational MUF.

Batchelor scale   Length scale at which turbulent concentration (or temperature) gradients in a fluid are damped out by molecular viscous effects; alternately, the length scale, at which the sharpening of the concentration (or temperature) gradients by the strain rate is balanced by the smoothing effect of molecular diffusion. This length depends on the kinematic viscosity ν, the molecular diffusion coefficient κ, and the dissipation rate of turbulent kinetic energy ϵ:

$$L_B = \left(\frac{\nu \kappa^2}{\epsilon} \right)^{1/4}$$

At scales smaller than L_B, scalar fluctuations disappear at a fast rate and subsequently the Batchelor spectrum drops sharply off. Most commonly the *Batchelor scales* are used for temperature ($\kappa = \kappa_T$) and salt ($\kappa = \kappa_S$), respectively. In oceanic turbulence studies, L_B is most commonly defined using κ_T, because temperature is easily measured. The Batchelor scales for temperature and salt (typical scales of mm and sub-mm in natural waters, respectively) are smaller than the *Kolmogorov scale* L_K, since in water molecular diffusivities are much smaller than viscosity ν ($D_T/\nu \sim 100$ and $D_S/\nu \sim 1000$ in natural waters). *See also* Kolmogorov scale.

Batchelor spectrum   Under isotropic and stationary conditions, the power spectrum of the one-dimensional temperature fluctuations follows

$$\phi_T(k_z) = \frac{\chi \kappa_T^{1/2} q^{3/2} \nu^{3/4}}{2 \varepsilon^{3/4}}$$
$$\left(\frac{\exp\left(-x^2\right)}{x} - \sqrt{\pi}(1 - erf(x)) \right)$$
$$\left[K^2 (\text{rad/m})^{-1} \right]$$

where k_z denotes the one-dimensional wavenumber [rad m^{-1}], $x = k_z q''^2 L_B = k_z \kappa_T^{1/2}$

$v^{1/4}q^{1/2} \epsilon^{-1/4}$ the nondimensional vertical wavenumber, v the *kinematic viscosity* of water, χ_T the *dissipation rate of temperature variance,* and ϵ the *dissipation rate of turbulent kinetic energy;* L_B is the Batchelor scale (Batchelor, 1959). The constancy of the turbulence parameter q, usually taken as $q = 3.4$ (Dillon and Caldwell, 1980), is subject to debate. The *Batchelor spectrum* of salt follows the same function with the molecular diffusivity κ_S of salt (instead of κ_T) and the *dissipation rate of salt variance* χ_S (instead of χ_T). For small k, the spectrum $\phi_T(k_z)$ drops as k^{-1} (*viscous-convective subrange*), and for k larger than the *Batchelor wavenumber* k_B, $\phi_T(k_z)$ rolls off at a much steeper rate (*viscous-diffusive subrange*). These two subranges are commonly referred to as Batchelor spectrum.

Batchelor wavenumber   The inverse of the Batchelor scale, expressing the wavenumber k_B, above which scalar fluctuations (such as those of temperature or salt) become eradicated by the smoothing effect of molecular diffusion (i.e., Batchelor spectrum drops off sharply above k_B). Two definitions are common: $k_B = (\varepsilon/v\kappa^2)^{1/4}$ [rad m^{-1}], or $k_B = (2\pi)^{-1}(\varepsilon/v\kappa^2)^{1/4}$ [cycles m^{-1}]. κ is the molecular diffusivity of the scalar ($\kappa = \kappa_T$ for temperature or $\kappa = \kappa_S$ for salt).

batholith   Large igneous body where molten rocks have solidified at depth in the ocean. A typical *batholith* has a thickness of a few kilometers and horizontal dimension of tens to hundreds of kilometers.

bathymetry   Measure of the depth of the sea floor below sea level.

b-boundary   (or bundle boundary) In relativity, assorted boundary points of the space-time manifold defined by equivalence classes of curves in the orthogonal frame bundle of the space-time (B. Schmidt). The goal of this work is to evade the difficulties of setting up a topology on a manifold with an indefinite metric by using the positive-definite measure on the frame bundle. Examples of cosmological space-times have been found which show that the notion of the *b-boundary* is not suitable for defining the

boundary of a space-time, due to topological problems.

beach cusps   Periodic (in space), transient, cuspate features which commonly appear in beach planforms. Wavelengths typically range from 1 to 50 meters, with amplitudes less than a wavelength.

beachface   The relatively planar portion of a beach in the vicinity of the still water line. Subject to swash by wave action; therefore, alternately wet and dry.

beach mining   The process of removing sand for some human activity, often industrial.

beach morphology   The study of the shape and form of a beach or coastal area.

beach nourishment   The process of placing sand on an eroding beach to advance the shoreline seaward. Sand is typically transported by hydraulic dredge via a pipeline, delivered by a split-hull barge, or transferred overland by truck. Typical project sand volumes are 10,000 to 10 million m^3.

beach profile   A cross-shore slice through a beach, illustrating how bathymetry varies in the shore-normal direction.

beach ridge   A linear ridge of sand that appears behind the modern beach and indicates a prograding beach. Multiple *beach ridges* are often visible. Referred to as "Cheniers" in Texas and Louisiana.

beam attenuation coefficient   The limit of the ratio of the spectral absorptance plus spectral scatterance to the distance of photon travel as that distance becomes vanishingly small [m^{-1}]; equal to the sum of the absorption and scattering coefficients.

beams, ion   Beam-like upwards flow of ions, usually O^+ (singly ionized oxygen), observed above the atmosphere during polar auroras. First seen in 1976 by the S3-3 satellite, ion beams are believed to be accelerated by the same parallel electric field that accelerates auroral electrons downwards, possibly a quasi-neutral electric field associated with upwards Birkeland currents. *See* conics.

beam spread function   The irradiance distribution on the inner surface of a sphere as generated by an initially collimated, narrow beam and normalized to the beam power $[m^{-2}]$; numerically equals the point spread function.

beam transmissometer   An instrument for measuring the fraction of a collimated beam lost by absorption and scattering per unit distance of photon travel; measures the beam attenuation coefficient.

Beaufort wind scale   A descriptive table of wind speeds developed by Admiral Sir Francis Beaufort in 1806. For practical use, in particular at sea, wind speeds are described by the states of the sea surface and the wave height H. The original scale considers 12 grades from calm to hurricane; in 1956, grade 12 was divided into grades from 12 to 17. The wind velocity is related to the Beaufort Force approximated by $v = B^{1.5}$, where v is the wind speed measured at 10 m height and B is the Beaufort Force. See tables on pages 39 and 40.

bed load   A term used in the study of sediment transport by moving fluids. *Bed load* denotes the fraction of sediment transported very close to the bed, such as the particles that bounce along the bed. The remainder of the sediment is suspended in the water column and is referred to as suspended load and wash load.

Beer's law   Radiation traveling in a certain direction in a scattering or absorbing medium is exponentially attenuated.

Belinda   Moon of Uranus also designated UXIV. Discovered by Voyager 2 in 1986, it is a small, irregular body, approximately 34 km in radius. Its orbit has an eccentricity of 0, an inclination of $0°$, a precession of $129°$ yr^{-1}, and a semimajor axis of 7.53×10^4 km. Its surface is very dark, with a geometric albedo of less than 0.1. Its mass has not been measured. It orbits Uranus once every 0.624 Earth days.

Benard cell   When a gas or plasma is heated uniformly from below, convection takes place in vertical cells, the *Benard cells*. The motion is upward in the center of the cell and downward at the cell boundaries. In laboratory experiments, the pattern of convection cells is regular and long-lived. Natural Benard cells can be observed in the terrestrial atmosphere under calm conditions or as granulation at the top of the solar convection zone.

Benioff zone   In a subduction zone the oceanic lithosphere descends into the mantle at velocities of about 10 cm per year. The upper boundary of the descending plate is a fault zone between the plate and overlying crustal and mantle wedge. The earthquakes on this fault define the *Benioff zone*. The zone has a typical dip of $30°$ to $45°$ and extends to depths of 670 km. The existence of these dipping zones of seismicity was one of the major arguments for plate tectonics.

benthic   The portion of the marine environment inhabited by marine organisms that live in or on the bottom of the ocean.

benthos   Bottom-dwelling marine organisms.

Bergen school   Meteorology. A school of analysis founded in 1918 by the Norwegian physicist Vihelm Bjerknes (1862–1951), his son Jacob Bjerknes (1897–1975), Halvor Solberg (1895–), and Tor Bergeron (1891–). V. Bjerknes began his career as a physicist. In the late 1890s, he turned his attention to the dynamics of atmosphere and oceans. The "circulation theorems" he developed during this period provided a theoretical basis for the basic concepts in the general circulation. During World War I, Bjerknes, as founding director of the Geophysical Institute at Bergen, was successful in convincing the Norwegian government to install a dense network of surface stations which provided data for investigating the surface wind field. These studies led to the concept of fronts and ultimately to models of the life cycle of frontal cyclones. In 1919, J. Bjerknes introduced the concept of warm, cold, and occluded fronts, and correctly explained their relationship to extratropical cyclones. By 1926, in collaboration with Solberg

Classic Beaufort Scale

B°	Name	Wind speed [knots]	Wind speed [m/s]	Sea surface characteristics	H [m]
0	Calm	< 1	0.0–0.2	Sea like a mirror	0
1	Light air	2	0.9	Ripples	0.1–0.2
2	Light breeze	5	2.5	Small wavelets	0.3–0.5
3	Gentle breeze	9	4.4	Large wavelets, crests begin to break	0.6–1.0
4	Moderate breeze	13	6.7	Small waves, white horses	1.5
5	Fresh breeze	18	9.3	Moderate waves, becoming longer	2.0
6	Strong breeze	24	12.3	White foam crests	3.5
7	Moderate gale	30	15.5	Sea heaps up; white foam from breaking waves blown in streaks	5.0
8	Fresh gale	37	19	Moderately high waves, greater length; edges of crests break into spindrift	
9	Strong gale	44	22.6	High waves, sea begins to roll	9.5
10	Whole gale	52	26.5	Sea-surface white from great patches of foam blown in dense streaks	12
11	Storm	60	30.6	Exceptionally high waves; sea covered by patches of foam; visibility reduced	15
12	Hurricane	68	34.8	Air filled with foam and spray; visibility greatly reduced	

and others, he had described the structure and life cycle of extratropical cyclones. Bergeron also made important contributions to the understanding of occluded fronts and the formation of precipitation.

Bergeron, Tor   (1891–) *See* Bergen school.

berm   A nearly horizontal ledge, ridge, or shelf of sediment which lies behind the beach face at a beach. More than one *berm* will sometimes be evident within the backshore region.

Bernoulli equation   For steady flow of a frictionless, incompressible fluid along a smooth line of flow known as a streamline, the total mechanical energy per unit weight is a constant that is the sum of the velocity head ($u^2/2g$), the elevation head (z), and the pressure head ($p/\rho g$): $u^2/2g + z + p/\rho g =$ constant, where u is the velocity, g is the acceleration of gravity, z is the elevation above some arbitrary datum, p is the fluid pressure, and ρ is the fluid density. The *Bernoulli equation* expresses all terms in the units of energy per unit weight or joules/newton, which reduces to meters. The Bernoulli equation therefore has the advantage of having all units in dimensions of length. The sum of these three factors is the hydraulic head h.

Besselian year   *See* year.

beta decay   Nuclear transition mediated by the weak force, in which the nuclear charge changes by one, either $Z \rightarrow Z+1$, $A \rightarrow A$ with the emission of an electron plus an antineutrino; or $Z \rightarrow Z-1$, $A \rightarrow A$ with the emission of a positron plus a neutrino. The paradigm is the free decay of the neutron:
$n \rightarrow p + e^- + \bar{\nu}$.

beta-effect   (β-effect) A combined effect of the rotation and curvature of the Earth which tends to produce ocean currents on the western boundaries of basins with speeds exceeding

Beaufort Wind Scale Limits of Wind Speed at 10 m

Force	Knots	m/sec	km/hr	mi/hr	Description of wind
0	<1	0-0.2	<1	<1	Calm
1	1-3	0.3-1.5	1-5	1-3	Light air
2	4-6	1.6-3.3	6-11	4-7	Light breeze
3	7-10	3.4-5.4	12-19	8-12	Gentle breeze
4	11-16	5.5-7.9	20-28	13-18	Moderate breeze
5	17-21	8.0-10.7	29-38	19-24	Fresh breeze
6	22-27	10.8-13.8	39-49	25-31	Strong breeze
7	28-33	13.9-17.1	50-61	32-38	Moderate gale
8	34-40	17.2-20.7	62-74	39-46	Fresh gale
9	41-47	20.8-24.4	75-88	47-54	Strong gale
10	48-55	24.5-28.4	89-102	55-63	Whole gale
11	56-63	28.5-32.6	103-117	64-72	Storm
12	64-71	32.7-36.9	118-133	73-82	Hurricane
13	72-80	37.0-41.4	134-149	83-92	Hurricane
14	81-89	41.5-46.1	150-166	93-103	Hurricane
15	90-99	46.2-50.9	167-183	104-114	Hurricane
16	100-108	51.0-56.0	184-201	115-125	Hurricane
17	109-118	56.1-61.2	202-220	126-136	Hurricane

those in the rest of the ocean basin. *See* beta-plane approximation.

Beta Lyrae systems   Binary stars in which the more massive star has recently filled its Roche Lobe so that material has begun flowing down onto the companion star (in a stream or accretion disk). This occurs very rapidly (on the Kelvin–Helmholtz time scale) until the ratio of masses has been reversed. The companion is often completely hidden by the disk, so that one sees eclipses of the mass donor by something that itself emits very little light. It is not absolutely certain that Beta Lyrae itself is actually a *Beta Lyrae system. See* Roche lobe.

beta-plane approximation   (Rossby et al., 1939) The effects of the Earth's sphericity are retained in the Cartesian metric by approximating the Coriolis parameter, f, with the linear function of y, which is the latidinal coordinate and is measured positive northward from the reference latitude. This approximation is called the *β-plane approximation.*

Bianca   Moon of Uranus also designated UVIII. Discovered by Voyager 2 in 1986, it is a small, irregular body, approximately 22 km in radius. Its orbit has an eccentricity of 0.001, an inclination of $0°$, a precession of $299°$ yr^{-1}, and

a semimajor axis of 5.92×10^4 km. Its surface is very dark, with a geometric albedo of less than 0.1. Its mass has not been measured. It orbits Uranus once every 0.435 Earth days.

Bianchi classification   A classification of three-parametric symmetries into inequivalent classes.

For three-dimensional spaces, the number of distinct classes is finite, and the complete list of all classes is called the *Bianchi classification* (after the Italian mathematician Luigi Bianchi). The total number of Bianchi's classes is 9, although some of the classes are themselves collections labeled by free parameters. This classification is useful in constructing mathematical models of the universe.

biased vacuum states (domain wall)   Domain walls are formed whenever a discrete symmetry is broken, and their existence is a consequence of the presence of degenerate potential minima in the broken phase. Domain walls in the universe would importantly alter standard cosmology. They are, however, produced in some particle physics theories, which are therefore highly constrained.

It is conceivable that the Higgs potential may be modified somehow, so that the above mentioned equivalent minima acquire different values. It would be possible then, with this *biased* potential, to have walls decaying sufficiently rapidly so that the cosmological problem they generate is cured. In fact, invoking quantum effects between the fields that form the walls and other particles in some models, one finds that the potential is slightly tilted (see figure) in such a way that the vacuum, defined as the minimum energy state, instead of being degenerate, becomes single-valued. This effect takes place after the phase transition has been completed so that domain walls actually form and subsequently decay. *See* cosmic topological defect, domain wall.

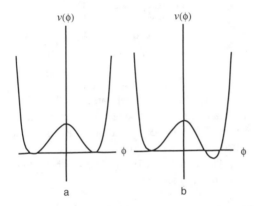

Possible self-interaction potentials: (a) the original potential, (b) the biased potential once the radiative corrections have been accounted for.

bias frame, bias overscan   CCD detectors contain a pedestal or "bias" level of counts. This is a DC offset that is added to the output signal of the CCD. Sometimes the CCD is "preflashed" to establish a bias before an observation is made. The function of this bias is to raise the number of counts for the data above some level where the detection of data is more accurate. The value of the bias can vary from pixel to pixel on a CCD, as well as vary over time or temperature. When processing observing data, one of the first steps is to remove the bias. This is a two-step process. The first step is to account for any bias changes that may vary on a frame-to-frame basis. This is usually accomplished by measuring the bias on a part of the CCD that is not exposed during the observation (this is called the "overscan re-

gion"), and subtracting this bias as a single value from the frame or fitting a function to the bias and subtracting it along columns and/or rows. The second step is to account for the pixel-to-pixel variations in the bias. This is often done by obtaining a number of bias frames or images that are essentially images with zero integration times with the shutter closed. Such frames contain only the bias, which is assumed to be representative of the bias on each data image. The average bias image is directly subtracted from each data image (including flat fields and other calibration images).

biasing parameter   Ratio between the amplitude of galaxy-galaxy and cluster-cluster correlation function. By extension, ratio between the amplitude in the matter power spectrum and those of galaxies and clusters. At each scale: $b^2(k) = P_{gal}(k)/P_{matter}$. If galaxies and clusters of galaxies form in high density environments, then galaxies are more correlated than the underline matter distribution (and clusters even more strongly than galaxies). It is usually assumed that the bias parameter is independent of scale. The current estimated value is: $b = 1.3 \pm 0.13$.

bidirectional reflection distribution function (BRDF)   The ratio of the radiance reflected by a surface into a specific localized direction to the collimated irradiance incident from a particular direction onto the surface [sr^{-1}].

Big Bang   The initial explosion that gave birth to the universe, and a standard model of the universe in which all matter, space, and time expands from an initial state of enormous density and pressure. All models of the universe constructed in the classical relativity theory (General Relativity) must take into account the fact that at present the universe is expanding (*see* expansion), i.e., that any two clusters of galaxies recede from one another. If the matter in the universe is either a dust or a perfect fluid, then Einstein's equations imply that at a certain moment in the past any given small portion of matter must have been squeezed to zero volume. This moment of infinite compression is called the *Big Bang*. It is a formal mathematical conclusion that only implies that every region of the universe must have been much denser and hotter in the past than it is now (and observations confirm this implication). The infinite compression implied by the models shows that classical General Relativity does not apply to such states of matter and must be replaced by a suitable generalization (nonexistent so far). Quantum gravity is a likely candidate to produce this generalization. The Robertson–Walker models imply that the Big Bang occurred simultaneously for all matter in the universe, but more general models exist in which there is a nonsimultaneous Big Bang. *See* inhomogeneous models, singularities.

Big Bang cosmology   Cosmological model that assumes that the whole observable universe has expanded from an earlier state of much higher density. It was originally proposed by George Gamow, Ralph Alpher, and Robert Hermann in the late 1940s long before the observations of cosmic microwave background radiation and abundances of light elements suggested the existence of very high densities and temperatures in the past.

The model is based on the cosmological principle. The dynamics are described by Einstein's general relativity theory. Contrary to the steady state model that requires new physics, the Big Bang model assumes the current laws of physics are locally valid everywhere. The dynamics are described by Friedmann models and only three parameters are necessary to specify the evolution of the universe: mean density, Hubble constant, and cosmological constant. The extrapolation of the atomic, nuclear, and particle physics to explain the behavior of the universe at early times has proven to be very successful: the present abundances of light elements and the existence of a cosmic microwave background radiation with a degree of anisotropy close to 10^{-5} have been confirmed by observations. The model has its own shortcomings. The existence of a particle horizon poses a difficulty as to how to explain the high degree of homogeneity of the observed universe. A solution to this problem has been proposed appealing to a period of rapid expansion. *See* inflation, Big Bang nucleosynthesis, Friedmann–Lemaître cosmological models, cosmic microwave background.

Big Bang flatness problem   Problem related to the need to fine-tune the initial value of the density parameter to allow the universe to survive 10^{17} s being close to flat. The present observational data indicate that the mean energy density ρ is, within one order of magnitude, close to the critical density ρ_c, so that the universe will (approximately) expand forever, and so that the time = constant hypersurfaces (the 3-spaces) are almost Euclidean. The success in predicting the abundances of light elements (*see* nucleosynthesis) strongly suggests that our understanding of the physical processes occurring at $t \sim 1$s is adequate. At that moment

$$\rho(1\sec) - \rho_c(1\sec) \sim 10^{-16} \, ,$$

and at earlier times the differences are even smaller. A larger difference would imply that our universe would either have recollapsed or would have become freely expanding, with negligible gravitational deceleration by now (*see* Friedmann models). The fact that it is in neither of these states indicates that either the universe is exactly flat or initial conditions tuned its initial value such that at 1s it differed from unity in the 16th decimal place.

Big Bang horizon problem   Problem related to the existence of particle horizons arising since fact that we can see, or affect, objects only as far away as light has traveled since the Big Bang. Causality poses a severe difficulty when explaining the large scale homogeneity and isotropy of the universe. In the standard Big-Bang model, homogeneity and isotropy are built-in (*see* cosmological principle). In this context, it is not surprising that, for example, the cosmic microwave background radiation from two opposite directions in the sky shows the same black body spectra. However, since those regions came within our horizon and have been visible only very recently, there has not been time since the Big Bang for light to travel from one to the other, so they have never been in casual contact with each other. Then how could they have acquired the same physical properties? If those regions had slightly different initial densities, the evolution of the universe will amplify those differences making the initial irregularities more evident. The horizon problem is therefore a statement about why the universe is so homogeneous. As a solution, a period of rapid expansion (*see* inflation) has been advocated that would have stretched the observed universe out of a horizon volume at early times.

Big Bang nucleosynthesis   The production of deuterium, ^{3}He and ^{4}He (the latter to about 25% mass fraction) in the first 500 to 1000 sec of the early universe. These light isotopes, plus measurable amounts of ^{7}Li and trace amounts of elements B, Be, are the result of non-equilibrium nuclear reactions as the universe cooled to about $10^8 K$. Heavier isotopes were produced in stellar nucleosynthesis.

big blue bump   A feature in the spectral energy distribution of active galactic nuclei, dominating the emission shortwards of 4000 Å and in the UV. The shape, luminosity, and spectral extension of the big blue bump are as yet uncertain since the big blue bump lies mostly in the unobserved far UV, with maximum emission probably right below the Lyman limit. Only the low and high energy tails of the big blue bump have actually been observed. The high energy end of the big blue bump appears to be for several objects in the soft X-ray domain, where a sharp rise toward lower energies, the so-called "soft X-ray excess" is observed. The current interpretation of the active galactic nuclei spectral energy distribution, still highly debated, ascribes this feature to thermal emission from a hot accretion disk surrounding a supermassive black hole. *See* active galactic nuclei, Lyman limit.

binaries, X-ray   *X-ray binaries* are close binaries in which one of the objects is either a black hole or neutron star, and the companion is either a star with a strong wind or a star overfilling its Roche-lobe. The material transfers from the companion to the compact object and forms an accretion disk. The gravitational energy released as the material accretes onto the disk powers the X-ray flux. Two classes of X-ray binaries exist: low-mass (neutron star) and high-mass (black hole) X-ray binaries. Low-mass X-ray binaries have companion star masses less than $5 M_\odot$, whereas high-mass X-ray binaries have more massive companions.

binary black holes   Two black holes orbiting each other, like stars in a binary system. A binary system of two black holes may radiate away significant orbital energy by emission of gravitational radiation, with efficiency that increases with the eccentricity of the orbit; orbital energy loss may lead to orbital decay, with the two black holes spiraling down toward each other and ultimately coalescing to form a single black hole. There is as yet no definitive evidence from observations of the actual occurrence of *binary black holes*. Binary black holes have been invoked to explain peculiar features of kpc-sized radio jets, of the photometric light curve of the blazar OJ 287, and of rarely observed peculiarities in the spectral line profiles of active galactic nuclei. Gravitational wave detectors may provide the first incontrovertible evidence of merging binary black holes. *See* black hole, black hole binary.

binary fraction   The fraction of stars in a stellar association which are binary stars. For many years it was believed that there were no binary stars among Population II stars. However, it now seems that there may be no difference in the relative number of binary stars between groups of stars of the same age, regardless of whether they are Population I or Population II.

binary pulsar   The pulsar PSR 1913+16, which is in orbit with another neutron star; both stars have a mass of approximately 1.4 solar masses. Orbital parameters are extracted from the almost periodic Doppler shifts of the pulses from the pulsar. The pulse repetition period is 59 milliseconds, and the orbital period is 2.79×10^4 sec. General relativistic effects, including periastron advance, are verified, and they, along with a general relativistic model of gravitational radiation from the orbiting neutron stars (period decrease of -2.4×10^{-12}), give excellent agreement with observations.

binary star   Two stars gravitationally bound together. There are numerous types of binaries including, among others, the following classes: Algol, RS Canus Venaticorum, W Ursa Majoris (*see* contact binary), β Lyrae, cataclysmic variables. Some binaries show eclipses as one star passes in front of the other. In such systems, the amount of light received as a function of time (*see* light-curve) varies. The binary appears brightest when neither star is eclipsed and is fainter when one star is eclipsed by the other.

binary star system A pair of stars in a gravitationally bound orbit around each other. Half or more of all stars in the sky are binaries, with the fraction varying from one cluster to another and one stellar population to another. The statistical properties of a binary population include the distribution of periods, total masses, mass ratios, and orbital eccentricities, and these also vary from one place to another (for reasons that must have to do with the conditions of *star formation* but which we cannot currently model). The range of orbit periods is less than an hour to millions of years; the range of separations from the sizes of the stars themselves is to at least 0.1 *parsec*; the range of masses is essentially the same as for single stars (*see* initial mass function); and the range of eccentricities is from 0 (circle) to at least 0.9, with short period systems generally in circular orbits.

Binary stars can be subdivided in several ways. Wide binaries are ones whose separation is larger than the maximum size either star will reach (as a red giant, super giant, or asymptotic giant branch star), so the stars evolve independently. Close binaries interact at some point in their lives, generally via gas flowing from the more rapidly evolving star (the more massive one) to its companion. They can also be subdivided by the method of detection, including visual binary, astrometric binary, eclipsing binary, spectroscopic binary, photometric binary, and spectrum binary systems.

Habitable planets are unlikely to be found in most binary systems because only orbits very close to one star or very far away from both will be stable. Most of our knowledge of the masses of stars comes from the analysis of binary systems. *See* initial mass function, parsec, star formation.

biogenic stratification In low saline water, biogeochemical processes, such as photosynthesis and subsequent settling of biogenic particles and mineralization by bacteria, may significantly change (generally enhance) vertical salinity gradients and thereby increase the *stability* N^2 of the water column. This phenomenon, common in many deep and eutrophic lakes (especially close to the sediment) is usually the cause of permanent density stratification (so-called meromixis). Ca^{2+} and HCO_3- are often the dominant ions for *biogenic stratification.*

bioluminescence Light produced by organisms as a result of conversion of chemical energy to radiant energy.

biomass The amount of living matter per unit of water surface or water volume (i.e., in the water column) [$kg\ m^{-2}$ or $kg\ m^{-3}$].

biosphere That part of the earth and its atmosphere that can support life. This extends from the upper atmosphere to underground into rock where living bacteria have been found. In more colloquial terms, it refers to the first hundred meters of the Earth's atmosphere, the oceans, and the soil down to bedrock, where easily recognizable organisms are found.

Biot–Savart law If the electromagnetic fields are time independent within a given region, then within the region it holds that the magnetic field for a given point in space is proportional to volume integral of the vectorial product of the current density times the unit vector in the direction of the relative position divided by the square of the relative position. In CGS units the constant of proportionality is the inverse of the speed of light. Named after Biot and Savart who in 1820 presented experimental evidence that established the law.

BIPM *See* Bureau International des Poids et Mesures.

bipolar flow The flow of material from a star in two streams in opposite directions. It is thought to be caused by the effects of stellar rotation on the mass outflow.

Birkeland current Field aligned electric current linking the Earth's ionosphere with more distant regions, often associated with the polar aurora and with substorms.

Birkhoff theorem (1923) The only spherically symmetric asymptotically flat vacuum space-time in general relativity with no cosmological term is the Schwarzschild space-time. Since the latter is static, this also rules out monopole gravitational waves. Further, the empty space-time inside a spherically symmetric source must be flat. Generalizations with the inclusion of a Maxwell field and cosmological constant have been given. *See* Schwarzschild solution.

Bjerknes *See* Bergen school.

Bjerknes circulation theorem　The rate of circulation change is due to either the baroclinicity or the change in the enclosed area projected in the equatorial plane:

$$\frac{DC}{Dt} = -\int \frac{dp}{\rho} - 2\Omega \frac{DA_e}{Dt}$$

where C is the relative circulation, p is the pressure, ρ is the density, Ω is the Earth's rotation rate and A_e is the enclosed area projection in the equatorial plane. D is the absolute derivative along the flow. In a barotropic fluid, the relative circulation for a closed chain of fluid particles will be changed if either the horizontal area enclosed by the loop changes or the latitude changes.

Bjerknes feedback　An ocean-atmospheric interaction mechanism first proposed by Jacob Bjerknes in 1969 to explain the El Nino/ Southern Oscillation phenomenon. Normally the easterly trade winds maintain a tilt of equatorial thermocline, shallow in the east and deep in the west. The equatorial upwelling induced by the trades brings cold upper thermocline water to the surface in the eastern equatorial Pacific. A slight relaxation in the trades weakens equatorial upwelling and depresses the thermocline in the east, both acting to warm the eastern Pacific. The warming in the east shifts the center of active atmospheric convection eastward and relaxes the easterly trades on the equator even more. This constitutes a positive feedback among the trade winds, thermocline depth, upwelling, and sea surface temperature.

black aurora　Name given to structured dark patches appearing on the background of bright aurora. Their origin is uncertain.

black-body radiation　The radiation from a hypothetical thermal radiating body with perfect emissivity. Practical black body sources consist of a heated cavity with a small exit aperture. Because radiation interacts repeatedly with the walls of the cavity before emerging, the emerging radiation is closely black body.

The spectral distribution of *black-body radiation* is given by Planck's formula:

$$\frac{d\omega}{d\lambda} = \frac{2\pi c^2 h}{\lambda^5} \left[\frac{1}{\exp^{hc/(\lambda kT)} - 1} \right]$$

(Joules per second per wavelength interval and per unit area of the emitter) in which $h = 6.62608 \times 10^{-27}$ erg sec is Planck's constant and $k = 1.3807 \times 10^{-16}$ erg/K is Boltzmann's constant; this was the first understood instance of a quantum phenomenon. At long wavelengths the spectral distribution is approximately

$$\frac{d\omega}{d\lambda} \sim 2\pi c \frac{kT}{\lambda^4} \, ,$$

which corresponds to earlier classical descriptions by Wien and others. The peak of the distribution obeys:

$$\lambda_{\text{Planck}} T = .2898 \text{ cm K} \, .$$

This is Wien's law.

blackbody temperature　The temperature at which the radiation distribution from an object can be characterized by Planck's blackbody equation. The distribution of radiation from most hot, compact astronomical objects is closely approximated by a *blackbody temperature,* yet not exactly. The energy spectrum of the sun, for example, has an energy distribution that is closely but not exactly described by a blackbody having a temperature $T_B = 6300$ K. The effective temperature of the sun, which takes into account the sun's surface area and total output power, is $T_{eff} = 5800$ K. *See* effective temperature, excitation temperature, color temperature.

black frost　Temperatures falling below freezing in air dry enough that white hoar frost does not form. Or, the blackening of vegetation due to water freezing within and disrupting their cells. *See* hoarfrost.

black hole A region of spacetime from which the escape velocity exceeds the velocity of light. In Newtonian gravity the escape velocity from the gravitational pull of a spherical star of mass M and radius R is

$$v_{esc} = \sqrt{\frac{2GM}{R}},$$

where G is Newton's constant. Adding mass to the star (increasing M), or compressing the star (reducing R) increases v_{esc}. When the escape velocity exceeds the speed of light c, even light cannot escape, and the star becomes a *black hole*. The required radius R_{BH} follows from setting v_{esc} equal to c:

$$R_{BH} = \frac{2GM}{c^2}.$$

For a solar mass black hole $M \sim 2 \times 10^{33}$ gm, and $R_{BH} \sim 3$ km. (An equivalent conclusion was first derived by P.S. Laplace in the 18th century, even though the notion of a black hole originated in relativity theory in the 1960s.) The so-called gravitational radius (or horizon radius) R_{BH} for the Earth is equal to about 0.88 cm. Thus, the Earth and the sun, squeezed to form black holes, would be extremely dense; such densities are not met anywhere in the real world. However, if $\bar{\rho}$ denotes the mean mass-density inside an object, then $M = \frac{1}{3}\pi\bar{\rho}r^3$, i.e., $r \propto M^{1/3}$ and $R_{BH} \propto M$, which means that with increasing mass R_{BH} grows much faster than r. The radius would be one astronomical unit for a black hole of $10^8 M_{\odot}$; a spherical object whose mean mass-density equals that of water (1g/cm^3), would become a black hole if its radius exceeded about $4.01 \cdot 10^8$ km $= 2.68$ astronomical units. This means that if the center of such an object were placed at the center of the sun, then its surface would be between the orbits of Mars and Jupiter. Hence, black holes might form under reasonable conditions.

In General Relativity for spherical black holes (Schwarzschild black holes), exactly the same expression R_{BH} holds for the surface of a black hole. The surface of a black hole at R_{BH} is a null surface, consisting of those photon trajectories (null rays) which just do not escape to infinity. This surface is also called the black hole horizon. Further, the gravitational redshift of radiation originating at or inside its horizon is infinite. Both mean that radiation emitted from inside the black hole can never be detected from outside. Hence, the surface at that radius is called a horizon. Material accreting from outside can get very hot, radiating copiously. Thus, black holes are associated with some of the most luminous objects known, including quasars and other active galaxies, some X-ray binaries, and gamma ray bursters. Besides collapsed astrophysical objects, primordial particles produced during the Big Bang are possible candidates as black holes.

If the black hole gains matter from outside (e.g., an accreting disc), it will increase in mass and its horizon will cover a larger portion of space. According to quantum mechanical computations pioneered by S. Hawking, a black hole can also radiate away energy via quantum effects, in which case its horizon contracts.

Black holes were first discovered as purely mathematical solutions of Einstein's field equations. This solution, the Schwarzschild black hole, is a nonlinear solution of the Einstein equations of General Relativity. It contains no matter, and exists forever in an asymptotically flat space-time. It is thus called an eternal black hole. Later these solutions were extended to represent the final stage in models of gravitational collapse when outward pressure does not balance self-gravity.

The general theory of relativity allows one to prove that only a small number of families of different black hole types can exist (no-hair theorems). They correspond to different mathematical vacuum solutions of Einstein's field equations which are related to the symmetry of the asymptotically flat space-time outside the horizon (domain of outer communication) and, equivalently, to the charges (conserved quantities in vacuum) of the black hole. The simplest case is given by the Schwarzschild metric, which represents a black hole fully characterized by its mass. If the black hole is electrically charged, then one has a Reissner–Nordström metric, and rotating cases are given by the (electrically neutral) Kerr metric and the (charged) Kerr–Newman metric. In these cases, the formula for the horizon location depends on the angular momentum (and charge) as well as the mass of the black hole. Further, the general properties of the above metrics have been widely investigated, leading to a set of laws very close in spirit to thermodynamics.

Rotating black holes, described by the Kerr solution, allow rotational energy to be extracted from a region just outside the horizon, especially if magnetic fields are present. *See* ADM mass, asymptotic flatness, black hole horizon, domain of outer communication, black hole horizon, Kerr black hole, Schwarzschild black hole.

black hole binary   An observed binary system of which apparently one member is a black hole. These are identified by their X-ray emission. Cyg-X1 is the prototype, with a compact object of mass $\approx 12 M_{\odot}$ (the putative black hole) accreting mass from a hot supergiant companion star through an accretion disk which thermally emits X-radiation.

black hole horizon   The future causal horizon that is the surface of a black hole: the boundary between light-rays which can reach infinity and those that do not. The description of horizons has been developed in terms of classical relativity, where one can prove, for instance, that the area of a black hole cannot decrease, so the horizon converges towards a final (bigger) future event horizon. However, Hawking radiation, a quantum phenomenon, leads to the eventual evaporation of an isolated black hole. If the black hole is eternal and does not change in time, the horizon is a true future event horizon and bounds a region of space which will never be experienced from outside the black hole. If the black hole evaporates away completely, its interior will eventually be seen from outside. If there is a limit to this effect, it must occur in the fully quantum limit, black hole mass $\sim 10^{-5}$ grams (the Planck Mass, $M_{Pl} = (\hbar c / G)^{1/2}$, associated length scale 1.6×10^{-33} cm). *See* apparent horizon, black hole, future/past causal horizon, future/past event horizon, Killing horizon.

black ice   Condition of aged ice on highway surfaces, which has a polished surface and so appears dark rather than bright.

blast wave shock   Shock created by a short, spatially limited energy release, such as an explosion. Therefore, the energy supply is limited, and the shock weakens and slows as it propagates outward. In models of the energetics and propagation of interplanetary traveling shocks, *blast wave shocks* are often used because their mathematical description is simple, and the assumptions about the shock can be stated more clearly. It is also speculated that blast wave shocks give rise to the metric type II radio bursts on the sun.

blazar   A class of active galactic nuclei which includes BL Lac objects and Optically Violently Variable (OVV) quasars, whose name derives from the contraction of the terms BL Lac and quasar. BL Lac and OVV quasars share several common properties, like high continuum polarization and large luminosity changes on relatively short time scales. All known *blazars* — a few hundred objects — are radio loud active galactic nuclei, and several of them have been revealed as strong γ-ray sources. Blazars are thought to be active galactic nuclei whose radio jets are oriented toward us, and whose nonthermal, synchrotron continuum is strongly amplified by Doppler beaming. *See* active galactic nuclei, BL Lacertae object.

blizzard   Winter storm characterized by winds exceeding 35 miles (56 km) per hour, temperatures below $20°$ F ($-7°$ C), and driving snow, reducing visibility to less than 1/4 mile (400 m) for 3 or more hours.

BL Lacertae   Prototype of extremely compact active galaxies [BL Lacertae objects, or Lacertids (also Blazar)] that closely resemble Seyfert and N galaxies, radiating in the radio, infrared, optical, and X-ray. *BL Lacertae,* of magnitude 14.5, is located at RA$22^h00^m40^s$, dec $42°02'$ and was observed in 1929 and incorrectly identified as a variable star but observed to be a radio source in 1969. About 40 Lacertids are presently known. They are characterized by a sharply defined and brilliant (starlight) nucleus that emits strong nonthermal radiation and whose continuous visible spectrum has no emission or absorption lines. Surrounding the bright nucleus is a faint halo resembling a typical elliptical galaxy from which redshifts can be measured. BL Lacertae is thus found to have a redshift $z = 0.07$, comparable to those of the nearer quasars; PKS 0215+015 is a BL Lac object with $z = 0.55$. BL Lacertae objects radiate most of their energy in the optical and infrared wavelengths and undergo rapid changes in brightness in visible light, the infrared, and X-rays brightening to the brightness of the brightest quasars. The central radio source is very small, consistent with the rapid fluctuations in brightness. X-ray brightness may vary in periods of several hours, suggesting that the emitting region is only a few light-hours across. It has been conjectured that the Lacertids, quasars, and radio galaxies are actually the same types of objects viewed from earth at different angles that either obscure or reveal the galaxy's central powerhouse of radiation, presumably powered by accretion onto a central black hole.

BL Lacertae object   An active galaxy very similar to quasars in appearance but with no emission lines, with a strong continuum stretching from rf through X-ray frequencies. There are no known radio quiet BL Lac objects. They can exhibit dramatic variability.

blocking   A persistent weather pattern where the mid-latitude westerly jet is blocked and diverted into a northern and a southern branch. The *blocking* is associated with a pair of anticyclonic (blocking high) circulation to the north and cyclonic circulation to the south. When a blocking occurs, extreme weather conditions can persist for a week or longer. Blockings often occur in winter over northwestern North Pacific off Alaska and northwestern North Atlantic off Europe.

blocking patterns   High-amplitude quasistationary wave disturbances in the extratropical atmosphere.

blue clearing (Mars)   The difference in albedo between Mars surface features is small in blue, except for the polar caps. Therefore, albedo features visible in green and red are not usually identified in blue. However, they are sometimes identified even in blue; this phenomenon is called *blue clearing*. In the first half of the twentieth century, it was thought that the Martian atmosphere was thick enough to hold a haze layer which absorbs light in blue, and that the albedo features are visible in blue only when the haze disappears and the sky clears up in blue, but the haze layer is now not believed to exist. The blue clearing has been observed most frequently around the opposition of Mars. The opposition effect may be one of the causes of the blue clearing. The degree of the opposition effect is larger in bright areas than in dark areas in all visible wavelengths, not only in red but also in blue. Therefore, albedo features are identified even in blue around the opposition. However, the degree of the opposition effect may depend on the Martian season, for the surface is covered with a thin dust layer and uncovered in a cycle of a Martian year.

blue ice   Old sea ice that has expelled impurities and appears a deep translucent blue in sunlight.

blue jet   A long-duration luminous structure observed directly above an active thundercloud, extending upwards from the cloud top for many tens of kilometers. The name is derived from their highly collimated blue beam of luminosity, which persists for several tenths of a second. Unlike sprites and elves, *blue jets* are relatively rare and consequently are poorly understood.

blue straggler   A star whose position on the HR diagram is hotter (bluer) and brighter than that allowed for stars of the age represented in the particular star cluster or other population under consideration. Such stars are sometimes interlopers from younger populations, but more often they are the products of evolution of binary stars, where material has been transferred from one star to another or two stars have merged, giving the recipient a larger mass (hence, higher luminosity and surface temperature) than single stars of the same age.

body waves   Earthquakes generate seismic waves that are responsible for the associated destruction. Seismic waves are either surface waves or *body waves*. The body waves are compressional p-waves and shear s-waves. These waves propagate through the interior of the Earth and are the first arrivals at a distant site.

Bogomol'nyi bound   The mass of a (magnetic) monopole depends on the constants that couple the Higgs field with itself and with the gauge fields forming the monopole configuration. The energy of a monopole configuration can be calculated, albeit numerically, for any value of these constants. It can, however, be seen that the mass always exceeds a critical value, first computed by E.B. Bogomol'nyi in 1976, which is

$$m_M \geq \eta/e \,,$$

where η is the mass scale at which the monopole is formed, and e is the coupling of the Higgs field to the gauge vectors A_μ^a (a being a gauge index). The bound is saturated, i.e., equality holds, when the ratio of the coupling constants vanishes, a limit known as the Prasad–Sommerfield limit. Since magnetic monopoles may be produced in the early universe, this limit is of relevance in cosmology. *See* cosmic topological defect, monopole, Prasad–Sommerfield limit, t'Hooft–Polyakov monopole.

Bohr's theory of atomic structure   Theory of the atomic structure of hydrogen postulated by Neils Bohr (1885–1962). The theory is based upon three postulates: 1. The electrons rotate about the nucleus without radiating energy in the form of electromagnetic radiation. 2. The electron orbits are such that the angular momentum of the electron about the nucleus is an integer of $h/2\pi$, where h is Planck's constant. 3. A jump of the electron from one orbit to another generates emission or absorption of a photon, whose energy is equal to the difference of energy between the two orbits. This theory has now been superseded by wave mechanics, which has shown that for the hydrogen atom spectrum, Bohr's theory is a good approximation.

bolide   A meteoroid that explodes or breaks up during its passage through the atmosphere as a meteor.

bolometric correction (B.C.)   The correction to an observed magnitude (luminosity) in a particular wavelength region to obtain the bolometric magnitude (luminosity). This correction is to account for the stellar flux that falls beyond the limits of the observed wavelength region.

bolometric magnitude/luminosity   The luminosity of a star integrated over all wavelengths the star radiates. This is a measure of the total energy output of a star.

Boltzmann's constant (k)   Constant of proportionality between the entropy of a system S and the natural logarithm of the number of all possible microstates of the system Ω, i.e., $S = k \ln(\Omega)$. When one applies statistical mechanics to the ideal or perfect gas, one finds that *Boltzmann's constant* is also the constant of proportionality of the equation of state of an ideal gas when the amount of gas is expressed in the number of molecules N rather than the number of moles, i.e., $PV = NkT$. $k = 1.3806503(24) \times 10^{-23}$ J K^{-1}. Named after Ludwig Boltzmann (1844–1906).

Bondi mass   In general relativity the function $m(r, t)$ equal to the total energy contained inside the sphere of radius r at the time t. If there is no loss or gain of energy at $r \rightarrow +\infty$, and the space is asymptotically flat, then

$$\lim_{r \rightarrow +\infty} m(r, t) = M_{ADM} \,,$$

where M_{ADM} is the ADM mass. *See* ADM mass, asymptotic flatness.

Bonnor symmetry   A stationary (i.e., rotating) axisymmetric vacuum space-time can be mapped to a static axisymmetric electrovacuum space-time. In this process, the twist potential must be extended in the complex plane. When the twist is proportional to a real parameter, the complex continuation of this parameter can be used to make the resulting twist potential real. *See* electrovacuum.

bore   A steep wave that moves up narrowing channels, produced either by regular tidal events, or as the result of a tsunami. A bore of this type, generated by an earthquake off the coast of Chile, destroyed Hilo on the island of Hawaii on May 22, 1960.

borrow site   A source of sediment for construction use. May be on land or offshore.

boson   An elementary particle of spin an integer multiple of the reduced Planck constant $\hbar$. A quantum of light (a photon) is a *boson.*

boson star   A theoretical construct in which the Klein–Gordon equation for a massive scalar field ϕ:

$$\Box\phi + m^2\phi = 0$$

is coupled to a description of gravity, either Newtonian gravity with ϕ_g the gravitational potential

$$\nabla^2\phi_g = 4\pi G\frac{1}{2}(\nabla\phi)^2$$

or General Relativistic gravitation:

$$G_{\mu v} = 8\pi G T_{\mu v}(\phi)$$

where $T_{\mu v}(\phi)$ is the scalar stress-energy tensor. In that case the wave operator applied to ϕ is the covariant one. $\Box\phi = \phi^\alpha{}_{;\alpha}$. Stable localized solutions, held together by gravity are called boson strings. Nonstationary and nonspherical *boson stars* may be found by numerical integration of the equations.

Bouguer correction   A correction made to gravity survey data that removes the effect of the mass between the elevation of the observation point and a reference elevation, such as the mean sea level (or geoid). It is one of the several steps to reduce the data to a common reference level. In Bouguer correction, the mass in consideration is approximated by a slab of infinite horizontal dimension with thickness h, which is the elevation of the observation point above the reference level, and average density ρ. The slab's gravitational force to be subtracted from the measured gravity value is then $\Delta g = 2\pi G\rho h$, where G is the gravitational constant.

Bouguer gravity anomaly   Much of the point-to-point variation in the Earth's gravity field can be attributed to the attractions of the mass in mountains and to the lack of attraction of the missing mass in valleys. When the attraction of near surface masses (topography) is used to correct the "free-air" gravity measurements, the result is a Bouguer-gravity map. However, major mountain belts (with widths greater than about 400 km) are "compensated". Bouguer gravity anomalies are caused by inhomogeneous lateral density distribution below the reference level, such as the sealevel, and are particularly useful in studying the internal mass distributions. Thus, the primary signal in Bouguer gravity maps is the negative density and the negative gravity anomalies of the crustal roots of mountain belts.

Bouguer–Lambert law (Bouguer's law) *See* Beer's law.

bounce motion   The back-and-forth motion of an ion or electron trapped in the Earth's magnetic field, between its mirror points. This motion is associated with the second periodicity of trapped particle motion ("bounce period") and with the longitudinal adiabatic invariant.

boundary conditions   Values or relations among the values of physical quantities that have to be specified at the boundary of a domain, in order to solve differential or difference equations throughout the domain. For instance, solving Laplace's equation $\nabla\phi = 0$ in a 3-dimensional volume requires specifying *a priori* known boundary values for ϕ or relations among boundary values of ϕ, such as specification of the normal derivative.

boundary layer pumping   Ekman pumping.

Boussinesq approximation   In the equation of motion, the density variation is neglected except in the gravity force.

Boussinesq assumption   The Boussinesq assumption is employed in the study of open channel flow and shallow water waves. It involves assumption of a linearly varying vertical velocity, zero at the channel bottom and maximum at the free surface.

Boussinesq equation   The general flow equation for two-dimensional unconfined flow in an aquifer is:

$$\frac{\partial}{\partial x}\left(h\frac{\partial h}{\partial x}\right) + \frac{\partial}{\partial y}\left(h\frac{\partial h}{\partial y}\right) = \frac{S_y}{K}\frac{\partial h}{\partial t}$$

where S_y is specific yield, K is the hydraulic conductivity, h is the saturated thickness of the aquifer, and t is the time interval. This equation is non-linear and difficult to solve analytically. If the drawdown in an aquifer is very small compared with the saturated thickness, then the saturated thickness h can be replaced with the average thickness b that is assumed to be constant over the aquifer, resulting in the linear equation

$$\frac{\partial^2 h}{\partial x^2} + \frac{\partial^2 h}{\partial y^2} = \frac{S_y}{Kb} \frac{\partial h}{\partial t} .$$

Bowen's ratio   The ratio between rate of heat loss/gain through the sea surface by conduction and rate of heat loss/gain by evaporation/condensation.

Boyle's law   For a given temperature, the volume of a given amount of gas is inversely proportional to the pressure of the gas. True for an ideal gas. Named after Robert Boyle (1627–1691).

Brackett series   The set of spectral lines in the far infrared region of the hydrogen spectrum with frequency obeying

$$v = cR \left(1/n_f^2 - 1/n_i^2 \right) ,$$

where c is the speed of light, R is the Rydberg constant, and n_f and n_i are the final and initial quantum numbers of the electron orbits, with $n_f = 4$ defining the frequencies of the spectral lines in the Brackett series. This frequency is associated with the energy differences of states in the hydrogen atom with different quantum numbers via $v = \Delta E / h$, where h is Planck's constant and where the energy levels of the hydrogen atom are:

$$E_n = hcR/n^2 .$$

Bragg angle   Angle θ which relates the angle of maximum X-ray scattering from a particular set of parallel crystalline planes:

$$\sin \theta = n\lambda/(2d)$$

where n is a positive integer, λ is the wavelength of the radiation, and d is the normal separation of the planes. θ is measured from the plane (not the normal).

Bragg Crystal Spectrometer (BCS)   *Bragg Crystal Spectrometers* have been flown on a number of solar space missions including the Solar Maximum Mission (1980–1989). Such a spectrometer is currently one of the instruments on the *Yohkoh* spacecraft. This instrument consists of a number of bent crystals, each of which enables a selected range of wavelengths to be sampled simultaneously, thus providing spectroscopic temperature discrimination. Typical wavelengths sampled by these detectors lie in the soft X-ray range (1 to 10 Å).

braided river   A river that includes several smaller, meandering channels within a broad main channel. The smaller channels intersect, yielding a braided appearance. Generally found at sites with steeper slopes.

Brans–Dicke theory   Or *scalar-tensor theory of gravitation*. A theory of gravitation (1961) which satisfies the weak equivalence principle but contains a gravitational coupling G_N, which behaves as (the inverse of) a scalar field ϕ. This ϕ field satisfies a wave equation with source given by the trace of the energy momentum tensor of all the matter fields. Thus G_N is not generally constant in space nor in time. The corresponding action for a space-time volume V can be written as

$$S = \frac{1}{16\pi}$$
$$\int_V d^4x \sqrt{-g} \left[\phi R + \omega \phi^{-1} \partial_\mu \phi \partial^\mu \phi + L_M \right] ,$$

where R is the scalar curvature associated to the metric tensor whose determinant is g, ω is a coupling constant, and L_M is the matter Lagrangian density.

In the limit $\omega \to \infty$ the scalar field decouples, and the theory tends to general relativity. For ω finite, instead, it predicts significant differences with respect to general relativity. However, from tests of the solar system dynamics, one deduces that $\omega > 500$ and the theory can lead to observable predictions only at the level of cosmology. In the Brans–Dicke theory the *strong* equivalence principle is violated, in that objects with different fractional gravitational binding energy typically fall at different rates.

A generalization of Brans–Dicke theory is given by the dilaton gravity theories, in which the scalar field ϕ is called the dilaton and couples directly to all matter fields, thus violating even the weak equivalence principle. *See* dilaton gravity.

Brazil current   A warm ocean current that travels southwestward along the central coast of South America.

breaker zone   The nearshore zone containing all breaking waves at a coast. Width of this zone will depend on the range of wave periods and heights in the wave train and on bathymetry.

breakwater   A man-made structure, often of rubble mound construction, intended to shelter the area behind (landward of) it. May be used for erosion control or to shelter a harbor entrance channel or other facility.

breccias   Composite rocks found on the moon, consisting of heterogeneous particles compacted and sintered together, typically of a light gray color.

Bremsstrahlung   (German for *braking radiation.*) Radiation emitted by a charged particle under acceleration; in particular, radiation caused by decelerations when passing through the field of atomic nuclei, as in X-ray tubes, where electrons with energies of tens of kilovolts are stopped in a metal anode. *Bremsstrahlung* is the most common source of solar flare radiation, where it is generated by deceleration of electrons by the Coulomb fields of ions.

bremstrahlung [thermal]   The emission of bremsstrahlung radiation from an ionized gas at local thermodynamic equilibrium. The electrons are the primary radiators since the relative accelerations are inversely proportional to masses, and the charges are roughly equal. Applying Maxwell's velocity distribution to the electrons, the amount of emitted radiation per time per volume per frequency is obtained for a given temperature and ionized gas density.

Brewster point   One of three points on the sky in a vertical line through the sun at which the polarization of skylight vanishes. Usually located at about $20°$ below the sun. *See* Arago point, Babinet point.

brightness   The luminosity of a source [J/sec] in the bandpass of interest. (For instance, if referred to visual observations, there are corrections for the spectral response of the eye.) Apparent *brightness* of a stellar object depends on its absolute luminosity and on its distance. The flux of energy through a unit area of detector (e.g., through a telescope objective) is $F = L/4\pi r^2$ [J/sec/m^2]. Stellar apparent magnitude gives a logarithmic measure of brightness: $m = -2.5 \log(F) + K$. The magnitude scale is calibrated by fixing the constant K using a number of fiducial stars. Absolute magnitude is defined as the apparent magnitude a source would have if seen at a distance of $10\,pc$ and is a measure of the absolute brightness of the source.

brightness temperature   The temperature of an equivalent black body with a given intensity and long wavelengths where the Rayleigh–Jeans approximation is valid. This temperature is used in radio and submillimeter astronomy and is given by

$$T_B = \frac{I_\nu c^2}{2k\nu^2}$$

where h and k are the Planck and Boltzmann constants, respectively, ν is the frequency in Hertz, and I_ν is the monochromatic specific intensity (i.e., the flux per frequency interval per solid angle) in units of erg s^{-1} cm^{-2} Hz^{-1} sr^{-1}. To relate an object's brightness temperature to its flux density in Janskys, the size of the source in stearadians (or the full width of the main beam of the radiotelescope) is used:

$$S_\nu = 2\pi k T_B \theta^2 4ln2/\lambda^2$$

which reduces to

$$S_\nu = 7.35 \times 10^{-4} \theta^2 T_B/\lambda^2$$

where θ is the width of the telescope's full beam at half maximum in arcseconds, T_B is the brightness temperature in Kelvins, λ is the center wavelength in centimeters, and S_ν is the flux density in Janskys.

bright point   A transient H-alpha brightening of flare intensity, less than 20 millionths of the solar hemisphere in area. Such a brightening when covering a larger area becomes classified as a solar flare.

brittle behavior (brittle fracture)   A phenomenological term describing the nature of material failure. According to the simplest or "classical" view, *brittle behavior (brittle fracture)* of a solid material is a discrete event in which the failure of the solid occurs, without significant prior deformation and without warning, at a particular stress. In general, material is described as "brittle" when fracture occurs with no preceding appreciable permanent deformation. Sometimes brittle behavior refers to failure following limited amounts of inelastic strain that are small compared with the elastic strain.

brittle-ductile transition   A temperature range over which the principal failure mechanism of a rock body (such as the continental crust) under deviatoric stress changes from brittle to ductile. Brittle failure takes place at low temperatures by developing fractures or slipping along pre-existing fractures. Ductile failure takes place at higher temperatures and, observed at the macroscopic scale, by continuous deformation. The transition occurs due to the activation of diffusion or dislocation creep at elevated temperatures and is affected by the rate of ductile deformation and the pore fluid pressure in the brittle region. In the Earth's continental crust, this transition is marked by the absence of earthquakes below a certain depth where the temperature is higher than about 300 to 400°C. In the oceanic crust and mantle, the transition occurs at higher temperatures.

brittle-plastic transition   *See* brittle-ductile transition.

broad line radio galaxies   Radio galaxies showing optical spectra very similar to, and in several cases almost indistinguishable from, those of Seyfert-1 galaxies. Broad line radio galaxies are type 1, low-luminosity, radio-loud active galactic nuclei; the radio-loud counterpart of Seyfert-1 galaxies. Differences between Seyfert-1 and Broad Line Radio galaxies encompass the morphology of the host galaxy (Seyfert-1 are mostly, albeit not exclusively, spirals, while broad line radio galaxies are hosted by ellipticals) and some features of the optical spectrum, like weaker singly ionized iron emission and larger internal absorption due to dust in broad line radio galaxies. *See* Seyfert galaxies.

broad line region   The region where the broad lines of active galactic nuclei are produced. The strongest lines observed in the optical and UV spectrum are the Balmer lines of hydrogen, the hydrogen Lyman α line, the line from the three times ionized carbon at 154.9 nm, and some recombination lines from singly ionized and neutral helium. Since no forbidden lines are observed, the *broad line region* is most likely a relatively high density region with particle density in the range 10^9 to 10^{13} ions per cubic centimeter. The broad line region is believed to be very close to the central source of radiating energy of the active galactic nucleus. Observations of variation of broad line profiles and fluxes suggest that the broad line regions line emitting gas is confined within 1 pc in Seyfert-1 galaxies. Models of the broad line regions invoke a large number of dense emitting clouds, rapidly rotating around a central illuminating source. Alternatively, it has been suggested that at least part of the emission of the broad line regions could come from the middle and outer region of the accretion disk suspected to be a universal constituent of the central engine of active galactic nuclei.

brown dwarf   A sphere of gas with the composition of a star (that is, roughly $\frac{3}{4}$ hydrogen, $\frac{1}{4}$ helium, and at most a few percent of heavier elements) but with a mass low enough that the center never, as the object contracts out of interstellar gas, gets hot enough for hydrogen burning to balance the energy being lost from the surface of the sphere. Very young *brown dwarfs* can be as bright as low-mass stars (10^{-4} to 10^{-5} of the solar luminosity), but they fade with time until they are too cool and faint to see. The number of brown dwarfs in our galaxy is not very well known. They are probably not common enough (at least in the galactic disk) to contribute much to the dark matter, but small numbers have been observed (a) in young clusters of stars, (b) as companions to somewhat more massive stars, and (c) in isolation in space. *See* dark matter, galactic disk, hydrogen burning, interstellar gas.

Brunt frequency *See* buoyancy frequency.

Brunt–Väisälä frequency *See* buoyancy frequency.

B star Star of spectral type B. Rigel and Spica are examples of *B stars*.

bulkhead A man-made wall, typically vertical, which holds back or protects sediment from the impact of water. May be held in place by driven piles or with anchors placed in the sediment behind the wall. *Bulkheads* are commonly made of treated wood or steel.

bulk modulus For a density ρ and a pressure P, the quantity $\rho\, dP/d\rho$ is called the *bulk modulus*. The reciprocal of compressibility. Two versions of this parameter are used. One has the temperature constant during the pressure change and is usually designated the *isothermal bulk modulus*, K_T. The other has the entropy constant during the pressure change and is called the *adiabatic bulk modulus*.

bulk parameters (**1.**) Quantities derived from spatial or temporal average values of other quantities. For example, non-dimensional scaling numbers, such as the Reynolds number or the Froude number, are often defined in terms of average values of velocity and length. These average values are sometimes also referred to as "characteristic scales" of the flow.

(**2.**) Quantities that are computed from spatial or temporal averages of field data. The data that constitute the *bulk parameter* are often sampled with a number of different instruments. It must be assumed that the data are statistically homogeneous and isotropic in the sampling interval.

buoyancy The upward force on a body equal to the weight of the fluid displaced by the body.

$$F_{\text{buoy}} = \rho V g$$

where ρ is the density, V is the volume of the fluid, and g is the acceleration of gravity.

buoyancy flux Turbulence in a stratified fluid affects the stratification. The buoyancy flux J_b [W kg^{-1}] quantifies (in differential form) the rate of change of the potential energy of the stratification. It is defined by $J_b = g\rho^{-1} < w'\rho' >= g\rho^{-1}F_\rho$, where g is the gravitational acceleration, ρ density, ρ' density fluctuations, w' vertical velocity fluctuations, and F_ρ the vertical density (mass) flux [kgm^{-2}s^{-1}]. Three principal cases can be distinguished:

(1) In a stable water column, the buoyancy flux J_b expresses the rate at which turbulent kinetic energy (source of J_b) is turned over to increase the potential energy of the water column by work against stratification: $J_b = -K_\rho N^2$, where K_ρ is the turbulent diapycnal diffusivity and N^2 is stability. This process is accompanied by dissipation of turbulent kinetic energy (ϵ). For the relation between J_b and ϵ, *see* mixing efficiency.

(2) In an unstable water column (*see* convective turbulence), potential energy (source of J_b) is released, which fuels kinetic energy into turbulence. If the source of the density change $\partial\rho/\partial t$ is at the surface or bottom of the mixed boundary layer, the boundary buoyancy flux is defined by, $J_b^o = \rho^{-1}h_{\text{mix}}g\partial\rho/\partial t$, where h_{mix} is the thickness of the mixed boundary layer.

(3) Under the conditional instability of double diffusion, the water column releases potential energy (source of J_b) from the water column to turbulent kinetic energy at a rate J_b, which is given by the so-called double-diffusive flux laws (Turner, 1974; McDougall and Taylor, 1984).

buoyancy frequency A measure of the degree of stability in the water column, defined by

$$N^2 = -\frac{g}{\rho}\frac{\partial\rho}{\partial z}$$

where g is the constant of gravity, ρ is the density, and z is the vertical coordinate (positive upwards). When N^2 is positive, the medium is stable, and N represents the frequency of oscillation of a water parcel in purely vertical motion. When N^2 is negative, the medium is unstable. In the large vertical gradient regions of the upper ocean, N typically ranges from 3 to 7 cycles per hour. In the deep ocean, N is between 0.5 and 1 cycles per hour. Also called the stability frequency, the Brunt–Väisäla frequency, the Brunt frequency, and Väisälä frequency.

buoyancy Reynolds number   The non-dimensional ratio $R_{eB} = \epsilon/(\nu N^2)$ expresses the intensity of turbulence, quantified by the dissipation of turbulent kinetic energy ϵ, relative to the turbulence-suppressing effect of stability N^2 and viscosity ν. If $R_{eB} > 20 - 30$, turbulence is considered as actively mixing despite the stratification (Rohr and Van Atta, 1987), but as long as $R_{eB} < 200$ turbulence is not expected to be isotropic (Gargett et al.; 1984).

buoyancy scale   Largest overturning length scale allowed by the ambient stratification. In stratified turbulent flows away from boundaries, the size of the largest overturns or eddies is limited by the work required to overcome buoyancy forces. The *buoyancy scale* can be estimated from

$$ L_N = 2\pi \left(\frac{\epsilon}{N^3} \right)^{1/2} $$

where N is the buoyancy frequency and ϵ is the dissipation rate of turbulent kinetic energy.

buoyancy scaling   In the interior of stratified natural water bodies, internal wave shear is usually the dominant source of turbulence. Because turbulence is limited by the presence of stratification, it depends on both the stability N^2 of the water column and the dissipation of turbulent kinetic energy ϵ. Dimensional analysis provides the relations for *buoyancy scaling* as a function of N and ϵ.

Length	$L_{bs} \sim (\epsilon N^{-3})^{1/2}$
Time	$\tau_{bs} \sim N^{-1}$
Velocity	$w_{bs} \sim (\epsilon N^{-1})^{1/2}$
Diffusivity	$K_{bs} \sim \epsilon N^{-2}$
Buoyancy flux	$J_{bs} \sim \epsilon$

buoyancy subrange   Range of the energy spectra, which is affected by buoyancy forces and where subsequently turbulence is expected to be anisotropic (*see* buoyancy Reynolds number). The relevant parameters to quantify turbulence in the *buoyancy subrange* are dissipation of turbulent kinetic energy ϵ and stability N^2. *See* buoyancy scaling.

Bureau International des Poids et Mesures (BIPM)   International bureau of weights and measures in Sevres, France, charged mainly with the production of International Atomic Time.

Burgers vector (b)   The dislocation displacement vector defined by the procedure first suggested by Frank (1951). When a dislocation line or loop is defined as the boundary between a slipped area and an unslipped area on a slip plane, it can be characterized by a unit vector expressing the amount and direction of slip caused by its propagation on the slip plane. This unit vector, which corresponds to unit motion of dislocations, is called *Burgers Vector* (**b**). Burgers vector must correspond to one of the unit vectors for a given crystal structure.

burst   In solar physics, a transient enhancement of the solar radio emission, usually associated with an active region or flare. In cosmic ray physics, a sudden flux of cosmic rays with a common origin.

bursty bulk flows   In the Earth's magnetosphere, intervals of rapid ion flows observed in the plasma sheet, lasting typically 10 min. Extreme velocity peaks within these intervals reach many hundreds of kilometers per second, last for a fraction of a minute, and are termed flow bursts. Flows closer to Earth than 25 R_E tend to be directed earthward; more distant ones are usually tailward.

Butcher–Oemler effect   The increase in the fraction of blue galaxies in distant clusters of galaxies. H. Butcher and A. Oemler, in papers published in 1978 and 1984, discussed an excess of galaxies with color index B-V significantly bluer than that of normal elliptical and S0 galaxies. This excess of blue galaxies was found in clusters of galaxies at redshift near 0.4: Butcher and Oemler discovered that 20% of galaxies in clusters at redshift larger than 0.2 were blue galaxies, while blue galaxies accounted for only 3% of all galaxies in nearby clusters. Blue color suggests strong star formation. The detection of a large number of blue galaxies indicates that significant galactic evolution is still occurring at a very recent epoch, about .9 of the present age of the universe.

butterfly diagram   A diagram showing the behavior of the observed latitude of a sunspot with time during the course of several solar cycles. The latitudinal drift of the sunspot emergence as the cycle proceeds (from high latitudes to the solar equator) resembles a butterfly's wings. The physical cause of this behavior is thought to be due to the generation of magnetic field by the solar dynamo.

Buys Ballots Law   In the Northern hemisphere, if one stands with one's back to the wind, the low pressure is to the left, and the high is to the right. The rule is opposite in the Southern hemisphere and arises because of the Coriolis force from the Earth's rotation. This is useful because bad weather is typically associated with low pressure regions.

b-value   One of the coefficients of the Guttenberg–Richter relation relating the frequency of occurrence to the magnitude of earthquakes.

Byerlee's law   A laboratory experimental result of static rock friction obtained by U.S. geophysicist J.D. Byerlee in 1978. The experiments show that, statistically, regardless of rock type, temperature, and confining pressure, the relation between normal stress σ_n and shear stress τ on the flat frictional surface between two rock specimens is:

$$\tau = 0.85\sigma_n \qquad (\sigma_n \leq 200 \text{ MPa})$$
$$\tau = 50 + 0.6\sigma_n \qquad (\sigma_n > 200 \text{ MPa}) \ .$$

Assuming that the top part of the Earth's lithosphere is pervasively fractured by faults of random orientations, *Byerlee's law* is frequently used to define the brittle strength of the lithosphere. Shear stresses along major geological faults have been inferred to be much lower than Byerlee's law predictions.

bypass   As used in the study of coastal geomorphology, the process by which sediment moves from one side of a tidal inlet to the other. Artificial bypassing schemes have also been used to transport sediment across a stabilized inlet or past a harbor entrance.

C

cabbeling Mixing mechanism resulting in buoyancy effects due to the non-linear relation between temperature, salinity, and density. If mixing occurs between two water masses with the same potential density but different salinities S and temperatures Θ, the resulting water will be denser than the original water masses due to variations of the thermal expansivity α and the saline contraction coefficient β with Θ and S. In the figure the two initial water types are represented by the labels A and B. The water type that results from mixing will have a density that lies on the straight line between A and B and, therefore, will be denser. The effect is particularly strong at low temperatures. Cabbeling leads to a diapycnal density (mass) flux, although the original mixing may be purely isopycnal.

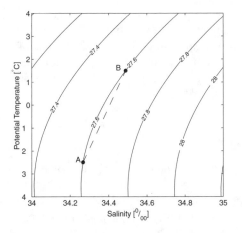

Contours of potential density σ_θ as a function of salinity and potential temperature.

cable A horizontal unit of length used in ocean navigation, equal to 100 fathoms, equal to 600 feet, equal to 182.88 meters; approximately one tenth of a Nautical mile. *See* nautical mile.

CAI *C*alcium-, *a*luminum-rich *i*nclusion found in chondritic meteorites. Usually of mil-limeter or centimeter size and roughly spherical, they are depleted in volatiles relative to solar composition and display a melt rim, which apparently formed in an episode of intense heating followed by rapid quenching.

caldera The central crater of a volcano. The *caldera* is formed when magma from a chamber in or under the volcanic structure is removed, either by eruption onto the surface or by withdrawal to another underground location. This removal of the magma removes the method of support for the overlaying crust, and the surface collapses to create the caldera. Volcanos often display multiple overlapping calderas, indicating that several episodes of magma injection and withdrawal have occurred over the volcano's history.

calendar A system of reckoning time in days and years enumerated according to cyclic patterns. Examples are the Julian calendar and the Gregorian calendar.

California current A cold ocean current flowing southeastward down the west coast of the United States and Baja California.

California nebula NGC1499: Very large nebula in the constellation Perseus, approximately 3° in length; the site of substantial star formation with strong hydrogen emission.

Callan–Rubakov effect Baryon violating processes have a very low probability of occurring at ordinary energies, since they require the exchange of very massive (grand unification scale) gauge particles. These particles are actually part of the structure of gauged monopoles, which therefore can act as seeds for enhancing processes such as baryogenesis.

For a monopole of mass m_M, the interaction probability with a proton say, would be proportional to m_M^{-2} which, for a grand unified model, is far too small to be cosmologically or astrophysically relevant. However, magnetic monopoles eventually exert an attractive force on all particles having a magnetic moment, and this enhances the interaction probability, as was discovered by C.G. Callan and V.A. Rubakov in the early 1980s. As it turns out, the resulting probability is as high as that characteristic of strong interactions and could, therefore, have a significant impact on cosmology. *See* cosmic topological defect, monopole, t'Hooft–Polyakov monopole.

Callisto   Moon of Jupiter, also designated JIV. Discovered by Galileo in 1610, it is one of the four Galilean satellites. Its orbit has an eccentricity of 0.007, an inclination of 0.51°, a precession of 0.643° yr^{-1}, and a semimajor axis of 1.883×10^6 km. Its radius is 2400 km, its mass, 1.07×10^{23} kg, and its density 1.85 g cm^{-3}. It has a geometric albedo of 0.20 and orbits Jupiter once every 16.69 Earth days.

calorie   The quantity of energy needed to raise the temperature of 1 g of water by 1°C. The exact equivalence to energy depends on the temperature and pressure at which the measurement is made. The preferred "thermochemical" or "defined" calorie is assigned a value of 4.1840 Joules.

Calypso   Moon of Saturn, also designated SXIV. Discovered by Pascu, Smith, and colleagues in 1980, it orbits Saturn in the same orbit as Tethys but trailing it by 60°. This is one of the two stable Lagrange points in the Saturn-Tethys system. Telesto orbits at the other. Its orbit has a semimajor axis of 2.95×10^5 km. Its size is $17 \times 11 \times 11$ km, but its mass has not been measured. It has a geometric albedo of 0.6 and orbits Saturn once every 1.888 Earth days.

Canary current   A cool ocean current flowing southward off the west coast of Portugal and along the northwest coast of Africa.

candela   SI unit of luminous intensity. The luminous intensity of a source that emits monochromatic radiation of frequency 540×10^{12} Hz with a radiant intensity of 1/683 watt per steradian.

canonical momentum   In classical mechanics, a generalized momentum in principle obtained from a Lagrangian, or if no Lagrangian exists, postulated as one of the independent phase space coordinates in a Hamiltonian system. The Lagrangian definition of the *canonical*

momentum p_α conjugate to the particular coordinate x^α is:

$$p_\alpha = \frac{\partial L}{\partial \left(\frac{dx^\alpha}{dt} \right)} \qquad \alpha = 1 \dots N$$

where t is the parameter describing the motion (typically the time).

If the Lagrangian is only a function of the parameter (time), the coordinates x^α, and their parameter derivatives (time derivatives) dx^β/dt, the Lagrange equations are:

$$\frac{d}{dt} \frac{\partial L}{\partial \left(\frac{dx^\alpha}{dt} \right)} = \frac{\partial L}{\partial x^\alpha} ,$$

which can be read:

$$\frac{dp_\alpha}{dt} = \frac{\partial L}{\partial x^\alpha} .$$

Notice that if the Lagrangian is quadratic in the velocities dx^α/dt, then p_α is linear in the dx^β/dt, and the relationship can be inverted to express dx^β/dt as a function of p_α. Only if the Lagrangian is homogeneous of degree one in the velocities is such an inversion impossible. *See* Lagrangian.

canonical transformation   In classical mechanics, transformation of variables and of the Hamiltonian, which preserves the form of the Hamilton equations. The Hamilton equations can be written

$$\begin{aligned} \dot{q}^i &= \partial H/\partial p_i \\ \dot{p}_j &= -\partial H/\partial q^j \end{aligned}$$

where $H = H(p_i, q^i, t)$ is the Hamiltonian. A *canonical transformation* introduces new coordinates (Q^i, P_j) on phase space and a new Hamiltonian $K(Q^i, P_j, t)$, such that the identical motion is described by

$$\begin{aligned} \dot{Q}^i &= \partial K/\partial P_i \\ \dot{P}_j &= -\partial K/\partial Q^j . \end{aligned}$$

The standard method to obtain a canonical transformation is to choose a generating function F, which is a function of one class of "old" variable ($\{q^j\}$ or $\{p_k\}$), one class of new variable ($\{Q^k\}$ or $\{P_m\}$), and time:

$$F = F\left(q^j, P_l, t \right) .$$

Here we mean that F is a function of all of the q^i, all of the P_l, and of time.

The equations connecting the old and new descriptions are then

$$p_l = \frac{\partial F}{\partial q^l}$$

$$Q_k = \frac{\partial F}{\partial P_k}$$

$$K = H + \frac{\partial F}{\partial t}.$$

Similar sets of equations can be derived by introducing different functions F. For instance, $F(p_i, Q^l, t)$ gives

$$q^k = -\partial F/\partial p_k$$

$$P_m = -\frac{\partial F}{\partial Q^m}$$

$$K = H + \frac{\partial F}{\partial t}.$$

For the remaining two possibilities, one has the following rules: $K = H + \partial F/\partial t$ always holds; derivatives with respect to the variables always come with fixed signs. That is, the other two sets of equations involve one of $\partial F/\partial q^l$ or $-\partial F/\partial p_k$ and one of $\partial F/\partial P_m$ and $-\partial F/\partial Q^l$ (note the signs).

Solution of one of these sets or its equivalent provides the coordinate transformation. The solution proceeds as follows: If, for instance $F = F(q^k, P_j, t)$, then $\partial F/\partial q^l$ is generally a function of q^k, P_j, and t. Hence $P_l = \partial F/\partial q^l$ can be inverted to solve for the new momenta P_l in terms of the old coordinates and momenta q^k, p_j, and t. Similarly, $\partial F/\partial P_k$ is a function of q^k, P_l, and t. Substituting in the expression for P_l in terms of q^k, p_l, t already found gives an explicit expression for Q^p in terms of q^l, p_k, t. Finally, the new Hamiltonian K must be expressed entirely in terms of the new variables Q^k, P_l, t using the relationships just found. *See* Hamilton–Jacobi Theory.

Canopus   -0.2 magnitude star, of spectral type FO at RA06^h 23^m 57.1^s dec $-52°41'44"$.

cap cloud   An orographic stratocumulus cloud mass that forms closely over the top of a mountain, arising from the condensation of

water in air as it rises to the peak; a particular case of *pileus.*

CAPE (Convective Available Potential Energy)   Provides a measure of the maximum possible kinetic energy that a statically unstable parcel can acquire, assuming that the parcel ascends without mixing with the environment and instantaneously adjusts to the local environmental pressure. *See* available potential energy.

Capella   0.08 magnitude binary star of spectral types G8+ FO at RA05^{h}16^{m}41.3^s, dec $+45°51'53"$.

capillarity correction   Due to the fact that mercury does not wet glass, in glass/mercury barometers the top surface of the mercury is convex. A correction for the curvature of the surface must be made; the mercury height is slightly less than the expected height for a given air pressure. The corrections are slight, on the order of 2 mm or less.

capillary fringe   The zone directly above the water table that is saturated (soil wetness = 1) and under tension (pressure < 0). The height of the capillary fringe (h_{cr}) is inversely proportional to the radius of the soil grains (r) and directly proportional to the surface tension (σ) and the cosine of the contact angle (θ_c):

$$h_{cr} = \frac{2\sigma \cos \theta_c}{\rho g r}$$

where ρg is the fluid weight density.

capillary wave   A water wave in which the primary restoring force is surface tension; waves with wavelength $\lesssim 1.7$ cm are considered capillary waves.

carbon-14 dating   A method to date a carbonaceous object by measuring the radioactivity of its carbon-14 content; this will determine how long ago the specimen was separated from equilibrium with the atmosphere-plant-animal cycle ("died"). Carbon-14 is continuously produced in the atmosphere by cosmic-ray bombardment and decays with a half-life typically described as 5568 years; dating is accomplished by comparing the carbon-14 activity per unit mass of the object with that in a contemporary sample. Estimates are fairly accurate, out to about 50,000 years.

carbonaceous chondrite   A chondritic meteorite that contains carbon and organic compounds in addition to the rocky minerals. They also contain water-bearing minerals, taken to be evidence of water moving slowly through their interiors not long after formation, and evidence of metamorphism of the meteorite by water and of possible cometary origin. It is believed that *carbonaceous chondrites* are the most primitive meteorites and that minerals they contain were the first minerals to crystallize during the formation of the solar system.

carbon burning   The set of nuclear reactions that convert carbon to oxygen, neon, and other heavier elements. *Carbon burning* occurs in hydrostatic equilibrium in the evolution of stars of more than about 10 solar masses (between helium burning and neon burning). It can also be ignited in degenerate matter, in which case the burning is explosive. This may happen in the cores of stars of about 8 solar masses. It definitely happens at the onset of a supernova of type Ia. Oxygen burning can also be either hydrostatic or explosive.

carbon cycle   a series of nuclear reactions that occur in the interior of stars, and in which carbon 12 acts as a sort of catalyst, not being consumed in the reaction. The chain is $^{12}C \rightarrow {}^{13}N \rightarrow {}^{14}O \rightarrow {}^{14}N \rightarrow {}^{15}O \rightarrow {}^{15}N \rightarrow {}^{12}C + {}^4He$. The two oxygen to nitrogen steps are beta decays and emit neutrinos. The energy release is 26.7MeV, of which 1.7MeV is in neutrinos. This cycle represents about 2% of the energy production in the sun (which is mostly the p-p cycle); however, its temperature dependence is approximately T^{15}, so it dominates in more massive stars.

carbon dioxide   Colorless, odorless trace component gas of the Earth's atmosphere (approximately 370 parts per million), CO_2. Over 99% of the Earth's carbon dioxide is found in oceans. *Carbon dioxide* is evolved in volcanos, and is also perhaps carried to Earth in cometary cores. Carbon dioxide is a major reaction product in combustion involving fossil fuels, all of which is hydrocarbon based. Coal, essentially pure carbon, produces much more CO_2 than does natural gas (CH_4) per unit of energy evolved. Carbon dioxide concentrations have been rising in the Earth's atmosphere throughout the last century. Carbon dioxide is a greenhouse gas, transparent in the visible, opaque in infrared. It is estimated to contribute more than half the current greenhouse warming of the Earth. Carbon dioxide is evolved by animals and sequestered into complex organic compounds by plants.

carbon monoxide   Colorless, odorless, and very toxic trace constituent of the air, CO. Formed in the incomplete combustion of carbon; also apparently produced by near surface ocean-biologic sources. A very tightly bound molecule, as a result of which it is the most common (apart from H_2) in the interstellar medium and, in the atmospheres of cool stars, it generally uses up all of the available C or O, whichever is less abundant, leaving only the other one to form additional detectable molecules.

carbon star   Cool, evolved star of relatively low mass in which sufficient carbon has been mixed to the surface for the carbon-oxygen ratio to exceed unity. As a result, carbon monoxide formation does not use up all the carbon, and other molecules, including CH, CN, and C_2, will appear in the star's spectrum. The stars are generally on the asymptotic giant branch and are seen at somewhat lower luminosity than might be expected from standard theories of convection and mixing. Other chemical peculiarities are common, and there are many subtypes. *See* asymptotic giant branch star.

Caribbean current   An extension of the North Equatorial current that flows northwestward into the Caribbean past the Yucatan peninsula and exits the Caribbean basin westward past the southern tip of Florida.

Carme   Moon of Jupiter, also designated JXI. Discovered by S. Nicholson in 1938, its orbit has an eccentricity of 0.207, an inclination of 164°, and a semimajor axis of 2.26×10^7 km. Its radius is approximately 20 km, its mass 9.5×10^{16} kg, and its density 2.8 g cm^{-3}. Its geometric albedo is not well determined, and it orbits Jupiter (retrograde) once every 692 Earth days.

Carnot cycle An ideal thermodynamic reversible cycle over a substance which consists of the following four processes: 1. isothermal expansion, in which the substance does work and there is inflow of heat Q_2 at a constant temperature T_2; 2. adiabatic expansion, in which the substance is thermally insulated and does work, there is no heat flow, and the temperature of the substance decreases; 3. isothermal compression, work is done on the system and there is outflow of heat Q_1 at a constant temperature T_1; 4. adiabatic compression, the substance is thermally insulated, and work is done on the substance, there is no heat flow, temperature of the substance increases, and at the end of this process the substance returns to its initial state. By applying the second law of thermodynamics to the *Carnot cycle,* one shows that the ratio of heat inflow and heat outflow Q_2/Q_1 for a substance undergoing a Carnot cycle depends only on the temperatures T_2 and T_1 and is independent of the substance. Applying the Carnot cycle to an ideal gas, one further shows that the ratio of heat outflow and heat inflow is equal to the ratio of the corresponding temperatures, i.e., $Q_1/Q_2 = T_1/T_2$. One can then represent an arbitrary reversible cyclic process by a series of Carnot cycles and show that $\oint_{rev} dQ/T = 0$ for any reversible process. This in turn leads to the definition (up to an arbitrary constant) of entropy $S = \int_{rev} dQ/T$ as a thermodynamic function of state. Named after Nicolas Léonard Sadi Carnot (1796–1832).

Carnot efficiency The ratio between the work done and the amount of heat introduced into a system going through a Carnot cycle. The *Carnot efficiency* is equal to the difference between the two temperatures of the isothermal steps of the cycle divided by the higher of the two temperatures.

Carnot engine An ideal heat engine whose working substance goes through the Carnot cycle. A heat engine receives heat at a given temperature, does work, and gives out heat at a lower temperature. The efficiency of a *Carnot engine* or Carnot efficiency is the maximum efficiency possible for a heat engine working between two given temperatures.

Carrington longitude A fixed meridian on the sun as measured from a specified standard meridian. Measured from east to west (0° to 360°) along the sun's equator. *Carrington longitude* rotates with the sun and is a particularly useful coordinate when studying long-lived features on the sun. When combined with a Carrington rotation number, the Carrington longitude is commonly used as an alternative to specifying a time.

Carrington rotation The period of time covering 360° of Carrington longitude. Used to provide a temporal reference frame where the time unit is a solar rotation period. For example, Carrington rotation 1917 corresponds to the time period 9 December 1996 to 5 January 1997, while Carrington rotation 1642 relates to the rotation between 28 May 1976 and 23 June 1976.

Carter–Peter model The dynamics of a superconducting cosmic string is macroscopically describable by means of the duality formalism. This formalism only requires the knowledge of a Lagrangian function $\mathcal{L}$ depending on a state parameter w, the latter being interpretable as the squared gradient of a phase ϕ, namely

$$w = \kappa_0 \gamma^{ab} \partial_a \phi \partial_b \phi \, ,$$

with κ_0 a normalization coefficient, γ^{ab} the inverse of the induced metric on the string surface, and the superscripts a, b representing coordinates on the worldsheet. In the *Carter–Peter* (1995) *model* for describing a conducting cosmic string of the Witten kind, the Lagrangian function involves two separate mass scales, m and $m_\star$ say, respectively describing the energy scale of cosmic string formation and that of current condensation; it takes the form

$$\mathcal{L} = -m^2 - \frac{m_\star^2}{2} \ln \left\{ 1 + \frac{w}{m_\star^2} \right\} .$$

This model implies the existence of a first order pole in the current, as is the case in realistic conducting string models taking into account the microscopic field structure. The figure shows

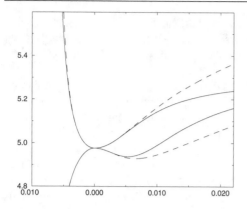

Energy per unit length U (upper curves) and tension T (lower curves) as functions of the sign-preserving square root of the state parameter: the full lines represent the actual values derived from the Witten microscopic model, the dashed line being the values obtained with the Carter–Peter macroscopic model. The fit is almost perfect up to the point where U and T both increase for positive values of the state parameter. This is satisfying since the macroscopic model there ceases to be valid because of instabilities.

the shape of the energy per unit length and tension as function of the state parameter $\sqrt{w}$ together with a comparison with the same functions in the Witten conducting string model. It is clear from the figure that the fit is valid in most of the parameter space except where the string itself is unstable. The string equation of state is different according to the timelike or spacelike character of the current. For the former it is

$$U = T + m_\star^2 \left[\exp\left(2\left(m^2 - T \right) / m_\star^2 \right) - 1 \right] ,$$

while for positive w we have

$$T = U - m_\star^2 \left[1 - \exp\left(-2\left(U - m^2 \right) / m_\star^2 \right) \right] .$$

See conducting string, cosmic string, current carrier (cosmic string), current generation (cosmic string), current instability (cosmic string), phase frequency threshold, summation convention, Witten conducting string, worldsheet geometry.

Cartesian coordinates   A coordinate system in any number of spatial dimensions, where the coordinates $\{x^i\}$ define orthogonal coordinate lines. In such a system, Pythagoris' theorem holds in its simplest form:

$$ds^2 = \delta_{ij} dx^i dx^j ,$$

where the summation convention is assumed for i and j over their range.

Or, in nonflat spaces or spacetimes, a system in which the coordinates have many of the properties of rectangular coordinates, but Pythagoris' theorem must be written as:

$$ds^2 = g_{ij}(x^k) dx^i dx^j ,$$

where $g_{ij}(x^k)$ is the coordinate dependent metric tensor. In this case the description is reserved for coordinates that all have an infinite range, and/or where the metric coefficients $g_{ij}(x^k)$ are "near" δ_{ij} everywhere.

Cartesian coordinates [in a plane]   A relationship between the points of the plane and pairs of ordered numbers called coordinates. The pair of numbers corresponding to each point of the plane is determined by the projection of the point on each of two straight lines or axes which are perpendicular to each other. *Cartesian coordinates* thus establish a nonsingular relationship between pairs of numbers and points in a plane. The point in which the two axes intersect is called the origin. The horizontal axis is called the x-axis, and the vertical axis is called the y-axis. Named after Rene Descartes (1596–1650).

Cartesian coordinates [in space]   A relationship that is established between the points of space and trios of ordered numbers called coordinates. *Cartesian coordinates* are a coordinate system in which the trio of numbers corresponding to each point of space is determined by the projection of the point on each of three straight lines or axes which are perpendicular to each other and intersect in a single point. The positive direction of the z-axis is generally set, such that the vectorial product of a non-null vector along the positive x-axis times a non-null vector along the positive y-axis generates a vector along the positive z-axis; this is called a right-handed coordinate system. Named after Rene Descartes (1596–1650).

Casagrande size classification   A classification of sediment by particle size (diameter). The basis for the Unified Soils Classification commonly used by engineers.

case 1 water   Water whose optical properties are determined primarily by phytoplankton and co-varying colored dissolved organic matter and detritus; not a synonym for open ocean waters.

case 2 water   Water whose optical properties are significantly influenced by colored dissolved organic matter, detritus, mineral particles, bubbles, or other substances whose concentrations do not co-vary with the phytoplankton concentration; not a synonym for coastal waters.

Cassegrainian   A type of reflecting telescope invented by G. Cassegrain, with a small convex secondary mirror mounted in front of the primary mirror, to reflect rays approaching a focus back through a hole in the primary mirror, where they are viewed using a magnifying lens (eyepiece) from behind the telescope.

Cassini   A spacecraft destined for Saturn that was launched on October 15, 1997, and is expected to arrive in July 2004. At this time, it will orbit Saturn for four years. It is a joint mission of NASA, the European Space Agency (ESA), and the Italian Space Agency. The spacecraft consists of an orbiter and ESA's Huygens Titan probe. The latter will be dropped through the atmosphere to the surface of Saturn's largest moon, Titan.

In total the spacecraft weighs 5650 kg. In order to get to Saturn in the nominal 6 years and 9 months, it was initially launched inward, not outward, and aimed toward Venus rather than Saturn, to provide a "gravity-assisted" trajectory. This consists of two Venus flybys, a flyby of Earth, and a flyby of Jupiter.

The mission is named in honor of the seventeenth-century, French-Italian astronomer Jean Dominique Cassini, who discovered the prominent gap in Saturn's main rings, as well as four icy moons. The Titan probe is named in honor of the Dutch scientist Christiaan Huygens, who discovered Titan in 1655, and realized that the strange Saturn "moons" seen by Galileo in 1610 were a ring system surrounding the planet.

Cassini's division   A gap, detectable by small telescope observation, between the A and B rings of Saturn, discovered by Cassini.

Cassiopeia A   A discrete strong radio source at RA $23^h21^m10^s$, dec $+58^o32'05''$ emitting 21-cm radiation. Cas A was apparently created in a supernova explosion in 1667. The constellation Cassiopeia is located almost directly opposite the Big Dipper across the north celestial pole. Five bright stars in the constellation form a rough W (or M) in the sky. (Tycho's supernova also appeared in the constellation in 1572 and disappeared in 1574.)

Castor   Double star (alpha Geminorum A, B). A is an A1 type 1.94 magnitude star located at RA $07^h34.4^m$, dec $+31°54'$; B is a type A2, 2.92 magnitude star located at RA $07^h34.5^m$, dec $+31°54'$.

Castor, John I.   Astrophysicist. In 1976, in collaboration with David Abbott and Richard Klein, developed the theory of winds in early type stars (CAK theory). Through hydrodynamic models and atomic data, they showed that the total line-radiation pressure is the probable mechanism that drives the wind in these systems, being able to account for the observed wind speeds, wind mass-loss rates, and general form of the ultraviolet P-Cygni line profiles through which the wind was originally detected.

cataclysmic variable (cataclysmic binary)
A star that suddenly and unpredictably brightens by several magnitudes, most likely by a transfer of stellar material from one star to its close companion. A few weeks or so after the eruptive event, the star returns to its original brightness, indicating that a permanent transformation or evolution of the system has not taken place. Stars exhibit some or all of the following characteristics: flat or blue optical spectral distribution, broad emission or absorption lines of hydrogen and helium, rapid variability, marked aperiodical changes in optical brightness, and low X-ray luminosity. Binary star in which one component is a white dwarf and one is a main sequence star or red giant, where the stars are close enough together that material flows onto the white dwarf from its companion, either through Roche Lobe overflow or in a stellar wind. Pairs without such transfer are called V471 Tauri stars. Mass transfer at a rate of 10^{-7-8} solar masses per year is relatively sta-

ble, leading to systems whose brightness varies rather little. These are nova-like variables and symbiotic stars. Another kind of cataclysmic variable star is the flare star, whose prototype is UV Ceti. Flare stars are intrinsically cool red stars (of type M or less commonly K) on the main sequence that will unpredictably brighten by up to two magnitudes over the course of a few seconds and then fade back to normal in 20 or 30 minutes. Most of these stars are known to have close companions. Mass transfer at 10^{-9} solar masses per year is unstable, with an accretion disk building up for a while and then dumping material onto the white dwarf rapidly. These events are the outbursts of the dwarf novae. In most cases, hydrogen gas builds up on the surface of the white dwarf until it is somewhat degenerate, at which point it burns explosively, producing a nova. If this happens frequently enough to have been seen twice or more in historic times, the system is called a recurrent nova. It is sometimes possible for similar accretion (perhaps of helium or heavier elements) to trigger degenerate ignition of the carbon-oxygen core of the white dwarf. This leads to a supernova explosion of Type Ia.

cataclysmic variable [binary models of]
Thought to be binary systems composed of a white dwarf (primary), a main sequence star (secondary), and characterized by accreting mass flowing from the secondary towards the primary. Typically the accreting mass forms a disk or accretion disk around the primary (with the exception of magnetic variables).

cataclysmic variable [galactic distribution]
Due to interstellar absorption, *cataclysmic variables* (which have luminosities on the order of solar luminosity) cannot be observed at a distance greater than 1 kpc. In spite of this, 380 CVs had been detected by 1976. Thus, it seems that CVs are relatively common astrophysical objects.

cataclysmic variable [outbursts]
Some CVs present "outbursts" or periodical increases in luminosity with respect to their usual quiescent state. Classical novae are *cataclysmic variables* that have been known to present only one outburst; recurrent novae are cataclysmic vari-

ables that present outbursts periodically every 10 years or more; dwarf novae are cataclysmic variables that present outbursts every few weeks or months.

cataclysmic variables [phenomenological classification of]
Cataclysmic variables are classified into five groups. Classical Novae: cataclysmic variables that have been known to present only one outburst. Recurrent Novae: cataclysmic variables that present outbursts every 10 years or more. Dwarf Novae: cataclysmic variables that present outbursts every few weeks or months. Nova-like: cataclysmic variables that have not been known to present outbursts but have the same spectroscopic characteristics as other cataclysmic variables when they are quiescent. Magnetic Variables: cataclysmic variables that present relatively strong magnetic fields.

cataclysmic variables [physical parameters]
The luminosities of *cataclysmic variables* vary between 0.001 and 10 solar luminosities. Binary orbital periods are typically between 0.7 hours and 1 day. Mass accretion rates vary from 10^{-10} to 10^{-7} solar masses per year. The primary star (white dwarf) typically has a mass of ~ 0.6 solar masses, varying between 0.4 and 0.9 solar masses, and a radius of ~ 0.01 solar radii. The secondary is a main sequence star with a mass lower than its companion star since the primary has evolved through the main sequence faster.

catastrophic formation of solar system
A theory attributing the formation of the solar system to a collision of another massive object (presumably another star) with the sun, which threw material out of the sun, or to a close encounter with another star — a tidal encounter. Now out of favor because it suggests solar systems are rare, since such encounters are rare, while recent observations provide evidence for planetary systems around a number of local stars and even around neutron stars.

Cauchy singularity
Any region of spacetime where violations of causality can occur because the deterministic evolution of physical systems from initial data is not preserved. *See* naked singularity, white hole.

causal boundary (Geroch, Kronheimer and Penrose, 1972). A boundary construction using only the causal properties of a space-time.

causal curve In relativity, a time-like or null curve, i.e., one that can be the history of a particle moving no faster than the speed of light. *See* null vector, time-like vector.

causal future/past The *causal future/past* $J^{\pm}(S)$ of a set S is defined as the union of all points that can be reached from S by a future/past-directed causal curve. *See* causal curve.

causality relations A trip from point x to point y of a causal space is an oriented curve with past endpoint x and future endpoint y consisting of future-oriented time-like geodesic segments. A causal trip is similarly defined, with the time-like geodesic segments replaced by causal geodesic segments.
1. Point x chronologically precedes point y (i.e., $x \ll y$) if and only if there exists a trip from x to y.
2. Point x causally precedes point y (i.e., $x \prec y$) if and only if there exists a causal trip from x to y. For an arbitrary point x of a space-time, the relation $x \prec x$ holds, since a causal trip may consist of a single point.

cavity, magnetic A region of weak magnetic field created by a relatively dense plasma expelling its field lines. Produced in some distant barium releases.

CCD Acronym of charge-coupled device, presently the most widely used detector in optical astronomy. The CCD is a two-dimensional detector, like a photographic film or plate. Each picture element (pixel) of a CCD is a photodiode where electrons, freed by the incoming radiation via photoelectric effect, are held in a positive potential for an arbitrary time (i.e., the exposure time). At readout time, an oscillating potential transfers the stored charges from pixel to pixel across each row of pixels to an output electrode where the charges are measured. Unlike photographic plates, CCD possess linearity of response (i.e., the number of electrons freed is proportional to the number of photons detected)

and detection quantum efficiency (i.e., the ratio between detected and incident photons) which is very high, close to 100% for red light. The pixel size can be as small as 15 μm × 15 μm; arrays of 2048 × 2048 pixels are among the largest available CCDs.

cD galaxies Luminous, large-size elliptical galaxies that are located at the center of dense clusters of galaxies. The notation "cD" indicates a cluster D galaxy in the Yerkes classification scheme. The photometric profile of a *cD galaxy* is different from that of other elliptical galaxies, since there is an excess of light at large radii over the prediction of the de Vaucouleurs law. cD galaxies possess a stellar halo that may extend up to 1 Mpc, exceptional mass and luminosity, and are thought to result from multiple merging of galaxies and from cannibalism of smaller galaxies belonging to the cluster. An example of a cD galaxy is Messier 87, located at the center of the Virgo cluster. *See* de Vaucouleurs' law, Yerkes classification scheme of galaxies.

Celaeno Magnitude 5.4 type B7 star at RA 03^{h}44^m, dec +24.17'; one of the "seven sisters" of the Pleiades.

celerity Phase speed; the speed of a wave deduced from tracking individual wave crests.

celerity The translational speed of a wave crest. Given by $C = L/T$, where C is *celerity* in units of length per time, L is wavelength, and T is period.

celestial equator Extension to the celestial sphere of the plane of the Earth's equator; the set of locations on the celestial sphere that can appear directly overhead at the Earth's equator. *See* nutation, precession.

celestial poles The points on the celestial sphere that are directly overhead the Earth's poles. *See* nutation, precession.

celestial sphere The imaginary sphere representing the appearance of the sky, in which all celestial objects are visualized at the same distance from the Earth and at different directions from the Earth. The *celestial sphere* thus surrounds the Earth, and locations on the sphere are given by the two angles necessary to define a given direction.

Celsius, Anders   Astronomer (1701–1744). Proposed the Centigrade temperature scale.

Celsius scale   Also called Centigrade Scale, scale for measuring temperature in which the melting point of ice is 0°, and the boiling point of water is 100°. This definition has been superseded by the International Temperature Scale 1968, which is expressed both in Kelvin and degrees Celsius. Named after Anders Celsius (1701–1744).

Centaur   An "outer planet crosser." A minor body whose heliocentric orbit is between Jupiter and Neptune and typically crosses the orbits of one of the other outer giant planets (Saturn, Uranus, Neptune). The orbits of the *Centaurs* are dynamically unstable due to interactions with the giant planets, so they must be transition objects from a larger reservoir of small bodies to potentially active inner solar system objects. The Kuiper belt is believed to be this source reservoir.

Centaurus A (Cen A)   Active galaxy at RA$13^h25^m28^s$, dec $-43°01'11''$ in constellation Centaurus. Also classified as NGC5128. Distance approximately 3.4Mpc. Large, elliptical galaxy with strong dust lanes seen in the visible and infrared, strong jets seen in radio (double lobed) and in the X-ray (single lobed). Also visible in the gamma ray range.

center of figure   To a first approximation the earth is a sphere. However, the rotation of the earth creates a flattening at the poles and an equatorial bulge. The shape of the earth can be represented as an oblate spheroid with equatorial radius larger than the polar radius by about a factor of 1/300. The center of the best fit spheroid to the actual shape of the earth is the *center of figure*. There is an offset between the center of figure and the center of mass of a few kilometers.

center of mass   In Newtonian mechanics, the "average" location $\mathbf{X}$ of the mass, given in com-

ponents by

$$X^a = \frac{\sum m_{(i)} x_{(i)}^a}{\sum m_{(i)}}, a = 1, 2, 3,$$

where the sums indicated by Σ are over all the masses, labeled (i). In relativistic mechanics there are many inequivalent formulations which all reproduce this nonrelativistic result in the limit of small velocities.

Centigrade scale   *See* Celsius scale.

centimeter burst   A transient solar emission of radiation at radio wavelengths $\sim$1 to 10 cm. *Centimeter bursts* provide a powerful diagnostic of energetic electrons in the solar atmosphere, especially during solar flares. The production mechanisms include thermal bremsstrahlung, gyrosynchrotron radiation, and collective plasma processes.

central meridian passage   The passage of a solar feature across the longitude meridian that passes through the apparent center of the solar disk. Useful for identifying a characteristic time during the transit of a solar feature (e.g., an active region) across the solar disk.

central peak   A mound of deformed and fractured rock found in the center of many impact craters. This material originally existed under the crater floor and was uplifted by the stresses associated with the impact event. *Central peaks* are believed to form by hydrodynamic flow during crater collapse. The target material behaves as a Bingham fluid, which displays properties of viscous fluids yet has a definite plastic yield stress. As the crater formed in this target material collapses, shear stresses cause material to be jetted up in the center of the crater. When the shear stresses fall below the cohesion of the target material, the motion of this central jet ceases, and the material freezes into a central peak. Central peaks are common features of complex craters but are associated with the smaller complex craters. As crater size increases, central peaks tend to be replaced first by craters with a ring of central peaks (called a peak ring), then by a combination of central peaks surrounded by a peak ring, and finally a multiple-ring basin.

central pit A depression found in the center of many impact craters. *Central pits* are particularly common on bodies where ice is (or believed to be) a major component of the upper crust, such as Jupiter's icy moons of Ganymede and Callisto and on Mars. The formation of central pits is not well understood but is believed to be related to the vaporization of crustal ice during crater formation.

centrifugal force The conservative force that arises when Newton's equations are applied in a rotating frame; the apparent force acting on a body of mass m that is rotating in a circle around a central point when observed from a reference system that is rotating with the body:

$$\mathbf{F} = m \, \boldsymbol{\omega} \times (\boldsymbol{\omega} \times \mathbf{r})$$

where $\boldsymbol{\omega}$ is the angular velocity vector, and $\mathbf{r}$ is the radius vector from the origin. The magnitude of the force can also be expressed in terms of the distance d from the axis and the velocity v in a circle or orthogonal to the axis: $F = mv^2/d$; the direction of this force is perpendicularly outward from the axis. These terms appear in dynamical equations when they are expressed in a rotating reference system or when they are expressed in cylindrical or spherical coordinates rather than Cartesian coordinates. These terms are generally treated as if they were actual forces.

centroid moment tensor (CMT) In seismology, a moment tensor obtained when a point source is put at a centroid of a source region. Dziewonski et al. began to determine CMT routinely in 1981 for earthquakes with $Ms \geq 5.5$. Recently, CMT has been used to estimate seismic moment and force systems on a source region of an earthquake, replacing a fault plane solution determined from polarity of P- and S-waves. Six independent components of the moment tensor and four point-source hypocentral parameters can be determined simultaneously through iterative inversion of a very long period ($T > 40$ s) body wave train between the P-wave arrival and the onset of the fundamental modes and mantle waves ($T > 135$ s). Not assuming a deviatoric source, some focal mechanisms have been found to have large non-double couple components.

Cepheid variable A giant or supergiant star crossing the instability strip in the HR diagram at spectral type F-G. The stars can be crossing either from blue to red as they first leave the main sequence or from red to blue in later evolutionary phases. In either case, the stars are unstable to radial pulsation because hydrogen at their surfaces is partially ionized on average and acts like a tap or faucet that turns the flow of outward radiation up or down, depending on its exact temperature. As a result, the stars change their size, brightness, and color (temperature) in very regular, periodic patterns 1 to 50 days, and with a change of 0.5 to 1 magnitude. Because the counterbalancing force is gravity, the pulsation period, P, is roughly equal to $(G\rho)^{-1/2}$, where ρ is the star's average density and G is Newton's constant of gravity. This, in turn, means that there is a relationship between period, luminosity, and color for whole populations of Cepheids. Thus, if you measure the period and color of a Cepheid, you know its real brightness and can, in turn, learn its distance from its apparent brightness. Cepheids can be singled out only in relatively nearby galaxies, and with the Hubble Space Telescope, they have been measured out to galaxies in the Virgo Cluster. Classical Cepheids are population I stars that are massive young objects. The W Virginis Cepheids belong to the older population II stars. The prototype star, δ Cephei, was discovered to vary in 1784. This period-luminosity relation was originally discovered for the Cepheids in the Large Magellanic Cloud by Henrietta S. Leavitt in the first decade of the 20th century. Once an independent measure of the distance to a few nearby Cepheids was accomplished (through spectroscopy and other methods), a period-luminosity relation was defined. The calibration of the period-luminosity relation is the underpinning of our distance measurements to the nearest galaxies (within 400 Mpc), which then calibrates the Hubble constant, the proportionality constant that relates red shift of distant galaxies to the expansion of the universe.

Čerenkov radiation The Čerenkov effect (in Russia: Vavilov–Čerenkov effect*) was first observed with the naked eye by P.A. Čerenkov in 1934 as a "feeble visible radiation" from fast electrons (Vavilov's interpretation, 1934) due to

γ-rays in pure water. In 1937 Čerenkov confirmed the theoretical prediction by Frank and Tamm about the sharp angular dependence of this radiation: Frank and Tamm earlier in the same year had given a complete classical explanation of the Čerenkov effect. Its quantum theory was given by Ginsburg in 1940. In 1958 Čerenkov, Tamm, and Frank were awarded the Nobel Prize.

Čerenkov radiation occurs when a charged particle is moving in a transparent medium with a velocity greater than that of light in this medium. This radiation is emitted with a conical front, the direction of motion of the particle being its axis making the angle θ with the cone's generatrix, while $\cos\theta = v/u$. Here v is the velocity of light in the medium and u, the velocity of the charged particle ($u > v$). This picture is very similar to that of waves on the surface of water when a boat is moving faster than the waves propagate. In the theory of Čerenkov's radiation, an important role is played by the medium's dispersion (the frequency dependence of the electric and magnetic properties of the medium, hence of its refraction index).

In fact, the classical theory of what we now call Čerenkov radiation, as well as of some similar phenomena involving the superluminal (in a vacuum or in refractive media), was developed as early as 1888 by Heaviside: the first publication in the *Electrician,* with subsequent publications in 1889 and 1892, and especially in 1912 (comprising more than 200 pages of the third volume of his Electromagnetic Theory). In 1904 and 1905 Sommerfeld published four papers on the same effect, though only in a vacuum; there he mentioned, at least once, the previous work of Heaviside, as did Th. des Coudres in 1900. Subsequently this early development was lost until 1979, when the references were again cited.

The Čerenkov effect is the basis of Čerenkov counters of ultrarelativistic charged particles which essentially consist of a (usually Plexiglas) transparent dielectric cylinder and some photomultipliers to detect the Čerenkov radiation emitted in the dielectric by these particles.

Ceres   The first observed asteroid, discovered by Giuseppe Piazzi in 1801. It has a diameter of 913 km. It orbits in the main asteroid belt. Its average distance to the sun agrees with the distance of the "missing planet" predicted by Bode's Law.

CGS (Centimeter-Gram-Second)   The system of measurement that uses these units for distance, mass, and time. This system incorporates the use of electrostatic units (esu) or electromagnetic units (emu) in the description of electromagnetic phenomena. The magnetic permeability (μ) and dielectric constant (ϵ) are dimensionless in the CGS system.

chalcophile   Elements that display an affinity for sulfur are called chalcophile elements. Such elements are readily soluble in iron monosulfide melts and tend to be found concentrated in sulfide ores.

Challenger Deep   Deepest part of the Marianas Trench, deepest ocean water. Located at approximately $142°15'E$, $11°20'N$ off the coast of Guam. Bottom depth approximately 11,000 m (36,100 ft); various "deepest" measurements range within 100 m of this value.

Chandler wobble   Although the Earth's pole of rotation (i.e., the axis about which the Earth is spinning) lies close to the axis of the solid Earth's largest principle moment of inertia, they are generally not quite coincident because the solid Earth has a small amount of angular momentum ΔL oriented perpendicular to the axis of the largest principle moment of inertia. There is a small angle between the two axes, and the action of ΔL as viewed by an observer on the planet is to cause the axis of rotation to swivel around the axis of the maximum moment of inertia. This is the free precession of the planet, as opposed to the forced precession associated with tidal couples due to the moon and sun, and is called the *Chandler wobble.* The theory of the rotation of solid bodies suggests that the period of the wobble should be approximately 300 days, but in fact the period is around 435 days.

---

*Čerenkov then was a student of S.I. Vavilov, performing a research in fluorescence as a part of his Ph.D. thesis under Vavilov's supervision.

This discrepancy is thought to be due to the fact that the Earth is deformable rather than rigid and also due to core-mantle and mantle-ocean coupling. The damping and excitation of the wobble are not entirely understood.

Chandrasekhar limit (or Chandrasekhar mass) The maximum mass of cold matter that can be supported by degeneracy pressure, especially of electrons. This sets the maximum possible masses for white dwarfs and for the cores of massive stars before they collapse (*see* supernovae, type II). The first correct calculations were done by Chandrasekhar in 1930–33, for which he eventually received a Nobel Prize. The maximum possible mass for a white dwarf is about 1.4 solar masses if it is made of helium, carbon, and oxygen, rather less if it is made of heavier elements. The corresponding limit for a neutron star is sometimes called the Oppenheimer–Volkoff limit.

chaos Property of system described by a set of nonlinear equations, such that for at least some initial data the deterministic solutions diverge exponentially from each other in time.

chaotic cosmology The suggestion by Charles Misner (1967) that the universe began in a highly chaotic state but smoothed out into the present ordered state in the course of time via dissipative processes.

chaotic system A complex system exhibiting chaos.

chaotic terrain Part of the highland-lowland boundary of Mars, located between longitudes 10° W and 50° W and latitudes 20° S and 10° N, in the old cratered terrain between Lunae Planum and Chryse Planitia. It regularly connects with the large outflow channels draining northwards and merges with Valles Marineris to the west. Crater counts indicate it formed between the early to late Hesperian.

By analogy with Earth, it has been proposed to be terrain in which the ground has collapsed, to produce jumbled arrays of blocks at lower elevations than the surroundings. The depth to which collapse occurred becomes less to the east, away from Valles Marineris. Thus, its for-

mation may in some way be related to the formation of Valles Marineris.

Chapman layer Simple formal description for an atmospheric layer such as ionospheric layers. Assumptions are the atmosphere consists of one atomic species only, and the incoming radiation is monochromatic. According to the barometric height formula, the atmospheric density decreases with increasing height. The intensity of the incoming ionizing electromagnic radiation increases with increasing height: it is maximal at the top of the atmosphere and then is absorbed according to Bougert–Lambert–Beer's law. Both effects combined lead to a Chapman profile: at a certain height, the ionization, and, therefore, also the charge density, is highest. Below, it decreases as the intensity of the ionizing radiation decreases. At higher altitudes, it decreases too because the number of particles available for ionization decreases.

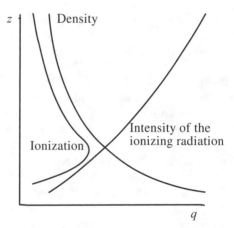

Chapman layer.

characteristic earthquake Major faults in active tectonic regions have large *characteristic earthquakes*. The 1906 San Francisco earthquake was the last characteristic earthquake on the northern section of the San Andreas fault. The southern section of the San Andreas fault had a characteristic earthquake in 1857.

characteristics, method of A mathematical technique, where a problem represented by a set of partial differential equations is converted to an ordinary differential equation problem. The solution may be obtained graphically or by numerical techniques. Used in the study of open channel flow and elsewhere.

charge exchange   An atomic process that may occur when a fast ion collides with a neutral atom, in which an electron is transferred from the atom to the ion. As a result, the atom becomes a slow ion, while the ion becomes an energetic neutral atom (ENA). It strongly depends on energy (also on the type of ion) and is much more pronounced around 5 keV than at 50 keV.

Charge exchange is believed to be the major mechanism by which ions produced in a magnetic storm are lost, leading to the decay of the enhanced ring current associated with the storm. Ions of that ring current are magnetically trapped by the Earth, but if they change into ENAs, the magnetic field no longer affects them, and they move in straight lines, making them a possible means of remotely sensing the ring current.

Charles' Law   At constant pressure, the volume of an ideal gas is directly proportional to the absolute temperature of that gas.

Charon   Moon of Pluto, although it is close enough in size to Pluto to be considered a double planet. It was discovered by Christy in 1978. Its orbit has an eccentricity of 0.0001, an inclination of 98.3°, and a semimajor axis of 1.96×10^4 km. Both Pluto and *Charon* rotate in such a way as to always keep the same face towards each other. Its radius is not well determined, with measurements giving between 590 and 630 km. Its mass and density are not well known. Its geometric albedo is 0.36, and it orbits every 6.387 Earth days.

chemically peculiar star   A star whose surface composition deviates from the standard set by the sun (3/4 hydrogen, 1/4 helium, a few percent of everything else in standard proportions: *See* cosmic abundance). This happens for one of three reasons: (1) products of nuclear reactions in the star itself have been mixed to the surface (as in carbon stars or R Coronae Borealis stars), (2) products of nuclear reactions in a companion star have been transferred to the observed, or (3) a combination of radiation pressure and

a very steady atmosphere has separated out unusual atoms, often in regions of strong magnetic fields. Europium and strontium are among the elements sometimes greatly overabundant in the third sort of star, called an Ap star (A for its spectral type, p for peculiar).

chenier   Beach ridge.

Chezy coefficient, Chezy equation   The *Chezy equation* is used to relate flow speed to flow geometry and other parameters. The equation is $V = C\sqrt{RS_f}$, where V is flow speed, C is the Chezy coefficient, $R = A/P =$ hydraulic radius, and S_f is friction slope (head loss per unit length). The *Chezy coefficient* is determined primarily by channel roughness but decreases with increasing roughness. An alternative to the Manning equation.

chlorinity   The concentration of chloride ions by mass in seawater (g/kg). Typical ocean *chlorinity* is of order 19 g/kg of seawater. There is an observed constant ratio of chlorinity to salinity: Salinity in parts per thousand = (1.80655) (chlorinity in parts per thousand). *See* salinity.

chlorophyll   Chemical compounds occurring in plants that enable radiant energy to be converted to chemical energy in the process of photosynthesis; there are several types (e.g., denoted as chl *a, b,* etc.) with chl *a* typically the most abundant.

chondrite   A relatively abundant stony meteorite characterized by the presence of small ($\sim$ 1 mm) glassy spherical inclusions called chondrules. *Chondrites* are further subdivided into ordinary, carbonaceous, and enstatite chondrites, depending on details of composition and thermal processing.

chondrite, carbonaceous   A form of chondrite with high abundances of water, organic, and other volatile materials. This, and the fact that the heavier elements are present in roughly solar proportions, indicates that these bodies have been little processed. These meteorites are further classified by their volatile content into CI (or C1), which are most nearly solar in composition, CII (CM) and CIII (CO or CV), which are successively poorer in volatile content. C1 chondrites do not contain chondrules but are chemically very similar to other chondrites.

chondrite, ordinary   The most common form of meteorite. Composed mostly of olivine with significant amounts of other metallic silicates. *Ordinary chondrites* are further classified according to their iron content into H (high iron), L (low iron), and LL (very low iron) chondrites. An additional subdivision, E (enstatite), is characterized by lower Mg/Si ratios, which give rise to a composition richer in enstatite and containing little olivine.

chondrule   Small spheroidal ($\sim$ 1 mm) glassy bead of silicates (olivine or enstatite) found within stony meteorites, called chondrites. *Chondrules* generally have a composition very similar to the matrix material of the chondrite in which they are embedded, but this material has undergone an episode of melting and rapid quenching early in the history of solar system formation. The actual process causing the melting is poorly understood.

Christoffel symbol   *See* affine connection, metricity of covariant derivative.

chromatic aberration   A situation arising in lens-based (refracting) optical instruments, in which light of different colors cannot be brought into focus at the same point, arising because glass or other lens material is dispersive and so produces different deviation of the path of light of different colors.

chromosphere   The layer in the sun or a cool star between the photosphere and corona. The layer is relatively thin and at a temperature of about 10,000 K in the sun. During a total solar eclipse, it is seen as a red ring (hence the name) around the moon's shadow. Outside eclipse, it contributes emission lines to the spectrum of the sun (or star). The intensity of the chromosphere varies through the solar cycle.

chromospheric evaporation   The upward flow of hot plasma in a solar flare resulting from the fast deposition of energy in the chro-

mosphere. In the *chromospheric evaporation* process, the energy released during a solar flare rapidly dissipates in the chromosphere (typically assumed to involve non-thermal electron beams or a thermal conduction front). The chromosphere is suddenly heated to coronal temperatures and subsequently expands upward. Evidence for these flows generally comes from blueshifts measured in spectral lines, such as those observed by a Bragg Crystal Spectrometer.

chromospheric heating   The process behind the enhanced temperature of the sun's chromosphere above that of the sun's surface (photosphere). Radiation is the bulk energy loss mechanism in the chromosphere, conductive losses being negligible, and both the photosphere and corona contribute to its heating. The amount of heating required to balance radiative losses in the chromosphere is about 4×10^3 Wm^{-2} for quiet sun and coronal hole regions. This rises to $\sim 2 \times 10^4$ Wm^{-2} for the chromosphere in active regions. The exact nature of the heating mechanism is uncertain.

chronological future/past   The *chronological future/past* $I^{\pm}(S)$ of a set S is defined as the union of all points that can be reached from S by a future/past-directed time-like curve. (A time-like curve must be of non-zero extent, thus isolated points are excluded.)

CHUMP (Charged Hypothetical Ultra Massive Particle)   In the framework of grand unification models, hypothetical *CHUMPs* are among the candidates to explain the missing mass problem in the universe (the fact that the observed luminous matter cannot account for the dynamical properties observed on large scales). As they carry an electric charge, they need to be very massive, otherwise they would already have been detected. An example of such a particle is the vorton, which would appear in theories having current-carrying cosmic strings. *See* dark matter, vorton.

cigar distribution   In plasma physics, a pitch angle distribution in which the highest intensities are field aligned, so that a contour of the flux density against polar angle θ resembles a cigar. *See* pancake distribution, conics.

circulation In fluid flow, the *circulation* of a closed loop (a circuit) C, which is the boundary of a topological disk S, is

$$\int_{C=\delta S} \mathbf{v} \cdot \mathbf{dx}$$

that is fixed in the fluid, i.e., moves with the fluid elements. According to Stokes' theorem, this is equal to

$$\int_S \nabla \times \mathbf{v} \cdot \mathbf{n} dS ,$$

where $\nabla \times \mathbf{v}$ is the curl of $\mathbf{v}$, also called the circulation of $\mathbf{v}$. Here $\mathbf{n}$ is the unit normal to the surface S. If the circulation vanishes, then $\mathbf{v} = \nabla \phi$ for some scalar potential ϕ.

cirrus *See* cloud classification.

CISK (Convective Instability of the Second Kind) An atmospheric instability responsible for hurricane and other intense "warm core" disturbances which form over the sea. In contrast to baroclinic instability, which feeds upon preexisting horizontal temperature gradients, these storms generate their own "warm core" by absorbing latent and sensible heat from the sea surface and converting latent heat into sensible heat in the rain areas near their centers.

civil time The time in use by a nation's civilian population, as legislated by its government. It approximates the standard time in its time zone, aside from daylight savings adjustments.

civil twilight *See* twilight.

Clapeyron–Clausius equation An equation that gives the relation between the vapor pressure of a liquid and its temperature. It is

$$\frac{dP}{dT} = \frac{L_{12}}{T(v_2 - v_1)}$$

where L_{12} is the latent heat absorbed for transferring from phase 1 to phase 2, v_1 and v_2 are the volumes of phase 1 and phase 2, respectively.

Applied to a water and water vapor system, it is

$$\frac{dE_{sw}}{dT} = \frac{L_v}{T(v_v - v_w)} \approx \frac{L_v}{T \cdot v_v}$$

where E_{sw} is saturated vapor pressure at the water surface, L_v is condensation latent heat, and v_v and v_w are the volumes of unit mass vapor and water. Thus

$$E_{sw} = E_s(T_0) \cdot \exp\left(\frac{L_v}{R_v T_0} - \frac{L_v}{R_v T}\right)$$

where R_v is the water vapor gas constant, and T_0 is reference temperature. If $T_0 = 273.15\,K$, then $E_s(T_0) = 6.11\,hPa$.

Applied to an ice and water vapor system, it is

$$\frac{dE_{si}}{dT} \approx \frac{L_s}{T \cdot v_v}$$

$$E_{si} = E_s(T_0) \cdot \exp\left(\frac{L_s}{R_v T_0} - \frac{L_s}{R_v T}\right)$$

where E_{si} is saturated vapor pressure at the ice surface, L_s is sublimation latent heat. In general 273.15 K is the reference temperature, then

$$E_{sw} = 6.11 \cdot \exp\left[\frac{L_v}{R_w}\left(\frac{1}{273.15} - \frac{1}{T}\right)\right]$$

$$E_{si} = 6.11 \cdot \exp\left[\frac{L_s}{R_w}\left(\frac{1}{273.15} - \frac{1}{T}\right)\right].$$

The unit of E_{sw} and E_{si} is hPa.

clapotis A term used in the study of water waves on a free surface, such as waves on the surface of the ocean. Denotes a complete standing wave — a wave which does not travel horizontally but instead has distinct nodes and antinodes. Results from complete reflection of waves from a structure; may also arise in other scenarios.

classical radius of electron The quantity $e^2/mc^2 = 2.82 \times 10^{-13}$ cm.

clathrate When water freezes, it can trap other, more volatile gases in "cages" formed by the crystal structure. As long as the water ice is solid, it will hold the trapped gas, even at temperatures where that gas would ordinarily sublimate. The ice-gas combination is called a *clathrate,* and the vapor pressure above such a clathrate generally has a value between that of the water and that of the gas.

clay A fine-grained sediment, with particle sizes less than 4 μm (definitions vary). Interparticle forces due to electrical charges are significant compared to particle weight.

clear air turbulence (CAT) Vigorous small-scale motions within the lowest kilometer of the atmosphere. Most clear air turbulence is the result of small-scale wavelike undulations that develop spontaneously when the vertical shear of the horizontal flow exceeds some critical value.

cleft, polar *See* cusp, polar.

climate regimes (Charney and DeVore, 1979). The equilibrium states in the extratropical circulation in the atmosphere.

climatic optimum Mid-Holocene event in which summer temperatures in the Northern hemisphere averaged 1 to 2°C above current temperatures. Dated as the period approximately 8000 to 5000 years ago.

closed magnetosphere A model of a magnetosphere (usually the Earth's), in which all planetary magnetic field lines are confined inside some surface, usually identified with the magnetopause. The Earth's magnetosphere resembles a *closed magnetosphere,* except for two details: (1) on the night side, field lines extend to great distances and no well-defined closure surface has been observed there; (2) indications exist that a small amount of interconnection usually exists. *See* open magnetosphere.

closed universe A model of the universe that is finite in total volume and age. It evolves from a "big bang" to a point of maximum expansion before contracting back to a "big crunch" of high density and temperature.

cloud classification A scheme of distinguishing and grouping clouds based on their appearance, elevation, and the physical processes generating them. The World Meteorological Organization classifies 10 genera in three major groups — cumulus, stratus, and cirrus, by criteria essentially according to the cloud formation processes. The major cloud genera are, cirrus (C_i), cirrocumulus (C_c), cirrostratus (C_s), altocumulus (A_c), altostratus (A_s), nimbostratus (N_s), stratocumulus (S_c), stratus (S_t), cumulus (C_u), and cumulonimbus (C_b). Clouds can also be referred to, according to their composition, as water clouds, ice clouds, or mixed clouds.

clump star A metal-rich star of a few solar masses that has begun helium burning. Because of the difference in opacity, such stars do not reach the giant branch but instead spend a long period of time in the red clump in the HR diagram, $B - V = 1mag$, and $M = 0mag$.

CNO cycle (or tricycle) A set of nuclear reactions in which hydrogen is fused to helium, with the liberation of about 7×10^{18} ergs per gram, 5 to 10% of which is lost in neutrinos, depending on the precise temperature at which the reactions occur, generally in the range 15–25 $\times 10^6$ K. These reactions, which require the presence of carbon, nitrogen, or oxygen as catalysts, are the primary source of energy for stars larger than about 2 solar masses during their main sequence lifetimes and for all stars during the red giant phase. They are the primary source of nitrogen in the universe and contribute to the production of ^{13}C, ^{15}N, ^{17}O, ^{18}O, and, through extensions to heavier catalysts at higher temperatures, probably also to ^{19}F, ^{21}Ne, ^{22}Ne, ^{23}Na, and perhaps ^{26}Al.

The primary reaction sequence is (see *nuclear reactions* for explanation of the symbols):

$$^{12}C(p, \gamma)^{13}N(e^+ v_e)^{13}C(p, \gamma)^{14}N(p, \gamma)$$
$$^{15}O(e^+ v_e)^{15}N(p, \alpha)^{12}C ,$$

thus the net effect is to convert four protons (hydrogen atoms) to one helium nucleus or alpha particle, returning the ^{12}C to its original form. Other closures include

$$^{15}N(p, \gamma)^{16}O(p, \gamma)^{17}F(e^+ v_e)^{17}O(p, \gamma)$$
$$^{18}F(e^+ v_e)^{18}O(p, \alpha)^{15}N .$$

See nuclear reactions.

cnoidal wave A nonlinear theory for periodic waves, attributed to Korteweg and Devries (1895), in which the water surface profile is expressed in terms of a Jacobian elliptical integral, cn(u).

coastal jet a coastal current system that intensifies toward the coast.

coastally trapped waves   Waves generated by wind blowing over a stratified ocean with a shelf topography. These are hybrid waves with characteristics of both Kelvin and continental shelf waves. In the northern hemisphere, waves propagate alongshore with the coast on the right.

coastal ocean   Shallow water generally situated over continental shelves; often, but not always, case 2 waters. *See* case 2 water.

coastal upwelling or downwelling   The resultant ocean volume transport in the Ekman layer is at right angles to and to the right of the wind direction in the northern hemisphere. If the wind blows parallel to the coast and is directed with the coast on the right/left in the northern/southern hemisphere, water in the Ekman layer will tend to move away from the coast and will be replaced by water moving upward from below the layer. This is called the *coastal upwelling*. Regions particularly noted for coastal upwelling are the coast of Peru, the U.S. West Coast, and the coasts of Northwest and Southwest Africa. There is also upwelling off Somalia and Arabia during the Southwest Monsoon.

Likewise, if the wind blows parallel to the coast and is directed with the coast on the left/right in the northern/southern hemisphere, water in the Ekman layer will tend to move toward the coast and will move downward to replace water below the layer. This is called the *coastal downwelling*.

Coastal Zone Color Scanner (CZCS)   A multi-spectral line scanner operational from October 1978 to June 1986, which had four channels devoted to ocean color, each of 20 nm bandwidth and centered at 443, 520, 550, and 670 nanometers.

cobble   A sediment particle between 64 and 250 mm in diameter (definitions vary). Generally rounded due to abrasive action.

coble creep (grain-boundary diffusion creep)   A deformation mechanism for the diffusion creep of a fine-grained polycrystal, in which the creep rate is controlled by diffusion along the grain boundary. It was first proposed by Coble

in 1963. For a spherical grain Coble found

$$\varepsilon = \frac{D_{GB}\delta\sigma\Omega}{\alpha d^3 kT}$$

where (ε) is the creep rate, D_{GB} is the grain boundary self-diffusion coefficient, δ is the thickness, σ is the differential stress, d is the grain size, Ω is the atomic volume, k is Boltzmann's constant, α is a numerical coefficient depending on the grain shape and the boundary conditions for σ, and T is the temperature. It is a linear rheology ($n = 1$, n is stress sensitivity of creep rate at steady-state stage) with a high sensitivity to grain-size (strain rate depends on the grain size as $\varepsilon \propto 1/d^3$).

cobpoint   Short for "connection of observer point", describing the point on a shock to which the observer is connected magnetically. *See* connection longitude.

coda wave   A wave group following P- and S-waves on a seismogram. For a near earthquake, there are many cases in which the directions of arrival of *coda waves* cannot be determined. The coda waves are considered to consist mainly of incoherent waves scattered in a small-scale heterogeneous structure in the Earth. The spectrum of coda waves hardly depends on hypocentral distances or ray paths of seismic waves. Coda Q as a measure of energy redistribution and magnitude of earthquakes can be determined from, respectively, decreasing rate of amplitude of coda waves and their duration time.

coefficient of thermal conductivity   A measure of the ability of a material to transport thermal energy. In solar physics the *coefficient of thermal conductivity,* usually denoted κ, is used in the calculation of heat flux transfer in a fully ionized medium. When a magnetic field is present, the coefficient of thermal conductivity depends on whether one is considering the heat transfer parallel ($\kappa_\parallel$) or perpendicular ($\kappa_\perp$) to the field. In the solar corona, the ratio of these two coefficients is $\kappa_\perp/\kappa_\parallel \simeq 10^{-13}$, and so for most situations, $\kappa_\perp$ is ignored and we have $\kappa = \kappa_0 = 9 \times 10^{-12}T^{5/2}$ W m^{-1} K^{-1} (the Spitzer conductivity). The $T^{5/2}$ dependence of κ_0 makes energy transport by thermal conduction negligible for low temperature plasmas, such as the lower chromosphere and photosphere.

coesite   A dense, heavy form of silica, formed at extremely high pressures and temperatures (40 kb at 700°C to 1700°C). *Coesite* is found naturally only in meteorite impact craters. Stishovite, an even higher temperature, higher pressure polymorph of quartz, is also found associated with impact craters.

cohesionless   Without cohesion; used to describe sediments, such as sand, that have little tendency to cling together when completely dry or wet (surface tension will hold moist particles together if partially wet).

cohesive sediment   Sediment in which particles cling together due to electrical charges on particle edges or sides. These forces are significant compared to particle weight. Clays represent the best example of *cohesive sediments.*

cold dark matter   *See* dark matter, cold.

cold front   The front line along which a wedge of cold sector air underruns and displaces the warm sector mass. The gradient of the upper surface of the cold air is steep, about 1/25 to 1/100. Along the *cold front* there is a strong instability that causes cumulonimbus clouds with rain and thunder. When the cold front passes, weather changes with significant temperature decrease, increasing pressure, and wind veering to northerly or north-westerly in the northern hemisphere.

cold plasma   Formal description of a plasma neglecting the thermal motion of the particles. Such *cold plasma* is the basic assumption in magnetohydrodynamics: it allows the description of a plasma as a fluid with all particles having the same speed, the bulk speed, and is sufficient to derive concepts, such as frozen-in fields, magnetic pressure and tension, and reconnection.

Coleman–Weinberg potential   Spontaneous symmetry breaking field-theory models predict that topological defects (magnetic monopoles, cosmic strings, domain walls) could survive from the early universe. However, they require that a scalar field ϕ evolves in a potential such that its minimum (which determines the actual vacuum) is reached for non-vanishing values of ϕ. *Ad hoc* fixing of such an arbitrary potential appears rather unnatural. However, Coleman and Weinberg realized in 1973 that in some cases, quantum radiative corrections to an otherwise more natural potential could generate the required shape for symmetry breaking. *See* Higgs mechanism, spontaneous symmetry breaking.

collision boundary   *See* convergent boundary.

collisionless shock   The momentum transport by collisions plays a crucial role in the formation of gas-dynamic shocks. In the rarefied plasmas of space, too few collisions occur between the constituents of the plasma to provide efficient momentum transport and allow for shock formation. Nonetheless, shocks do exist in space plasmas. These shocks are called *collisionless shocks;* here the collective behavior of the plasma is not guaranteed by collisions but by the collective effects of the electrical and magnetic properties of the plasma, which allow for frequent interactions between the particles and, consequently, also for the formation of a shock wave.

color-color diagram   A plot of one color index vs. another color index of a collection of stars. In the Johnson photometry system, the commonly used color-color diagram plots the U-B color vs. the B-V color. The placement of a star on this diagram is a function of its intrinsic colors (which are a measure of the star's temperature) and the strength of the interstellar reddening. The effect of reddening is to move stars along a line of increasing U-B and B-V values.

colored dissolved organic matter (CDOM) High-molecular-weight organic compounds (humic and fulvic acids) formed from the decomposition of plant tissue; they strongly absorb light at the blue end of the spectrum and can give water a yellowish color at high concentrations.

color excess   A measure of the reddening of starlight due to small intervening interstellar dust grains. It has been found for most interstellar dust clouds that the extinction of light is proportional to the amount of reddening that a given interstellar cloud causes.

$$\frac{A_V}{E} = 3.2 \ .$$

A_V is the visual extinction (obscuration) of light in magnitudes, and E is the color excess of B and V magnitudes, given by the intrinsic color index subtracted from the measured color index

$$E = (B - V) - (B - V)_o$$

where (B−V) and (B−V)$_o$ are the measured and expected color indices, respectively, of a star of a known temperature (*see* color index). Since the ratio of extinction to reddening is known, then the luminosity of the star can be derived and hence the distance to the star.

color index   The difference between a star's apparent magnitude at some given wavelength to that at another (longer) wavelength. A typical *color index* used is B–V, where B and V are the apparent blue and visible magnitudes measured with standard filters at 4200 and 5400, respectively. The color index is independent of distance and gives a measure of a star's color temperature, that temperature which approximates the radiation distribution of the star as a black body (*see* color temperature). For any given stellar type, there is an expected color index, calculated by assuming a black body temperature for the star. Any difference between the observed color index and the measured color index of a star is due to interstellar reddening. *See* color excess.

color-magnitude diagram   *See* HR (Hertzsprung–Russell) diagram.

color temperature   The temperature that describes a black body radiation distribution based on the intensity ratios at two or more wavelength intervals (or "colors"). One reason that the radiation distribution from stars deviates from that of a black body is that atoms and molecules in the stellar atmosphere deplete radiation from the continuum for spectral line formation. Then a least squares fit of a Planck curve to the stellar spectrum will underestimate the radiation temperature of the star. However, since the shape of a black body curve for any temperature is the same, the ratio of the intensity at two wavelength intervals judiciously chosen to avoid strong absorption features will give a temperature that more closely describes the radiation distribution that the star is actually putting out.

coma   In a comet, the region of heated gas and dust surrounding the nucleus of the comet, distinct from the tails which are generated by nuclear and coma material driven off by solar wind and solar radiation pressure. The gas in the *coma* may be ionized, and its composition can vary with distance from the comet as the different molecules undergo photo-dissociation with time. The coma can reach many millions of kilometers in radius. In optics, an image defect in which the image of a point consists of an off-center bright spot in a larger, fainter "coma".

comet   A small solar system object composed substantially of volatile material, which is distinctive because of its rapid motion across the sky and its long tail(s) of gas and dust emitted from the body. An accurate description of a *comet* is a "dirty snowball", "dirty iceberg" since comets contain some rocky material but are primarily composed of ice. The rocky-icy center of a comet is called the nucleus. Comets usually orbit the sun in highly elliptical orbits. As the comet nears the sun, the increased temperature causes the ice in the nucleus to sublimate and form a gaseous halo around the nucleus, called the coma. Comets often possess two tails: a dust tail that lies in the orbit behind the comet generated by surface activity, and a brighter, ionized gas tail, that points away from the sun, driven by solar wind. Long period comets (periods greater than 200 years) are thought to originate in the Oort cloud, a spherical shell surrounding the Earth at distances exceeding 50,000 AU. They are perturbed by the planets (especially Jupiter) to fall in toward the sun. Their orbits typically have random inclinations and very large eccentricities; some hyperbolic orbits have been observed. Short period comets apparently arise in the Kuiper Belt, in the zone from 20 to 50 AU. Their orbits typically have small eccentricities. Both cometary reservoirs are thought to represent primordial solar system material. Comets are distinguished observationally by the emission of gas. As a result, a comet with a dust coating on its surface that inhibits gas production might be classified as an asteroid. Because of this ambiguity, objects such as Chiron, a Centaur asteroid, have now been reclassified as comets. Kuiper belt objects, which are expected to be composed mostly of ice, are classified as comets. Since Pluto and Charon are also composed mainly of ice and are thought to have originally been Kuiper belt objects, they too may be thought of as comets. Triton, although almost certainly a captured Kuiper belt object, orbits a planet and is identified as an icy satellite rather than a comet. Comet nuclei have radii which are typically in the range of 1 to 100 km, although they may be considerably larger. Each time the comet passes the sun, it loses more ice and dirt (if the Earth later passes through this debris, we see a meteor shower) until eventually the comet no longer displays a coma and/or tail. A comet discovered at this point of its evolution is often identified as an asteroid, so some asteroids are probably dead comets. *See* Oort cloud, Kuiper belt.

comet(s): chemical composition of The volatile material of comets is primarily amorphous water ice but also contains, with some variation in quantity, other simple molecules including a few percent (relative to water) of carbon dioxide (CO_2), carbon monoxide (CO), formaldehyde (H_2CO), methanol (CH_3OH), and methane (CH_4). At a lower level many other molecules have been detected including NH_3, HCN, HNC, C_2H_6, C_2H_2, H_2S, SO_2, and sometimes OCS (Hicks and Fink, 1997). Larger molecules include: HC_3N, NH_2CHO, $HCOOH$, CH_3CN, $HNCO$, and possibly $HCOOCH_3$. The solid materials are probably amorphous silicates, minerals commonly observed in meteorites, such as olivines and pyroxines, and small grains of circumstellar or interstellar origin seen in interplanetary dust particles, such as graphitic or diamond-like carbon grains and silicon carbide. Repeated evaporation of volatiles from successive passages through the inner solar system may result in a surface layer of complex non-volatile organics and rocky material.

comet(s): dirty iceball or snowball model
Comets are composed of a mixture of volatile and rocky material, what Whipple described as a dirty snowball. It has been estimated that comet Shoemaker–Levy 9, which impacted Jupiter, had a low density and little physical cohesion, suggesting that comets are loosely packed light snowballs with void spaces. Greenberg advocates comets as aggregates of remnant microscopic interstellar ice grains that have experienced little heating and suggests that comets, at least originally, were very homogeneous and similar to one another. Abundances and deuterium enrichments of molecules in the coma and tail of Hale–Bopp and Hyakutake are similar to interstellar values, and the ortho to para ratio (the quantum spin state of the hydrogen atoms — a measure of the conditions the ice has experienced) of the water from comet Hale–Bopp implies that it formed at and was never warmed above ∼ 25 K. Taken together these support the contention that Hyakutake and Hale–Bopp (and by extension other comets) have interstellar heritage and have not experienced much heating. However, comets are notoriously unpredictable in their behavior, which could imply heterogeneity. Weidenschilling has pointed out that comets display complex variations (i.e., outbursts and jetting), and believes that this points to compositional inhomogeneities on the order of tens of meters, indicating that these comet nuclei either formed heterogenously or later became differentiated.

comet(s): missions to Deep Space 1 (http://nmp.jpl.nasa.gov/ds1/) launched in October 1998 and will fly by Comet West–Kohoutek–Ikemura in June 2000, although that is not its primary target. The Stardust mission (http://stardust.jpl.nasa.gov), which launched in early 1999, will use aerogel to collect dust from Comet Wild 2 in 2004 and return the sample to Earth in 2006. The Deep Space-4, Champollion mission (http://nmp.jpl.nasa.gov/st4/) will launch in 2003, meet and land on Comet Tempel 1 in 2006, and return a sample in 2010. The Rosetta Mission (http://www.esoc.esa.de/external/mso/projects-index.html) will launch in 2003, rendezvous with Comet Wirtanen in 2112-13, and RoLand (the lander) will make further measurements on the surface of the comet.

comet, artificial *See* barium release.

comminution The breaking up and fragmentation of a rock or other solid. The Earth's crust is comminuted on a wide range of scales by tectonic process. In California the crust has been fragmented into blocks on scales of microns to tens or hundreds of kilometers. Fault gouge is an example of *comminution* on the smallest scale.

common envelope A binary star system enters a *common envelope* phase when one, or both, of the stars in the binary overfills its Roche-lobe and the cores orbit within one combined stellar envelope. Common envelope phases occur in close binaries where the Roche-lobe overflowing star expands too rapidly for the accreting star to incorporate the accreting material. Most common envelope phases occur when a star moves off the main sequence and expands toward its giant phase (either in Case B or Case C mass transfer phases). The standard formation scenarios of many short-period binary systems (e.g., low-mass X-ray binaries, double neutron stars) require a common envelope phase which tightens the orbital separation and ejects the common envelope.

common envelope binary Binary stars that are so close to one another that both fill their respective Roche surfaces, resulting in a common envelope that surrounds both stars.

comoving frame In general relativity coordinates are just labels for space-time points and have no *a priori* physical meaning. It is, however, possible to associate those labels to part of the matter present in the universe, in which case one has a material realization of a reference frame. Since such coordinates follow the matter in its motion, the corresponding frame is said to be comoving. A fundamental requisite is that the trajectories of the objects considered do not cross at any point; otherwise the map of coordinates would become singular in such points. A useful, but not necessary, property is that those objects interact only gravitation-ally, so as to track geodesics in the background spacetime.

compact group of galaxies Isolated group of galaxies for which the separation between the galaxies is comparable to the size of the galaxies themselves; groups of galaxies isolated by Hickson from the Palomar Observatory Sky Survey according to three criteria: (1) there are at least four members whose magnitudes differ by less than three magnitudes from the magnitude of the brightest member; (2) if R_G is the radius of the circle on the sky containing all group members, then the distance to the nearest galaxy outside the group must be larger than 3 R_G (in other words, the group must be reasonably isolated and not an obvious part of a larger structure); (3) the mean surface brightness within R_G should be brighter than 26 mag per square second of arc, i.e., the group must not contain vast empty sky areas and hence should be "compact".

compaction As rocks are buried to a greater depth in a sedimentary basin, the "lithostatic" pressure increases. This causes the rock to compact. The void space, or "porosity" of the rock, decreases with increasing depth, and the density increases.

compact steep spectrum radio sources A class of radio sources which includes radio galaxies and quasars unresolved at resolution ≈ 2 arcsecs. They are differentiated from other core-dominated radio sources by showing a steep radio spectrum. Observations at higher resolution show that compact steep spectrum radio sources are either classical lobe dominated sources whose lobe size is less than the size of the galaxy or quasars with a core single-side jet morphology. In both cases, the radio morphology appears often to be disrupted and irregular. *Compact steep spectrum radio sources* are thought to be young radiogalaxies which are expanding or, alternatively, radio sources in which the expansion of the radio plasma is hampered by interstellar or intergalactic medium.

compensation In geophysics, the positive mass of major mountain belts is compensated by the negative mass of crustal mountain roots. The crustal rocks are lighter than the mantle rocks beneath. In a compensated mountain belt, the total mass in a vertical column of rock is equal to the total mass in the adjacent lowlands. A mountain belt behaves like a block of wood floating on water.

In cosmological scenarios where topological defects are suddenly formed, both the geometry of spacetime and ordinary matter and radiation fluids are perturbed in such a way as to satisfy theoretical conservation laws imposed by general relativity theory. Matching conditions between times before and after the relevant phase transition during which defects are generated require that cosmic fluid perturbations *compensate* for the energy density inhomogeneity of space in the presence of the defect. *See* cosmic phase transition, cosmic topological defect, Kibble mechanism.

compensation depth   The depth of the Earth at which the overlying rocks are assumed to exert a constant pressure. Below this depth lateral variations in density are assumed to be small.

composite volcanos   *Composite volcanos,* also called stratovolcanos, are steep-sided volcanic cones consisting of alternating layers of lava flows and ash deposits. Eruptions from composite volcanos are often very explosive and deadly, as evidenced by the eruptions of volcanos such as Vesuvius in Italy, Mt. St. Helens in the United States, and Mt. Pinatubo in the Philippines. Composite volcanos are found near subduction zones on Earth and are produced by silica-rich magmas moving upwards from the subducted plate. The high silica content of the magmas allows them to be very viscous and retain much gas, which leads to the explosive nature of the eruptions.

compound channel   A channel or river in which the equation for flow area vs. depth exhibits a discontinuity.

compressibility   The ratio of the fractional change in volume dV/V (volumetric strain) in response to a change in pressure dp, that is,

$$\beta = \frac{1}{V}\frac{dV}{dp} = -\frac{1}{\rho}\frac{d\rho}{dp}$$

where ρ is density.   Isothermal or adiabatic changes of V and p yield slightly different values of β. β is a modulus of elasticity, and its reciprocal is the bulk modulus.

Compton, Arthur H.   Physicist (1892–1962). His studies in X-rays led him to discover the Compton effect, that is the change in wavelength of a photon when it is scattered by a free electron. The discovery of the Compton effect confirmed that electromagnetic waves had both wave and particle properties.

Compton cooling   The reduction of energy of a free electron, due to its interaction with a photon. If the kinetic energy of the electron is sufficiently high compared to the incoming photon, the energy of the incoming photon plus part of the kinetic energy of the electron is redirected as a photon of higher energy. Named after Arthur H. Compton (1892–1962).

Compton reflection   The Compton scattering of hard X-ray radiation by a layer of dense and thick matter, such as the surface of a star or of an accretion disk. Hard X-ray radiation is scattered off the surface of the layer after having lost part of its energy. *Compton reflection* creates a distinguishing spectral feature, an enhancement in the spectral energy distribution between 10 and 50 keV. Such features have been detected in the spectra of several Seyfert-1 galaxies and of a galactic object, the black hole candidate Cyg X-1. *See* Seyfert galaxies, Cygnus X1.

Compton scattering   The inelastic scattering of high energy photons by charged particles, typically electrons, where energy is lost by the photon because of the particle recoil. A photon carries momentum, part of which is exchanged between the photon and the particle. Conservation of energy and momentum yields an increase in the photon wavelength (and hence a decrease in photon energy) as measured in the initial rest frame of the electron equal to

$$\lambda - \lambda_0 = \lambda_C (1 - \cos\theta)\,,$$

where λ_0 is the wavelength of the incident photon, θ is the angle between the initial and final direction of propagation of the photon, and λ_C is a constant, called the Compton wavelength, and defined by $\lambda_C = h/mc$, where h is the Planck constant, m the particle mass, and c the speed of light. In the case of scattering by electrons, $\lambda_C = 0.02426$ Å. If $\lambda \gg \lambda_C$, then the energy exchange is irrelevant, and the scattering is elastic (Thomson scattering). *Compton scattering* by electrons occurs for photons in the X-ray domain. *See* inverse Compton effect, Thomson scattering.

computational relativity   Numerical relativity.

conditional unstability   The atmosphere is said to be conditionally unstable when the lapse rate is between the adiabatic lapse rate and moist lapse rate. In this case, parcels displaced downward will be restored, whereas saturated parcels displaced upward will continue to move upward.

conducting string   In cosmology, possible topological defects include conducting cosmic strings. In generic grand unified models, one may have couplings between the cosmic string-forming Higgs field and fermionic fields, and the vanishing of Higgs-generated fermionic mass terms in the core of the defect allows the existence of fermionic zero modes carrying currents along the string. Alternately, bosonic conductivity arises when charged boson fields acquire non-zero expectation values in the string cores. These currents are persistent, and the vortex defects containing them are called *conducting cosmic strings*. They are also often referred to as superconducting strings, for it can be seen that the electric (or other) current they carry is dissipationless.

The production of equilibrium current-carrying string loop configurations called vortons may contribute to the dark-matter density of the universe. Such loops might also serve as seeds for the generation of primordial magnetic fields. *See* current carrier (cosmic string), current generation (cosmic string), fermionic zero mode, vorton, Yukawa coupling, Witten conducting string.

conduction   (**1.**) Transport of electric current. (**2.**) Transfer of heat without the flow of particles from one part of a medium to another by the transfer of energy from one particle to the next

and by lattice oscillations (phonons) in a solid. This flow of heat is directed by temperature gradients in the medium. In the solar corona, the conductive flux, F_{cond}, is directly proportional to the temperature gradient for classical Spitzer conduction, via $F_{cond} = \kappa_0 \nabla T$, where κ_0 is the coefficient of thermal conductivity parallel to the magnetic field.

conductive heat transfer   Transfer of heat due purely to a temperature difference. Heat conduction is a diffusive process, in which molecules transmit kinetic energy to other molecules by colliding with them. Fourier's law of heat conduction states

$$q = -\lambda \cdot \nabla T$$

where q is the heat flux vector, λ is the thermal conductivity (tensor), and ∇T is the temperature gradient. The minus sign indicates that heat is transferred by conduction from higher temperature to lower temperature regions.

conductivity of water   The ionic content of water enhances the electric conductivity of water (strongly temperature-dependent). Since most of the dissolved solids in natural waters are present in the form of ions, the easy-to-measure conductivity is a practical way to estimate salinity. Instead of salinity, often conductivity normalized to T = 20 or 25°C is used in lakes to express the concentration of ionic content. (Conductivity in ocean and fresh water is in the range of $\sim 50 - 70$ mS/cm and $\sim 50 - 500$ μS/cm, respectively.)

conformal infinity   In relativity there exists a conformal isometry of Minkowski spacetime with a region of an Einstein universe. *Conformal infinity* $\mathcal{I}$ is the boundary of this region. The boundary $\mathcal{I}$ has the topology of a light cone, with the vertex points $i^{\pm}$ (timelike future/past infinity) missing. Future/past conformal infinity $\mathcal{I}^{\pm}$ is the set of future/past endpoints of null geodesics in $\mathcal{I}$. *See* null infinity.

conformal tensor   Weyl tensor.

congruence   A family of curves at least in some small region, such that one and only one curve passes through each point in the region.

This implies that the tangent vectors to the *congruence* form a vector field and, further, every smooth vector field generates a congruence. *See* vector field.

conical spacetime   *See* deficit angle (cosmic string).

conics   Pitch angle distributions whose peak intensity is neither along the field line ("cigar") nor perpendicular to it ("pancake") but at some intermediate angle. Ion conics (generally of O^+ ions) are often observed on magnetic field lines above the auroral zone, at altitudes of the order of 5000 km, associated with the aurora. It is believed that they arise from wave-ion interactions which energize the ions at an altitude below the one where the conic is observed. Such an interaction preferentially increases the ion's velocity components perpendicular to the magnetic field, producing a pancake distribution. By the conservation of magnetic moment, the pancake transforms into a cone at higher altitudes.

conic section   Any of the plane figures obtained by intersecting a circular cone with a plane. If the plane is parallel to the base of the cone, the figure is a circle. As the tilt angle β of the plane increases, $0 < \beta < \alpha$, where α is the apex half angle of the cone, the figure is an ellipse. If $\beta = \alpha$ the figure is a parabola, and if $\beta > \alpha$, the figure is a hyperbola.

In Newtonian physics, gravitational motion is an orbit that is a conic section with the sun at one focus.

conjugate depths   The two water depths that appear on either side of a hydraulic jump in open channel flow. Also referred to as sequent depths. They differ from alternate depths in that the specific energy is not the same for *conjugate depths* due to energy loss in the hydraulic jump.

conjunction   Orientation of planets so that the angle planet-Earth-sun equals zero. For outer planets, the planets are as far from Earth as possible in their orbits, because they are then on opposite sides of the sun. For inner planets, conjunction is either a closest approach to Earth (both planets on the same side of the sun), or the most distant (the inner planet on the opposite side of the sun from Earth).

connection   *See* affine connection.

connection longitude   Heliographic longitude to which a spacecraft is connected magnetically. The *connection longitude* ϕ_{conn} can be determined from the "offset" $\Delta\phi$ of the footpoint of the archimedian magnetic field spiral through the observer with respect to the observer's heliolongitude ϕ_0 according to

$$\phi_{conn} = \phi_0 + \Delta\phi = \phi_0 + \frac{r\omega_\odot}{v_{sowi}}$$

with r being the radial distance of the observer from the sun, $\omega_\odot$ the sun's angular speed, and v_{sowi} the solar wind speed. The connection longitude is important in the study of energetic particles because, in contrast to the electromagnetic radiation, particles do not propagate radially but along the magnetic field line. For average solar wind conditions, the Earth, or an earth-bound spacecraft, magnetically is connected to a position around W 58 on the sun. Connection longitudes also can be established to shocks. *See* cobpoint.

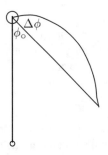

Heliolongitude and connection longitude.

Conrad discontinuity   Seismic discontinuity at a depth of around 20 km between the upper and the lower crusts beneath a continent and an island arc. The *Conrad discontinuity* is not a sharp discontinuity like Mohorovičić's discontinuity. The Conrad discontinuity was originally thought to represent a boundary between the lower crust consisting of basaltic rocks and the upper crust consisting of granitic rocks, deducing from velocity of Conrad head waves (P^*).

But now it is thought to be a thermodynamical interface or a rheological boundary.

conservation of angular momentum   In the absence of external torques, the total angular momentum **L** of a system is a constant:

$$\mathbf{L} = \sum_a \mathbf{L}_{(a)} = \text{constant pseudovector.}$$

Here a sums over the subsystems comprising the system.

conservation of energy   The observation that energy cannot be created or destroyed but can only be changed in form.
    Forms of energy include

1. potential energy (the ability to do work),

2. kinetic energy (the energy of motion),

3. and relativistically, the energy equivalent of mass: $E = mc^2$.

Since an understanding of special relativity implies that mass and energy are interconvertable, the conservation law is often posed as conservation of mass and energy. In normal macroscopic laboratory physics, mass is very closely conserved; the deviations are of order $p/\rho c^2$. In astrophysical settings, the interconversion of mass and energy via $E = mc^2$ is very important, describing the internal energy source of stars via nuclear fusion, which releases energy because of an overall conversion of the mass of the constituents to energy.

conservation of momentum   In any isolated system, the total momentum **P** is constant:

$$\mathbf{P} = \sum \mathbf{p}_a, \text{ a constant vector.}$$

In situations where constituents are separated "before" and "after" (in a collision, for instance), Newton's second law guarantees that each individual $\mathbf{p}_{(a)}$ will be constant except during the collision. *Conservation of momentum* then provides (three) relations between the "before" and "after" momenta.

conservative system   In mechanics, a system in which there is a potential energy function

that depends only on position, and such that the sum of the kinetic energy and the potential energy, for any particular particle motion, remains constant. Simple examples are mass point motion in a fixed Newtonian gravity field, motion of charged particles in electrostatic fields (neglecting radiation reaction), and the exchange between kinetic and potential energy in the motion of a mass-on-spring, harmonic oscillator or in the motion of simple pendulum is a uniform gravitational field. In all of these cases, the total energy E is unchanged during the motion, and $E =$ kinetic energy plus potential energy, so computing one determines the other. In Lagrangian mechanics, a *conservative system* is one in which the Lagrangian $L = T - V$, the kinetic energy T is quadratic in velocities, the potential V is a function only of coordinates, and the Lagrangian has no explicit time dependence. Such a Lagrangian is related through a Legendre transformation to a Hamiltonian H, which is a function of the momenta and the velocities, is constant in time, and $H = T + V$. The idea may be extended more generally to a time independent Hamiltonian, even if H is not simply $T + V$.

constellation   An apparent association of stars in the sky which are given symbolic or mythic significance by associating figures to their pattern on the sky. According to current practice, the sky is divided into 88 such areas of association.

constitutive law of frictional sliding   *See* fault constitutive law.

constraint equations   In general relativity, an ADM formulation separates the spacetime into a 3 + 1 decomposition (3 space plus 1 time). Einstein's equation can then be written with the 3-metric g_{ij} and the extrinsic curvature k_{ij} (both symmetric 3-dimensional 2-tensors) as the fundamental variables. K_{ij} appears as a momentum conjugate to g_{ij}, and six of the ten 2-order Einstein equations become a hyperbolic set of equations giving the time derivative of g_{ij} and of K_{ij} (twelve first-order equations). However, there are in addition four equations (the $G_{\sigma 0}$ and the $G_{\sigma i}$ components of the Einstein tensor), which are equated to the corresponding component of the matter stress-energy tensor, or to zero for an empty (vacuum) solution. Those four equations, the *constraint equations,* contain no time derivative of either g_{ij} or of K_{ij}; hence, they must be satisfied at the particular time in question. They can be written as one equation, which is a nonlinear generalization of Newton's gravitational equation including the contribution of the gravitational field as a source; and three components of a tensor equation which is a transversality condition on the momentum. It is easy to show that these four equations put real restrictions on the g_{ij} and the K_{ij} that can be chosen for initial data (hence *constraints*). However, it can further be shown that the evolution via the hyperbolic equations preserves the solution of the constraint equations after their solution at the initial time. These equations are elliptic, and a standard procedure has been developed to solve this. *See* ADM form of the Einstein–Hilbert action.

contact binary   A binary star in which the two components are touching. Main-sequence contact binaries are also called W Ursa Majoris (abbreviated as W UMa) and β Lyrae binaries, depending on the mass of the primary.

continent   Regions of substantial area on the surface of the Earth that rise above sea level. The *continents* are made up of thick, light continental crust. The granitic continental crust has a typical thickness of 35 km, whereas the basaltic oceanic crust has a typical thickness of 6 km. The continents do not participate in the plate tectonic cycle but move and float on the surface of the mantle so that the continents are much older than the oceanic crust, about 2 billion years vs. 100 million years.

continental climate   The climatic type describing the climate character within a great land. Comparing to the marine climate, the main features of *continental climate* are strong annual temperature variations, which cause colder winters and hotter summers, and strong diurnal variations.

continental collision   When subduction in an ocean basin dominates over sea-floor spreading, the ocean closes and the continents on the two sides of the ocean collide. This is happening today in the Himalayas where the Indian subcontinent has collided with Eurasia.

continental drift   Refers to the movement of continents with respect to each other. The remarkable similarity between the east coast of the Americas and the west coast of Europe and Africa has been recognized for several centuries and led to the hypothesis that these continents were once attached into a single supercontinent, which then broke apart and created the distribution of land masses we see today, called *continental drift.* The idea was popularized by German meteorologist Alfred Wegener in 1922, although many other people from 1620 onward contributed to the idea. The remarkable similarity extends beyond the similarity of continent boundaries to the distribution of geologic features, such as mountain ranges and glacial remains and fossil types, and the divergence in the evolution of flora and fauna after a certain point in time. It was initially rejected by geophysicists who could not envision that the solid mantle of the earth could flow.

All the continents were combined into a single supercontinent called Pangea about 300 million years ago. Pangea broke apart into a northern continent (Laurasia, composed of North America, Europe, Asia, and Greenland) and a southern continent (Gondwana, composed of South America, Africa, Australia, Antarctica, and India) about 250 million years ago, and those two continents subsequently broke apart into the continents we see today. The original idea of continental drift suggested the continents moved over the ocean floors. In the 1960s, studies of the magnetic polarity of ocean crust revealed that the seafloor was also moving. Combining the ideas of continental drift and seafloor spreading led to the development of the theory of plate tectonics to explain the geologic activity on the Earth's surface. Direct ranging to geodetic satellites accurately measures the rate of continental drift, with speeds of up to 15 cm/yr.

continental margin   The boundary between the continents, including continental shelves, and the ocean basins. There is no unique definition of the margin, it is usually taken as an arbitrary water depth, say 1 km. There are two types of *continental margins,* active and passive. An active margin is also a plate boundary, usually a subduction zone. An example is the west coast of South America. A passive margin is not an active plate boundary. An example is the east coast of the United States.

continental plate   In geophysics, the large pieces of material comprising the outer surface of the lithosphere. Plates are about 50 km in thickness and support the continents, which float on the denser viscous-fluid mantle below them. Current knowledge defines 15 major plates: Pacific, Philippine, Juan De Fuca, Cocos, Nazca, Antarctic, Scotia, South American, North American, Caribbean, Arabian, Indian, African, Eurasian, and Australian.

continental shelf   Shallow oceanic margins underlain by continental crust. The water depth is usually taken to be less than 1 km.

continental shelf waves   The waves that are generated by the sloping continental shelf topography. In the Northern Hemisphere, the *continental shelf wave* propagates along a constant topography line with the coast on the right.

continuum   In mechanics, a description that ignores the granular or quantum nature of matter. In spectroscopy, the part of a spectrum without apparent lines, arising from solid, liquid, or gaseous sources, in which atomic lifetimes are too short to produce specific line emission or absorption.

contravariant vector   *See* vector.

convection   Large-scale flow of gases (or other fluids) in stars (or elsewhere) that carries heat energy from one place to another. In the stellar context, hot currents rise, cool ones fall, and the solar granulation is direct evidence for the occurrence of *convection* near the surface of the sun. Convection sets in whenever a stellar temperature gradient is steeper than the adiabatic one:

$$-\frac{dT}{dr} > -\left(1 - \frac{1}{\gamma}\right)\frac{T}{p}\frac{dp}{dr}$$

where γ is the ratio of specific heats. This can happen either when gas is quite opaque to radiation (as it is near the surfaces of cool stars, including the sun, where hydrogen is in the process of being ionized) or when a nuclear reaction depends on a very high power of temperature, as does the CNO cycle at the cores of massive stars. Stars are fully convective during early phases of their formation (guaranteeing initial chemical homogeneity) and throughout their lives for stars of less than 0.3 solar masses. Convection is the primary way that material inside stars is mixed from one zone to another. In relatively dense regions of stars, convection will carry all the available energy, and the temperature gradient will be very nearly adiabatic. At lower densities, convection becomes inefficient, radiation carries much of the energy, and the temperature gradient can be much steeper.

In the absence of any adequate theory of convection (or other turbulent processes in fluids), convection is often treated in the Mixing Length Approximation, in which gas is assumed to rise or fall through a fixed fraction of the pressure scale height (half is typical) and then come into temperature equilibrium with its surroundings, depositing its extra heat or soaking up its deficit. The approximation is more than 50 years old, and modern numerical computations of gas flow processes are just beginning to replace it in standard computations of stellar structure and evolution. The absence of an adequate theory of convective energy transport is one of the major remaining uncertainties in our understanding of stellar physics. *See* CNO cycle, solar granulation.

convection zone   A layer in a star in which convection currents provide the main energy transport mechanism. In the sun, a convection zone extends from just below the photosphere down to about 0.7 $R_\odot$.

convective adjustment   One way to parameterize the physical process of convection in climate modeling. For example, in a simple version, one first examines the relative humidity and lapse rate in each grid column at the end of each time step of integration; if the lapse rate is superadiabatic, the temperature profile is adjusted to dry static neutrality in a manner that conserves energy; if the column is conditionally unstable and the humidity exceeds a specified value, the column is adjusted to moist static neutrality.

convective cloud   A type of cloud that is generated by convective activities in the atmosphere. The main feature of it is its strongly vertical development. Strong vertically developed clouds are also called heap clouds.

convective heat transfer   Transfer of heat by mass movement, due to free or forced convection. The latter case is also referred to as advective heat transfer.

convective instability   Stratification instability caused by convective activities, i.e., the lower layer has higher moisture and becomes saturated first when being lifted, and hence cools thereafter at a rate slower than does the upper, drier portion, until the lapse rate of the whole layer becomes equal to the saturation adiabatic and any further lifting results in instability. In general, use $\frac{\partial \theta_{sw}}{\partial Z} < 0$ or $\frac{\partial \theta_{se}}{\partial Z} < 0$ as the criterion of the *convective instability*, where θ_{sw} and θ_{se} are pseudo-wet-bulb potential temperature and pseudo-equivalent potential temperature, respectively.

convective scaling   In the boundary layer, pure convective turbulence depends only on the thickness H_{con} of the convectively well-mixed layer and the boundary buoyancy flux J_b^o [Monin and Obukhov, 1954]. Dimensional analysis provides the scaling relations as a function of H_{con} and J_b^o by:

Length	$L_{cs} \sim H_{con}$
Time	$\tau_{cs} \sim H_{con}/w_{cs}$
Velocity	$w_{cs} \sim (H_{con}J_b^o)^{1/3}$
Diffusivity	$K_{cs} \sim w_{cs}H_{con}$
Dissipation of turbulent kinetic energy	$\varepsilon_{cs} \sim J_b^o$

If convection is driven purely by heat fluxes (i.e., $J_b^o = g\alpha F_{th}/(c_p\rho)$), further scaling relations for temperature are as shown in the table on page 86.

convective turbulence   If density increases at the top of a fluid (i.e., cooling, evaporation of salty water at the surface, etc.) or if density decreases at the bottom of a fluid (i.e., warming from below, etc.), convective plumes (thermals for temperature) will set in and mix a progressively thicker boundary layer. Convective scaling allows the quantification of the relevant physical parameters of convection as a function of the boundary buoyancy flux J_b^o and the thickness H_{con} of the convectively unstable layer. *See also* penetrative convection.

convergent boundary   Plates are destroyed or severely deformed at *convergent boundaries.* Two types of convergent boundaries exist: subduction and collision. A subduction convergent boundary occurs when two plates composed of oceanic crust (thin, basalt composition) or an ocean plate and a continental plate (thick, more silicic composition) meet. At the ocean-ocean boundary, one of the oceanic plates dives down under the other plate. At an ocean-continent boundary, the oceanic plate is always subducted under the continental plate. The location where the first plate subducts under the second plate is characterized by a deep trough, called a trench. As the oceanic plate dives deeper into the Earth's interior, the temperature rises and sediments which accumulated on the ocean floor begin to melt. This magma rises towards the surface and erupts on the overriding plate, creating the very explosive volcanos called stratovolcanos (or composite volcanos). Subduction boundaries are characterized by this explosive volcanism and earthquakes from a variety of depths (down to about 200 km). Japan is an example of an ocean-ocean subduction boundary, while the Cascade volcanos (including Mt. St. Helens) in the northwestern U.S. are an example of the ocean-continent subduction boundary. If both plates are composed of continental crust, neither plate is subducted. Instead the two plates crumple to form high mountain ranges, such as the Himalayas. This type of convergent boundary is called a collision boundary and is characterized by earthquakes but no volcanism. Convergent boundaries are believed to occur over the descending portions of convection cells within the Earth's asthenosphere.

convergent plate boundaries   *See* convergent boundary.

Temperature (fluctuations)	$\Theta_{cs} \sim F_{th}/(c_p\rho w_{cs}) \sim (g\alpha)^{-1}(J_b^{o2}/H_{con})^{1/3}$
Dissipation of temperature variance	$\chi_{cs} \sim T_{cs}^2 w_{cs}/H_{con} \sim (g\alpha)^{-2}(J_b^{o5}/H_{con}^4)^{1/3}$
Temperature gradient	$\partial\Theta_{cs}/\partial z \sim F_{th}/(c_p\rho K_{cs}) \sim (g\alpha)^{-1}(J_b^{o2}/H_{con}^4)^{1/3}$

conversion efficiency   *See* energy conversion efficiency.

cooling flow   In cosmology: Clusters of galaxies may contain of order 1000 visible galaxies, which contribute only 5 to 10% of the cluster mass, some baryonic gas observable in its X-ray emission, which constitutes $\sim 30\%$ of the mass needed to bind the cluster; and an amount $\sim 60\%$ of the mass in currently unknown dark matter. In the outer rarified portions of the cluster, the baryon gas cools inefficiently, but toward the center of the cluster, the higher density leads to rapid cooling of this gas. The gas accordingly loses pressure support and falls into the center at a typical rate $\sim 100M\odot$/year. All clusters observed with cooling flows have a giant elliptical at their center, suggesting that the inflow has persisted for cosmological times (roughly one Hubble time) to form this central galaxy, further suggesting that cooling flows are generic in large clusters. However, cooling flows do not exist in interacting clusters (the result of cluster mergers), and it is a current active topic of research to understand the mechanism of disruption and the timescale for re-establishment of cooling flows.

Coordinated   Universal   Time   (UTC) Starting in 1972, Greenwich Mean Solar Time was split into *Coordinated Universal Time (UTC),* the basis of civil timekeeping ever since, and Universal Time (UT1), which is a measure of Earth's rotation.   UTC advances in step with International Atomic Time (TAI), except that at leap seconds it is adjusted to remain within 0.9 s of UT1.   Therefore, to compute the true elapsed seconds between any two events defined in UTC since 1972, a table of leap seconds is required. For example, because the leap seconds totaled 24 during all of 1989, and 29 during all of 1995 (five seconds having been inserted during the interim), when calculating the time interval between any date and time in 1989 and 1995, it is necessary to add 5 seconds to what would be calculated

on the basis of equal 24 hour days, each hour having 60 minutes comprised of 60 seconds. In other words, on five distinct occasions between January 1, 1989, and December 31, 1995, there was an hour whose final minute had 61 seconds.

coordinate singularity   A location in a space (or spacetime, in relativity) where description of physical fields is impossible because the coordinates do not correctly map to (a region of) a rectangular coordinate chart.   An example is the origin in spherical coordinates, where angular directions have no meaning. In general relativity, the situation is more difficult because the metric itself can change, and recognizing a coordinate singularity as such requires subtlety: A coordinate singularity is a singularity which is removable by (singular) transformation to a frame in which all components of physical objects remain bounded. In general relativity, the surface of a spherical black hole, the surface $r = 2M$ (where r is defined so that $4\pi r^2$ measures the area of a sphere) is a singularity in the original coordinates used to describe it.   (*See* Schwarzschild black hole.)   A transformation found by Kruskal and Szekeres removes the coordinate singularity and shows that all geometrical measurable quantities are finite.

coordinate system   A way of assigning a set of labels to each point in a space (or, by extension, to each event in spacetime).   Since common experience suggests that space is 3-dimensional, one assigns three independent functions, (e.g., x, y, z) to label points in space. The $x = 0$ surface, for instance, consists of all those points in space where the function x (as a function of position) vanishes. For purposes of physical description, the coordinate functions are taken to be continuous functions of the space points.   In spacetime, one introduces a fourth coordinate, time.   In Newtonian physics, time is a universal function, known and measurable by any observer. In special and in general relativity, time is a function of the motion of the observer (at least), and different observers use different space and time coordinate functions. Although the notation $\{x, y, z\}$ suggests rectangular coordinates, the constant $x-$ surfaces can in fact be curved, for instance, if x is really the radius from the origin in a spherical coordinate system. In general, the coordinate functions can lead to curved constant-coordinate surfaces (curvilinear coordinates). Then the four spacetime coordinates form a set of functions, say $\phi^{\alpha}(P)$ where $\phi^{\alpha}, \alpha = 0, 1, 2, 3$ correspond to the time and the three spatial coordinate functions; P represents a point in spacetime.

A coordinate system is closely related to a reference frame. For instance, one can align basis vectors (which constitute the frame) along the intersection of constant-coordinate surfaces, with some rule for assigning length or magnitude of the basis vectors.

coordinate time　Time defined relative to an inertial (in particular, nonrotating) reference frame, whose relationship with time measured on the surface of the Earth can be calculated using relativity (not to be confused with "coordinated time" like Coordinated Universal Time). General uses are "Geocentric Coordinate Time" (TCG) and "Barycentric Coordinate Time" (TCB), the latter referring to the solar system barycenter.

coordinate transformation in special relativity　The transformation of space-time coordinates between two reference systems that are moving uniformly with respect to each other. Classically, any physical system is composed of particles, and a full description of the system is obtained if all the positions of each particle are known for any given time. The position of each particle is represented by a trio of numbers whose value depends on the location of the reference system. Thus if two observers used two different systems of reference, then a coordinate transformation is needed in order to compare the observations. Classically time is considered to be an absolute variable; that is its value is the same regardless of the reference system. Thus, classical coordinate transformations transform an arbitrary trio of spatial coordinates at any given time. Special relativity states that the velocity of light is constant for two reference systems that are moving at constant velocities with respect to each other. In order for this to hold time can no longer be an absolute variable, and its value must depend on the reference system. The universal character of the speed of light plus the assumption that space is homogeneous and isotropic leads to the Lorentz transformation of space-time coordinates. The Lorentz transformations transform an arbitrary foursome of coordinates (three spatial coordinates plus time) from one system of reference to another that is moving uniformly with respect to the first.

Copernicus, Nicholas　Astronomer (1473–1543). Proposed that the sun, rather than the Earth, was the center of the solar system.

coplanarity theorem　In magnetohydrodynamics, the coplanarity theorem

$$\mathbf{n} \cdot (\mathbf{B}_d \times \mathbf{B}_u) = 0$$

states that the shock normal $\mathbf{n}$ and the magnetic fields $\mathbf{B}_u$ and $\mathbf{B}_d$ in the upstream and downstream medium all lie in the same plane. The coplanarity theorem is a consequence of the jump conditions for the electromagnetic field at the shock as described by the Rankine–Hugoniot equations. *See* Rankine–Hugoniot relations.

Cordelia　Moon of Uranus also designated UV. Discovered by Voyager 2 in 1986, it is a small, irregular, body, approximately 13 km in radius. Its orbit has an eccentricity of 0, an inclination of $0.1°$, a precession of $550°$ yr^{-1}, and a semimajor axis of 4.98×10^4 km. It is the inner shepherding satellite for Uranus' epsilon ring. Its surface is very dark, with a geometric albedo of less than 0.1. Its mass has not been measured. It orbits Uranus once every 0.335 Earth days.

cordillera　An extensive chain of parallel mountains or mountain ranges, especially the principal mountain chain of a continent. The term was originally used to describe the parallel chains of mountains in South America (las Cordilleras de los Andes).

core　Differentiated central volume of the Earth and (some) other planets. *Cores* vary in composition, size, and physical state among the different solar system bodies. In geophysics, the Earth has a core with a radius of 3480 km, compared to the Earth's radius of approximately 6370 km. The core is much denser than the mantle above and is composed primarily of iron with some other alloying elements; most of the terrestrial planets are believed to have cores composed of iron, based on their high densities. Although most of the terrestrial planets are believed to have only one core of either solid or liquid iron, the Earth (due to its large size) has two cores, an inner solid iron core with a radius approximately 600 km and an outer liquid iron core. The Earth's core formed early in the evolution of the Earth as a large fraction of heavy iron gravitationally segregated from the silicic components. Latent heat of fusion released in the freezing of the Earth's liquid core supplies the energy to drive mantle convection and to support plate tectonics. The cores of the lower-density Jovian planets are believed to be composed of rock and/or ice. The presence or absence of a core is best determined from seismology, although the moment of inertia of the planet also provides information on how centrally condensed the body is. In astronomy, "core" is used to describe the central flat density region of some star clusters and galaxies. It is also used to describe the central homogeneous region of a chemically differentiated star.

core collapse   The beginning of the end for massive stars, which have built up cores of iron from silicon burning, when the mass of the core reaches the Chandrasekhar limit. The collapse happens in a few seconds (after millions of years of evolution of the star) and releases a total energy of about 10^{53} ergs, the gravitational binding energy of the neutron star left behind. Much of this is radiated in neutrinos (and perhaps gravitational radiation), about 1% appears as kinetic energy of the expanding supernova remnant, some is radiated as visible light (so that we see a supernova of type II), and some is stored in the rotation and magnetic field of the neutron star or pulsar left behind at the center. The total available is GM^2/R where M and R are the mass and radius of the neutron star. Core collapse may sometimes continue on past the neutron star stage and leave a black hole.

core convection   It is generally understood that the Earth's magnetic field arises predominantly from electrical instabilities associated with the flow of conducting fluid in the Earth's core. The common viewpoint is that the energy source for the motions that generate the field is convection in the Earth's core associated with heat loss from the core to the mantle. If there is radioactive heating of the core from isotopes such as potassium 40, then it is possible that the temperature of the core might stay roughly constant with time, but it is usually thought that there is little radioactivity in the core and the heat loss from the core is associated with the overall secular cooling of the planet as a whole and increase in size of the solid inner core. Convection arises when density decreases with depth, either because fluid cooled near the core-mantle boundary becomes more dense than underlying fluid, or because as the core cools, the inner core freezes out, excluding light elements and releasing latent heat and hence generating buoyant fluid at the inner core boundary. This latter case may lead to "compositional convection" related to the chemical makeup of the buoyant fluid rather than its temperature (i.e., "thermal convection"), which is energetically favorable for maintaining magnetic field but can have the seemingly perverse effect of transporting heat against a thermal gradient.

core-dominated quasars   High luminosity, radio-loud active nuclei whose radio morphology is characterized by a luminous core which dominates the source emission. Mapped at milliarcsecond resolution, the core becomes partly resolved into a one-sided jet.   Many core-dominated radio quasars exhibit radio knots with superluminal motion, indicative of ejection of plasma at a velocity very close to the speed of light. The quasars 3C 273 and 3C 120 (whose name means that they were identified as radio sources 273 and 120 in the third Cambridge radio survey) are two of the brightest quasars in the sky and prototypical core-dominated superluminal sources. In the framework of the unification schemes of active galactic nuclei, core- and lobe-dominated quasars are basically the same objects: core-dominated objects are observed with the radio axis oriented at a small angle with respect to the line of sight, while the radio axis and the line of sight form a larger angle in lobe-dominated objects. The jet one-sidedness suggests that radiation is boosted by relativistic beaming: If the emitting particles are moving at a velocity close to the speed of light, the detection of the jet on the approaching side is strongly favored. In this case, a very large dynamical range is needed to detect the radio-lobes, which, seen pole-on, may appear as a faint fuzz surrounding the core.

core flow   Magnetic field may be generated in a conducting fluid by fluid flow, as described by the induction equation of magnetohydrody-namics. Such flow in the Earth's core is thought to be the generating mechanism for the bulk of the Earth's magnetic field. The motions and field within the core cannot be directly calcu-lated from observations of the magnetic field at the Earth's surface (although it may be possible to indirectly infer certain parts of the internal flow and field). However, observations of the surface field can be used to calculate the field at the core-mantle boundary by assuming that the electrical currents in the mantle are negligi-ble, in which case the flow at the surface of the core is constrained using the radial component of the induction equation of magnetohydrody-namics in the frozen-flux limit (i.e., assuming that magnetic diffusion within the core may be neglected):

$$\frac{\partial B_r}{\partial t} = \nabla_H \cdot (B_r \mathbf{u})$$

where B_r is the radial field and $\mathbf{u}$ the veloc-ity vector. Since the flow does not penetrate the core-mantle boundary, there are two compo-nents of the velocity but only one constraining equation, which means that although the above equation can be used to invert time varying mod-els of B_r for possible flows, they will not be uniquely determined. Extra constraints on the flow have been used to alleviate the nonunique-ness. *See* nonuniqueness.

core-mantle boundary   At around 3480 km from the center of the Earth, the material compo-sition is thought to change from molten iron plus dissolved impurities (the outer core) to crys-talline silicate rock (the mantle). This *core-mantle boundary* is in terms of absolute den-sity contrast the Earth's major transition, with a jump from 5.6 g/cm^3 at the base of the man-tle to 9.9 g/cm^3 at the top of the core. There is also a significant contrast in viscosity (al-though the viscosities of both sides are poorly constrained) and also, quite possibly, conduc-tivity (although it has been proposed that the conductivity of the base of the mantle is highly elevated). The degree to which chemical ex-change occurs across the core-mantle boundary is a frequent topic of study. Analysis of seismic data has indicated significant lateral variation in seismic wave speeds at the base of the mantle and also anisotropy, and there have also been claims of seismic observations of topography of the boundary itself. *See* core-mantle coupling.

core-mantle coupling   This term may be used to refer either to the exchange of material across the core-mantle boundary (e.g., via chemical reaction), or the exchange of momentum be-tween parts of the core and the mantle. The mechanism for the latter type of coupling is dis-puted, as is the degree to which the former type of coupling occurs at all. Changes in the ro-tation rate and direction of the Earth's mantle on timescales of decades are usually attributed to momentum exchange between the core and the mantle, and there have been claims that the core is important on even shorter timescales, even though on subannual timescales the at-mosphere is the predominant driving force for changes in Earth's rotation. There are several possibilities for how momentum exchange may occur: viscous coupling, electromagnetic cou-pling, topographic coupling, and gravitational coupling. Viscous coupling is usually ruled out as the cause of the observed changes in Earth ro-tation because the viscosity of the core is usually regarded as very small. Electromagnetic cou-pling would occur via magnetic linkage between the liquid metal of the core and conductivity in the mantle, in which electrical currents would be induced by changes in the magnetic field. Topo-graphic coupling would occur through the resis-tance of topography on the core-mantle bound-ary to flow at the top of the core, while gravi-tational coupling would happen if lateral varia-tions in the density of the mantle and core are arranged so as to yield a torque between the two. *See* core-mantle boundary.

Coriolis   A term used to refer to the force or acceleration that results due to the rotation of a coordinate system, such as a system fixed to the rotating Earth. Increases in importance as angular velocity or length or time scale of the problem increase.

Coriolis effect (Coriolis force)   A nonconservative effective inertial force contributing to the deviation from simple trajectories when a mechanical system is described in a rotating coordinate system. It affects the motion of bodies on the Earth and in molecular spectroscopy leads to an important interaction between the rotational and vibrational motions. The effect is described by an additional term in the equations of motion, called the *Coriolis force, $F_{cor} = 2m\omega \times \mathbf{v}$*, where ω is the angular velocity, and $\mathbf{v}$ is the velocity measured in the rotating frame. In meteorology, it is an apparent force acting on a moving mass of air that results from the Earth's rotation. Coriolis force causes atmospheric currents to be deflected to the right in the northern hemisphere and to the left in the southern hemisphere. It is proportional to the speed and latitude of the wind currents, and, therefore, varies from zero at the equator to a maximum at the poles. Coriolis force is very important to large-scale dynamics. To a unit mass fluid, on Earth it is expressed as

$$f = -2\Omega \times \vec{V} = -2\left[\vec{i}(w\Omega cos\Phi\right.$$
$$\left. - v\Omega sin\Phi) + \vec{j}(u\Omega sin\Phi) + \vec{k}(-u\Omega cos\Phi)\right]$$

where Ω is the angular velocity of Earth, Φ is latitude, and u, v, w are components of wind. Coriolis force is always perpendicular to the motion direction. Thus it affects only the direction of wind and never affects wind speed. (Gustave Coriolis, 1835.) *See* centrifugal force.

Coriolis, Gaspard Gustave de   Physicist (1792–1843). Presented mathematical studies on the effects of Earth's rotations on atmospheric motions.

Coriolis parameter   The local vertical (or radial) component of twice the Earth's angular velocity, that is,

$$f = 2\Omega \sin(\Theta)$$

where Θ is the geographical latitude (equator: $\Theta = 0$; poles: $\Theta = \pi/2$), and Ω is the Earth's angular frequency ($\Omega = 2\pi/86'00$ s^{-1}). It expresses the Coriolis force by the momentum equation $\partial v/\partial t = -f_u$ and $\partial u/\partial t = f_v$. The current components u, v are directed towards east and north, respectively.

corner frequency   On Fourier amplitude spectrum for seismic displacement waveforms, amplitudes are almost constant on the lower frequency side of a frequency, whereas amplitudes become small with increasing frequency on the higher frequency side. The border frequency between them is called the *corner frequency.* The larger an earthquake is and the slower the rupture associated with an earthquake is, the lower the corner frequency becomes. From a finite line source model with unilateral rupture propagation, it is expected that the corner frequency is related to finiteness concerning apparent duration time of rupture in the direction of the length of the fault plane and finiteness concerning rise time of source time function, resulting in amplitude decrease on the higher frequency side, inversely proportional to the square of frequency. The flat amplitudes on the lower frequency side represent a pulse area of displacement waveforms.

corona   In astronomy, the tenuous outer atmosphere of the sun or other star, characterized by low densities and high temperatures ($> 10^6 K$). Its structure is controlled by solar magnetic fields, which form the *corona* into features called coronal streamers. The solar corona has a total visible brightness about equal to a full moon. Hence, since it is near the sun (~ 2 to 4 solar radii), it is normally invisible, but can be observed with a coronagraph or during a total solar eclipse. This visible portion of the corona consists of two components: the F-corona, the portion which is caused by sunlight scattered or reflected by solid particles (dust), and the K-corona, which is caused by sunlight scattered by electrons in the extended hot outer atmosphere of the sun. Stellar coronae are sources of X-rays and radio emission, and their intensity varies with the period of the stellar activity cycle, about 11 years for the sun.

In planetary physics, a corona is a large structure of combined volcanic and tectonic origin. Most coronae are found on Venus, although the term is also used to describe tectonic features on the Uranian moon of Miranda. The Venusian coronae typically consist of an inner circular plateau surrounded first by a raised ridge and then an annulus of troughs. Most of the interior features of the corona are typical volcanic structures, including calderas, small shield volcanos, and lava flows. Coronae tend to be very large structures, often 300 km or more in diameter. Planetary scientists believe they form when a large blob of hot magma from Venus's interior rises close to the surface, causing the crust to bulge and crack. The magma then sinks back into the interior, causing the dome to collapse and leaving the ring. Rising and collapsing diapirs of material have also been proposed to explain the coronae on Miranda.

coronagraph   A telescope designed to observe the outer portions of the solar atmosphere. The bright emission of the solar disk is blocked out in *coronagraphs* by means of an occulting disk, bringing the faint outer corona into view. Coronagraphs typically view the corona in white light, though filters can be used to achieve specific wavelength observations. Because of the need for an occulting disk, they only observe the corona above the solar limb, projected onto the plane of the sky. Modern coronagraphs, such as the one on the SOHO spacecraft, can observe the corona between 1.1 and 30 solar radii. Coronagraphs can also be operated from the ground as long as the air column above the coronagraph is thin enough to reduce atmospheric scattering sufficiently. The first coronagraph was operated by B. Lyot from the Pic du Midi in the Pyrenees at an altitude above 2900 m. Coronagraphs provide the most startling observations of coronal mass ejections, helmet streamers, and prominences.

Coronal Diagnostic Spectrometer (CDS)   A Wolter II grazing incidence telescope equipped with both a normal incidence and a grazing incidence spectrometer flown on board the SOHO spacecraft. This instrument is designed to measure absolute and relative intensities of selected EUV lines (150 to 800 Å) to determine temperatures and densities of various coronal structures.

coronal dimming   During an eruptive event such as coronal mass ejection or a long duration flare, a large mass of plasma is ejected from the solar corona. When observed in soft X-ray wavelengths, the expulsion of million degree plasma is called *coronal dimming*. This coronal dimming relates the removal of hot material from the low corona to the higher, cooler material commonly associated with a coronal mass ejection, as seen in white light.

coronal heating   The temperature of the solar atmosphere increases dramatically from the photosphere, through the chromosphere and transition region, to the corona with temperatures in the corona varying from 2 to 3 million degrees Kelvin in the quiet diffuse corona to as much as 5 to 6 million degrees in active regions. The reason why the corona is so hot remains a mystery although it is now clear that the sun's magnetic field plays a crucial role in the transport and dissipation of the energy required to heat the corona. The total energy losses in the corona by radiation, conduction, and advection are approximately 3×10^{21} J or about 500 W m^{-2}. Balancing these losses requires only about 1 part in 100,000 of the sun's total energy output.

coronal hole   A low density extended region of the corona associated with unipolar magnetic field regions in the photosphere, appearing dark at X-ray and ultraviolet wavelengths. The magnetic field lines in a *coronal hole* extend high into the corona, where they couple to the solar wind and are advected into space. The corona, the outermost gravitationally bound layer of the solar atmosphere, is a very hot plasma (temperatures in the range of 1 to 2×10^6 K). The large-scale structure of the coronal gas consists of relatively dense regions whose magnetic field lines are "closed" (anchored at two points in the photosphere) and lower-density regions (the coronal holes), whose magnetic field lines are "open" (anchored at a single point in the photosphere and extending outward indefinitely). The solar wind emerges along these open field lines.

Except possibly for the periods of highest solar activity, the largest coronal holes are located at relatively high heliographic latitude, often with irregularly shaped extensions to lower latitude, sometimes into the opposite hemisphere. During maximum activity periods, equatorial coronal holes can appear and last for several solar rotations.

coronal lines   Forbidden spectral emission lines emitted from highly ionized atomic species, in a high temperature, dilute medium where collision between ions and electrons dominates excitation and ionization, as in the solar corona. In such plasma the temperature (1 to 2×10^6 K in the solar corona) and hence the kinetic energy of ions and electrons is so high that collisions have sufficient energy to ionize atoms. The first coronal emission line was identified at 530.3 nm during the total solar eclipse of 1869. Only in the 1940s were most of the *coronal lines* identified as forbidden transitions from elements such as iron, nickel, and calcium in very high ionization stages. Ratios of coronal line fluxes, similarly to ratios of nebular lines, are used as diagnostics of temperature and density. *See* forbidden lines, nebular lines.

coronal loops   The solar corona is comprised primarily of magnetic loop-like structures which are evident at all scales in the corona and are thought to trace out the magnetic field. Loops are seen at soft X-ray, EUV, and optical wavelengths. Typical configurations of loops occur in active regions, where many bright compact loop structures are associated with strong surface magnetic fields, and in arcades spanning a magnetic neutral line and often overlaying a filament or filament channel. The interaction and reconfiguration of these structures often accompany the dynamic eruptive phenomena on small scales in solar flares and very large scales in coronal mass ejections.

coronal mass ejection (CME)   An ejection of material from the sun into interplanetary space, as a result of an eruption in the lower corona. This material may sometimes have higher speeds, densities, and magnetic field strengths relative to the background solar wind and may produce shocks in the plasma. The fastest CMEs can have speeds of 2000 km s^{-1} compared with normal solar wind speeds closer to 400 km s^{-1}. CMEs are more common at solar maximum, when three per day can be seen, than solar minimum, when one may be seen in five days. If the material is directed towards the earth, then the CME may cause a disturbance to the Earth's geomagnetic field and ionosphere. *See* solar wind.

coronal rain   Cool plasma flowing down along curved paths at the solar free-fall speed of 50 to 100 km s^{-1}; material condensing in the corona and falling under gravity to the chromosphere. Typically observed in Hα at the solar limb above strong sunspots.

coronal transients   A general term for short-time-scale changes in the corona but principally used to describe outward-moving plasma clouds. Erupting prominences are accompanied by *coronal transients,* which represent outward moving loops or clouds originating in the low corona above the prominence. As many as one coronal transient per day is observed to occur during the declining phase of the solar cycle and are most commonly associated with erupting filaments.

coronal trap   The region of the corona in which charged particles are trapped between two areas of converging magnetic field, i.e., a magnetic bottle. The converging field causes a strengthening of the field and consequently a strengthening of the Lorentz force felt by a charge particle of velocity v. The particle's pitch-angle, $\theta = cos^{-1}(v_z/v)$ where z is the direction parallel to the field direction, increases as the particle moves into the region of increasing field strength until all of the particle's momentum is converted into transverse momentum ($\theta = 90°$). This location is known as the mirroring point because the particles cannot pass into a region of greater field strength and therefore become trapped. When collisions and wave-particle interactions are ignored, the conditions for a particle to be trapped are defined by the equation $sin\theta/B = sin\theta_0/B_0$ where θ_0 is the particle's initial pitch angle and B_0 is the coronal field. Note that for a prescribed field convergence B/B_0, particles with initial pitch angles

$\theta < \theta_0$ will escape the trap, i.e., they thermalize in the ambient plasma before they bounce.

corotating interaction region   *See* heliospheric stream structure.

correlation length   In phase transitions, topological defects may arise when growing spatial domains with different orientations (phases) of the correlated field fail to match smoothly. Hence, fluctuations in the phase of the field (determined by local physics only) will be uncorrelated on scales larger than a given *correlation length* ξ, whose details depend on the transition taking place.

The tendency of the field configuration after the transition will be to homogenize, and thus ξ will grow in time. Causality imposes an upper limit, as information cannot propagate faster than light. Hence, in cosmology the correlation length must be smaller than the distance signals can have traveled since the Big Bang, which for both radiation- and matter-dominated eras implies $\xi \lesssim t$, with t the cosmic time.

The correlation length is of utmost importance for the subsequent evolution of a cosmic defect network. In fact, the initial length scale of the network will be determined by the probability of defect formation out of the coalescence of different domains.

coseismic deformation   Displacement such as uplift and subsidence that occurs during an earthquake. The term is used particularly to describe static displacement and not vibration associated with seismic wave propagation. Depending on the means of measurement, the term may represent displacements that occur within a fraction of a second to a period of several years.

cosine collector   A radiant energy detector whose effective light collection area is proportional to the cosine of the angle between the incident light and the normal to the detector surface; used to measure plane irradiances.

cosmic abundance   The relative abundance of elements in the universe. Hydrogen provides approximately 75% of the mass density of the universe. ^4He provides about 24%. Lithium, beryllium and boron are each at the 10^{-12} to 10^{-10} level. These elements are thought to have been produced in the Big Bang.

Heavier elements were produced in stars or supernovae. Carbon, nitrogen, oxygen, and neon are present at parts greater than 10^{-4}. Silicon and iron are abundant at the 10^{-4} level. Elements with atomic number exceeding approximately 30 are present at the 10^{-10} to 10^{-11} level. There is strong "odd-even" effect; even atomic numbers (numbers of protons) or even numbers of neutrons make the isotope much more abundant than nearby isotopes.

cosmic censorship   The conjecture put forward by R. Penrose that the formation of naked singularities (singularities visible from infinitely far away) is evaded in nature because singularities in space-time are always surrounded by an event horizon which prevents them from being observed and from influencing the outside world. In this simple formulation the hypothesis was proven false by counterexamples, i.e., models of spacetime whose metrics obey Einstein's equations, but in which naked singularities exist. According to proponents of *cosmic censorship,* these examples are not generic. Some of the spacetimes in question are highly symmetric or require tuning of parameters. In others, the gravitational fields in the neighborhood of those singularities are, in a well-defined mathematical sense, too weak (i.e., produce too weak tidal forces) to be considered genuinely singular. There is as yet no well-formulated statement of cosmic censorship or a proof of its holding in general relativity.

cosmic microwave background   Single component of cosmic origin that dominates the electromagnetic background at wavelengths in the millimeters to centimeters range. It was serendipitously discovered by Penzias and Wilson in 1965. The *cosmic microwave background* radiation has the spectrum of a black body at temperature $T_o = 2.728 \pm 0.002$ K. It can be detected in any direction of the sky. Its high degree of isotropy is an observational evidence that on the largest scale, the universe is homogeneous and isotropic.

The Planckian spectrum of the cosmic microwave background is a strong vindication of the Big Bang picture. Since the universe at present is transparent to radiation (radiogalaxies at redshifts $z < 1$ are observed at microwave frequencies), a thermal spectrum could not have been produced recently, i.e., at redshifts smaller than unity as would be required by the steady state cosmological model. In the Big Bang model, the effect of the expansion is to decrease the cosmic microwave background temperature, so in the past the universe was hotter. At redshift z, the temperature would be $T = T_o(1+z)$. At $z \geq 1000$ matter and radiation would have achieved thermal equilibrium. In the 1940s, Gamow, Alpher, Herman, and Follin predicted that, as the universe expands and temperature drops, the interactions that kept matter and radiation in thermal equilibrium cease to exist. The radiation that would then propagate freely is the one observed today.

On scales up to a few tens of Mpc, the universe is not homogeneous. The presence of inhomogeneities induces temperature anisotropies on the background radiation. Irregularities in the matter distribution at the moment of recombination, our peculiar motions with respect to the Hubble flow, the effect of hot plasma on clusters of galaxies (*see* Sunyaev–Zel'dovich effect), and several other contributions, induced anisotropies at the level of one part in 10^3 (dipole) and at 10^5 on smaller angular scales. These effects convert the CBR into an excellent probe of the history of structure formation (galaxies, clusters of galaxies) in the evolving universe. *See* cosmic microwave background, dipole; quadrupole; temperature fluctuations; spectral distortions.

cosmic microwave background, dipole component   Dipole variation in the thermodynamic temperature as a function of direction. It is the largest anisotropy present in the *cosmic microwave background* radiation. The motion of an observer with velocity v with respect to a reference frame where a radiation field (of temperature T_o) is isotropic produces a Doppler-shifted temperature $T(\theta) = T_o(1 + (v/c)^2)^{1/2}/(1-(v/c)\cos(\theta))$ where θ is the angle between the direction of observation and the direction of motion, and c is the speed of light. Immediately after the discovery of the cosmic microwave background, the search started for the Doppler anisotropy described above, and

the first results were obtained at the end of the 1960s. The best-fit of the dipole amplitude is 3.358 ± 0.023 mK in the direction $(l, b) = (264°.31 \pm 0°.16, 48°.05 \pm 0.09)$ in galactic coordinates. The current understanding is that the largest contribution to the dipole anisotropy comes from the motion of the Earth. All other contributions are negligible. Under this assumption, the data quoted above corresponds to a sun velocity, with respect to the cosmic microwave background, of $v_\odot = 369.0 \pm 2.5$ km/s towards the constellation Leo, and the velocity of the local group is $v_{LG} = 627 \pm 22$ km/s in the direction $(l, b) = (276° \pm 3, 30° \pm 2)$. *See* peculiar motion.

cosmic microwave background, quadrupole component   Quadrupole variation of the temperature pattern of the cosmic microwave background across the sky. It was first measured in 1992 by the Differential Microwave Radiometer (DMR) experiment on board the COBE satellite, launched by NASA in 1989. The cosmic microwave background temperature fluctuations were measured at an angular resolution of $7°$ at frequencies of 31.5, 53, and 90 GHz. The r.m.s. quadrupole anisotropy amplitude is defined through $Q_{rms}^2/T_o^2 = \sum_m |a_{2m}|^2/4\pi$, with T_o the cosmic microwave background temperature and a_{2m} the five ($l = 2$) multipoles of the spherical harmonic expansion of the temperature pattern (*see* cosmic microwave background temperature anisotropies, Sachs–Wolfe effect). The observed cosmic microwave background quadrupole amplitude is $Q_{rms} = 10.7 \pm 3.6 \pm 7.1 \mu$ K, where the quoted errors reflect the 68% confidence uncertainties from statistical errors and systematic errors associated to the modeling of the galactic contribution, respectively.

More interesting for cosmological purposes is the quadrupole obtained from a power law fit to the entire radiation power spectrum, Q_{rms-PS}. The data indicates that the spectral index of matter density perturbations is $n = 1.2 \pm 0.3$ and the quadrupole normalization $Q_{rms-PS} = 15.3^{+3.8}_{-2.8} \mu$ K. For $n = 1$, the best-fit normalization is $Q_{rms-PS}|_{(n=1)} = 18 \pm 1.6 \mu$ K. The difference between the two definitions reflects the statistical uncertainty associated with the large sampling variance of

Q_{rms} since it is obtained from only five independent measurements.

cosmic microwave background, spectral distortions   The cosmic microwave background is well characterized by a 2.728 ± 0.002 K black body spectrum over more than three decades in frequency. *Spectral distortions* could have been produced by energy released by decaying of unstable dark matter particles or other mechanisms. Free-free processes (bremsstrahlung and free-free absorption) become ineffective to thermalize the radiation below redshift $z_{ff} \simeq 10^5 (\Omega_B h^2)^{-6/5}$, where Ω_B is the baryon fraction of the total density in units of the critical density, and h is the dimensionless Hubble constant. Any processes releasing energy later than z_{ff} will leave a distinctive imprint on the spectrum of the cosmic microwave background. After z_{ff}, Compton scattering between the radiation and electron gas is the only process than can redistribute the photon energy density, but as it conserves photon number, it does not lead to a Planckian spectrum. The lack of any distortion on the cosmic microwave background spectrum sets a very strict upper limit on the fractional energy released in the early universe: $\Delta E / E_{cmb} < 2 \times 10^{-4}$ for redshifts between 5×10^6 and recombination. The only distortion detected up to now is a temperature deficit in the direction of clusters of galaxies due to inverse Compton scattering of cosmic microwave background photons by hot electrons in the cluster atmosphere. *See* Sunyaev–Zeldovich effect.

cosmic microwave background, temperature fluctuations   Variation on the cosmic microwave background temperature across the sky. Current observations show that the cosmic microwave background has a dipole anisotropy at the 10^{-3} level and smaller scale anisotropies at the 10^{-5} level in agreement with the expectations of the most widely accepted models of structure formation. It is customary to express the cosmic microwave background temperature anisotropies on the sky in a spherical harmonic expansion,

$$\frac{\Delta T}{T}(\theta, \phi) = \sum_{lm} a_{lm} Y_{lm}(\theta, \phi) \,.$$

The dipole ($l = 1$) is dominated by the Doppler shift caused by the Earth's motion relative to the nearly isotropic blackbody field (*see* cosmic microwave background dipole component). The lower order multipoles ($2 \leq l \leq 30$), corresponding to angular scales larger than the horizon at recombination, are dominated by variations in the gravitational potential across the last scattering surface (*see* Sachs–Wolfe effect). On smaller angular scales, peculiar motions associated with the oscillation in the baryon-photon plasma dominate the contribution, giving rise to variations in power between $l \sim 100$ and $l \sim 1000$ known as *Doppler peaks*. Together, other physical processes can contribute to increase the intrinsic anisotropies along the photon trajectory such as integrated Sachs–Wolfe, Sunyaev–Zeldovich, Vishniac or Rees–Sciama effects. The pattern of temperature anisotropies and the location and relative amplitude of the different Doppler peaks depend on several cosmological parameters: Hubble constant, baryon fraction, dark matter and cosmological constant contributions to the total energy density, geometry of the universe, spectral index of matter density perturbations at large scales, existence of a background of gravitational waves, etc. In this respect, the cosmic microwave background is an excellent cosmological probe and a useful test of models of galaxy and structure formation. The character of the fluctuations is usually described by the best fitting index n and Q_{rms-PS}, the mean r.m.s. temperature fluctuations expected in the quadrupole component of the anisotropy averaged over all cosmic observers.

A cosmological model does not predict the exact cosmic microwave background temperature that would be observed in our sky, but rather predicts a statistical distribution of anisotropy parameters, such as spherical harmonic amplitudes: $C_l = < |a_{lm}|^2 >$ where the average is over all cosmic observers. In the context of these models, the true cosmic microwave background temperature observed in our sky is only a single realization from a statistical distribution. If the statistical distribution is Gaussian, and the spectral index of matter density perturbation spectrum is $n = 1$, as favored by inflation, then $C_l = 6C_2/l(l+1)$. The figure displays the mean temperature offset $\delta T_l = (l(l+1)C_l/2\pi)^{1/2}T_o$ of several experiments carried out to measure temperature anisotropies on all angular scales. The data indicate a plateau at $l \simeq 20$, suggesting a spectral index close to $n = 1$ and a rise from $l = 30$ to 200, as it would correspond to the first Doppler peak if $\Omega = 1$. Error bars in the vertical direction give 68% confidence uncertainty. In the horizontal direction indicate the experiments angular sensitivity. We have supersoped the predictions of three flat models: $\Omega_{cdm} = 0.95 \ h = 0.65$, $\Omega_{cdm} = 0.95$, $\Omega_\Lambda = 0.7$ and $h = 0.7$, and $\Omega_{hdm} = 0.95$ $h = 0.65$ to show the agreement between data and theory. In both cases the baryon fraction was $\Omega_B = 0.05$, in units of the critical density. Ω_{cdm} represents the fraction of cold dark matter, Ω_{hdm} of hot dark matter, and Ω_Λ the contribution of the cosmological constant to the total energy density.

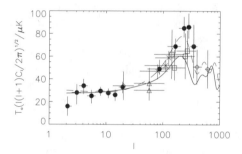

Experimental results as to January 1999. Measurements of anisotropies have been converted into temperature offsets. The predictions of two models are superposed with: cold dark matter ($\Omega_{CDM} = 0.95$, $\Omega_B h^2 = 0.02$, $h = 0.65$) and cold dark matter with cosmological constant ($t_o = 14 \times 10^9$ yr, $\Omega_\Lambda = 0.7$, $\Omega_{CDM} = 0.25$, $\Omega_B h^2 = 0.02$).

cosmic nucleosynthesis   Big Bang nucleosynthesis.

cosmic phase transition   The idea that phase transitions would occur in the early universe, originally borrowed from condensed matter and statistical physics. Examples of such cosmic transitions are the quark to hadron (confinement) phase transition, which quantum chromodynamics (QCD, the theory of strongly interacting particles) predicts at an energy around 1 GeV, and the electroweak phase transition at

about 250 GeV. Within grand unified theories, aiming to describe the physics beyond the standard model, other phase transitions are predicted to occur at energies of order 10^{15} GeV.

In cosmic phase transitions, the system is the expanding, cooling down universe, and the role of the order parameter is commonly assigned to the vacuum expectation value $\langle |\phi| \rangle$ of hypothetical scalar Higgs fields (denoted ϕ in what follows), which characterize the ground state of the theory. In the transition the expected value of the field goes from zero (the high temperature symmetric phase) to a nonvanishing value (in the low temperature broken symmetry phase, which does not display all the symmetries of the Hamiltonian).

The evolution of the order parameter (ϕ in our case) can be a continuous process (for second-order transitions). It can also proceed by bubble nucleation or by spinoidal decomposition (first-order transition), in which case ϕ changes from zero to its low temperature value abruptly. Typical effective potentials are shown on the figures.

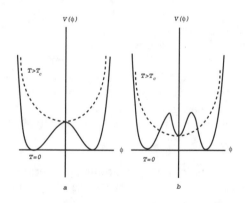

Effective potentials used to describe cosmic phase transitions: (a) for a second-order phase transition and (b) for a first-order phase transition. In both cases, the potential has a minimum value for $\phi = 0$ at high temperatures (i.e., $T \gg T_c$, dotted lines), while its shape is modified at low temperatures (solid lines). First-order transitions are characterized by the fact that there exist two qualitatively distinct types of minima at low temperature, the symmetric phase ($\phi = 0$) being metastable instead of unstable as is the case for a second-order transition.

cosmic rays   High energy subatomic particles. Cosmic-ray primaries are mostly protons and hydrogen nuclei, but they also contain heavier nuclei. Their energies range from 1 to 2 billion electron volts to perhaps 10^{16} eV, though the higher energies are rare. On colliding with atmospheric particles, they produce many different kinds of lower energy secondary cosmic radiation. Many cosmic rays are produced in the sun, but the highest-energy rays are produced outside the solar system, perhaps in shocks associated with supernova explosions.

cosmic spring   Cosmic strings formed in the early universe might have the ability to carry currents. These currents were at some stage believed to have the capability to locally remove the string tension: the energy carried by the current indeed tends to balance the string tension, so that the effective tension could be made to vanish, or even become negative, hence turning a string into a *spring* (i.e., a tensionless string). Detailed numerical investigations revealed that the maximum allowed current was, in the case of spacelike currents, not enough for this mechanism to take place. The possibility that timelike currents could make up springs is still open, although quite unlikely, since reasonable equations of state show the phase space where it could happen to be very small.

Another possibility is that of static electromagnetically supported string loops (some authors adopt this as a definition of a spring). Here it is not the current inertia that balances the string tension, but the long-range electromagnetic field generated by the current that would support the whole configuration. In this particular case, the string loop would be required to be unnaturally large, due to the slow (logarithmic) growth of the electromagnetic support around the string core. Non-static but still stationary (rotating) configurations are now believed to have a much more important role to play in cosmology. *See* Carter–Peter model, current saturation (cosmic string), magnetic regime (cosmic string), tension (cosmic string), vorton.

cosmic string   A type of cosmic topological defect that arises from symmetry breaking schemes when the low temperature minimum of the potential has a phase structure: $\phi = \phi_0 e^{i\varphi}$,

where the φ is an arbitrary real phase, all values of φ having the same (zero) energy. Then, at large distances from the string, the phase can continuously increase around the string, forcing a high energy region along the line describing the string.

Both local (which have an associated gauge vector field that compensates much of the string energy), and global (which have no such gauge vector) strings may be formed depending on whether the broken group is a gauge or a rigid symmetry of the system before the transition, respectively. *See* Abelian string, cosmic topological defect, deficit angle (cosmic string), global topological defect, homotopy group, local topological defect.

cosmic texture   Cosmic structures in which multicomponent fields provide large scale matter sources. Their dynamics can generate local energy concentrations which act to seed subsequent formation of structure (super cluster of galaxies, etc.) in the universe. *See* cosmic topological defect.

cosmic topological defect   Current understanding of the physics of the early universe is based in part on the spontaneous breaking of fundamental symmetries. These symmetry breaking processes take place during phase transitions, and many of these transitions might have occurred at grand unified energy scales. At these scales spacetime gets "oriented" by the presence of a hypothetical field called generically the "Higgs field", pervading all the space. Different models for the Higgs field lead to the formation of a whole variety of topological defects with very different characteristics and dimensions.

Some of the proposed theories have symmetry breaking patterns leading to the formation of "domain walls" (mirror reflection discrete symmetry): incredibly thin (thickness comparable to a Compton wavelength associated with particle energy $\sim 10^{15}$ GeV) planar surfaces, trapping enormous concentrations of mass-energy, which separate domains of conflicting field orientations, similar to two-dimensional sheet-like structures found in ferromagnets.

In other theories, cosmological fields are distributed in such a way that the old (symmetric) phase gets confined into a finite region of space surrounded completely by the new (non-symmetric) phase. This situation leads to the generation of defects with linear geometry called "cosmic strings". Theoretical reasons require that these strings (vortex lines) not have any loose ends in order that the two phases stay separated. This leaves infinite strings and closed loops as the only possible alternatives for these defects to manifest themselves in the early universe. Point-like defects known as (magnetic) "monopoles" do arise in other particular symmetry breaking schemes. These are extremely important configurations, since their formation is predicted in virtually all grand unified theories whose low energy limit matches the standard model of particle interactions. *See* cosmic phase transition, cosmic string, cosmic texture, domain wall, Kibble mechanism, monopole, spontaneous symmetry breaking, texture.

cosmochemistry   The study of the chemical make-up of solar system bodies, how the chemistry of these bodies has evolved (through radioactive decay, cooling temperatures, etc.), and the chemical reactions that occur between different regions of a body (such as surface-atmosphere interactions). *Cosmochemistry* describes which elements will condense from the solar nebula at various temperatures and pressures, which explains why the inner terrestrial planets are composed of dense refractory elements while the outer Jovian planets and their moons are composed of more volatile gases and ices. Cosmochemical reactions tend to move to a state of equilibrium, which explains why certain molecules are found on planetary surfaces and others are found in the planetary atmospheres.

cosmogenic nuclides   Nuclides produced by the interaction of cosmic rays with the atmosphere. For instance, if thermal neutrons are captured by atmospheric nitrogen ^{14}N, a proton is emitted and the cosmogenic nuclide ^{14}C results. Formally, this reaction can be written as ^{14}N(n,p)^{14}C. The spallation of atmospheric nitrogen or oxygen, due to the capture of fast protons or neutrons, produces nuclides such as ^{10}Be under emission of nucleons or smaller fragments such as ^{2}H or ^{4}He. *Cosmogenic nuclides* are produced in the lower stratosphere and can

be transported down to the troposphere. Some cosmogenic nuclides, such as ^{10}Be, are washed out by rain, and their traces are stored in the polar ice sheets; others, such as ^{14}C, are assimilated by living matter and stored, for instance in trees. The production rate of cosmogenic nuclides depends on the intensity of the galactic cosmic radiation and thus varies during the solar cycle as the galactic cosmic radiation is modulated. The isotopes mentioned above frequently are used in palaeoclimatology because their records provide an indirect measure for solar activity. On longer time-scales, the intensity of the galactic cosmic radiation in the atmosphere and, therefore, of the cosmogenic nuclides is also modulated by the geomagnetic field. Thus, the long-term trend (some thousand years) in the cosmogenic nuclides can be used as a measure for the strength of the geomagnetic field. *See* modulation of galactic cosmic rays.

cosmological constant   A constant (usually denoted Λ) that measures the curvature of an empty space devoid of gravitational fields. In the real universe, where gravitational fields exist throughout the whole space, this curvature would provide a tiny background (approx. 10^{-50} cm^{-2}) to the total curvature, but its effects on the evolution of the universe could be profound. Depending on the sign of Λ, a Friedmann–Lemaître model with positive spatial curvature could go on evolving forever, or the one with negative spatial curvature could recollapse. In the first version of his relativity theory, Einstein did not use Λ (i.e., effectively he assumed that $\Lambda = 0$). Then it turned out that, contrary to everybody's expectations, the theory implied that the universe cannot be static (i.e., unchanging in time) if it is spatially homogeneous (*see* homogeneity). Consequently, Einstein modified his theory to allow for $\Lambda > 0$, and in the modified theory a model of a static universe existed (*see* Einstein universe). Later, E.P. Hubble discovered that the real universe is nonstatic indeed (*see* expansion of the universe). When Einstein realized how close he was to a prediction of this discovery (14 years in advance), he called the introduction of the *cosmological constant* "the biggest blunder of my life". Nevertheless, the constant is routinely taken into account in solving Einstein's equations, and in some modern theories of the early universe it must be nonzero. (*See* inflation.) A positive cosmological constant implies a universal repulsive force acting on all objects in the universe. The force is very weak at short distances, but can outbalance the gravitational force at large distances, hence the existence of the static Einstein universe. The Einstein equations written with the cosmological constant are

$$R_{\mu\nu} - \frac{1}{2}g_{\mu\nu}R - \Lambda g_{\mu\nu} = 8\pi G T_{\mu\nu}$$

where $R_{\mu\nu}$ is the Ricci tensor, $g_{\mu\nu}$ is the metric tensor, R is the curvature scalar, and $T_{\mu\nu}$ is the stress-energy tensor.

cosmological constant problem   In quantum field theory, all the fields contribute to the vacuum energy density, that is, to the cosmological constant and, therefore, the theoretical value for it is quite big. From dimensional analysis one expects the density of the cosmological constant in the universe to be of the order of (mass of heaviest particle)^4, and very heavy particles have been detected, for instance the top-quark with mass about 175 GeV. At the same time, the observable value of cosmological constant is zero, or at least it is very small. The astronomical observations do not give any definite lower bounds for the cosmological constant, but instead put the upper bound for this density corresponding to a mass of about 10^{-47} GeV. Thus, there is an explicit discrepancy between the theoretically based matter field's contributions to the induced cosmological constant and observations. To eliminate this discrepancy one has to introduce the vacuum cosmological constant, which must cancel the induced one with tremendous precision. The exactness of this cancellation is the *cosmological constant problem*.

Suggestions for the solution of the cosmological constant problem typically reduce the fine-tuning of the vacuum cosmological constant to the fine-tuning of some other quantities like the potential for a cosmic scalar field. Other suggestions are based on supersymmetry, which can prevent the contributions of matter fields to the induced cosmological constant via the cancellation of the contributions between bosons and fermions. However, supersymmetry, if it exists in nature, is believed to be broken at the energies

of the order M_F, the Fermi scale, $\approx (250)^4$ GeV. Thus, there is no acceptable and reliable solution of the cosmological constant problem at the moment. A solution has been postulated in terms of an anthropic many-world hypothesis. According to this hypothesis our universe is just a single one among many others, and we live in it because the small cosmological constant lets us do so. Most of the other universes are strongly compactified because of the large cosmological constant.

One can also mention that some cosmological and astrophysical theories require small but non-zero cosmological constant. The density of the cosmological constant may, in such theories, serve as a dark matter and, in particular, provide a desirable age of the universe. Recent observations of Ia supernovae in distant galaxies suggest the existence of dark matter in the form of a very small, repulsive cosmological constant. This small cosmological constant has apparently little to do with the cosmological constant problem because its possible existence can reduce the necessary exactness of the cancellation between the induced and vacuum cosmological constants by at most one order of magnitude. *See* cosmological constant, induced gravity, spontaneous symmetry breaking.

cosmological model   A solution of Einstein's equations (*see also* metric) that can be used to describe the geometry and large-scale evolution of matter in the universe. The term is sometimes misused (for various reasons) to denote such models of spacetime that do not apply to the real universe. *See* Bianchi classification, homogeneity, inhomogeneous models, perturbative solution.

cosmological principle   An assumption that says that every observer in the universe would see the same large-scale distribution of matter around him (*see* homogeneity), and that for a fixed observer, the large-scale distribution of matter would be the same along every direction (*see* isotropy). Philosophically, the *cosmological principle* is the extreme opposite of the pre-Copernican view that the Earth was the center of the universe; according to the cosmological principle all positions in the universe are equivalent. The cosmological principle clearly does not apply at small (e.g., galactic) scales, and several explicit general relativistic models of the universe exist in which the cosmological principle does not apply (*see* inhomogeneous models, microwave background radiation). A strengthened version of the cosmological principle called a "perfect cosmological principle" gave rise to the steady-state models.

cosmology   The science that investigates the whole universe as a single physical system. It combines mathematics (heavily used to find cosmological models from Einstein's equations or other theories of gravity), physics (that guides the theoretical research on observable effects in cosmological models), and astronomy (that provides observational support or negation of various theoretical results).

The universe is usually assumed to be a continuous medium (usually a perfect fluid or dust) to which the laws of hydrodynamics and thermodynamics known from laboratory apply. In the very early universe, seconds after the Big Bang, the continuous medium is a mixture of elementary particles, and then a plasma. The medium resides in a spacetime, whose metric is that of a cosmological model. This spacetime is an arena in which geodesics (i.e., trajectories of bodies moving under the influence of gravitational fields) can be investigated. Among the geodesics are light-rays that lie on light-cones; they are trajectories along which astronomers receive their observational data from distant galaxies and quasars. In particular, the microwave background radiation consists of photons that travel along light-like geodesics; it brings information from an era shortly after the Big Bang. These notions provide the foundation on which theoretical *cosmology* is built. Observational cosmology deals mainly with the spectrum and temperature of the microwave background radiation, in particular the dependence of its temperature on the direction of observation, with the formation of the light elements in the early universe, with spatial distribution of matter in the large scale, with properties and evolution of galaxies and quasars, also with gravitational lenses. Cosmology seeks to explain, among other things, the creation and evolution of the large-scale matter distribution in the universe,

formation of galaxies, and the sequence of physical processes following the Big Bang.

coude focus   An optical system that directs the beam of light by bending the path at an "elbow"; ("coude" = "elbow") from the primary mirror of a reflecting telescope down the hollow polar axis of the instrument to a remote fixed focal position.

coulomb   Standard international unit of electric charge, equal to the charge that passes through any cross-section of a conductor in 1 sec during a constant current flow of 1 A. *See* ampere.

Coulomb collisions   In a plasma, collisions are mediated through long-range electrostatic (Coulomb) forces between electrons and protons. The dynamics of a particle in the plasma are governed by the electrostatic interactions with all other particles in the plasma. The Coulomb interaction serves to slow down the incident particle, which releases energy in the form of a photon of a given wavelength (e.g., via bremsstrahlung). X-ray production is dominated by close encounters of electrons with protons, while long range *Coulomb collisions* are primarily responsible for radio bremsstrahlung.

Coulter counter®   One of a class of instruments that measures particle size distribution from the change in electrical conductivity as particles flow through a small orifice; originally developed by Coulter Electronics.

Courant number   A dimensionless number used to assess the numerical stability of a numerical solution scheme. Commonly used in the study of computational fluid mechanics.

covariant derivative   A differential operator defined on the tensors of an arbitrary rank; the map ∇ from tensor fields of type (j, k) to tensor fields of type $(j, k + 1)$ on a manifold; i.e., it produces a tensor of one higher covariant rank and the same contravariant rank. The covariant derivative satisfies the properties of derivative operators: (i) linearity, (ii) Leibnitz rule for derivatives of products, and (iii) commuting with the operation of contraction. In the theory of general relativity, the covariant derivative is required to be (iv) torsion-free, (i.e., when acting on a differentiable function f (a type-(0,0) tensor), the commutation property holds: $\nabla_{[a}\nabla_{b]}f = 0$) and (v) metric preserving, $\nabla g = 0$. Then a covariant derivative acting on a scalar is identical to partial derivative of the same scalar. For vectors or tensors of a higher rank, the covariant derivative is always a sum of the partial derivative and some linear combination of the initial tensor (or vector) with the coefficients of affine connection. In the general case the commutator of two covariant derivatives is given by the curvature tensor. *See* affine connection, curvature tensor.

covariant vector *See* one-form.

Cox number For a given background scalar gradient, the *Cox number* is a measure of the amount of turbulent activity in a stratified water column. For an arbitrary scalar θ, it is defined as the ratio of mean square gradient to the square of the mean gradient, i.e.,

$$\text{Cox} = \frac{\overline{(\nabla\theta)^2}}{(\overline{\nabla\theta})^2}$$

where ∇ denotes the gradient operator. In practical oceanic or atmospheric applications, temperature is most easily measured, and so for a one-dimensional vertical profile of temperature, the one-dimensional Cox number is given by

$$\text{Cox}_T = \frac{\overline{(\partial T'/\partial z)^2}}{(\partial\overline{T}/\partial z)^2}$$

where T' is the fluctuating temperature and $\overline{T}$ is the mean temperature. Under isotropic conditions, this one-dimensional estimate is equal to one-third of the total value of the Cox number.

The physical interpretation of the Cox number is that turbulent eddies tend to homogenize a background temperature gradient by moving fluid parcels against the ambient stratification. This displacement of fluid parcels leads to a certain amount of variance in the observed vertical gradient of the temperature fluctuations.

CP problem In quantum field theoretical descriptions of matter, the observation that strong (nuclear) interactions conserve the symmetries of parity and CP (simultaneous reversal of charge and parity) in spite of the fact that weak nuclear interactions violate those symmetries. Thus far unobserved particles called axions are postulated to accommodate the descrepancy.

Crab Nebula Nebula of visual brightness 8.4 (mag), apparent dimension 6×4 (arc min), right ascension $05^h34.5^m$, dec $+22°01'$, and a distance of 2 kpc. The supernova remnant resulting from the Type II supernova seen by Chinese, Japanese, Korean, Native American, and Arab astronomers in the year 1054. It was one of the first sources of radio waves and X-rays outside the solar system to be identified. It has a pulsar at the center and is expanding about 2000 km/sec.

Crab Pulsar Pulsar (neutron star) remnant of a supernova explosion which was witnessed in 1054 AD by Chinese and Japanese astronomers and apparently Native American observers at right ascension $05^h34.5^m$, dec $+22°01'$, and a distance of 2 kpc. The pulsar has a repetition period of 33ms, which is increasing at 4.2×10^{-13} sec/sec. The *Crab Pulsar* radiates two pulses per revolution: this double pulse profile is similar at all radio frequencies from 30 MHz upwards, and in the optical, X-ray and gamma ray parts of the spectrum. M1 in the Messier classification.

CRAND (Cosmic Ray Albedo Neutron Decay) Source of radiation belt particles. Cosmic rays interacting with the Earth's atmosphere create neutrons, as well as other particles and cosmogenic nuclides. Because neutrons are neutrals, their motion is not influenced by the geomagnetic field. If a neutron decays in the radiation belt regions, a proton and an electron are added to the radiation belt population.

crater A generic term used to describe any approximately circular depression. The term was first used in a geologic context by Galileo to describe the circular depressions he saw on the moon through his telescope. *Craters* can be produced by several processes, including impact, volcanism, and subsidence. Impact craters are produced when debris from space collides with a surface. Small impact craters, called simple craters, are usually bowl-shaped in appearance with little subsequent modification. Larger craters, called complex craters, show a range of features, including shallow depths, central peaks or central pits, and wall terracing. Extremely large (generally 500 km in diameter) impact craters are called basins. Craters on the planets are generally attributed to the impacts of planetary bodies, meteorites, and asteroids. Craters are a primary surface feature on Mars, Venus, Mercury, and the moon. Erosion has destroyed evidence of older craters on the Earth, but a significant number of younger craters have been recognized. Volcanic craters, now preferentially called calderas, are produced by the withdrawal of magma from a volcanic region. Removal of any underlying support, such as by removal of groundwater or the dissolution of rocks such as limestone, can create craters by subsidence. Volcanic and subsidence craters are not usually as circular as impact craters.

crater depth-diameter plots   Graphical representation of a crater's depth and width. Comparison of the plots from different planetary bodies shows that gravity is the factor controlling crater depth. They illustrate that there is an inverse relation between the depth of complex craters and the acceleration due to gravity; that shallower craters form for a given diameter when the gravity field of a planet's surface is higher; and that there is a smaller transition diameter from simple to complex craters in a higher gravity field.

cratering rates   The number of impacts per time unit on a planetary surface in the course of its geological history. *Cratering rates* are required to determine the absolute ages of planetary bodies using crater counting studies. A fairly accurate picture of the cratering rate on the moon has been compiled from radiometric dating of returned samples, but not for the other planetary bodies, at this time. In the latter cases, attempts have been made to determine it theoretically, first by analyzing the distribution of objects within the solar system and predicting how the cratering rates at the planetary body in question may be related to those on the moon, and second by comparing the crater statistics for

the planetary body with those of the moon and making assumptions about how the two are related. Both approaches cause large uncertainties that will be best reduced through sample return missions.

cratering record   The retention of impact craters by a planetary surface provides a record of how the frequency and size of impact craters have changed over time. The number of craters on a surface is related to the age of the surface; hence, older surfaces have greater crater frequencies than younger surfaces. Analysis of the number of craters in the ancient highlands of the moon indicate that the cratering rate was much higher early in solar system history. This period of higher cratering rates is called the period of heavy bombardment. The material responsible for cratering during the heavy bombardment period was leftover material from the formation of the planets. The formation of large impact basins was common during this time since the frequency of larger debris was greater. Analysis of the lunar cratering record indicates that the heavy bombardment period ceased about 3.8×10^9 years ago in the Earth-moon system. The cratering rate since that time has been much lower and is due to the impacts of asteroids and comets. Study of the *cratering record* can provide important information on the sources of the impacting material as well as the geologic evolution of the different solid-surface bodies in the solar system.

crater number: index of age   Older surfaces accumulate more impact craters than young ones. On planetary surfaces where erosion rates are low (such as the moon and Mars), crater densities depend on the age of the surface and the cratering rate only, such that relative ages and chronological relations can be defined fairly reliably. Determining absolute ages is more difficult in the absence of age-dated samples, etc. Absolute age determination has been attempted using impact-rate models (all assume that the impact rate has remained unchanged over the last 3 billion years) along with size-distribution curves. *See* cumulative size-frequency curves.

crater production function   The size frequency of craters expected if there was no crater erosion or destruction. It is determined by either examining the size distribution of interplanetary debris and calculating what size distribution of craters would result, or by analyzing areas of a planet that are sparsely cratered such that there is no crater overlap, there are distinguishable secondary craters, and there is little to no crater degradation.

crater saturation   When equilibrium is reached between the number of new craters that form on a planetary surface and the number of old craters that are destroyed. In this case, destruction of old craters is by the formation of new craters only. In other words, for a specific crater size range, a time will be reached when no more craters of that size range can be accommodated on the fresh surface without obliterating pre-existing craters. At this point the number of craters at that specific size range will no longer increase with time. A saturated surface has proportionately fewer small craters than a production surface because small craters are preferentially destroyed.

craton   Very old stable part of the continental crust; a major structural unit of the Earth's crust that is no longer affected by orogenic activity. *Cratons* are generally of igneous or metamorphic origin sometimes covered with a thin layer of sedimenary rocks or shallow water. They are comparatively rigid and stable areas of the Earth's crust, with relatively subdued seismic activity in their interior. Cratons are composed of shields and any adjacent platforms. Shields are the oldest regions, usually marked by exposed crystalline rocks. Platforms are thinly-mantled bordering areas.

creep   Continental drift and mantle convection require that the solid Earth's mantle behaves like a fluid on geological time scales. This behavior is known as creep. The fact that a crystalline solid can behave as a fluid is easily documented by the creeping flow of glaciers.

Cressida   Moon of Uranus also designated UIX. Discovered by Voyager 2 in 1986, it is a small, irregular body, approximately 33 km in radius. Its orbit has an eccentricity of 0, an inclination of $0°$, a precession of $257°$ yr^{-1}, and

a semimajor axis of 6.18×10^4 km. Its surface is very dark, with a geometric albedo of less than 0.1. Its mass has not been measured. It orbits Uranus once every 0.464 Earth days.

critical current (cosmic string)   *See* current saturation (cosmic string).

critical density   The cosmological density which is just sufficient to eventually halt the universal expansion, through its gravitation attraction:

$$\rho_{critical} = \frac{3H_0}{8\pi G}$$

where H_0 is the current Hubble parameter and G is the Newtonian gravitational constant.

critical depth   In hydraulics, depth of flow in an open channel, such as a river or canal, corresponding to a Froude number, $V/\sqrt{gD}$, of unity, where V is the flow speed, g is the acceleration of gravity, and D is the hydraulic depth. In oceanography, the depth in seawater below which respiratory carbon loss by phytoplankton exceeds photosynthetic carbon gain; no net phytoplankton production occurs.

critical flow   Flow in an open channel at critical depth.

critical frequency   If an ionospheric layer posseses a distinct maximum in ionization, a radio frequency capable of just penetrating to this height is called the *critical frequency* of the layer. It is the greatest frequency that can be reflected vertically from the layer. Strictly speaking, this frequency is not reflected by the layer but is infinitely retarded and absorbed; the radio wave slows down (retardation) as it approaches the critical frequency and absorption is increasingly effective as the pathlength changes (hence, deviative absorption). It is also identified as the frequency to which the virtual height vs. frequency curve becomes asymptotic. The ordinary ray critical frequency for a given ionospheric layer is denoted by the symbol "fo" plus the name of the layer (e.g., foE, foF1, foF2). Similarly the extraordinary ray critical frequency is denoted by fx. While these parameters refer to the layer electron density, normally they are recorded in units of megahertz since they are recorded from ionograms with a frequency, rather than electron density, scale. *See* ionogram, plasma frequency.

critical level   A level at which the mean flow speed is equal to the wave speed.

critical phenomena in gravitational collapse Discrete scaled-similar, or continuous scaled-similar behavior in the gravitational collapse of physical fields, near the threshold of black hole formation. Discovered by Choptuik in 1993.

critical point   In an expanding flow, the location of a transition from subsonic to supersonic flow. The concept of the *critical point* plays a central role in theories of the expansion of the solar wind. The term is frequently generalized to refer to a point of transition through the Alfvén speed or the fast or slow magnetoacoustic wave speed.

critical temperature   In thermodynamics, the upper limiting temperature at which a gas can be forced to condense (to a liquid or a solid) by compression at constant temperature. For water, the *critical temperature* is 374°C. The corresponding pressure is 221 bar. In superconductors, it is the temperature above which the superconducting behavior disappears. In a permanent magnet, it is the temperature above which the magnetism disappears. Also called Curie point. In symmetry breaking particle theories relevant to early universe cosmology, the critical temperature T_c for a particular model separates temperature zones in which the effective potential has very different qualitative features. For temperatures above T_c, simple potentials will have the form of a generalized paraboloid, with its minimum attained for a vanishing field. However, at temperatures below T_c, a degenerate space of minima develops (e.g., as in the "Mexican-hat" potential) and the field, in trying to minimize its potential energy, will select one of these minima, hence breaking the symmetry previously possessed by the system. *See* cosmic phase transition, Kibble mechanism, spontaneous symmetry breaking.

critical velocity   The velocity of the flow in an open channel corresponding to a Froude number of unity. *See* critical depth.

cross helicity   One of the quadratic invariants occurring in the theory of hydromagnetic turbulence. The *cross helicity* within a volume V is defined as

$$H_C = \int \mathbf{V} \cdot \mathbf{B} d^3 x \, ,$$

where $\mathbf{V}$ is the velocity, $\mathbf{B}$ is the magnetic field, and the integral is taken over V. In an incompressible dissipation-free fluid, for suitable boundary conditions, H_C is conserved. In fully developed three-dimensional dissipative hydromagnetic turbulence, the cross helicity, as well as the energy, cascades from large eddies down to smaller eddies where dissipation can occur. *See* helicity, hydromagnetic turbulence, magnetic helicity.

cross-section   Any of several quantities with units of area, which describe the interaction of an object with an incident flux of particles or of radiation.

cross-shore   Perpendicular to the general trend of a coastline. A beach profile represents a slice through a beach in the cross-shore direction.

cross slip   Refers to the manner of dislocation motion. With Burgers vector parallel to the dislocation line, screw dislocations are not tied to specific slip planes, and they can move from one plane to another intersecting plane which contains the same Burgers vector. This is called *cross slip*. Cross slip is an important mechanism of deformation at high temperature, and it is involved in the way recovery takes places.

cross waves   Waves in an open channel which propagate across the channel, normal or nearly normal to the direction of flow.

crust   The outermost layer of a differentiated solid-surfaced body; on Earth, the outer layer of the solid earth, above the Mohorovicic discontinuity. Its thickness on Earth averages about 35 km on the continents and about 7 km below the ocean floor. The *crust* is composed of the low-density, low-melting-point materials which floated to the top during differentiation of the body. The crusts of the terrestrial planets are composed primarily of silicate-rich rocks, such as granite (Earth's continental crust), basalt (Earth's oceanic crust; and volcanic plains on the moon, Mars, and Venus), and anorthosite (lunar highlands).

crustal deformation   Crustal movement.

crustal movement   In geophysics, crustal movement refers to present-day crustal displacements and deformation identified by geodetic measurement.   Changes in components such as vertical movement, horizontal movement, tilt, extension, and contraction can be obtained from repeated measurement and continuous observations of leveling and trilateration surveys, electro-optical distance measurement, GPS (Global Positioning System), VLBI (Very Long Baseline Interferometry), SLR (Satellite Laser Ranging), SAR (Synthetic Aperture Radar), tiltmeter, extensometer, volume-strainmeter, and tide gage. Periodic changes due to the Earth's tide and seasonal factors, deformation accompanied by plate motion, and deformation associated with large earthquakes and volcanic eruptions are considered causes of *crustal movement.*

crystal   A solid formed by the systematic arrangement or packing of atoms, ions, or molecules. The repetitive nature of this packing may cause smooth surfaces called crystal faces to develop on the *crystal* during the course of its unobstructed growth. A crystal is called anhedral, subhedral, or euhedral if it, respectively, lacks such faces, is only partly bounded by them, or is completely bounded.

crystallization age   The *crystallization age* of a rock tells when it solidified from its parent magma. It is usually obtained from the radioactive decay of elements within the rock's minerals. The concentrations of the parent and daughter elements are obtained from several minerals within the rock and plotted on an isochron diagram. The slope of the line, together with knowledge of the half-life of the radioisotope, allow the crystallization age to be determined.

CTD   A standard oceanographic or limnic instrument (profiler) to cast profiles of conductivity (C) and temperature (T) as a function of depth (D). Most instruments carry additional sensors such as O_2, transmissivity, pH, etc.

CTRS   "Conventional Terrestrial Reference System," a geographic reference frame, fixed in the mean crust of the Earth, and defined by international agreement under supervision of the IUGG (The International Union of Geodesy and Geophysics).   The definition is based on agreed coordinates and velocities (due to continental drift) of numerous observing stations and accepted methods for interpolating between them.   See http://hpiers.obspm.fr/webiers/general/syframes/convent/UGGI91.html

cumulative size-frequency curves   Distribution curves in which the logarithm of the cumulative number of impact craters (above a certain diameter) is plotted against the logarithm of the diameter. It shows the number of craters on a planetary surface generally increases with decreasing crater size. Size-frequency slopes may be affected if secondary craters have been included in the counts; by slopes on the planetary surface (surface craters are preserved more readily on level terrain); and if crater saturation has been reached.

Incremental plots are also used, for which the number of craters within a specific size increment is plotted against the diameter. *See* crater saturation.

cumulus   A sharply outlined cloud with vertical development. The summit of the cloud, in general dome-shaped, shows rounded bulges, and its base is usually horizontal. The mean lower level of the cumulus is 600 to 2000 m. Cumulus clouds consist of water droplets and are the product of water vapor condensation in convective activity. Labeled with four subtypes, based on the shape of the cloud: (1) cumulus fractus — Cu fra; (2) cumulus humilis — Cu hum; (3) cumulus congestus — Cu con; (4) an intermediate type between Cu hum and Cu con — Cu med.

Curie (Ci)   The special unit of radioactivity equal to 2.2×10^{12} disintegrations per minute.

Curie point   The temperature above which magnetization is eliminated in a material that is ferromagnetic at low temperatures.

Curie point survey   A method to estimate depth distribution of temperature at which spontaneous magnetism of ferromagnet is lost (Curie point, about 840 K) from measurement of the geomagnetic field by magnetic exploration in the air. Assuming that the average geothermal gradient is 30 K/km, the depth of Curie point in the crust is about 20 km. *Curie point surveys* are used for geothermal prospecting.

curl   An antisymmetric derivative of a vector, defined as a (pseudo)vector: $(curl\mathbf{V})_x = V_{y,z} - V_{z,y}$, where $\mathbf{V}$ is a vector, and the notation $_{,y}$ means partial derivative with respect to y. This expression gives the x-component of the curl, and the other components are given by cyclic permutation of the component indices.

current carrier (cosmic string)   A particle physics model for a topological defect in general, and that of a cosmic string in particular, may contain various couplings between the Higgs field and other particle fields. Whenever these particles are charged and have the possibility to move along the defect, they will induce a current, and are, therefore, called *current carriers*. See cosmic topological defect, Witten conducting string.

current, curvature   In a guiding center plasma, the part of the electric current due to the curvature drift.

current generation (cosmic string)   A *cosmic string* is a topological defect of the vacuum where a Higgs field is responsible for the existence of a large energy density confined in a linelike configuration. In grand unified models, this Higgs field is often coupled to many other fields whose dynamics will be different depending on whether the background space is the core of the string or the outside space. In particular, they might condense, i.e., acquire a net nonzero value inside the core.

For complex fields, the core condensate will have a well-defined phase. At the time of condensation, this phase will take random values along the string; any mismatch between those values will generate a gradient: it is precisely the gradient in the phase of one of these (charged) fields that powers the currents. *See* conducting string, current carrier (cosmic string), Witten conducting string.

current, gradient   In a guiding center plasma, the part of the electric current due to the gradient drift.

current instability (cosmic string)   In many grand unified models, cosmic strings have the ability to carry persistent currents. These currents could have drastic effects on the evolution of a network of strings as they could, for instance, imply the existence of vortons. However, not all such currents are completely stable for at least two reasons: other phase transitions subsequent to that leading to the formation of strings and currents, and classical and quantum perturbations could in fact destabilize the currents, leaving just an ordinary, nonconducting string. *See* conducting string, cosmic string, current carrier (cosmic string), vorton.

current, magnetization   In a guiding center plasma, the part of the electric current due to $\nabla \times \mathbf{M}$, with $\mathbf{M}$ being the magnetization term. Intuitively, this is the current associated with variations in density across the plasma, e.g., the current at the edge of a finite body of plasma.

current, polarization   In a guiding center plasma, the current associated with the change of the electric field, $d\mathbf{E}/dt$. If the electric field $\mathbf{E}$ changes, the energy associated with the electric drift $\mathbf{E} \times \mathbf{B}/B^2$ changes, and the polarization current transmits that energy.

current quenching   *See* current saturation (cosmic string).

current saturation (cosmic string)   The mechanism of *current saturation* (or quenching) in cosmic strings takes place when the magnitude of the current flowing along the core becomes comparable (in appropriate units) with the mass measured outside the string of the particle which is the current carrier. Then the fields acquire a nonvanishing probability to jump off the string and into a free particle state. There exists then a maximum current, which in the case of fermionic carriers of mass and charge qm is given by $J_{max} = qm/2\pi$, beyond which the current will fall drastically, showing saturation. *See* Carter–Peter model, phase frequency threshold, Witten conducting string.

current screening (cosmic string)   Conducting cosmic strings moving at a supersonic velocity v in a plasma of density ρ induce a magnetic shock since the charged particles in the plasma cannot penetrate in regions too close to the string core. The distance r_s from the string core to the shock is given by

$$r_s = \frac{J}{cv\sqrt{\rho}},$$

with c the velocity of light and J the current flowing along the string. At distances larger than r_s, the plasma therefore does not see the current on the string, so that it is said to be screened. *See* conducting string.

current sheet   A non-propagating boundary between two plasmas with the magnetic field tangential to the boundary, i.e., a tangential discontinuity without any flow. *Current sheets* are thought to exist in the solar corona with thicknesses much smaller than typical coronal length scales. Current sheets are required to separate magnetic fields of opposite polarity. Examples are the heliospheric current sheet separating the two hemispheres of opposing polarity in the interplanetary magnetic field, the tail current separating the northern and southern part of the magnetosphere's tail, or the tips of helmet streamers. The large current densities inherent to current sheets may have an important role in heating the corona, producing solar flares and prominence formation. Current sheets are topological regions which are most suitable for reconnection to occur. *See* reconnection.

curvature   A geometrical tensor describing how much a given space differs from the flat space of the same number of dimensions. Calculationally the components of the curvature tensor (the Riemann tensor) involve the connection coefficients and their first derivatives, and the structure coefficients describing the behavior of the basis used (simple example: spherical coordinates, where the structure coefficients vanish vs. unit vectors parallel to the spherical coordinate lines, structure constants nonzero). In Einstein's theory of gravity (general relativity), the curvature is an expression of the gravitational field. Flat space (zero curvature) is described by special relativity. The components of the curvature can be measured by repeated experiments involving the acceleration of separation of geodesics moving in various planes. That this involves relative acceleration shows that the effects of the curvature-tensor are a second-order derivation from flatness. *See* connection, structure coefficients.

curvature invariant   Invariant functions of the curvature tensor. There exist two types:

1. Polynomial invariants. In general relativity, the Ricci scalar R is pointwise determined by the matter distribution. Two quadratic and two cubic invariants exist. An example of these is the Kretschmann scalar $I = R_{abcd}R^{abcd}$.

2. Ratios of various tensor functions of the curvature tensor of like type.

See Riemann tensor.

curvature tensor   *See* Riemann tensor.

curve   A $1 - 1$ mapping from the real numbers to a space; for instance, the mapping from the real numbers Z to the four-space given in Minkowski coordinates: $\{Z\} : \{Z \longrightarrow X^\alpha(Z), \alpha = 0, 1, 2, 3, \}$. It is expedient to consider continuous and differentiable mappings (hence curves) of this type.

curved space-time   A general term for the pseudo-Riemannian manifold with a non-zero curvature metric. This is a concept that proves very useful for the construction of quantum field theory in *curved space-time* and especially in the models of induced gravity. Any theory of gravity should have two basic components: equations for the matter fields and particles and equations for the gravity itself. In quantum (and classical) field theory, it is sometimes useful to formulate the fields and their interactions on the background of an arbitrary metric which is not *a priori* a solution of the concrete field equations. First, one can achieve many results working on general curved space-time without specifying the equations for the metric itself. These results include, in particular, the covariant formulation of matter fields and their renormalization, calculation of the leading terms in the quantum corrections to the (undefined) classical action of gravity, and some applications of these. Second, the originally unspecified geometry of the space-time may be defined by the quantum effects of matter fields (or string). In this case, the action of gravity does not have a classical part, and gravitational interaction is induced by quantum effects of other fields. *See* general relativity (which defines a curved spacetime as a solution to Einstein's equations). *See* induced gravity, nonminimal coupling, quantum field theory in curved spacetime, quantum gravity.

curve of growth   A curve giving the optical density of a spectral line as a function of the atomic column density. Thus, densities can be estimated from the strength of emission or absorption lines.

curvilinear coordinates   A coordinate system in which the coordinate lines are not straight, and in which the metric tensor expressing Pythagorean theorem

$$ds^2 = g_{ij}(x^k)dx^i dx^j$$

has significantly nonconstant metric tensor components $g_{ij}(x^k)$.

curvilinear coordinates [in a plane]   A coordinate system is a relationship that is established between the points of the plane and pairs of ordered numbers called coordinates. *Curvilinear coordinates* are a coordinate system that is not Cartesian (*see* Cartesian coordinates). While Cartesian coordinates may be visualized as the determination of points in a plane by the intersection of two straight lines perpendicular to each other (each line corresponding to one of the pair of numbers or coordinates), curvilinear coordinates may be visualized as the determination of points by the intersection of two curves

in general (each curve corresponding to one of the coordinates). The two curves are not necessarily perpendicular to each other. In curvilinear coordinates, the relationship between points in a plane and pairs of numbers is not necessarily nonsingular at all points. A frequently used curvilinear coordinate is the polar coordinates (r, θ) defined by the expressions: $x = r\cos(\theta)$ and $y = r\sin(\theta)$. Polar coordinates can be visualized as the determination of points by the intersection of circles of radius r with rays starting at the origin of the coordinate system and extending outward at an angle θ with respect to the x-axis. They are, however singular at $r = 0$, and in principle, a different coordinate system has to be used at this point.

curvilinear coordinates [in space]   A coordinate system is a relationship that is established between the points in space and trios of ordered numbers called coordinates. *Curvilinear coordinates* are a coordinate system that is not Cartesian (*see* Cartesian coordinates), while Cartesian coordinates may be visualized as the determination of points in space by the intersection of three planes perpendicular to each other (each plane corresponding to one of the trio of numbers or coordinates). Curvilinear coordinates may be visualized as the determination of points by the intersection of three surfaces in general (each surface would correspond to one of the coordinates). Although the relation between curvilinear and rectangular coordinates is required to be nonsingular, in typical cases there are isolated points (e.g., at $r = 0$) where the relation is singular. In this case, in principle, a different coordinatization should be used.

cusp (cosmic string)   Small localized regions on cosmic strings which attain velocities close to that of light. These regions are highly energetic with enormous string curvature that may favor the emission of Higgs constituent particles (i.e., the scalar field making up the string) away from the string core as well as gravitational radiation. *See* cosmic string.

cusp, polar   One of two points or regions in the magnetosphere where, in the noon-midnight meridional surface, field lines swept back into the tail part company with the ones closing near

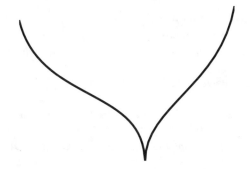

Generic shape of a string segment at the location of a cusp.

the sunward boundary. In simple idealized models, the cusps contain neutral points, located on the magnetopause. The models predict that at those points field lines intersect, and the field intensity drops to zero. In the actual magnetosphere, the cusps are observed as regions of weak varying field.

The term *polar cusp* is also applied to the funnel-shaped regions extending from the above weak-field regions near the magnetopause to their footprints on the ionosphere (those regions are also sometimes referred to as the cusps). The polar cleft is the name applied to the same region by researchers who propose its shape to be slit-like rather than funnel-like.

cutoff energy   Of cosmic ray protons at a given point P on Earth, the energy below which such protons can no longer reach P. For heavier particles, the *cutoff energy* can be derived from that of protons.

Proton orbits arriving at P can be trapped in the Earth's magnetic field (as those of radiation belt protons are), or they can extend beyond the Earth's magnetic field. All low energy proton orbits at P are trapped, while at sufficiently high energy, none are. The cutoff energy at P may be viewed as the energy below which all orbits at P are trapped.

Actually at any point P, the transition between trapped and non-trapped orbits is somewhat irregular and depends on direction. At high energy E, protons can arrive from any direction. At lower energies, some orbits arriving from the east are trapped and, therefore, empty of cosmic ray protons. As E decreases, the trapped/free boundary (its structure is com-

plex) expands across zenith until it reaches the western horizon, at the value of E below which all access is cut off.

cut-off rigidity   Cosmic rays below the *cut-off rigidity* cannot penetrate down to the Earth's surface but are reflected back towards space (*see* Störmer orbits). Since this shielding is due to the geomagnetic field, the cut-off rigidity depends on the geomagnetic latitude and the altitude of the observer: for normally incident protons the cut-off rigidity at the geomagnetic equator is about 15 GeV, that is all particles below 15 GeV are unable to reach the Earth, while the cut-off rigidity is only about 1.4 GeV at 58° geomagnetic latitude at sea level. In general, the cut-off rigidity P_{cutoff} is related to the geomagnetic latitude Φ_c by

$$P_{cutoff} = 14.9 \ \text{GeV} \cos^4 \Phi_c \,. \qquad (1)$$

Thus, the magnetosphere acts as a giant mass spectrometer.

Historically, the term cut-off rigidity refers to the rigidity at which the lower part of the spectrum of galactic cosmic rays is cut off. The cut-off rigidity is also used to characterize the energy threshold of a neutron monitor.

cyclic coordinate   In classical mechanics, a coordinate that does not explicitly appear in the Lagrangian (or in the Hamiltonian) for a system, though its velocity may be present. Via Lagrange equations, if $p_a = \partial L/\partial \dot{q}^a$ (where the coordinate q^a does not appear in the Lagrangian though $\dot{q}^a$ does), then

$$\frac{d}{dt} p_a = \frac{\partial L}{\partial q^a} = 0 \,.$$

Hence, momenta conjugate to *cyclic coordinates* are constants and provide first integrals for the system. *See* first integral.

cyclone   In meteorology, a 3-dimensional depression vortex system with closed cells and low central pressure. Its horizontal scale is from 200 to 3000 km. A *cyclone* has a characteristic pattern of wind circulation (counterclockwise in the northern hemisphere, clockwise in the southern). Mid-latitude cyclones are associated with the convergence of polar and tropical air masses along fronts. According to the structure of temperature and pressure, a cyclone can be classified as a barotropic cyclone (cold cyclone and warm cyclone), a baroclinic cyclone, or a neutral cyclone. Based on the distribution of geography, it can be classified as an extratropical cyclone, subtropic cyclone, or tropical cyclone. Since its lower level convergence can cause ascending air motion, cyclones are often accompanied by clouds and rain.

cyclongenesis   A development of synoptic-scale weather disturbances.

cyclotron damping and instability   In a collisionless plasma, damping or instability associated with the $n = \pm 1$ resonance, of importance in space physics and astrophysics as a mechanism of pitch-angle scattering of charged particles. *See* resonant damping and instability.

cyclotron frequency   *See* Larmor frequency.

cyclotron radius   *See* Larmor radius.

Cygnus A   Nearby (z = .056, about 200Mpc distant) active galaxy (3C 405) at RA19^{h}59.4^m, dec+40°43′ which is a strong radio source, with radio jets that extend for about 50kpc in either direction. The closest and second strongest radio galaxy. Recently detected to have a small quasar-like core.

Cygnus Loop   Supernova remnant at RA20^{h}49^{m}00^s dec+30°30′, 230′ × 160′ in extent. In the optical, a large filamentary loop, also strong in radio, visible in X-ray. Age of remnant is estimated at 20,000 years.

Cygnus X1 (Cyg-X1)   A binary system at RA 19^{h}56^{m}22^s.0, dec +35°03′36″, at distance 2.5 kpc, consisting of the O9.7 supergiant HDE 226868 (Gies and Bolton 1986) and a compact object with an orbital period of 5.6 days. The mass of the unseen companion is significantly larger than $5M_\odot$, probably as large as $12M_\odot$. *Cygnus X1* is the second brightest X-ray source in the sky, with X-ray emission exhibiting strong variability at time scales from milliseconds to years. It also radiates in γ-radiation. Because no theoretical models exist of compact degenerate (neutron) stars of mass exceeding $5M_\odot$, the system is taken as the prototype of a black hole binary. In this model, the compact object of mass $\approx 12M_\odot$ is the putative black hole, accreting mass from the hot supergiant companion star which has overflowed its Roche lobe through an accretion disk which thermally emits X-radiation. Cygnus X1 is also a radio source, with radio flux correlated to X-ray output. Also called 4U 1956+35.

cynthion   Of or pertaining to the moon.

Cytherean   Venusian, referring to the planet Venus.

D

D'' layer   The bottom layer of the Earth's mantle, approximately 150 km thick. The name derives from K.E. Bullen's assignment of letters to various layers in the Earth and the subdivision of these layers. D'' has a distinct seismic signature that distinguishes it from the overlying mantle. Since the mantle convects, and at least enough heat is extracted from the core to power the geodynamo, it seems reasonable to suppose that D'' is a thermal boundary layer between the bulk of the mantle and the core. However, it is also possible that it is chemically distinct from the bulk mantle, either through differentiation allowed by the elevated temperatures and perhaps partial melting at the bottom of the layer, chemical core-mantle coupling, or perhaps because all or part of D'' is composed of the dregs of dense plate that was once subducted from the Earth's surface. There is seismic evidence of lateral variations in the properties of D'', and consequently there has been speculation that it is a significantly heterogeneous layer perhaps composed of "crypto-continents" involved in some inverted form of tectonics at the base of the mantle. It is often suggested as the place of origin for some or all of the plumes which rise to the Earth's surface and are manifested as hot spots such as Hawaii.

Dalton's Law   The additivity of partial pressures. Accurate for ideal gases, approximate law for mixtures of real gases. The total pressure is Σp_i, where $p_i = kN_iT/V$ and N_i is the number of molecules of type i, k is Boltzmann's constant, T is the temperature, and V is the volume of the container of gas mixtures. Discovered by John Dalton, 1766–1844.

Darcy's law   An empirical law that governs the macroscopic behavior of fluid flow in porous media. It is credited to H. Darcy, who conducted experiments on the flow of water through sands in 1856. The law ignores the details of tortuous paths of individual fluid particles and defines

average flow rate per unit area (Darcy velocity) q as being related to fluid pressure p in the form of mass diffusion,

$$q = -\frac{\kappa}{\mu} \cdot \nabla p$$

where κ is the permeability (tensor) of the porous medium, and μ is the viscosity of the fluid. In application to groundwater hydrology, *Darcy's law* is modified as follows to include the effect of Earth's gravity:

$$q = -\frac{\kappa}{\mu} \cdot (\nabla p - \rho g)$$

where g is the gravitational acceleration vector, and ρ is fluid density. Darcy's law applies only when flow is laminar within a porous medium and flow velocities are low enough that inertial forces are negligible and breaks down at very high flow velocities and possibly very low permeabilities. *See* Darcy velocity.

Darcy velocity (or specific discharge)   The volume of fluid flow per unit time through a unit area of a porous medium. *Darcy velocity* q equals the average flow velocity v of fluid particles times porosity n. *See* Darcy's law.

Darcy–Weisbach friction factor   A dimensionless friction factor intended primarily for determination of head loss (energy loss per unit weight) of a flow in a full conduit, such as a pipe. The friction factor is a function of the Reynolds number for the flow and the relative roughness of the conduit.

dark cloud   A part of the interstellar medium that emits little or no light at visible wavelengths and is composed of dust and gas that strongly absorb the light of stars. Most of the gas is in molecular form and the densities are of the order of 10^3 to 10^4 particles cm^{-3} with masses of 10^2 to 10^4 solar masses and sizes of a few parsecs. *Dark cloud* temperatures range from 10 to 20 K. *See* interstellar medium.

dark matter   Matter component that does not radiate in the electromagnetic spectrum and, therefore, is not detected by means of telescopes. The first evidence of existence of large fractions of non-luminous matter came from the study of clusters of galaxies by Zwicky in the early 1930s. If clusters of galaxies form bound systems, the velocities of the member galaxies within the clusters are characteristic of cluster mass. This turns out to be about an order of magnitude larger than the sum of the luminous masses observed within the galaxies themselves. Since the 1970s it was known that there is a similar situation in the outer parts of spiral galaxies and in some elliptical galaxies. At that time it was assumed that the dark mass was ordinary (baryonic) matter in some not readily detectable form such as gas, low mass stars, planets, or stellar remnants (white dwarfs, neutron stars, and black holes). However, the nucleosynthesis bound limits $\Omega_B \leq 0.1$, while dynamical measurements suggest $\Omega_{matter} \leq 0.3$. To explain the discrepancy, the existence of a more exotic and yet undetected form of matter has been postulated. A wide class of *dark matter* candidates fall into two categories depending on their mean kinetic energy at high redshifts: cold and hot. The main difference lays in the behavior of the post-recombination power spectrum at galactic scales. *See* matter density perturbations, dark matter, cold and dark matter, hot.

dark matter, cold   Dark matter made of particles with negligible random velocities. The standard *cold dark matter* model of structure formation assumes the dark matter particle contribution makes the universe flat. The primordial Gaussian density field is characterized by a power spectrum with a spectral index $n = 1$. Density perturbations that come within the horizon before matter-radiation equality (*see* thermal history of the universe) are frozen. As a result, the post-recombination power spectrum is modified and bends gently from $n = -3$ on subgalactic scales to the initial $n = 1$. The model had severe observational difficulties and several variants have been proposed: spectral index $n \leq 1$, spectral index $n \geq 1$, a 20% fraction of hot dark matter, and a 70% contribution of a cosmological constant, etc. Figure on page 113.

dark matter, hot   Dark matter made of particles that are highly relativistic at early times. As the universe cools, their momentum is redshifted away and becomes non-relativistic. The standard example would be massive neutrinos.

Several extensions of the standard model of particle interactions suggest the existence of neutrinos with masses up to $m_\nu \sim 90$ eV, enough to equate the mean density to the critical density. The standard *hot dark matter* model, like its cold counterpart, assumes the primordial Gaussian density field is characterized by a power spectrum with a spectral index $n = 1$. The post-recombination spectrum is rather different. Perturbations that came within the horizon while the neutrinos were relativistic (of wavelength $30h^{-1}$ Mpc and smaller) are erased. Due to their large velocity, neutrinos diffuse out of the perturbation in a Hubble time. The final result is that while on large scales the power spectrum retains its original $n = 1$ spectral index, on galaxy and cluster scales it is exponentially damped.

The standard hot dark matter model is ruled out by observations. However, neutrinos could still have a smaller mass and give a contribution of 20 to 30%, the rest being made of cold dark matter and baryons. This model, termed *mixed dark matter,* seems to fit observations of large scale structure acceptably well. Figure on page 113.

dark nebula   A nebula that can be seen because the dust within it obscures the light coming from stars or bright nebulae behind it.

dart leader   In a lightning flash, a second flow of current from the cloud along the channel already opened by the stepped leader and the return stroke. The *dart leader* does not step; it rapidly and smoothly flows along the channel about 50 ms after the first flash.

data assimulation   A process in which the observational data are modified in a dynamically consistent fashion in order to obtain a suitable data set for numerical model initialization in the weather predication and climate modeling.

Davidson current   An ocean countercurrent flowing northward during the winter months between the California Current and the coasts of northern California, Oregon, and Washington.

day   An interval of 86,400 seconds approximating one rotation of the Earth relative to the

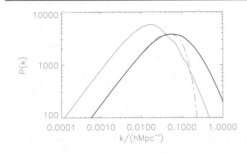

Theoretical power spectra for three models with critical density. Thick solid line: cold dark matter ($\Omega_{cdm}=0.95, \Omega_B h^2=0.02, h=0.65$); thin solid line: hot dark matter ($\Omega_{hdm}=0.95, \Omega_B h^2=0.02, h=0.65$) and dash dotted line: cold dark matter with cosmological constant ($t_o=14\times10^9 yr$, $\Omega_B=0.7$, $\Omega_B h^2=0.02$). All models are normalized to produce the same quadrupole anisotropy on the Cosmic Microwave Background.

sun (apparent or mean solar time), the fixed stars (sidereal time), or an atomic clock (atomic time).

daylight savings time An adjustment frequently adopted by nations to their civil time, specifically a subtraction of 1 hour from standard time during a regular period approximating summer in their hemisphere. In the U.S., the changeover currently occurs at 2:00 AM, beginning the first Sunday in April and ending the last Sunday in October.

DDO classification scheme A variant of the Hubble classification scheme for galaxies, named after the David Dunlop Observatory (DDO) where it was developed. The emphasis is on the prominence and length of the spiral arms: The DDO scheme identifies a new class of spirals, the anemic spirals (indicated by the letter A), which are intermediate in terms of arm prominence between the S0 galaxies and the grand-design, or gas-rich, spirals. Other labels are as in Hubble's scheme. The original DDO scheme has undergone a major revision. The revised DDO type includes a luminosity class in addition to the morphological description. The luminosity class is indicated with a Roman numeral and ranges from I to V, in order of decreasing luminosity. For example, Messier 31 is of type Sb I-II according to the revised DDO scheme. The luminosity class subdivision refines the separation into the three classes S0, A, and S, since a good correlation is found between the degree of spiral arm development and luminosity class, and it is therefore possible to assign a luminosity class on the basis of the appearance of spiral arms.

dead zone An area within a flow field that has very low velocities and thus will trap any contaminant that enters it.

Debye length In a plasma, the maximum length scale for which substantial deviation from charge neutrality can occur. This length scale is of order v_{th}/ω_p, where v_{th} is the electron thermal speed, and ω_p is the plasma frequency. The *Debye length* can be interpreted as the maximal radius of a sphere which in a two-component plasma might be depleted of electrons due to their thermal motion. On spatial scales small compared to the Debye length, the quasi-neutrality of a plasma is likely to be violated, while on larger scales the plasma is quasi-neutral: the kinetic energy contained in the thermal motion is not large enough to disturb the particle distribution over a range wider than the Debye length. For instance, if an unbalanced (but insulated) electric charge is placed in a plasma or electrolyte, ions and electrons near it will shift their average positions in response to its electric field. That creates a secondary field which cancels the charge's field further away than the Debye length.

With k_B as Boltzmann constant, T as temperature, n_e as electron density, and ω_{pe} as electron plasma frequency, the Debye length can be written as

$$\lambda_D = \sqrt{\frac{\epsilon_0 k_B T}{2e^2 n_e}} = \sqrt{\frac{k_B T}{m_e}} \cdot \frac{1}{\omega_{pe}} .$$

In a plasma of absolute temperature T and density n cm^{-3}, D $= 743$ cm $(T/n)^{1/2}$. The Debye length is also important in measuring plasma parameters: within the Debye length the electrons are influenced by the presence of a test charge, such as a satellite in a space plasma, while at larger distances the test charge goes unnoticed. Thus, in order not to influence the measurement, plasma instruments have to be mounted on sufficiently long booms.

decelleration parameter $q_0 = -\ddot{R}/(R(\dot{R})^2)$, where $R(t)$ is the length scale of the universe and · indicates the time derivative. This definition assumes an approximately isotropic universal expansion (as is observed). The quantity $\dot{R}$ is related to the Hubble parameter: $H_0 = \dot{R}/R$.

decibel   A dimensionless measure of the ratio of two powers, P1 and P2, that is equal to 10 times the logarithm to the base 10 of the ratio of two powers (P1/P2). The units expressed this way are one-tenth of a bel and are referred to as *decibels*. The power P2 may be some reference power. For instance, in electricity, the reference power is sometimes taken as 1 milliwatt (abbreviated to dBm).

declination   In terrestrial magnetism, at any given location, the angle between the geographical meridian and the magnetic meridian; that is, the angle between true north and magnetic north is the *declination*. Declination is measured either east or west as the compass needle points to the east or west of the geographical meridian. East is taken as the positive direction. Lines of constant declination are called isogonic lines, and the one of zero declination is called the agonic line. *See* dip, magnetic. In astronomy, an angle coordinate on the celestial sphere corresponding to latitude, measured in degrees north or south of the celestial equator.

decollement   A near horizontal detachment zone between distinct bodies of rocks.

deep(-focus) earthquake   Earthquakes at depths ranging from about 300 to 700 km that occur along the Wadati–Benioff zone, which is inclined from a trench toward a continental side beneath the subduction zone of an oceanic plate. Fault plane solutions with down-dip compression are dominant. Since a *deep(-focus) earthquake* takes place under high-pressure conditions, where friction is large, it is difficult to explain its generation mechanism by frictional sliding processes such as those for a shallow earthquake. Recent laboratory experiments indicate that shear melting, a self-feedback system of phase transformations of olivine and fault growth, and brittle fracturing due to pore pressure are possible generation mechanisms of deep(-focus) earthquakes.

Deep Space 1 (DS1)   A New Millennium spacecraft launched October 24, 1998. It is the first mission under NASA's New Millennium Program to test new technologies for use on future science missions. Its objective is to test 12 advanced technologies in deep space to lower the cost and risk to future science-driven missions that use them for the first time. Among these technologies are a xenon ion propulsion system (performing beyond expectations), autonomous navigation, a high-efficiency solar array, and a miniature camera/spectrometer. By December 1, 1998, DS1 had accomplished enough testing to satisfy the technology validation aspects of the minimum mission success criteria and is well on its way toward meeting maximum criteria.

It carried out a flyby of the near-Earth asteroid 1992 KD on July 28, 1998 at an altitude of 10 km. The primary mission ended on September 18, 1999. It is now on a new trajectory to encounter Comets Wilson–Harrington and Borrelly.

Deep Space 2   Two microprobes that were onboard the Mars Polar Lander spacecraft launched on January 3, 1999 and lost in the landing on Mars on December 3, 1999. The primary purpose of the Mars Microprobe Mission was to demonstrate key technologies for future planetary exploration while collecting meaningful science data (thus, it was named *Deep Space 2*). In this case, the scientific objectives were to determine if ice is present below the Martian surface; to characterize the thermal properties of the Martian subsurface soil; to characterize the atmospheric density profile; to characterize the hardness of the soil and the presence of any layering at a depth of 10 cm to 1 m. *See* Mars Microprobe, Deep Space 1.

Deep Space Network (DSN)   The NASA *Deep Space Network* is a world-wide network of large antennas with the principal function of maintaining communications with spacecraft beyond the moon's orbit. The three main tracking complexes are in Goldstone, California (U.S.), near Canberra (Australia), and Madrid (Spain).

deep water wave   A wave in water that has a depth at least half of one wavelength. Then $c = \sqrt{gL/(2\pi)}$, where c is the wave speed, g is the acceleration of gravity, and L is the wavelength. *See* shallow water wave.

defect   *See* cosmic topological defect.

deferent   In the Ptolemaic theory of the Earth-centered universe, the large orbital circle around a point between the Earth and the equant, followed by the center of a planet's circular epicycle.

deficit angle (cosmic string)   Cosmic strings are inhomogeneities in the energy density field that form during cosmic phase transitions. In the limit of energy below the Planck energy (10^{19} GeV), strings can be well described in the weak gravity and thin string limit. The first condition allows simplification of the relevant equations, while the second assumes that dimensions transversal to the string are effectively negligible compared to the length of the string. In this limit, the Einstein equations predict a metric with the usual Minkowski aspect in cylindrical coordinates

$$ds^2 = dt^2 - dz^2 - dr^2 - r^2 d\theta^2$$

but where the azimuthal angle θ varies between 0 and $2\pi(1 - 4GU)$, with G and U being Newton's constant and the energy per unit length of the string, respectively. While space-time looks locally flat around the string, globally, however, it is non-Euclidean due to the existence of this missing angle $\delta\theta \equiv 8\pi GU$ (called the *deficit angle*) which, in usual simple models, is small, of order 10^{-5} radian. This peculiar feature implies that constant-time surfaces perpendicular to different segments of a string will have the shape of a cone. *See* Abelian string, cosmic string, cosmic topological defect.

deflation   A term used to denote the reduction in elevation of a beach or other area subject to sediment transport, due to transport of sediments by wind.

deformation radius   *See* Rossby radius of deformation.

degeneracy   The condition in which some fermion (a particle with angular momentum = $1/2\hbar$ and subject to the Pauli exclusion principle) is packed as tightly as quantum mechanical considerations permit. Electrons are degenerate in white dwarfs and neutrons in neutron stars and pulsars. Neutrinos, if they have nonzero rest mass and contribute to hot dark matter, may be degenerate in dwarf galaxies. The ignition of a nuclear reaction in degenerate matter leads to an explosion because the reaction heats the gas. The gas does not expand (because the pressure in degenerate matter depends only on the density, not on the temperature), and so it cannot cool. The reaction goes faster at the higher temperature and releases more energy. The gas gets hotter, and so forth, until finally it is no longer degenerate, and it expands explosively.

degree (temperature)   On the Celsius thermometer scale, under standard atmosphere pressure, the freezing point of water is 0 degrees, and the boiling point of water is 100 degrees. The space between these two temperature points is separated into 100 parts. Each part represents 1 *degree*, i.e., 1° C. On the Fahrenheit thermometer scale, under standard atmosphere pressure, the freezing point of water is 32 degrees, and the boiling point is 212 degrees. Thus,

$$\text{degree}°F = \frac{9}{5}\ \text{degree}°C + 32$$

or

$$\text{degree}°C = \frac{5}{9}\left(\text{degree}°F - 32\right)\ .$$

The thermodynamic temperature scale, also called Kelvin temperature scale or absolute temperature scale, is an ideal temperature scale based on Carnot cycle theory. It was chosen as the basic temperature scale in 1927. Additionally, the international practical temperature scale has been established, based on the thermodynamic temperature scale. Currently the standard international practical temperature scale in use is the International Practical Temperature Scale 1968, IPTS-68. Symbol is T_{68} and unit is K. The relation between international practical Celsius temperature scale (t_{68}, unit is °C) and international practical temperature scale is

$$t_{68} = T_{68} - 273.15$$

de Hoffman–Teller frame   Frame of reference in which a magnetohydrodynamic shock is at rest. In contrast to the normal incidence frame, the shock rest frame most commonly used. In the *de Hoffman–Teller frame* the plasma flow is parallel to the magnetic field on both sides of the shock and the $\mathbf{v} \times \mathbf{B}$ induction field in the shock front vanishes. Compared to the normal incidence frame, this frame moves parallel to the shock front with the de Hoffman–Teller speed $\mathbf{v}_{HT} \times \mathbf{B} = -\mathbf{E}$.

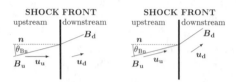

de Hoffman–Teller frame (right) and normal incidence frame (left).

Deimos   Moon of Mars, also designated MII. It was discovered by A. Hall in 1877. Its orbit has an eccentricity of 0.0005, an inclination of $0.9 - 2.7°$, a precession of $6.614°$ yr^{-1}, and a semimajor axis of 2.35×10^4 km. Its size is $7.5 \times 6.1 \times 5.5$ km, its mass is 1.8×10^{15} kg, and its density is 1.7 g cm^{-3}. Its geometric albedo is 0.07, and its surface is similar in reflectivity to C-type asteroids. It may be a member of that group that was captured in the past. *Deimos* orbits Mars once every 1.262 Earth days.

delta   An alluvial deposit where a river meets a larger body of water or near the mouth of a tidal inlet. A *flood delta* is deposited inshore of an inlet by flood tidal currents; an *ebb delta* is located seaward of the inlet throat and deposited by ebb tidal currents.

Delta Scuti stars   Main sequence stars, generally of spectral type A, located within the instability strip on the HR diagram. They are subject to pulsational instabilities driven by hydrogen ionization, but these are generally quite subtle, amounting to brightness changes of 10% or less and with a number of modes (with periods of hours) excited simultaneously. Both the amplitudes and frequencies of the modes can change over a period of years.

Delta surface approximation   (Yang, 1987) The effects of the Earth's sphericity are retained by a quadratic function of y, the meridinal co-ordinate measured positive northward from the reference latitude.

density   Mass per unit volume.

density current   A flow that is driven by density variations within a fluid. Typically a result of temperature or salinity gradients within a body of water.

density inversion   Situation in which fluid density decreases with depth. The ocean is normally stably stratified, and the water density increases monotonically with depth. Turbulence created by surface wind stress, internal waves, or tidal flow can disrupt this density profile by mechanical displacement of water parcels. This can lead to situations in which density locally decreases with depth. *Density inversions* can also be created by the local loss of buoyancy of water at the surface caused by loss, e.g., nighttime cooling of the ocean surface, or by intrusive flows, such as the outflow of saline Mediterranean water into the Atlantic at the Strait of Gibraltar.

Since the density of sea water is a (non-linear) function of temperature and salinity, density inversion is usually accompanied by inversions of the temperature and salinity profiles. Normally, temperature decreases with depth, and salinity increases with depth. However, the existence of either a temperature or salinity inversion alone does not necessarily create a density inversion.

depleted mantle   Mantle that has been depleted of its lightest basaltic components through processes such as partial melting. The residue after extraction of crust.

depletion layer   A region adjacent to the magnetopause but outside it, where plasma density has become abnormally low. Usually found during times of northward IMF, it is caused by the compression of magnetic flux tubes as they are pushed against the magnetopause, squeezing out their plasma and replacing its pressure by that of the magnetic field.

depth of compensation   The depth over which the density of the Earth varies in order to "compensate" the excess mass of topography.

Derrick theorem   Let ϕ be a scalar field subject to the generalized Klein–Gordon equation

$$\left(\Delta - \partial_t^2\right) \phi = \frac{1}{2} V'(\phi) ,$$

with $V(\phi)$ the potential. The theorem states that there is no stable, time-independent and localized (i.e., with finite energy) solution of this equation in more than one dimension. The theorem seems to imply the non-existence of topological defects; however, this is not the case, as one usually assumes some gauge fields which render the energy finite. On the other hand, if no gauge field is present, as for the global cosmic string in three space dimensions, we find that the energy is indeed (mildly) divergent, in agreement with the theorem. Notice that gravitational fields can stabilize such structures, even on a small scale. Thus, boson stars can be stable objects. *See* boson star, cosmic string, cosmic texture, global topological defect.

Descartes ray   The ray traveling through water droplets producing a rainbow, which yields the limiting (minimum) deflection. Hence, there is a concentration of light near the *Descartes ray;* the colors of the rainbow are due to different refraction of different colors in the droplets, giving slightly different return angles for the Descartes rays corresponding to the different colors.

Desdemona   Moon of Uranus also designated UX. Discovered by Voyager 2 in 1986, it is a small, irregular body, approximately 29 km in radius. Its orbit has an eccentricity of 0, an inclination of 0.2°, a precession of 245° yr^{-1}, and a semimajor axis of 6.27×10^4 km. Its surface is very dark, with a geometric albedo of less than 0.1. Its mass has not been measured. It orbits Uranus once every 0.474 Earth days.

deSitter Universe   A solution of Einstein's equations that describes a spacetime devoid of any matter and gravitational fields in the presence of the cosmological constant. When Einstein introduced the cosmological constant, he hoped that it would have a second consequence (in addition to providing a static model of the universe; *see* Einstein Universe): that a solution of Einstein's equations corresponding to such a completely empty space would not exist. This expectation resulted from his belief in Mach's principle that said, in effect, that gravitation is induced by matter and would not exist in the absence of matter. Einstein expected that his theory of relativity would automatically obey Mach's principle. It turned out not to be so, and the *deSitter Universe* was the first counterexample. The word "universe" applied to this particular solution of Einstein's equations is justified by tradition only; according to today's criteria, deSitter's metric is not a model of the actual universe.

Despina   Moon of Neptune also designated NV. Discovered by Voyager 2 in 1989, it is a small, roughly spherical body approximately 74 km in radius. It is very dark, with a geometric albedo of 0.059. Its orbit has an eccentricity of 0.00014, an inclination of 0.066°, a precession of 466° yr^{-1}, and a semimajor axis of 5.25×10^4 km. Its mass has not been measured. It orbits Neptune once every 0.335 Earth days.

Dessler–Parker–Sckopke (DPS) theorem
An approximate theorem in magnetospheric physics, by which the change ΔB in the surface magnetic field at the magnetic equator (in a magnetic storm) is proportional to the energy E of the additional ring current due to the storm. If B_e is the field of the Earth's dipole at 1 R_E and $U_e = B_e^2 R_E^3/3$ is the magnetic energy of the dipole field above 1 R_E, by the DPS theorem $\Delta B/B_e = (2/3)E/U_e$. The theorem was derived for a specific simple model by Dessler and Parker in 1959 and was extended by Sckopke in 1966: further work suggested that the actual dependence was nonlinear, and that for large magnetic storms it overestimated E by a factor 1.5 to 3. Still, the theorem helps estimate the energy deposited in the inner magnetosphere by magnetic storms. The magnitude of ΔB may be obtained from the Dst index.

detritus   The particulate decomposition or disintegration products of plankton, including dead cells, cell fragments, fecal pellets, shells, and skeletons, and sometimes mineral particles in coastal waters.

deuterium ^2H; the isotope of hydrogen whose nucleus consists of 1 proton and 1 neutron. The nucleus is called a *deuteron*.

deuterium burning   The capture of a proton by a deuteron to produce ^3He (which then easily burns through to ^4He; *see* proton-proton chain). *Deuterium burning* occurs at the lowest temperature of any important nuclear reaction, about 10^6K. Thus, its onset marks the end of the protostellar collapse stage of star formation (*see* proto-star). It is the only reaction that occurs in the lowest mass stars, near 0.08 solar masses (*see* brown dwarf), and it has long ago destroyed all the deuterium in the sun and other normal stars. Thus, the presence of deuterium in the Earth and other planets means they are made of material that was never inside the sun. Deuterium is produced only in the early universe (*see* Big Bang nucleosynthesis).

de Vaucouleurs' classification scheme   A classification scheme that refines and extends the Hubble scheme of classification for galaxies, introduced by G. de Vaucouleurs in 1959. de Vaucouleurs' scheme attempts to account for the variety of morphologies observed for each Hubble type. His scheme employs three main parameters: (1) a refined Hubble type, where several intermediate stages are added to the Hubble sequence, notably E^+, $S0^-$, $S0^+$, which account for some lenticular features in ellipticals or weak arms in S0 galaxies, and Sd, Sm, and Im, which more closely detail the transition from Sc to Magellanic irregulars; (2) a parameter describing the spiral design, as ring shaped (r) or s-shaped (s), or intermediate (rs); and (3) a parameter designating barred galaxies (SB), non-barred (SA), and intermediate (SAB), for galaxies where the bar is less developed than in classical Hubble's barred galaxies. In addition, the presence of an outer ring or of a ring-like feature formed by joining spiral arms is indicated with an uppercase R preceding all other labels. The *de Vaucouleurs classification scheme* has been extensively used in the three editions of the Reference Catalogue of Galaxies, where de Vaucouleurs' types are given for several thousand galaxies. For example, Messier 31, the spiral galaxy nearest to the Galaxy, and the nearby spiral Messier 101 are classified as Sb and as Sc according to Hubble, and as SA(s)b and as SAB(rs)cd according to de Vaucouleurs.

de Vaucouleurs' law   Empirical law describing the brightness profile of an elliptical galaxy. The surface brightness Σ of typical giant elliptical galaxies, apart from constants, decreases with radius as

$$\log \Sigma(r) \propto (r/r_e)^{-1/4} \; ,$$

where r_e denotes a scaling parameter, the effective radius, within which half the light of the galaxy is emitted. *de Vaucouleurs' law* applies more frequently to giant elliptical galaxies; dwarf elliptical galaxies are often better fitted by other laws. *See* elliptical galaxies.

deviative absorption   *Deviative absorption* of a radio wave occurs near the point of reflection in the ionosphere. *See* critical frequency.

deviatoric strain   A state of strain defined by subtracting one-third of the volumetric strain from the total strain. If ε_{ij} is the total strain tensor, the deviatoric strain tensor is $\varepsilon'_{ij} = \varepsilon_{ij} - \theta\delta_{ij}/3$, where $\theta = \varepsilon_{ii} = \varepsilon_{11} + \varepsilon_{22} + \varepsilon_{33}$, and δ_{ij} is the Kronecker delta.

deviatoric stress   A state of stress with pressure subtracted. If σ_{ij} is the total stress tensor, the *deviatoric stress* tensor is $\sigma'_{ij} = \sigma_{ij} - p\delta_{ij}$, where $p = \sigma_{ii}/3 = (\sigma_{11} + \sigma_{22} + \sigma_{33})/3$, and δ_{ij} is the Kronecker delta.

dew   During night or early morning, due to radiational cooling, water condensed onto objects near the ground whose temperature has fallen below the dew point of adjacent air but still above freezing. *Dew* often appears in warm seasons with clear weather and light winds. Although the amount of water is small, in dry season and regions it amounts to a significant contribution to plant growth requirements.

dew point   The temperature at which liquid water reaches saturation with respect to a plane water surface as the temperature of a gas mixture is lowered at constant pressure in the presence of nucleation centers so supercooling is excluded. In meteorology, it is the temperature at which moisture begins to condense on a surface in contact with the air.

dex   Decimal exponent.

dextral fault   An alternative term for a right-lateral, strike-slip fault.

diapir   A body of light material, for example salt, that moves upwards in the Earth's crust due to buoyant forces; originally meant a form of dome or anticline folding structure produced by intrusion of fluid-like material such as rock salt and mudstone. A *diapir* is thought to be formed by the processes that low density rock relative to the ambient country rock flows viscously due to its buoyancy and ascends with a mushroom shape, deforming the upper layer with higher density. Recently, the mechanism of ascent of diapir due to inversion of density has been applied to material circulation of igneous rocks and to the whole mantle.

diapycnal   Motion or transport directed across surfaces of constant density (isopycnals).

diapycnal flux   In hydrology, flux occurring in the direction perpendicular to surfaces of equal density. Every diapycnal exchange alters the density structure and thereby affects the potential energy of the water column (sink or source of energy). In lakes, vertical and diapycnal can usually be used synonymously. In polar regions of the ocean, diapycnal and vertical deviate substantially. *See also* buoyancy flux, cabbeling.

dichotomy of Mars   The surface division of Mars into a lightly cratered third of the planet in the middle to upper latitudes of the northern hemisphere, and a remaining heavily cratered two-thirds. The lightly cratered third stands approximately 3 km lower than the heavily cratered surface.

diel   Pertains to occurrences on a 24-hour cycle; any periodic diurnal or nocturnal (i.e., day or night) cycle.

dielectric strength   For an insulating material, the maximum electric field strength (volts per meter) that it can withstand intrinsically without breaking down. For a given configuration of dielectric material and electrodes, the minimum potential difference that produces breakdown.

differential charging   One cause for spacecraft damage in space by substorms and magnetic storms. Since electrons are much lighter than positive ions, in a plasma with equal densities and comparable particle energies, the electron flux is larger. This causes the spacecraft to be hit each second by many more electrons than ions, so that it becomes negatively charged. The charging of an entire spacecraft is hard to prevent and need not cause any damage (except for distorting particle observations); but when different parts charge up differently, damage-causing discharges are possible. Electrically conducting paint is one way of dealing with the problem.

differential diffusion   If turbulent diffusion is almost negligible (such as in an extremely stable water column and weak turbulence), molecular diffusion becomes the dominant flux. Since the molecular diffusivities of ions, which are present in natural waters, have slightly different values, the fluxes will be slightly different even for equal gradients. Over a long time period, the ionic composition of the water column will change (Sanderson et al., 1986).

differential emission measure   A measure of the emitting power of a volume of plasma over a given range of temperatures $T \rightarrow T + dT$. The *differential emission measure* is often employed to draw conclusions from spectral observations made of a number of distinct emission lines at a variety of temperatures. The differential emission measure, $Q(T)$, is defined as $\int Q(T) dT = \int n_e^2 dV$, where n_e is the electron density, and V is the plasma volume being considered.

119

differential heating/cooling   Due to different exposure (to the atmosphere; solar radiation, wind, etc.), changing water transmissivity (algae, kelp forests, etc.), or varying depth, neighboring water bodies may experience different rates of cooling or heating. *Differential heating/cooling* leads subsequently to lateral density gradient and may drive lateral convection.

differential rotation   Change of the rotation speed with distance from the rotation axis.

1. Change of the solar rotation speed with heliographic longitude: the sun rotates faster close to the equator and slower towards the poles. Typical values for the sidereal rotation period of the photosphere at different latitudes are given in the table.

Sidereal Rotation Periods of the Sun

Latitude	Period [days]
equator	26.8
30°	28.2
60°	30.8
75°	31.8

In addition, there is also a differential rotation in radius in the convection zone. In particular at latitudes above about 30°, the rotation speed of the sun increases towards the bottom of the convection zone. At lower latitudes there is a rather small decrease in rotation speed while at latitudes around 30° there is no differential rotation in radius. Observed in the photosphere, low latitudes rotate at a faster angular rate (approximately 14° per day) than do high latitudes (approximately 12° per day). This pronounced latitude dependence appears to be the result of convective flows driven radially by the buoyancy force and deflected horizontally by the Coriolis force due to solar rotation. *Differential rotation* plays a crucial role in the understanding of solar activity and the solar dynamo.

2. *Differential rotation* can also be observed in systems of stars when the rotation speed varies with distance from the center of rotation.

differentiation   In geophysics, the process of separation of different materials, in geological or astrophysical settings, via different physical properties. For instance, the atmosphere of the Earth is differentiated in that hydrogen is very rare, having escaped from the upper atmosphere, while oxygen and nitrogen have not. Their larger molecular weight gives them lower speeds at comparable temperatures, so they do not achieve escape velocity at the top of the atmosphere.

Similarly, internal heating (caused by accretion, radioactive decay, tidal heating, etc.) causes the interior of a solid body to become partially or completely molten. Materials making up the body's interior separate, depending primarily on their densities — denser material (such as iron) sinks to the center while less dense materials (such as silicates) float to the surface. This *differentiation* thus creates the layered structure of the crust, mantle, and core suspected in most of the larger solar system bodies (those with diameters > 1000 km).

In mathematics, the linear operation ($\prime$) obeying Leibnitz' law: $(ab)' = a'b + b'a$ which retrieves the slope of a function if it exists, in some suitable generalization, if not.

diffraction   A term relating to the spread of wave energy in a direction lateral to the dominant direction of wave propagation. Used in the description of light and water waves. Results in interference patterns (constructive and destructive) and accounts for the spreading of energy into areas that would otherwise be shadowed.

diffraction grating   A light analyzer, used to disperse different chromatic elements in a light beam, based on the principle of light diffraction and interference by a series of parallel slits. A typical *diffraction grating* consists of a large number of equally spaced, tilted grooves ($\sim$ 100 to 1000 grooves per millimeter). The spectral resolving power increases with the number of grooves per millimeter for a fixed incident bandpass. A diffraction grating can either reflect or transmit light. To avoid loss of light because of a maximum of the diffraction pattern at zero order, the grooves are tilted to shift the maximum of the diffraction pattern toward the first or, occasionally, toward the second or higher diffraction orders. In this case, a grating is said to be blazed.

diffuse absorption coefficient   For downwelling (upwelling) irradiance: the ratio of the absorption coefficient to the mean cosine of the downward (upward) radiance. *See* absorption coefficient.

diffuse attenuation coefficient   For downwelling (upwelling) irradiance: the ratio of the sum of the absorption coefficient and the scattering coefficient to the mean cosine of the downward (upward) radiance.

diffuse aurora   A weak diffuse glow of the upper atmosphere in the auroral zone, caused by collisions with the upper atmosphere of electrons with energies around 1 keV. It is believed that these electrons leak out from the plasma sheet of the magnetosphere, where they are trapped magnetically. The *diffuse aurora* is not conspicuous to the eye, but imagers aboard satellites in space see it as a "ring of fire" around the magnetic pole. Discovered by ISIS-1 in 1972, its size, intensity, and variations — in particular, its intensifications and motions in substorms — are important clues to the state of the magnetosphere.

diffuse galactic light   The diffuse glow observed across the Milky Way. A large part of the brightness of the Milky Way, which is the disk of our galaxy seen from the inside, can be resolved into stellar sources. The *diffuse galactic light* is a truly diffuse glow which accounts for the remaining 25% of the luminosity and, by definition, is unresolved even if observed with large telescopes. The diffuse galactic light is due to light emitted within our galaxy and scattered by dust grains, and it is not to be confused with light coming from extended sources like reflection or emission nebulae. The brightness close to the galactic equator due to diffuse galactic light is equivalent to 50 stars of 10th magnitude per square degree; for comparison, the total star background is 170 10th magnitude stars per square degree, and the zodiacal light 80.

diffuse interstellar bands (DIBs)   A series of interstellar absorption features recorded on photographic plates in the early 1900s. They were labeled "diffuse" because they arise from electronic transitions in molecules, so they are broad in comparison to atomic lines. There are now well over 100 such bands known in the UV, visible, and near IR regions of the spectrum arising in interstellar clouds. DIBs must be molecular, given the complexity of the absorption lines. DIBs are easily seen when observing spectra of hot, fast rotating stars whose spectrum has a strong continuum. Even with very high resolution spectroscopy, the *diffuse interstellar bands* continue to show blended structures. DIBs show considerable scatter in strength vs. the amount of stellar reddening suggesting inhomogeneous variation of chemistry and dust-to-gas ratio. This may arise because the molecule(s) in the volume may be able to add hydrogen to the molecular structure in certain circumstances, as is known for some carbon compounds. Identifying the carriers of these absorptions has become perhaps the classic astrophysical spectroscopic problem of the 20th century, and numerous molecules have been put forth as the source of these features. Recently most attention has focused on carbon rich molecules such as fullerenes and polycyclic aromatic hydrocarbons.

diffuse scattering coefficient for downwelling (upwelling) irradiance   The ratio of the scattering coefficient to the mean cosine of the downward (upward) radiance.

diffusion   The gradual mixing of a quantity (commonly a pollutant) into a fluid by random molecular motions and turbulence.

diffusion-convection equation   Transport equation for energetic charged particles in interplanetary space considering the effects of spatial diffusion and convection of particles with the solar wind. The transport equation can be derived from the equation of continuity by supplementing the streaming with the convective streaming $\mathbf{v}f$, yielding

$$\frac{\partial f}{\partial t} + \nabla \cdot (\mathbf{v}_{\text{sowi}} f) - \nabla \cdot (D \nabla f) = 0$$

with f being the phase space density, D the (spatial) diffusion coefficient, and $\mathbf{v}_{\text{sowi}}$ the solar wind speed. The terms then give the convection with the solar wind and spatial scattering. The *diffusion-convection equation* can be used to model the transport of galactic cosmic rays or the transport of solar energetic particles beyond the orbit of Earth.

If **v** and D are independent of the spatial coordinate, the solution of the diffusion-convection equation for a δ-injection in the radial-symmetric case, such as the explosive release of energetic particles in a solar flare, reads

$$f(r, t) = \frac{N_0}{\sqrt{(4\pi Dt)^3}} \exp\left\{-\frac{(r - vt)^2}{4Dt}\right\}.$$

This latter equation is a handy tool to estimate the particle mean free path from intensity-time profiles of solar energetic particle events observed in interplanetary space, although for careful studies of propagation conditions numerical solutions of the complete transport equation should be used.

diffusion creep   When macroscopic strain is caused by diffusion transport of matter between surfaces of crystals differently oriented with respect to differential stress, it is called *diffusion creep*. Nabarro–Herring creep, where vacancies diffuse through the grain between areas of its boundary, and Coble creep, where diffusion takes place along the grain boundary, are two examples. Diffusion creep results dominantly from the motions (diffusion) of species and defects (vacancies and interstitial). It is characterized by (1) a linear dependence of strain rate on stress ($n = 1$, n is stress sensitivity of creep rate at steady-state stage); (2) high grain-size sensitivity; (3) the rate-controlling species is the lowest diffusion species along the fastest diffusion path; (4) no lattice preferred orientation formed; (5) little transient creep; (6) deformation is stable and homogeneous.

diffusion, in momentum space   Momentum transfer between particles can be due to collisions as well as due to wave-particle interaction. If these collisions/interactions lead to energy changes distributed stochastically, and the energy changes in individual collisions are small compared to the particle's energy, the process can be described as *diffusion in momentum*

space. Instead of the particle flow considered in spatial diffusion, a streaming S_p in momentum results

$$S_p = -D_{pp}\frac{\partial f}{\partial p}$$

with p being the momentum, f the phase space density, and D_{pp} the diffusion coefficient in momentum.

Also if non-diffusive changes in momentum, e.g., due to ionization, can happen, the streaming in momentum can be written as

$$S_p = -D_{pp}\frac{\partial f}{\partial p} + \frac{dp}{dt}f$$

with the second term corresponding to the convective term in spatial diffusion. *See* diffusion-convection equation.

diffusion, in pitch angle space   Wave-particle interactions lead to changes in the particle's pitch angle. When these changes are small and distributed stochastically, *diffusion in pitch-angle space* results. The scattering term can be derived strictly analogous to the one in spatial diffusion by just replacing the spatial derivative by the derivative in pitch-angle μ:

$$\frac{\partial}{\partial \mu}\left(\kappa(\mu)\frac{\partial f}{\partial \mu}\right)$$

with f being the phase space density and $\kappa(\mu)$ being the pitch-angle diffusion coefficient. Note that κ depends on μ, that is scattering is different for different pitch-angles, depending on the waves available for wave-particle interaction. *See* resonance scattering, slab model.

diffusively stable   A water column is called *diffusively stable* if both vertical gradients of salinity ($\beta\partial S/\partial z$) and of temperature ($\alpha\partial\Theta/\partial z$) enhance the stability N^2 of the water column. The practical implication is that molecular diffusions of salt and temperature cannot produce local instabilities, as they can in double diffusion.

diffusive regime   The *diffusive regime* is one of two possibilities, which allow double diffusion to occur. Under diffusive regimes, salinity stabilizes and temperature destabilizes the water column in such a way that the resulting vertical density profile is stable, i.e.,

$R\rho = (\beta\partial S/\partial z)/(\alpha\partial\Theta/\partial z) > 1$ (and stability $N^2 > 0$).

diffusive shock acceleration   Acceleration due to repeated reflection of particles in the plasmas converging at the shock front, also called Fermi-acceleration. *Diffusive shock acceleration* is the dominant acceleration mechanism at quasi-parallel shocks because here the electric induction field in the shock front is small, and therefore shock drift acceleration is inefficient. In diffusive shock acceleration, the scattering on both sides of the shock front is the crucial process. This scattering occurs at scatter centers frozen-in into the plasma, thus particle scattering back and forth across the shock can be understood as repeated reflection between converging scattering centers (first order Fermi acceleration).

SHOCK FRONT

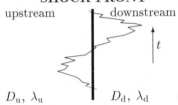

upstream   downstream

D_u, λ_u D_d, λ_d

Particle trajectory in diffusive shock acceleration.

With f being the phase space density, $\mathbf{U}$ the plasma bulk speed, D the diffusion tensor, p the particle momentum, and T a loss time, the transport equation for diffusive shock acceleration can be written as

$$\frac{\partial f}{\partial t} + \mathbf{U}\nabla f - \nabla(D\nabla f) - \frac{\nabla\mathbf{U}}{3}p\frac{\partial f}{\partial p} + \frac{f}{T}$$

$$+\frac{1}{p^2}\frac{\partial}{\partial p}\left(p^2\left(\frac{dp}{dt}\right)f\right) = Q(r, p, t)$$

with $Q(r, p, t)$ describing an injection into the acceleration process. The terms from left to right give the convection of particles with the plasma flow, spatial diffusion, diffusion in momentum space (acceleration), losses due to particle escape from the acceleration site, and convection in momentum space due to processes that affect all particles, such as ionization or Coulomb losses.

In a first-order approximation, the last two terms on the right-hand side (losses from the acceleration site and convection in momentum space) can be neglected. In addition, if we limit ourselves to steady state, some predictions can be made from this equation:

1. *Characteristic acceleration time.* With the indices u and d denoting the properties of the upstream and downstream medium, the time required to accelerate particles from momentum p_0 to p can be written as

$$t = \frac{3}{u_u - u_d}\int_{p_0}^{p}\frac{dp}{p}\cdot\left(\frac{D_u}{u_u} + \frac{D_d}{u_d}\right) .$$

Here D denotes the diffusion coefficient. Alternatively, a characteristic acceleration time τ_a can be given as

$$\tau_a = \frac{3r}{r-1}\frac{D_u}{u_u^2}$$

with $r = u_u/u_d$ being the ratio of the flow speeds in the shock rest frame. For a parallel shock, r equals the compression ratio. τ_a then gives the time the shock needs to increase the particle momentum by a factor of e. Note that here the properties of the downstream medium have been neglected: It is tacitly assumed that the passage of the shock has created so much turbulence in the downstream medium that scattering is very strong and therefore the term D_d/u_d is small compared to the term D_u/u_u.

2. *Energy spectrum.* In steady state, diffusive shock acceleration leads to a power law spectrum in energy $J(E) = J_0 \cdot E^{-\gamma}$. Here the spectral index γ depends on the ratio $r = u_u/u_d$ of the flow speeds only:

$$\gamma = \frac{1}{2}\frac{r+2}{r-1}$$

in the non-relativistic case, or $\gamma_{rel} = 2\gamma$ in the relativistic case.

3. *Intensity increase upstream of the shock.* The spatial variation of the intensity around the shock front can be described as

$$f(x, p) = f(x, 0)\exp\{-\beta|x|\}$$

with $\beta = u_u/D_u$. If β is spatially constant, an exponential intensity increase towards the shock results. Because the particle mean free path λ increases with energy, the ramp is steeper for lower energies than for higher ones. In addition, the intensity at the shock front is higher for lower energies, reflecting the power-law spectrum. In the study of particle events, the upstream intensity increase is often used to determine the scattering conditions upstream of the shock.

4. *Self-generated turbulence.* A crucial parameter for the acceleration time is the strength of the interplanetary scattering, as can be seen from the equation for the acceleration time. Downstream of the shock turbulence is high because the disturbance shock has just passed by. Thus upstream scattering is the limiting factor. For typical conditions in interplanetary space, the Fermi process would develop too slowly to reach MeV energies during the time it takes the shock to travel from the sun to Earth. Nonetheless, these particles are observed. It is assumed that *self-generated turbulence* allows for more efficient scattering in the plasma upstream of the shock: at first, particles are accelerated to low energies only. As these particles propagate away from the shock, they generate and amplify Alfvén waves in resonance with the field parallel motion of the particles. These waves grow in response to the intensity gradient of the energetic particles and scatter particles back to the shock. These particles therefore interact again with the shock, gaining higher energy and, as they stream away from the shock front, generating waves with longer wavelength. This process repeats itself with the faster particles, and as acceleration on the shock continues, the particles acquire higher and higher energies and a turbulent region develops upstream of the shock. Such turbulent foreshock regions have been observed at traveling interplanetary shocks (proton energies up to some 100 keV and the waves in resonance with these particles) and at the quasi-parallel portion of the terrestrial bow shock (proton energies up to some 10 keV and waves in resonance with these particles). *See* resonance scattering.

dike   A crack through which magma flows, the magma subsequently solidifying to form a thin planar igneous body.

dilatancy model   A model to explain processes of earthquake generation, connecting with phenomena of anelastic volumetric expansion of rocks (dilatancy). At the beginning of the 1970s, C.H. Scholz proposed the model, dividing processes from strain accumulation to generation of a large earthquake into five stages. According to the dilatancy model, with increase of underground stresses, many cracks are formed, and pore water flows into the cracks. Then, pore pressure decreases, causing dilatancy hardening. Subsequently, the pore pressure gradually increases due to water supply from the ambient region, reaching main rupture. Land uplift preceding a large earthquake, temporal change in *P*-wave velocity, advent of seismic gap, and activity of foreshocks might be better explained by the model. However, since actual earthquakes take place on planes with mechanical defects in the crust, dilatancy does not necessarily develop sufficiently to explain a large earthquake. Reliability of observation for temporal change in *P*-wave velocity is also suspect. Therefore, the *dilatancy model* has attracted little attention in recent years.

dilatation of time-Lorentz transformation
The increase in the time interval of an event when measured in a uniformly moving reference system rather than in the reference system of the event, as calculated by the Lorentz Transformations in the Special Theory of Relativity. In special relativity time is not an absolute variable, and it therefore varies for different reference systems. *See also* coordinate transformation in special relativity. *See* time dilatation.

dilaton   A scalar component of gravity which emerges in the low energy limit of string theory. *See* dilaton gravity.

dilaton gravity   In the framework of string theory the field equations of general relativity are obtained as an approximation which is valid only for distances larger than the typical (microscopic) string length (low energy). Further, since string theory is a theory of extended objects (including *p*-dimensional branes, with $p \geq 2$ and integer), one expects to have (non-local) corrections to Einstein's field equations. The simplest corrections are extra fields, among which the *dilaton* plays the role of a scalar component of gravity and can also be thought of as a variable (in both space and time) Newton's "constant" (*see* Brans–Dicke theory). The dilaton couples to matter fields, e.g., Yang–Mills $SU(N)$ fields and Maxwell's $U(1)$ electromagnetic field. Because of this, the presence of a non-trivial dilaton in general spoils the principle of equivalence. *See* dilaton.

dilatonic black hole In field theory applied to cosmology, a dilaton is an additional scalar field associated with gravity. A few dilatonic gravity solutions are known in four space-time dimensions which represent black holes with a non-trivial dilaton field. When the latter field couples to the electromagnetic tensor, the black holes must be electrically charged. One exact solution is given by the analog of the spherically symmetric general relativistic black hole with charge, the Reissner–Nordström metric. When angular momentum is present (*see* Kerr–Newman metric) only perturbative solutions are known for either small angular momentum or small electric charge. All the solutions have in common a dilaton which decreases and vanishes at a large distance from the center of the hole. However, near the event horizon the dilaton is nonzero and can possibly affect the scattering of passing radiation.

Other solutions have been found when the dilaton couples to Yang–Mills $SU(N)$ fields giving rise to richer structures. *See* black hole, dilaton gravity, future/past event horizon, Kerr–Newman metric, Reissner–Nordström metric.

dilution Defined as the total volume of a sample divided by the volume of effluent (contaminant) contained in the sample.

dimension A statement of the number of independent parameters necessary to uniquely define a point in the space under consideration. Everyday experience with space indicates that it is 3-dimensional, hence specifying the x, y, z labels (for instance) of a point uniquely defines the point. If it is wished to specify the locations of two mass points, it is convenient to introduce a 6-dimensional space, giving, for instance, the x, y, z labels of the location of each of the mass points. In special and general rela-

tivity, time is considered a separate dimension, and so events are given by specifying x, y, z, and t (time); thus, spacetime is 4-dimensional. In many mathematical operations, e.g., in integration, the dimension of the space enters explicitly and as a consequence the solution to standard equations depends on the dimension of the space in which the solution is found. It is found effective theoretically to allow noninteger dimensions in those cases. Also, since the ratio of the "volume" to the "surface" depends on the dimension, this concept has been generalized to a "fractal dimension" which is defined in terms of the ratio of these quantities in some suitable sense. *See* fractal.

dimensional analysis In usual physical descriptions, quantities are assigned units, e.g., centimeter for length, gram for mass, second for time. In *dimensional analysis,* one constructs a combination of known quantities which has the dimension of the desired answer. Because specific systems have typical values for dimensional quantities, results constructed in this way are usually close to the correctly computed result. Typically dimensional analysis omits factors of order one, or of order π; the results then differ from exactness by less than an order of magnitude. For instance, the typical length associated with a sphere is its radius r (measured in centimeters, say). Its volume by dimensional analysis is then r^3 (cm^3), whereas the exact value including π and factors of order unity is

$$4\pi/3r^3 \sim 4.18r^3 \, .$$

dimensional transmutation (Coleman Weinberg, 1973) In field theory, *dimensional transmutation* occurs in a first or second order phase transition in an originally massless theory. The classical potential of the massless scalar field φ has single minima at $\varphi = 0$ of a particular shape. Due to the quantum effects, the potential acquires the second minima at φ_c. In turn, the existence of the critical point φ_c leads to spontaneous symmetry breaking (the field is nonzero, even though the underlying theory does not pick out a nonzero value for φ: it could have ended up in the other minimum). Additionally, quantum effects can modify the shape of the potential at the new minima, so that the field now corresponds to massive particles — this is dimensional transmutation. Such effects are important at the level of cosmic defects in cosmology, and also more fundamentally because, for instance, Newton's constant of gravity G may have arisen by a similar process from a "simpler" theory with $G = 0$. *See* induced gravity, spontaneous symmetry breaking.

Dione   Moon of Saturn, also designated SIV. It was discovered by Cassini in 1684. Its orbit has an eccentricity of 0.002, an inclination of $0.02°$, a semimajor axis of 3.77×10^5 km, and a precession of $30.85°$ yr^{-1}. Its radius is 560 km, its mass 1.05×10^{21} kg, and its density is 1.43 g cm^{-3}. Its geometric albedo is 0.7, and it orbits Saturn in a synchronous orbit once every 2.737 Earth days.

dip-equator   Isocline with inclination $I = 0$, geomagnetic equator.

dip, magnetic   The local angle observed between the horizontal plane and the direction of the Earth's magnetic field. Also known as magnetic inclination.

dip slip fault   A fault upon which the motion on the fault is vertical.

disappearing filament   In solar physics, disappearance of a filament/protuberance without the typical emission of electromagnetic radiation related to a flare. The *disappearing filament* might be observed as a coronal mass ejection. Compared to coronal mass ejections accompanied by a flare, these are normally rather small and slow coronal mass ejections, too slow to drive a shock wave.

discharge coefficient   A calibration coefficient employed to relate the flowrate over a weir to the depth (head) on the weir and other geometric properties.

disconnection event   In solar physics, reconnection in the tip of a helmet streamer leads to the expulsion of coronal plasma. In contrast to a typical coronal mass ejection, the disconnected magnetic field configuration is open, and the image in the coronograph is a jet- or fan-like structure. Observations suggest *disconnection events* are related to the emergence of new magnetic flux on the sun. The accompanying disturbance of the coronal structure then "squeezes" the opposing fields in the tip of a helmet streamer, eventually causing reconnection and the expulsion of magnetic flux and matter. Because of the open magnetic field structure, the remainders of such disconnection events do not form magnetic clouds and consequently are more difficult to identify in interplanetary space.

discordant redshift   Redshift of a galaxy significantly differing from the redshift of other galaxies belonging to an apparently interacting system of galaxies, such as a group. For example, in the case of the Stephan Quintet, a group of five galaxies, four galaxies have redshifts around 6000 km/s, and one has a *discordant redshift* of only 800 km/s. Several other intriguing examples exist, in which two galaxies with large redshift difference are connected by bright filaments. Galaxies with discordant redshift can be explained as due to casual superposition of foreground and background galaxies, or in some cases (at small angular separation) on gravitational lensing of background galaxies. However, H. Arp and collaborators have suggested the existence of "non-velocity" redshifts. They concluded that redshift is not a reliable distance indicator, questioning the validity of Hubble's law and especially of quasar distances deduced from their redshift. This is a distinctly non-standard interpretation, at variance with the opinion of most of the astronomical community.

discrete aurora   A term used for aurora appearing in well-defined formations, e.g., arcs and ribbons, distinguishing it from the diffuse aurora. Auroral arcs visible to the eye are of this kind. *Discrete auroras* are caused by electrons of up to 5 to 15 keV, accelerated earthward along magnetic field lines. Often these same field lines also carry upward-directed currents, suggesting that the electrons are carriers of that current, and the accelerating electric field helps achieve the current intensity demanded by the sources of that current, which are more distant. While those sources provide the energy, the acceleration process occurs mainly within about 8000 km of Earth.

discrimination (seismic)   The use of seismology to determine whether a seismic event was caused by the test of an explosive device (such as the relative slip of rock on either side of a fault, while explosions generate waves through a more isotropic compression of the surrounding rock). The difference in the source mechanisms and their relative durations (explosions tend to be very rapid) means that the seismic signals generated differ in the geometrical pattern of the radiation, in the proportion of different types of seismic wave emitted, and in frequency content. Large explosions that can be detected at long distances may be reliably discriminated through study of the seismic signals, although smaller explosions may be more difficult both to detect and to distinguish from earthquakes. Potential complications include the geological setting of the event (which may not be well known) and whether an effort has been made to decouple the explosion from the surrounding rock by setting it off in a cavity.

disk warp   A deviation from planarity of the disk of a galaxy, in which the outer parts of a galactic disk gradually tilt. More precisely, one can think of the disk of a galaxy as a sequence of concentric, adjacent rings: If the disk is flat, the rings are coplanar; if the disk is warped, the inclination and the position angle of the intersection line between rings (line of nodes) varies continuously from ring to ring. A *disk warp* is more frequently observed in the gaseous than in the stellar component of a galaxy. The distribution of the 21-cm hydrogen emission line often reveals a warp outside the boundaries of the optical disk. Small warps are probably present in the outer regions of most disk galaxies (including the galaxy); strong warps, where the disk plane tilts by 40 to 50°, as in the case of NGC 660, are of rare occurrence.

dislocation climb   Dislocation motion when an edge dislocation moves along the direction perpendicular to its slip plane under the action of a shear stress. The climb of dislocation involves transport of matter. For the dislocation line to climb one interatomic distance up or down, a line of atoms along the edge of the extra-half plane has to be removed or added according to the position of the extra-half plane, which is completed by diffusion.

dislocation creep   A continuous, usually slow deformation of solid crystalline materials resulting from the motion (glide, climb, and cross slip) of dislocations under the action of shear stress. The resistance to dislocation motion includes: (1) intrinsic resistance (the Peierls stress); (2) impurities, and (3) mutual interaction of dislocations. In general, the largest resistance force controls the rate of deformation. *Dislocation creep* is characterized by (1) a non-linear dependence of strain rate on stress ($n = 3 - 5$, n is stress sensitivity of creep rate at the steady-state stage); (2) crystals deform on specific crystallographic planes along certain orientations so that it can produce strong lattice preferred orientation; (3) significant transient creep is expected.

dislocation energy (self-energy of dislocation)   The energy stored in a solid by the existence of a unit length of dislocation line. The magnitude of *dislocation energy* depends on the nature of particular dislocations such as the type and configuration. For example, the Peierls energy is the energy associated with the Peierls dislocation, which can be divided into two portions: the elastic energy stored in the two half-crystals, and the misfit energy (core energy) associated with the distorted bond.

dislocation glide   The manner of dislocation motion when a dislocation moves on its slip plane under the action of a shear stress. Glide of dislocation involves no transport of matter by diffusion: atomic bonds are shifted, but there is no need for atoms to be brought in or evacuated. The resistance to the *dislocation glide* is intrinsic, mainly from the Peierls stress.

disparition brusque   The sudden disappearance, within the space of a few hours, of a solar filament. This disappearance begins with a slow rising motion at a few kms^{-1} and is typically associated with a brightening in X-rays and occasionally by the appearance of Hα flare ribbons.

dispersion A phenomenon in which wave velocity (phase velocity, group velocity) changes with its wavelength. In seismology, for a layered structure, phase velocity becomes closer to S-wave velocity of the lower and upper layers for longer and shorter wavelengths, respectively. Phenomena in which phase velocity increases and decreases with increasing wavelength are respectively referred to as normal dispersion and reverse dispersion. (This is the case for visible light in glass and is the usual case in seismology.) In seismology, curves representing the relation between surface wave velocity and its wavelength are called dispersion curves, from which the velocity structure of the crust and the mantle can be estimated.

dispersionless injection A sudden rise in the intensity of energetic ions in the Earth's nightside equatorial magnetosphere, in general at or beyond synchronous orbit, occurring simultaneously over a wide range of energies. It is widely held that such particles must have been accelerated locally because if their acceleration occurred some distance away, the faster ones would have arrived first.

dispersion measure (DM) The integral along the line of sight distant source of the electron number density.

The pulse arrival time for two different frequencies f_2, f_1 is related by

$$\Delta t = e^2 / (2\pi m_e c) \, \Delta \left(f^{-2} \right) DM \, .$$

dispersive Tending to spread out or scatter; having phase and group velocities that depend on wavelength. Used to describe both physical and numerical processes.

displacement vector In a Euclidean space, the difference vector between position vectors to two points. The *displacement vector* is often thought of as the difference in position of a particular object at two different times.

dissipation In thermodynamics, the conversion of ordered mechanical energy into heat. In computational science, deliberately added to differential equations to suppress short wave-

length oscillations that appear in finite representations of differential equations, but have no analog in the differential equations themselves.

dissipation of fields One of the basic concepts of magnetohydrodynamics. In the case of a finite conductivity, the temporal change of the magnetic flux in a plasma can be written as

$$\frac{\partial \mathbf{B}}{\partial t} = \frac{c^2}{4\pi\sigma} \nabla^2 \mathbf{B}$$

with $\mathbf{B}$ being the magnetic flux, σ the conductivity, and c the speed of light. Formally, this is equivalent to a heat-conduction equation, thus by analogy we can interpret the equation as describing the temporal change of magnetic field strength while the magnetic field lines are transported away by a process that depends on conductivity: the field dissipates. Note that while the magnetic flux through a given plane stays constant, the magnetic energy decreases because the field-generating currents are associated with ohmic losses. Magnetic field dissipation seems to be important in reconnection. *See* reconnection.

Aside from the conductivity, the temporal scale for field dissipation depends on the spatial scale of the field. With τ being the characteristic time scale during which the magnetic field decreases to $1/e$ and L being the characteristic spatial scale of the field, the dissipation time can be approximated as

$$\tau \approx \frac{4\pi\sigma}{c^2} L^2 \, .$$

Thus, the dissipation depends on the square of the characteristic scale length of the field: Smaller fields dissipate faster than larger ones. Thus in a turbulent medium, such as the photosphere, where the field lines are shuffled around and therefore a polarity pattern on very small spatial scales results, the field dissipates rather quickly. Or in other words, turbulence can accelerate magnetic field dissipation.

Note that for infinite conductivity the dissipation time becomes infinite as well, leading to frozen-in fields.

dissipation of temperature variance For scales smaller than the Batchelor scale temperature fluctuations T' become extinguished by the smoothing effect of molecular thermal diffusivity κ_T. In natural waters, χ is mostly estimated by profilers that measure the temperature fluctuations T' at a high rate (at least such, as to resolve the structures to the Batchelor scale). If small-scale turbulence is isotropic, the rate of *dissipation of temperature variance* $\chi\,[\mathrm{K^2 s^{-1}}]$ is estimated from the temperature gradient spectra by $\chi = 6\kappa_T < (\partial T'/\partial z)^2 >$, where z is the profiling direction. The vertical turbulent diffusivity κ_v^t can be estimated by the relation $\kappa_v^t = \chi[2(\partial T/\partial z)^2]^{-1}$, a procedure often referred to as *Osborn–Cox method* (Osborn and Cox, 1972). *See* Cox number.

dissipation of turbulent kinetic energy
The rate at which turbulent kinetic energy $\epsilon\,[\mathrm{W\ kg^{-1}}]$ is transformed to heat by internal friction caused by viscosity v. Dissipation is given by $\epsilon = 2v\sum_{ij}[0.5 \cdot (\partial u_i/\partial x_j + \partial u_j/\partial x_i)]^2$ (*see* strain rate). In natural waters, ϵ is mostly estimated by profilers that measure one velocity component u' in the direction perpendicular to the motion of the measuring profiler (direction z). For isotropic small-scale turbulence, the above nine terms in the summation collapse to the simple form $\epsilon = 7.5v < (\partial u'/\partial z)^2 >$ ($<>$ indicates the average over a length scale, chosen typically 0.5 to several meters). An alternative estimate can be determined by the inertial dissipation method.

dissipation profiler   Instrument for the measurement of oceanic turbulence levels. It is equipped with fast sampling airfoil probes and thermistors to resolve velocity and temperature fluctuations in the dissipation range. Generally, these instruments are designed to either fall or rise vertically in the water in order to produce a profile of the turbulent activity in the water column. For certain applications, the instruments are towed horizontally.

dissipation range   The range of length scales or wavenumbers at which velocity fluctuations in turbulent flows are damped out (dissipated) by molecular viscosity. *See* Kolmogorov scale, turbulent cascade.

dissipation rate   Rate at which turbulent kinetic energy (TKE) is removed from turbu-

lent flows at length scales defined by the Kolmogorov scale. The *dissipation rate* enters the turbulent kinetic energy equation as the term

$$\epsilon = 2v\overline{e_{ij}e_{ij}}$$

where v is the viscosity and e_{ij} is the fluctuating strain rate tensor defined by

$$e_{ij} \equiv \frac{1}{2}\left(\frac{\partial u_i}{\partial x_j} + \frac{\partial u_j}{\partial x_i}\right)$$

where u_i and x_i are the velocity and position, respectively, for the spatial directions $i = 1, 2, 3$. In oceanic turbulence studies, ϵ can be estimated from measurements of velocity gradients. Under the assumption of isotropic and steady turbulence, ϵ is estimated from

$$\epsilon = \frac{15}{2}v\overline{\left(\frac{\partial u_2}{\partial x_1}\right)^2}$$

where x_1 is the direction along which the measurement device measures the perpendicular turbulent velocity fluctuations u_2. *See also* airfoil probe.

dissolved organic matter   *See* colored dissolved organic matter.

distance indicator   *See* standard candle.

distance modulus   The distance to an object can be derived by comparing its apparent magnitude (m) and absolute magnitude M, where M is defined to be the flux of the star at a standard distance of $(10pc^2)$. Using the standard definition for magnitude, we have

$$m - M = -2.5\log\left(\frac{L/4\pi R^2}{L/4\pi(10pc^2)}\right)$$

where R is the distance to the star and L is the star's intrinsic brightness (luminosity). This reduces to

$$m - M = 5\log(R) - 5\,.$$

diurnal   Due to the daily variation of the solar radiation received at the Earth's surface, meteorological quantities, such as temperature, pressure, atmospheric pollution, wind speed and direction, etc. have daily variations. *Diurnal* variation is a periodic variation and does not contain the non-periodic variation caused by synoptic situations (such as advection processes). The strength of diurnal variation is related to location, stronger in continental areas and weaker in maritime areas, e.g., the strongest diurnal variation region is the Tibetan Plateau. The diurnal range in equatorial areas exceeds the annual variation in average temperature. More strictly, *diurnal* pertains to occurrences during the day, as opposed to nocturnal occurrences; diel occurrences are those that happen once per day.

diurnal motion   Apparent motion of objects on the celestial sphere due to the rotation of Earth from west to east, which causes objects to appear to rise in the east and set in the west daily.

divergence   Branching off or moving in different directions. Also denotes a mathematical operation; the divergence of a vector $\vec{u}$ is denoted by $\nabla \cdot \vec{u}$.

divergence law for irradiance   *See* Gershun's law.

divergence theorem   Also called Gauss' theorem:

$$\int_A \vec{B} \cdot \vec{n}\, \mathrm{d}A = \int_V \nabla \cdot \vec{B}\, \mathrm{d}V$$

with V being a volume enclosed by surface A and $\vec{n}$ is a unit normal to A. Used, for instance, to convert between integral and differential forms of Maxwell's equations.

divergent boundary   In tectonics, two plates move apart from each other at a *divergent boundary*. Magma moves up from the Earth's asthenosphere at the divergent boundary. The magma rises to the surface, where it cools and solidifies as the volcanic rock basalt. Continued magma ascent forces the newly formed basaltic crust to move to the sides as the process repeats. Thus, divergent boundaries are areas where new crust is formed and where the plates move apart. Divergent boundaries are characterized by volcanism (quiet eruptions with very fluid lavas) and shallow earthquakes. Divergent boundaries are believed to occur over the uprising portions of convection cells within the Earth's asthenosphere. The Mid-Atlantic Rift and the East African Rift Valley are examples of divergent boundaries.

divergent plate boundary   *See* divergent boundary, seafloor spreading.

diversity reception   Often, a radio circuit will have many different possible paths between the transmitter and receiver. There may be differences in the quality of the service on these paths and the quality may vary in time, space, and frequency. Diversity methods seek to exploit these differences. In the simplest form, signals on two different paths may be received and the maximum signal selected. More complex diversity systems may make use of redundancy between signals collected from several paths. *See* ionospheric radio propagation path.

divided bar   The name of an apparatus used to measure thermal conductivities of disk-shaped rock samples. Two metal, usually brass, heads with circular cross-sections are maintained at different constant temperatures. Rock samples with known and unknown thermal conductivities are sandwiched in between. The heat flux q through the axis (z) of the system is determined from the measured temperature gradient dT/dz across the rock sample of known thermal conductivity λ_k from Fourier's law of heat conduction. The thermal conductivity of the other sample is then given by this heat flux and the measured temperature gradient across it. *See* conductive heat transfer.

D-layer   The D-layer is the lowest ionospheric layer at heights between 60 and 85 km. In contrast to the other ionospheric layers, it is still inside the mesosphere. Ionization is primarily due to energetic particles of solar and galactic origin and some UV lines which penetrate deep into the atmosphere, such as the Lyman-α-line. The main constituent is NO, which results from a combination of atomar oxygen O and molecular nitrogen N_2. The layer can be described very well by a Chapman profile. Due to the high densities, free electrons are rare and ion clusters and negative ions form instead. Polar cap absoptions (PCAs) and sudden ionospheric disturbances (SID) strongly modify this layer.

dog days   The time period (July 3 through August 11) during the northern summer when Sirius is high in the daytime sky, supposedly adding to the summer sun's heat.

doldrums   The equatorial zone, characterized by high temperatures with small seasonal and diurnal change (and heavy rainfall) and light winds, so that sailing ships have difficulty sailing through the region.

domain of dependence   Let S be an achronal set. The *future/past domain of dependence* $D^\pm(S)$ of S is the set of points x such that an arbitrary past/future endless trip containing x intersects S. The *domain of dependence* of S is $D(S) = D^+(S) \cap D^-(S)$. *See* achronal set, causality relations.

domain of outer communication   The region outside of all black hole surfaces (horizons). The region of spacetime that is visible from infinity.

domain wall   Cosmological topological defects arising in phase transitions for which a discrete symmetry is spontaneously broken: the phase transition $G \rightarrow H$ induces a discrete family of equivalent vacuum states and domains having a different value for the Higgs field responsible for the symmetry breaking will form, separated by a correlation length. At the intersections between these domains, the Higgs field will, by continuity, not be able to lie in the true vacuum (the minimum of the potential), and thus the region (a wall) will contain an enormous amount of energy. Any theory predicting such walls contradicts observational cosmology as, for instance, they would contribute to the total energy density of the universe as well as to the anisotropies of the cosmic microwave background at a level well over that observed. *See* cosmic topological defect, homotopy group, spontaneous symmetry breaking.

dominant energy condition   For all future-directed time-like vectors ξ^a, the vector $-T^a_b \xi^b$ is a future-directed time-like or null vector. Here T^a_b is the stress tensor of the matter. This condition expresses the requirement that the speed of energy flow is less than the speed of light.

Doodson number   A set of six integers in the notation $d_1 d_2 d_3 . d_4 d_5 d_6$ defined by A.T. Doodson in 1921 to uniquely classify tide components. In this scheme, the tidal band is decomposed into integer multiples of six astronomical functions: τ, the mean lunar time; s, the mean longitude of the moon; h, the mean longitude of the sun; p, the mean longitude of the lunar perigee; N', the negative mean longitude of the ascending lunar mode; and p_s, the mean longitude of the solar perigee. Thus, for each component j, the tidal frequency, ω_j, and phase, β_j, are decomposed as:

$$\omega_j + \beta_j =$$
$$d_1 \tau + (d_2 - 5)\,s + (d_3 - 5)\,h +$$
$$(d_4 - 5)\,p + (d_5 - 5)\,N' + (d_6 - 5)\,p_s$$

The first digit, d_1 defines the tidal species and is always equal to the order of the spherical harmonic component of the tide potential from which it originates. Long-period tides have $d_1 = 0$, diurnal tides have $d_1 = 1$, and semi-diurnal tides have $d_1 = 2$. The combination of the first two digits $d_1 d_2$ defines the tidal group number, and the first three digits $d_1 d_2 d_3$ define the tidal constituent number. For example, the largest semi-diurnal tide, M_2, has a *Doodson number* of 255.555, while the largest diurnal tide, K_1, has a Doodson number of 165.555.

Doppler beaming   Beaming of radiation due to the rapid, i.e., close to the speed of light, motion of an emitting source with respect to an observer. Light emitted isotropically in the rest frame of a source is observed greatly enhanced if the source is moving toward the observer: For a radiating particle moving at a velocity close to the speed of light, corresponding to a Lorentz factor $\gamma \gg 1$ (γ is equal to $1/\sqrt{(1 - (v/c)^2)}$, where v is the velocity of the radiating matter, and c is the speed of light), the observer would see most light concentrated in a narrow beacon of half-opening angle $1/\gamma$ radians, and enhanced (or "Doppler boosted") by a factor that can be proportional to a large power (3 to 4) of γ. *Doppler beaming* is relevant whenever there are charges moving at a velocity close to the speed of light (for example, if $v = 0.95c$ then $\gamma = 3$), as in the case of radio jets in radio galaxies and quasars.

Doppler broadening   The broadening of a spectral line in a gas because of red and blue shifts associated with thermal motion in the gas.

Doppler dimming   A means by which to determine outflow plasma velocities from coronagraph measurements of the solar corona, where the outflows are perpendicular to the line-of-sight. In the solar wind of the outer corona, *Doppler dimming* observed in Lyman α at 1216 Å provides a diagnostic tool for determining the outflow speed. At heights of interest, $\sim 4R_\odot$, densities are too small for significant coronal Lyman α emission, so the observed radiation is produced by resonance scattering of chromospheric Lyman α off the H_I locally present in the corona. The intensity of the detected radiation is determined both by the chromospheric intensity and by the systematic velocity of the scattering H_I. This scattering is largest when the velocity is zero and Doppler dimming occurs when the velocity, relative to the chromosphere, increases.

Doppler effect   The observed frequency of the wave signal from a standard source is a function of the motion of the emitter and the detector. In Newtonian physics with a universal time t, for waves moving at a given speed v_p with respect to a medium (e.g., for sound waves in air):

$$\text{received frequency } f' = \frac{1 \pm v_o/v_p}{1 \mp v_s/v_p} f$$

where f is the source frequency, measured at rest near the source, v_s is the source speed, v_o is the observer speed relative to the medium, both measured along the line joining source and observer; the upper sign is for motion reducing the relative separation. In Newtonian physics only the component along the line between the source and observer contributes to the *Doppler effect*.

When considering the Doppler effect of light or other electromagnetic signals, one uses the special relativistic formula. Here there is no medium, so the concept of motion relative to the medium is meaningless. The relativistic formula is:

$$f' = \frac{1 - v/c \cos \theta}{\sqrt{(1 - v^2/c^2)}} f \;,$$

where **v** is the velocity of relative motion of the source and observer, θ is the angle between the direction of propagation of the photon and the velocity of the observer, measured in the rest frame of the emitter. ($\theta = 0$ corresponds to motion increasing the separation.) Notice that in relativistic systems there is a transverse Doppler shift arising from the denominator, even when there is no motion along the line between source and observer.

Doppler shift   Doppler effect.

double couple   A seismological model that a focal mechanism is produced by release of double-couple torques with mutually opposite direction at a hypocenter. In contrast, a model that a focal mechanism consists of a single torque is referred to as single couple. Although radiation patterns for initial motion of *P*-waves are the same for both models, those for *S*-waves become of four quadrant type for *double couple,* whereas they are of two quadrant type for single couple. Radiation patterns for Love wave and Rayleigh wave and their amplitude ratios are also different between the two models. At the beginning of the 1960s, a point source model equivalent to lateral faulting was theoretically proved to be double couple. The force and moment are zero for a double couple, and the amount of one of the two torques is called seismic moment.

double diffusion   Mixing process resulting from differential ("double") diffusion of salt and heat between two water masses. In the absence of mechanical stirring caused by shear currents, this type of mixing occurs in two layer situations, in which two water types of different heat and salt composition are stacked vertically. A necessary condition for *double diffusion* is that the gradients of temperature and salinity have the same sign. *See also* stability ratio.

Two scenarios are possible: in the first, warmer and saltier water is above colder, fresher water. Since the rate of molecular diffusion for heat is about 100 times larger than that of salt, the upper water loses heat to the lower water faster than it loses salt. This results in a loss of buoyancy of the upper water in the vicinity of the interface. If the initial density difference between the layers is small, saltier water will sink into the lower layer in the form of thin columns (or fingers, hence, "salt fingering"). Likewise, the water below the interface gains heat faster than it gains salt, resulting in penetration of fresher, colder water into the upper layer.

On the other hand, if a layer of cold, fresh water is above a layer of warm, salty water, the water just above the interface gains buoyancy from the heat transfer through the interface and tends to rise. At the same time, the water below the interface loses buoyancy and tends to sink. This fluid motion leads to the development of strata of homogeneous convective layers separated by thin regions of high vertical gradients of temperature and salinity. This phenomenon is often called layering.

The degree of double diffusion is usually expressed by the stability ratio R_ρ, which is defined by

$$R_\rho = \frac{\text{stability due to the stabilizing component}}{\text{instability due to the destabilizing component}}.$$

Two cases of double-diffusion occur:

$$R_\rho = (\alpha \partial \Theta / \partial z)/(\beta \partial S / \partial z) > 1$$
$$\text{for the finger regime},$$
$$R_\rho = (\beta \partial S / \partial z)/(\alpha \partial \Theta / \partial z) > 1$$
$$\text{for the diffusive regime}.$$

Double diffusion is potentially active when $1 < R_\rho < (\kappa_T/\kappa_s) \approx 70$ (temperature-dependent, *see* Lewis number), where κ_T, κ_s are the molecular diffusivities of temperature (heat) and salt, respectively. In fact, double diffusion becomes manifest by convectively well-mixed layers, if $1 < R\rho <\sim 4$ and is most intense for $R_\rho \longrightarrow 1$.

double probe   A pair of DC antennas extended in opposite directions from a satellite. By observing the voltage difference between their ends, the component of the electric field **E** in that direction can be estimated.

Double probes extending 5 to 10 m are often used in low Earth orbit to observe electric fields in the auroral and polar regions of the upper atmosphere. Their readings must be corrected for the voltage due to the spacecraft motion, shadow effects, and other problems, but because of the high ambient ion density, they tend to work well since they are much larger than the Debye length. Some have been used in more distant regions, where boom lengths of the order of 100 m are needed.

double star   Roughly synonymous with binary star, but generally restricted to mean those pairs where the stars are far enough apart that they will complete their evolution without affecting each other. Also, frequently, binary stars that can be resolved into two separate points of light on the sky. *See* binary star.

downward (upward) irradiance diffuse attenuation coefficient   The negative of the derivative with respect to depth of the natural logarithm of the downward (upward) plane irradiance [m^{-1}].

downward (upward) radiance mean cosine   The average cosine of the nadir (zenith) angle of all downward (upward) traveling photons; it equals the ratio of the plane irradiance to the scalar irradiance for the downward (upward) direction.

downward (upward) scalar irradiance diffuse attenuation coefficient   The negative of the derivative with respect to depth of the natural logarithm of the downward (upward) scalar irradiance [m^{-1}].

downwelling   The process of accumulation and sinking of warm surface waters along a coastline. The primary *downwelling* regions are in the subtropical ocean waters. Downwelling may bring waters rich in oxygen to deeper layers of oceans or lakes.

draconitic month   *See* month.

drag coefficient   Coefficient linking the current velocity squared to the drag force. In natural waters, the stress τ (drag force per unit area) onto the boundary is usually parameterized by $\tau = \rho C_h u_h^2$ [Nm^{-2}]. The index h indicates at which level above bottom (often $h = 1m$) the velocity u_h is considered. For smooth bottom boundaries (*see* roughness length), C_h is typically $(2.2 \pm 0.6) \cdot 10^{-3}$ and is equal to the skin friction. If bottom irregularities are present (*see* rough flow) a so-called form drag adds to the skin friction. Also in this case, the same parameterization for τ is customary but with a larger value for C_h. At the surface boundary, roughness is due to viscosity at very low wind speed ($W_{10} < 2ms^{-1}$) and due to waves at wind speed $W_{10} > \sim 6ms^{-1}$. Therefore, C_{10} has a minimum at about $W_{10} = 3 - 5ms^{-1}$ of $C_{10} \approx 1 \cdot 10^{-3}$ and increases to 10^{-3} outside the maximum. The index 10 on W_{10} refers to the standard height for wind measurements of $h = 10m$ above water level.

dragging phenomenon   *See* frame dragging.

dredge   A machine that removes sediment from an area for deposition elsewhere. Common designs include the cutterhead dredge, which functions much like a giant drill with a hollow shaft for transporting sediment, and the clamshell dredge, which is essentially a crane with a clamshell-shaped bucket for grabbing and moving sediments.

D region   The mid- and low-latitude *D region* is the portion of the ionosphere formed below 90 km above the surface of the Earth. It lies below the E region. Above 70 km the D layer is mainly due to ionization of the trace constituent NO by solar Lyman-α radiation, while at lower altitudes ionization by galactic cosmic rays is important. The D region has a marked diurnal variation, with daytime maxima just after local noon (typically 10^8 to 10^9 electrons m^{-3}), and disappears at night. The ionization is highest in summer. In winter, on some days the D-region ionization can increase above normal levels. This winter anomaly is thought to be due to meteorological effects on the lower ionosphere together with post (geomagnetic) storm particle precipitation. The D region is responsible for MF and most HF radio wave absorption. The nature of the D region depends on collisions between free electrons in the layer and the atmosphere. Collisions are relatively more important in the lower ionosphere than in the higher layers. Absorption results from radio waves passing through the D region and losing energy to the local atmosphere as a result of these collisions. At LF/VLF/ELF the ionization in the D region may be high enough to reflect the radio waves. At high latitudes particle precipitation can strongly enhance D-region ionization leading to increased absorption. *See* ionosphere, ionospheric absorption, polar cap absorption.

drift   The gradual motion by ions and electrons undergoing guiding center motion. It is generally defined as the net motion, perpendicular to the magnetic field, of the guiding center itself.

drift, curvature   The guiding center drift $\mathbf{v}_c$ associated with the curvature of magnetic field lines. For a particle of mass m and charge e, with velocities in m/sec and the velocity component $v_{||}$ along the magnetic field $\mathbf{B}$, the non-relativistic curvature drift velocity is $\mathbf{v}_c = 2\Omega v_{||}^2[\mathbf{B} \times (\mathbf{B} \cdot \nabla \mathbf{B})/B^3]$ where ∇ is in m^{-1} and Ω is the gyration frequency. In a field of B nT (nanotesla), for protons $\Omega = 0.096$ B, for electrons $\Omega = 176$ B.

drift, electric   The guiding center drift $\mathbf{u}$ imposed on a charged particle in a magnetic field $\mathbf{B}$ by an electric field $\mathbf{E}$. In MKS units, given $\mathbf{E}$ in V/m, $\mathbf{B}$ in Tesla, $\mathbf{u} = \mathbf{E} \times \mathbf{B}/B^2$ m/sec. Unlike other drifts, $\mathbf{u}$ acts equally on positive and negative charges, and therefore does not produce an electric current, but rather a bulk flow of the plasma: convection in the Earth's magnetosphere, for instance, is generally viewed as associated with *electric drifts*.

Intuitively, the electric drift can be viewed as an effect of the voltage drop across the circular orbit of guiding-center motion in a magnetic field. Positive ions accelerate in the direction of $\mathbf{E}$, and being fastest at the part of the circle furthest in that direction, their motion there curves less than at the opposite end of the circle. That causes their guiding centers to move sideways, in a direction perpendicular to both $\mathbf{E}$ and $\mathbf{B}$. Negative electrons are accelerated in the opposite direction, but since they also circle the guiding center in the opposite sense, their drift velocity $\mathbf{u}$ is the same.

drift, gradient   The guiding center drift $\mathbf{v}_g$ associated with the gradient of the magnetic field intensity B. For a particle of mass m and charge e and with velocity component $v_\perp$ perpendicular to the magnetic field, the non-relativistic *gradient drift* velocity is $\mathbf{v}_g = \Omega v_\perp{}^2 (\mathbf{B} \times \nabla\mathbf{B})/B^2$, where distances are in meters and where in a field of B nT (nanotesla), for protons $\Omega = 0.096$ B, for electrons $\Omega = 176$ B.

drift, magnetization The guiding center drift implied by the magnetization current.

driftmeter An instrument for measuring the bulk flow of plasma in space; by ascribing this flow to electric guiding center drift, two components of the ambient electric field $\mathbf{E}$ can be deduced. *Driftmeters* generally consist of two matched plasma detectors observing diametrically opposed directions: any bulk motion is expressed as a difference in the observed plasma flux. Driftmeters are widely used in research satellites as an alternative to double probes.

drift, polarization The guiding center drift implied by the polarization current.

driven shock Shock building in front of an object moving with a speed higher than the local sound speed. Examples are the shock in front of a super-sonic aircraft or the shock in front of a coronal mass ejection. Since *driven shocks* are connected to a moving obstacle, they are also traveling shocks.

drizzle Rain in the form of very small drops (less than .5 mm) that fall slowly to the earth. Typically formed in thin stratus cloud layers, so that the drops have little time to enlarge as they fall.

drumlins Streamlined hills of unconsolidated, unsorted material deposited by a glacier. They are parallel to the direction of movement of the glacier. *Drumlins* are often 25 to 50 m high and about 1 km in length. They can be formed by ice erosion of an earlier glacial deposit or by shaping of the accumulated debris under an active glacier.

drunkards walk Descriptive term for the zig-zag motion of a particle in spatial diffusion.

dry-adiabatic lapse When dry or unsaturated air ascends or descends adiabatically, the temperature change is given by the *dry-adiabatic lapse* rate. In static equilibrium conditions,

$$\gamma_d = -\frac{dT}{dz} = \frac{Ag}{C_p}$$

where γ_d is dry-adiabatic lapse rate, A is mechanical equivalent of heat, g is gravitational acceleration, C_p is specific heat at constant pressure of dry air. At standard conditions the dry-adiabatic lapse rate is 9.8° C/km.

dry freeze A weather event in which the temperature drops below freezing, thus killing vegetation, but the humidity is so low that no hoarfrost forms.

duality in elastic string models A cosmic string may carry a current that can be either time-like or spacelike. Such a string is then called an elastic string, and would in fact be the relativistic analog of an ordinary violin-like string. In order to model the presence of a current in the string core, a single parameter is required, called the state parameter w, whose sign reflects the nature of the current: $w > 0$ for a spacelike current, and $w < 0$ for a timelike current. This parameter w is interpretable as the square of the gradient (with respect to the string worldsheet internal coordinates) of the phase of a current carrier field. Given the state parameter, the string motion is governed by a Lagrangian function $\mathcal{L}(w)$ whose integrated form over the string worldsheet provides the action. Defining $\mathcal{K} = -2d\mathcal{L}/dw$, it can be shown that the square χ of the current is given by

$$\chi = \frac{w}{\mathcal{K}^2} \, .$$

Define $\Lambda(\chi)$:

$$\Lambda = \mathcal{L} + \mathcal{K}\chi$$

Depending on the particular regime (electric or magnetic) one considers, the two functions Λ and $\mathcal{L}$ may play the roles of energy per unit length or tension so that the knowledge of either of them together with the duality transformation is equivalent to the equation of state. The equivalence between both formalisms [either in terms of $\mathcal{L}(w)$ or in terms of $\Lambda(\chi)$] can also be understood as interchanging the roles of the variables of space and time along the string. It was used to reproduce the basic features of the Witten conducting string in the Carter–Peter model. *See* Carter–Peter model, cosmic string, cosmic topological defect, current carrier (cosmic string), elastic string model, electric regime (cosmic string), energy per unit length (cosmic string), equation of state (cosmic string), tension (cosmic string).

ductile behavior   The phenomenological macroscopic nature of solid material deformation. In general *ductile behavior* refers to the capability for substantial change of shape without gross fracturing. The deformation processes that may enter into ductile behavior can be distinguished as cataclastic flow, crystal plasticity, and diffusion flow/grain-sliding. Which combination of processes will be involved in forming *ductile behavior* depends greatly on the properties of the material itself and the deformation conditions such as pressure, temperature, and strain rate.

dune   Any deposit of sand-sized (1/16 to 2 mm in diameter) windblown material. *Dunes* are common on bodies with an atmosphere, having been seen on Earth, Mars, and Venus. Dunes typically form in areas with a prevailing wind direction and an abundant source of sand. Dunes often migrate, as wind blows the sand from one deposit into another area to build up a new deposit. Vegetation and other obstructions cause dunes to stabilize and cease their migration. There are three primary classes of dunes: (1) longitudinal dunes, which are oriented parallel to the prevailing wind direction; (2) transverse dunes, which are oriented perpendicular to the wind direction; and (3) parabolic dunes, which are U-shaped with the dune pointing in the downwind direction. Many subclasses of dunes occur, depending on differences in wind, sand supply, and vegetation.

dusk   The part of morning or evening twilight between complete darkness and civil twilight. *See* twilight, civil twilight, nautical twilight, astronomical twilight.

dust   In astrophysics, grains formed in the envelopes of cool stars, and blown outward via stellar winds and radiation pressure, or formed in supernova explosions. Interstellar *dust* has typical radii of order $a \sim 1\ \mu m$ to $10\ \mu m$ and an exponential size distribution $n(a) \propto a^{\delta}$, where n is the number density. The exponent δ is of order 3.5. Dust plays a substantial part in star forming regions and in the foundation of planetary systems in accretion onto young stars, since the dust has a strong influence on radiative effects.

In geophysics, aerosols (or potential aerosols) of order $1\ \mu m$ in size which have a large effect in the atmospheric energy balance, as well as have dramatic effects on local climate.

In relativity, a continuous medium having a well-defined mass-density and velocity, but whose pressure is equal to zero. It follows from Einstein's equations that each point of dust moves on a geodesic. Dust is an approximate model of the matter distribution in the universe at the current stage of its evolution (in the early phase of the evolution, pressure cannot be neglected).

dust storm   Wind and elevated suspended dust that visibly reduces atmospheric transparency.

dust storm (Mars)   In southern summer of Mars, global dust storms break out sometimes in mid-latitudes of the southern hemisphere. They were photographically observed in 1956, 1971, 1973, and 1977. The dust is wafted on the easterlies, and encircles Mars along a latitude line. At the same time the Hadley cell carries the dust to the northern hemisphere. After about two weeks the whole of Mars, except polar regions, is obscured by the dust. Although albedo features become visible after a month or so, it takes a Martian year for the atmosphere to return to the clear state of before the *dust storm*. Global dust storms have not yet been observed in the northern summer when Mars is near aphelion. Local dust storms break out in all seasons in both hemispheres, especially in low latitudes and near polar caps. Dust clouds seem to be yellowish to visual observers. However, the true color is reddish: dust clouds are bright in red, but not identified in blue. Some observers report that dust clouds in their initial state, in a day or two after their outbreaks, are white visually and bright in blue photographically, but not in red.

dwarf galaxy   A galaxy having a small mass and low luminosity. They are categorized into: *D Ir* dwarf irregular; *D Sp* dwarf spiral; *D El* dwarf elliptical; and *D Sph* dwarf spheroidal (and nucleated dwarf spheroidal).

dwarf nova   The subset of cataclysmic variables in which hydrogen accreted on the white dwarf has ignited while degenerate and so burned explosively. The system survives and the explosions repeat in periods from about 30 years (recurrent novae) up to 10^{4-5} years (classical novae). Most or all of the accreted hydrogen and its burning products are expelled at speeds of about 1000 km/sec, leading to a visible nebula called a nova remnant. These remain detectable for decades until the gas (only about 10^{-5} solar masses) dissipates. A few novae are found in the Milky Way each year, and the total number is probably 20 to 50, many of which are hidden behind galactic dust. The ejecta make some contribution to the galactic inventory of carbon and of rare isotopes like nitrogen-15 and (probably) aluminum-26.

dwarf spheroidal galaxies   Low mass ellipsoidal galaxies, which differ from other dwarf galaxies because of their lower surface brightness and lower luminosity. More quantitatively, a *dwarf spheroidal* (dSph) *galaxy* can be defined as a galaxy with absolute blue magnitude fainter than -14 (-8.5 for the faintest dwarf spheroidal known), surface brightness lower than 22 visual magnitudes per square arcsec (a surface brightness comparable to that of the night sky at new moon), and no nucleus. dSph galaxies have been discovered in the local group, including the first ever, the sculptor system, discovered in 1937 by H. Shapley. The local group dSph galaxies appear as a collection of faint stars, with no diffuse light coming from a background of unresolved, less luminous stars. In addition, dSph galaxies are extremely poor in atomic gas.

dynamical friction   The retarding effect on a massive body as it moves through a cloud of lighter bodies, arising from the fact that momentum is transferred to the lighter bodies by their gravitational interaction with the massive body. Hence, heavier objects tend to sink to the center of clusters (and some of the lighter objects are expelled in the interactions).

dynamical time (DT)   A family of atomic times introduced in 1984 to replace ephemeris time. DT incorporates a relativistic correction for the change in the frequency of a clock due to the effect of its different, or changing, gravitational potential. The heavier the clock, the slower it runs. The second of DT is scaled from the second of International Atomic Time. "Terrestrial Dynamical Time" (TDT) is measured on the Earth's geoid and is meant to approximate Terrestrial Time used in solar motion theories. "Barycentric Dynamical Time" (TDB) is referred to as the solar system barycenter. TDB differs from TDT only by periodic variations. Planetary ephemerides are computed using TDB, but published referred to TDT. *See* Barycentric Dynamical Time, Terrestrial Dynamical Time.

dynamic height   Height measured in units of dynamical meters.

dynamic meter   A unit of gravity potential used to represent the amount of work performed in lifting a unit mass vertically 1 m and defined as 10^5 dyn-cm/gm or 10 J/kg. *Dynamic meters* can be related to a depth, D, in the ocean as $D = gh/10$, where g is the gravitational acceleration and h is the geometric depth. One dynamic meter corresponds roughly to 1.02 geometric meters at the surface of the Earth where $g = 9.8 \text{m/sec}^2$.

dynamic pressure   The inferred momentum flux ρv^2 of the solar wind (ρ density, v velocity) before it strikes the magnetosphere. It is widely used as a parameter in theoretical or empirical models of the magnetopause and bow shock, and equals the pressure on a plate perpendicular to the flow of the solar wind, far ahead of Earth, assuming it absorbs all solar wind particles that hit it. A typical value is 2 nP (nanopascal). The size of the magnetosphere shrinks (expands) with growing (diminishing) dynamic pressure.

dynamic recrystallization   Deformation-induced reworking of grain size, shape, or orientation in a crystal with little or no chemical change. An important mechanism of ridding the crystal of tangled dislocations, in which the new dislocation-free crystals nucleate and grow through the deformed or partially recovered structure. *Dynamic recrystallization* will result in a new undeformed polycrystalline state with high-angle grain boundary (i.e., no particular orientation relationship between grains).

dynamics   The study of motion arising from interaction between bodies, and of bodies under the influence of external fields (gravitational, electric, and magnetic).

dynamic viscosity   The coefficient of viscosity, usually denoted μ equal to the ratio of the shearing stress to the shear of the motion. *See* kinematic viscosity, eddy viscosity, absolute viscosity.

dynamo   The mechanism whereby strong magnetic fields are produced from small field fluctuations through distortion of field lines by motion of conducting fluids. A *dynamo* effect is responsible for the Earth's magnetic field. Numerical simulations of the Earth's internal dynamo can, in fact, reproduce the field reversals that have taken place in the Earth's field.

E

Eady, E.T.   (1915–1966) English mathematician and meteorologist. Worked virtually alone in developing the theory of baroclinic instability during World War II.

Eady model   A baroclinic geophysical fluid dynamic model that clearly illustrates the baroclinic instability process.

EAL   *See* International Atomic Time.

Earth   Third planet in the solar system from the sun, orbital semimajor axis 1.496×10^8 km, eccentricity 0.0167, polar inclination to the ecliptic $23°.45$; Mass 5.9736×10^{24} kg, radius 6126 km. Home of the human race. A planet with active plate tectonics, and frequently (by geologic standards) reversing magnetic field. The Earth has a dense but transparent atmosphere (approximately 1.013 bar at the surface). The atmosphere consists of 77% nitrogen, 21% oxygen, with traces of argon, carbon dioxide, and water and produces a greenhouse effect contributing about 40°C to its average temperature of approximately 14°C. Oceans cover 71% of the Earth's surface, which acts as a thermal reservoir, and CO_2 buffer. Earth is the only planet with a known active biosphere, extending from as much as 1 km into the Earth to 30 km above it. Photosynthetic organisms produce copious amounts of oxygen (O_2) and remove CO_2 from the atmosphere, using energy from the sun (G2 dwarf).

earth ellipsoid and flattening factor   The figure of the Earth is approximately an oblate spheroid, symmetrical about its North-South axis. A spheroid is a special case of an ellipsoid, with two of its three principal axes (A, B, C) equal, viz: $A = B > C$. The flattening factor $f = (A - C)/A$, now approximately 1/298, slowly decreases, as tidal dissipation slows the Earth's rotation. The "weighted mean axis" is $(2A + C)/3$. The "GCT Sphere" is used in mapping transformations that do not accommodate oblateness. Most models were determined by fitting a spheroid to surveys. Ground-based surveying methods depend on the plumb bob or its equivalent to find the local vertical, yielding not the spheroid, but the geoid. Some models were determined dynamically, by measuring the Earth's gravity field through the analysis of spacecraft orbits. Among these, cases such as GEM-9 and GEM-10, with identical axes, were derived from different parent models, whose gravity fields differ in detail.

Earth Orbiter (EO-1)   A spacecraft belonging to the New Millennium Program, intended to validate revolutionary technologies for future land imaging missions. Launched November 21, 2000, with three advanced land imaging instruments that collect multispectral and hyperspectral scenes in coordination with the Enhanced Thematic Mapper (ETM+) on Landsat-7. EO-1 flies in a 705-km circular, sun-synchronous orbit at a 98.7° inclination allowing it to match within 1 min the Landsat-7 orbit and collect identical images for later comparison.

earth orientation parameters (EOP)   Coordinates of the Earth's rotational pole as measured along the Greenwich meridian and the meridian 90° west. They are determined geodetically by very long baseline radio interferometry or satellite or lunar laser ranging.

earthquake   The sudden movement of the ground caused by the release of stress along a fracture (or fault) within the lithosphere. *Earthquakes* usually occur along plate boundaries, although intraplate earthquakes do occur and probably represent the release of excess stress. When too much stress builds up within the planet's lithosphere, the material will fracture and elastic waves carry the resulting energy throughout the planet. The actual location of the initial slip along the fault is called the focus of the earthquake; it is usually located at some depth below the surface. The location on the surface directly above the focus is called the epicenter of the earthquake. Energy released by an earthquake is measured in terms of magnitude on the Richter scale — an increase of one magnitude represents a 32-fold increase in the energy released by the earthquake. The study of earthquakes is called seismology. Similar releases of energy are seen on other bodies, resulting in moonquakes, Marsquakes, etc.

earthquake intensity   A measure of the local perceived strength of an earthquake based on the local damage that is done.

earthquake magnitude   The Richter magnitude is a measure of the intensity of an earthquake. It is a logarithmic scale with an increase of one in magnitude corresponding to about a factor of 32 increase in energy (a factor of 10 in amplitude recorded on a seismogram). *Earthquake magnitudes* are instrumental in that they are obtained from earthquake seismographs. The table on page 141 relates the earthquake magnitude to earthquake intensity and the associated damage.

earthquake moment   The *earthquake moment M* is defined by the relation

$$M = md\,A$$

where m is the shear modulus in the rock where the earthquake occurs, d is the mean displacement across the fault during an earthquake, and A is the rupture area. The earthquake moment is empirically related to the earthquake magnitude.

earthquake precursor   Any phenomenon that occurs prior to an earthquake that warns of its future occurrence. Examples are foreshocks, ground motion (uplift, tilt), electromagnetic signals, emission of gases (for example, radon), changes in the water table, and animal behavior. Although *earthquake precursors* have been documented in some cases, no reliable precursors to earthquakes have been found.

earthquake prediction   Predictions of earthquakes can be divided into two classes: (1) earthquake hazard assessment and (2) prediction of a particular earthquake. Hazard assessment provides an estimate of probability that an earthquake of a specified magnitude will occur in a specified region in a specified time interval. Hazard assessments are based on a number of observations including the number of smaller earthquakes that occur in a region. Reasonably accurate hazard assessments are available. No reliable short-term predictions of actual earthquakes have been documented.

earthquake swarm   A swarm of earthquakes that takes place, concentrating spatially and temporally. The relation among foreshocks, mainshock, and aftershocks is obscure, and an *earthquake swarm* does not include a remarkably large event (mainshock). Earthquake swarms tend to occur in markedly heterogeneous crustal structures such as a volcanic region and crush zone. Possible generation mechanisms are (1) region of accumulated elastic strain is divided into blocks, and earthquakes take place in the respective subregions, or (2) earthquakes occur at the same place repeatedly during short period due to fast supply of energy. A remarkable example of earthquake swarm is the Matsushiro earthquake swarm in Japan (Maximum magnitude M 5.4), for which more than 60,000 felt earthquakes were observed for several years from 1965.

earth radius (R_E)   A widely used unit in measuring distances in the Earth's magnetosphere, usually measured from the center of the Earth. (In other planetary magnetospheres, the planet's radius is often used in a similar way.) The mean *earth radius* is 6,371,315 m. The average distance to the subsolar point of the magnetopause is about 11 R_E, to that of the bow shock 14 R_E, to the synchronous orbit 6.6 R_E, to the moon about 60 R_E and to the L1 and L2 Lagrangian points 236 R_E.

earth rotation parameters (ERP)   Earth orientation parameters and Universal Time.

earthshine   Reflected light from the dayside of the Earth that illuminates the part of the moon that is not directly lit by sunlight.

East Australian current   An ocean current flowing southward along the east coast of Australia.

East Greenland current   An ocean current flowing southward along the east coast of Greenland.

Magnitude distance (km)	Felt	Intensity	Damage
2	0	I	Not felt
		II	Felt by a few people
3	15	III	Hanging objects sway
		IV	Windows and doors rattle
4	80	V	Sleepers awaken
5	150	VI	Windows and glassware broken
		VII	Difficult to stand
6	220	VIII	Branches broken from trees
7	400	IX	Cracks in ground – general panic
		X	Large landslides – most masonry structures destroyed
8	600	XI	Nearly total destruction

east-west effect   An east-west anisotropy in the arrival of cosmic ray particles. At equal inclinations to the vertical, a higher flux of particles is observed from the west than from the east because of an asymmetry in the distribution of trapped orbits (*see* cutoff energy). The direction of the asymmetry (more from west than from east) shows that primary cosmic ray particles have a positive electric charge.

easy access region   Prior to the observations of the spacecraft Ulysses, the topology of the heliosphere was assumed to be similar to the one of the magnetosphere. In particular, above the solar poles cusp-like regions were expected where the cosmic radiation should have an easy access to the inner heliosphere, leading to higher fluxes over the poles compared to the ones at the same radial distance in the equatorial plane. Although Ulysses' findings show an increase in the intensity of the galactic cosmic radiation over the sun's poles, this increase is much smaller than the one expected in the picture of an easy access.

Two physical mechanisms seem to contribute to this lack of easy access: (a) an unexpected high level of magnetic field turbulence, leading to an enhanced scattering and thus preventing a relatively large number of particles from penetrating deep into the heliosphere, and (b) a more peanut-like shape of the heliosphere with a wider extent over the poles owing to the fast solar wind flowing out of the polar coronal holes.

ebb current   The tidal current that results when the water in bays and estuaries is higher than that in the adjoining sea. The opposite is referred to as a flood current.

ebb delta   A deposit of sediments immediately offshore a tidal inlet.

ebb shoal   *See* ebb delta.

ebb tide   Used essentially interchangeably with ebb current.

eccentric   Not centered, or not circular. Refers to elliptical orbits, where eccentricity is defined as the distance between the foci divided by the major axis, or equivalently

$$\epsilon^2 = 1 - \left(\frac{b}{a}\right)^2 ,$$

where ϵ is the eccentricity, a is the semimajor axis, and b is the semiminor axis. The eccentricity of a circle is zero. For parabolas eccentricity = 1; for hyperbolic orbits, eccentricity exceeds 1. *See* eccentricity.

eccentric dipole   An approximation to the internal magnetic field of the Earth. It replaces that field with the field of a magnetic dipole, suitably oriented, but achieves additional accuracy by displacing that dipole from the center of the Earth in such a way that the quadrupole harmonic terms (terms which diminish as $1/r^4$) are reduced as much as possible.

eccentricity   A characterization of conic sections, which are also solutions to the Newtonian equations of motion for a mass in the field of a central mass, thus applicable to planetary orbits. Eccentricity ϵ = zero for circles, eccentricity $\epsilon < 1$ corresponds to an ellipse, $\epsilon = 1$ to a parabola, and $\epsilon > 1$ to a hyperbola. *See* conic section, Kepler's laws.

echelle spectrograph   A grating spectrograph designed to achieve high spectral resolution, employed as an analyzer of optical and UV radiation. To increase resolution, the *echelle spectrograph* works with high diffraction orders (10 to 100). The light diffracted by the echelle grating is made of several high order spectra, covering adjacent narrow spectral ranges. They would overlap spatially if they were not separated by a cross-disperser, i.e., a grating with the grooves aligned perpendicularly to those of the echelle grating. The final echelle spectrum is a sequence of spatially displaced spectra of increasing order, and must be recorded on a two-dimensional detector, such as a CCD or a photographic plate. With echelle spectrographs, a spectral resolving power of several 10^4 can be achieved with a compact design. *See* diffraction grating, grating spectrograph.

echo sounder   A device that sends an acoustic signal into a water column and records the travel time for the sound to be reflected off of an object and returned to the sender. Generally used from a boat to measure water depth to the seafloor.

eclipse   The obscuration of one astronomical body by another which moves between the first body and the observer.

eclipse year   About 346.62 days; because of the precession of the nodes of the moon's orbit (the moon must be near the node for a solar eclipse to occur) alignment of the nodal line with the Earth-sun direction recurs every eclipse year.

eclipsing binary   A pair of stars whose orbiting around each other is revealed because one periodically passes in front of (eclipses) the other from our point of view. This is probable only when the stars are quite close together or the stars are very large. The detailed shape of the light curve can be analyzed to reveal the physical sizes of the two stars, the ratio of their surface temperatures, and the angle the orbit plane makes with our line of sight. The bigger the stars, the longer the eclipse lasts; the hotter the star in back, the deeper the eclipse; and angles different from 90° tend to round the corners of the dips in brightness (though illumination of one star by the other can have somewhat similar effects and the analysis can be complicated).

ecliptic   The apparent path of the sun through (actually in front of) different stellar constellations as seen from the Earth; the extension on the celestial sphere of the plane of the orbit of the Earth. The orbits of the visible planets are close to the *ecliptic* (deviating by $\approx 7°$ for Mercury). This path crosses through the zodiac constellations. *See* celestial sphere.

Eddington approximation   In radiative transfer, the approximation that the radiative flux is constant over direction in the upper hemisphere, and separately constant over the lower hemisphere.

Eddington limit   The maximum luminosity, or accretion rate, beyond which the spherical infall of matter on a massive body stops because the infalling matter is pushed outward by radiation pressure. In the case of spherical accretion, i.e., matter falling radially and uniformly onto a body, the gravitational force is given by: $F_{\text{gravity}} = GM_{f\,\text{source}}M_{\text{cloud}}/D^2$ where D is the distance of the cloud from the source. The radiation force is determined by assuming that the cloud is optically thin and the photons are traveling radially from an isotropic source. Then, each photon absorbed imparts its entire momentum to the cloud ($p_\gamma = E_\gamma/c$). The radiation force is then given by: $F_{\text{radiation}} = \kappa M_{\text{cloud}}\frac{L_{\text{source}}}{4\pi D^2 c}$ where κ is the opacity of the cloud. The Eddington luminosity is independent of distance: $L_{\text{source}}^{\text{Eddington}} = 4\pi GM_{\text{source}}c/\kappa$. Then the Eddington luminosity can be written as

$$L_{Edd} = 1.3 \times 10^{38}(M/M_\odot)\text{ergs } s^{-1} \,,$$

A consequence of the *Eddington limit* is that central black holes need to be very massive to radiate at $L \sim 10^{45}$ to 10^{47} erg, the typical luminosity of quasars. Since the accretion luminosity can be written as $L = \eta \dot{M}c^2$, i.e., as the fraction η of the rest mass falling onto the black hole per unit time, $(\dot{M})$, that is converted into radiating energy, a limiting accretion rate is associated to the Eddington luminosity. *See* quasar.

Eddington luminosity   *See* Eddington limit.

Eddington ratio   The ratio between the bolometric luminosity of a source, and the Eddington luminosity. The *Eddington ratio* can be equivalently defined from the accretion rate. The Eddington ratio is a parameter expected to influence the structure and the radiating properties of an accretion disk in a fundamental way: If the Eddington ratio is $\lesssim 1$ a geometrically thin disk is expected to form, while if $\gtrsim 1$ the accretion disk may inflate to form a torus whose thickness is supported by radiation pressure. *See* Eddington limit.

eddy   A current that runs in a direction other than that of the main current; generally "spins off" from a larger flow and defines a circular path.

eddy correlation method   Method to directly compute turbulent fluxes of scalars. Turbulent velocity fluctuations cause a net transport of scalar properties of the fluid. The turbulent flux is then given by the time or space averaged product $\overline{v'\theta'}$ of the velocity v and the scalar θ, where the primes denote the fluctuations (*see* Reynolds decomposition) and the over-bar denotes the average. For example, the fluctuations of vertical velocity w and temperature T may be combined to give the vertical flux of heat, $\overline{Cw'T'}$, where C is the specific heat.

eddy diffusivity   An analog to molecular diffusivity, used to model diffusion in a turbulent flow.

eddy flux   The flux of chemical properties, momentum, energy, heat, etc. via the eddies in turbulent motion.

eddy-resolving   *Eddy-resolving* models are able to describe the turbulent flow down to a resolution including all scales on which viscosity is not dominant.

eddy viscosity   An analog to molecular viscosity, used to describe shear stresses in a turbulent flow. A coefficient of proportionality to relate shear stress to rate of strain (velocity gradient) in turbulent flow.

edge wave   Waves that are trapped at a coast by refraction. Waves strike the shore and some energy is reflected, and then turned by refraction. Depending on the incident angle and bathymetry, some of this energy will be trapped at the coast. The trapped wave moves in the longshore direction as a progressive wave.

effective charge   A somewhat obsolete expression in field theory for the renormalized quantities which have logarithmic dependence of the scale parameter μ. "Effective charge" is sometimes used instead of the more common "effective coupling constant," "effective mass," "effective parameter," etc. *See* effective couplings.

effective couplings   The values of the coupling constants in quantum field theory depend on the dimensional parameter μ, which measures the typical scale of the energy of the interaction. This dependence is governed by the renormalization group and it has, in general, logarithmic form. Thus, instead of constant coupling, one finds some function of μ. For example, in quantum electrodynamics (QED) the charge of the electron is running (in the one-loop approximation) as

$$\frac{e(\mu)}{e} = \left[1 - \frac{2}{3}\frac{e^2}{(4\pi)^2}\ln\frac{\mu}{\mu_0}\right]^{-1}$$

where $e = e(\mu_0)$ and μ_0 corresponds to some fixed scale. The minus sign in the bracket indicates the lack of asymptotic freedom in QED. This leads to the well-known formal problem (problem of charge zero, or Landau zero) because for some μ corresponding to a very high energy the bracket becomes zero and the *effective coupling* infinite. This in turn makes the perturbation theory inapplicable and addresses serious questions about the fundamental validity of the theory. Indeed, the resolution of this problem is beyond the scope of QED. It is supposed that at these very high energies QED is not an independent theory but part of some nonabelian unified (Grand Unification) theory in which the asymptotic freedom (or finiteness) takes place. At the energies comparable to the Planck energy the local quantum field theory should be (presumably) abandoned and instead one has to consider a more fundamental string or superstring theory. *See* effective charge.

effective pressure   The pressure term of effective stress in a porous medium. If p is the total pressure, the *effective pressure* is defined as $p' = p - \alpha p_f$, where p_f is the pore fluid pressure. The parameter α is defined as $\alpha = 1 - K/K_s$, where K and K_s are the bulk moduli of the matrix frame and the solid grains that constitute the matrix. In most practical cases, $K_s \gg K$ and $\alpha = 1$. *See* effective stress.

effective stress   In a porous medium, the pressure term of the total stress σ_{ij} is partially sustained by the pore fluid. The stress tensor with the effect of pore fluid subtracted is called the *effective stress*. It is the effective stress that determines the deformation and failure of the solid component of the porous medium. The effective stress is defined as

$$\sigma'_{ij} = \sigma_{ij} - \alpha p_f \delta_{ij}$$

where p_f is the pore fluid pressure. The parameter α is defined as $\alpha = 1 - K/K_s$, where K and K_s are the bulk moduli of the matrix frame and the solid grains that constitute the matrix. In most practical cases, $K_s \gg K$ and $\alpha = 1$.

effective temperature   The *effective temperature* of a blackbody is that temperature which characterizes the energy flux (total power output) at the surface of an object. The energy per second emitted by an object at a given frequency over a unit area is called the surface flux. It is found by integrating the blackbody equation over all solid angles and all frequencies. The surface integral gives

$$F_\nu = \oint B_\nu(\nu, T) \cos\theta \, d\phi \, d\theta = \pi B_\nu(\nu, T)$$

in units of erg cm^{-2} s^{-1} Hz^{-1}, where θ is the angle between the normal to the surface and the path of an emitted photon, ϕ is the azimuthal angle, and B_ν is the Planck blackbody intensity in units of erg s^{-1} cm^{-2} Hz^{-1} sr^{-1}. The total energy emitted per second per unit area is the surface flux (above) integrated over all frequencies:

$$F = \int \pi B_\nu(\nu, T) d\nu = \frac{2\pi^5 k^4}{15 h^3 c^3} T_{eff}^4 = \sigma T_{eff}^4$$

where σ is the Stefan–Boltzmann constant. The total energy flux of a blackbody is related to its luminosity by,

$$F = L/4\pi r^2 = \sigma T_{eff}^4 \,.$$

In astronomy, one computes the effective temperature of a star (or the sum) from its luminosity by this formula. The effective temperature then, is the characteristic temperature that relates a star's total output power to its size.

effluent   Something that is discharged; commonly used to refer to the discharge from a sewer or factory outfall lying in a river or coastal waters.

eigenray   The integral curve of a principal direction of the Killing bivector $\nabla_{[a} K_{b]}$. By the Killing equation, the symmetrized derivative of the Killing vector $\mathbf{K}$ vanishes. In a spinorial notation, the null eigendirection is given by the solution α_A of the eigenvalue problem $\phi^{AB} \alpha_B = \lambda \alpha^A$ where ϕ^{AB} is the spinor representation of the Killing bivector. *See* Killing vector.

eigenvalue   An allowed value of the constant a in the equation $Au = au$, where A is an operator acting on a function u (which is called an eigenfunction). Also called characteristic value.

eikonal approximation   The approximation to a wave equation which assumes the wave function is of the form $exp[i\omega t + i\mathbf{kx}]$ where ω and $\mathbf{k}$ are large. This replaces the second order derivatives of the wave function by terms proportional to the square of ω or of $\mathbf{k}$.

einstein   One mole of photons (6.023×10^{23} photons).

Einstein–Cartan gravity   An important particular case of gravity with torsion. The action of *Einstein–Cartan gravity* is a direct generalization of the Einstein–Hilbert action of General Relativity:

$$S_{EC} = -\frac{1}{16\pi G} \int d^4x \sqrt{-g}\, g^{\mu\nu} \tilde{R}_{\mu\nu}$$

where $\tilde{R}_{\mu\nu}$ is the (nonsymmetric) Ricci tensor with torsion

$$\tilde{R}_{\mu\nu} = \partial_\lambda \tilde{\Gamma}^\lambda_{\mu\nu} - \partial_\nu \tilde{\Gamma}^\lambda_{\mu\lambda} + \tilde{\Gamma}^\lambda_{\mu\nu}\tilde{\Gamma}^\tau_{\lambda\tau} - \tilde{\Gamma}^\tau_{\mu\lambda}\tilde{\Gamma}^\lambda_{\tau\nu}$$

and $\tilde{\Gamma}^\lambda_{\mu\mu}$ is a non-symmetric affine connection with torsion, which satisfies the metricity condition $\tilde{\nabla}_\mu g_{\alpha\beta} = 0$. The above action can be rewritten as a sum of the Einstein–Hilbert action and the torsion terms, but those terms are not dynamical since the only one derivative of the torsion tensor appears in the action in a surface term. As a result, for pure Einstein–Cartan gravity the equation for torsion is $T^\lambda_{\mu\nu} = 0$, and the theory is dynamically equivalent to General Relativity. If the matter fields provide an external current for torsion, the Einstein-Cartan gravity describes contact interaction between those currents. *See* metricity of covariant derivative, torsion.

Einstein equations The set of differential equations that connect the metric to the distribution of matter in the spacetime. The features of matter that enter the equations are the stress-energy tensor $T_{\mu\nu}$, containing its mass-density, momentum (i.e., mass multiplied by velocity) per unit volume and internal stresses (pressure in fluids and gases). The *Einstein equations* are very complicated second-order partial differential equations (10 in general, unless in special cases some of them are fulfilled identically) in which the unknown functions (10 components of the metric $g_{\mu\nu}$) depend, in general, on 4 variables (3 space coordinates and the time):

$$\begin{aligned} G_{\mu\nu} &\equiv R_{\mu\nu} - \frac{1}{2}g_{\mu\nu}R \\ &= \frac{8\pi G}{c^2} T_{\mu\nu} \end{aligned}$$

Here $G_{\mu\nu}$ is the Einstein tensor, $R_{\mu\nu}$ the Ricci tensor and R its trace, and G is Newton's constant. The factor c^{-2} on the right assumes a choice of dimensions for $G_{\mu\nu}$ of $[\text{length}]^{-2}$ and

for $T_{\mu\nu}$ of $[\text{mass/length}^3]$. It has been common since Einstein's introduction of the cosmological constant Λ to add $\Lambda g_{\mu\nu}$ to the left-hand side. Positive Λ produces a repulsive effect at large distances. Modern theory holds that such effects arise from some other (quantum) field coupled to gravity, and thus arise through the stress-energy tensor. A solution of these equations is a model of the spacetime corresponding to various astronomical situations, e.g., a single star in an otherwise empty space, the whole universe (*see* cosmological models), a black hole. Unrealistic objects that are interesting for academic reasons only are also considered (e.g., infinitely long cylinders filled with a fluid). In full generality, the Einstein equations are very difficult to handle, but a large literature exists in which their implications are discussed without solving them. A very large number of solutions has been found under simplifying assumptions, most often about symmetries of the spacetime. Progress is also being made in computational solutions for general situations. Through the Einstein equations, a given matter-distribution influences the geometry of spacetime, and a given metric determines the distribution of matter, its stresses and motions. *See* constraint equations, gauge, signature.

Einstein–Rosen bridge A construction by A. Einstein and N. Rosen, based on the Schwarzschild solution, wherein at one instant, two copies of the Schwarzschild spacetime with a black hole of mass M outside the horizon were joined smoothly at the horizon. The resulting 3-space connects two distant universes (i.e., it has two spatial infinities) each containing a gravitating mass M, connected by a wormhole. Subsequent study showed the wormhole was dynamic and would collapse before any communication was possible through it. However, recent work shows that certain kinds of exotic matter can stabilize wormholes against such collapse.

Einstein summation convention *See* summation convention.

Einstein tensor The symmetric tensor

$$G_{ab} = R_{ab} - \frac{1}{2}g_{ab}R$$

where R_{ab} is the Ricci tensor, g_{ab} is the metric tensor, and $R = R_a^a$ is the Ricci scalar. The divergence of the *Einstein tensor* vanishes identically. *See* gravitational equations, Ricci tensor.

Einstein Universe　　The historically first cosmological model derived by Einstein himself from his relativity theory. In this model, the universe is homogeneous, isotropic, and static, i.e., unchanging in time (*see* homogeneity, isotropy). This last property is a consequence of the long-range repulsion implied by the cosmological constant which balances gravitational attraction. This balance was later proved, by A.S. Eddington, to be unstable: any small departure from it would make the universe expand or collapse away from the initial state, and the evolution of the perturbed model would follow the Friedmann–Lemaître cosmological models. Because of its being static, the *Einstein Universe* is not in fact an acceptable model of the actual universe, which is now known to be expanding (*see* expansion of the universe). However, the Einstein Universe played an important role in the early development of theoretical cosmology — it provided evidence that relativity theory is a useful device to investigate properties of the universe as a whole.

ejecta　　In impact cratering, material that is tossed out during the excavation of an impact crater. The ejected material is derived from the top 1/3 of the crater. Some of the ejected material falls back onto the floor of the crater, but much is tossed outside the crater rim to form an *ejecta* blanket. Ejecta blankets on bodies with dry surface materials and no atmosphere (like the moon and Mars) tend to display a pattern with strings of secondary craters (craters produced by material ejected from the primary crater) radiating outward from the main crater. Glassy material incorporated into the ejecta can appear as bright streaks, called rays. A radial ejecta blanket is typically very rough within one crater radii of the rim — few individual secondary craters can be discerned. Beyond one crater radii, the ejecta blanket fans out into radial strings of secondary craters. On bodies with a thick atmosphere (like Venus) or with subsurface ice (like Mars and Jupiter's moons of Ganymede and Callisto), impact craters are typically surrounded by a more fluidized ejecta pattern, apparently caused by the ejecta being entrained in gas from either the atmosphere or produced during vaporization of the ice by the impact. The extent of the fluidized ejecta blankets varies depending on the state/viscosity of the volatiles and the environmental conditions. In supernova physics, the ejecta are the material blown from the stars as a result of the explosion.

Ekman convergence　　The stress on the Earth's surface varies from place to place and hence so does the Ekman transport. This leads to convergence of mass in some places, and hence to expulsion of fluid from the boundary layer, called the *Ekman convergence*. In other places, the Ekman transport is horizontally divergent, i.e., mass is being lost across the sides of a given area, so fluid must be "sucked" vertically into the boundary layer to replace that which is lost across the sides, called Ekman suction. This effect is called the Ekman pumping.

Ekman layer　　(Ekman, 1905) The top or bottom layer in which the (surface or bottom) stress acts. The velocity that was driven by the stress is called the Ekman velocity. Typically, the atmospheric boundary *Eckman layer* is 1 km thick, whereas the oceanic boundary Eckman layer is 10 to 100 m thick.

Ekman mass transport　　The mass transport by the Ekman velocity within the boundary Ekman layer is called the *Ekman mass transport* or Ekman transport. In steady conditions, the Ekman transport is directed at right angles to the surface stress. In the atmosphere, the transport is to the left in the northern hemisphere relative to the surface stress. In the ocean, the transport is to the right in the northern hemisphere relative to the surface stress.

Ekman pumping　　*See* Ekman convergence.

Elara　　Moon of Jupiter, also designated JVII. Discovered by C. Perrine in 1905, its orbit has an eccentricity of 0.207, an inclination of 24.77°, and a semimajor axis of 1.174×10^7 km. Its radius is approximately 38 km, its mass 7.77×10^{17} kg, and its density 3.4 g cm^{-3}. It has a geometric albedo of 0.03 and orbits Jupiter once every 259.7 Earth days.

elastic deformation   The reversible deformation of a material in response to a force. In a macroscopic sense, if the strain is proportional to stress (Hooke's law) and falls to zero when the stress is removed, the deformation is defined as elastic deformation. In a microscopic sense, if atomic displacement is small compared to the mean atomic spacing and atoms climb up the potential hill only to a small extent they will always be subjected to a restoring force, and upon the removal of a force, they will come back to the original position. Thus, elastic deformation is characterized by (i) reversible displacement (strain) and (ii) instantaneous response.

elastic limit   The greatest stress that a material is capable of sustaining without any permanent strain remaining after complete release of the stress.

elastic lithosphere   That fraction of the lithosphere that retains elastic strength over geological time scales. The *elastic lithosphere* is responsible for holding up ocean islands and volcanos, and it is responsible for the flexural shape of the lithosphere at ocean trenches and at many sedimentary basins. A typical elastic lithosphere thickness is 30 km. Because of this thickness many sedimentary basins have a width of about 200 km.

elastic modulus   *See* Young's modulus.

elastic rebound   The process that generates an earthquake. A fault is locked and tectonic deformation builds up the elastic stress in the surrounding rock. When this stress exceeds the rupture strength of the fault, slip occurs on the fault. The surrounding rock *elastically rebounds* generating seismic waves.

elastic string model   Linear topological defects known as cosmic strings can carry currents and their dynamics are described by an equation of state relating the energy per unit length to the tension. This equation of state can be classified according to whether the characteristic propagation speed of longitudinal (sound type) perturbations is smaller (supersonic), greater (subsonic), or equal (transonic) to that of transverse perturbations. Corresponding models bear the same names. *See* Carter–Peter model, current instability (cosmic string), equation of state (cosmic string).

E layer   The lowest thermospheric layer of the ionosphere at heights between 85 and 140 km. Ionization is primarily due to solar X-rays and EUVs, and to a much smaller amount, also due to particles with rather low energies (some keV) and to an even smaller extent, due to meteorite dust. Dominant ionized particle species is molecular oxygen O_2^+. Formally, the *E layer* can be described by a Chapman profile.

E-layer screening   The process whereby the E layer prevents radio signal propagation from taking place by a higher layer, normally the F region. Generally, *E-layer screening* occurs when the E layer MUF is greater than the operating frequency. There are two ways in which this can occur. During much of the daytime, a radio wave that would normally have propagated by the F layer will be screened, or prevented, from reaching the F layer because it is first reflected back to the Earth by the intervening E layer. While the signal may be propagated by successive reflections from the E region, the additional hops will lead to sufficient attenuation to prevent the signal from being useful. It is also possible that a signal propagating away from the nighttime terminator will, after being reflected by the F region then encounter an increasing E region MUF and be reflected back from the top of the E region towards the F region. In this case, the signal is prevented from reaching the receiver, but may be effectively ducted along between the E and F regions until it again reaches the nighttime terminator and can propagate to ground level. In both cases, the E-layer screening effect can be enhanced by the presence of sporadic E. On long (> 5000 km) multi-hop paths E-layer screening enhanced by sporadic E, rather than absorption, is the low frequency limiting factor. *See* ionospheric radio propagation path.

Electra Magnitude 3.8 type B5 star at RA 03^h44^m, dec +24.06'; one of the "seven sisters" of the Pleiades.

electric drift *See* drift, electric.

electric field, parallel *See* parallel electric field.

electric regime (cosmic string) It has been known since the seminal work of E. Witten in 1985 that a current can build up in a cosmic string. As current generation proceeds via random choices of the phase of the current carrier, the resulting current can be of two distinct kinds, timelike or spacelike, depending on whether its time component is greater or smaller than its space component (there is also the possibility that they are equal, leading to a so-called lightlike current, but this is very rare at the time of current formation and would only occur through string intercommutation). Explicitly, setting the phase of the current carrier as a function of the time t and the string coordinate z in the form

$$\varphi = \omega t - kz ,$$

with ω and k arbitrary parameters, one can define a state parameter through

$$w = k^2 - \omega^2 ,$$

the case $w < 0$ (respectively, $w > 0$) corresponding to a timelike (respectively, spacelike) current.

In the timelike case, the configuration is said to be in the *electric regime,* whereas for spacelike currents it is in the magnetic regime. The reason for these particular names stems from the possibility that the current is electromagnetic in nature, which means coupled with electric and magnetic fields. Then, for the electric regime, one can always find a way to locally remove the magnetic field and thus one is led to describe solely an electric field surrounding a cosmic string. The same is true in the magnetic regime, where this time it is the electric field that can be removed and only a magnetic field remains. *See* current carrier (cosmic string), current generation (cosmic string), intercommutation (cosmic string).

electroglow A light emitting process in the upper atmospheres of Jupiter, Saturn, Uranus, and Saturn's satellite Titan. Sunlight dissociates some H_2 and ionizes the hydrogen; the electrons are accelerated and interact with H_2, producing the glow.

electromagnetic current meter A device that uses Faraday's Law of magnetic induction to measure flow velocities. The current meter head establishes a magnetic field. A moving conductor (water) creates an electrical potential that is measured by the instrument. The electrical potential is proportional to the speed of the current. Often used for one- or two-dimensional velocity measurements.

electromagnetic induction This is the generation of currents in a conductor by a change in magnetic flux linkage, which produces a magnetic field that opposes the change in the inducing magnetic field, as described by Faraday's laws and Lenz's law. In geophysics, this can be used to study the conductivity of the mantle through measurement of its response to magnetic fluctuations originating in the ionosphere and/or magnetosphere. Maxwell's equations can be used to show that in the appropriate limit, a magnetic field obeys a diffusion equation:

$$\frac{\partial \mathbf{B}}{\partial t} = -\nabla \times \left(\frac{1}{\mu_0 \sigma} \nabla \times \mathbf{B} \right)$$

where σ is the local conductivity. Fluctuating external fields of frequency ω diffuse into a layer at the top of the mantle with a skin depth $\sqrt{2/\mu_0 \sigma \omega}$. The induced field is then essentially a reflection of this external field from the skin-depth layer, and it has a phase relative to the external field that relates the conductivity structure of the layer to the frequency of the signal (essentially, long period signals penetrate deeper and therefore depend more on deeper conductivity than do short period signals).

electromagnetic radiation Radiation arising from the motion of electric charges, consisting of variations in the electric and magnetic fields, and propagating at the speed of light. Depending on the wavelength, observable as radio, infrared, visible light, ultraviolet, X-rays, or gamma rays.

electromagnetism The study of the relationships between electric and magnetic fields, their causes, and their dynamical effects.

electromagnetism, Lorentz transformation and An important property of Lorentz transformations is that the equations that describe the relationship between electric and magnetic field, their causes and effects, are Lorentz invariant. It then follows that special relativity presents a consistent theoretical groundwork for both mechanics and electromagnetism.

electron An elementary particle in the family of leptons, with mass of $9.1093897 \times 10^{-31}$ kg, negative charge and spin of 1/2.

electron precipitation In solar physics, the transport of an accelerated population of electrons from the corona to the chromosphere where they deposit the bulk of their energy. The electrons can precipitate directly, immediately following their production, or after a delay during which they are trapped in the corona by converging magnetic field. The pitch-angle of the electrons, the nature of the field convergence, and the collision rate in the corona are all important in deciding which electrons precipitate and when.

electron temperature The kinetic temperature of electrons in an ionized gas.

electron volt The energy an electron (or proton) gains when accelerated through an electric potential difference of 1 volt. Equal to 1.602177×10^{-19} J.

electrostatic unit A unit of charge defined so that two equal such charges exert a mutual electrostatic force of 1 dyne, when separated by 1 cm.

electrovacuum A space-time in general relativity containing a source-free electromagnetic field interacting with the gravitational field.

element A group of atoms, each with the same number of protons, defines an *element*. The number of protons in a given atom is called the atomic number. There are 92 naturally occurring elements. Elements with over 100 protons can be synthesized. Although an atom consists of protons and neutrons in the nucleus surrounded by clouds of electrons, it is only the number of protons that defines the atom as belonging to a particular element. Atoms of the same element that have different numbers of neutrons are known as isotopes of the element. Atoms of the same element that have different numbers of electrons are called ions.

elemental abundances: general The bulk elemental composition of most solar system objects are roughly the solar system average, but contain distinct sub-components that are presolar. Isotopic anomalies represent the only unequivocal signatures of the survival of circumstellar and interstellar materials within extraterrestrial objects such as meteorites and interplanetary dust particles (IDPs).

elemental abundances: in minerals Isotopic anomalies in a variety of elements have been used to identify a number of circumstellar mineralogical species in meteorites. These isotopic anomalies are thought to have a nucleosynthetic origin, i.e., their carrier grains were formed in the circumstellar environments of various kinds of stars, survived transport through the interstellar medium, and incorporation into our solar system.

elemental abundances: in organics Significant deuterium enrichments have been seen in the organic components of both meteorites and interplanetary dust particles. Unlike the isotopic signatures in the circumstellar materials, denterium anomalies do *not* have a nucleosynthetic origin. Instead, it has been proposed that they result from reactions at low temperatures in the interstellar medium. In a few cases, the D/H ratios of specific classes of organic compounds have been measured, for example in PAHs in meteorites and IDPs.

elemental abundances: of comet(s) The overall elemental composition of Comet Halley resembled the solar system average, but abundances and deuterium enrichments of molecules in cometary tails and comae, and the types of solid materials observed in interplanetary dust particles have been interpreted as an indication of the primitive, pre-solar nature of comets, or at least their sub-components. For example, the ratios of D/H in water from comets Halley, Hyakutake, and Hale–Bopp were all around 3×10^{-4}, similar to what is seen in certain interstellar environments and approximately a factor of two over ocean water, and DCN/HCN ratios up to 2.5×10^{-2} were measured in Hale–Bopp.

elevation   The vertical distance measured from the geoid, from the earth ellipsoid, or from the local terrain. When interpreting *elevation* data, care is needed to ascertain which of the three surfaces mentioned is the reference one. When no specification is given, the context must be examined. For geophysics, hydrology, and cartography, the geoid is the usual reference surface because most elevation data are produced in surveys based on the local horizontal, as found by the spirit level or the normal to the plumb line. For construction or related work, the local mean terrain (earth surface) is the more likely reference. The ellipsoid is rarely the reference surface, but it may be so in space science contexts.

elevation head (sometimes gravitational head) (z) The potential energy per unit fluid weight is given by the elevation above an arbitrary horizontal datum. The elevation head has units of length and is a component of the hydraulic head.

Eliassen–Palm (EP) flux   A vector in the meridional (y,z) plane, which has the eddy momentum flux and eddy heat flux as its horizontal and vertical components. Its convergence is directly related to the eddy forcing on the zonal mean flow.

elliptical galaxies   Galaxies of regular, ellipsoidal appearance, and of rather reddish colors. The photometric profiles of most elliptical galaxies are described by empirical laws in which the surface brightness decreases smoothly as a function of the distance from the galaxy center. *Elliptical galaxies* do not show features such as bars, spiral arms, or tails. Only a minority of them show ripples, shells, or asymmetric radial distribution of surface brightness.

Elliptical galaxies are characterized by the absence of significant neutral or molecular gas, and hence the absence of star formation, and by a stellar content mostly made of old stars belonging to stellar population II. They account for about 1/3 of all observed galaxies, and are the majority of galaxies in dense cluster environments. They cover a wide range of masses, from $\sim 10^6$ to $\sim 10^{11}$ solar masses, the most massive being located at the center of clusters of galaxies (cD galaxies), the less massive being dwarf elliptical galaxies. *See* cD galaxies, dwarf spheroidal galaxies.

Ellison scale   Like the Thorpe scale, the *Ellison scale* is a quantity to estimate the overturning eddy size. The Ellison scale L_E is based on density ρ instead of temperature T, and the definition $L_E = <\rho^2>/(\partial\rho/\partial z)$ deviates slightly from the procedure for the Thorpe scale L_T estimation. Both scales, L_E and L_T are considered adequate measures for the overturning eddy size and generally agree well with the Ozmidov scale.

El Niño   The warm phase of the Southern Oscillation beginning at about Christmas time (hence the name "El Niño", Spanish for "Christ child") and lasting 12 to 18 months. Characterized by warming of sea surface temperatures in the central and eastern equatorial Pacific Ocean. The anomalously warm water causes the sardine population to die off the Peru coast. A series of effects arise, including an increased westerly wind and a shift in Pacific ocean circulation. This warming occurs in the entire tropics and causes drought in Indonesia and Australia. Enhanced North-South temperature differences transport energy into the atmosphere, modifying global atmospherics flow, causing warm dry weather in Northern U.S., and wet cool weather in the Southern U.S. *See* Southern Oscillation Index, La Niña.

elongation   The angle between the sun and the observer, measured at the object being observed.

elongation   The apparent angular separation between the sun and a solar system object as viewed by a distant observer, i.e., the sun-observer-object angle.

elves   Transient air glow events observed near 90 km, nearly simultaneously with a strong cloud-to-ground lightning stroke. They often precede sprites, which may occur at lower altitudes a few milliseconds later. It is currently believed that *elves* are the result of wave heating by very low frequency (VLF) radio pulses emitted by the lightning discharge current. *See* sprites.

Elysium Province   The second most pronounced region of central vent volcanism on Mars. It is 5 km high, ≈ 2000 km in diameter, and is centered on latitude 25° N and longitude 210° W. It is considerably smaller than Tharsis and has only three shield volcanoes of appreciable size (namely, Albor Tholus, Hecates Tholus, and Elysium Mons). However, the volcanoes are still large compared to those on Earth, with Hecates Tholus standing ≈ 6 km above the plains and Elysium Mons standing ≈ 9 km above the plains. All of them are greater than 150 km in diameter. The volcanoes are considered to be older than those at Tharsis formed during the late Noachian to early Hesperian. They all present a diverse range of volcanic morphologies, all of which indicate more pyroclastic activity compared to the Tharsis volcanoes. Confined outflow channels exist NW of Elysium Mons and Hecates Tholus. They drain north-west and extend ≈ 1000 km into the plains, and may be of volcanic origin.

Similarities of the *Elysium Province* to Tharsis include the volcanic and tectonic history, a broad free air gravity anomaly, and Phlegra Montes which is assumed to be an island of old cratered terrain on the northern flank of Elysium Mons. If the favored proposal for subsurface thermal activity at Tharsis is applicable to Elysium, then it appears that there is a bimodal distribution of hotspots on Mars.

embedded defect   Cosmic topological defects are predicted to form at phase transitions when the symmetry G of a system is broken down to a smaller symmetry group H and there exists a nontrivial homotopy group of the vacuum manifold $\mathcal{M} \sim G/H$. In the case where all the homotopy groups of $\mathcal{M}$ are trivial, no topological defect can form but there might still be defect-like solutions. These arise from the existence of subgroups $g \subset G$ and $h \subset H$ that are also broken through the scheme $g \to h$ during the breaking $G \to H$. For the same reasons as with ordinary topological defects, if there is a non-trivial structure of the vacuum submanifold g/h, then defects might form. However, here stability is not ensured by topology, but by dynamical arguments.

Even these unstable defects could have a role in cosmology, as for instance the so-called $Z-$strings, associated with the total breaking of a $U(1)$ subgroup of $SU(2)$ in the scheme $SU(2) \times U(1) \to U(1)$ of the electroweak model (the actual group that is broken is essentially $SU(2)$, which contains a $U(1)$ subgroup, and it is the breaking of this subgroup that gives rise to the defect-like solutions), and that could be responsible for the primordial baryogenesis. Other *embedded defects* are observed in other branches of physics such as condensed matter experiments. *See* cosmic phase transition, cosmic topological defect, homotopy group, vacuum manifold.

emerging flux region   New bipolar active regions emerge from below the solar photosphere in a characteristic pattern known as an emerging flux region. A flux loop brought to the surface by magnetic buoyancy intersects the surface to form a bipole. As the loop emerges, the opposite poles move apart, the preceding spot moving ~ 1 km s^{-1} and the following spot less than 0.3 km s^{-1}. In Hα the phenomenon is characterized by arch filaments which appear to trace the rising flux loops.

emission line   A feature in the spectrum of the light produced by a medium that emits light by a quantum transition, thus increasing the intensity of the light at certain wavelengths. *See* spectrum. *Compare with* absorption line.

emission lines: interstellar and cometary Extraterrestrial molecules, fragments, and atoms can be detected or characterized based on emission, the emanation of radiation as a result of excitation such as resulting from collisions or the absorption of photons. Interstellar examples include Lyman and Balmer series of molecular and atomic hydrogen, respectively, in the UV/Vis, and in the infrared, the fluorescence of aromatic hydrocarbons pumped by UV photons. Similarly, emission from both atoms (i.e., sodium D lines) and fluorescence from molecules (i.e., 3.4 μm, methanol) and fragments (i.e., Swan bands of C_2*) excited by sunlight have been observed from cometary comae.

emissivity   An indirect measure of reflectivity, where emissivity = 1 − reflectivity and reflectivity is the reflected energy measured from a surface by a single energy bounce. Mirrored surfaces reflect large amounts of energy (around 98%) but absorb very little (around 2%). A blackbody surface, on the other hand, reverses the ratio, absorbing 98% of the energy and reflecting only 2%. Emissivity is affected by the geometric shape of the object, the electrical properties of the radiating surface, and the measuring wavelength.

A Magellan radiometer experiment observed the 12.6-cm-wavelength radiothermal emissivity on Venus. The nominal pattern of radiothermal *emissivity* shows high mountain summits display abnormally low emissivity and plains regions high emissivity. One explanation for the low emissivity of mountaintops is the presence of electrically conductive minerals, produced by weathering, embedded in the surface rocks.

empirical model of the magnetosphere   A mathematical representation of the global magnetic field of the magnetosphere, whose coefficients are fitted to data. It is a convenient tool for predicting the magnetic field vector or field line linkage to the ionosphere which a satellite at some point P in space is most likely to find. The construction of empirical fields is also the best way of extracting global information from magnetic field data.

The parameters of modern empirical models are generally derived from flexible representations of the fields of the different magnetospheric current systems, e.g., those of the magnetopause, tail, ring current, and the Birkeland current circuit. The field predicted at P depends not only on the location but also on the tilt angle ψ, on the dynamic pressure of the solar wind, on the interplanetary magnetic field and on geomagnetic activity indices such as Dst and AE.

Enceladus   Moon of Saturn, also designated SII. Discovered by Herschel in 1789 its surface has the highest albedo of any solar system body. It also displays evidence of resurfacing, possibly as a result of water volcanism. Since it is too small to retain radioactive heat, tidal heating aided by a 1:2 resonance with Dione may provide the required energy. Its orbit has an eccentricity of 0.0045, an inclination of 0.00°, a precession of 156.2° yr^{-1}, and a semimajor axis of 2.38×10^5 km. Its radius is 250 km, its mass, 8.40×10^{19} kg, and its density 1.28 g cm^{-3}. It has a geometric albedo of 1.0, and orbits Saturn once every 1.370 Earth days.

Encke's comet   Comet with the shortest known orbital period: 3.30 years. Its orbit has semimajor axis 2.21 AU and perihelion distance 0,338 AU.

endothermic   A process that absorbs heat as it proceeds. Opposite of exothermic.

energetic particles   Supra-thermal particles; that is all particles with speeds large compared to the thermal plasma speed.

energetic particles in interplanetary space
Energetic particles in interplanetary space can be observed with energies ranging from the supra-thermal up to 10^{20} eV. The main constituents are protons, α-particles, and electrons; heavier particles up to iron can be found in substantially smaller numbers. The particle populations originate in different sources, all having their typical spectrum, temporal development, and spatial extent.
1. Galactic cosmic rays (GCR) are the high-energy background with energies extending up to 10^{20} eV. They are incident upon the heliosphere uniformly and isotropically. In the inner heliosphere, the galactic cosmic radiation is modulated by solar activity: the intensity of GCRs is highest during solar minimum and reduced during solar maximum conditions. *See* modulation of galactic cosmic rays.
2. Anomalous galactic cosmic rays (AGCR), also called anomalous component, energetically connect to the lower end of the galactic cosmic rays but differ from them with respect to composition, charge states, spectrum, and variation

Energetic Particle Population in the Heliosphere

Population	temporal scales	spatial scales	energy range	acceleration mechanism
GCR	continuous	global	GeV to > TeV	diffusive shock
AGCR	continuous	global	10 – 100 MeV	shock?
SEP	?	?	keV – 100 MeV selective heating, shock	reconnection, stochastic
ESP	days	extended	keV – 10 MeV shock-drift, stochastic	diffusive shock,
RII	27 days	extended	keV – 10 MeV	diffusive shock
PBSP	continuous	local	keV – MeV	diffusive shock, shock drift

with the solar cycle. The anomalous component stems from originally neutral particles which became ionized as they traveled through interplanetary space towards the sun. The now charged particles are then convected outwards with the solar wind and are accelerated at the termination shock, the outer boundary of the heliosphere.

3. Solar energetic particles (SEP) are accelerated in solar flares, their injection therefore is point-like in space and time. Energies extend up to about 100 MeV, occasionally even into the GeV range. In this case, particles can also be observed from the ground (*see* ground level event). Properties of solar energetic particles differ, depending on whether the parent flare was gradual or impulsive. *See* gradual flare, impulsive flare.

4. Energetic storm particles (ESP) are accelerated at interplanetary shocks. Originally, ESPs were thought to be particle enhancements related to the passage of an interplanetary shock. The name was chosen to reflect their association with the magnetic storm observed as the shock hits the Earth's magnetosphere. Today, we understand the particle acceleration at the shock, their escape and the subsequent propagation through interplanetary space as a continuous process lasting for days to weeks until the shock finally stops accelerating particles. *See* energetic storm particles.

5. Recurrent intensity increases (RII) are due to particles accelerated at the shocks around co-rotating interaction regions (CIRs). The energetic particles can even be observed remote from these co-rotating shocks at distances where the shocks have not yet been formed or at higher solar latitudes when a spacecraft is well above the streamer belt where the CIR form. *See* corotating interaction region.

6. Planetary bow shock particles (PBSP) Particles accelerated at a planetary bow shock are a local particle component with energies extending up to some 10 keV. An exception is the Jovian magnetosphere where electrons are accelerated up to about 10 MeV. With a suitable magnetic connection between Earth and Jupiter, these Jovian electrons can be observed even at Earth's orbit.

energetic storm particles   Particles accelerated at an interplanetary shock. The name stems from the first observations of shock-accelerated particles: around the time of shock passage at the Earth when the interaction between the shock and the magnetosphere caused a geomagnetic storm, an increase in particle intensities could be observed, which was termed *energetic storm particles*.

Originally, the term referred to a bump in the intensity on the decaying flank of a solar energetic particle event. In protons up to energies of a few hundred keV, such a bump lasted for some hours around the time of shock passage. Energetic storm particles are observed only at quasi-parallel shocks, where they are accelerated by diffusive shock acceleration. At quasi-perpendicular shocks, on the other hand, short shock-spikes, lasting only for some 10 min, are observed at the time of shock passage. Thus, the appearance of the shock accelerated particles strongly depends on the local angle θ_{Bn} between the magnetic field direction and the shock normal and the dominant acceleration mechanism related to this local geometry.

Today, the term *energetic storm particles* is often is used in a broader context and basically refers to the fact that part or all of the observed particles are accelerated at a traveling interplanetary shock. Such intensity increases can be observed in electrons as well as nuclei with electron increases predominately up to energies of some 10 keV or occasionally up to a few MeV and proton increases up to some MeV, at strong shocks even up to about 100 MeV.

In contrast to energies up to some hundred keV, in these higher energies no obvious dependence of the intensity time profile on the local geometry, that is the angle θ_{Bn}, can be seen and therefore no distinction in shock spikes and shock bumps related to quasi-perpendicular and quasi-parallel shocks can be observed. In the MeV range, particles are much faster than the shock and therefore easily escape from the shock front. Thus, the intensity time profile at the observer's site does not reflect the local properties of the shock and the associated particles but samples particles accelerated at the shock during its propagation from the sun to the observer. Thus, the intensity profile can be interpreted as a superposition of particles continuously accelerated at the outward propagating shock and their subsequent propagation through interplanetary space. Since the acceleration efficiency of the shock changes as it propagates outward and the magnetic connection (see cobpoint) of the observer to the shock moves eastward along the shock front into regions of different acceleration efficiency, for different locations of the observer with respect to the shock different intensity profiles result. Therefore, in the MeV range, the appearance of the energetic particle event does not depend on local geometry but on the location of the observer relative to the shock. This dependence, of course, is modified by the characteristics of the shock, in particular its ability to accelerate particles and the radial and azimuthal variations of this acceleration efficiency.

energy   Work, or the ability to do a particular amount of work. *Energy* is usually categorized as either kinetic (the energy of actual mass motion), or potential (stored energy, which can be used eventually to cause mass motion). Thermal energy is recognized as kinetic energy of molecular or atomic structures, i.e., it is motion on the small scale. Thermal energy can be extracted from heated systems provided there is a difference of the average molecular or atomic kinetic energy. Such a difference is equivalent to a difference in temperature. Potential energy exists in many forms: gravitational (a rock poised at the top of a hill), electrical (an electron situated at the negative terminal of the battery). In classical electromagnetism, there is an energy associated with static electric and magnetic fields in vacuum: Energy $= \frac{1}{2}\epsilon_0 E^2 + \frac{1}{2}B^2/\mu_0$, where ϵ_0 is the permittivity and μ_0 is the magnetic susceptibility. The presence of polarizable materials modifies these expressions by replacing the vacuum ϵ_0 by a permittivity ϵ specific to the material, and by replacing the vacuum μ_0 by a value μ specific to the material. The units of energy (and work) in metric systems are ergs: 1 erg $= 1$ dyne $\cdot$ cm (the work done by moving a force of 1 dyne through a distance of 1 cm), and Joules: 1 Joule $= 10^7$ ergs. Since the unit of power is watts: 1 watt $= 1$J/sec, derived units of energy are often in use. For instance, the kilowatt hour $= 10^3$ watts $\times$ 3600 sec $= 3.6 \times 10^6$ J. Another energy unit is the calorie, defined as the energy required to raise the temperature of 1 gm of water $1°$K from a temperature of $15°$C $= 288.15$K; 1 calorie $= 4.18674$ J. The dieting "Calorie," properly written capitalized, is 1,000 calories. In the British system, the basic unit of work is the foot-pound. The British system unit analogous to the calorie is the British thermal unit (Btu), approximately the energy to raise 1 lb of water $1°$F. The precise relation is 1 Btu $= 251.996$ calories.

energy-containing scale   As the energy content at scales larger than the system and below the Kolmogorov scale vanishes, the turbulent kinetic energy spectrum has a maximum. The range of the spectrum that contains the significant part of the turbulent energy is referred to as the *energy-containing scale*. The aim of eddy-resolving models is to describe the turbulent flow down to the resolution of the energy-containing scales in order to capture most parts of the kinetic energy of the system.

energy conversion efficiency   In oceanography, the rate of chemical energy accumulation per unit volume divided by the rate of absorption of light energy by phytoplankton per unit volume; it is linearly related to quantum yield.

energy grade line A visual representation of the energy in a flow. Indicates the sum of the velocity head, $V^2/2g$, elevation, and pressure head, p/γ, for the flow, where V is the flow speed, g is the acceleration of gravity, p is pressure, and γ is the unit weight (weight per unit volume) of the fluid.

energy-momentum relations — special relativity In special relativity physical laws must be the same in reference systems moving uniformly with respect to each other; that is, they must be invariant under Lorentz Transformations. In special relativity time is not an absolute variable, and therefore special relativity is mathematically described through a four-dimensional spacetime with time as the first coordinate in addition to the three spatial coordinates. In order for the Lorentz electromagnetic force to be incorporated into a law of mechanics that is invariant under Lorentz transformations, and for the mechanics law to also reduce to Newton's law at low velocities, a four-component relativistic momentum vector is defined such that the first component equals the energy of a given particle and the other three components equal the momentum components of the particle. Since the four-vector scalar product is invariant under Lorentz transformations, one obtains the relativistic relationship between the energy and the momentum of a particle by calculating the four-vector scalar product of the four-vector momentum with itself.

energy per unit length (cosmic string) In the framework of a cosmological model with the generation of topological defects, a cosmic string is an approximation of a vacuum vortex defect in terms of a line-like structure, confined to a two-dimensional world sheet. For a complete macroscopic description (as opposed to microscopic, in terms of relevant fields like the Higgs field and other microscopic fields coupled to it) we need to know quantities such as the string tension T and the *energy per unit length* U (often denoted μ in the literature). For a microscopic model, specified by its Lagrangian, we can compute its energy-momentum tensor $T^{\mu\nu}$ by standard methods. Given the cylindrical symmetry of the string configuration, the energy per unit length is calculated as

$$U = 2\pi \int r\, dr\, T^{tt} ,$$

where T^{tt} is the time-time component of the energy-momentum tensor. *See* equation of state (cosmic string), Goto–Nambu string, tension (cosmic string), wiggle (cosmic string).

enthalpy An extensive thermodynamic potential H given by

$$H = U - PV ,$$

where U is the internal energy, P is the pressure, and V is the volume of the system. The change of the *enthalpy* is the maximum work that can be extracted from a reversible closed system at constant P. For a reversible process at constant S and P, work stored as enthalpy can be recovered completely.

entrainment Jets and plumes, moving through fluid at rest, have the tendency to entrain ambient fluid into the flow. The rate $\partial Q/\partial x$ [m^{2}s^{-1}], at which the volume flow Q [m^{3}s^{-1}] increases per unit distance x [m], is called the *entrainment* rate. The most customary parameterization of the entrainment rate is $\partial Q/\partial x = E R u$ (Morton, 1959), where R and u are circumference and velocity of the flow Q. The non-dimensional proportionality factor E is called the entrainment coefficient, which increases as the gradient Richardson number Ri decreases. This implies that entrainment is more efficient for large velocity differences and small density differences. Often entrainment involves the transport of one substance in another, such as suspended particles in a current, or parcels of moist air in dry winds.

entropy That thermodynamic potential defined by the exact differential

$$dS = dQ/T ,$$

where dQ is a reversible transfer of heat in a system, and T is the temperature at which the transfer occurs. *Entropy* is conceptually associated with disorder; the greater the entropy the less ordered energy is available.

environmental lapse rate   To a cloud or a rising parcel of air, the actual variation rate with height of temperature and moisture conditions in the atmosphere surrounding it. The overall average rate is a decrease of about 6.5 K/km, but the rate varies greatly from region to region, air stream to air stream, season to season. An inverse environmental lapse rate is a negative environmental lapse rate (temperature increase with height).

eolian   Refers to wind-related processes or features. Wind is created when different regions of an atmosphere experience temperature and/or pressure differences, causing the atmosphere to move in an attempt to average out these differences. Wind is a major agent of erosion, reducing rocks to sand and dust and transporting the material between locations. Eolian erosional features include yardangs and ventifacts. Eolian depositional features include dunes, ergs, and loess. Eolian material can be transported by saltation, suspension, traction, or impact creep.

eon   Often used in place of 1 billion (10^9) years.

EOP   *See* earth orientation parameters.

Eötvös experiment   Baron Loránd Eötvös measured the dependence of gravitational acceleration on the chemical composition of matter. He used a torsion pendulum to measure the difference δg of the acceleration of samples at the end of the two arms. By 1922, these measurements established to the precision $\delta g/g \sim 10^{-9}$ that the acceleration is independent of the chemical composition. (More recent measurements by Dicke and Braginski have improved this precision to $\sim 10^{-12}$). The Eötvös experiment provides the experimental support for the geometrical nature of the gravitational interaction, as predicted by the theory of general relativity.

epeiric sea   An inland sea with limited connection to the open ocean that is shallow, typically with depths less than 250 m.

ephemeris   A book or set of tables predicting the location in the sky of planets and satellites.

ephemeris time (ET)   Uniform time based on seconds having the duration they did on 0.5 January 1900, since seconds based upon the Earth's rotation generally lengthen with time (*see* Universal Time). Prior to 1972, accurate measurements of the motions of the moon and planets provided the best measure of time, superior to any laboratory clock. *Ephemeris Time* was thus defined as a continuous measure of time based on these motions, although, technically, its definition rested on Simon Newcomb's theory of the motion of the sun. Ephemeris Time, which came into usage in 1956, was replaced by Dynamical Time in 1984. The two systems coexisted from 1977 to 1984. Ephemeris Time was used, in the obvious way, as the basis of the ephemeris second, which is the direct ancestor of the SI second; i.e., the SI second was matched as well as possible to the ephemeris second.

epicenter   Point on the Earth's surface above the initial earthquake rupture.

epicentral distance   The distance of an observation point at the Earth's surface from the epicenter of an earthquake. It is expressed either as a length S along the great circle path between the observation point and the epicenter or as an angle $\Delta = S/R$, where R is the average radius of the earth.

epicycle   Secondary circle along which the planet moves in the Ptolemaic (geocentric) description of the solar system. The *epicycle* center moves along a larger circle called the deferent. This combination can explain the retrograde motion of the planets. *See* deferent, geocentric, retrograde motion.

Epimetheus   Moon of Saturn, also designated SXI. Discovered by Walker, Larson, and Fountain in 1978 and confirmed in 1980 by Voyager 1. It is a co-orbital partner of Janus. Its orbit has an eccentricity of 0.009, an inclination of $0.34°$, and a semimajor axis of 1.51×10^5 km. Its size is $72 \times 54 \times 49$ km, its mass, 5.6×10^{17} kg, and its density 0.7 g cm^{-3}. It has a geometric albedo of 0.8, and orbits Saturn once every 0.694 Earth days.

epoch   An arbitrary fixed instant of time used as a chronological reference for calendars, celestial reference systems, star catalogs, or orbital motions. Prior to 1984, star catalog coordinates were commonly referred to the mean equator and equinox of the beginning of the Besselian year (*see* year). Thereafter, the Julian year (365.25 days) has been used.

equant   A point in the Ptolemaic description of orbits about which the center of the epicycle moved at constant angular velocity as it moved along the deferent. The *equant* was equidistant on the opposite side of the center of deferent from the Earth. *See* deferent, epicycle.

equation of continuity   Any of a class of equations that express the fact that some quantity (mass, charge, energy, etc.) cannot be created or destroyed. Such equations typically specify that the rate of increase of the quantity in a given region of space equals the net current of the quantity flowing into the region. Thus, the change of a property ϵ (e.g., mass) inside a volume element results from the convergence of a flux $\vec{C}(\epsilon)$ across the boundaries of the volume and from sources and sinks $S(\epsilon)$ inside the volume:

$$\frac{\partial \epsilon}{\partial t} + \nabla \cdot \vec{C}(\epsilon) = S(\epsilon) .$$

The most common application is the conservation of mass

$$\frac{\partial \rho}{\partial t} = -\nabla \cdot (\rho \vec{u}) = -\nabla \cdot \vec{j}$$

with ρ being the density and $\vec{j} = \rho \vec{u}$ the mass current. Using the total derivative instead of the partial one, this can be written as

$$\frac{d\rho}{dt} = \frac{\partial \rho}{\partial t} s + \vec{u} \cdot \nabla \rho = -\rho \cdot \nabla \vec{u} .$$

equation of state   An equation relating the pressure, volume, and temperature (or some other set of thermodynamical variables) of a substance or system.

equation of state (cosmic string)   For an accurate macroscopic description of a vacuum vortex defect as a cosmic string, one needs the string

equations of state, which give both the string energy per unit length U and its tension T (in general both are variable), and also the relation between them. The simplest example of *equation of state* is $T = U =$ const, corresponding to the Goto–Nambu string.

Also conducting cosmic strings can be described by fitting analytical expressions to the numerical solutions of the classical equations of motion for the microscopic fields. The best analytical expression available for the Witten model is known as the Carter–Peter equation of state. *See* Carter–Peter model, conducting string, energy per unit length (cosmic string), Goto–Nambu string, tension (cosmic string), wiggle (cosmic string).

equation of state of pure water   Potential density of pure water is a function of temperature only. For actual values see Chen and Millero (1986).

equation of time   The hour angle of the true (observed) sun minus the hour angle of the (fictitious) mean sun of mean solar time.

equator   The locus of a point traveling around the globe that is equidistant from the north and south poles forms the *equator*.

equator, geomagnetic   (**1.**) The line around the Earth, along which the vertical component of the magnetic field is zero. Approximately equidistant from both magnetic poles, but distorted from that shape by non-dipole harmonics of the Earth's magnetic field.

(**2.**) A term sometimes applied to the equatorial surface of the magnetosphere.

equatorial anomaly   Although processes in the ionospheric F region are complex, it is still reasonable to expect the maximum daytime ionization to fall roughly near to the location of the overhead sun. However, in the equatorial region, and most noticeable near equinox, the daytime maximum in ionization splits forming crests 20° to 30° to the north and south of the dip equator, leaving an ionization trough at the dip equator. This is called the *equatorial anomaly*. It is formed by a process analogous to a fountain, ionization near the dip equator being driven upward by the local E-region horizontal dynamo electric fields and the local horizontal magnetic field. As the ionization is lifted to higher altitudes, it encounters magnetic field lines connected to higher latitudes and falls back down, forming the crests. Generally, the anomaly is most pronounced around 1400 LT and then declines steadily until it disappears around 2000 LT. However, this behavior can change greatly with season, phase of the solar cycle, and diurnally. *See* equatorial electrojet, geomagnetic dip equator.

equatorial bulge   The rotation of the earth causes the equatorial radius to be larger than the polar radius by about 21 km. This results in an *equatorial bulge.*

equatorial cold tongue   A strip of cold surface water that extends like a tongue along the equator from the southeastern boundaries of the Pacific and Atlantic Oceans. The southeasterly trade winds are its direct cause, pushing the thermocline close to the surface and inducing upwelling that pumps cold thermocline water to the surface. The contrast between the warm western and cold eastern oceans in turn enhances the easterly trade winds on the equator. The surface water temperatures in the cold tongue peak in March-April and reach the minimum in July-September, despite negligible seasonal change in solar radiation at the top of the atmosphere on the equator. The water in the *equatorial cold tongue* is rich in nutrients and carbon dioxide, making it the largest natural source of atmospheric CO_2.

equatorial convergence zone   Also called intertropical convergence zone, i.e., ITCZ. Convergence zone at the tropospheric low level wind fields over tropics region, usually from $5°S - 10°S$ to $5°N - 10°N$. From cloud pictures, ITCZ appears as a long cloud band and consists of many cloud clusters with west–east direction and almost surrounds the Earth. This cloud band width is about 200 to 300 km. In some regions and some times there are two cloud bands located on the south and the north sides of the equator, the so-called "double ITCZ."

On average, in January, the ITCZ is located at about the equator and 15°S; in April at about

4°N and 5°S; in July at 20°N and 2°S; and in November at 10°N and 5°S. The ITCZ has both active and inactive periods of short term variation. During the active period, it is most distant from the equator and has strong convective activity, with a series of large scale cloud clusters and frequent cyclonic vortices. During the inactive period, it is near the equator and has weakly developed clouds, which are mainly smaller scale scattered cumulus.

The ITCZ is the main region with ascending air in tropics, in which the deep cumulus convective activities are strong and frequent, especially during the active period. At times some of the cumulus can reach to the tropopause. In strong convergence regions, tropical disturbances are often created in the ICTZ. These tropical disturbances may develop into tropical cyclones in suitable environmental conditions. About 70 to 80% of tropical cyclones come from tropical disturbances developed in the ITCZ.

equatorial easterlies   At the south side of the subtropical high in the high level of the troposphere, there are strong easterlies located at about $5 - 20°N$, $30 - 130°E$. The maximum speed is often located at $5 - 10°N$ in the Arabian Sea region at about the 150 hPa level and is about 35 to 50 m/s. Along the Arabian Sea to the west, the *equatorial easterlies* move south gradually and reach the equator at North Africa. Over its entrance region (east of 90°E) there is a direct meridional circulation, while over its exit region (west of 60°E) there is an indirect meridional circulation. Through these meridional circulations, equatorial easterlies are connected to the southwest monsoon system at lower troposphere.

equatorial electrojet   A current flowing above the Earth's magnetic equator, inferred from magnetic disturbances on the ground. It flows on the sunlit side of Earth, from west to east.

equatorial Kelvin wave   Oceanic Kelvin wave that propagates along the equator. For the first baroclinic mode in the ocean, a Kelvin wave would take about 2 months to cross the Pacific from New Guinea to South America.

equatorial surface of the magnetosphere   A surface with respect to which the Earth's magnetospheric field and its trapped plasma exhibit approximate north-south symmetry. It is usually defined as a collection of the points of weakest magnetic intensity $|\mathbf{B}|$ on "closed" field lines which start and end on the surface of Earth. In the plasma sheet the surface is often called the "neutral sheet" and it undergoes a periodic deformation ("warping") due to the daily and annual variation of the geomagnetic tilt angle ψ. *See* minimum-B surface.

equatorial undercurrent   A strong narrow eastward current found in the region of strong density gradient below the ocean mixed layer and with its core close to the equator. Its vertical thickness is around 100 m and its half-width is a degree of latitude. The maximum current is typically 1 ms^{-1}. This eastward current is a major feature of the equatorial ocean circulation, particularly in the Pacific and Atlantic Oceans.

equatorial upwelling   The resultant volume transport in the Ekman layer is at right angles to and to the right of the wind direction in the northern hemisphere. If the wind is easterly, water in the Ekman layer will tend to move away from the equator on both sides of the equator and will be replaced by water moving upward from below the layer. This is called *equatorial upwelling*. If the wind is westerly, water in the Ekman layer will tend to move toward the equator in both sides of the equator and repel the water below the layer. This is called equatorial downwelling.

equatorial waveguide   An oceanic or atmospheric wave that is confined to propagate near the equator due to the vanishing of the Coriolis force at the equator. While the conditions for geostrophic balance theoretically fail at the equator, in practice any mass crossing the equator will be influenced by the Coriolis force on either side. The force turns such motion back towards the equator, thus creating a trap or a waveguide.

equatorial waves   A class of equatorially trapped wave solutions first obtained by Taroh Matsuno in 1966, consisting of the Rossby, in-

ertial gravity, mixed Rossby-gravity, and Kelvin waves. The Rossby and inertial gravity waves are waves of rotating fluid, having their counterparts in off-equatorial regions. The Kelvin and mixed Rossby-gravity waves are unique to the equatorial waveguide arising from the singularity of the equator where the Coriolis parameter $f = 2\Omega \sin\theta$ vanishes. Here Ω is the angular velocity of the Earth's rotation around the north pole and θ is latitude.

equilibrium   In mechanics, a configuration in which the total force on a system vanishes, so if placed in such a configuration the system remains in it.

equilibrium beach profile   A theoretical beach profile shape that results if wave conditions and water level are held constant for an indefinite period. It has a monotonic, concave-up shape.

equilibrium range   High-wavenumber part of the turbulent kinetic energy spectrum that includes the inertial subrange and the dissipation range. The turbulence at these wavenumbers is nearly isotropic and the shape of the spectrum at these wavenumbers does not depend on the amount of kinetic energy present at larger scales or the size of these energy containing scales. Kinetic energy is merely transferred by inertial forces through the inertial subrange until the energy is dissipated into heat at the Kolmogorov microscale.

equilibrium space-times   Stationary electrovacuum space-times describing the external fields of an arbitrary array of electrically charged, massive sources. Equilibrium is achieved by the balance of gravitational and electromagnetic forces. The static *equilibrium space-times* were found by S.D. Majumdar and A. Papapetrou (1947). *See* Israel–Wilson–Perjés space-times.

equilibrium tide (gravitational tide)   A hypothetical ocean tide that responds instantly to tide producing forces, forming an equilibrium surface. The effects of friction, inertia, and the irregular distribution of land mass are ignored.

equilibrium vapor pressure   The vapor pressure in an equilibrium state system with two or more phases of water. If the system consists of vapor and pure water (ice), the *equilibrium vapor pressure* is the water (ice) surface saturated vapor pressure, and it is a function of temperature only. If the system consists of vapor and waterdrop or solution, due to curvature effects or solution effects, the equilibrium vapor pressure will be higher (vapor-waterdrop system) or lower (vapor-solution system) than the equilibrium vapor pressure in vapor-water system.

equinox   Dates on which the day and night are of equal length. Dates on which the sun is at (one of two) locations of the intersection of the ecliptic and the celestial equator. Because the Earth poles are inclined by $23°27'$ to its orbital plane, Northern and Southern hemispheres typically receive different daily periods of sunlight. At the *equinoxes* the location of the planet in its orbit is such that the sun strikes "broadside," equally illuminating the two hemispheres. *See* autumnal equinox, vernal equinox.

equivalence principle   A principle expressing the universality of gravitational interactions. It can be expressed in various degrees of "strength".

Weak equivalence principle: The motion of any freely falling test particle is independent of its composition or structure. (Here "test particle" means that the particle's gravitational binding and gravitational field are negligible.)

Medium strong equivalence principle: Near every event in spacetime, in a sufficiently small neighborhood, in every freely falling reference frame all nongravitational phenomena are exactly as they are in the absence of gravity.

Strong equivalence principle: Near every event in spacetime, in a sufficiently small neighborhood, in every freely falling reference frame all phenomena (including gravitational ones) are exactly as they are in the absence of external gravitational sources.

Various candidate descriptions of gravity obey one or the other of these laws. General relativity, which is very well verified in terms of large scale phenomena, obeys the strong equivalence principle. *See* Eötvös experiment.

Eratosthenes   Mathematician, astronomer, geographer, and poet (third century BC). Measured the circumference of the Earth with considerable accuracy ($\sim 15\%$ above actual value) by determining astronomically the difference in latitude between the cities of Syene (now Aswan) and Alexandria, Egypt. He also compiled a star catalog and measured the obliquity of the ecliptic.

E region   The portion of the ionosphere formed between approximately 95 and 130 km above the surface of the Earth. It lies between the D and F regions. The *E region* is directly controlled by the sun's ionizing radiation and the amount of ionization present in the E region is directly related to the amount of radiation present. Thus, the E region has a maximum in ionization at local noon ($\sim 10^{11}$ electrons m^{-3}), disappears shortly after sunset at E-region heights, and forms again at E-layer sunrise. It has a readily predicted behavior globally, diurnally, and seasonally that can be described by the sun's solar zenith angle. There is also a small solar cycle variation in the daytime electron densities. At night, a small amount of ionization is found at E-region heights and is thought to be due to solar radiation scattered around from the daylight hemisphere together with a contribution due to cosmic rays. *See* ionosphere, sporadic E, spread E.

erg   A cgs unit of energy equal to work done by a force of 1 dyne acting over a distance of 1 cm. 1 erg $= 10^{-7}$J; 10^7 erg s$^{-1} \equiv 1$ Watt.

ergodic motion   In mechanics, motion such that the trajectory of any given initial condition passes through every point of the surface in phase space having the same total energy as the initial condition.

ergoregion   In general relativity, in spinning black hole spacetimes (Kerr black holes) the region in which particle kinematics allows rotational energy to be extracted from the black hole into energy of orbits that can reach infinity. *See* Kerr black hole.

ergs   A large region covered by sand-sized (1/16 to 2 mm in diameter) material. *Ergs are* sometimes called sand seas since they represent vast regions of sand. Most ergs are located in desert basins downwind from terrain which is experiencing high amounts of erosion. Various types of dunes occur within the ergs. Ergs are found on Earth and around the north polar cap on Mars.

Ernst equation (1967)   The complex partial differential equation

$$\mathcal{R}eE\Delta E = \nabla E \cdot \nabla E$$

in a Euclidean 3-space introduced by F.J. Ernst. The axially symmetric solutions $E = E(x^1, x^2)$ represent the metric potentials of stationary, axisymmetric vacuum gravitational fields. A second form

$$(\bar{\xi}\xi - 1)\Delta\xi = 2\bar{\xi}\nabla\xi \cdot \nabla\xi$$

is obtained by the substitution $E = (\xi - 1)/(\xi + 1)$. The *Ernst equation* has been found to describe additionally various physical systems, such as colliding plane waves in general relativity, monopoles in SU(2) gauge theory, and states of Heisenberg ferromagnets. The symmetries of the equation and solution generating techniques have been extensively studied.

erosion   Rock is destroyed by weathering. Mechanical weathering breaks up rock into small particles which can be transported hydrologically. In chemical weathering the rock is dissolved by water. In either case, the removal of the rock is *erosion*.

ERP   Earth rotation parameters.

Ertel potential vorticity   *See* potential vorticity.

eruption   Volcanos are subject to periodic *eruptions* during which molten rock (magma) and/or volcanic ash flows to the Earth's surface. Eruptions can occur for months between periods of quiescence.

eruptive prominence   Solar prominences can become activated and exhibit several types of large-scale motion. The prominence may grow and become brighter with a corresponding in-

crease in helical motion or flow along the prominence axis. This type of activation can lead to an eruption, especially if it exceeds a height of 50,000 km. In this case, it rises as an *erupting prominence* and eventually disappears. The eruption of a quiescent prominence (disparition brusque) is a slow process lasting several hours. The eruption of an active region prominence is much more rapid taking about 30 min.

escape velocity   In Newtonian physics, under gravitational interactions, the gravitational field contributes negatively to the total energy of a particle. The *escape velocity* is that speed giving a zero total energy (kinetic + gravitational potential). A particle moving without dissipation at or exceeding the escape velocity will escape to infinity.

eskers   Long narrow ridges of sand and gravel deposited in the middle of glaciers. They have stream-like shapes and are believed to form from meltwater streams flowing in tunnels beneath a melting glacier. Sinuous ridges in the high latitude regions of Mars have been suggested to be *eskers*.

estuary   A sheltered region that lies inshore of a sea and is subject to tidal action. Generally has a reduced salinity, compared to the adjoining sea, as a result of river and other freshwater inputs.

ET   *See* ephemeris time (ET).

eternal black hole   A black hole that was not formed in the collapse of matter, but was present *ab initio* as a stable topological structure in spacetime.

Eudoxos of Cnidus   Greek geometer and astronomer, born in Cnidus, Asia Minor approximately 408 B.C., died about 353 B.C. Eudoxos established the foundation of geometrical principles developed by Euclid, and applied the subject to the study of the moon and the planets.

Euler equations   The equations of motion for an inviscid fluid. Equivalent to the Navier–Stokes equations if viscosity is taken as zero.

Represents Newton's 2nd Law for an inviscid fluid.

Eulerian   A term used to denote a description of fluid behavior which involves description of fluid flow parameters (particularly velocity) at fixed points in space. The alternative is a Lagrangian description, which essentially involves describing the behavior of selected fluid particles as they move through space.

Eulerian coordinates   In hydrodynamics, physical parameters such as pressure, fluid velocity, and density can be expressed as functions of positions in space and time; thus, a coordinate system fixed to an external reference frame, in which physical phenomena, for instance hydrodynamical flow, move through the hydrodynamical grid. Named after Leonhard Euler (1707–1783). *See* Lagrangian coordinates.

Eulerian representation   Description of a phenomenon relative to a framework fixed in space. Measurements in moorings are typical applications of Eulerian-type observations. *See* Lagrangian representation.

Eulerian velocity   Velocity measured from a fixed point or set of points (i.e., by a moored current meter). *See also* Lagrangian velocity.

Euler, Leonhard   Mathematician (1707–1783). Made contributions in the areas of algebra, theory of equations, trigonometry, analytic geometry, calculus, calculus of variations, number theory, and complex numbers. He also made contributions to astronomy, mechanics, optics, and acoustics.

Euler pole   Euler's theorem states that a movement on a sphere may be represented as a rotation about an axis. This holds not only for the movement of a point on a sphere but also the movement of a continuous undeformable patch on a sphere. This turns out to be important for the theory of plate tectonics, which models the surface of the planet as a set of rigid plates in relative motion. Therefore, an axis may be found for each plate relative to some given reference frame. These axes are known as *Euler poles*. In particular, by considering the motion of one

plate relative to another plate, one may find the pole of relative rotation. If two plates meet at a mid-ocean ridge, then the ridge will often be fractured by transform faults oriented perpendicular to a line running toward the pole of rotation between the two plates. Because velocity about the pole depends on distance from the pole, the velocity between adjoining plates will vary along the boundary between the plates unless the boundary happens to lie on a cylinder centered on the pole. This means that, for example, subduction may be much faster at one end of a trench than the other.

Euler potentials   Two scalar functions of position (α, β) used for describing a magnetic field **B** and satisfying $\mathbf{B} = \nabla\alpha \times \nabla\beta$. Their use is equivalent to that of a vector potential $\mathbf{A} = \alpha\nabla\beta$ (of no particular gauge, non-covariant) and is useful because the points of any magnetic field line share the same value of α and β. As long as the field inside Earth is excluded, (α, β) in the Earth's vicinity are usually unique to one field line (not valid in more general geometries) and this allows using them in mapping field lines and as variables in the theory of plasma convection by electric fields in the magnetosphere. Field line motion can also be described by *Euler potentials*. In toroidal geometries, (α, β) are generally multiply valued.

euphotic depth   In oceanography, the depth to which significant phytoplankton photosynthesis can take place; typically taken to be the depth at which photosynthetically available radiation falls to 1% of its value just below the surface [m].

euphotic zone   In oceanography, the water layer in which significant phytoplankton photosynthesis can take place; typically taken to be the layer down to which photosynthetically available radiation falls to 1% of its value just below the sea surface.

euphotic zone midpoint   In oceanography, the layer at which photosynthetically available radiation falls to 10% of its value just below the sea surface.

Europa   Moon of Jupiter, also designated JII. Discovered by Galileo in 1610, it is one of the four Galilean satellites. Its orbit has an eccentricity of 0.009, an inclination of 0.47°, a precession of 12.0° yr^{-1}, and a semimajor axis of 6.71×10^5 km. Its radius is 1569 km, its mass 4.75×10^{22} kg, and its density 2.94 g cm^{-3}. It has a geometric albedo of 0.64, and orbits Jupiter once every 3.551 Earth days. *Europa's* surface is covered with ice, and it appears to have liquid water under the ice.

eustatic sea-level   The worldwide change of sea level elevation with time. Specifically, *eustatic sea-level* describes global sea level variations due to absolute changes in the total quantity of water present in the oceans. These changes are distinct from redistributions of water from circulation, tides, and density changes. Diastrophic eustatism is a change of sea level resulting from variation in capacity of the ocean basins, through the deeping of basins or raising of continents. Glacial eustatism refers to changes in sea level produced by withdrawal or return of water to the oceans.

eutrophic water   Water with high phytoplankton biomass; chlorophyll *a* concentration exceeds 10 mg m^{-3}.

evaporation (*E*)   The process by which water moves from the liquid state to the vapor state follows Fick's first law and can be written in finite difference form as $E = K_E u_a(e_s - e_a)$, which shows that E increases with increasing wind speed (u_a) and with the difference between the vapor pressure of the evaporating surface (e_s) and the overlying air (e_a); K_E is a coefficient that reflects the efficiency of vertical transport of water vapor by turbulent eddies of the wind. Sublimation is the process by which water moves from the solid state (snow and ice) to the vapor state.

evapotranspiration (*ET*)   The sum of all processes by which water changes phase to vapor and is returned to the atmosphere, including sublimation from snow and ice, evaporation from lakes and ponds, and transpiration by vegetation. Actual evapotranspiration (AET) is the amount of *evapotranspiration* that actually occurs under given climatic and soil-moisture conditions. Potential evapotranspiration (PET) is the amount of evapotranspiration that would occur under given climatic conditions if soil moisture was not limiting. Moisture deficit is PET-AET and is the amount of irrigated water added to maximize crop yield.

evening cloud (Mars)   *Evening clouds of Mars* appear near the evening limb or terminator in spring to summer of the northern hemisphere. Similar to morning clouds, evening clouds show a diurnal variation. In their most active period from late spring to mid summer, evening clouds appear in early to mid afternoon, increasing their brightness as they approach the evening limb. They form the equatorial cloud belt together with morning clouds. *See* morning cloud, afternoon cloud.

event horizon   A null hypersurface in space-time separating observable events from those that an observer can never see. A uniformly accelerated observer in Minkowski space-time experiences an event horizon. An *absolute event horizon* in a weakly asymptotically simple space-time is the boundary of the causal past of future null infinity. A black hole is a domain of space-time separated from null infinity by an absolute event horizon, which is the surface of the region surrounding a black hole where the gravitational field is strong enough to prevent the escape of anything, even light. It is a one-way membrane in the sense that no light signal can be transmitted across it from the interior to the external world. *See* black hole.

Evershed effect   The strong horizontal outflow observed in the penumbra of sunspots, extending several sunspot diameters from the umbral edge. This outflow has speeds of ~ 2 to 6 kms^{-1} as detected in photospheric and chromospheric lines. The Evershed flow is concentrated within dark penumbral filaments that are thought to be surface manifestations of convective rolls aligned with nearly horizontal field in the penumbra.

evolutionary track   As a star ages and moves off the zero age main sequence, its effective temperature (T_{eff}) and luminosity (L) change.

The path of a star in the T_{eff}, L plane (the Hertzsprung–Russell, HR, diagram) as those physical parameters change is called the star's *evolutionary track*. This track is primarily a function of the star's initial mass. Collections of these tracks derived from theoretical stellar evolution models are often plotted on an HR diagram along with data from a collection of stars to estimate initial masses and ages.

exact solution   (of Einstein's equations) A solution of Einstein's equations, i.e., a metric. The term *exact solution* is used only when it is necessary to stress that the metric was derived from Einstein's equations by strict mathematical procedures. The opposites of *exact* are an approximate (i.e., a perturbative solution) and a numerical solution. This opposition is irrelevant for several purposes; for example, all observational tests of the relativity theory necessarily involve approximations, and numerical calculations are often performed on top of exact solutions to obtain quantitative results. However, some theoretical problems (among them cosmic censorship and the existence of singularities) have been extensively studied within the framework of exact solutions.

excitation temperature   In a given object, the constituent atoms and molecules are in quantum mechanically excited states, and the ratio of their numbers in the ground state to the excited state is characterized by an *excitation temperature,* such that

$$\frac{N_u}{N_l} = \frac{g_l}{g_u} e^{-h\nu/kT_{ex}}$$

where N_u and N_l are the number per volume in the upper and lower states, respectively, and g_l and g_u are the statistical weights. The excitation temperature T_{ex} characterizes this distribution for a given upper and lower state. If the atoms and molecules are in thermodynamic equilibrium (collision dominated), then the kinetic temperature, which describes a distribution in speed of the atoms and molecules, is equal to the excitation temperature.

exosphere   The outermost part of the Earth's atmosphere, beginning at about 500 to 1000 km above the surface. It is characterized by densities so low that air molecules can escape into outer space.

exotic terrane   Continental deformation is partly achieved through folding, which preserves the juxtaposition of stratigraphic layers, but also through faulting. A terrane is a block of continent bounded by faults with a stratigraphy that is internally coherent but distinct from that of surrounding blocks. Mountain chains may be composed of several such blocks. A "suspect terrane" is a terrane whose geology appears to be unrelated to the surrounding rock. If it can be shown (through paleomagnetic methods, for example) that the rock in the terrane originated far away from the surrounding rock, then the terrane is known as an *exotic terrane*. This may happen if part of a continent is sliced away by a long strike-slip fault (such as the San Andreas fault) and is later attached to the continent or another continent elsewhere, or if an island arc or other oceanic landform collides with and is engulfed by a continent.

expansion   The mathematical quantity (a scalar) that determines the rate of change of volume of a given small portion of a continuous medium (*see also* acceleration, kinematical invariants, rotation, shear). An observer placed in a medium in which acceleration, rotation, and shear all vanish while *expansion* is nonzero would see every point of the medium recede from (or collapse toward) him/her with the velocity v proportional to its distance from him/her, l, so that $v = \frac{1}{3}\theta l$. The factor θ is the expansion ($\theta > 0$ for receding motion, $\theta < 0$ for collapse). (In astronomy, $H = \frac{1}{3}\theta$ is called the Hubble parameter.) A medium with the above-mentioned properties is used as a model of the universe in the Friedmann–Lemaître cosmological models. In these models, the expansion is uniform, i.e., at any given time it is the same at all points of space and can only vary with time. In general (in particular in inhomogeneous models of the universe) the expansion can be different at every point of space. In a perfect fluid or dust such a nonuniform expansion must be accompanied by shear.

expansion of the universe   The recession of distant galaxies from our galaxy. The evidence for the expansion of the universe is the redshift of light received from distant sources, first discovered by E.P. Hubble in 1929 (*see* Hubble parameter). Expansion, in turn, is the evidence that the universe is evolving (*see* steady-state models) from a state of higher density toward lower density. Extrapolation of this phenomenon backward in time leads to the notion of the Big Bang. The rate of expansion depends on the value of the cosmological constant Λ. With $\Lambda = 0$, the rate of expansion is constantly being decreased by gravitational attraction between any two massive objects. The Friedmann–Lemaître cosmological models predict the following possible scenarios for the future fate of the universe when $\Lambda = 0$. If the mean matter-density in the universe, $\bar{\rho}$, is at present lower than the critical density $\rho_0 \approx 10^{-29} \text{g/cm}^3$, then the decelerating force is small and the universe will go on expanding forever. If $\bar{\rho} > \rho_0$, then the expansion will be completely halted at a certain moment in the future and followed by collapse toward the final singularity; this collapse is the time-reverse of the Big Bang. The determination of the present value of $\bar{\rho}$ is thus of crucial importance and is currently one of the main (unsolved) problems of cosmology. Certain vigorously advertised theories are critically dependent on $\bar{\rho}$ being equal to ρ_0, but unquestionable observations imply only that $\bar{\rho} \geq 0.2\rho_0$.

experimental craters　A means of observing craters as they form, during which the experimental conditions are changed in order to determine the effects of different parameters. Results show that the impact process consists of three formation stages: compression, excavation, and modification, and that the strength of the impact target influences crater shape and diameter. For example, a low target strength reduces the threshold diameter for the formation of complex craters.

Experimental studies are usually of two types. In the first, a small projectile is fired (at several kilometers per second) from a light gas gun toward a target in a vacuum chamber. Very high-speed cameras photograph the craters as they form. Scaling over several orders of magnitude is then applied to extrapolate the results (this introduces some error since crater formation is dependent on size). In the second method,

explosive charges (such as nuclear or chemical explosives) are used. This method reduces the scaling problem but introduces a gas acceleration stage that does not exist in natural impacts and causes a slightly different final impact morphology.

extended object　Astronomical source that has an angular diameter larger than the resolving power of the instrument used to observe it.

extensions of space-times　The omission of some region from a space-time may be detected by the appearance of incomplete geodesics. These are geodesics for which the affine parameter does not assume all real values. Thus, an observer moving along an incomplete time-like geodesic will reach the boundary in a finite time. The extension of a space-time will produce a spacetime that contains the original spacetime (is isometric to it where they overlap). In the presence of singularities, inextendable geodesics will occur. *See* maximal extension of a space-time.

extinction and reddening　Extinction of light equally at all visible wavelengths can be caused by dust in interstellar clouds between the observer and the continuum source. Grains of all sizes contribute to absorption and scattering of light, resulting in a general obscuration of the continuum light. However, most interstellar dust grains are slightly smaller than visible wavelengths. Reddening is a selective extinction of bluer wavelengths due to scattering off of these small dust grains. The extinction of the bluer light causes the object to appear redder than it should compared with the predicted temperature based on its spectral type. Reddening increases the color index of a star (*see* color index, color excess).

extratropical storm　A storm that occurs when cold, denser air moving toward the equator meets the warmer, more humid air moving away from the equator. A circulation results that can cause high winds over a large area. It is more common in winter when temperature gradients between the equatorial regions and the higher latitudes are stronger.

extreme ultraviolet (EUV)   The portion of the electromagnetic spectrum in the wavelength range from approximately 100 to 1000 Å.

extreme ultraviolet imaging telescope (EIT) Telescope aboard the SOHO spacecraft which obtains images of the sun at extreme-ultraviolet wavelengths. The EIT is able to image the solar transition region and inner corona in four selected bandpasses in the EUV, namely, Fe IX/X at 171 Å, Fe XII at 195 Å, Fe XV at 284 Å, and He II at 304 Å. The EIT is used to image active regions, filaments and prominences, coronal holes, coronal "bright points", polar plumes, and a variety of other solar features using either full-disk or sub-field images.

eye (of a storm)   In strong tropical cyclones, the central region of the storm is often relatively calm, sometimes with a clear cylindrical appearance with blue sky overhead and storm clouds rising all around it. Winds are generally 10 knots or less, and no rain occurs. Sizes may range from 6 to 60 km in diameter, with most frequent size of order 20 to 40 km.

F

F1-layer   Relatively unstable ionospheric layer at a height of between about 140 and 200 km which fills the gap between the E-layer and the F2-layer. Ionization is due to EUV radiation. Formally, the *F1-layer* can be described by a Chapman profile.

F2-layer   Uppermost ionospheric layer at heights of between 200 and about 1000 km. In addition to local ionization and recombination, large-scale transport processes influence the chemistry and charge density. In addition, vertical transport leads to a separation of different particle species and charge states. Despite already reduced particle densities, electron densities are higher than in the other ionospheric layers. The dominant particle species is ionized atomic oxygen O^+. N^+ and N_2^+ are dominant in the lower parts, while in the upper part of the F2-layer ionized helium He^+ and atomic hydrogen H^+ become dominant.

Faber–Jackson law   An empirical relationship between the total luminosity and the central velocity dispersion of elliptical galaxies: $L \propto \sigma^4$, i.e., the galaxy luminosity is proportional to the fourth power of the velocity dispersion σ. This law was first discussed by S.M. Faber and R.E. Jackson in 1976. Since the velocity dispersion can be measured from the broadening of absorption lines in the galaxy spectrum, the *Faber–Jackson law* can, in principle, be used to determine the luminosity and, once the apparent magnitude of the galaxy is measured, to derive the distance of the galaxy. The Faber–Jackson law is analogous to the Tully–Fisher law for spiral galaxies: The stellar velocity dispersion substitutes the H I rotational width, since there is little atomic gas in early-type galaxies. *See* Tully–Fisher law, velocity dispersion.

facula   A bright region of the photosphere seen in white light at the solar limb. *Faculae* are the manifestations of the supergranule boundaries, seen at the limb.

fading   *Fading* is a common characteristic of radio propagation and is typified by aperiodic changes in the received signal amplitude and other signal characteristics. The depth of fading can be described by an amplitude probability distribution and the rapidity of the fading may be described by autocorrelation functions. These statistics are required to make allowances for fading in system design. Understanding the properties of the fading may also be important in constructing systems to capitalize or minimize the effects. Fading on ionospheric propagation paths can arise from sources that alter the properties of a single path (e.g., absorption, ionization changes resulting in pathlength changes or skip distance changes, ionization irregularities near the reflection point) and more commonly from multiple propagation paths (e.g., interference fading, polarization fading). Understanding the source of the fading may allow some control over its effects (e.g., skip fading may mean using a lower operating frequency; absorption fading may mean a higher frequency is required). *See* flutter fading, ionospheric radio propagation path, multipath fading.

failed arm   *See* aulacogen.

faint young sun paradox   As the sun ages, the fusion processes at its core gradually intensify, and models predict that the sun has increased brightness by about 35% over the last 4 billion years. The paradox is that one expects the Earth to have been completely icebound, but geological evidence shows liquid water at least that far into the past. A resolution has been proposed in terms of intense greenhouse effect, driven by ammonia and hydrocarbon haze, which could have maintained the surface above freezing.

Falkland current   An ocean current flowing northward along the coast of Argentina, between Argentina and the Falkland Islands.

fallout   Term describing airborne dangerous particles that deposit from the atmosphere after days, weeks, or longer in suspension in the air. Usually referring to radioactive particles which can be lofted into the air by nuclear explosions, or via leaks from nuclear power plants, or by fire associated with catastrophic failure in a nuclear installation.

fall speed (velocity)   The equilibrium speed reached by a sediment particle falling in a column of still water. Governed by the size, shape, and density of the particle, as well as any cohesive forces between particles, and the density and viscosity (and thus temperature) of the fluid. Used for classification of particle size and assessment of the mobility of the sediment.

Fanaroff–Riley (FR) class I and II radio galaxies   Lobe-dominated radio galaxies whose luminosity at 178 MHz is below (class I) or above (class II) 5×10^{32} ergs $s^{-1} Hz^{-1}$. In 1974, B.L. Fanaroff and J.M. Riley noted a dichotomy in the radio morphology of radio galaxies with a sharp threshold luminosity: FR I type galaxies, of lower luminosity, show smooth two-sided and poorly collimated jets and edge-darkened lobe structures. FR II type galaxies, of higher luminosity, show edge-brightened lobes, often with prominent hot spots at the inner end of the lobes, connected to the nucleus by pencil-like jets that are usually very faint, and in several sources not visible at all.

Faraday Cup   A detector for low-energy plasma, absorbing incident ions or electrons and measuring the rate at which their electric charge is deposited.

Faraday effect   The rotation of the plane of plane-polarized light by a medium placed in a magnetic field parallel to the direction of the light beam. The effect can be observed in solids, liquids, and gasses.

Faraday rotation   When propagating through the ionosphere, a linearly polarized radio wave will suffer a gradual rotation of its plane of polarization due to its interaction with the ionization in the ionosphere in the presence of the Earth's magnetic field. The magnitude of rotation depends on the electron density and the inverse square of the frequency. Typically, for a total electron column of 10^{16} electrons m^{-2} and a frequency of 1 GHz the *Faraday rotation* is

0.01 radian or $0.57°$ and for a total electron column of 10^{18} electrons m^{-2} the Faraday rotation is 1 radian or $57°$.

fast magnetohydrodynamic shock   A *fast magnetohydrodynamic shock* forms when a fast magnetohydrodynamic wave steepens. The magnetic field increases from the upstream to the downstream medium and is bent away from the shock because the normal component of the field is constant. In contrast, in a slow magnetohydrodynamic shock, the magnetic field is bent toward the shock normal. The normal component of the upstream (downstream) flow speed is larger (smaller) than the propagation speed of fast magnetohydrodynamic waves and both upstream and downstream flow speeds exceed the Alfvén speed. Traveling interplanetary shocks in general and planetary bow shocks always are fast magnetohydrodynamic shocks.

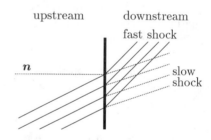

Change in magnetic field direction across a fast and a slow shock.

fast shock wave   *See* hydromagnetic shock wave.

fast solar wind   Solar wind escaping from the open magnetic field regions of the coronal holes. Plasma speeds range from 400 km/s to 800 km/s, densities are low with about 3 ions/cm^3 at Earth's orbit. Particles are mostly hydrogen; helium amounts to about 4%. Electron and proton temperatures are 1×10^5 K and 2×10^5 K, respectively. Fast solar wind streams are remarkably stable and variations from stream to stream are small. Despite the differences in individual parameters, the average momentum and total energy flux are similar in fast and slow solar wind streams.

Fata Morgana   A complex mirage made up of multiple images as of cliffs and buildings, distorted and magnified, which occasionally gives the impression of elaborate castles floating in the air. A Fata Morgana can only occur where there are alternating warm and cold layers of air near the ground or surface of water. Named after the effect seen in the Straits of Medina, and attributed to the sorcery of King Arthur's half-sister, Morgan Le Fay.

fathom   A nautical unit of depth, equal to 6 ft. It is also sometimes used for horizontal measure; a related unit is the cable, equal to 100 fathoms. Also, a nautical mile is approximately 1000 fathoms. *See* nautical mile.

fault   A seismic fracture across which lateral motion occurs. In some cases a fracture is simply a crack in rocks where slip has occurred. Major *faults* may be broad zones of granulated rock (fault gouge) accommodating lateral motions of 1000 km or more. Faults can be classified as thrust faults, normal faults, or strike-slip faults.

fault constitutive law   A representation of fault mechanical properties using relations among stress, strain rate, and displacement on fault planes.   A variety of constitutive laws have been proposed such as the slip weakening model which represents the relation between shear stress and slip, and the rate- and state-dependent friction law which represents the relation between friction and slip accompanied by changes in slip rate. These are empirical laws derived from laboratory experiments of rock friction. There are two modes for friction: stable sliding and unstable sliding. It is thought that aseismic slips such as fault creep and slow earthquakes correspond to the former, while usual earthquakes correspond to the latter. Numerical simulations are being carried out to tackle problems of earthquake cycle and seismic nucleation process of large earthquakes, using the *fault constitutive laws.*

fault gouge   The granulated material on a fault that has been generated during the many earthquakes that have occurred on the fault.

fault parameter   Parameter that characterizes faulting. As a geometrical quantity, there is the strike of a fault plane, dip angle and slip direction, while as quantities which represent fault size, there is fault area (length times width) and the amount of slip. Furthermore, rupture velocity, rise time, and slip rate are physical parameters which represent source processes. From these *fault parameters,* seismic moment and stress drop can be calculated. These fault parameters can be obtained from radiation patterns of seismic waves, waveforms, and aftershock distributions.

fault plane solution   An earthquake is generated by the relative motion of rocks across a fault. The movement has to be parallel to the fault plane, i.e., a vector representing the motion of one side with respect to the other will lie on the fault plane, and therefore there will be a plane normal to this vector (termed the "auxiliary plane") that will be perpendicular to the fault plane. For the simplest type of earthquake — simple movement along a flat fault plane, known as a "double-couple" earthquake — the pattern of radiation is divided into four lobes separated by the fault and auxiliary planes. The reason for this is that these planes divide the rock into areas in which the released stress is compressive or extensional, and thereby directions in which the first motions of the propagated radiation are compressions or rarefactions. With sufficient seismic data from around the world, it is possible to reconstruct the pattern of radiation for these first motions, and so to pick out the fault and auxiliary planes. This is known as the *fault plane solution.* Without other information it is difficult to tell which plane is the fault plane and which is the auxiliary plane, but local observations of the fault where the earthquake occurred or knowledge of the tectonic setting can help to determine this. The fault plane solution is invaluable in discriminating between different types of fault (e.g., thrust faults and transcurrent faults).

fault scarp   When there is vertical movement on a fault during an earthquake, one side of the fault is elevated relative to the other. This is a *fault scarp.*

F corona   The Fraunhofer, or F, solar corona is generated by the diffraction of photospheric radiation by slow moving dust in the interplanetary medium. The contribution of the *F corona* to the total coronal white light emission becomes increasingly important beyond $\sim 2R_\odot$. The Fraunhofer lines are clearly visible in the spectrum of this near-sun enhanced zodiacal light.

feeder beach   A sacrificial beach; a region of placed sand intended to be eroded to benefit an adjoining area.

feldspar   Metal aluminosilicate rocks with approximate specific gravity in the range of 2.5 to 2.76. The principal types are $KAlSi_3O_8$ (orthoclase), $NaAlSi_3O_8$ (albite), $CaAl_2Si_2O_8$ (anorthite).

Fermat's principle   A ray of light or other radiation follows the path that requires the least time to travel from one point to another, including reflections and refractions that may occur. The similar statement holds for sound expressed in terms of ray acoustics.

fermion   An elementary particle with internal angular momentum (spin) equal to an odd multiple of $\frac{1}{2}\hbar$ where $\hbar$ is Planck's reduced constant, $\hbar = 1.054571596 x 10^{-34} Js$. *Fermions* obey the exclusion principle; their wavefunction is antisymmetric under interchange of particle position. Electrons, neutrinos, protons, and neutrons are fermions.

fermionic zero mode   In field theory, the mass of fermions (spin 1/2 particles) is proportional to the value of a field (the Higgs field) to which they couple. A topological defect is a region of space inside which the Higgs field has a vanishing value. As a result, such fermions have effectively vanishing masses inside the core of the defect, and can travel at the velocity of light. Such fermions are called zero modes. Such string behavior may be very important in the early development of structure in the universe. *See* Higgs mechanism, Witten conducting string, Yukawa coupling.

Ferrel cell   Eddy-driven midlatitude zonal mean circulation cell in the atmosphere. *See* Hadley circulation.

fertile mantle   Mantle rock that includes a basaltic component.

fetch   A linear distance across a body of water over which the wind speed and direction may be assumed to be reasonably constant.

FG Sagittae star   Member of a very small class of variable stars which has just experienced the last flash of helium burning on the asymptotic giant branch. This causes the star to change its color and brightness very quickly. In addition, the last hydrogen is likely to be lost from the surface and carbon mixed into the visible atmosphere. *FG Sagittae stars* probably evolve to R Coronae Borealis stars.

fibril   A linear pattern in the Hα chromosphere of the sun occurring near strong sunspots and plage or in filament channels. *Fibrils* are similar in appearance to spicules except that they are bent over and extend along the solar surface at a height of about 4000 km rather than protruding radially outwards. They have lengths typically ~ 15000 km with a thickness of ~ 2000 km and exhibit axial proper motions ~ 20 to 30 kms^{-1}.

Fickian flux (Fick's law)   The statement that the flux **J** of a diffusing substance is proportional to the concentration gradient, i.e., $\mathbf{J} = -D(\nabla_x C)$ where D is called the diffusion coefficient. Often written in 1-dimensional form: $J_C = -D_C \partial C/\partial x$. While the first Fickian law is well suited for molecular diffusivities D_C, it is in fact often invoked as the first-order closure scheme for turbulent fluxes. The turbulent flux for this so-called eddy-formulation is $\mathbf{J}_i = -K_i \partial C/\partial x_i$, where K_i is the turbulent diffusivity in direction i. This first-order closure scheme collapses if the eddies get very large and the local gradients are too small (non-local diffusion, see Stull, 1988).

Fick's second law   The rate of change of a property C is given by the divergence of all the fluxes ($\mathbf{J}_C$) and all sinks and sources (λ_C)

of C. The conservation equation $\partial C / \partial t = -\mathrm{div}(\mathbf{J}_C) + \lambda_C$ is often referred to as the second Fickian law.

field   A mathematical construct that represents physical interactions as spread through space and time. Any quantity that can be defined at every point of (a region of) space (or spacetime) can be defined to be a *field*. Classical examples include the electromagnetic and the gravitational field.

field capacity (θ_{fc})   The maximum amount of water in the unsaturated zone of the soil that can be held against the pull of gravity.

field line motion   A theoretical formalism that helps visualize the effects of electric fields and varying magnetic fields on the plasma they permeate. In a magnetic field $\mathbf{B}$, it ascribes to each point a velocity $\mathbf{v}$ that satisfies $\partial \mathbf{B} / \partial t - \nabla \times (\mathbf{v} \times \mathbf{B}) = 0$.

If the field is embedded in a highly conducting fluid (e.g., the molten metal in the Earth's core) or in a collision-free plasma (such as is found in space around Earth), the bulk flow velocity $\mathbf{v}$ of the fluid or plasma in general comes close to satisfying the **MHD condition** $\mathbf{E} = -\mathbf{v} \times \mathbf{B}$, where $\mathbf{E}$ is the ambient electric field. By Maxwell's equations, the curl of this is the equation defining field line motion; hence, the fluid or plasma moves with the field lines.

In a collisionless plasma the MHD condition is related to the electric drift of the plasma. In all such cases, two particles of the fluid or plasma which initially share the same field line continue doing so as time advances. It should, however, be noted that this is only the motion perpendicular to field lines: in addition, the fluid or plasma may also slide along field lines.

Field line motion helps intuitive understanding of plasma motions. In limiting cases where the motion dominates the magnetic field — e.g., the solar wind which is known to move (more or less) radially — field line motion provides a shortcut to calculating the magnetic field $\mathbf{B}$, given the sources of $\mathbf{B}$ on the sun. On the other hand, if $\mathbf{B}$ completely dominates the plasma (e.g., where it is very rarefied), if we know the way $\mathbf{B}$ changes (i.e., $\partial \mathbf{B} / \partial t$), the observed field

line structure can help derive the bulk flow velocity $\mathbf{v}$.

field of view   The angular size of the full image formed in an optical instrument; the angular separation of two points that lie at the edges of the optical field.

field ordering   In condensed matter physics, when the temperature of a ferromagnet goes below the critical point, T_c, a non-zero magnetization $\vec{M}$ develops. The rotational symmetry previously possessed by the system is then broken due to the presence of a preferred direction, the one fixed by $\vec{M}$. The value of $\vec{M}$, zero for the high-temperature phase and non-zero for temperatures below T_c, plays the role of the order parameter of the phase transition.

In cosmological transitions, the role of the order parameter is played by the vacuum expectation value of the Higgs field (here denoted ϕ). Standard topological defects (like monopoles, walls, strings) involve regions in space where this order parameter remains in the high-temperature symmetric phase (vanishing ϕ). In this case the field potential energy (the false vacuum trapped inside the defect) is the main source of the energy associated with the defects.

There are, however, other types of defects where the bulk of energy is concentrated not as potential energy but in spatial gradients. Cosmic textures are one example of this. They have the property that the broken-symmetry phase vacuum manifold $\mathcal{M}$ of the order parameter ϕ has the same dimension as space (equal to three, for cosmological applications), and this allows ϕ to always stay on $\mathcal{M}$, regardless of the location considered. Hence, possessing no potential energy, all the relevant dynamics comes from the ordering of this field ϕ, that is from the tendency to minimize its gradients.

Texture knots will shrink (instability to collapse) as explained by the Derrick theorem. This will result in the ordering field becoming increasingly tightly wound in the vacuum manifold. Then the spatial gradients (kinetic energy terms) in the configuration will eventually become so high as to be able to exceed the energy of the symmetric state and unwind the knot.

See cosmic topological defect, Derrick theorem, spontaneous symmetry breaking.

figure of the Earth   The shape of the Earth. To a first approximation it is an oblate spheroid with a polar radius of 6357 km and an equator radius of 6378 km.

filament   A structure in the corona consisting of cool (~ 7000 K), dense ($\simeq 10^{12}$ cm^{-3}) plasma supported by magnetic fields and seen as dark lines threaded over the solar disk. When seen in emission at the solar limb, a *filament* appears as a protuberance: a bright arc of matter extending high above the photosphere, spanning latitudes of up to some $10°$. Their density is about a factor of ten higher than the ambient density (therefore the arc is bright when seen above the solar limb), and the filament can extend up to about 100 times the scale height in the corona. Filaments are aligned along the separation of opposing magnetic field patches in the photosphere. Their existence therefore is related to solar activity with only a few filaments observed during solar minimum and a much larger number during solar maximum. Filaments can have very long lifetimes, lasting 2 to 3 solar rotations in some cases. They are found, preferentially, in two latitude belts on the sun; in a strip at high latitudes known as the polar crown and in active mid-latitudes. Typical magnetic fields are ~ 5 to 10 G in quiescent filaments and may be as high as 200 G in active region filaments fields.

The cold end dense matter of the filament/protuberance is held against gravity by magnetic tension in the anchoring magnetic field lines. Two configurations can be distinguished by comparing the photospheric magnetic field pattern with that of the filament:

1. Kippenhahn–Schlüter configuration, also called normal configuration: the magnetic field inside the filament has the same direction as the photospheric field below it.

2. Raadu–Kuperus configuration, also called inverse configuration: the magnetic field inside the filament is directed opposite to the one in the photosphere. This is possible only because the anchoring field lines have a neutral point below the filament. In particular, in large and high-rising filaments, which tend to give rise to coronal mass ejections, the Raadu–Kuperus configuration seems to be the dominant one. This is attributed to the X-point below the filament where reconnection is likely to occur, leading to the expulsion of the filament (coronal mass ejection) and the generation of electromagnetic emission due to the accelerated electrons (the flare). *See* coronal mass ejection, reconnection.

filament channel   A broad pattern of fibrils in the chromosphere, marking where a filament may soon form or where a filament recently disappeared.

filter layer   In civil or coastal engineering, denotes a layer of material (typically stone, gravel, or sand, possibly combined with a geotextile fabric) intended to prevent migration of fine material into or out of a structure. May also be employed to reduce settlement. As an example, breakwaters are often built with a *filter layer* in the base to help prevent scouring and settlement.

finestructure   In atmospheric dynamics, structures with scales from tens of meters to decimeters, covering the processes of internal waves and intrusions. *Finestructure* processes are intermediary between the large-scale (overall stratification) structures and the small-scale (turbulence).

fine structure constant   Dimensionless constant relating to strength of electromagnetic interactions:

$$\alpha = e^2 / (4\pi\epsilon_0\hbar c) = 7.297352533(27)\times 10^{-3}.$$

fingering   *See* double diffusion.

finger regime   The double-diffusive regime, where temperature stabilizes and salinity destabilizes (i.e., warmer and saltier water is on top of cooler and fresher water). In this case, the non-dimensional stability ratio is defined as $R_\rho = (\alpha\partial\Theta/\partial z)/(\beta\partial S/\partial z)$. The term "finger" refers to the saltier finger-like plumes driving the double-diffusive convection. Classical locations of the *finger regime* are under the Mediterranean Outflow, in the East North-Atlantic and the C-SALT field east of Barbados.

finite amplitude wave   A wave of finite amplitude. In oceanography, a wave of finite height. All waves are finite amplitude, but the simplest water wave theory (linear wave theory) assumes waves of infinitesimal wave height. A variety of finite amplitude (higher order) wave theories are available for water and other wave phenomena.

finite difference   A method of approximating functions by specifying values on a specified grid of points, and approximating derivatives by taking algebraic differences of the values on the points. For instance, the expression

$$[f(x + \delta) - f(x - \delta)]/(2\delta)$$

is an approximation of the first partial derivative with respect to the coordinate x, where the points along the x axis are separated by distance δ.

finite element method   A numerical method used to solve partial differential equations. In the *finite element method,* the spatial domain is divided into a set of non-overlapping elements. Neighboring elements share nodal points along their common boundaries. By approximation, an unknown variable inside an element is interpolated from values at the nodal points of the element using a set of interpolation functions. A numerical solution subject to given initial and boundary conditions is obtained by solving for all nodal values in the domain.

fireball   An aerial display associated with a meteor that reaches a brightness greater than that of the full moon. Applied often to any bright meteor.

firehose instability   Instability of a thermally anisotropic collisionless plasma, associated with an excess of pressure parallel to the magnetic field. The nonrelativistic condition for instability is (cgs units)

$$1 + \frac{4\pi}{B^2}\left(P_\perp - P_\parallel - \Pi\right) < 0\,,$$

where $\mathbf{B}$ is the mean magnetic field strength, $P_\perp$ and $P_\parallel$ are, respectively, the pressures transverse and parallel to the mean magnetic field, and

$$\Pi = \frac{1}{\rho}\sum_\alpha \rho_\alpha \left(\Delta\mathbf{V}_\alpha\right)^2\,.$$

ρ_α is the mass density of charge species α and $\Delta\mathbf{V}_\alpha$ is its relative velocity of streaming relative to the plasma.

Π is effectively an enhancement of $P_\parallel$ due to intraspecies streaming.

first fundamental form   The metric $\mathbf{g}$ on a space or spacetime $\mathcal{M}$. (Perhaps as induced from a higher dimensional space $\mathcal{N}$ if $\mathcal{M}$ is embedded in $\mathcal{N}$.)

first integral   A quantity that is constant for a particular motion (depending on the initial conditions) because the differential equations describing a system allow a partial analytic integration.

first law of thermodynamics   The conservation of energy:

$$dQ = dU + p\,dV$$

where dQ is an amount of heat input, dU is a change in the internal energy of the system, and $p\,dV$ is the work done.

first order phase transitions   For first-order transitions in cosmology, which may for instance lead to cosmic defects (strings, monopoles, domain walls ...) at very high energies the symmetry breaking potential has $\phi = 0$ as the only vacuum state, i.e., the only minimum of the potential. When the temperature goes down to below the critical temperature T_c, a set of different minima develops, a potential barrier separating the old (false) and the lower energy new (true) vacua. Provided the barrier at this small temperature is high enough, compared to the thermal energy present in the system, the field ϕ will remain trapped in the false vacuum state even for small ($< T_c$) temperatures.

Classically, this is the complete picture. However, quantum tunneling effects can liberate the field from the old vacuum state, at least in some regions of space: there is a probability per unit time and volume in space that at a point $\vec{x}$ a bubble of true vacuum will nucleate. The result is the formation of bubbles of true vacuum with the value of the Higgs field in each bubble being independent of the value of the field in all other bubbles. This leads to the formation of domains (the bubbles) where the fields are correlated, whereas no correlation exists between fields belonging to different domains. After its creation, a bubble will expand at the speed of light surrounded by a "sea" of false vacuum domains. As opposed to second-order phase transitions, here the nucleation process is extremely inhomogeneous and $\phi(\vec{x})$ is an abruptly changing function of time. *See also* Ginzburg temperature, GeV, Kibble mechanism, QCD, spontaneous symmetry breaking.

first-order wave theory   Also known as linear wave theory or Airy wave theory.   The derivation involves assumption that terms of second and higher order are negligible.

fjord   (also fiord). A long, narrow, typically deep inlet, connected to a sea. Generally found in mountainous regions at high latitudes (Norway, Alaska, New Zealand).

FK Comae stars   A rapidly rotating, red giant. Because conservation of angular momentum during the expansion of a star from the main sequence to red giant stage should slow its surface rotation, the phenomenon must arise from the interaction between binary stars, possibly from the mergers of W Ursa Majoris binaries.

flare   A pronounced, transient increase in the sun's global output related to sunspots and solar activity. *Flares* are generally manifested by a rapid (≈ 1 min) brightening in the red hydrogen line, emitted in the sun's chromosphere, and a very sudden increase (a few seconds) in X-ray emission from that region. Energies up to 10^{25} J can be released over a time period of some minutes though much shorter events are known. White-light flares, that is a brightening of the sun in visible light, are rare because even for the largest flares the brightness is less than 1% of the total luminosity of the photosphere. In certain frequency ranges, e.g., at the wings of the black body, the intensity of the electromagnetic radiation can increase by orders of magnitude. Although the flare is defined as an outburst in electromagnetic radiation, large flares are associated with large fluxes of energetic ions from the sun, at times at energies of 1 GeV and more (*see* energetic particles) and the eruption of filaments, coronal mass ejections (*see* coronal mass ejec-

tions). Depending on the time-scales and the occurrence of coronal mass ejections, flares can be classified as impulsive or gradual. The energy for these rapid events is taken from reconnection in the magnetic fields of the sunspot region near which the flare occurs. Particle events in interplanetary space caused by these different kinds of flares also show distinct properties. *See* gradual flare, impulsive flare.

Flares can be classified according to the size of the Hα emitting region (cf. table), the peak intensity in X-ray emission (SXR class), or the radio flux at 5000 MHz (given in solar flux units sfu).

Flare Classification

Class	Area* (Hα)	SXR Class	Radio Flux
S	≤ 200	C2	5
1	200–500	M3	30
2	500–1200	X1	300
3	1200–2400	X5	3000
3+	>2400	X9	30000

* In millionth of sun's area with 1 millionth equal to $6.08 \cdot 10^6$ km^2.

flare electromagnetic radiation   The electromagnetic radiation released in a solar flare shows typical time profiles in different frequency ranges. In a large flare, the flare emission can be divided into three phases: (a) a precursor (also called preflare phase) lasting for some minutes to some 10 minutes, visible as a weak brightening of the flare region in Hα and soft X-rays, indicating the heating of the flare site. (b) The impulsive or flash phase in which electromagnetic radiation is emitted over the entire frequency range from γ-rays over X-rays, UV, the visible, IR up to radio waves. In this phase, the hardest part of the electromagnetic emission is most abundant, which indicates the acceleration of particles in addition to just a heating of the flare site. The impulsive phase can be observed in most flares, lasting for some minutes. In larger flares it can be followed by (c) a gradual phase lasting for some 10 minutes to some hours during which the emission mainly occurs in Hα and soft X-rays. This emission stems from the heated plasma at the flare site, and does not indicate the presence of accelerated particles. Microwave and radio emission can also continue, evidence for energetic electrons trapped in closed magnetic loops.

1. Soft X-rays and Hα originate as thermal emission in a plasma with temperatures of about 10^7 K. Most of the emission is continuum, lines of highly ionized O, Ca, and Fe are observed as well.

2. Hard X-rays are photons with energies between a few tens of keV and a few hundred keV generated as bremsstrahlung of electrons with slightly higher energies. Only a very small amount of the total electron energy, about 1 out of 10^5, is converted into hard X-rays.

3. Microwaves are generated by the same electron population that also generates the hard X-rays as can be deduced from the similarities, in particular multiple spikes, in both intensity time profiles. Microwave emission is gyro-synchrotron emission of accelerated electrons.

4. γ-rays are the best indicators for the presence of energetic particles. The spectrum can be divided into three parts: (a) bremsstrahlung of electrons and, to a lesser extent, the Doppler broadening of closely neighbored γ-ray lines leads to a γ-ray continuum. (b) Nuclear radiation of excited CNO-nuclei leads to a γ-ray line spectrum in the MeV range. The most important lines are the 2.23 MeV line due to neutron capture in the photosphere and the 4.43 MeV line from the relaxation of excited ^{12}C nuclei. These reactions require particle energies of some 10 MeV/nucl, thus the γ-ray line emission indicates the presence of very energetic nuclei at the flare site. (c) Decaying pions lead to γ-ray continuum emission above 25 MeV.

5. Radio emission results from electron streams in the corona, exciting Langmuir oscillations. Frequencies are in the meter range (*see* metric radio emission). According to their frequency drift, radio bursts are classified as type I to type V metric bursts. *See* type i radio burst, $i = I \dots V.$

flare star A star whose brightness increases detectably for a few hours at irregular intervals due to flare analogous to those in the sun (solar flares), but much brighter. They are a signature of a star that is rotating rapidly and has a strong magnetic field, either because it is young or because it is in a close binary system. The most conspicuous *flare stars* are of spectral type MV, both because the stars are intrinsically faint and because the convection zones are deep, producing stronger fields and brighter flares.

flaring angle On the magnetopause at a given point P on it — the angle χ (sometimes also denoted α) between the magnetopause surface and the flow direction of the distant solar wind (or else, $90° - χ$ is the angle between flow of the distant solar wind and the normal to the magnetopause at P). By the Newtonian approximation, the perpendicular pressure of the solar wind on the magnetopause at P equals $p \sin χ$, where p is the dynamic pressure of the solar wind.

flat field, sky flat, dome flat A CCD's pixel-to-pixel variation in sensitivity is called the flat field function. To make accurate measurements of the number of counts from objects on an image, this variation (which is a multiplicative factor) must be removed. This is done by making a flat field image, and dividing this image into all the data images.

There are a number of methods for obtaining a flat field image, and they are very dependent on the instrument, type of data, and filters used. Different flat field images must be obtained for each filter, as filters can modify the illumination across a CCD, and narrowband filters can produce interference fringes. A common method is to get a number of exposures of a blank screen in the telescope dome and average them together; these are called *dome flats*. One may also get images of the twilight sky, or combine many data images. In these latter cases, one must be sure that no actual objects (stars, etc.) appear on the flats. If these *sky flats* are not of the same point in the sky, it is possible to combine them with a median filter and obtain a good *flat field* frame. Each data image is divided by an average flat field image of the same filter, etc.

flattening of the Earth Because of its rotation the polar radius of the Earth is smaller than the equatorial radius by about 21 km. This leads to polar flattening and an equatorial bulge.

flat universe   A model of the universe in which the average density is exactly equal to the critical density at which it neither expands freely nor collapses again. It is called flat because at any instant the global geometry of space is Euclidean; space is flat rather than curved.

flexure   Bending of a plate. In tectonics, the Earth's lithosphere is subjected to flexure in a variety of geological settings. Examples include the bending of the lithosphere prior to subduction at an ocean trench, bending under the load of a volcanic island, for example the Hawaiian Islands, and the bending associated with sedimentary basins.

floating breakwater   A breakwater which does not rest on the seafloor, but is instead anchored in place and has a buoyant force equal to its weight.

flocculation   The process in which suspended soil forms lumps during the mixing of freshwater with seawater, especially in estuaries. *Flocculation* occurs as the result of molecular attractive forces known as the van der Waals forces. These forces are weak, and are only significant as clay particles are brought close together, such as during mixing turbulence. Normally, in fresh water clay minerals carry a negative charge which repels particles from each other. In saline waters, these charges are neutralized, and the attractive van der Waals forces dominate.

flood current   The tidal current that results when the water in bays and estuaries is lower than that in the adjoining sea. The opposite is referred to as an ebb current.

flood delta   A deposit of sediments immediately inshore of a tidal inlet, deposited by flood tidal currents.

flood shoal   *See* flood delta.

flood tide   *See* flood current.

Flora   Eighth asteroid to be discovered, in 1847. Orbit: semimajor axis 2.2015AU, eccentricity 0.1562, inclination to the ecliptic 5°.8858, period 3.27 years.

Florida current   Ocean current flowing northward along the south-east coast of the U.S. from the Florida Straits to Cape Hatteras, feeding into the Gulf Stream. This current transports approximately 26 million cubic meters per second through the Florida Straits.

flow regime   A categorization of open channel flow based on flow speed and bed form. The lower flow regime corresponds to lower flow speeds and ripples or dunes on the bed. A transitional flow is somewhat faster and will yield dunes, a planar bed, or antidunes. The upper flow regime is faster and yields a planar bed, antidunes, or more complicated flow and bed patterns.

fluence   The time integrated flux.

fluorescence   The inelastic scattering process in which a photon is absorbed by a molecule and shortly thereafter (10^{-11} to 10^{-8} s) another photon of greater wavelengh is emitted; the emitted radiance is unpolarized.

flutter fading   Very rapid fading of radio signals in an ionospheric circuit (5 to 10 fades per second, or 5 to 10 Hz fades, and faster). Called *flutter fading* because of the audible effect it has on signals. Fades can be deep, dropping below the local noise levels and resulting in the signal being drowned in the noise. It is a common problem at low latitudes where it is associated with equatorial spread F. Flutter fading is also observed at high latitudes on non-great circle paths, suggesting large ionization gradients, and near the auroral oval where it has been linked with slant-type sporadic E. *See* fading.

fluvial   Refers to processes or features produced by the flow of some liquid, usually water. River channels are among the most common features produced by fluvial processes. These features can be produced by surface flow or by groundwater flow. On Mars, large outflow channels have been produced by catastrophic surface floods while the smaller valley networks have been created by the sapping process (removal of groundwater, causing the overlying terrain to collapse).

flux   The rate of flow of fluid, particles, or energy through a given surface.

flux density   In radiation, a measure of flux (power per unit area) per wavelength or frequency interval. The unit of *flux density* is called a Jansky and is defined as 10^{-23} Joule s^{-1} m^{-2} Hz^{-1}.

fluxgate   A type of magnetometer widely used both for geophysical studies and aboard spacecraft. It utilizes the property of some magnetic materials to saturate abruptly at a well-defined value of the magnetic intensity.

flux, magnetic   The magnetic flux crossing a small area dA equals B_ndA, where B_n is the field component perpendicular to dA. The flux crossing a finite area is accordingly the integral $\int B \cdot dA$ over the area, and flux density is *magnetic flux* per unit area. The magnetic flux through a surface is sometimes referred to informally as "the number of magnetic field lines crossing it".

flux, particle   Total flux is the number of particles per unit area crossing a given surface each second. Differential flux is the flow of particles per steradian crossing a given surface from a particular direction.

flux Richardson number   The non-dimensional ratio $R_f = J_b/J_R$ expresses the rate of storage of potential energy in the stratification of the water column (J_b, buoyancy flux) relative to the rate of production J_R of turbulent kinetic energy by Reynolds stress. The *flux Richardson number* and the mixing efficiency γ_{mix} are related by $R_f = \gamma_{\mathrm{mix}}(1 + \gamma_{\mathrm{mix}})$.

fluxrope   A cylindrical body of magnetized plasma with twisted fields. *Fluxrope* topologies observed in interplanetary space have come to be known as magnetic clouds and are thought to be the interplanetary counterparts to coronal mass ejections.

fluxtube   The volume enclosed by a set of magnetic field lines which intersect a simple closed curve. A way of visualizing magnetic fields, a tube whose surface is formed by field lines. The strength of a *fluxtube, F,* is often defined as the amount of flux crossing an areal section, S, via $F = \int_S \mathbf{B} \cdot \mathbf{dS}$ where $\mathbf{dS}$ is taken in the same sense as $\mathbf{B}$ to yield $F \geq 0$. Fluxtube properties include: (i) the strength of a fluxtube remains constant along its length, (ii) the mean field strength of a fluxtube varies inversely with its cross-sectional area, (iii) a compression of a fluxtube increases the field and gas density in the same proportion, (iv) an extension of a fluxtube without compression increases the field strength. Examples of fluxtubes in the solar atmosphere are sunspots, erupting prominences and coronal loops, the Io/Jupiter system where Io injects ions into the Jupiter magnetosphere. Because of the localized source, these ions fill up tubes following the Jovian field lines to the poles, where they induce bright isolated auroral displays.

flux tube   *See* fluxtube.

flux tube volume   *See* specific volume.

fly's eye   A telescope adapted to wide angle sky searches, consisting of a set of fixed mirrors with detectors at the focus of each mirror. Directional information is obtained from the field of view of each reflector, but typically no attempt is made to obtain an image from each reflector.

flywheel effect   An exchange of momentum between the neutral and charged components of the ionosphere. If some magnetospheric source creates a strong electric field $\mathbf{E}$ in a region in the ionosphere, the local plasma flows with the electric drift velocity $\mathbf{v}$, and collisions with neutral gas also impel that gas to move with some fraction of $\mathbf{v}$. Like a flywheel, the neutral gas stores some of its momentum and later, when $\mathbf{E}$ is decaying, its collisions help the ions maintain some of their flow, creating in the process a secondary electric field.

focal length   The distance behind a converging lens (or the distance in front of a converging mirror) that rays arriving parallel from infinity come together in a close approximation to a point (positive focal length); *or* the distance of the point in front of a diverging lens (or the distance behind a diverging mirror) from which rays arriving parallel from infinity appear to diverge (negative focal length).

focal mechanisms   Earthquakes send out compressional P-waves that are either compressional or tensional. The azimuthal variation in the first arrivals defines the focal mechanism. This variation provides information on whether the fault displacement was thrust, normal, or strike-slip and information on the orientation of the fault.

focus   In optics, the point where light rays converge and/or from which light rays (appear to) diverge.

In geometry, in a circle, the center of the circle. In an ellipse, each of the two points on the major axis located at $a\epsilon$ from the center, where a is the semimajor axis, and ϵ is the eccentricity. In a hyperbola, the point located inside the hyperbola, on the axis a distance of $a(\epsilon - 1)$ from the point the orbit crosses the axis, such that the transverse distance to the orbit at that point is $a(\epsilon^2 - 1)$. Here a is called the semiaxis of the hyperbola, and ϵ is $1/(\cos\alpha)$ with α the slope of the asymptote to the hyperbola.

In a parabola, the point inside the parabola on the axis of the parabola a distance $p/2$ from the point the orbit crosses the axis, such that the transverse distance to the orbit at that point is 2p.

In Newtonian physics, gravitational motion is an orbit that is a conic section (circle, ellipse, parabola, hyperbola) with the sun at one focus.

focused transport equation   Model suggested by Roelof (1969) to describe the interplanetary transport of charged energetic particles in terms of field-parallel propagation, focusing on the diverging interplanetary magnetic field, and pitch-angle scattering:

$$\frac{\partial f}{\partial t} + \mu v \frac{1 - \mu^2}{2\zeta} v \frac{\partial f}{\partial \mu} - \frac{\partial}{\partial \mu}\left(\kappa(\mu)\frac{\partial f}{\partial \mu}\right) = Q(s_0, \mu, t).$$

Here f is the phase space density, v the particle speed, μ the particle's pitch-angle, $\kappa(\mu)$ the pitch-angle diffusion coefficient (see diffusion, in pitch-angle space, slab model), s the spatial coordinate along the Archimedian spiral, and

$\zeta = -B(s)/(\partial B/\partial s)$ the focusing length (see focusing). The term Q on the right-hand side describes a particle source.

focusing   In plasma physics, reduction of a particle's pitch angle as it propagates outward in a slowly diverging magnetic field such as the interplanetary magnetic field. *Focusing* can be characterized by a focusing length ζ

$$\zeta = -\frac{B(s)}{\partial B/\partial s}$$

with s being the spatial scale along the magnetic field B. In interplanetary space, the divergence of the magnetic field would reduce a particle's pitch-angle from nearly 90° on the sun to about 0.7° at the orbit of Earth.

Focusing is a direct consequence of the constancy of the magnetic moment (first adiabatic invariant). Its reverse effect is the increase of the particle's pitch angle in a convergent magnetic field, eventually leading to mirroring. *See* adiabatic invariant, interplanetary propagation, magnetic mirror.

foehn   Hot winds on the down-slope side of a mountain on which background winds impinge. There are two basic types of *foehns* of thermodynamic and dynamic causes, respectively. Thermodynamic foehns occur when the lower atmosphere is humid and the uplift by the mountain forces water vapor to condense on the upstream side, causing temperature increases on the down-slope side. Dynamic foehns occur when the atmosphere is strongly stratified on the up-slope side and the air near the surface cannot flow over the mountain and is blocked. As the upper air is forced to descend downslope, adiabatic compression raises its temperature. The second-type of foehn occurs when the Froude number U/Nh is below a certain critical value, where U is the background wind speed, N the Brunt–Väisälä frequency, and h the height of the mountain.

fog   A situation in which clouds form at ground level which reduces visibility below 1 km; consisting of water droplets too small to fall out of suspension.

folds   In geophysics, when the continental crust is deformed under compression the result is often a near periodic, sinusoidal structure. These are folds, and they can occur on scales from a meter or less to hundreds of kilometers. Folding is usually associated with layered sedimentary rocks with some layers being rigid and others more ductile.

following spot   A sunspot or sunspot group lying on the eastern side of an active region complex, i.e., following in the direction of solar rotation. *Following spots* tend to be smaller and greater in number than preceding (or "leader") spots. During the evolution of an active region, the following spots tend to remain at the same longitude or move backwards (relative to the direction of solar rotation).

Fomalhaut   1.16 magnitude star of spectral type A at RA22$^\mathrm{h}$ 57$^\mathrm{m}$ 38.9$^\mathrm{s}$, dec 19°37′20″.

forbidden lines   Spectral emission lines violating quantum mechanics selection rules for electric dipole emission. In both permitted and forbidden transitions, the photon of a spectral line is emitted when an electron moves from an upper to a lower energy level. The photon energy is equal to the difference between the energy of the two levels. In the case of *forbidden lines*, the probability of a spontaneous transition between the upper and lower energy level is very small, and the electron remains a much longer time in the excited state than in the case of a permitted transition. In this case, the upper level is said to be metastable. Forbidden lines in several astronomical sources are collisionally excited, i.e., the electron bound to an ion is brought to a higher, metastable level via the collision with a free electron or with another ion. At densities typical of the terrestrial environment, subsequent collisions would quickly de-excite the atom without emission of radiation. At electron densities $n_e \lesssim 10^3 - 10^7$ electrons cm^{-3}, the probability of a collision is much lower, and the electron can decay to a lower level with the emission of a photon of the forbidden line. Forbidden lines are, therefore, very sensitive indicators of density in several gaseous nebulæ, for example H$\textsc{i}$ regions, or planetary nebulæ.

forbidden orbits   *See* Störmer orbits.

forbidden region   A term used in the study of cosmic ray ions or particles convecting into the Earth's magnetic field. It is a region in space, or among the directions of arrival, which (for some stated energy) contains no orbits that connect to infinity, but only trapped orbits. Since the cosmic ray ions, or the convecting ions, come from great distances and are not trapped, the *forbidden region* contains none of them. Forbidden regions for convection particles are responsible for Alfvén layers.

Forbush decrease   The sudden dramatic reduction in the flux of cosmic rays detected at the Earth (e.g., with a neutron monitor) due to the interaction of solar flare induced shock waves with the interplanetary medium. The shock waves sweep across the geomagnetic field lines and deflect the galactic cosmic ray particles. One to three days after a strong flare counting rates decrease sharply by a few percent. This decrease coincides with the passage of the shock. Sometimes the decrease occurs in two steps, the first step coinciding with shock passage, the second with the arrival of the magnetic cloud. Although the decrease is sharp, the recovery is slow, lasting for some days.

force   In Newtonian physics, a vector quantity describing the external influence on an object which tends to accelerate the object. *Force* may be measured by its effect in compressing a spring:

$$\mathbf{F} = -k\mathbf{x},$$

where k is a constant, and $\mathbf{x}$ is the amount of displacement of the spring from its resting position. Newton's second law relates force to acceleration:

$$\mathbf{F} = m\mathbf{a},$$

where m is the mass, and $\mathbf{a}$ is the acceleration. Thus, force is the rate of change of momentum with time.

Units: dynes = gm cm/sec^2; Newtons = kg m/sec^2.

In relativity, a number of formulations of force have been given, all of which reduce to the Newtonian form in the limit of small velocities.

force balance   A term used in the study of plasmas in space, referring to the requirement (inherent in Newton's laws of motion) that the sum of all forces on any volume element of plasma equals its mass times its acceleration. In collision-free plasmas, mass is small and the acceleration term can generally be neglected, leading to the requirement that all forces on the element balance. In general, gravity is negligible, and the two main terms are the magnetic force $\mathbf{j} \times \mathbf{B}$ and the pressure gradient ∇p (or $\nabla \cdot \mathbf{P}$, if pressure is a tensor $\mathbf{P}$). This equation (non-linear in $\mathbf{B}$, since $\mathbf{j} = \nabla \times \mathbf{B}$) and its modifications are among the conditions imposed by MHD.

force-free magnetic field   A magnetic field and associated current system for which the Lorentz force vanishes. Equivalently, the current density and magnetic field must be parallel, i.e., $curl \mathbf{B} = \alpha \mathbf{B}$, where $\mathbf{B}$ is the magnetic field and α is a scalar quantity such that $\mathbf{B} \cdot \nabla \alpha = 0$. The force-free approximation is often appropriate for plasmas whose fluid pressure is small in comparison with the magnetic pressure (a "low-beta" plasma), and may be regarded as intermediate between the potential field (zero current density) approximation and a full magnetohydrodynamic description.

Forchhammer's Principle   Principle of Constant Proportions: the ratio of major salts in samples of seawater from various locations is invariant.

forearc basin   A sedimentary basin on the trench side of a volcanic arc at a subduction zone.

forearc sliver   At a subduction zone where plate convergence is oblique (not normal to strike), there is often a strike-slip fault or shear zone in the upper plate trenchward of the volcanic arc. The narrow block of the upper plate between the subduction thrust fault and the shear zone is called the *forearc sliver*. Forearc slivers are observed to move in the tangential component of the direction of subduction and, as a result, the relative motion of the subducting plate with respect to the sliver is less oblique. This phenomenon is called slip partitioning.

forecasting (wave)   The process of predicting ocean wave conditions from anticipated wind and weather conditions. May refer to the prediction of bulk quantities, such as wave height or period, or of wave energy spectra.

foreshock   The region ahead of the Earth's bow shock, linked to the shock by magnetic field lines. Although the velocity of the solar wind is super-Alfvénic, preventing the shock from affecting the approaching solar wind by Alfvén waves, fast ions and electrons can travel upstream and can affect the *foreshock* region, creating a noisy plasma regime.

In seismology, a preceding earthquake which has smaller magnitude than the main earthquake. Unfortunately, a foreshock cannot be identified as a foreshock until a subsequent, perhaps larger earthquake occurs, so the concept has no predictive ability.

foreshore   The part of a beach that lies nearest the sea, from the low-tide line to the high-tide line; often taken as synonymous with the term beachface. May be used to indicate the beachface plus a short region of the beach profile offshore of the beachface.

Foucault pendulum   A pendulum constructed for little dissipation, and suspended so that it can swing freely in any plane. Over a period of (up to) days, the plane of the swing relative to the Earth turns at a rate:

$$\Omega = 2\omega_\oplus \sin\theta \,,$$

where θ is the latitude and $\omega_\oplus$ is the rotational rate of the Earth. The motion is clockwise looking down at the pendulum in the northern hemisphere, vanishes at the equator, and is counterclockwise in the southern hemisphere.

four-velocity   In relativity, the coordinates x, y, z, t of a series of events along a world line are considered functions of the proper time measured by an observer traveling along that world line. The *four-velocity* is the four-vector obtained by differentiating the four-coordinates with respect to the proper time τ associated with the orbit:

$$u^\alpha = dx^\alpha/d\tau, \alpha = 0, 1, 2, 3 \,.$$

Thus, four-velocity is the tangent four-vector $u^\alpha = dx^\alpha/ds$ of an affinely parametrized timelike curve $\gamma = \{x^\alpha(s)\}$ in the space-time. The norm is $g(u, u) = -1$. *See* signature, spacelike vector, timelike vector.

f-plane approximation   In calculating motions on the Earth's surface, the effects of the Earth's sphericity are retained by approximating the value of the Coriolis parameter, f, as a constant.

fractal   A geometrical object that is self-similar under a change of scale; i.e., that appears similar at all levels of magnification. *Fractals* can be considered to have fractional dimensionality. A statistical distribution is fractal if the number of "objects" N has a fractional inverse power dependence on the linear dimension of the objects r.

$$N \sim r^{-D}$$

where D is the fractal dimension. Examples occur in diverse fields such as geography (rivers and shorelines), biology (trees), geology, and geophysics (the number-size distribution of fragments often satisfies this fractal relation; the Guttenberg–Richter frequency-magnitude relation for earthquakes is fractal), and solid state physics (amorphous materials).

fractionation   In geophysics, separation of different minerals during the melting of rocks, and refreezing of a solid; applied to the geophysical processes modifying rocks.

fracture   In geophysics, the Earth's crust is filled with fractures on a range of scales from centimeters to thousands of kilometers. The term *fracture* covers both joints and faults. If no lateral displacement has occurred across a fracture, it is classified as a joint; if a lateral displacement has occurred, it is classified as a fault.

fracture zone   In geophysics, deep valleys caused by faults on the ocean floor. The ocean floor on the two sides of a *fracture zone* can be of very different ages and this leads to differential elevations and subsidence.

fragmentation   The breaking up of a mass into fragments. Applied in astrophysics to discuss collisions of minor planets, and in geophysics to describe rock processing where *fragmentation* takes place on a wide range of scales and occurs on joints and faults. On the largest scale the plates of plate tectonics are fragments that are the result of fragmentation.

frame dragging   The phenomenon in relativistic theories of gravitation, particularly in general relativity in which the motion of matter, e.g., translating or rotating matter "drags" the inertial frame, meaning the inertial frames near the moving matter are set in motion with respect to the distant stars, in the direction of the matter motion. For instance, the plane of a circular polar orbit around a rotating primary rotates in the direction of the central rotating body. In the case of an orbit around the Earth, the gravitational fields are weak, and the effect was first calculated by Lense and Thining in 1918. The plane of the polar orbit rotates in this case at a rate:

$$\dot{\Omega} = \frac{2GJ}{c^2 r^3} .$$

Here c is the speed of light, G is Newton's gravitational constant, and J is the angular momentum of the isolated rotating planet. For an orbit close to the surface of the Earth, this rate is approximately 220 milliarc sec/year.

This result actually applies to the dragging of the line of nodes of any orbit, with the following modification due to eccentricity e:

$$\dot{\Omega} = \frac{2J}{a^3(1 - e^2)^{3/2}} .$$

The pointing direction of a gyroscope near a rotating object is also affected by the frame dragging. The precession rate due to this effect for a gyroscope in circular orbit is:

$$\dot{\Omega}_\delta = \frac{3\hat{\mathbf{r}}(\mathbf{J} \cdot \hat{\mathbf{r}}) - J}{r^3} ,$$

where $\hat{r}$ is the unit vector to the gyroscope position. For a gyroscope in orbit at about 650 km altitude above this rate is approximately 42 milliarc sec/year. A second relativistic effect, the deSitter precession, constitutes

$$\frac{3}{2} \frac{M}{r^2} (\hat{\mathbf{r}} \times \mathbf{v}) ,$$

where M is the mass of the central body, and $\mathbf{v}$ is the velocity of the body in its orbit. For a 650-km altitude orbit around the Earth, this contribution is of order 6.6 arcsec/year.

Proposed space experiments would measure the orbital dragging (LAGEOS-III experiment) and the gyroscope precession (Gravity Probe B, GPB).

In strong field situations, as in close orbit around (but still outside) a rotating black hole dragging is so strong that no observer can remain at rest with respect to distant stars.

Fraunhofer lines   Absorption lines in the solar spectrum (first observed by Fraunhofer in 1814).

free-air correction   A correction made to gravity survey data that removes the effect of the elevation difference between the observation point and a reference level such as mean sealevel. It is one of several steps taken to reduce the data to a common reference level. The *free-air correction* accounts only for the different distances of the two elevations from the center of the earth, ignoring the mass between the two elevations. The commonly used formula for free-air correction for elevation difference h is $\Delta g = 2h\, g_h/R$, where g_h is the measured gravity value, and R is the average radius of the Earth.

free-air gravity anomaly   The difference between the measured gravity field and the reference field for a spheroidal Earth; the correction to the value of the gravitational acceleration (g) which includes the height of the measuring instrument above the geoid. It is called "free air" because no masses between the instrument and the geoid are included in the correction. Inclusion of the excess gravitational pull due to masses between the measuring instrument and the geoid gives rise to the Bouguer anomaly.

free atmosphere   The atmosphere above the level frictionally coupled to the surface, usually taken as about 500 m.

free bodies   A physical body in which there are no forces applied. Such a body will maintain a uniform motion (until a force acts upon

it) according to the first law of Newtonian Mechanics.

free-bound continuum emission   Radiation produced when the interaction of a free electron with an ion results in the capture of the electron onto the ion (e.g., recombination emission). The most energetic spectral lines of importance in solar physics belong to the Lyman series of Fe XXVI. The continuum edge for this series lies at 9.2 keV, so that all radiation above this energy must be in the form of continua, both free-bound and free-free. As the energy increases, the contribution of the free-bound emission to the total emission falls off relative to the free-free emission.

free convection   Flow of a fluid driven purely by buoyancy. In the case of thermal free convection, the necessary condition for its onset is that the vertical thermal gradient must be greater than the adiabatic gradient. Whether *free convection* actually occurs depends on geometrical constraints. Parts of the Earth's mantle and core are believed to be freely convecting, but little is known about the patterns of convection.

free-free continuum emission   Radiation produced when the interaction of a free electron with an ion leaves the electron free (e.g., bremsstrahlung).

free oscillations   When a great earthquake occurs the entire Earth vibrates. These vibrations are known as *free oscillations*.

freeze   A meteorological condition in which the temperature at ground level falls below 0°C.

F region   The *F region* is the part of the ionosphere existing between approximately 160 and 500 km above the surface of the Earth. During daytime, at middle and low latitudes the F region may form into two layers, called the F_1 and F_2 layers. The F_1 layer exists from about 160 to 250 km above the surface of the Earth. Though fairly regular in its characteristics, it is not observable everywhere or on all days. The F_1 layer has approximately 5×10^5 e/cm^3 (free electrons per cubic centimeter) at noontime and minimum sunspot activity, and increases to roughly 2×10^6 e/cm^3 during maximum sunspot activity. The density falls off to below 10^4 e/cm^3 at night. The F_1 layer merges into the F_2 layer at night. The F_2 layer exists from about 250 to 400 km above the surface of the Earth. The F_2 layer is the principal reflecting layer for HF communications during both day and night. The longest distance for one-hop F_2 propagation is usually around 4000 km. The F_2 layer has about 10^6 e/cm^3 and is thus usually the most densely ionized ionospheric layer. However, variations are usually large, irregular, and particularly pronounced during ionospheric storms. During some ionospheric storms, the F region ionization can reduce sufficiently so that the F_2 region peak electron density is less than the F_1 region peak density. *See* ionosphere, ionogram, ionospheric storm, spread F, traveling ionospheric disturbance, winter anomaly.

frequency of optimum traffic (FOT)　*See* Optimum Working Frequency (OWF).

Fresnel reflectance　The fraction of radiant energy in a narrow beam that is reflected from a surface at which there is an index of refraction mismatch.

Fresnel zone　Any one of the array of concentric surfaces in space between transmitter and receiver over which the increase in distance over the straight line path is equal to some integer multiple of one-half wavelength. A simplification allowing approximate calculation of diffraction.

fretted channels　One channel type considered to indicate fluvial activity on Mars. *Fretted channels* are a channel type affecting much of the fretted terrain, thus straddling the highland–lowland boundary of Mars. Their formation is restricted to two latitude belts centered on 40°N and 45°S and spanning $\approx 25°$ wide. They are most extensive between longitude 280°W to 350°W. They extend from far in the uplands down to the lowland plains and represent broad, flat-floored channels, in which flow lines are a common feature. *See* fretted terrain.

fretted terrain　Part of the highland–lowland boundary region of Mars, lying along a great

circle having a pole at $\approx 145°$W and 55°N. The terrain also exists around the high-standing terrain retained in the northern lowland plains. It is characterized by flat-topped outliers of cratered uplands, termed plateaux, mesas, buttes, and knobs depending on their size. The differences in size are thought to reflect different extents of fracturing and subsequent modification, whereby greater fracturing and modification created the smaller landforms.

Plateaux, mesas, and buttes are generally considered to have formed during creation of the relief difference. Alternatively, it has been proposed that they represent flat-topped table mountains that formed due to the interaction of basaltic lava and ground water at low eruption rates. The final form of the *fretted terrain* has also been accounted for as representing the shorelines of a sea and by scarp retreat owing to mass wasting and sublimation of volatiles, or ground-water sapping.

friction　The process whereby motion of one object past another is impeded, or the force producing this impediment. *Friction* is caused by microscopic interference between moving surfaces, and/or by microscopic fusion between the surfaces, which must be broken to continue motion. Friction is dissipative, producing heat from ordered kinetic energy. The frictional shear stress on a surface S'_f is a fraction of the normal stress on the surface S_n so that

$$S'_f = f S_n$$

where f is the coefficient of friction. This is known as Amontons's law and f has a typical value of 0.6 for common surfaces in contact, and in geophysical processes. Friction controls the behavior of geophysical faults.

friction factor　A coefficient that indicates the resistance to a flow or movement. Generally empirical and dimensionless (e.g., Darcy–Weisbach friction factor).

friction slope　Energy loss, expressed as head (energy loss per unit weight of fluid) per unit length of flow. Appears in the Manning Equation.

friction velocity   In geophysical fluid dynamics, the ratio

$$u_* = \sqrt{\frac{\tau}{\rho}},$$

often used to express the shear stress τ at the boundary as a parameter that has the dimensions of velocity. It represents the turbulent velocity in the boundary layer, which is forced by the surface shear stress. *See* law-of-the-wall layer.

Friedmann, Aleksandr Aleksandrovich (July 16, 1888–September 16, 1925) Russian mathematician/physicist/meteorologist/aviation pioneer who was the first man to derive realistic cosmological models from Einstein's relativity theory (in 1922 and 1924). His models, later refined by Lemaître and others, still are the backbone of modern cosmology.

Friedmann–Lemaître cosmological models   The cosmological models that incorporate the cosmological principle and in which the cosmic matter is a dust. (For models with the same geometrical properties, but with more general kinds of matter, *see* Robertson–Walker cosmological models.) Because of the assumed homogeneity and isotropy, all scalar quantities in these models, like matter-density or expansion, are independent of position in the space at any given time, and depend only on time. All the components of the metric depend on just one unknown function of one variable (time) that is determined by one of the 10 Einstein's equations — the only one that is not fulfilled identically with such a high symmetry. The only quantity that may be calculated from these models and (in principle) directly compared with observations is the time-evolution of the average matter-density in the universe. However, these models have played a profound role in cosmology through theoretical considerations that they inspired. The most important of these were: cosmical synthesis of elements in the hot early phase of the evolution, the emission of the microwave background radiation, theoretical investigation of conditions under which singularities exist or do not exist in spacetime, and observational tests of homogeneity and isotropy of the universe that led to important advances in the knowledge about the large-scale matter-distribution. The *Friedmann–Lemaître models* come in the same three varieties as the Robertson–Walker cosmological models. When the cosmological constant Λ is equal to zero, the sign of the spatial curvature k determines the future evolution of a Friedmann–Lemaître model. With $k \leq 0$, the model will go on expanding forever, with the rate of expansion constantly decelerated (asymptotically to zero velocity when $k = 0$). With $k > 0$, the model will stop expanding at a definite moment and will afterward collapse until it reaches the final singularity. In the Friedmann–Lemaître models with $\Lambda = 0$, the sign of k is the same as the sign of $(\bar{\rho} - \rho_0)$, where ρ_0 is called the critical density. *See* expansion of the universe. With $\Lambda \neq 0$, this simple connection between the sign of k and the sign of $(\bar{\rho} - \rho_0)$ (and so with the longevity of the model) no longer exists. The Friedmann–Lemaître models are named after their first discoverer, *A.A. Friedmann* (who derived the $k > 0$ model in 1922 and the $k < 0$ model in 1924), and *G. Lemaître* (who rediscovered the $k > 0$ model in 1927 and discussed it in connection with the observed redshift in the light coming from other galaxies).

fringe (interference)   In interferometry, electromagnetic radiation traversing different paths is compared. If the radiation is coherent, interference will be observed. Classically the detection area had sufficient transverse extent that geometric factors varied the path lengths, and some parts of the field exhibited bright fringes while other parts exhibited dark fringes.

front   In meteorology, a surface of discontinuity between two different and adjacent air masses which have different temperature and density. Through the front region, the horizontal cyclonic shear is very strong, and the gradients of temperature and moisture are very large. *Fronts* can have lengths from hundreds to thousands of kilometers. The width of a front is about tens of kilometers. As the height increases, the front tilts to the cold air side, with warm air above cold air. In a frontal region, temperature advection, vorticity advection and thermal wind vorticity advection are very strong leading to the development of complicated weather systems, increasingly with the strength of the front. Different types of fronts are distinguished according to the nature of the air masses separated by the front, the direction of the front's advance, and stage of development. The term was first devised by Professor V. Bjerknes and his colleagues in Norway during World War I.

frontogenesis   Processes that generate fronts which mostly occur in association with the developing baroclinic waves, which in turn are concentrated in the time-mean jet streams.

frost point   The temperature to which air must be cooled at constant pressure and constant mixing ratio to reach saturation with respect to a plane ice surface.

Froude number   A non-dimensional scaling number that describes dynamic similarity in flows with a free surface, where gravity forces must be taken into consideration (e.g., problems dealing with ship motion or open-channel flows). The *Froude number* is defined by

$$Fr \equiv \frac{U}{\sqrt{gl}}$$

where g is the constant of gravity, U is the characteristic velocity, and l is the characteristic length scale of the flow. Even away from a free surface, gravity can be an important role in density stratified fluids. For a continuously stratified fluid with buoyancy frequency N, it is possible to define an internal Froude number

$$Fr_i \equiv \frac{U}{Nl} .$$

However, in flows where buoyancy effects are important, it is more common to use the Richardson number $Ri = 1/Fr_i^2$.

frozen field approximation   If the turbulent structure changes slowly compared to the time scale of the advective ("mean") flow, the turbulence passing past sensors can be regarded as "frozen" during a short observation interval. Taylor's (1938) *frozen field approximation* implies practically that turbulence measurements as a function of time translate to their corresponding measurements in space, by applying $k = \omega/u$, where k is wave number [rad m^{-1}], ω is measurement frequency [rad s^{-1}], and u is advective velocity [m s^{-1}]. The spectra transform by $\phi(k)dk = u\phi(\omega)d\omega$ from the frequency to the wavenumber domain.

frozen flux   If the conductivity of a fluid threaded by magnetic flux is sufficiently high that diffusion of magnetic field within the fluid may be neglected, then magnetic field lines behave as if they are frozen to the fluid, i.e., they deform in exactly the same way as an imaginary line within the fluid that moves with the fluid. Magnetic forces are still at work: forces such as magnetic tension are still communicated to the fluid. With the diffusion term neglected, the induction equation of magnetohydrodynamics is:

$$\frac{\partial \mathbf{B}}{\partial t} = \nabla \times (\mathbf{u} \times \mathbf{B}) .$$

By integrating magnetic flux over a patch that deforms with the fluid and using the above equation, it can be shown that the flux through the patch does not vary in time. *Frozen flux* helps to explain how helical fluid motions may stretch new loops into the magnetic field, but for Earth-like dynamos some diffusion is required for the dynamo process to work. The process of calculating flows at the surface of the core from models of the magnetic field and its time variation generally require a frozen flux assumption, or similar assumptions concerning the role of diffusion. *See* core flow.

"frozen-in" magnetic field   A property of magnetic fields in fluids of infinite electrical conductivity, often loosely summarized by a statement to the effect that magnetic flux tubes move with the fluid, or are "frozen in" to the fluid. An equivalent statement is that any set of mass points threaded by a single common magnetic field line at time $t = 0$ will be threaded by a single common field line at all subsequent times t.

It is a quite general consequence of Maxwell's equations that for any closed contour C that co-moves with a fluid, the rate of change of magnetic flux Φ contained within C is (cgs units)

$$\frac{d\Phi}{dt} = \oint \left(\mathbf{E} + \frac{1}{c}\mathbf{V} \times \mathbf{B} \right) \cdot d\mathbf{x} ,$$

where **E** and **B** are the electric and magnetic fields, **V** is the fluid velocity, c is the (vacuum) speed of light, and the integral is taken about the contour C. In fluids of infinite electrical conductivity (and for that matter, to an excellent approximation in most situations involving collisionless plasmas)

$$\mathbf{E} + \frac{1}{c}\mathbf{V}\times\mathbf{B} = 0 \, ,$$

so that

$$\frac{d\Phi}{dt} = 0 \, .$$

When this condition is satisfied, the magnetic flux through any closed contour C that co-moves with a fluid is constant; it is in this sense that magnetic flux tubes may be considered to "move with a fluid". This result is sometimes called Alfvén's theorem.

F star   Star of spectral type F. Canopus and Procyon are *F stars*.

Fukushima's theorem   In two articles in 1969 and 1976, Naoshi Fukushima showed that the main contributions to the magnetic field, observed on the ground from field aligned currents which flow in and out of the Earth's ionosphere, tended to cancel (the mathematical principle might have been known before, but was not applied to the ionosphere). He showed that with a uniformly conducting spherical ionosphere, linked to infinity by straight conducting filaments (an idealization of the actual geometry), there is no magnetic effect below the ionosphere. It explained why the main magnetic effect seen on the ground comes from secondary currents, the Hall currents which form the auroral electrojets.

fully arisen sea   A sea condition whereby continued energy input by wind will not increase wave energy.

fully rough flow   Hydrodynamic flow near a boundary in which the Reynolds number computed using the typical surface irregularity scale ϵ as the length exceeds approximately 100:

$$u\epsilon/\nu \geq 100$$

where u is the velocity and ν is the kinematical viscosity. *See* kinematic viscosity, Reynolds number.

fulvic substance   In oceanography, high molecular weight organic compounds resulting from plant decay, especially phytoplankton. *See* colored dissolved organic matter.

fundamental tensors of a worldsheet   A cosmic string is a type of cosmic topological defect which may play a role in producing the structures and features we see in our present universe. It can be conveniently described as long and infinitely thin, but a line evolving in space and time and describing a two-dimensional "surface" called a worldsheet, whose evolution is known provided one knows its dynamics and its geometry. Fundamental tensors provide the knowledge of the worldsheet geometry, while the equation of state allows one to compute its dynamics.

The position of the worldsheet in spacetime requires knowledge of its coordinates $x^\mu(\ell, \tau)$ depending on two variables internal to the worldsheet: a curvilinear space coordinate ℓ and a time τ, denoted collectively as ξ_a. One can define a 2×2 induced metric γ_{ab} on the string worldsheet, using the spacetime metric $g_{\mu\nu}$, through

$$\gamma_{ab} = g_{\mu\nu}\frac{\partial x^\mu}{\partial\xi^a}\frac{\partial x^\nu}{\partial\xi^b} \, ,$$

with inverse γ^{ab}. Then the *first fundamental tensor* of the worldsheet is

$$\eta^{\mu\nu} = \gamma^{ab}\frac{\partial x^\mu}{\partial\xi^a}\frac{\partial x^\nu}{\partial\xi^b} \, .$$

The surface spanned by the string in spacetime is curved in general. This is quantified by the *second fundamental tensor* $K_{\mu\nu}{}^\rho$ which describes how the 2-dimensional sheet is embedded in spacetime, and is calculable by means of the covariant derivative in spacetime ∇_μ

$$K_{\mu\nu}{}^\rho = \eta_\mu^\sigma\eta_\nu^\alpha\nabla_\alpha\eta_\sigma^\rho \, ,$$

with symmetry

$$K_{\mu\nu}{}^\rho = K_{\nu\mu}{}^\rho \, , \text{ and trace}$$
$$K^\mu = K^\alpha{}_\alpha{}^\mu \, .$$

One interesting particular case is the Goto–Nambu string, for which we simply have $K^\mu = 0$ as the sole equation. *See* covariant derivative, equation of state (cosmic string), Goto–Nambu string, metric, summation convention, Witten conducting string.

future/past causal horizon   The boundary that separates two disconnected regions I and O of space-time such that: **1.** no physical signal originating from I can ever reach (at least) a particular class of observers in O (*future horizon*) or **2.** (at least) a particular class of observers in O cannot send signals into I (*past horizon*). By *causal horizon* (with no further specifications) case **1.** is usually meant.

future/past event horizon   A *future event horizon* is the future causal horizon which separates an eternal black hole B from the disconnected external region O such that no physical signal originating from inside B can ever reach any observer located in O. On the other hand, it is possible to send signals from O into B (a mechanical analog would be a one-way permeable membrane).

Any external observer (in O) not falling in the gravitational field of the black hole sees a signal sent towards the black hole approach the horizon but never cross it. An observer co-moving with the signal, as any locally inertial observer, would instead cross the horizon at a finite (proper) time without noticing any local effect (without experiencing any particular discomfort).

It can be proven that an event horizon is a light-like surface which necessarily shares the symmetry of O. This leads to a classification of all possible families of black holes in terms of the ADM mass and a few other parameters *see* No-hair theorems.

A *past event horizon* is the past causal horizon circumventing a white hole. *See* ADM mass, black hole, black hole horizon, future/past causal horizon, Rindler observer, white hole.

G

gabbro   A coarse-grained igneous rock with the composition of basalt, composed of calcic plagioclase, a ferromagnesium silicate, and other minerals. Classified as plutonic because the large crystal size indicates formation by slow crystallization, typical of crystallization at great depth in the Earth.

galactic bulge   Many spiral galaxies have a central core region which is roughly spherical with radius greater than the disk thickness. This is the *galactic bulge.*

galactic cluster   *See* open cluster.

galactic coordinates   Coordinates measuring angular positions on the sky, taking the location of Sagittarius (RA $12^h 49^m$, dec $27°24'$) as the origin of galatic latitude (measured north or south of the galactic plane), and longitude, measured increasing eastward. The north galactic pole is taken in the same hemisphere as the pole north of the celestial equator. The tilt of the galactic plane to the celestial equator is $62°36'$. Transformations between galactic and celestial coordinates are:

$$\cos b^{II} \cos(l^{II} - 33°)$$
$$= \cos \delta \cos(\alpha - 282.25°) ,$$
$$\cos b^{II} \sin(l^{II} - 33°)$$
$$= \cos \delta \sin(\alpha - 282.25°) \cos 62.6°$$
$$+ \sin \delta \sin 62.6° ,$$

$$\sin b^{II} = \sin \delta \cos 62.6°$$
$$- \cos \delta \sin(\alpha - 282.25°) \sin 62.6° ,$$
$$\cos \delta \sin(\alpha - 282.25°)$$
$$= \cos b^{II} \sin(l^{II} - 33°) \cos 62.6°$$
$$- \sin b^{II} \sin 62.6° ,$$
$$\sin \delta = \cos b^{II} \sin(l^{II} - 33°) \sin 62.6°$$
$$+ \sin b^{II} \cos 62.6° ,$$

$l^{II} =$ new galactic longitude ,
$b^{II} =$ new galactic latitude ,
$\alpha =$ right descension (1950.0) ,
$\delta =$ declination (1950.0) ,

For, $l^{II} = b^{II} = 0$:

$$\alpha = 17^h 42^m 4, \ \delta = -28°55'(1950.0) ;$$
$$b^{II} = +90.0, \ \text{galactic north pole:}$$
$$\alpha = 12^h 49^m , \ \delta = +27°4(1950.0) .$$

galactic disk   A relatively thin (400 to 1000 pc) component of disk galaxies extending out to 15 to 25 kpc, usually exhibiting a spiral structure of bright stars. Stellar population is mostly Population I (young) stars. In the Milky Way Galaxy, the oldest Population I stars have 0.1 times the metal abundance of the sun and have slightly elliptical orbits rising up to 1000 pc from the disk plane. Stars like the sun (0.5 to 1 times solar metal abundance) can be found up to 300 pc from the plane. Stars with metals 1 to 2 solar extend to 200 pc above the disk. Star forming regions in the spiral arms (containing O and B stars) have circular orbits and are within 100 pc of the disk. Metal abundances there are 1 to 2.5 solar. *See* metalicity, galactic bulge, spiral arm.

galactic globular cluster   Associations of 100,000 to a million stars which have extremely high stellar densities. These clusters are composed of stars that all have the same chemical composition and age. The stars all formed from the same proto-Galactic fragment of gas, which accounts for the fact that the stars have the same composition. Moreover, all of the stars in a globular cluster are the same age because 20 to 30 million years after the first stars formed, the first supernovae would have detonated. These supernovae drove all of the remaining gas out of the cluster, thereby ending star formation. As the oldest stellar populations in the Milky Way Galaxy, *globular clusters* provide a firm lower limit to the age of the galaxy, and therefore the universe.

galactic noise   Galactic radio noise originates in the galactic center. However, it is common to include in this any contributions from noise sources outside the Earth's atmosphere. Because the source is above the Earth's ionosphere, some *galactic noise* will be reflected back into space and only contributions from above the local peak F region electron density will be observed at the ground. Consequently, this source of noise is only important for radio circuits operating above the local foF2. *See* atmospheric noise.

galactic wind Large-scale outflow of gaseous matter from a galaxy. Evidence of *galactic winds* is provided by the morphology of X-ray emitting regions, elongated along an axis perpendicular to the major axis of a highly inclined disk galaxy. More rarely, it is possible to observe, as in the case of the spiral galaxy NGC 1808 or of the prototype Starburst galaxy M82, the presence of optical filaments suggesting outflow from the inner disk. A galactic wind is currently explained as due to an intense, concentrated burst of star formation, possibly induced by gravitational interaction with a second galaxy. As the frequency of supernova blasts increases following the production of massive stars, supernova ejecta provide mechanical energy for the outflow and produce tenuous hot gas, which is seen in the X-ray images. Most extreme galactic winds, denoted super-winds, could create a bubble of very hot gas able to escape from the potential well of the galaxy and diffuse into the intergalactic medium. Super-winds are thought to be rare in present-day universe, but may have played an important role in the formation and evolution of elliptical galaxies, and in the structure of the medium within clusters of galaxies. *See* starburst galaxy.

Galatea Moon of Neptune also designated NVI. Discovered by Voyager 2 in 1989, it is a small, roughly spherical body approximately 79 km in radius. It is very dark, with a geometric albedo of 0.063. Its orbit has an eccentricity of 0.00012, an inclination of 0.054°, a precession of $261°$ yr^{-1}, and a semimajor axis of 6.20×10^4 km. Its mass has not been measured. It orbits Neptune once every 0.429 Earth days.

galaxies, classification of A scheme, the most prominent of which is due to Hubble, to classify galaxies according to their morphology. In the Hubble scheme, elliptical galaxies are classified $E0$ (spherical) to $E7$ (highly flattened). The number indicator is computed as $10(a - b)/a$ where a is the semimajor axis and b is the semiminor axis of the galaxy. Spiral galaxies are divided into two subclasses: ordinary, denoted S, and barred, denoted SB. *S0* galaxies are disk galaxies without spiral structure, and overlap $E7$ galaxies. Spirals are further classified as a, b, or c, progressing from tight, almost circular, to more open spirals. An additional class is Irregular, which are typically small galaxies, with no symmetrical (elliptical or spiral) structure.

Other classification systems are those of de-Vaucouleurs, of Morgan, and the DDO system.

galaxy (**1.**) The Milky Way galaxy. A spiral galaxy with a central bulge, of about 1000 pc in extent, a disk component of order 1000 pc in thickness and extending about 15 to 20 kpc from the center, and a roughly spherical halo extending 50000 kpc. The total mass is about $1.5 \times 10^{11} M_\odot$ and the number of stars in the *galaxy* is of order 2×10^{11}; the total luminosity is of order $10^{11} L_\odot$. The sun is located in a spiral arm about 10 kpc from the center.

(**2.**) A large gravitational aggregation of stars, dust, and gas. Galaxies are classified into spirals, ellipticals, irregular, and peculiar. Sizes can range from only a few thousand stars (dwarf irregulars) to $10^{13} M_\odot$ stars in giant ellipticals. Elliptical galaxies are spherical or elliptical in appearance. Spiral galaxies range from *S0*, the lenticular galaxies, to *Sb*, which have a bar across the nucleus, to *Sc* galaxies which have strong spiral arms. Spirals always have a central nucleus that resembles an elliptical galaxy. In total count, ellipticals amount to 13%, *S0* to 22%, *Sa, b, c* galaxies to 61%, irregulars to 3.5% and peculiars to 0.9%. There is a morphological separation: Ellipticals are most common in clusters of galaxies, and typically the center of a cluster is occupied by a giant elliptical. Spirals are most common in the "field", i.e., not in clusters.

Galilean invariance The invariance of physical expressions under the Galilean transformation from one coordinate system to another coordinate system which is moving uniformly with respect to the first. Newton's Laws of Mechanics are Galilean Invariant. Named after Galileo

Galilei (1564–1642). *See* Galilean transformation.

Galilean relativity   The variance of physical expressions under the Galilean transformation from one coordinate system to another coordinate system which is moving uniformly with respect to the first. Position and velocity are relative or variant under Galilean transformation, but the Laws of Mechanics are invariant under such transformations. Named after Galileo Galilei (1564–1642). *See* Galilean transformation.

Galilean transformation   The transformation of spatial coordinates from one reference system to another reference system, moving uniformly with respect to the first, according to the following expression: $\vec{r}' = \vec{r} - \vec{v}t$, where $\vec{v}$ is the relative velocity between the two reference systems. Under *Galilean transformation* time is considered an absolute variable; that is, its value is the same for all reference systems. In special relativity, the Galilean transformations are superseded by the Lorentz transformations. Named after Galileo Galilei (1564–1642).

Galilei, Galileo   Physicist and astronomer (1564–1642). One of the first to use the telescope for astronomical observations, and improved its design. Discovered sunspots, lunar mountains, and valleys, the four largest satellites of Jupiter now known as the Galilean Satellites, and the phases of Venus. In physics he discovered the laws of falling bodies and the law of the pendulum.

Galileo spacecraft   A spacecraft launched in 1989 that arrived at Jupiter on December 7, 1995. Galileo's atmospheric probe plunged into the Jovian atmosphere on the same day, to relay information on its structure and composition, including major cloud decks and lightning. The spacecraft's orbiter studied the giant planet, its rings and its moons, and the magnetic environment. The primary mission ended in December 1997. A two-year extended mission was then concentrated on Europa and ended with two close flybys of Io in December 1999. A mosaic of images of Europa show several terrain types that provide evidence for the existence of a liquid ocean under the surface. Galileo is now continuing its mission under another extension, studying Jupiter's largest moon, Ganymede.

NASA's Galileo project is managed by JPL. The Galileo probe development and operations are the responsibility of NASA's Ames Research Center.

Galiliean satellite   One of the four Jovian satellites discovered via telescope observations by Galileo: Io, Europa, Ganymede, Callisto, in order of distance from Jupiter.

gallium   A silver-white rare metallic element having the symbol Ga, the atomic number 31, atomic weight of 69.72, melting point of 29.78°C, and a boiling point of 2403°C, soft enough to cut with a knife. Specific gravity of solid (29.6°C), 5.904; specific gravity of liquid (29.8°C), 6.095. *Gallium* compounds, especially gallium arsenide, are used as semiconductors. The metal is used as a substitute for mercury in high temperature thermometers.

gamma   A unit of magnetic field intensity employed in geophysics, equal to one nanotesla (1.0×10^{-9} tesla) and 1.0×10^{-5} gauss. This unit is used because of its convenient size for expressing fluctuations that occur in the Earth's field, due to solar influence and geomagnetic storm phenomena. *See* geomagnetic field.

γ-ray   Electromagnetic radiation (photons) with energy greater than about 0.1 MeV (wavelength less than about 0.2 Å).

gamma ray burst (GRB)   Tremendous flashes of radiation first detected in the gamma-ray region of the spectrum. GRBs were discovered about 30 years ago with instruments intended to monitor nuclear test ban treaties. Presently thought to be the most powerful explosions in nature (after the Big Bang), their sources have only recently been localized by observations of associated afterglows in X-rays, visible light, and radio waves, delayed in that order. In gamma-rays, GRBs last from a millisecond to hours, although optical and radio afterglows may go on for weeks. Several afterglows attributed to GRBs have redshifted spectral lines indicating occurrence at several billion light-years distant. At this distance, the energy released is similar to the rest-mass energy of the sun. Afterglow features match many of the theoretical predictions of an expanding "fireball," but the source mechanism itself remains unknown. Other transients, seemingly similar and found with instrumentation designed to study gamma ray bursts, have in recent years been confirmed as distinctly separate phenomena, coming from neutron stars located in this galaxy or its Magellanic cloud satellite galaxies. *See* soft gamma repeaters, March 5th event.

gamma-ray burst, black hole accretion disks
Under the gamma-ray burst "fireball" paradigm (*see* gamma-ray bursts, fireball), any gamma-ray burst engine must produce high energies without ejecting too much baryonic matter. A class of models, all of which produce rapidly accreting tori around black holes, provide natural explanations for the high energies, but low ejecta mass. The gravitational potential energy is converted to energy in a pair/plasma fireball either from the neutrino annihilation or magnetic field energy mechanisms, both of which require an asymmetry in the mass accretion. The energy is deposited along the disk rotation axis, producing a beamed jet. This beaming lessens the burst energy requirements (most papers quote gamma-ray burst energies assuming isotropic explosions) and avoids excessive baryon contamination. This class of models includes mergers of double neutron star systems, mergers of black hole and neutron star binaries, mergers of black hole and white dwarf binaries, collapsars, and helium core mergers.

gamma-ray burst, classical   Classical gamma-ray bursts make up the bulk of the observed gamma-ray bursts. They do not appear to repeat, and have hard spectra (95% of energy emitted by photons with energy greater than 50 keV). However, beyond these characteristics, classical gamma-ray bursts represent a very heterogeneous set of objects. Burst durations range from 0.01 to 300 s, during which time the burst may be chaotic, exhibiting many luminosity peaks, or it may vary smoothly. They are distributed isotropically and are thought to originate outside of the galaxy (cosmological bursts). This translates to burst energies in the range: $\sim 10^{48}$

to 10^{53} ergs, in some cases the most energetic explosions in the universe since the Big Bang.

gamma-ray burst, classification   Gamma-ray bursts have been separated into two major classes: soft gamma-ray repeaters and "classical" gamma-ray bursts. Soft gamma-ray repeaters repeat, have average photon energies of 30 to 50 keV, burst durations of ~ 0.1 s, smooth light-curves, and lie in the galactic plane. Classical bursts do not seem to repeat, emit most of their energy $\gtrsim 50$ keV, have a range of durations (0.01 to 300 s) and light-curve profiles. Classical bursts are distributed isotropically and are thought to originate at cosmological distances outside of the galaxy or galactic halo. Soft gamma-ray repeaters are thought to be caused by accretion, magnetic field readjustment, or quakes in neutron stars in the galactic disk.

gamma-ray burst, cosmological mechanisms
Spectra of the optical counterpart of GRB970508 reveal many red-shifted ($z = 0.835$) absorption lines, confirming that at least some gamma-ray bursts are cosmological; that is, they occur outside of the Milky Way. Cosmological models provide a simple explanation for the isotropic spatial distribution of gamma-ray bursts, but require energies in excess of 10^{51} ergs. Most cosmological models rely upon massive accretion events upon compact objects (e.g., the merger of two neutron stars). Several mechanisms ultimately produce accretion disks around black holes where the gravitational energy released can be more readily converted into relativistic jets. These black-hole accretion disk models provide an ideal geometry for facilitating the potential energy conversion via the neutrino annihilation or magnetic field mechanisms into beamed gamma-ray burst jets.

gamma-ray burst, fireball   The gamma-ray burst fireball refers to a mechanism by which the energy produced near the black hole or neutron star source is converted into the observed burst. Due to photon-photon scattering and electron scattering opacities at the source of the burst, the burst photons are initially trapped. As the "fireball" expands adiabatically, the optical depth decreases, but so does the temperature. In the current scenario, the fireball does not become optically thin until much of the internal energy has been converted into kinetic energy through expansion. The observed gamma-ray burst spectrum is then produced by internal shocks in the fireball or by shocks created as the fireball sweeps up the interstellar medium.

gamma-ray burst, galactic mechanisms
Prior to the launch of the Burst and Transient Source Experiment (BATSE), galactic models were the "favored" mechanisms for gamma-ray bursts. The bulk of these models involve sudden accretion events onto neutron stars or some sort of glitch in the neutron star (e.g., neutron star quakes). However, the data from BATSE revealed that the bursts are isotropically distributed in the sky, limiting galactic models to those that occur in the galactic halo. The advantage of such mechanisms is that the energy requirements for galactic explosions is nearly 8 orders of magnitude lower than their cosmological counterparts ($\sim 10^{43}$ ergs). Recent observations of the optical afterglow of gamma-ray bursts, specifically the detection of redshifted lines in GRB970508, place at least some of the gamma-ray bursts at cosmological redshifts.

gamma-ray burst, GRB 970508   The May 8, 1997 gamma-ray burst is the first gamma-ray burst where a reliable optical counterpart revealed identifiable metal absorption lines. The lines are probably caused by some intervening material between our galaxy and the gamma-ray burst source. A redshift of the absorbing medium of $z = 0.835$ was inferred from these 8 lines, setting the minimum redshift of the gamma-ray burst (Metzger et al., 1997). Other evidence indicates that the absorbing material was part of the host galaxy of the gamma-ray burst, placing the gamma-ray burst at a redshift $z = 0.835$. This gamma-ray burst provides indisputable evidence that at least some gamma-ray bursts are cosmological.

gamma-ray burst, hypernova   The hypothetical explosion produced by a collapsar, the collapse of a rotating massive star into a black hole, which would produce a very large γ-ray burst. *See* gamma-ray burst models, collapsar.

gamma-ray burst, magnetic fields   The generation and stretching of magnetic field lines have been proposed as mechanisms to convert the energy of material accreting onto a black hole in active galactic nuclei and gamma-ray bursts alike. The "standard" mechanism is that described by Blandford and Znajek (1977) which uses magnetic field interactions in the disk to extract the rotational energy of the black hole. Other mechanisms exist which extract the potential energy of the accreting matter. Gamma-ray bursts require magnetic field strengths in excess of $\sim 10^{15}$ Gauss, which is roughly 10% of the disk equipartion energy.

gamma-ray burst, mechanisms   Over 100 proposed distinct mechanisms for classical gamma-ray bursts exist which can be grouped roughly into three categories based on their location: solar neighborhood GRBs, galactic GRBs, and cosmological GRBs. Solar neighborhood GRB mechanisms are the least likely and calculations of the proposed mechanisms do not match the observations. Galactic models have the advantage that they require much less energy than cosmological models. However, the isotropic distribution of bursts require that these models be in the galactic halo. In addition, absorption lines in the spectra of the optical counterparts of GRBs, namely GRB970508, indicate that some bursts must be cosmological. Because of this evidence, cosmological models are the favored class of models, despite their high energy requirements (up to 10^{53} ergs).

The major constraint of any mechanism is that it must produce sufficient energy ($\gtrsim 10^{51}$ ergs for an isotropic cosmological burst) with relatively little contamination from baryons. The low mass in the ejecta is required to achieve the high relativistic velocities ($\Gamma = 1 + E_{burst}/M_{ejecta}$). Beaming of the burst reduces both the energy requirement (the energies generally quoted in papers assume an isotropic explosion) and limit the baryonic contamination. This beaming is predicted by most of the viable gamma-ray burst models.

gamma-ray burst models, collapsar   The cores of stars with masses above $\sim 10 M_\odot$ overcome electron degeneracy pressure and collapse. For stars with masses $\lesssim 25$ to $50 M_\odot$, this collapse eventually drives a supernova explosion and the formation of a neutron star. However, more massive stars eventually collapse into black holes. If those stars are rotating, they may drive a gamma-ray burst by forming an accretion disk around the collapsed core via the black hole accretion disk paradigm. A hypernova is the observed emission from a collapsar, just as a Type II supernova is the observed emission from a core-collapse supernova.

gamma-ray burst models, helium merger
X-ray binaries are powered by the accretion of material onto a neutron star or black hole from its close-binary companion. In some cases, the star may continue to expand until it engulfs the compact object. The compact object then spirals into the hydrogen envelope of this companion, releasing orbital energy which may eject the envelope. However, if there is insufficient energy to eject the envelope, the neutron star or black hole will merge with the core, accreting rapidly via neutrino emission (*see* accretion, super-Eddington). The compact object quickly collapses to a black hole (if it is not already) and the angular momentum of the orbit produces an accretion disk around the black hole. This system may power a gamma-ray burst under the black hole accretion disk paradigm.

gamma-ray burst models, merging compact objects   Merging compact objects (double neutron star binaries, black hole and neutron star binaries, and black hole and white dwarf binaries) evolve into black hole accretion disk systems which then produce the pair/plasma jet which may ultimately power the burst. The actual merger of two neutron stars ejects too much material and it is not until a black hole forms with a $\sim 0.1 M_\odot$ accretion disk around it that a viable gamma-ray burst might be powered via the black hole accretion disk paradigm. In the other systems, the compact companion of the black hole is shredded into an accretion disk, again forming a black hole accretion disk system.

gamma-ray burst, soft gamma-ray repeaters
Gamma ray bursts are characterized by the short-duration burst of photons with energies greater than 10 keV. A very small subset of bursts (three

of hundreds) have been observed to repeat. Soft gamma-ray repeaters have softer spectra (photon energies ~ 30 to 50 keV vs. ~ 0.1 to 1 MeV for classical bursts). The three observed soft gamma-ray repeaters also have durations which are on the low end of the gamma-ray burst duration (~ 0.1 s vs. the 0.01 to 300 s for the classical bursts). All three soft gamma-ray repeaters lie in the galactic plane and are thought to be the result of accreting neutron stars.

Ganymede   Moon of Jupiter, also designated JIII. Discovered by Galileo in 1610, it is one of the four Galilean satellites. Its orbit has an eccentricity of 0.002, an inclination of $0.21°$, a precession of $2.63°$ yr^{-1}, and a semimajor axis of 1.07×10^6 km. Its radius is 2631 km, its mass 1.48×10^{23} kg, and its density 1.94 g cm^{-3}. It has a geometric albedo of 0.42 and orbits Jupiter once every 7.154 Earth days. *Ganymede* is the largest satellite in the solar system.

garden hose angle   Angle between the interplanetary magnetic field line and a radius vector from the sun. *See* Archimedian spiral.

gas constant   The quantity $= 8.314472$ J mol^{-1} K^{-1} appearing in the ideal gas law:

$$R = PV/(nT) ,$$

where P is the pressure, V is the volume, T is the temperature, and n is the quantity of the gas, measured in gram-moles.

gaseous shocks   Abrupt compression and heating of gas, caused by matter moving at velocity larger than the sound speed of the surrounding medium. Since material is moving supersonically, the surrounding gas has no time to adjust smoothly to the change and a shock front, i.e., a thin region where density and temperature change discontinuously, develops. Shocks can form in any supersonic flow: in astronomy, in the case of supernovae, flare stars, and in stellar winds. Since heating causes emission of radiation, the excitation and chemical composition of the gas can be diagnosed by studying the emitted spectrum.

gas thermometer   A thermometer that utilizes the thermal properties of an (almost) ideal gas. It is either a constant volume thermometer, in which the pressure is measured, or a constant pressure thermometer in which the volume is measured. The measurements are based on the ideal gas law:

$$nTR = PV \, ,$$

(where P is the pressure, V is the volume, T is the temperature, R is the gas constant, and n is the quantity of the gas, measured in gram-moles), or on some calibration using the actual real gas.

gauge   In general relativity, a statement of the behavior of the coordinates in use. In the ADM form, these may be explicity statements of the metric components α and β^i. Or a differential equation solved by these quantities. *See* ADM form of the Einstein–Hilbert action.

gauge pressure   The pressure in excess of the ambiant (typically local atmospheric) pressure. A tire gauge reads *gauge pressure.*

Gauss   The unit of magnetic induction in the cgs system of units. 1 Gauss $\equiv 10^{-4}$ Tesla.

Gauss–Bonnet topological invariant   A quantity formed by integrating over all space time a certain skew-square of the Riemann tensor, which in four dimensions is independent of variations in the metric. Even in higher dimensional theories, it does not contribute to the propagator of the gravitational perturbations. Thus, despite the fact that the Gauss–Bonnet invariant contains fourth derivatives, it does not give rise to the massive spin-2 ghosts and therefore does not spoil the unitarity of (quantum) gravity and thus can be regarded as an acceptable ingredient in quantum gravity. *See* higher derivative theories, massive ghost.

Gauss coefficients   In an insulator, the magnetic field may be written in terms of a potential that satisfies Laplace's equation (i.e., $\mathbf{B} = \nabla \Phi$ where $\nabla^2 \Phi = 0$). For the Earth's magnetic field, because the planet is nearly spherical it is convenient to expand Φ in terms of the spherical harmonic solutions to Laplace's equation:

$$\Phi = r_e \sum_{n=1}^{\infty} \sum_{m=0}^{n} \left(\frac{r_e}{r} \right)^{n+1}$$
$$\left[g_n^m \cos m\phi + h_n^m \sin m\phi \right] P_n^m (\cos \theta)$$
$$+ r_e \sum_{n=1}^{\infty} \sum_{m=0}^{n} \left(\frac{r}{r_e} \right)^{n}$$
$$\left[q_n^m \cos m\phi + s_n^m \sin m\phi \right] P_n^m (\cos \theta)$$

where r is the distance from the center of the Earth, r_e is the radius of the Earth, θ is the geocentric colatitude, ϕ is the east longitude, and the P_n^m are Schmidt partially normalized associated Legendre polynomials. The g_n^m, h_n^m, q_n^m, and s_n^m are termed the Gauss coefficients. g_n^m and h_n^m represented fields of internal origin, as their associated radial functions diverge as $r \rightarrow 0$, while q_n^m and s_n^m represent external fields.

gegenschein   A band of diffuse light seen along the ecliptic 180° from the sun immediately after sunset or before sunrise. It is created by sunlight reflecting off the interplanetary dust particles, which are concentrated along the ecliptic plane. To see the *gegenschein,* you must have very dark skies; under the best conditions it rivals the Milky Way in terms of brightness. The material within the gegenschein is slowly spiraling inward towards the sun due to the Poynting–Robertson Effect; hence, it must be continuously replaced by asteroid collisions and the debris constituting comet tails. The gegenschein is the name given to the band of light seen 180° from the sun while zodiacal light is the term applied to the same band of light located close to the sun.

gelbstoff   *See* colored dissolved organic matter.

general circulation model (GCM)   A set of mathematical equations describing the motion of the atmosphere (oceans) and budgets of heat and dynamically active constituents like water vapor in the atmosphere and salinity in the oceans. These equations are highly nonlinear and generally solved by discrete numerical methods. Due to computer resource limits, grid size in a GCM typically measures on the order of 100 km. Therefore, many sub-grid processes such as turbulence mixing and precipitation have to be parameterized based on physical considerations and/or empirical relations. Atmospheric GCMs have been successfully applied to weather forecasts for the past three decades. Efforts are being made to couple the ocean and atmospheric GCMs together to form coupled climate models for the purpose of predicting the variability and trends of the climate.

general relativity   Description of gravity discovered by Einstein in which the curvature of four-dimensional spacetime arises from the distribution of matter in the system and the motion of matter is influenced by the curvature of spacetime. Described mathematically by

$$G_{\mu\nu} = \frac{8\pi G}{c^2} T_{\mu\nu} \, ,$$

where $G_{\mu\nu}$ is the Einstein tensor, constructed from the Ricci tensor, and has dimensions of inverse length, and $T_{\mu\nu}$ is the four-dimensional stress-energy tensor and has dimensions of mass per unit volume. *General relativity* is the theory of mechanical, gravitational, and electromagnetic phenomena occurring in strong gravitational fields and involving velocities large compared to the velocity of light. General relativity generalizes special relativity by allowing that the spacetime has nonzero curvature. The observable manifestation of curvature is the gravitational field. In the language of general relativity every object moving freely (i.e., under the influence of gravitation only) through space follows a geodesic in the spacetime. For weak gravitational fields and for objects moving with velocities small compared to c, general relativity reproduces all the results of Newton's theory of gravitation to good approximation. How good the approximation is depends on the experiment in question. For example, relativistic effects in the gravitational field of the sun are detectable at the sun's surface (*see* light deflection) and out to Mercury's orbit (*see* perihelion shift). Farther out, they are currently measurable to the orbit of Mars, but are nonsignificant for most astronomical observations. They are important for very accurate determination of the orbits of Earth satellites, and in the reduction of data from astrometric satellites near the Earth.

General relativity differs markedly in its predictions from Newton's gravitation theory in two situations: in strong gravitational fields and in modeling the whole universe. The first situation typically occurs for neutron stars (*see* binary pulsar) and black holes. In the second situation, the nonflat geometry becomes relevant because of the great distances involved. This can be explained by an analogy to the surface of the Earth. On small scales, the Earth's surface is flat to a satisfactory precision, e.g., a flat map is perfectly sufficient for hikers exploring small areas on foot. However, a navigator in a plane or on a ship crossing an ocean must calculate his/her route and determine his/her position using the spherical coordinates. A flat map is completely inadequate for this, and may be misleading. *See* separate definitions of notions used in general relativity: black hole, Brans–Dicke theory, cosmic censorship, cosmological constant, cosmology, curvature, de Sitter Universe, Einstein equations, Einstein Universe, exact solution, Friedmann–Lemaître cosmological models, geodesic, gravitational lenses, horizon, light cone, light deflection, mass-defect, metric, naked singularity, nonsimultaneous Big Bang, relativistic time-delay, scale factor, singularities, spacetime, topology of space, white hole, wormhole.

GEO   Laser interferometer gravitational wave detection being constructed near Hanover Germany; a joint British/German project. An interferometer is a L-configuration with 600-m arm length. Despite its smaller size compared to LIGO, it is expected that *GEO* will have similar sensitivity, due to the incorporation of high quality optical and superior components, which are deferred to the upgrade phase (around 2003) for LIGO. *See* LIGO.

geocentric   Centered on the Earth, as in the Ptolemaic model of the solar system in which planets orbited on circles (deferents) with the Earth near the center of the deferent.

geocentric coordinate time   *See* coordinate time.

geocentric latitude   Latitude is a measure of angular distance north or south of the Earth's equator. While geodetic latitude is used for most mapping, *geocentric latitude* is useful for describing the orbits of spacecraft and other bodies near the Earth. The geocentric latitude ϕ' for any point P is defined as the angle between the line $\overline{OP}$ from Earth's center O to the point, and Earth's equatorial plane, counted positive northward and negative southward. *See* the mathematical relationships under latitude. *Geocentric latitude* ϕ' is the complement of the usual spherical polar coordinate in spherical geometry. Thus, if L is the longitude, and r the distance from Earth's center, then right-handed, earth-centered rectangular coordinates (X, Y, Z), with Z along the north, and X intersecting the Greenwich meridian at the equator are given by

$$
\begin{aligned}
X &= r \cos\left(\phi'\right) \cos\left(L\right) \\
Y &= r \cos\left(\phi'\right) \sin\left(L\right) \\
Z &= r \sin\left(\phi'\right)
\end{aligned}
$$

geocorona The outermost layer of the exosphere, consisting mostly of hydrogen, which can be observed (e.g., from the moon) in the ultraviolet glow of the Lyman α line. The hydrogen of the *geocorona* plays an essential role in the removal of ring current particles by charge exchange following a magnetic storm and in ENA phenomena.

geodesic The curve along which the distance measured from a point p to a point q in a space of $n \geq 2$ dimensions is shortest or longest in the collection of nearby curves. Whether the *geodesic* segment is the shortest or the longest arc from p to q depends on the metric of the space, and in some spaces (notably in the spacetime of the relativity theory) on the relation between p and q. If the shortest path exists, then the longest one does not exist (i.e., formally its length is infinite), and vice versa (in the latter case, quite formally, the "shortest path" would have the "length" of minus infinity). Examples of geodesics on 2-dimensional surfaces are a straight line on a plane, a great circle on a sphere, a screw-line on a cylinder (in this last case, the screw-line may degenerate to a straight line when p and q lie on the same generator of the cylinder, or to a circle when they lie in

the same plane perpendicular to the generators). In the spacetime of relativity theory, the points (called events) p and q are said to be in a time-like relation if it is possible to send a spacecraft from p to q or from q to p that would move all the way with a velocity smaller than c (the velocity of light). Example: A light signal sent from Earth can reach Jupiter after a time between 30-odd minutes and nearly 50 minutes, depending on the positions of Earth and Jupiter in their orbits. Hence, in order to redirect a camera on a spacecraft orbiting Jupiter in 10 minutes from now, a signal faster than light would be needed. The two events: "now" on Earth, and "now + 10 minutes" close to Jupiter are not in a timelike relation. For events p and q that are in a timelike relation, the geodesic segment joining p and q is a possible path of a free journey between p and q. ("Free" means under the influence of gravitational forces only. This is in fact how each spacecraft makes its journey: A rocket accelerates to a sufficiently large initial velocity at Earth, and then it continues on a geodesic in our space time to the vicinity of its destination, where it is slowed down by the rocket.) The length of the geodesic arc is, in this case, the lapse of time that a clock carried by the observer would show for the whole journey (*see* relativity theory, time dilatation, proper time, twin paradox), and the geodesic arc has a greater length than any nearby trajectory. The events p and q are in a light-like (also called null) relation if a free light-signal can be sent from p to q or from q to p. (Here "free" means the same as before, i.e., mirrors that would redirect the ray are not allowed.) In this case the length of the geodesic arc is equal to zero, and this arc is the path which a light ray would follow when going between p and q. The zero length means that if it was possible to send an observer with a clock along the ray, i.e., with the speed of light, then the observer's clock would show zero time-lapse. If p and q are neither in a timelike nor in a light-like relation, then they are said to be in a spacelike relation. Then, an observer exists who would see, on the clock that he/she carries along with him/her, the events p and q to occur simultaneously, and the length of the geodesic arc between p and q would be smaller than the length of arc of any other curve between p and q. Given a manifold with metric g, the equations of the geodesics arise by variation of the action $S = \int g(\dot{x}, \dot{x}) ds$ and are

$$\ddot{x}^c + \Gamma^c_{ab} \dot{x}^a \dot{x}^b = \lambda(x) \dot{x}^c$$

where $\dot{x}^a = dx^a/ds$ is the tangent of the curve $x = x(s)$. The parameter s chosen such that $\lambda = 0$ is called an affine parameter. The proper time is an affine parameter.

geodesic completeness   A geodesic $\gamma(\tau)$ with affine parameter τ is complete if τ can take all real values. A space-time is geodesically complete if all geodesics are complete. *Geodesic completeness* is an indicator of the absence of space-time singularities.

geodesy   The area of geophysics concerned with determining the detailed shape and mass distribution of a body such as Earth.

geodetic latitude   The *geodetic latitude ϕ* of a point P is defined as the angle of the outward directed normal from P to the Earth ellipsoid with the Earth's equatorial plane, counted positive northward and negative southward. It is zero on the equator, 90° at the North Pole, and $-90°$ at the South Pole. When no other modifier is used, the word latitude normally means geodetic latitude. *See* equations under latitude.

geodynamics   The area of geophysics that studies the movement of planetary materials. Plate tectonics is the major subdiscipline of geodynamics, although the study of how rocks deform and how the interior flows with time also fall under the geodynamics category.

geodynamo   The interaction of motions in the liquid, iron rich outer core of the Earth with the magnetic field, that generate and maintain the Earth's magnetic field. Similar mechanisms may (may have) exist(ed) on other planetary bodies.

Geographus   1620 Geographus, an Earth-crossing asteroid. Discovered September 14, 1951. It is named in honor of the National Geographic Society, which funded the survey that found it. It is a very elongated object, with dimensions approximately 5.1×1.8 km. Its mass is estimated at 4×10^{13} gm. Its rotation period

has been measured as 5.222 hours. Its orbital period is 1.39 years, and its orbital parameters are semimajor axis 1.246 AU, eccentricity 0.3354, inclination to ecliptic 13.34°.

geoid   A reference equipotential surface around a planet where the gravitational potential energy is defined to be zero. On Earth, the *geoid* is further defined to be sea level.

geoid anomalies   The difference in height between the Earth's geoid and the reference spheroidal geoid. The maximum height of *geoid anomalies* is about 100 m.

geomagnetic activity   Routine observations of the Earth's geomagnetic field exhibit regular and irregular daily variations. The regular variations are due to currents flowing in the upper E region resulting from neutral winds due to solar heating effects. The regular variations show repetitive behavior from one day to the next and have repeatable seasonal behavior. The small differences observed in the regular behavior can be used to learn more about the nature of the upper atmosphere. The irregular variations (daily to hourly) in the geomagnetic field are called *geomagnetic activity* and are due to interactions of the geomagnetic field and the magnetosphere with the solar wind. While these variations show no regular daily patterns, there is usually a global pattern so that common geomagnetic latitudes will show similar levels of disturbance. This behavior is described by Ap index, geomagnetic indices, geomagnetic storm, K index, Kp index, magnetosphere.

geomagnetic dip equator   The locus of points about the Earth where the geomagnetic field is horizontal to the surface of the Earth. It is where the geomagnetic inclination is zero. The *geomagnetic dip equator,* often referred to as the dip equator, is offset with respect to the geographic equator because the Earth's magnetic field is best described by a tilted dipole approximation. *See* geomagnetic field.

geomagnetic disturbance   Any type of rapidly varying perturbation to Earth's magnetic field induced by variations in the solar magnetic field and its interaction with the magnetic field of the Earth. These interactions can generate large current sheets in the magnetosphere and ionosphere, whose associated magnetic fields can have direct effect on power transmission systems because of the continental scale of those systems.

geomagnetic elements   The geomagnetic field intensity, **F**, can be characterized at any point by its *geomagnetic elements.* These may be expressed as the magnitudes of three perpendicular components, or by some other set of three independent parameters. Several different sets are used, depending on context. Generally, the preferred set is (X, Y, Z) where X is the component of the geomagnetic field measured to the geographic north direction; Y is the component of the geomagnetic field measured to the geographic east direction; and Z is the component of the geomagnetic field measured in the vertical, with the positive sign directed downwards (so the geomagnetic Z component is positive in the northern hemisphere). Other elements are H, D, and I where H is the horizontal component of the geomagnetic field measured in the direction of the north geomagnetic pole; D, the magnetic declination, is the angle between the direction of the geomagnetic field, the direction a compass needle points, and true geographic north, reckoned positive to the east; and the inclination, I, or dip angle, is the angle that the geomagnetic field dips below the horizontal. *See* geomagnetic field, nanotesla.

geomagnetic field   The magnetic field intensity measured in and near the Earth. This is a vector, with the direction defined such that the north-seeking pole of a compass points toward the geomagnetic north pole and in the direction defined as positive field. Thus, the magnetic field lines emerge from the Earth's South Pole and point into the North Pole. The surface intensity of the Earth's magnetic field is approximately 0.32×10^{-4} tesla (T) at the equator and 0.62×10^{-4} T at the North Pole. Above the Earth's surface, the field has the approximate form of a magnetic dipole with dipole moment 7.9×10^{15} Tm^{-3}. Many earth-radii away, this dipole is distorted into a teardrop with its tail pointing anti-sunward by the magnetized solar wind plasma flow. The principal sources of the geomagnetic field are convective motions of the Earth's electrically conducting fluid core, magnetization of the crust, ionospheric currents, and solar wind perturbations of the geomagnetic neighborhood.

geomagnetic indices   *Geomagnetic indices* give a very useful descriptive estimate of the extent of geomagnetic disturbances. Most common indices of the Earth's magnetic field are based on direct measurements made at magnetic observatories. Local indices are calculated from these measurements and then global indices are constructed using the local indices from selected, standard locations. These global indices are often referred to as planetary indices. Commonly used indices are the K index (a local index) and the Kp index (the associated planetary K index, which depends on a specific number of magnetic observatories); the A index (essentially a linear local index) and Ap index (the associated planetary index). Other important indices are the AE index, which gives a measure of the currents flowing in the auroral region, and the Dst index, which gives a measure of the currents flowing in the Earth's magnetosphere. *See* Ap, K, Kp indices.

geomagnetic jerk   It has been observed that there are on occasion impulses in the third time derivative of the geomagnetic field (i.e., "jerks" in the field, which appear as an abrupt change in the gradient of the secular variation). These appear to be global phenomena of internal origin, with recent examples around 1969, 1978, and 1991, and claims of several earlier examples. The effect appears to be most notable in the eastward component of the magnetic field, and the phenomenon appears to take place over a timescale of perhaps 2 years. Unless the mantle behaves as a complicated filter causing slow changes in the core to be revealed as relatively rapid variation at the Earth's surface, the sharpness of this phenomenon places a constraint on mantle conductivity: it must be sufficiently low so that the magnetic diffusion timescale of the mantle (over which sharp changes in the field at the core's surface would be smoothed out when observed at the Earth's surface) is at most the same order of magnitude as that of the jerk. It is not entirely clear what causes *geomagnetic* *jerks,* although one proposal has been sudden unknotting of field loops within the core, and there have been claims that they are associated with jumps in the Chandler wobble.

geomagnetic potential   An auxiliary function, denoted γ or V, used for describing the internally generated magnetic field of the Earth. *See* harmonic model.

geomagnetic storm   A *geomagnetic storm* is said to occur when the geomagnetic indices exceed certain thresholds. At these times, a worldwide disturbance of the Earth's magnetic field is in progress. There are categories of geomagnetic storm. In one form of classification, a minor geomagnetic storm occurs when the Ap index is greater than 29 and less than 50; a major geomagnetic storm occurs when the Ap index is greater than 49 and less than 100; and a severe geomagnetic storm occurs when the Ap index exceeds 100. A geomagnetic storm arises from the Earth's magnetic and plasma environment response to changes in the solar wind plasma properties. A key factor is a change in the solar wind interplanetary magnetic field from north to south, facilitating easier access to the Earth's magnetic domain. The geomagnetic storm can be described by various phases. During the initial phase of a storm there may be an increase in the observed middle-latitude magnetic field as the solar wind plasma compresses the Earth's magnetosphere. This is then followed by the main phase of the storm when the magnetic field at middle latitudes decreases below normal levels and can exhibit large, sometimes rapid, changes. This phase is associated with the formation of large-scale current systems in the magnetosphere caused by the solar wind interactions. Finally, during the recovery phase the geomagnetic field gradually returns to normal levels. A storm may last for one or more days. Statistically, there is a greater likelihood of larger geomagnetic storms during the equinoxes. *See also* coronal mass ejection, geomagnetic indices, gradual commencement storm, solar wind.

geometrodynamics   A particular interpretation of general relativity in which the geometry of the 3-space (*see* ADM form of the Einstein–Hilbert action) is treated as a dynamical object (along with matter fields) which evolves according to the spatial components of Einstein equations and satisfies the Hamiltonian and (super-) momentum constraints. The full set of physical fields forms the so-called superspace with a finite number of degrees of freedom at each space point. It can also be restricted so as to have a finite total number of degrees of freedom. *See* minisuperspace, Hamiltonian and momentum constraints in general relativity, superspace.

geophysics   The subdiscipline of geology which deals with the application of physics to geologic problems. Some of the major areas of geophysical analysis are vulcanology, petrology, hydrology, ionospheric and magnetospheric physics, geochemistry, meteorology, physical oceanography, seismology, heat flow, magnetism, potential theory, geodesy, and geodynamics. The aims of *geophysics* are to determine a body's interior composition and structure, and the nature of the processes that produce the observed features on the body's surface.

geopotential   The sum of the Earth's gravitational potential and the centrifugal potential associated with the Earth's rotation. *Geopotential* is sometimes given in units of the geopotential meter (gpm) defined by

$$1 \text{ gpm } = 9.8 \text{m}^2\text{s}^{-2} = 9.8 \text{ Jkg}^{-1} ,$$

so that the value of the geopotential in geopotential meters is close to the height in meters. Alternatively, the geopotential height or dynamic height is defined by the geopotential divided by acceleration due to gravity, so that the geopotential height in meters is numerically the same as the geopotential in geopotential meters.

geopotential surface   If the sea were at rest, its surface would coincide with the *geopotential surface.* This geopotential surface is called sea level and is defined as zero gravitational potential.

geospace   One of several terms used to describe the totality of the solar-terrestrial environment. It is the domain of sun-Earth interactions and comprises the particles, magnetic and electric fields, and radiation environment that extend from the sun to Earth and includes the Earth's space plasma environment and upper atmosphere. *Geospace* is considered to be the fourth physical geosphere (after solid earth, oceans, and atmosphere).

geostrophic adjustment   A process in which the pressure and flow fields adjust toward the geostrophic balance. The upper panel shows such an example, where the shallow water system is initially at rest but its surface level has a discontinuous jump. In a non-rotating system, the water surface will become flat at the average of the initial levels (middle). In a rotating system, in contrast, the water level difference is sustained by flows in geostrophic balance with the surface slope (lower panel). The discontinuous jump in water level deforms into a smooth slope whose horizontal scale is the radius of deformation.

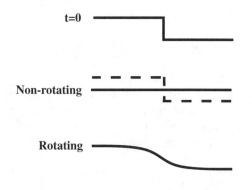

geostrophic approximation   The assumption of geostrophic balance.

geostrophic balance   A balance between the pressure gradient and Coriolis forces: $f\mathbf{u} \times \mathbf{k} = -\nabla p/\rho$, where f is the Coriolis parameter, $\mathbf{u}$ the horizontal velocity, $\mathbf{k}$ the unit vertical vector, p the pressure, and ρ the water density. For large-scale ocean-atmospheric motion, it provides a good approximate relation between flow and pressure fields. In the Northern (Southern) Hemisphere, pressure increases toward the right (left) for an observer facing downstream in a geostrophic flow. Thus, geostrophic winds associated with a circular low pressure system rotate counter-clockwise in the Northern Hemisphere. The resulting flow from this balance is called the geostrophic flow and the resulting transport from this balance is called the geostrophic transport.

geostrophy   *See* geostrophic balance.

Geosynchronous Operational Environmental Satellites (GOES)   A series of meteorology observing satellites operated by the National Oceanic and Atmospheric Administration (NOAA). GOES also monitors space weather via its onboard Space Environment Monitor (SEM) system. The three main components of space weather monitored by GOES at 35,000 km altitude are X-rays, energetic particles, and magnetic field.

geosynchronous orbit   A prograde circular Earth-satellite orbit at a radius of 42,300 km above the equator. At this distance from Earth, the orbital period of a satellite is 1 day, i.e., equal to the rotational period of Earth. The satellite thus remains above the same point on the ground. Such orbits have extreme commercial value for broadcast and communications.

geotherm   Because the Earth is losing heat to its surface, the interior of the Earth is hot. The temperature of the Earth's interior as a function of depth is the geotherm. Typically the increase in temperature with depth is about 25 K/km.

geothermal energy   The energy contained in heat below the surface of the Earth or the development of this energy for commercial purposes. The dominant form of use is natural hot water from subsurface formations, but other approaches, such as pumping water underground to heat it, can be carried out. Though in some places (Iceland) hot geothermal water is used to heat buildings, in most cases it is used to generate electricity or in dual (cogeneration) use. The worldwide installed geothermal electrical production in 1998 was about 8000 MW electrical, equivalent to several large nuclear plants.

geothermal gradient   The vertical gradient of temperature in the Earth. Downward increase (about 25 K/km near the surface) is usually taken to be positive.

Geroch group   In general relativity, an infinite-parameter symmetry group of stationary axisymmetric vacuum space-times. It is generated by two noncommuting 3-parameter symmetry groups; one is the group of constant linear combinations of the space-like and time-like Killing vectors. The other generating group (called the Ehlers group) acts on the gravitational potentials and is isomorphic to O(2,1).

Geroch–Hansen moments (1974)   Gravitational multipole moments of stationary and axisymmetric isolated sources in the theory of general relativity. They form a complex infinite series. The real parts are the mass (or gravielectric) moments and the imaginary parts are the current or gravimagnetic moments. The Schwarzschild spacetime has a mass monopole moment equal to the mass m, and all other moments vanish. The nth moment of the Kerr spacetime with rotation parameter a is $m(ia)^n$. *See* Kerr black hole, Schwarzschild black hole.

Gershun's law   The conservation of energy equation obtained by integrating the monochromatic radiative transfer equation (for a medium with no inelastic scattering or other sources) over all directions; it states that the depth derivative of the net plane irradiance equals the negative of the absorption coefficient multiplied by the scalar irradiance.

Gershun tube   A tube used to limit the field of view of a radiometer to a small solid angle; used in measuring radiances.

GeV   Giga electronvolt; a unit of energy equal to a billion (10^9) electronvolts.

giant branch   The collection of stars in which a contracting helium core heats an overlying hydrogen shell, accelerating hydrogen burning. Stars develop a large hydrogen envelope and occupy a specific region in the HR diagram, with K to M spectral type ($B - V = 1.0 - -1.8$, absolute magnitude 2 to -2). The Red Giant phase ends when the temperature of the core reaches

$10^8 K$, and helium ignites explosively in its core (the "helium flash").

giant cells   The largest of the discrete scales governing convective motions on the sun; the other three being granulation, mesogranulation, and supergranulation. The existence of giant cells is less conclusive than the other scales of convection. However, they are believed to have a dimension comparable with that of the depth of the convection zone ($\sim 300,000$ km), surface velocities of as little as 0.03 to 0.1 kms^{-1} and lifetimes of ~ 14 months.

giant planet   One of Jupiter, Saturn, Uranus, and Neptune in the solar system, or a similar extrasolar planet. Typically much more massive than terrestrial planets, giant planets are composed of volatiles (hydrogen, helium, methane, ammonia) with small solid cores, if any.

Gibbs free energy (Gibbs Potential)   An extensive thermodynamic potential H given by

$$H = U - ST - PV,$$

where U is the internal energy, S is the entropy, T is the temperature, P is the pressure, and V is the volume of the system. For a reversible process at constant T and P, work stored as *Gibbs Potential* can be recovered completely.

gilvin   *See* colored dissolved organic matter.

Ginzburg temperature   In quantum (or classical) systems with multiple minima in some potential, separated by a barrier, the minimum temperature needed to induce thermal jumps from one minimum of the potential to another. Applications include magnetism, superconductivity, and the mechanism of formation of topological defects in the early universe. *See* correlation length, cosmic topological defect, Kibble mechanism.

glacial   Referring to processes or features produced by ice.

glaciation   A geologic epoch when 30% of the Earth's land surface is covered with moving ice. The most recent *glaciation* began in the

Pleistocene, about 2 million years ago, ending only about 10,000 years ago.

glacier   A mass of moving ice formed by accumulation and compactification of snow, and flowing from an accumulation source to an edge where it is ablated. *Glaciers* are found on Earth and are suspected to have occurred on Mars in the past. Ground ice is present on both Earth and Mars. Some of the icy moons of the outer solar system show evidence of ice moving on top of liquid water oceans, either presently (such as Europa) or in the past (such as Ganymede). Glaciers contained within mountain valleys are called valley or alpine glaciers, while those spread out laterally over large areas are called continental glaciers, found in Greenland and Antarctica. Glaciers are very efficient at eroding the underlying material and transporting this material to other locations where it is deposited as eskers, moraines, and drumlins. Ground ice also creates a number of geologic features, primarily thermokarst features produced by the collapse of overlying material as subterranean ice is removed by heating.

global loop oscillation   Coronal loops act as high quality resonance cavities for hydromagnetic waves and, consequently, display a large-scale frequency response sharply peaked at the global resonant frequency of the loop, defined by $v_{\text{Alfvén}} = 2L/\text{period}$. This global mode is crucial to wave heating models of the solar corona since, in order to be efficient, the resonant absorption of waves requires a close matching to the length of the coronal loop through the condition of global-mode resonance.

globally hyperbolic space-time   (Leray, 1952.) A space-time $(\mathcal{M}, g)$ is globally hyperbolic if

1. $(\mathcal{M}, g)$ is strongly causal.

2. For any two points $p, q \in \mathcal{M}$, the intersection of the causal sets $J^+(p) \cap J^-(q)$ is compact where $J^+(p)$ is the causal future set of p, i.e., those points that can be influenced by p, and $J^-(q)$ is the causal past set of q. In a globally hyperbolic space-time, the wave equation with source term $\delta(p)$ has a unique solution which vanishes outside the causal future set $J^+(p)$.

global positioning system (GPS)   A U.S. Department of Defense system of 24 satellites used as timing standards and for navigation to monitor positions on the Earth's surface. Satellites carrying precise cesium or rubidium atomic clocks which are synchronized with Coordinated Universal Time broadcast coded data streams giving their time and their position. By observing four satellites, all four parameters (x, y, z, time) describing the location and time at a near Earth receiver can be extracted. Precise mode (a military classified mode) allows locations to less than 1 m. Commercial devices are available that will provide a location of the device with an accuracy of 10s of meters using signals from these satellites. Differential GPS and interferometric applications can reduce errors to millimeters.

global seismology   A research field of seismology aimed at elucidation of the Earth's deep structure, dynamics, and its driving forces through an international exchange of information and technique in seismology. In the 1980s, through analyses of digital data of global seismograph networks, three-dimensional velocity structural models of the Earth were proposed, and research on global seismology made remarkable progress. In order to know more about the structures and dynamics of the Earth, a seismograph network with broadband and high dynamic range was equipped, and many countries started to participate in a global seismograph network. As representative networks, there are GEOSCOPE (France), IRIS (U.S.), CANDIS (Canada), CDSN (China), FKPE (Germany), DRFEUS (Europe), and POSEIDON (Japan).

global thermohaline circulation   A vertical overturning circulation in which cold water sinks in localized polar/subpolar regions, spreads to and rises to the surface in the rest of the world ocean. At low temperatures close to the freezing point, sea water density becomes less sensitive to temperature than to salinity differences. So the formation of deep water is determined by salinity differences among the oceans. Currently, most of the world ocean's deep water sinks in the northern North Atlantic, and flows southward along the western boundary into the Southern Ocean. Riding on the

Antarctic circumpolar current, the North Atlantic Deep Water spreads into the Indian and Pacific basins, along with the Antarctic Bottom Water. Compensating the deep outflow, warm surface water flows back into the North Atlantic. Because of a large temperature difference between the surface ($10°C$) and deep ($0°C$) waters, this overturning circulation transports a huge amount of heat into the North Atlantic, making winter much warmer in northern Europe than at the same latitudes in North America.

global topological defect   Topological defects that may be important in early universe cosmology, formed from the breakdown of a rigid (or global) symmetry that does not have "compensating" gauge fields associated to it. Long-range interactions between the defects in the network and energy stored in gradients of the field cannot be compensated far away from the defect; in general, these constitute divergent-energy configurations. A cutoff for the energy is physically given by other relevant scales of the problem under study, such as the mean distance between two arbitrary defects in the network, or the characteristic size of the loop (in the case of cosmic string loops), for example.

globular cluster   A dense spherical cluster of stars of Population II, typically of low mass ($\approx 0.5 M_\odot$). Diameter of order 100 pc; containing up to 10^5 stars. *Globular clusters* are a component of the halo of the galaxy.

GMT   *See* Universal Time (UT or UT1).

gnomon   A vertical rod whose shadow in sunlight is studied to measure the angular position of the sun.

Goddard, Robert H.   Rocket engineer (1882–1945). Designed the liquid-propellant rocket.

Goldberg–Sachs theorem (1962)   A vacuum spacetime in general relativity is algebraically special if and only if it contains a shear-free geodesic null congruence. *See* congruence, Petrov types.

Goldstone boson   *See* Goldstone model.

Goldstone model   A quantum field model of a scalar field with a nonlinear self-interaction (Goldstone, 1961). A theory in which the symmetry of the Lagrangian is not shared by the ground state (the vacuum, i.e., the lowest energy) solution. The Lagrangian density reads

$$\mathcal{L} = (\partial_\mu \phi)^* (\partial^\mu \phi) - V(\phi)$$

with ϕ a complex scalar field (* means complex conjugate) and the symmetry breaking potential $V(\phi)$ has the "Mexican hat" form $V = \frac{1}{4}\lambda(\phi^*\phi - \eta^2)^2$, with λ and η positive constants.

$\mathcal{L}$ is invariant under the global transformation $\phi \rightarrow e^{i\Lambda}\phi$, with Λ a constant in spacetime. This model has a local potential maximum at $\phi = 0$ and the minima occur when the absolute value of ϕ equals η. Hence, the minima can be expressed as $\phi = \eta e^{i\theta}$ and the phase θ can take any of the equivalent values between zero and 2π. Once one of these phases is chosen (and we have, say, $\phi_{\text{vac}} = \eta e^{i\theta_{\text{vac}}}$ as our vacuum state) the original symmetry possessed by the model is lost (broken). To see this, the original $U(1)$ transformation $\phi \rightarrow e^{i\Lambda}\phi$ will now change θ_{vac} by $\theta_{\text{vac}} + \Lambda$: the model is no longer invariant under the original symmetry.

By further analyzing this model in the vicinity of the new vacuum state, one deduces that the original Lagrangian can be written in terms of massive and massless scalar fields, plus other uninteresting interaction terms. It is the particle associated with this massless field (whose degree of freedom is related to motion around the equal energy circle of minima of the potential V) that became known as the Goldstone boson.

In cosmology, global strings may arise from configurations of the Goldstone field; these can have important implications for the structure of the universe. *See* Goldstone theorem, Higgs mechanism.

Goldstone theorem   Any spontaneous breaking of a continuous symmetry leads to the existence of a massless particle. This theorem shows that when the Lagrangian of the theory is invariant under a group of symmetries G, but the ground state is only invariant under a subgroup H, there will be a number of massless (Goldstone) particles equal to the dimension of the quotient space $\mathcal{M} \sim G/H$, and hence equal to the difference between the number of generators of G and H, or said otherwise, to the number of generators of G that are not generators of H. Actually, the theorem is more general, also applying in the case the symmetry is broken, but not spontaneously.

A particularly simple example is the Goldstone model. There we have a breaking scheme $U(1) \rightarrow \{1\}$ and the production of one Goldstone particle. *See* Goldstone model, spontaneous symmetry breaking.

Gondwanaland   Continent in existence prior to the opening of the South Atlantic Ocean, consisting of South America, Africa, and Australia.

Goto–Nambu string   Structureless vacuum vortex defects that can be macroscopically described by a simple cosmic string equation of state $T = U = $ const., where U is the string energy per unit length and T is its tension.

The motion is derived by finding ξ^0 and ξ^1 that minimize the action:

$$S = -U \int \sqrt{-\gamma} d^2 \xi$$

where, the string being a one-dimensional object, its history can be represented by a surface in spacetime $x^\mu(\xi^a)$ with internal worldsheet coordinates $\xi^0 = t$, $\xi^1 = \ell$. γ is the determinant of the induced metric on the two-dimensional worldsheet of the string evolving in time. The integrand essentially represents the element of surface area swept out by the string in spacetime.

One can then say that the string evolves in such a way as to extremize the area of its spacetime worldsheet. *See* cosmic string, cusp (cosmic string), energy per unit length (cosmic string), fundamental tensors of a worldsheet, Higgs mechanism.

GPS   *See* global positioning system.

graben   Common features on most planetary surfaces and often synonymous with the term rift valley. They are long, narrow troughs bounded by two (sometimes more) parallel normal faults, giving rise to a downdropped valley surrounded by high plateaus. *Graben* are created when two blocks of rock are pulled apart by tensional tec-

tonic forces. Divergent boundaries are usually characterized by graben.

GRACE (Gravity Recovery and Climate Experiment)   An approved Earth System Science Pathfinder NASA space mission which will employ a satellite-to-satellite microwave tracking system between two spacecraft to measure the Earth's gravity field and its time variability. The mission is scheduled to be launched in 2001 and will operate over five years.

gradient   For a scalar function f of coordinates, the collection of partial derivatives with respect to the coordinates: $\{f_{,i}\}$ (the $_{,i}$ indicating partial derivative with respect to coordinate i). Or, more formally, the $1 - form$ constructed by

$$\Sigma f_{,i} dx^i .$$

See 1-form, partial derivative.

gradient drift   *See* drift, gradient.

gradient Richardson number Ri   The non-dimensional ratio $Ri = N^2/(\partial u/\partial z)^2$ between the stability N^2 and the vertical shear $\partial u/\partial z$ of a stably stratified water column, which is subject to vertical gradients of the horizontal currents u. A necessary condition for turbulence to occur (typically Kelvin–Helmholtz instabilities) is the condition $Ri < 0.25$. In natural systems (atmosphere, oceans, lakes, etc.) the experimental determination of Ri depends on the scale over which N^2 and $(\partial u/\partial z)$ are determined. The rate of mixing, generally increasing as Ri decreases, is not a unique function of Ri.

gradual commencement storm   Geomagnetic storms may commence gradually, the storm having no well-defined onset time. These storms may persist for many days, geomagnetic activity increasing irregularly, sometimes to major storm levels. Geomagnetic activity associated with *gradual commencement storms* may recur every 27 days, consistent with the solar rotation. They are thought to be due to high-speed solar wind streams arising in long-lived coronal hole regions on the sun. *See* geomagnetic storm.

gradual flare   Solar flare with long-lasting electromagnetic radiation, originating high in the corona. In more detail, the properties of *gradual flares* are: (a) the soft X-ray emission lasts for more than 1 h, (b) the decay constant of the soft X-ray emission is larger than 10 min, (c) the duration in hard X-rays is at least 10 min, (d) the duration in microwaves is at least 5 min, (e) gradual flares are always accompanied by metric type II and metric type IV bursts, metric type III bursts are observed in half of the gradual flares, (f) the height in the corona is about $5 \cdot 10^4$ km, (g) the flare volume is between 10^{28} cm^3 and 10^{29} cm^3, (h) the energy density is low, (i) the size in Hα is large, and (j) gradual flares are always accompanied by coronal mass ejections.

If a gradual flare gives rise to an energetic particle event observed in interplanetary space, this event has properties different from particle events caused by impulsive flares. Many of these differences can be attributed to the interplanetary shock caused by the coronal mass ejection in gradual flares. In detail, the properties of a particle event originating in a gradual flare are: (a) the event is proton-rich with H/He being about 100, (b) the ^{3}He/^{4}He ratio is about 0.0005 which is the ratio observed in the solar wind, (c) the Fe/O ratio is 0.155, which is also close to the ratio in the solar wind, (d) the charge state of iron is about 14, which suggests rather low temperatures in the acceleration region and therefore acceleration out of the corona or the solar wind, (e) the particle event in interplanetary space at the orbit of Earth lasts for some days, (f) particles can be observed over a longitude cone of about 180°, and (g) an interplanetary shock is observed. Occurrence of such events is about 10 per year.

The distinct feature of the gradual flare is the coronal mass ejection and consequently the interplanetary shock. Some instability below a filament causes reconnection at the anchoring field lines which leads to heating and the generation of electromagnetic radiation (the flare) and to a disconnection and expulsion of the filament (the coronal mass ejection).

gradually varied flow   A description of flow in an open channel (such as a canal or river) which assumes that the rate of change of depth

is small (i.e., $\Delta h/L$ is small, where Δh is the change in depth, and L is the distance over which the depth changes).

Grad–Vasyliunas theorem   An equation used in the study of guiding center plasmas in the Earth's magnetic field, an equivalent formulation of force balance. It assumes that under the influence of convection electric fields, the plasma of each flux tube moves together, from one tube to another, that its particle motion has no preferred direction (isotropy), that a scalar (rather than tensor) pressure p can be used, and that Euler potentials (α, β) of the magnetic field are given. Under those assumptions, it relates the field aligned current density j$_{||}$ with the distributions of pressure p(α, β) and of specific volume V(α, β), as $j_{||} = |\nabla V(\alpha, \beta) \times \nabla p(\alpha, \beta)|$. Since the above assumptions are best met in the Earth's plasma sheet, it has been widely used to deal with plasma convection there. Also known as Vasyliunas theorem.

grain-boundary migration   To help minimize the energy of the solid materials in response to changing P-T conditions, the atoms forming the contacts between individual grains self rearrange to a more stable configuration. To a large extent this is achieved by the process of *grain-boundary migration*. This involves movement at a high angle to the plane of the grain boundary. Such a process usually produces regular interfaces and a polygonal aggregate of grain.

grain-boundary sliding   Movement of grains is limited within the plane of the grain boundary. *Grain-boundary sliding* can be envisaged in terms of the physical movement of individual grains past each other under an applied shear stress. Grain-boundary sliding is recognized as one of the most important deformation mechanisms in fine-grained polycrystalline aggregates, and is considered one of the main deformation mechanisms responsible for superplasticity of polycrystalline materials.

grain chemistry: dense interstellar   In dense interstellar media (i.e., visual extinction $A_v \geq 5$ magnitudes, or $n_H \geq 10^4$ cm^{-3}), where the radiation field is attenuated and the average temperature is ~ 10 K everything except H_2, He, and Ne freezes out and ice mantles form on grains. In regions where $H < H_2$ the less abundant heavier atoms such as O, C, and N can react with one another to form oxidized species such as CO, CO_2, and presumably O_2 and N_2. Where $H > H_2$ the heavy atoms are reduced and the ice mantles are dominated by molecules such as H_2O, CH_4, and NH_3. When these ices are exposed to cosmic rays and UV photons, bonds are broken and the resulting unstable species react to form more complex materials, in some cases organic molecules of the type seen in meteorites and interplanetary dust particles.

grain chemistry: diffuse interstellar In the diffuse interstellar medium (i.e., visual extinction $A_v \leq 2$ magnitudes, or $n_H \leq \sim 10^1$ cm^{-3}), where ice is not stable, the dust is essentially bare and grain surfaces are available as catalysts and reactants. Since hydrogen is the most abundant atom, H_2 formation from hydrogen atoms is by far the most common surface reaction.

grains: in other galaxies Extragalactic extinction curves are comparable to those within the Milky Way galaxy, consistent with grains in other galaxies being of similar size and composition to those in our own. Most spiral galaxies seem to have similar dust-to-gas ratios as in our own, although there are some exceptions.

grains, interstellar: destruction and formation Grains presumably form in the outflows from cool evolved (red giant) stars, novae and supernovae ejecta and grow as they mix with interstellar molecules, accreting ice mantles and sticking to one another. Such *interstellar grains* may be fragmented or destroyed by shocks, collisions, sputtering, or incorporation into forming stars. A typical grain lifetime is thought to be approximately 10^8 years.

grains, interstellar: size and composition A wide-spread *interstellar grain size* is inferred from optical, IR, and UV extinction curves throughout our galaxy. The grains are usually deduced to have sizes in the range of 0.01 to 0.2 μm, are probably of irregular (perhaps fractal) shape, and composed of (presumably amorphous) silicates, carbon, SiC, or metal ox-

ides. The position and profile of the 2175 Å bump is consistent with some of the carbon grains being graphitic, or composed from amorphous or diamond-like carbon containing graphitic (π-bonded) domains. In the diffuse interstellar medium where radiation is abundant, these grains are probably bare, but in dense molecular clouds, where the radiation is attenuated and the average temperature low, ice mantles condense on the surfaces of these grains.

grain size A measure of particle size in a sediment. Often refers to the median *grain size* in a sample. Typically determined by sieving.

grain size analysis A process of determining sediment grain size. Most commonly done using a stack of sieves of varying size.

grand potential The thermodynamic potential Ω equal to

$$\Omega = U - TS - \mu N ,$$

with U the internal energy, S the entropy, μ the chemical potential, and N the number of particles in the system.

grand unification Modern particle physics describes interactions through the exchange of so-called gauge particles, whose existence is due to an underlying symmetry. For instance, the gauge particle we know as the photon exists because of the invariance of the physics under local phase variation, which is a $U(1)$ symmetry. When the universe had a temperature corresponding to a typical photon energy of roughly 250 GeV, the symmetry of particle physics interactions is believed to have been that of the group product $SU(3) \times SU(2) \times U(1)$ describing the strong and electroweak interactions.

Whenever a symmetry is exact, as is the case for electromagnetism, the corresponding gauge particle is massless (as is the photon), while if it is broken, the gauge particle is massive and the interaction is much weaker (short range). This is in particular the case for the weak interaction. It is currently expected that the product symmetry we have at the electroweak level is just part of a larger (simple and compact) symmetry group that unifies forces and interaction — hence the name *grand unification*. This larger group could have been broken at extremely high energies. Thus, many gauge particles would be very massive, explaining why we have not yet observed them. *See* spontaneous symmetry breaking.

granulation   Cellular structure of the photosphere visible at high spatial resolution. The characteristic scale of the *granulation* is $\sim$ 1000 km, ranging widely from $\sim$ 300 km to over 2000 km. Typical velocities present in the granules are horizontal outflows $\sim$ 1.4 km s^{-1} and upflows in the center of the granules $\sim$ 1.8 km s^{-1}. Identifiable granules live for $\sim$ 5 to 10 min and at any one time $\sim$ 3 to 4 $\times$ 10^6 granules cover the surface of the sun.

grating spectrograph   An instrument for the analysis of radiation at different wavelengths, in which light is dispersed by a diffraction grating. In a typical *grating spectrograph* design, light focused on the focal plane of the telescope is collimated (i.e., the rays of the beam are made parallel) on a blazed diffraction grating. Light of different wavelengths is thus diffracted along different directions, and it is then re-focused on a detector, for example a photographic plate or a CCD, by a lens or a mirror. Several designs exist, based on different choices of the focusing and collimating elements, or on the use of a reflection or a transmission grating. The spectral resolving power, i.e., the ability to separate two close spectral lines spaced by $\Delta\lambda$ at wavelength λ is usually $\lambda/\Delta\lambda \lesssim 10^4$. *See* diffraction grating.

gravitation   The interaction between objects that in Newtonian theory depend only on their distances and masses:

$$\mathbf{F} = Gm_1m_2\hat{r}r^{-2},$$

where $\mathbf{F}$ is the attractive force, m_1 and m_2 are the masses, r is their separation, $\hat{r}$ is a unit vector between the masses, and G is Newton's gravitational constant.

In the more accurate description in general relativity, the gravitational field is a symmetric rank two tensor in four dimensions, which obeys Einstein's equations, which have as a source the stress-energy tensor of the matter. General relativity reduces to the Newtonian description in weak-field, quasi stationary situations. *See* Newtonian gravity, general relativity, Newtonian gravitational constant.

gravitational collapse   One of several possible final episodes of stellar evolution. Stars are supported against collapse under their own weight by gas pressure resulting from high temperature. The equilibrium of cold matter can be supported by the pressure of degenerate electron gas. The equilibrium of white dwarf stars is thought to be due to this. Chandrasekhar showed in 1939 that the maximal mass of a white dwarf is 1.4 solar mass and that stars more massive than this will collapse (to a neutron star or to a black hole).

gravitational constant G   In Newtonian physics, the acceleration of a particle towards the Earth is Gm/r^2 where m and r are the mass and radius of the Earth, respectively and

$$G = 6.670 \times 10^{-8} \text{ cm}^3/g \cdot \sec^2$$

is the gravitational constant.

gravitational deflection of light   The presence of a central mass (e.g., the sun) causes local curvature of spacetime, and trajectories of photons (the quanta of light) are deflected (attracted to the mass). In the case of a spherical central body, for small angle deflection, the angle deflected is

$$\theta = 4GM/dc^2,$$

where M is the central mass, G is Newton's constant, c is the speed of light, and d is the impact parameter of the light past the mass. If the deflection were as if the photon were a particle in Newtonian gravity traveling at the speed of light, the deflection would be half the relativistic result. Deflection of light has been verified to parts in 1000 by observations of the direction to extragalactic radio sources, as the sun passes near their position in the sky. In the case of distant sources, deflection by intervening galaxies or clusters of galaxies causes lensing, leading to the appearance of multiple images, and of rings or arcs of distorted images. *See* gravitational lens, light deflection.

gravitational equations   The field equations of the gravitational interaction. The basic tenet of Einstein's general relativity theory is that the stress-energy tensor $T_{\alpha\beta}$ of a matter distribution is proportional to the Einstein tensor $G_{\alpha\beta}$:

$$G_{\alpha\beta} = \frac{8\pi G}{c^2} T_{\alpha\beta}$$

Here G is Einstein's gravitational constant and c is the speed of light. If a nonzero cosmological constant is considered, $+\Lambda g_{\alpha\beta}$ is added to the left side of the equation (Λ is the cosmological constant). *See* Einstein tensor, stress energy tensor.

gravitational instability   In meteorology, an instability caused by Earth's gravity, under the unstable stratification lapse rate condition $\partial\theta/\partial z < 0$, in which θ is potential temperature, and z is height. It can be seen that with such lapse rate, any infinitesimal displacement not along θ surfaces is unstable. If instability is only caused by such conditions, it is pure gravitational instability, and Earth's rotation has no effect. From the vertical motion equation,

$$\frac{Dw}{Dt} = \frac{g}{\theta}(\theta' - \theta)$$

where θ' is the potential temperature of the moving air parcel, θ is the potential temperature of the environment, w is vertical velocity, t is time, and g is the gravitational acceleration. If $\partial\theta/\partial z < 0$, the value of an air parcel's vertical speed will always be accelerated, whether it is moving upward or downward.

gravitational lens   A gravitating object (in observational practice usually a quasar, galaxy, or cluster of galaxies) whose gravitational field deflects the light rays emitted by another object (usually a galaxy or a quasar) so strongly that some of the rays intersect behind the deflector (*see also* light deflection). An observer placed in the region where the rays intersect sees either multiple images of the source or a magnified single image. Unlike optical lenses that have well-defined focal points, *gravitational lenses* do not produce easily recognizable images of the light sources. For a spherically symmetric lens, the angle of deflection is approximately given by the Einstein formula $\Delta\phi = 4GM/(c^2 r)$, where G is the gravitational constant, M is the mass of the light-deflector, c is the velocity of light, and

r is the smallest distance between the path of the ray and the deflector's center. The formula applies only when the ray passes outside the deflector and $r \gg r_g$, where $r_g = 2GM/c^2$ is the gravitational radius of the deflector. Hence, rays passing closer to the lens are deflected by a larger angle, and so an image of a point-source is smeared out throughout a 3-dimensional region in which the rays intersect. Qualitatively, gravitational lenses are just a manifestation of the gravitational light deflection, confirmed in 1919 by A.S. Eddington. For nonspherical or transparent lenses, more complicated formulae apply. In spite of the distortion of the image, given the mass-distribution in the deflector and the light-intensity-distribution in the light-source, it is possible to calculate the patterns of light intensity seen at the observer's position. Comparing the patterns calculated for various typical situations with the patterns observed, it is possible to extract information about the mass- and light-intensity-distribution in the actual source and the actual deflector. *See* gravitational radius.

gravitational multipole moments   Coefficients of the asymptotic expansion of gravitational fields, introduced by Thorne, Epstein, and Wagoner. There exist two infinite series of gravitational multipole moments. The 0th mass moment or monopole moment characterizes the total mass of the source distribution. The odd mass moments are absent — a manifestation of the attractive nature of the interaction. The current moments are odd, beginning with the current dipole moment generated by the rotation of matter. A theory of exact gravitational multipole moments has been developed by Geroch and Hansen for axially symmetric and stationary fields.

gravitational perturbations   Small deviations from a given space-time with metric g_{ab}^0. The metric of the perturbed space-time has the form $g_{ab} = g_{ab}^0 + h_{ab}$ where the quantities h_{ab} and their derivatives are infinitesimal. These techniques are used in the theory of cosmological perturbations, gravitational radiation theory, and in quantization schemes. *See* linearized gravitation.

gravitational potential   The gravitational potential energy per unit mass. The gravity field of the Earth is the gradient of the *gravitational potential* (usually with a minus sign).

gravitational radius   The radius, also called Schwarzschild radius, at which gravitational attraction of a body becomes so strong that not even photons can escape. In classical Newtonian mechanics, if we set equal the potential energy of a body of unit mass at a distance r in the gravitational field of a mass M, GM/r, to its kinetic energy if moving at the speed of light, $c^2/2$, we find $r = R_g$ with the gravitational radius $R_g = 2GM/c^2$, where G is the gravitational constant, and M is the mass of the attracting body. An identical expression is found solving Einstein's equation for the gravitational field due to a non-rotating, massive body. The *gravitational radius* is ≈ 3 km for the sun. (The relation is not so simple for rotation in nonstationary black holes.) The gravitational radius defines the "size" of a black hole, and a region that cannot be causally connected with our universe, since no signal emitted within the gravitational radius of a black hole can reach a distant observer. *See* black hole.

gravitational redshift   Frequency or wavelength shift of photons due to the energy loss needed to escape from a gravitational field, for example from the field at the surface of a star, to reach a distant observer. Since the energy of a photon is proportional to its frequency and to the inverse of its wavelength, a lower energy photon has lower frequency and longer wavelength. The *gravitational redshift* is a consequence of Einstein's law of equivalence of mass and energy: even a massless particle like the photon, but with energy associated to it, is subject to the gravitational field. The shift increases with the mass of a body generating a gravitational field, and with the inverse of the distance from the body. A photon will be subject to a tiny frequency shift at the surface of a star like the sun, but to a shift that can be of the order of the unshifted frequency if it is emitted on the surface of a compact body like a neutron star.

gravitational wave   In general relativity, the propagating, varying parts of the curvature tensor. The notion is in general approximate, since the nonlinearity of the gravitational equations prevents a unique decomposition of the curvature to a background and to a wave part.

General relativity allows for a description of gravitational waves which travel through space with the velocity of light. Exact solutions of the field equations representing plane-fronted gravitational waves are known. The detection of typical astrophysical gravitational waves is extremely difficult because laboratory or terrestrial sources are very weak. Calculations show that even violent phenomena occurring at the surface of the Earth produce undetectably weak gravitational waves. For example, a meteorite of mass $2 \cdot 10^7$ kg hitting the Earth with the velocity of 11 km/sec and penetrating Earth's surface to the depth of 200 m emits $2 \cdot 10^{-19}$ ergs of energy in the form of gravitational waves. This would be sufficient to raise one hydrogen atom from the surface of the Earth to the height of 120 cm (or to raise one flu virus to the height of $0.29 \cdot 10^{-9}$ cm, which is somewhat less than $\frac{1}{3}$ of the diameter of a hydrogen atom). Hope for detection is offered by astronomical objects, such as binary systems or exploding supernovae that radiate strongly gravitational waves. The strength of gravitational waves is measured by the expected relative change of distance l between two test masses induced by a gravitational wave passing through the system, $A = \Delta l/l$, where l is the initial distance and Δl is the change in distance. For known astronomical objects that are supposed to be sources of gravitational waves, this parameter is contained between 10^{-23} and 10^{-19}. Measuring such tiny changes is an extreme challenge for technology. Several detectors are under construction at present. If the gravitational waves are detected, then they will become very important tools for observational astronomy, giving the observers access to previously unexplored ranges of phenomena, such as collisions of black holes or the evolution of binary systems of black holes or neutron stars. *See* GEO, LIGO, LISA, pp-waves, Virgo.

graviton   The putative quantum of the gravitational interaction. There exists no precise description of this term and there is no experimental evidence or consensus about the nature of the *graviton*. General relativity theory provides a picture in the weak-field limit in which the graviton is the quantum mode of a polarized weak gravitational plane wave. This particle then is a spin-2 and zero-mass quantum, bearing a close kinship with the photon. However, the perturbational approach to quantum gravitation is non-renormalizable, and approaches to quantization from a nonlinear viewpoint, or by embedding relativity in a larger theory (e.g., string theory), are being sought.

gravity   According to Newton's law of gravity, any mass m exerts a gravitational attraction g on any other mass which is given by

$$g = \frac{Gmm'}{r^2}$$

where r is the distance between the masses and the universal constant of gravity G is a fundamental constant in physics. The gravitational attraction of the mass of the Earth is responsible for the acceleration of gravity at its surface. In general relativity, *gravity* is the expression of the intrinsic curvature of the 4-dimensional spacetime.

gravity anomaly   To a first approximation the earth is a sphere. However, the rotation of the earth causes it to deform into an oblate spheroid. The result is polar flattening and an equatorial bulge. The equatorial radius is larger than the polar radius by about 3 parts in 1000. Thus, the reference gravitational field of the Earth is taken to be that of an oblate spheroid. *Gravity anomalies* are the differences between the actual gravity field and the reference field.

gravity assist   A way of changing the energy of a spacecraft by an encounter with a planet. When distant from the planet, the spacecraft is essentially in orbit around the sun. As it approaches the planet, it is best to consider the interaction in the planet's frame of reference. In this case the spacecraft is approaching the planet at some velocity, and leaves in a different direction at the same speed. However, from the point of view of the solar system, the energy of the spacecraft has changed by the addition of the planet's velocity. For instance, a satellite in an extreme elliptical orbit (essentially only radial motion) around the sun has the minimum energy possible to achieve a particular aphelion, and the negative binding energy is twice that of a circular orbit at that radius. If the satellite encounters a planet at the radius, it is essentially overtaken by that planet, and a parabolic orbit reverses the satellite's motion with respect to the planet. Hence, after encounter, the satellite speed is twice that of the planet in its orbit, and the satellite velocity is $\sqrt{2}V_{escape}$. Thus, after the one encounter the satellite can escape the solar system.

gravity wave   A water wave that has gravity as its primary restoring force (as opposed to surface tension); waves with wavelength $\gtrsim 1.7$ cm are considered *gravity waves*.

grazing incidence optics   An important means by which to image radiation at X-ray wavelengths. X-ray radiation entering a telescope is reflected at a very shallow angle (grazing incidence), by a special combination of hyperboloid and paraboloid mirrors, before being brought to a focus.

Great Attractor   An as yet unidentified and somewhat hypothetical entity, probably a massive super-cluster of galaxies, whose existence is suggested by a large flow of galaxies toward an apex at galactic longitude $\approx 307°$, galactic latitude $\approx 9°$, and recessional velocity 4500 km s^{-1}. The Local Cluster, as well as the Virgo Cluster and the Hydra-Centaurus cluster, are thought to be falling toward the putative *Great Attractor,* located at a distance from the galaxy which should be 3 to 4 times the distance to the Virgo cluster. The structure associated to the Great Attractor has yet to be identified, not least because the center of the Great Attractor probably lies close to the galactic plane, where gas and dust heavily obscure any extragalactic object. The very existence of the Great Attractor is debatable, since there is no consensus of the observation of infall motions toward the Great Attractor of galaxies located beyond it. *See* galactic coordinates.

Great Red Spot   A long-lived atmospheric phenomenon in the southern hemisphere of the planet Jupiter with a distinctive orange-red color. It is an approximate ellipse approximately

14,000 × 40,000 km across. The *Great Red Spot* has been in existence for at least 300 years; it was probably first observed by Jean Dominique Cossinin (1625–1712). The Great Red Spot is an anticyclone with counter-clockwise winds of up to 400 km/h.

great salinity anomaly (GSA)   A low salinity event that first appeared in the late 1960s northeast of Iceland, then propagated counter-clockwise in the northern North Atlantic and returned to the Norwegian, Greenland, and Iceland Seas in the late 1970s. The low salinity cap reduced the formation of various deep water masses in the North Atlantic. There is evidence that similar low salinity events — of smaller magnitude though — occurred both before and after the one in the 1970s.

Greenhouse Effect   The enhanced warming of a planet's surface temperature caused by the trapping of heat in the atmosphere by certain types of gases (called greenhouse gases; primarily carbon dioxide, water vapor, methane, and chlorofluorocarbons). Visible light from the sun passes through most atmospheres and is absorbed by the body's surface. The surface reradiates this energy as longer-wavelength infrared radiation (heat). If any of the greenhouse gases are present in the body's troposphere, the atmosphere is transparent to the visible but opaque to the infrared, and the infrared radiation will be trapped close to the surface and will cause the temperature close to the surface to be warmer than it would be from solar heating alone. In the case of Venus, the extensive amount of atmospheric carbon dioxide has created a runaway *Greenhouse Effect* with a surface temperature of 740 K. For the Earth, the net warming is about 17 K above the expected radiative equilibrium temperature of 271 K. Most climate models suggest that as more carbon dioxide is introduced into the atmosphere from the burning of fossil fuels, there is an enhanced Greenhouse Effect leading to global warming. This strongly suggests that there will be anthropogenically driven global warming becoming apparent in the next few decades, which would affect the Earth's climate, and agricultural practices throughout the world. Mechanisms that take up carbon dioxide, by the oceans (where it is transformed into

carbonate rocks) and by plants (particularly the tropical rain forests), apparently have longer timescales than the current timescale for adding greenhouse gases to the atmosphere.

greenhouse gases   The gases that can cause the greenhouse effect. Water vapor, carbon dioxide, ozone, methane, chlorofluorocarbons, and some oxides of nitrogen are *greenhouse gases*.

green line   A coronal line observed at 5303 Å resulting from a forbidden transition in highly ionized iron atoms (Fe XIV). Important for the study of coronal structures at temperatures of order 2 MK.

Green's theorem   An integral of the divergence theorem. For two scalar fields, ϕ and ψ,

$$\nabla \cdot (\phi \nabla \psi) = \phi \nabla^2 \psi + \nabla \phi \cdot \nabla \psi ,$$

and

$$\phi \nabla \psi \cdot \mathbf{n} = \phi \partial \psi / \partial n .$$

Inserting this into the divergence theorem, constructing the similar expression with ψ and ϕ exchanged, and subtracting the two yields

$$\int_V \left(\phi \nabla^2 \psi - \psi \nabla^2 \phi \right) d^3 x$$

$$= \int_S [\phi \partial \psi / \partial n - \psi \partial \phi / \partial n] \, da ,$$

where S is the closed surface bounding the volume V, and a is the area element on S.

Greenwich mean (solar) time (GMT)   A time system based on the angle from the Greenwich meridian to the "mean sun", an artificial construct moving on the celestial equator at a constant rate with one revolution per year. The sun actually traverses the ecliptic, not the celestial equator, during the year, and does that at a nonconstant rate, as well, because of the eccentricity of the Earth's orbit, and because the inclination of the ecliptic causes further variation in the motion of the projection of the sun's position on the celestial equator. Greenwich mean solar noon is when this "mean sun" is on the meridian at Greenwich, England. Other times of the day were derived from noon using clocks or astronomical observations. GMT was split into Coordinated Universal Time (UTC) and Universal

Time (UT1) on January 1, 1972. In the United Kingdom, GMT is sometimes used to refer to Coordinated Universal Time (UTC).

Greenwich sidereal date (GSD)   The number of days of sidereal time elapsed at Greenwich, England since the beginning of the Greenwich sidereal day that was in progress at Julian date 0.0.

Gregorian   The type of reflecting telescope invented by James Gregory, with a small concave secondary mirror mounted beyond focus in front of the primary mirror, to reflect rays back through the primary mirror, where they are viewed using a magnifying lens (eyepiece) from behind the telescope.

Gregorian calendar   The calendar introduced by Pope Gregory XIII in 1582 to replace the Julian calendar and currently in civil use in most countries. Every year therein consists of 365 days, except that every year exactly divisible by 4 is a leap year aside from centurial years, which are leap years only if they are also exactly divisible by 400.

Greisen–Kuzmin–Zatsepin cutoff   An energy limit at about 5×10^{19} eV per nucleon, beyond which, theoretically, cosmic rays have great difficulty in penetrating the 2.75° cosmic microwave background radiation (q.v.). In 1966, Kenneth Greisen (U.S.) and, independently, Vadem Kuzmin and Georgi Zatsepin (U.S.S.R.), pointed out that cosmic rays of very high energy would interact with the photons of the microwave background. This happens because, at their very high velocity, the cosmic rays find the photons colliding from in front (the most likely direction, of course) blue-shifted by the Doppler effect up to gamma ray energies. As a result, cosmic ray protons undergo pi-meson production, while other cosmic ray nuclides are photo-disintegrated. The largest distance from which cosmic rays of this high energy could presumably originate is 50 Mpc, quite local on the cosmic scale. Recent observations with the "Fly's Eye" (Utah, U.S.), AGASA (Japan), and Haverah Park (England) detectors have, puzzlingly, established that the cosmic ray spectrum extends, basically unchanged, to energies

higher than 10^{20} eV. There are very few plausible sources within 50 Mpc, and the arrival directions seem isotropic, as well, deepening the puzzle.

grism   (After the contraction of grating and prism.)   A diffraction grating coated onto a prism. *Grisms* are instrumental to the design of highly efficient spectrographs devoted to the observations of faint objects. For example, in the Faint Object Spectrograph at the 3.6 meter William Herschel telescope on La Palma (Canary Islands), there is no collimator, and light allowed into the spectrograph by the slit directly illuminates a grism which is mounted on the corrector plate of a Schmidt camera to minimize optical elements, and consequently, light losses. The prism acts as a cross disperser, separating the first and second order spectrum produced by the diffraction grating, in a similar way as obtained with a second diffraction grating employed as a cross-disperser in an Echelle spectrograph. *See* diffraction grating, echelle spectrograph.

groin   (also groyne) A structure built perpendicular to the trend of the coastline, lying partially in and partially out of the water. May be built of stone or other materials (wood, steel sheet piles, sand bags). Generally placed in order to impede longshore transport of sediment.

gross photosynthetic rate   The total rate of carbon dioxide fixation with no allowance for the CO_2 simultaneously lost in respiration. [μmoles CO_2 (or O_2) (mg chl)$^{-1}$ h^{-1}] or [mg C (carbon) (mg chl)$^{-1}$ h^{-1}].

ground level event (GLE)   In sufficiently strong solar flares, particles can be accelerated up to GeV energies. Given a suitable magnetic connection (*see* connection longitude) these particles can be detected on Earth as an increase in neutron monitor counting rates a few hours after the flare. *Ground level events* are rare and even during solar maximum only a few GLEs can be detected each year.

ground stroke   A lightning stroke that flashes between a cloud and the ground, usually from negative charges at the bottom of the cloud to positive charges on the ground.

groundwater   Water within the pores of a soil. May be still or moving, and have a pressure corresponding to hydrostatic conditions, pressure flow (as in an artesian acquifer), or below atmospheric pressure, due to the effects of surface tension.

group velocity   The speed at which the profile of a wave form (and its energy) propagates. Specifically, for waves in water, the group velocity reduces to one-half the wave phase speed in deep water; wave phase speed and group velocity are the same in shallow water ($\sqrt{gh}$ per linear water wave theory, where g is acceleration of gravity and h is water depth).

growing season   Generally, and vaguely, the interval between the last killing frost in Spring and the first killing frost in Autumn; or some suitable historical average of this period.

GSD   *See* Greenwich sidereal date.

G star   Star of spectral type G. Capella and our sun are *G stars*.

Guiana current   An ocean current flowing northwestward along the northeast coast of South America.

guiding center   The concept of the *guiding center* was introduced in the 1940s by H. Alfvén to describe the motion of particles in electromagnetic fields. It separates the motion $\mathbf{v}$ of a particle into motions $\mathbf{v}_\parallel$ parallel and $\mathbf{v}_\perp$ perpendicular to the field. The latter can consist of a drift $\mathbf{v}_D$ and a gyration ω around the magnetic field line:

$$\mathbf{v} = \mathbf{v}_\parallel + \mathbf{v}_\perp = \mathbf{v}_\parallel + \mathbf{v}_D + \omega = \mathbf{v}_{gc} + \omega \, .$$

If we follow the particle for a longer time period, for instance from the sun to the orbit of Earth, the gyration itself is of minor importance. In some sense the gyration can be averaged out. Then we can describe the particle motion by the motion $\mathbf{v}_{gc}$ of its guiding center, consisting of a field-parallel motion and a drift. The particle then is always within a gyro-radius of this position.

guiding center approximation (GCA)   A regime of plasma behavior which applies to most plasmas in the Earth's magnetosphere and in interplanetary space. In this regime, the motion of a particle can be separated (to a good approximation) into a rotation around a guiding field line of the magnetic field ("gyration") and a sliding motion (like a bead on a wire) along that line. Modifications of the motion are introduced as added corrections, e.g., as guiding center drifts such as the gradient drift and the curvature drift. The instantaneous center of rotation is known as the guiding center and the instantaneous radius as the gyroradius.

Alfvén's criterion states that the GCA is applicable if the gyroradius is much smaller than the scale distance over which the magnetic field **B** varies. The GCA does not apply to cosmic rays in the Earth's field because their gyration radius is too big, nor to dense plasmas dominated by particle collisions, where collisions usually intervene before a particle manages to complete even one circle.

Guiding center motion is also known as adiabatic motion, because particles undergoing it in the Earth's magnetic field conserve one or more adiabatic invariants, each associated with a nearly periodic component of the motion. The magnetic moment is associated with gyration around the guiding field line, the second invariant or longitudinal invariant is associated with the bounce motion between mirror point, and the third invariant is associated with drift around the Earth. The term "nonadiabatic motion" is frequently applied to motion near neutral points, lines, or sheets, where for at least part of the orbit the magnetic field is so weak (and the gyration radius so big) that Alfvén's criterion no longer holds.

gulf stream   Warm ocean current that flows up the eastern coast of North America. It turns eastward at about 40°N and flows toward Europe.

Gunn–Peterson bound   Limit on the abundance of neutral hydrogen in the intergalactic medium. Photons of wavelength 1216 Å present in the radiation emitted by a quasar at redshift z would be absorbed if neutral hydrogen is present in the interstellar medium. The quasar spectra would show a dip at frequencies higher than the Lyman-α emission line. The absence of such dip in the spectra of a quasar sample with mean redshift $z \simeq 2.6$ leads to the upper limit $n_H(z = 2.64) \leq 8.4 \times 10^{-12} h \text{cm}^{-3}$ for a flat universe. This strict upper limit implies that very little neutral hydrogen exists and what remains in the interstellar medium must be ionized. The physical mechanism that provided the energy necessary to reheat the interstellar medium is not known. Here h is the Hubble parameter $H_0/(100\text{km/sec/M}_{pc})$.

GUT Grand Unified Theory. *See* grand unification.

Guttenberg–Richter relation
The Guttenberg–Richter relation is

$$N = -b \log m + a$$

where N is the number of earthquakes in a specified area and time interval with a magnitude greater than m. This relation is applicable both regionally and globally. b-values are generally close to 0.9.

guyot In oceanographic geophysics, an isolated swell with a relatively large flat top. Believed to be of volcanic origin.

gyrofrequency The natural (Larmor) frequency f of a charged particle in a magnetic field. For nonrelativistic motion, $f = qB/(2\pi m)$, where q is the charge, B is the magnitude of the magnetic field, and m is the mass of the particle. In relativistic situations, the frequency depends on the particle's velocity through the Lorentz factor γ which enters the definition of the momentum. Applied to electrons in the Earth's magnetic field. *See* Larmor frequency.

gyroradius *See* Larmor radius.

H

Hadley (cell) circulation A circulation in the meridional plane known to exist in the tropics due to the ascending warm air near the equator and descending cold air in high latitudes.

Hadley cell Convection cells within the atmosphere of a body. On planets where most of the atmospheric heating is produced by the sun (such as Earth, Venus, and Mars), air over the equatorial regions will be hotter than air over the poles. This hotter air is less dense than cooler air and thus rises, eventually losing heat as it moves toward the polar regions. Over the poles, the air becomes colder and more dense, thus sinking towards the surface. The cooler air moves back along the planet's surface towards the equator, where it warms up and the cycle repeats. This basic cycle of warm air rising over the equator and cooler air sinking over the poles is called Hadley Circulation. Rapid rotation and variations in surface temperature (caused, for example, by oceans vs. continents) complicate this basic pattern. *Hadley cells* fairly accurately describe the atmospheric circulation only for Venus, although they form the basics for physical studies of other planetary atmospheres.

hadron Any particle that interacts with the strong nuclear force. *Hadrons* are divided into two groups: baryons ("heavy ones," consisting of three quarks), which are fermions and obey the exclusion principle, and mesons which are bosons, and consist of a quark anti-quark pair. *See* fermion, boson, quark.

hail Large frozen pellets (greater than 5 mm in diameter) of water that occur in thunderstorms, in which updrafts keep the hail suspended at an altitude with freezing temperatures for long periods of time, growing the hailstone until it finally falls out of the cloud.

hailstone A single unit of hail.

Hale cycle The observation of sunspot numbers alone reveals an 11-year cycle. In contrast, the *Hale cycle* is a 22-year cycle which in addition to sunspot numbers also considers polarity patterns. While after 11 years the polarity of the sun is reversed, the original polarity pattern is restored only after 22 years. The Hale cycle therefore is also called the magnetic cycle of the sun.

Hale–Nicholson Polarity Law In a given solar cycle, examination of solar magnetograms reveals a distinctive alternation of positive and negative polarities in active regions. In the sun's northern hemisphere the positive polarity is located in the "preceding" (westerly) part of the active region and the negative polarity is located in the "following" (easterly) part. The sense is reversed in the southern hemisphere. The hemispherical polarity patterns alternate with each successive activity cycle. This behavior of alternating active region magnetic polarities is known as the *Hale–Nicholson Polarity Law.*

Halley's comet A comet with a period of 74 to 79 years which was identified with several historical passages (including a visit coincident with the defeat of King Harold in 1066) of bright comets by Edmund Halley (1656–1742), validated when the comet reappeared after Halley's death in 1758. The most recent perihelion passage of Halley's comet occurred on February 9, 1986.

Halley's identification of the comet and suggestion of perturbations on its orbit provided an explanation of comets in the context of Newtonian mechanics and Newtonian gravity.

halo Arcs or spots of light in the sky, under suitable conditions even a bright circle around the sun or the moon. *Halos* are caused by the refraction of sunlight on ice crystals in the atmosphere; thus, halos can be observed best in cold climates where ice crystals also form in the lower (and denser) troposphere. The angular extend of a halo is always about 22°; however, depending on crystal shape and orientation, additional arcs and even a wider ring can form.

halocline The region of large vertical gradient of density due to salinity in oceans.

Hα condensation   The downflow of Hα emitting material in the chromospheric portions of solar flares. Typical downflow velocities are of the order $50\,\mathrm{kms}^{-1}$ and are observed as redshifts in Hα line profiles.

Hα radiation   An absorption line of neutral hydrogen (Balmer α) which lies in the red part of the visible spectrum at 6563 Å. At this wavelength, H_α is an ideal line for observations of the solar chromosphere. In H_α, active regions appear as bright plages while filaments appear as dark ribbons.

Hamiltonian   In simple cases, a function of the coordinates x^α, the canonical momenta p_σ conjugate to x^σ, and the parameter (time) t.

$$
\begin{aligned}
H &= H\left(x^\alpha, p_\beta, t\right) \\
&= p_\alpha \frac{dx^\alpha}{dt} - L\left(x^\delta, \frac{dx^\gamma}{dt}, t\right)
\end{aligned}
$$

where

$$
p_\alpha = \frac{\partial L}{\partial\left(\frac{dx^\alpha}{dt}\right)},
$$

and L is the Lagrangian. This relation is inverted to express the right side of the equation in terms of x^α, p_β, and t. Such a transformation is called a Legendre transformation. This can be done only if L is *not* homogeneous of degree 1 in $\frac{dx^\alpha}{dt}$; special treatments are needed in that case.

The standard action principle, written in terms of the *Hamiltonian,* provides the equations of motion.

$$
I = \int_{t_1}^{t_2} \left[p_\alpha \dot{x}^\alpha - H\left(x^\gamma, p_\delta, t\right)\right] dt
$$

is extremized, subject to x^α being fixed at the endpoints. This yields

$$
-\dot{p}_\alpha - \frac{\partial H}{\partial x^\alpha} = 0
$$

$$
\dot{x}^\alpha - \frac{\partial H}{\partial p_\alpha} = 0.
$$

From these can be formed an immediate implication:

$$
\begin{aligned}
\frac{dH}{dt} &= \frac{\partial H}{\partial t} + \frac{\partial H}{\partial p_\sigma}\frac{dp_\sigma}{dt} + \frac{\partial H}{\partial x^\mu}\frac{dx^\mu}{dt} \\
&= \frac{\partial H}{\partial t}
\end{aligned}
$$

since the last two terms on the right side of the equation cancel, in view of the equations of motion. Hence $H = \mathrm{constant}$ if H is not an *explicit* function of t. Also, notice from the equations of motion that conserved quantities are easily found. If H is independent of x^σ then p_σ is a constant of the motion. *See* Lagrangian.

Hamiltonian and momentum constraints in general relativity   The Einstein field equations, as derived by varying the Einstein–Hilbert action S_{EH}, are a set of 10 partial differential equations for the metric tensor **g**. Since the theory is invariant under general coordinate transformations, one expects the number of these equations as well as the number of components of **g** to be redundant with respect to the physical degrees of freedom. Once put in the ADM form, S_{EH} shows no dependence on the time derivatives of the lapse (α) and shift (β^i, $i = 1, 2, 3$) functions; their conjugated momenta π_α and π_β^i vanish (primary constraints). This reflects the independence of true dynamics from rescaling the time variable t and relabeling space coordinates on the space-like hypersurfaces Σ_t of the 3+1 slicing of space-time.

We denote by π^{ij} the momenta conjugate to the 3-metric components γ_{ij}. Once primary constraints are satisfied, the canonical Hamiltonian then reads

$$
H_G + H_M =
$$

$$
\int_{\Sigma_t} d^3x \left[\alpha\left(\mathcal{H}_G + \mathcal{H}_M\right) + \beta_i\left(\mathcal{H}_G^i + \mathcal{H}_M^i\right)\right],
$$

where the gravitational Hamiltonian density is

$$
\mathcal{H}_G =
$$

$$
8\pi G \gamma^{-1/2}\left(\gamma_{ik}\gamma_{jl} + \gamma_{il}\gamma_{jk} - \gamma_{ij}\gamma_{kl}\right)\pi^{ij}\pi^{kl} - \frac{1}{16\pi G}\gamma^{1/2\,(3)}R,
$$

and the gravitational momentum densities are

$$
\mathcal{H}_G^i =
$$

$$
-2\pi_{\;|j}^{ij} = -\frac{1}{16\pi G}\gamma^{il}\left(2\gamma_{jl,k} - \gamma_{jk,l}\right)\pi^{jk}.
$$

The corresponding quantities for matter have been denoted by the subscript M in place of G and their explicit expressions depend on the particular choice of matter fields that one wishes to consider.

Conservation of the primary constraints, namely arbitrariness of the momentum associated with α and β^i, leads to the vanishing of the Poisson brackets between the canonical Hamiltonian and, respectively, the lapse and shift functions. These are the secondary constraints

$$\mathcal{H}_G + \mathcal{H}_M = 0$$
$$\mathcal{H}_G^i + \mathcal{H}_M^i = 0 ,$$

the first of which is known as the Hamiltonian constraint, the second ones as the momentum constraints. One finds that, when the space-time coordinates are $(t, \vec{x})$ defined by the $3+1$ splitting, the *Hamiltonian constraint and the momentum constraints* are equivalent to the Einstein field equations, respectively, for the G_{00} and the G_{0i} Einstein tensor. *See* ADM form of the Einstein–Hilbert action, ADM mass, Einstein equations, initial data, tensor.

Hamilton–Jacobi Theory   In classical mechanics, a method of solution of Hamiltonian systems that makes use of the fact that the Hamiltonian is the generator of infinitesimal canonical transformations in time. Consider a canonical transformation to a set of variables $\{Q^l, P_n\}$ where the new Hamiltonian K is identically zero. Then Q^l, P_n are constants (because the right side of Hamilton's equations is zero). $K = 0$ implies the equation

$$0 = \frac{\partial F}{\partial t} + H .$$

Thus H, which is known to generate infinitesimal canonical transformations in time, is integrated up in time to produce the generating function F. F therefore represents a canonical transformation from the current phase space coordinates q^k, p_l evolving in time, to (functions of) the constants of the motion Q^l, P_k given by

$$p_l = \frac{\partial S}{\partial q^l}$$
$$P_k = -\frac{\partial S}{\partial Q^k}$$

where S (called the action) is the solution to the first equation. In order to solve the first equation, we write: $F = F(q^l, Q^k, t)$; $H = H(q^l, p_n, t) \equiv H(q^l, \frac{\partial F}{\partial q^n}, t)$.

Thus, the first equation is a (generally nonlinear) partial differential equation for S. There are $n+1$ derivatives of F appearing in the first equation, so $n+1$ integration constants, but because only the derivatives of S are ever used, there are n significant constants that we take to be the Q^k, functions of the constants of the notion, i.e., of the initial data. Once the first equation is solved for $S(q^l, Q^k, t)$, then the third equation gives a connection between the constant P_k, the constant Q^k, and q^l and t, which gives the explicit time evolution of q^l. Since the right side of the second equation is a function of $q^l(t)$, and of constants Q^k, P_k, the second equation gives the explicit time evolution of p_l. A variant of this method writes $f = f(p_k, Q^k, t)$; then the transformation equations are:

$$q_l = -\frac{\partial S}{\partial p_l}$$
$$P_k = -\frac{\partial S}{\partial Q_k} ,$$

and the *Hamilton–Jacobi* equation is written with the substitution $H(q^l, p_m, t) \equiv H(-\frac{\partial F}{\partial q^l}, p_m, t)$ as

$$\frac{\partial F}{\partial t} + H\left(-\frac{\partial F}{\partial q^l}, p_m, t\right) = 0 .$$

The analysis follows analogously to the first one.

hard freeze   A freeze in which surface vegetation is destroyed, and water in puddles and the Earth itself are frozen solid.

hard radiation   Ionizing radiation.

Hard X-ray Telescope (HXT)   Telescope on the Yohkoh spacecraft designed to provide imaging of solar flares at hard X-ray energies. Consists of four distinct energy channels: L (13.9 – 22.7 keV), M1 (22.7 – 32.7 keV), M2 (32.7 – 52.7 keV), H (52.7 – 92.8 keV).

harmonic analysis   A technique for identification of the relative strength of harmonic components in a tidal signal. Involves assuming the tide signal to be the sum of a series of sinusoids and determining the amplitude of each component.

harmonic model   The representation of a magnetic or gravitational field by a scalar potential V(x,y,z) satisfying Laplace's equation $\nabla^2 V = 0$, making V "harmonic". Because of the spherical geometry of the Earth, both its gravity field and magnetic field are customarily expanded in spherical harmonics, which naturally group the expressions that make up V into monopole, dipole, quadrupole, octopole (etc.) terms, decreasing with radial distance r as $1/r$, $1/r^2$, $1/r^{3\prime}$, $1/r^4$, etc. The rate at which corresponding field components decrease is larger by one power of r, i.e., these decrease as $1/r^2$, $1/r^{3\prime}$, $1/r^4$, $1/r^5$, etc.

The Earth's gravity field is dominated by its monopole term, but the axisymmetric terms of higher order m, up to $m = 6$ (terms whose potential decreases like $1/r^{m+1}$) are also needed in accurate calculation of satellite orbits and careful satellite studies give terms up to $m \approx 20$. The magnetic field **B** of the Earth inherently lacks the monopole term and its leading term, which dominates it, is the dipole term. Higher orders can also be fairly important near Earth, while far from Earth additional field sources need to be considered (*see* empirical models).

Since the Earth's magnetic field gradually changes with time ("secular variation"), scientists periodically extract from magnetic surveys and observations of each epoch (usually 10 years) an International Geomagnetic Reference Field (IGRF), a harmonic model meant to give (for that epoch) the best available representation of the internal magnetic field and its rates of change, expressed by a given set of spherical harmonic coefficients and their time derivatives. The magnetic fields of other planets have also been represented in this manner, but because of the scarcity of observations, their harmonic models have a much lower accuracy.

heat capacity   The thermodynamic quantity dQ/dT, where dQ is an increment in heat energy, and dT is the corresponding increment in temperature. Always specified with some thermodynamic variable held fixed, as heat capacity at constant volume, or heat capacity at constant pressure.

heat flow   The study of how bodies generate interior heat and transport this heat to their surfaces; a subdiscipline of the field of geophysics. Most planetary bodies begin with substantial amounts of interior heat. Among the sources of such heat are the heat leftover from the formation of the body (accretion), heat produced by differentiation, heat produced by radioactive decay, tidal heating, and solar electromagnetic induction. This heat can melt interior materials, producing magma which can later erupt onto the body's surface as volcanism. The heat contained in the body's interior is transported to the surface of the body, where it escapes to space. The three ways in which this energy can be transported through the interior are by radiation (the absorption and reemission of energy by atoms), convection (physical movement of material, with hot material rising and cool material sinking), and conduction (transfer of energy by collisions between atoms). Larger bodies are more efficient at retaining their interior heat, which translates to a longer period of geologic activity. The thermal evolution of a body can be estimated by determining what mechanisms are responsible for its heating, how the body transports that energy to the surface, and how long the body can retain its internal heat.

heat flow density   *See* heat flux.

heat flux   The flow of heat energy per unit area and per unit time. It is often called heat flow density or heat flow in geophysics.

Heaviside, Oliver (1850–1925)   Physicist and mathematician. Developed the modern vector form of Maxwell's equations and understanding of the classical electrodynamics (via fundamental physical effects predicted and evaluated by him). He also developed the vector and operational calculi, the ideas and applications of δ- and step-functions, as well as many practical applications of Maxwell's theory in telephony and electromagnetic waves propagation in the atmosphere (the ionospheric layer, thus long-range radio communications). *Heaviside* wrote in the telegraph equation and analyzed its technological consequences in 1887, predicted Čerenkov radiation in 1888, was the first to introduce the Lorentz force (in 1889, three years before H.A. Lorentz), and he predicted the existence of the Heaviside–Kennelly ionized atmospheric layer in 1902. His highly informal and heuristic approach to mathematics led to general disregard for his results and innovations on the part of formally thinking arbiters of the mathematical and theoretical fashion of his time.

From 1870 to 1874 Heaviside worked as a telegraphist in Newcastle. He began his own experimental research in electricity in 1868, publishing his first paper in 1872. From 1874 Heaviside never had any employment, performing all his research privately at home in London. He lived his later years in need and even in poverty. Heaviside was elected a Fellow of the Royal Society (1891) and honorary member of the American Academy of Arts and Sciences (1899), then awarded Doctor Honoris Causa of the Göttingen University (Germany) in 1905, honorary memberships of the Institute of Electrical Engineers (1908) and the American Institute of Electrical Engineers (1919), and awarded the Faraday Medal of the Institute of Electrical Engineers in 1921.

heavy minerals   A generic term used to denote beach sediments that have a specific gravity significantly greater than that of the common quartz and feldspar components of many beach sands. Includes hornblende, garnet, and magnetite, among others, and often appears as dark bands on a beach.

Hebe   Sixth asteroid to be discovered, in 1847. Orbit: semimajor axis 2.4246AU, eccentricity 0.2021, inclination to the ecliptic $14°.76835$, and period 3.78 years.

hedgehog configuration   A configuration frequently encountered in various condensed matter phase transitions where molecules point outward away from a pointlike topological defect. It is also used to describe monopoles produced during cosmic phase transitions. *See also* monopole.

Helene   Moon of Saturn, also designated SXII. It was discovered by Laques and Lecacheux in 1980. Its orbit has an eccentricity of 0.005, an inclination of 0°, a semimajor axis of 3.77×10^5 km, and it orbits Saturn at the leading Lagrange point in Dione's orbit. Its size is $18 \times 16 \times 15$ km, but its mass is not known.

Its geometric albedo is 0.7, and it orbits Saturn once every 2.737 Earth days.

heliacal rising   The first visibility of an astronomical object (star) in the predawn sky, after months of being invisible by virtue of being up in day.

helicity   In plasma physics, the sense and amount of twist of magnetic fields characterized by several different parameters. The density of *magnetic helicity* is $\mathbf{H}_m = \mathbf{A} \cdot \mathbf{B}$, where $\mathbf{A}$ is the magnetic vector potential of the magnetic field, $\mathbf{B}$. $\mathbf{H}_m$ determines the number of linkages of magnetic field lines. The density of *current helicity*, $\mathbf{H}_c$, is $\mathbf{B} \cdot \nabla \times \mathbf{B}$ which varies in a way similar to $\mathbf{H}_m$ and describes the linkage of electric currents. In hydrodynamics, one of two quadratic invariants (the other is energy) occurring in the theory of three-dimensional incompressible Navier–Stokes turbulence. The helicity within a volume $\mathcal{V}$ is defined as

$$H_{NS} = \int \boldsymbol{\omega} \cdot \mathbf{V} d^3 x \, ,$$

where $\mathbf{V}$ is the velocity, $\boldsymbol{\omega} =$ curl $\mathbf{V}$ is the vorticity, and the integral is taken over $\mathcal{V}$. In a dissipation-free fluid, for suitable boundary conditions, H_{NS} is conserved. In fully developed three-dimensional turbulence with viscous dissipation, the helicity, as well as the energy, cascades from large eddies down to smaller eddies where dissipation can occur. *See* cross helicity, hydromagnetic turbulence, magnetic helicity.

heliocentric   Centered on the sun, as in the Copernican model of the solar system.

heliopause   The boundary between the heliosphere and local interstellar medium. *See* heliosphere, solar wind.

helioseismology   The science of studying wave oscillations in the sun. Temperature, composition, and motions deep in the sun influence the oscillation periods. As a result, *helioseismology* yields insights into conditions in the solar interior.

heliosheath   Shocked solar wind plasma, bounded on the inside by the heliospheric termination shock, and on the outside by the heliopause. *See* heliosphere, solar wind.

Helios mission   German–US satellite mission to study the inner heliosphere. The instrumentation includes plasma, field, particle, and dust instruments. Two identical satellites, Helios 1 and 2, were launched into highly elliptical orbits with a perihelion at 0.3 AU and an aphelion at 0.98 AU. The combination of the two satellites allowed the study of radial and azimuthal variations; the mission lasted from 1974 to 1986 (Helios 1) and 1976 to 1980 (Helios 2).

heliosphere   The cavity or bubble in the local interstellar medium due to the presence of the solar wind. The size of the *heliosphere* is not yet established, but typical length scales must be of order one to several hundred astronomical units. Heliospheric plasma and magnetic field are of solar origin, although galactic cosmic rays and neutral interstellar atoms do penetrate into the heliosphere.

The heliosphere is thought to comprise two large regions; the interior region is the hypersonic solar wind, separated from the exterior shocked-plasma (heliosheath) region by the heliospheric termination shock. The boundary between the heliosheath and local interstellar medium is called the heliopause.

If the flow of the local interstellar medium is supersonic with respect to the *heliosphere,* a termination shock will be formed in the interstellar gas as it is deflected by the heliosphere. Because of substantial temporal variations in the solar wind (and, possibly, in the local interstellar medium) it is likely that the termination shock and heliopause are never static, but undergo some sort of irregular inward and outward motions.

heliospheric current sheet (HCS)   The current sheet that separates magnetic field lines of opposite polarity which fill the northern and southern halves of interplanetary space ("the heliosphere"). The sun's rotation, combined with the stretching action of the solar wind, gives the HCS the appearance of a sheet with spiral waves spreading from its middle. The spiral structures are responsible for the interplanetary sectors observed near the Earth's orbit. *See* interplanetary magnetic sector.

heliospheric magnetic field   The magnetic field that fills the heliosphere. Because coronal and heliospheric plasmas are excellent electrical conductors, the magnetic field is "frozen into" the expanding gas. Solar wind gas, once it has accelerated away from its coronal-hole origin, is hypersonic and hyper-Alfvénic, so its kinetic energy exceeds its magnetic energy and the field is passively carried along by the wind. The field lines are anchored in a rotating solar source, but carried along by a wind flowing in the outward radial direction, and may be idealized as lying on cones of constant heliographic latitude within which they are twisted to form a global spiral pattern.

The angle ψ between the field and the radial direction (called the Parker spiral angle) is given by $\tan\psi = r\Omega\cos\lambda / V$, where r is heliocentric distance, Ω is the angular velocity of rotation of the sun, λ is heliographic latitude, and V is the solar wind flow speed. In the ecliptic plane at 1 AU, ψ is of order $45°$, and as $r \rightarrow \infty$ the field is transverse to the flow direction. *In situ* observations of the heliospheric field are in accord with this idealized global picture, although modest quantitative deviations have been reported.

Although the magnetic field at the sun's surface is very complicated, there is an underlying dipole pattern except perhaps for brief periods near the time of maximum sunspot activity, when the solar magnetic dynamo reverses its polarity. When the underlying dipole component is present, the polarity of the heliospheric magnetic field at high northern or southern heliographic latitudes coincides with the polarity of the magnetic field in the corresponding high-latitude regions of the solar surface. For example, the Ulysses spacecraft's mid-1990s observations of high-latitude fields show outward polarity at northern latitudes and inward polarity at southern latitudes; these polarities will be reversed in the next sunspot cycle.

Thus, the *heliospheric magnetic field* may be characterized in the first approximation as consisting of two hemispheres consisting of oppositely directed spiraling field lines. These two hemispheres are separated by a thin current sheet, called the heliospheric current sheet.

Now the heliospheric field is not perfectly aligned with the solar poles and equator (equivalently, the underlying solar dipole is tilted with respect to the rotation axis), so that the heliospheric current sheet is tilted with respect to the heliographic equator. Moreover, the solar surface field is always far from being a perfect dipole, and the heliospheric current sheet itself is warped.

The sun rotates, and the heliospheric magnetic field configuration must rotate as well. Hence, a fixed observer near the equatorial plane will be immersed first in field of one polarity (inward, say), then in field of outward polarity, then inward again, etc. The net result is that the magnetic record shows two or more sectors of opposite polarity, and this pattern is usually approximately repeated in the next solar rotation. The phenomenon is ubiquitous at low heliographic latitudes, and in particular at each planet the magnetic field in the solar wind will exhibit alternating polarity. This organization of the low-latitude heliospheric field into sectors of opposite polarity is known as the interplanetary magnetic sector structure. The heliospheric current continues to exist at the largest heliocentric distances where measurements have been made, and presumably extends out to the heliospheric current shock.

The heliospheric magnetic field, like all quantities observed in the solar wind, also exhibits a rich variety of transitory variations. The large-scale morphology described above is a background on which waves, turbulence and other transient structures are superposed.

heliospheric stream structure   Longitudinal organization of the solar wind into faster and slower streams (also called interplanetary stream structure). Apart from the occasional violent outbursts associated with coronal mass ejections, the fastest solar wind comes from coronal holes. Coronal holes most often are found at relatively high solar latitudes, so that on average the high-latitude wind is faster than the wind at equatorial latitudes (velocity 800 km/s as against 400 km/s). However, the polar coronal holes often have equatorward extensions for a substantial range of longitudes, so that there is a substantial longitudinal variation in the equatorial solar wind speed.

For much of the solar activity cycle this longitudinal structure is long-lived enough that the wind forms stream patterns that approximately corotate with the sun. It should be stressed that this corotation is wave-like in the sense that although the pattern may corotate out to large heliocentric distances, the plasma itself, which is subject only to relatively weak magnetic torques, does not corotate. In fact, beyond a few solar radii the flow must be essentially radial, so the rotation of the sun sets up a situation in which fast wind will overtake slow wind from below.

The interaction between fast and slow wind occurs over several astronomical units in heliocentric distance. By 1 a.u. there is substantial compression near the stream interfaces. By 5 a.u. the interaction has proceeded to the point that regions of compressed plasma bounded by shocks, the so-called corotating interaction regions or CIRs, are common; the amplitude of the stream-structure velocity variation at 5 a.u. is substantially smaller than at 1 a.u. This erosion of the velocity structure continues as the wind flows farther out, and beyond about 10 a.u. the CIRs are no longer apparent.

heliospheric termination shock   The shock wave associated with the transition from supersonic to subsonic flow in the solar wind; *see* solar wind, heliosphere. The location of the termination shock is estimated to be at 100 astronomical units from the sun. Outbound heliospheric spacecraft have passed 60 astronomical units heliocentric distance, but have not yet encountered the termination shock. There is optimism that Voyagers 1 and 2 will encounter the shock within the next decade or two.

helium   (From Greek. helios, the sun.) An odorles colorless gas; the second lightest (after hydrogen) gas, atomic number: 2. Consisting of two isotopes: ^{3}He, and ^{4}He. ^{4}He is the second most abundant element in the universe (presumably formed early on in the Big Bang), but is very rare on Earth (partial pressure $\approx 10^{-5}$ atm) at the surface of the Earth. The helium content of the atmosphere is about 1 part in 200,000.

Helium was first detected in 1868 in the solar spectrum and in 1895 in uranium-containing minerals. In 1907, alpha particles were demonstrated to be ^{4}He. *Helium* is extracted from natural gas, produced by alpha decay of heavy elements in the rock. It is the fusion of hydrogen to helium which produces the energy of all main sequence stars, like the sun. The fusion of hydrogen into helium provides the energy of the hydrogen bomb.

Helium is widely used in cryogenic research as its boiling point is close to absolute zero and can be controlled by pumping the helium vapor; its use in the study of superconductivity is vital. The specific heat of helium gas is unusually high. Helium exhibits superfluidity at low temperatures. Liquid helium (He4) exists in two forms: He4I and He4II, with a sharp transition point at 2.174 K. He4I (above this temperature) is a normal liquid, but its superfluid form He4II (below 2.174 K) is unlike any other known substance. It expands on cooling; its conductivity for heat is enormous; and neither its heat conduction nor viscosity obeys normal rules. It has other peculiar properties.

Helium is the only liquid that cannot be solidified by lowering the temperature at ordinary pressures. It can be solidified by increasing the pressure.

helium burning The set of nuclear reactions that converts helium to carbon and oxygen. *See* triple-alpha process.

helium flash The onset of helium burning in a star of less than about 1.5 solar masses. The event is sufficiently explosive to change the structure of the star suddenly from a red giant to a horizontal branch or clump star. The reason for the explosion is that the helium fuel is degenerate when it ignites. This means that energy generation, though it increases the local temperature, does not immediately increase the pressure of the gas, so the core does not expand and cool. Rather, the fusion reaction goes faster and faster until the gas is so hot that it is no longer degenerate and the core of the star expands and cools, slowing down the nuclear reaction rate. *See* clump star, degeneracy, helium burning, horizontal branch star, red giant.

helmet streamer Structure of closed coronal loops overlaid by open field lines which extend outwards to larger radial distances like the

feather on a medieval helmet. Helmet streamers are associated with solar activity: they are much more frequent around solar maximum than during solar minimum, their closed loops are often the anchoring field lines of filaments or protuberances, and reconnection in the tip of a helmet streamer might lead to the expulsion of matter and magnetic flux. *See* disconnection event.

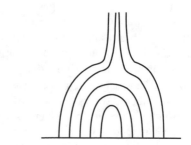

Helmet streamer.

Helmholtz free energy (free energy) An extensive thermodynamic potential H given by

$$H = U - ST \, ,$$

where U is the internal energy, S is the entropy, and T is the temperature of the system. For a reversible process at constant T and V, work stored as free energy can be recovered completely.

Helmholtz theorem A general continuous and first differentiable vector field $\mathbf{A}$ is the sum of two vectors: $\mathbf{A} = \mathbf{B} + \mathbf{C}$, where $\nabla \times \mathbf{B} = 0$ ($\mathbf{B}$ is irrotational, has no vorticity); and $\nabla \cdot \mathbf{C} = 0$, $\mathbf{C}$ is solenoidal, has no divergence.

Henyey–Greenstein scattering phase function An approximate phase function that is parameterized by only the mean cosine of the scattering angle.

Herbig Ae/Be star The analog of T Tauri stars when the object is about 3 to 10 solar masses. They are hotter, brighter, and shorter lived than the lower-mass T Tauri stars. Characterized also by excess infrared emission from circumstellar dust, perhaps arranged in a disk. *See* T Tauri star.

Herbig–Haro object   An object that is apparently a forming star, surrounded by still accreting matter. These objects have strong stellar winds, and mass loss including jets. Herbig–Haro objects are powerful emiters of infrared radiation; the spectrum consists of strong emission lines presumably produced by internal shocks within protostellar jets or winds.

herringbone burst   *See* type II radio burst.

Hertz (Hz)   Unit of frequency equal to s^{-1}.

Hertzsprung Gap   The region on a Hertzsprung–Russell Diagram between the main sequence and red giant branches in which relatively few stars are found. The reason is that stars evolve across the gap (at roughly constant luminosity) very quickly because the hydrogen fuel in their cores has been exhausted and the hydrogen around the core is not yet hot enough to fuse. Thus, the only energy source is gravitational potential energy from core contraction, some of which is used up in simultaneously expanding the outer layers of the star. The gap is particularly conspicuous in the HR diagram of a young open cluster of stars, where all the stars are the same age. *See* Hertzsprung–Russell diagram, main sequence star, open cluster, red giant.

Hertzsprung–Russell diagram   *See* HR diagram.

Hesperian   Geophysical epoch on the planet Mars, 1.8 to 3.5 Gy BP. Channels on Mars give evidence of large volumes of water flow at the end of the Hesperian and the beginning of the Amazonian epoch.

H_I 21-cm line   Spectral line emitted in the radio domain, at a wavelength of 21 cm (corresponding to a frequency of 1420 GHz), due to the hyperfine transition between two energy states in the ground level of the hydrogen atom, the lower state with electron spin and proton spin anti-parallel, the higher state with the two spins parallel. The H_I 21-cm emission line was first detected in 1951; since then it has been used to map the distribution of neutral hydrogen within the galaxy, and in external galaxies, with radio telescopes and interferometers. *See* forbidden lines.

Higgs mechanism   Mechanism responsible for the existence of massive particles in the standard model of quantum field theory. The mechanism arises in a spontaneous breakdown of symmetry. For instance, far before a gauge $U(1)$ symmetry the symmetry breakdown (i.e., at high temperatures), we have a complex scalar field ϕ and a massless gauge field A^μ (like the photon, for example) with two degrees of freedom (two polarizations). As the temperature lowers, ϕ selects one possible phase in the range $[0 - 2\pi]$. This is called spontaneous symmetry breakdown. After the breakdown, ϕ represents a massive scalar field and A^μ becomes a massive gauge field (three degrees of freedom), totaling again exactly four degrees of freedom as before the transition. Such events were presumably important in the very early universe. *See* Abelian Higgs model, cosmic topological defect, Goldstone model, spontaneous symmetry breaking.

higher derivative theories   Theories of gravity in which the equations of motion contain higher than second derivatives of the basic variables (the potentials) of the theory.

highlands, lunar   *See* lunar highlands.

high-pressure and high-temperature experiment   An experiment to measure density of rocks and minerals, velocity of elastic waves and elastic modulus, and to investigate materials, phases, and melting, realizing the high-pressure and high-temperature state of the Earth's interior. There are two kinds of high-pressure apparatuses: a dynamic high-pressure apparatus utilizing shock waves and a static high-pressure apparatus applying constant load. For the former, it is possible to produce pressure of several hundred gigaPascals (GPa) instantaneously, corresponding to the pressure at the Earth's core. However, because of the short reaction time and inaccuracies of the estimate of temperature, other approaches are being used, such as piston cylinder apparatus which is able to produce pressure and temperature characteristic of the uppermost mantle (up to 6 GPa and 2,800 K), multi-anvil type apparatus which can produce pressure and temperature characteristic of the upper-most part of the lower mantle (up to 30 GPa and 2,300 K), and diamond-anvil apparatus which produces pressure up to 300 GPa and temperature over several thousand K. A quench method has been used to identify the phases for phase equilibrium experiments, and *in situ* observation under high-pressure and high-temperature state is now being done employing X-rays.

H$_I$ region   A part of the interstellar medium where hydrogen atoms and most other atoms remain neutral. Most H$_I$ regions have important concentrations of molecules and dust. The main chemical constituents are hydrogen atoms and molecules. The temperature can vary from 10^2 to 10^3 K while the density is of the order of 100 or less particles cm^{-3}. *See* interstellar medium.

H$_{II}$ region   A part of the interstellar medium where all hydrogen is ionized by ultraviolet light from one or many main sequence stars of effective temperature between 3×10^4 to 5×10^4 K. Most other elements are also singly ionized. H$_{II}$ regions can have varying amounts of mass up to several thousand solar masses and temperatures between 6×10^3 to 1.5×10^4 K. The density varies from 10 to 10^4 particles cm^3, and up to 10^6 particles cm^{-3} in ultracompact H$_{II}$ regions. Their shapes are often irregular, and determined by the surrounding medium, and are generally associated with molecular clouds. The mass typically varies from 0.1 to 10^4 solar masses with sizes from 0.01 to 10 parsecs. They are located mostly in the spiral arms of galaxies. Many of them, associated with particular phases of stellar evolution, are also called supernova remnants, planetary nebulae, and nova remnants. *See* interstellar medium. *Compare with* planetary nebula.

Himalia   Moon of Jupiter, also designated JVI. Discovered by C. Perrine in 1904, its orbit has an eccentricity of 0.158, an inclination of 27.63°, and a semimajor axis of 1.148×10^7 km. Its radius is 93 km, its mass 9.56×10^{18} kg, and its density 2.83 g cm^{-3}. It has a geometric albedo of 0.03, and orbits Jupiter once every 250.6 Earth days.

hindcasting (wave)   A procedure of determining wave conditions which existed at a prior time, using observations of atmospheric pressure or wind measurements.

historical climate   The accumulation of weather records analyzed for long-term trends over periods of years. Climate records kept beginning about 2000 years ago, until very recently, were over land only.

hoarfrost   Frost that crystallizes onto vegetation and surfaces by direct deposition from saturated air during a hard freeze. *See* hard freeze.

Holmberg radius   The length of the semi-major axis of a galaxy, either expressed in angular or linear units, measured from the center to a minimum surface brightness of 26.5 photographic magnitudes per square second of arc (approximately 1.5% the surface brightness of the night sky).

homogeneity   A system is homogeneous if all points in it are equivalent with regard to relevant properties.

In cosmology, independence of position in space. The Euclidean space, used as a background for Newtonian physics, is homogeneous even though its contents are not. The space-time describing a realistic cosmology may be spatially homogeneous, i.e., its geometry may be independent of position in each space of constant cosmological time, but in a correct (general relativistic) description, this requires the matter stress tensor to be homogeneous under the same symmetry that describes the homogeneity of the space. It cannot be fully homogeneous as a space-time because homogeneity in four dimensions would imply independence of time, while there is observational evidence for the expansion of the universe. Whether our actual spacetime is even approximately spatially homogeneous is a question that can, in principle, be answered by observations. However, the observations are difficult and, thus far, inconclusive (*see* cosmological principle). At small scales (of clusters of galaxies at least) the universe is obviously inhomogeneous. If our spacetime is not spatially homogeneous, even on long scale average, then it must be modeled by inhomogeneous models.

A space that is isotropic with respect to all of its points is also homogeneous, but the converse is not true. *See* isotropy, Killing vector.

homologous flares   Solar flares that occur repetitively in the same active region, with essentially the same position and with a common pattern of development.

homologous temperature (T/T_m)   The temperature (T), normalized by the "melting temperature", T_m.

homothety   A one-parameter family of maps of space-times with the property

$$\mathcal{L}_\xi g_{ik} = \sigma g_{ik} \, .$$

Here ξ is the generator and g_{ik} the metric. $\mathcal{L}$ denotes the Lie derivative along the vector ξ. *Homothety* is a special one-parameter conformal group when σ is a constant.

homotopy group   Mathematical structure that allows classification of topological defects formed during phase transitions. Denoting $\mathcal{M}$ as the vacuum manifold, one can define the zeroth homotopy group π_0 as a counter of its disjoint parts if it is disconnected. For a connected manifold, it reduces to the identity element.

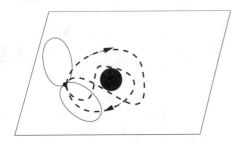

Inequivalent paths on a plane with a hole. The closed curves in full line, not encircling the hole, are all equivalent, and so are those in dashed line that wind once around the hole. However, because of the hole, it is not possible to smoothly transform full line curves into dashed lines. The first homotopy group of the plane with a hole is $\mathbb{Z}$, the group of integers, because the curves can wind an integer a number of times around the hole, in one way or another (hence there is a positive or negative sign associated to each curve).

The first, or fundamental, *homotopy group* π_1 counts the inequivalent closed paths on $\mathcal{M}$: for a plane with a hole for instance, all paths encircling the hole are said to be equivalent depending on the number of times they pass around it (see figure). As they can also wind clockwise or counter-clockwise inequivalently one has, for a plane with one hole, $\pi_1 \sim \mathbb{Z}$, the group of integers.

Similarly, one can define the *n*th homotopy group π_n of $\mathcal{M}$ by means of the inequivalent *n*-dimensional closed hypersurfaces that exist on $\mathcal{M}$.

The topological defect classification stems from these homotopy groups: when $\pi_0(\mathcal{M})$ is non-trivial, then domain walls must form, while strings, monopoles, and textures will exist, respectively, if $\pi_1(\mathcal{M})$, $\pi_2(\mathcal{M})$, and $\pi_3(\mathcal{M})$ are non-trivial. *See* cosmic phase transition, cosmic topological defect, spontaneous symmetry breaking.

hopper dredge   A dredge that includes a hopper or bin for storage of sediment that will later be pumped out. Used to remove sand from channel entrances and harbors.

horizon   On the Earth, or other planet, the distance at which a line from the viewpoint above the surface tangentially touches the surface of the planet. Hence, the maximum distance that can be seen from that viewpoint. The distance to the *horizon* depends on the height of the viewpoint; for 2 m above sea level the distance on Earth is approximately 5 km. In general relativity one defines an *event horizon,* which bounds parts of a spacetime which can never be observed from outside the horizon; and *particle horizon,* which delimits the boundary visible up to a particular time at a particular location in space. *See* apparent horizon, black hole horizon, event horizon, particle horizon.

horizontal branch star   A star of low mass ($\lesssim$ 1.5 solar masses) and low heavy-element content (0.2%) in the evolutionary phase where the two main sources of energy are helium burning in the core and CNO-cycle hydrogen burning in a thin shell. The star is about 30 times as bright as it was on the main sequence, and the phase lasts about 3% as long. The name comes from the loci of the stars in an HR diagram, which is roughly horizontal, at least for some choices of temperature indicator. RR Lyrae variables, one of the common distance indicators, are *horizontal branch stars*. The corresponding phase for stars of higher metallicity is called the clump phase, while more massive stars pass through the phase without abrupt changes in their locations on the HR diagram. *See* clump star, CNO cycle, helium burning, HR diagram, metallicity.

horse latitudes   The latitudes approximately 30°N and 30°S which are regions of calm winds and generally high pressure.

horst   An elevated block of crustal material forming a ridge or plateau, bounded by parallel normal or reverse faults. A *horst* is usually produced by compressional tectonic forces.

hot dark matter   *See* dark matter, hot.

hot spot   (**1.**) A bright, compact component observed in the radio lobes of powerful radio sources, such as radio galaxies and quasars. *Hot spots* are frequently found in radio source of Fanaroff–Riley class II, which includes the brightest lobe dominated radio source. Hot spots are $\sim$ 1 kpc in size, and they appear unresolved when observed at moderate resolution. Their radio spectrum is described by a power-law over the frequency and a spectral index ≈ -0.5, suggesting that radiation is emitted via a synchrotron process. Their location, often at the outer end of the jet, suggests that hot spots are the site of impact between the high speed particles of the jet and the lobes. *See* Fanaroff–Riley.

(**2.**) In geophysics, areas of enhanced volcanic activity not directly associated with plate boundary volcanism, either ocean ridges or the volcanic lines behind subduction zones. *Hot spots* represent areas where magma from the mantle erupts directly onto the surface, creating features associated with fluid lavas such as flood basalts and shield volcanos. The Hawaiian islands show the passage of the Pacific Plate over a hot spot. Iceland is considered to be a hot spot because of its large volcanic flux. Hot spot volcanism is a common process on other plan-

ets, particularly Mars and Venus where many of the large volcanos are the result of this process.

hot towers   The cores of large cumulonimbus clouds in the intertropical convergence zone. *See* intertropical convergence zone.

hour angle   The angular distance on the celestial sphere measured westward along the celestial equator from the vernal equinox to the hour circle passing through the celestial object in question. It is equal to 360° minus the right ascension.

HR (Hertzsprung–Russell) or color-magnitude diagram   A plot of brightness or luminosity vs. some indicator of effective temperature for observed stars, calculated models of stars, or a comparison between the two. Such diagrams have been a primary tool for displaying and testing ideas of stellar structure and evolution since the early 20th century. Such a plot was first attempted for stars in the Pleiades in 1910 by Hans Rosenberg, following advice from Karl Schwarzschild, while Einar Hertzsprung and Henry Norris Russell soon after considered stars of known distances in the solar neighborhood.

The vertical axis can be luminosity in solar units or absolute magnitude or even apparent magnitude for a cluster of stars known to be all the same distance from us, either over a full range of wavelengths (bolometric luminosity) or over some restricted range of wavelengths (U, B, V ,R, I, etc., UBV colors). The horizontal axis can be effective temperature (especially in the case of models), a color index, or spectral type.

Real stars do not populate such a diagram at random. Ninety percent (in a distance-limited sample) will fall on a diagonal from Hot-Bright (upper left) to Cool-Faint (lower right) called the main sequence. The scattering of stars above the main sequence are called red giants and supergiants; those fainter are sub-dwarfs and white dwarfs. A cluster of stars all with the same age and initial chemical composition has a characteristic HR diagram with signatures for age and metallicity. An HR diagram representing observed stars normally shows individual symbols for each star.

When calculations are displayed, there can be points for stars of particular masses, compositions, and evolutionary phases, or, more often, continuous trajectories representing the temporal evolution of a particular model (evolutionary track) or the loci of a number of different star masses at the same time (isochrones). Comparisons between theory and observation are made by plotting both real stars and tracks or isochrones. The problem of converting calibration from observers' quantities (like spectral type and absolute visual magnitude) to theorists' quantities (bolometric luminosity and effective temperature) remains a serious one.

Major phases of stellar evolution are frequently named for the loci of the stars in an HR diagram, including the main sequence, subgiant branch, Hertzsprung gap, red giant branch, horizontal branch, and asymptotic giant branch. Each of these corresponds to a fairly well-defined nuclear source of stellar energy and to a fairly definite interior structure in temperature, density, and chemical composition vs. radius. Both observations and models show that the primary determinant of what a star will do throughout its life is its initial mass. *See* absolute magnitude, apparent magnitude, asymptotic giant branch star, bolometric magnitude luminosity, color index, effective temperature, Hertzsprung Gap, horizontal branch star, main sequence star, metallicity, red giant.

Hubble deep field   A high galactic latitude sky field (of width $\approx$ 2 arcmin), intensively observed in four colors with the Wide Field Planetary Camera mounted on board the Hubble Space Telescope in December 1995. Three sets of observations for a total of 35-h exposure time were obtained with broad band filter centered at 450, 606, and 814 nm, and a set of 50 h exposure time with a near UV filter centered at 300 nm. The *Hubble deep field* was chosen in an area of low H$_I$ column density, small far-IR flux, with no radio source brighter than 1 mJy, no bright stars, and no nearby galaxy clusters. These selection criteria were aimed at making possible the identification and morphological study of a large number of faint, field galaxies. The observations allowed counting the number of galaxies in the field down to a magnitude $\lesssim$ 29, an unprecedented achievement. *See* Jansky (J_y).

Hubble diagram   A diagram plotting observed redshift of a galaxy on one axis vs. some measure of the distance of the galaxy on the other. Used to verify or to detect deviations from Hubble's law: $\mathbf{v} = H_0 d$, where d is the distance to the galaxy.

Hubble parameter (Hubble constant)   The proportionality factor H_0 in the Hubble law $v = H_0 l$, where v is the velocity with which a given galaxy G recedes from our galaxy, a measure of the present expansion rate of the universe, and l is the distance between G and the observer in our galaxy. H_0 is traditionally measured in km/(sec $\cdot$ Mpc), so that v comes out in kilometers per second when l is measured in megaparsecs (Mpc). The actual numerical value of H is subject to many random and systematic errors. In the 1930s H_0 was believed to be of the order of 500 km/(sec $\cdot$ Mpc). Currently it is believed to be between 50 and 100 km/(sec $\cdot$ Mpc). In determining the Hubble constant, H_0, it is the distance determination that introduces uncertainty since the recession is directly measured by the redshift. This uncertainty in the measured value of H is usually taken into account by incorporating the parameter h (called the dimensionless Hubble parameter) into all formulae that depend on H; by definition $h = H/(100 \text{ km/s} \cdot \text{Mpc})$ so that $h = 1$ when $H = 100$ km/s $\cdot$ Mpc. The value of H determines the age of the universe (i.e., the time since the Big Bang up to now) and the distances between galaxy clusters, so knowing it is of fundamental importance for cosmology. The *Hubble parameter* is colloquially called "Hubble constant", but in the currently accepted cosmological models it is a decreasing function of time and "the value of H" actually means the current value H_0. In inhomogeneous models the Hubble law does not apply globally; in them, v is a nonlinear function of l, a different one for every observer. If our universe is inhomogeneous, then the Hubble law applies only as a first approximation to the actual function $v(l)$ in a sufficiently small neighborhood of every observer, and the local value of the Hubble parameter depends on the position of the observer and on the direction of observation. Reference: E.P. Hubble, *The Realm of the Nebulae,* [republication by] Yale University

Press 1982; *Mon. Not. Roy. Astr. Soc.*, **113**, 658, (1953).

Hubble radius   Distance at which the recession velocity of a galaxy will be the speed of light: $cH_o^{-1} = 3000\ h^{-1}$ Mpc. Galaxies located at larger distances will not be in casual contact with us. Hence, it gives the scalelength of the particle horizon. *See* Hubble parameter.

Hubble–Reynolds law   Empirical law describing the brightness profile of an elliptical galaxy, introduced by J.H. Reynolds in 1913. According to *Hubble–Reynolds law,* the surface brightness depends upon the distance from the galaxy center, r, as $\Sigma(r) = \Sigma_0/(1 + r/r_0)$, where the scaling parameter r_0 is the radius at which the surface brightness falls to one quarter its central value Σ_0. This law, remarkable because of its simplicity, predicts a deficit of light close to the center and more light in the outer envelope of a galaxy with respect to de Vaucouleurs' law. *See* de Vaucouleurs' law.

Hubble sequence   A classification scheme of galaxies created by E. Hubble. In the *Hubble sequence,* galaxies are subdivided into elliptical galaxies, S0 galaxies (galaxies showing evidence of an amorphous disk and a bulge, but no spiral arm, also called lenticular galaxies), spiral galaxies, either barred or non barred, and irregular galaxies. An elliptical galaxy is conventionally indicated with the uppercase letter E and an integer number ranging from 0 to 7, increasing with the apparent flattening, and defined as the integer part of $10 \times (1 - (b/a))$, where (b/a) is the axial ratio measured on a photograph or digital image. Spiral galaxies are farther subdivided along the sequence in S0, Sa, Sb, Sc according to three criteria: (1) decreasing bulge prominence with respect to disk; (2) spiral arms less tightly wound; (3) appearance of arms more resolved. At the end of the sequence, irregular galaxies (subdivided into Magellanic and amorphous or M82-type) do not show regularly decreasing surface brightness nor spiral arms and are of patchy appearance. Elliptical and S0 galaxies are collectively referred to as "early morphological types", and Sc and irregulars as "late type" galaxies. Hubble attributed to these terms an evolutionary meaning, i.e., he thought that a spiral could be an evolved elliptical galaxy. This view is not considered appropriate anymore, since the angular momentum per unit mass, a constant for an isolated galaxy, increases along the sequence from the elliptical to the most flattened galaxies (Sc). Nevertheless, gravitational interaction between galaxies can affect the morphology of spiral galaxies to the point of changing their Hubble type. *See* elliptical galaxies, spiral galaxy.

Hubble's law   Relation between the recession velocity of a galaxy v and our distance to it d: $v = H_o d$ where H_o is called Hubble parameter to honor Edwin Hubble who discovered this relation in 1928. In an expanding universe, this direct proportionality is a consequence of the homogeneity and isotropy (*see* cosmological principle). The law is only exact on the average. Local irregularities in the matter distribution create small deviations. *See* peculiar motion.

Hubble space telescope (HST)   A space-based telescope of 2.4 m aperture, launched in 1990 and orbiting in a low terrestrial orbit. Although the optical design of HST is similar to that of a mid-sized ground-based telescope, the absence of atmosphere allows the telescope to operate at a resolution close to the diffraction limit (0.03 arcsec at 3000 Å), and to detect UV light which is absorbed by the terrestrial atmosphere. Currently available instruments on board HST include two imaging cameras, a long-slit spectrograph, and a camera and spectrometer operating in the near infrared. The Wide Field Planetary Camera, which is composed of three CCD detectors in an L-shape configuration plus a single, smaller CCD detector at the center of field, has limiting magnitude of 28 (with 1-h exposure time and S/N ratio 5), and reaches a resolution of 0.053 sec of arc. The highest resolution, 0.042 sec of arc, is achieved with the Faint Object Camera, which has a much smaller field of view, 7×7 square arcsec. (Traditional ground-based optical telescopes have resolution limited by atmospheric seeing to about 1 arcsec, although modern adaptive optics techniques can improve ground-based resolution in some wavelengths to well below 1 arcsec.) The Space Telescope Imaging Spectrometer, which operates between 115 and 1100 nm, offers spectral resolving power ranging from 150 to 100000 and long slit capabilities.

Hubble time   Inverse of Hubble constant; $H_o^{-1} = 9.78h^{-1} \times 10^9$ years corresponding to the time it would take the universe to reach its present size expanding at constant rate H_o. Here $h = H_0$ (100 km/sec/Mpc). *See* Hubble radius.

humic substance   High molecular weight organic compounds resulting from plant decay, especially terrestrial plants; water-soluble soil humic substance imparts a yellow color to water. *See also* colored dissolved organic matter.

hurricane   Intense tropical cyclone of the Atlantic Ocean, Caribbean region, and on the northeastern coast of Australia. Similar types of tropical cyclone appearing over the western Pacific are called typhoons. A *hurricane* is a circulatory strong wind system (180 km/h or so) with low central pressure (below 950 mb), and is confined to a few hundred kilometers. In the northern hemisphere the wind rotation is counterclockwise.

Huygens principle   In radiation fields, the phase front at any instant can be thought of as the locus of radiating sources, and the phase front at the next instant is obtained by a phase coherent addition of the radiation from all those sources.

Hyades   An open cluster of about 100 stars visible to the naked eye in the constellation of Taurus (45 parsecs from the sun, at right ascension 4^h 27^m, declination $+16°0'$). The *Hyades* cluster is about 600 million years old.

hybrid topological defect   Different symmetry breaking schemes in quantum field theory allow for the possibility of formation of various kinds of topological defects, like monopoles, cosmic strings, and domain walls, in the early universe. Realistic grand unified models generally involve several phase transitions and therefore one type may become attached to another, and the system will end up, say, with walls bounded by strings or strings connecting monopoles. Such mixed configurations are called *hybrid topological defects* and are dynam-

ically unstable since no topological constraint can be applied on them. *See* cosmic phase transition, cosmic topological defect, grand unification, Langacker–Pi mechanism.

hydraulic head (piezometric head)   The equivalent height of a fluid required to maintain a pressure hydrostatically. The *hydraulic head* is of practical meaning only when defined for a uniform fluid density. For a flow field with variable fluid density, a reference head is often defined using a reference density. Given density ρ, the hydraulic head h is defined as

$$h = \frac{p}{\rho g} + z$$

where p is fluid pressure, g is gravitational acceleration, and z is elevation measured from an arbitrary reference point. In groundwater hydrology, the hydraulic head is also called the piezometric head.

hydraulic conductivity (K)   The ability of a porous medium to transmit fluid is dependent on both the properties of the fluid and the medium:

$$K = \kappa \left[\frac{\rho g}{\mu} \right]$$

where κ is the intrinsic permeability of the porous medium, ρg is the fluid weight, and μ is the viscosity of the fluid. *Hydraulic conductivity* is generally a tensor because of κ. It has the dimension of velocity. *See* Darcy's law. The hydraulic conductivity of natural substances ranges over many orders of magnitude, from 10^{-21} m s^{-1} for unfractured crystalline rock to 10^{-2} m s^{-1} for karst limestone and gravel. The hydraulic conductivity decreases very rapidly as the medium becomes unsaturated, because the larger pores of the subsurface become airfilled first, causing flow to occur in smaller pores which conduct water at much lower flow rates according to Poiseuille's law. We therefore write the hydraulic conductivity in unsaturated systems as $K(\theta)$, indicating that it is a function of the volumetric moisture content.

hydraulic depth   The cross-sectional area of a free-surface flow (such as a river) divided by the top width of the flow.

hydraulic-fracturing method   A mensuration of crustal stress. The only method applicable to the measurement of crustal stress deeper than a depth of 100 m. Confining the upper and the lower sides of a portion of a drilled borehole with expansive stoppers, the wall of the portion of the borehole is broken, applying water pressure to the portion. The direction of resultant cracks becomes parallel to axis of maximum horizontal compressive stress. The amount of maximum and minimum horizontal compressive stresses can also be obtained from the water pressure when the cracks open and close.

hydraulic gradient (dh/dl)   A dimensionless number, the driving force in flow through porous media, equal to the change in hydraulic head (dh) with a change in distance in a given direction (dl).

hydraulic head (h)   The *hydraulic head* has units of length and is the total mechanical energy per unit weight of fluid, calculated as the sum of the pressure head ($p/\rho g$) and the elevation head (z), or $h = p/\rho g + z$, where p is the fluid pressure, ρg is the fluid weight or weight density (γ), and z is the elevation head. Assumptions are that the fluid is both homogeneous and incompressible, and that kinetic energy is negligible. In unsaturated flow, the pressure head is negative and the hydraulic head is then equal to the tension head ψ plus the elevation head, or $h = \psi + z$.

hydraulic jump   A rapid change of the flow in a river or other open channel where the velocity drops and depth increases over a short distance. Involves a change in the Froude Number, $Fr = V/\sqrt{gD}$, where V is the velocity, g is acceleration of gravity, and D is hydraulic depth, from > 1 (supercritical flow) to < 1 (subcritical flow).

hydraulic radius   The cross-sectional area of a flow divided by the wetted perimeter for the flow.

hydraulic routing   A technique for studying the propagation of a flood through a section of river. Involves the solution of the continuity and momentum equations for a moving fluid. Compare to hydrologic routing, which is a simplified method for investigation of similar problems.

hydraulic transmissivity   A parameter used to describe the ability of a confined aquifer to transmit water along it, assuming flow across the aquifer is negligible. For aquifer thickness $h(x, y)$, the transmissivity is defined as

$$T(x, y) = \int_0^{h(x,y)} K(x, y, z)\, dz$$

where K is the hydraulic conductivity, and z is the coordinate in the direction across the aquifer. The principal hydraulic conductivity in the z direction is zero, and therefore the transmissivity is a tensor defined in the aquifer (x, y) plane.

hydrodynamic   Relating to the flow or movement of water or other fluid.

hydrodynamic instability   A mean flow field is said to be hydrodynamically unstable if a small disturbance introduced into the mean flow grows spontaneously, drawing energy from the mean flow. There are two kinds of fluid instabilities. One is parcel instability such as the convective instability, inertial instability, and more generally, symmetric instability. The other kind is called wave instability, which is associated with wave propagation.

hydrogen   Colorless, explosive, flammible gas; H. The most common element in the universe. Naturally occurring atomic mass 1.00794. Constituent of water. Melting point 14.01 K, boiling point 20.28 K. Two natural isotopes: ^{1}H, 99.9885%, ^{2}H (deuterium), 0.0115%. One known radioactive isotope, ^{3}H (tritium), which undergoes β-decay to ^{3}He with a half life of 12.32 years.

hydrogen burning   The fusion of four hydrogen atoms (protons) to make one helium atom (alpha particle), via either the proton-proton chain or the CNO cycle. Because four hydrogen atoms are more massive than one helium atom by about 0.8%, *hydrogen burning* releases more energy than any other nuclear reaction in stars, $6-7 \times 10^{18}$ erg/gram (depending on lifetimes), as main sequence stars and red giants. Hydrogen burning was proposed as the main source of energy for stars as early as 1925 by Sir Arthur Eddington and Hans Bethe and others. The interaction between two protons or a proton and a heavier nucleus is a quantum mechanical process in which barrier penetration must occur. As a result, hydrogen burning occurs at a temperature of 10 to 20 million K in stars of low to high mass (0.085 to 100 solar masses). Lower mass configurations do not reach this temperature range, derive little or no energy from hydrogen burning, and are called brown dwarfs. *See* brown dwarf, CNO cycle, main sequence star, proton-proton chain, red giant.

hydrograph   A plot showing discharge vs. time. Commonly used to illustrate flood events on rivers or creeks.

hydrographic survey   A term often used synonymously with bathymetric survey. A survey of the geometry of the seafloor. Typically involves one system for establishment of horizontal position and another for vertical position or water depth. Traditionally performed from a boat, but amphibious vehicles, sleds, and helicopters have also been used.

hydrologic equation   A water balance equation often used in catchment hydrology based on the law of mass conservation that states that basin outputs are related to basin inputs plus or minus changes in storage, and written in simplified form as $Q = P - ET \pm \Delta S$, where Q is surface water runoff at the basin outlet, P is precipitation input to the basin, ET are evapotranspiration losses of water from the basin, and ΔS is the change in storage of water in the basin.

hydromagnetic   Pertaining to the macroscopic behavior of a magnetized electrically conducting fluid or plasma. *Hydromagnetic* phenomena are generally associated with large length scales compared with the Larmor radii and long time scales compared with the Larmor periods of the particles that comprise the fluid. Theoretical descriptions of such phenomena may be based on magnetohydrodynamics or kinetic theory. *See* magnetohydrodynamics, Vlasov–Maxwell equations.

hydromagnetics   *See* magnetohydrodynamics.

hydromagnetic shock wave   An abrupt transition between two regions in a magnetized plasma, analogous to the acoustic shock wave in air. *Hydromagnetic shock waves* propagate relative to the plasma, and may be regarded as the short-wavelength limit of large-amplitude magnetoacoustic waves. There are fast and slow shock waves that correspond, respectively, to the fast and slow magnetoacoustic wave modes; a large-amplitude wave of either mode may steepen to form a shock.

In a reference frame in which a shock is at rest, the flow speed of the plasma upstream of the shock is greater than the (fast or slow) magnetoacoustic wave speed, and downstream this inequality is reversed. As in the case of shocks in an ordinary gas, plasma flowing through a hydromagnetic shock undergoes heating and associated entropy production. In ideal magnetohydrodynamics a shock is infinitely thin; real hydromagnetic shocks have a structured transition region of finite thickness, comparable to the ion Larmor radius in many situations.

Hydromagnetic shock waves are routinely observed *in situ* in collisionless plasmas in space. The high-speed solar wind passes through a bow shock when it encounters a planetary or cometary obstacle. From time to time material ejected from the solar corona drives transient shock waves into the heliosphere; if such a shock encounters a planetary magnetic field, it may produce magnetic storms and aurorae in the planetary magnetosphere and upper atmosphere. Shock waves can also be produced internally in the solar wind as fast streams catch up with slower streams; these shocks are frequently observed from about 3 to about 10 astronomical units from the sun. *See* hydromagnetic wave.

hydromagnetic turbulence   The turbulent, irregular fluctuations that can arise in a magnetized fluid or plasma. Just as turbulence arises at high Reynolds number in an ordinary unmagnetized fluid, *hydromagnetic turbulence* is associated with high magnetic Reynolds number (the ordinary Reynolds number with kinematic viscosity replaced by electrical resistivity). Magnetized cosmic plasmas are likely sites of hydromagnetic turbulence, due to their large size and high electrical conductivity. Such turbulence, in turn, may be of importance in the physics of astrophysical systems, playing a role in energy or angular momentum transport, or even support against gravitational collapse.

Large-amplitude, apparently random fluctuations of magnetic field and other physical quantities are commonly observed *in situ* in space plasmas, especially the solar wind, which is often regarded as our best laboratory for the study of hydromagnetic turbulence. There is also strong indirect observational evidence for the existence of hydromagnetic turbulence in distant astrophysical systems.

The theory of hydromagnetic turbulence is not yet well understood, despite extensive theoretical and computational efforts.

Observationally, it is well established that there are extensive regions within the solar wind stream structure that appear turbulent in the sense that their fluctuations seem random and exhibit turbulence-like spectra; but often these same fluctuations also exhibit the properties expected for Alfvén waves. These latter properties include constant density and magnetic field strength and an alignment between velocity and magnetic field corresponding to maximum cross helicity. In the limit of perfect Alfvénic behavior, it can be shown that no turbulent energy cascade can occur. Thus there is a debate, still unsettled, about whether the interplanetary Alfvénic fluctuations are generated dynamically as part of the solar wind flow process, or are rather the remanent fossil of turbulent processes that occur near the sun and are passively convected out through the heliosphere. *See* Alfvén wave, Alfvénic fluctuation, cross helicity, helicity, hydromagnetic wave, magnetic helicity.

hydromagnetic wave Any long-wavelength, low-frequency wave that can propagate in a magnetized plasma; the analogous phenomenon in an ordinary gas is the sound wave. The standard constraints on the wavelength λ and frequency ω of hydromagnetic waves are that $\lambda >>$ the typical Larmor radius and $\omega <<<<$ the Larmor frequency for particles of every charge species. The theory of *hydromagnetic waves* can be developed in the framework of ei-

ther magnetohydrodynamics or kinetic theory; the former approach has the advantage of greater mathematical tractability, but cannot adequately describe some phenomena involving dissipation or instability that may occur in a collisionless plasma.

Either approach to the theory of hydromagnetic waves predicts several kinds of propagating waves. These may be either compressive (magnetoacoustic waves) or noncompressive (Alfvén waves). Magnetoacoustic waves can steepen to form shock waves, and in a hot collisionless plasma are subject to strong Landau damping. Alfvén waves do not steepen or Landau-damp. The discontinuity that corresponds to the short-wavelength limit of an Alfvén wave is called the rotational discontinuity.

In addition to the propagating modes, there is a one-dimensional static (nonpropagating) structure, the tangential pressure balance. In this structure, all gradients are normal to the local magnetic field, and the total transverse pressure $P_\perp + B^2/8\pi$ must be constant throughout the structure. The short-wavelength limit of the tangential pressure balance is called the tangential discontinuity.

Another nonpropagating structure that formally comes from magnetohydrodynamics is the "entropy wave", which is a simple balance of gas pressure alone. The entropy wave is a useful concept in gas dynamics, but strictly speaking it cannot exist in a collisionless plasma because of rapid particle mixing across the structure. However, the entropy wave, and especially its short-wavelength limit, the contact discontinuity, is a helpful concept in space physics and astrophysics in situations involving strong shocks, in which the magnetic field may be of secondary importance and gas dynamics are a fairly good approximation.

Most kinds of hydromagnetic waves and analogous nonpropagating structures, as well as their associated "discontinuities", have been identified in solar wind plasma and magnetic observations. *See* Alfvén wave, hydromagnetic shock wave, hydromagnetic turbulence, magnetoacoustic wave.

hydrosphere The water content of the Earth. The water on the Earth resides primarily in the oceans, and also in other surface waters in underground waters, and in glacial deposits. The term *hydrosphere* should also include water involved in the hydrologic cycle, i.e., water vapor in the air and droplets in clouds.

hydrostatic   Related to still water. For example, hydrostatic pressure is the pressure resulting from a stationary column of water.

hydrostatic approximation   An approximation in which the pressure is assumed to be equal to the weight of the unit cross-section column of fluid or air overlying the point.

hydrostatic equation   The variation of pressure (p) with depth (d) in a fluid at rest is one of the fundamental relationships in fluid mechanics: $dp = -\rho g \, dz$. The change in pressure (dp) of the fluid is equal to the unit fluid weight (ρg) over some vertical distance (dz), which we can integrate from the bottom of a fluid column ($z = -d$) to the surface ($z = 0$):

$$\int_p^{p_s} dp = -\rho g \int_{-d}^0 dz$$

where p_s is the pressure at the surface. Integrating, $p_s - p = -\rho g[0 - (-d)]$, or $p - p_s = \rho g d$, which can be used to calculate the absolute pressure at any point in a static fluid. Commonly we take the fluid pressure at the surface to be zero gage pressure ($p_s = 0$), or the fluid pressure relative to atmospheric pressure. The absolute pressure is then the sum of gage pressure and atmospheric pressure. If p is the gage pressure and the fluid is static with a constant density (ρ), then pressure increases linearly with depth: $p = \rho g d$.

hydrostatic equilibrium   The condition in a star or other object where the inward force of gravity is precisely balanced by the outward force due to the gradient of pressure, that is

$$\frac{dP}{dr} = -\frac{GM(r)\rho}{r} \, ,$$

where P and ρ are the local values of pressure and density at radius r, $M(r)$ is the mass interior to this point, and G is Newton's constant of gravity. In the absence of hydrostatic equilibrium, a star will expand or contract on the free-fall time scale (about 1 hour for the sun). A slight imbalance can lead to stellar pulsation (*see,* e.g., Cepheid variables). A large imbalance leads to either catastrophic collapse or violent explosion, or both (*see* supernova, type II). In general relativity in spherical systems the equation is modified to require larger $|dP/dr|$ to maintain equilibrium; as a result, such systems are less stable when analyzed in full general relativity then as predicted from Newtonian theory.

hydrostatic pressure   The hydrostatic pressure p_h is the pressure in water, say the oceans, as a function of depth h,

$$p_h = \rho_w g h$$

where ρ_w is the density of water and g is the acceleration of gravity.

Hygiea   Tenth asteroid to be discovered, in 1849. Diameter 430 km. Orbit: semimajor axis 3.1384 AU, eccentricity 0.1195, inclination to the ecliptic $3°.847$, period 5.56 years.

hygrometer   A device to measure relative humidity.

Hyperion   Moon of Saturn, also designated SVII. Discovered by Bond and Lassell in 1848, it is one of the largest non-spherical bodies in the solar system. Its orbit has an eccentricity of 0.104, an inclination of $0.43°$, and a semimajor axis of 1.48×10^6 km. Its size is $205 \times 130 \times 110$ km, its mass, 1.77×10^{19} kg, and its density 1.47 g cm^{-3}. It has a geometric albedo of 0.3, and orbits Saturn once every 21.28 Earth days.

hypersonic   Pertaining to speeds or flows with Mach number exceeding 5. In hypersonic motion of a body through a fluid, the shock wave starts a finite distance from the body. *See* Mach number.

hysteresis   The property whereby a dependent variable can have different values according to whether the independent variable is increasing or decreasing. In hydrologic systems, a loop-like curve develops that relates pairs of hydraulic properties of an unsaturated porous medium because volumetric moisture content

(θ), tension head (ψ), and hydraulic conductivity (K) covary along different curves depending on whether the soil is wetting or drying.

I

Iapetus   Moon of Saturn, also designated SVIII. Discovered by Cassini in 1671, it is noted for a surface where the leading edge has an albedo of 0.02 to 0.05, and the trailing edge has an albedo of 0.5. It has been conjectured that the difference is due to the leading edge accreting dark material (possibly from Phoebe) as it orbits. Its orbit has an eccentricity of 0.028, an inclination of 14.72°, and a semimajor axis of 3.56×10^6 km. Its radius is 730 km, its mass is 1.88×10^{21} kg, and its density is 1.15 g cm^{-3}. It orbits Saturn once every 79.33 Earth days.

Icarus   An Earth-crossing Apollo asteroid with perihelion 0.187 AU, aphelion 1.969 AU, approaching within 0.04 AU of the Earth's orbit. Its orbital inclination is 22.9° and its orbital period is about 1.12 years. Based on its brightness, its diameter is 1 to 2.5 km.

ice   A generic term referring to those substances of intermediate volatility, which are sometimes solid and sometimes gaseous. Examples of ices that are commonly encountered in astrophysical contexts are H_2O, NH_3, CH_4, and CO_2. Organic materials with intermediate volatilities are sometimes called *organic ices, CHON* (i.e., substances composed mainly of the elements C, H, O, and N) or simply organics.

ice age   A period of geologic history during which considerable portions of the Earth were covered with glacial ice. There have been many *ice ages* in the geological history of the Earth; during some of them, the ice sheets were situated in the polar regions and some of them were in equatorial regions. In the last ice age, the Pleistocene, about one-fifth of the Earth's surface was covered by glacial ice. The climatic feature of ice ages is a very large temperature decrease. Over mid-latitude region, temperature can decrease more than 10° C and ice and snow can cover 20 to 30% of the Earth's surface. The reasons for the ice ages may include the long-term variations of solar radiation, orbit parameters of the Earth, and variations of climatic systems and processes of the rock lithosphere (including nonlinear feedback processes).

ICM-ISM stripping   An important consequence of the interactions between a member galaxy in a cluster and the intracluster medium (ICM), resulting in stripping the galaxy of its interstellar medium (ISM) due to the ram pressure of ICM (Gunn and Gott, 1972). Suspected sites of the ongoing *ICM-ISM stripping* include some of the ellipticals and spirals in the Virgo cluster, and a few spirals in the cluster A1367. Observed signatures for the ICM-ISM stripping rather than other internal processes such as starbursts, cooling flows, or gravitational interactions between member galaxies, consist in a relatively undisturbed stellar disk, together with a selectively disturbed HII region containing abruptly truncated H_α disk and one-sided filamentary structures above the outer disk plane. The ISM-ICM stripping is consistent with observational facts about spiral galaxies in clusters: many of them are deficient in gaseous disks, those with truncated gaseous disks have relatively undisturbed stellar components, they sustain vigorous star formation in the inner disks but very little in the outer disks. The degree of severity of HII disk truncation signifies the decline in star formation rates and the time elapsed since the ISM-ICM stripping began in cluster spirals.

ideal gas   A theoretical gas consisting of perfectly elastic particles in which the forces between them are zero or negligible.

ideal gas [equation of state]   The thermodynamic relationship between pressure, volume, number of particles, and temperature for an ideal gas. It is mathematically expressed by the equation $PV = NkT$, where P is the pressure, V is the volume, N is the number of particles, k is Boltzmann's constant, and T is the absolute temperature.

igneous   Referring to rock that has been in a totally molten state. *Igneous* rocks are classified as volcanic, or extrusive, if they have been extruded from the Earth in volcanic flow. Examples are rhyolite, andesite, and basalt. Igneous rocks are classified as plutonic (intrusive igneous) if they formed totally subsurface. Granite, diorite, gabbro, and peridotite are plutonic. Igneous rocks are also classified as felsic, intermediate, mafic, and ultramafic, depending on the silica content. The types listed above are in order felsic to mafic (periodotite is classified ultramafic, with silica content below 45%).

igneous rocks   Rocks that cooled and solidified from a magma.

illuminance (lux)   The luminous power (measured in lumens) incident per unit area of a surface. One lux = one lumen per square meter.

immersed weight   The weight of an object (or sediment particle) when underwater. Particle weight minus buoyant force.

impact basin   Large craters ($\geq$ 300 km in diameter) that exhibit many of the characteristics found in craters greater than a few tens of kilometers in diameter.

Terrestrial circular *impact basins* often display concentric rings of mountains (rising to several kilometers) and lack a central peak. Concentric ring structures are most common on the moon. Morphological differences in impact basins include the degree to which they appear to have been eroded, or the amount they have been modified by later processes such as volcanism. Examples of lunar basins include Imbrium and Orientale (greater than 900 km in diameter), and martian basins include Hellas (2000 km in diameter) and Argyre (1200 km in diameter). Because impact basins and their associated ejecta deposits are so large, they dominate the surface of the moon and other planetary bodies. *See* crater.

impact crater   *See* crater.

impact crater ejecta   Debris ejected from an impact crater that surrounds the crater rim. Cumulatively ejecta fragments form a curtain of material.

Ejecta deposited near the crater rim is coarser grained than ejecta far from the rim. In this way, deep-seated (coarse) materials that only just make it over the crater rim form the rim deposits, while originally shallow materials are deposited at increasingly large radial distances. This produces an inverted stratigraphy in the ejecta deposits relative to the stratigraphy of the original target.

On the moon, ejecta patterns most often comprise continuous deposits, discontinuous deposits, and finally rays at the greatest distance from the crater rim. On Mars, ejecta is regularly lobate, implying it was emplaced by flow across the surface, possibly having been fluidized by near-surface ice. *See* impact excavation phase.

impact creep   The method of moving material along a surface by the transfer of energy from saltating particles. A particle lying on a surface can gather sufficient energy to move if it is struck by a bouncing (saltating) grain. The energy from a saltating particle is transferred to the stationary particle, causing it to move by traction.

impact excavation phase   The second phase of impact crater formation following compression and preceding modification. It is the phase during which the main mass of material is ejected from the crater as the impact shock waves expand via a hemispherical shell away from the point of impact. Material is thus thrown upwards and outwards at progressively lower levels and velocities, and forms a conical curtain of debris around the growing crater that leans outwards at angles that are approximately 50° from the horizontal. Toward the end of excavation, when the upward velocities are too low to launch excavated rock, the crater rim is formed by uplift and overturning of the crater edge. Close to the rim, the debris curtain is deposited at fairly low velocities and angles as continuous ejecta, whereas further out higher angles and velocities pursue such that debris from the curtain may strike the surface to produce discontinuous ejecta and secondary craters.

impact polarization   The polarization of radiation from an emitted resonance line caused by excitation of the upper level of the relevant transition with a flux of charged particles. Used as a diagnostic of energetic particle beams in solar flares.

impact velocities A factor in crater formation that determines crater size and is controlled by the thickness of the atmosphere around the planet (assuming cosmic velocities are constant throughout the solar system for the same sized objects). Thus, the moon has a higher impact velocity than Mars. For example, a typical asteroid may hit Mars at 10 km/sec and the moon at 14 km/sec, causing the crater formed on Mars to be smaller by 0.66. By recognizing that smaller craters are produced on Mars, size-frequency curves developed there can be displaced to make them directly comparable with those on the moon, etc. This is also applicable to the other solar system planets.

impermeable Does not allow penetration by a fluid.

impulsive flare Solar flares with short duration electromagnetic radiation occurring low in the corona in compact volumina. In contrast to gradual flares, impulsive flares are not accompanied by coronal mass ejections. In more detail, the properties of an impulsive flare are: (a) the soft X-ray emission lasts less than one hour, (b) the decay constant of the soft X-ray emission is less than 10 minutes, (c) the duration of the hard X-ray emission is less than 10 minutes, (d) the duration of the microwave emission is less than 5 minutes, (e) impulsive flares are always accompanied by metric type III bursts, about 75% of them are also accompanied by metric type II bursts, but metric type IV emission is observed only rarely, (f) the height in the corona is less than 10^4 km, (g) the flare occupies a volume between 10^{26} cm^3 and 10^{27} cm^3, (h) the energy density is high, (i) the size in Hα is small, and (j) impulsive flares are rarely accompanied by coronal mass ejections. If coronal mass ejections are observed in *impulsive flares,* they tend to be small and slow with speeds well below the combined solar wind and Alfven speed which is required to drive a shock.

If an impulsive flare gives rise to a particle event in interplanetary space, this event has properties different from the ones of particle events caused by gradual flares. These properties include: (a) the event is electron-rich and the H/He ratio is about 10, (b) the ^{3}He/^{4}He ratio is of the order of 1, which is enhanced by a factor of 2000 compared to abundances in the corona and in the solar wind, (c) the Fe/O ratio is about 1.2, an enhancement by a factor of 8 compared to the corona or the solar wind, (d) the charge states of iron are about 20, indicative for the acceleration out of a very hot environment with temperatures of about 10 million K, (e) the particles event lasts (at the orbit of Earth) for some hours up to about 1 day, (f) particles can be observed only in a relatively narrow cone of $\pm 30°$ around the source region, and (g) there is no interplanetary shock accompanying the event. Event rates are up to about 1000 per year during solar maximum.

In impulsive flares, particles are accelerated in a hot but confined volume low in the corona. The enrichment in ^{3}He and in heavier elements such as Fe can be explained by the process of selective heating: Particles are accelerated inside a close loop, giving rise to electromagnetic emission (the flare). The loop is very stable, preventing particles from escaping into interplanetary space. As the particles bounce back and forth in the closed loop, they excite electromagnetic waves which can propagate into all directions, in particular also perpendicular to the magnetic field lines. When these waves interact with the ambient plasma, they can accelerate particles. If these "secondary" particles are accelerated on open field lines, they can escape into interplanetary space. Because the acceleration requires particles and waves to be in resonance, different particles are accelerated by different types of waves. If a particle species is common in the corona, such as H and ^{4}He, the corresponding waves are absorbed more or less immediately; thus, these particles are predominately accelerated inside the closed loop and therefore do not escape into interplanetary space. Other waves, however, can travel larger distances before being absorbed by the minor constituents and therefore are more likely to accelerate these species on open field lines. Since the escaping particle component is selectively enriched in these minor species, the acceleration process is called selective heating.

impulsive flare Flares displaying impulsive spikes or bursts in their hard X-ray time profiles. These flares are generally confined in long (10^4 km) sheared loops and are considered as non-thermal X-ray sources.

inclination   In geophysics, the angle at which the magnetic field dips into the interior of the Earth; another term for geomagnetic dip angle; *see also* declination.

In astronomy, the angle the plane of the Earth's equator makes with the ecliptic, i.e., the angle the pole of the Earth makes with the normal to the ecliptic; or in general the angle the pole of a planet makes with the normal to its orbit, or the angle a planetary orbit makes with the ecliptic; or the angle the normal to the plane of a binary star's orbit makes with the line to the Earth. Most solar system planetary orbits have low *inclinations* (Pluto's is the largest at 17°) while asteroids and comets display larger inclination ranges.

incompatible element   An element that enters the molten fraction when the degree of partial melting is low.

index, geomagnetic activity   Disturbances of the geomagnetic field are quantitatively characterized by numerous indices of magnitude. The International Association of Geomagnetism and Aeronomy (IAGA) has formally adopted 19 such indices. An oft-used index of general activity, the index K measures irregular variations of standard magnetograms recorded at a given observatory. After average regular daily variations are subtracted, the deviation of the most disturbed horizontal magnetic field component in each 3-hour interval is derived. Auroral activity near the poles is the dominant source of these deviations, so an adjustment is made for observatory location in geomagnetic latitude, and the deviation is converted to a roughly logarithmic scale (ranging from 0 to 9 for deviations from 0 to 1500 nT). The most widely used of all geomagnetic activity indices is probably Kp (the"p" denoting planetary), intended as a worldwide average level of activity. The K indices from 12 stations between 48° and 63° geomagnetic latitude, at a wide range of longitudes, are combined to compute Kp. Detailed explanations of Kp and other indices may be found in *Handbook of Geophysics and the Space Environment,* A.S. Jursa, Ed., 1985. *See* geomagnetic field.

index of refraction (n)   The ratio of the speed of light in a material to the speed ($c =$ 299792458ms^{-1}) in a vacuum. The *index of refraction* enters essentially in computing the angel of refraction via Snell's law. Typical indices of refraction are between 1 and 2. The index of refraction depends on the wavelength of the radiation being considered. *See* Snell's law.

indices, geomagnetic   Quantitative variables derived from observations, giving the state of the magnetosphere. They include the Dst index, which gauges the intensity of the ring current; the Kp index, which gives the variability of the field; and the AE, AU, and AL indices, which provide information on the auroral electrojet and through that, indirectly, on the flow of Birkeland currents. For details, see individual items.

induced Compton scattering   Compton scattering induced by an extremely intense radiation field, as found in compact radio sources and pulsars. *Induced Compton scattering* is a form of stimulated emission, in which a photon of a given frequency stimulates the emission of a second photon of identical frequency, phase, direction of motion, and polarization. Unlike stimulated emission due to the transition of an electron between two-bound states of an atom or ion, induced Compton effect transfers energy to electrons.

induced gravity   (Sakharov, 1967) An alternative concept to the theories of quantum gravity in which the gravitational interaction is not fundamental, but is defined by the quantum effects of the matter fields. The main mathematical problem arising in this approach is the calculability of the dimensional parameters: cosmological and gravitational constants. Only one of them may be defined from the quantum field theory of matter fields, which leaves the room to the fine-tuning for the value of a cosmological constant. A form of *induced gravity* has been achieved in string theory. In this case, neither gravity nor other observable interactions are fundamental. All the interactions including gravity appear to be induced ones due to the quantum effects of (super)string. In particular, the Einstein equations arise as a consistency condition for the string effective action. *See* curved space-time, quantum field theory in curved spacetime, quantum gravity.

inelastic collision Collision between two or more bodies in which there is loss of kinetic energy.

inelastic scattering of radiation Scattering in which the wavelength of radiation changes because radiant energy is transferred to the scatterer.

inertia In Newtonian physics, the tendency of a material object to remain at rest, or in a state of uniform motion.

inertia coefficient Also referred to as a mass coefficient; appears in the Morison Equation for description of wave-induced force on a vertical pile or cylinder. Denotes the force that arises due to the acceleration of the fluid around the cylinder. A second term denotes the force due to the square of the instantaneous velocity and includes a drag coefficient.

inertial-convective subrange For wave numbers k well below $L_B{}^{-1}$, the molecular diffusivity of heat or salt does not influence the spectrum very much, and so the spectra are similar to the velocity spectrum $E(k)$, which falls off proportional to $k^{-5/3}$ (inertial-convective subrange). For smaller scales, velocity fluctuations are reduced progressively by viscosity, but the diffusivity of heat or salt is not yet effective (viscous-convective subrange).

inertial coordinate system An unaccelerated coordinate system in which the laws of Newton and the laws of special relativity hold without correction. In the absence of gravity, this coordinate system can be extended to arbitrarily large distances. In the presence of gravity, such a coordinate system can be erected locally, but cannot be extended beyond lengths corresponding to the typical tidal scale of $r = c/\sqrt{G\rho}$, where c is the speed of light, G is Newton's gravitational constant, and ρ is a measure of the matter density.

inertial frequency When waves are long compared with the Rossby radius, the frequency is approximately constant and equal to the Coriolis parameter, f, or twice the Earth's rotation rate. In this limit, gravity has no effect, so fluid particles are moving under their own inertia. Thus, f is often called the *inertial frequency*. The corresponding motion is called the inertial motion or inertial oscillation; the paths are called inertial circles. Likewise the wave with this frequency is known as the inertial wave.

inertial instability The instability that occurs when a parcel of fluid is displaced radially in an axisymmetric vortex with negative (positive) absolute vorticity (planetary vorticity plus relative vorticity) in the northern (southern) hemisphere.

inertial mass The mass that opposes motion. In general, for small accelerations, $\mathbf{a} = \mathbf{F}/m$, where m is the inertial mass. The *inertial mass* is usually contrasted to the passive gravitational mass, which is a factor in the Newtonian force law, and to the active gravitational mass, which generates the gravitational field. In Newtonian gravitation, and in general relativity, all these masses are proportional, and are set equal by convention.

inertial oscillation A fluid particle with an initial velocity v but free of force in the Northern Hemisphere will be bent by the inertial Coriolis force of magnitude $2\Omega \sin\theta$ to its right, where Ω is the angular velocity of Earth rotation around the North Pole and θ is the local latitude. In the absence of background currents, the particle's trajectory is a circle with a radius of $v/2\Omega \sin\theta$. The particle returns to its original position in $1/(2\sin\theta)$ days (inertial period). *Inertial oscillations* are often observed in the ocean after strong wind events like hurricanes.

inertial subrange Part of the turbulent kinetic energy spectrum where turbulent kinetic energy is neither produced nor dissipated by molecular diffusion, but only transferred from larger to shorter length scales by initial forces (*see also* turbulence cascade). Wavenumbers in this part of the spectrum are much larger than the energy containing scales of turbulence and shorter than the Kolmogorov wavenumber $k_\eta = (\epsilon/\nu^3)^{1/4}$ at which kinetic energy is dissipated into heat. For sufficiently high Reynolds numbers, this part of the spectrum is nearly isotropic and is independent of molecular viscosity. The shape of the energy spectrum (see the figure) in the *inertial subrange* is given by Kolmogorov's "$k^{-5/3}$" law

$$\Phi(k) = A\epsilon^{2/3}k^{-5/3}$$

where k is the wavenumber and ϵ is the dissipation rate of turbulent kinetic energy. The parameter A is a universal constant that is valid for all turbulent flows, and $A \simeq 0.3$ for the streamwise components of $\Phi(k)$ and $A \simeq 1.5$ for the cross-stream components.

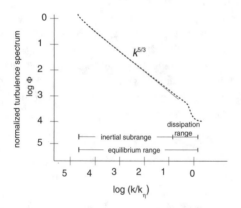

Typical wavenumber spectrum observed in the ocean plotted against the Kolmogorov wavenumber.

inferior conjunction   *See* conjunction.

inferior mirage   A spurious image of an object formed below its true position by atmospheric refraction when temperature decreases strongly with height. *See* superior mirage.

infinitesimal canonical transformation   In classical mechanics, a canonical transformation in which the change between old canonical variables $\{q^k, p_l\}$ with Hamiltonian $H(q^k, p_l, t)$ and new canonical variables $\{Q^k, P_m\}$ with Hamiltonian $K(Q^m, P_n, t)$ is infinitesimal, so that squares of differences can be neglected. A basic form of canonical transformation postulates a generating function $F(q^k, P_l, t)$ and

then solves

$$p_k = \frac{\partial F}{\partial q^k}$$

$$Q^l = \frac{\partial F}{\partial P_l}$$

$$K = \frac{\partial F}{\partial t} + H$$

Notice that $F = q^k P_k$ (summed on k) generates the identity transformation: $Q^k = q^k$, $P_l = p_l$ via the above equation.

The generator for an *infinitesimal canonical transformation* is then:

$$F = q^k P_k + \epsilon G\left(q^k, P_m, t\right)$$

where ϵ is a small parameter (so that ϵ^2 terms are ignored). From the previous equations,

$$p_k = P_k + \epsilon \frac{\partial G}{\partial q^k}$$

$$Q^l = q^l + \epsilon \frac{\partial G}{\partial P_l}.$$

The terms on the right involve derivatives of G, which are functions of q^k, P_l, and t. However, because of the small parameter ϵ, $G(q^k, P_l, t)$ differs only at first order in ϵ from $G(q^k, p_l, t)$ (the same function, evaluated at a slightly different argument). Hence we can write, to first order in ϵ,

$$P_k = p_k - \epsilon \frac{\partial}{\partial q^k}\left(G\left(q^l, p_n, t\right)\right).$$

This is an infinitesimal canonical transformation, and G is the generator of the infinitesimal canonical transformation. Compared to the general form, it has the great advantage of being explicit. A full nonlinear transformation can be carried out by integrating infinitesimal transformations. *See* canonical transformation, Hamilton–Jacobi Theory.

inflation   In cosmology, a period of rapid universal expansion driven by a matter source whose energy density falls off slowly in time or not at all. The simplest example appears in the behavior of the scale $a(t)$ for an isotropic, homogeneous universe (a cosmology modeled by a Robertson–Walker cosmology):

$$3\left(\frac{\dot{a}}{a}\right)^2 = \frac{8\pi G\rho}{c^2} - \frac{k}{a^2} + \Lambda$$

(G = Newton's constant; c = speed of light; dot indicates proper time derivative) where ρ is energy density associated with "ordinary" matter (thus ρ decreases as the universe expands), $-\frac{k}{a^2}$ reflects the global topology (k is a parameter, $k = -1, 0, 1$); Λ is the cosmological constant, which does not depend on the size scale, a, of the universe. As the universe expands, Λ dominates the right-hand side of this equation for $k = 0, -1$; if Λ is large enough it also dominates the solution for late times for $k = +1$. In these cases at late times, one has

$$ a \sim \exp\left(\sqrt{\frac{\Lambda}{3}}t\right). $$

This exponential growth is called *inflation*. As originally posited by Einstein, Λ is precisely constant, and observationally at present, neither Λ nor the term $\frac{k}{a^2}$ is greater than order of magnitude of $\frac{8\pi G\rho}{c^2}$. In modern theories of quantum fields, however, those fields can produce long-lived states with essentially constant (and very large) energy density, mimicking a Λ. In 1979 Guth noted that those states can drive long periods of inflation in the very early universe, then undergo a transition to more normal matter behavior. (Hence, $\Lambda \rightarrow \frac{8\pi G\rho}{c^2}$ in the transition called "reheating".) If the inflation exceeds $\sim$60 e-foldings, then the ratio of ρ to the $\frac{-k}{a^2}$ term after reheating becomes consistent with an evolution to the current observed state. This explains why the universe appears nearly "flat" ($k = 0$) now, solving the "flatness problem". Similarly, inflation blows the size of some very small causally connected region up to a size larger than the current observable universe. This has the effect of suppressing large initial inhomogeneities. As similar by-products of the accelerated expansion, all topological defects (and monopoles in particular) formed before the inflationary phase are diluted in such a way that their remnant density, instead of overfilling the universe, becomes negligible. At the same time, small-scale quantum fluctuations are spread out by this inflation. With correct choice of the inflation parameters, one can achieve reasonable consistency with the fluctuations necessary to create the observed large scale structure of the universe. This restriction of parameters could be viewed as restricting the physics at very early times in the universe. With typical such parameter theories, the period of inflation occurred during the first 10^{-35} sec after the Big Bang at temperatures corresponding to very high particle energies, $\sim 10^{14}$ GeV. *See also* monopole excess problem, Robertson–Walker cosmological models.

infragravity wave   A water gravity wave with a period in the range of 20 sec to 5 min.

infrared   Referring to that invisible part of the electromagnetic spectrum with radiation of wavelength slightly longer than red-colored visible light.

infrasound   "Sound" waves of frequency below 20 Hz (hence inaudible to humans).

inherent optical property (IOP)   In oceanography, any optical quantity that depends only on the properties of the water and is independent of the ambient light field; examples include the absorption coefficient, the scattering coefficient, and the beam attenuation coefficient; apparent optical properties become inherent optical properties if the radiance distribution is asymptotic.

inhomogeneous models (of the universe)
Cosmological models that do not obey the cosmological principle. Several such models have been derived from Einstein's equations, starting as early as 1933 with the solutions found by Lemaître (often called the "Tolman–Bondi" model) and McVittie. The research in this direction was partly motivated by the intellectual challenge to explore general relativity and to go beyond the very simple Robertson–Walker cosmological models, but also by the problems offered by observational astronomy. The latter include the creation and evolution of the large-scale matter-distribution (in particular the voids), the recently discovered anisotropies in the microwave background radiation, also gravitational lenses that cannot exist in the Robertson–Walker models. The more interesting effects predicted by inhomogeneous models are:

1. The Big Bang is, in general, not a single event in spacetime, but a process extended in time. Different regions of the universe may be of different age.

2. The curvature index k of the Robertson–Walker models is, in general, not a constant, but a function of position. Hence, some parts of the universe may go on expanding forever, while some regions may collapse in a finite time to the final singularity.

3. The spatially homogeneous and isotropic models (*see* homogeneity, isotropy) are unstable with respect to the formation of condensations and voids. Hence, the formation of clusters of galaxies and voids is a natural phenomenon rather than a problem.

4. An arbitrarily small electric charge will prevent the collapse to the final singularity. This also applies to the initial singularity — the model with charge, extended backward in time, has no Big Bang. The charge may be spread over all matter (then it has to be sufficiently small compared to the matter density) or concentrated in a small volume; the result holds in both cases. Reference: A. Krasiński, *Inhomogeneous Cosmological Models.* Cambridge University Press, 1997.

initial condition   A boundary condition applied to hyperbolic systems at an initial instant of time; or for ordinary differential equations (one independent variable) applied at one end of the domain. For more complicated situations, for instance, in theories like general relativity one may specify the state of the gravitational field and its derivative (which must also satisfy some consistency conditions) at one instant of time. These are the *initial conditions.* One obtains the later behavior by integrating forward in time.

initial data   In general relativity, *see* constraint equations.

initial mass function (IMF)   The distribution of newly formed stars as a function of mass. The *initial mass function* is estimated from the photometric and spectroscopic properties of stars in open clusters and associations of stars. Ideally, the IMF can be measured counting the stars of each spectral type in an association of stars so young that the shortest-lived massive stars are still in the main sequence. The initial mass function is usually assumed to be of the form

$\Psi(m) \propto m^{-\Gamma}$, where m is the mass of the star. The index Γ may vary for different mass ranges, but it is always positive, implying that high mass stars are formed less frequently than low mass stars. According to E.E. Salpeter, $\Psi(m) \propto m^{-2.35}$, for all masses. From this law, we expect that for one star of 20 solar masses ($M_\odot$) ≈ 1000 stars of $1 M_\odot$ are formed. According to G.E. Miller and J.M. Scalo, the IMF valid for the solar neighborhood can be approximated as $\Psi(m) \propto m^{-1.4}$, for $0.1 \lesssim m \lesssim 1 M_\odot$, $\Psi(m) \propto m^{-2.5}$, for $1 \lesssim m \lesssim 10 M_\odot$, and $\Psi(m) \propto m^{-3.5}$, for $m \gtrsim 10 M_\odot$. This law predicts fewer high mass stars ($m \gtrsim 10 M_\odot$) for a given number of solar mass stars than Salpeter's law.

injection boundary   A line in the nightside equatorial magnetosphere along which dispersionless plasma injections appear to be generated.

inner core   The Earth's *inner core* has a radius of 1215 km, and is solid. The inner core is primarily iron and its radius is growing as the Earth cools. As the outer core solidifies to join the inner core, elements dissolved in the outer core are exsolved. The heat of fusion from the solidification of the inner core provides a temperature gradient, and convection, and the ascent of these light exsolved components are major sources of energy to drive the geodynamo.

inner radiation belt   A region of trapped protons in the Earth's magnetosphere, typically crossing the equator at a geocentric distance of 1.2 to 1.8 R_E (Earth radii), with energies around 5 to 50 MeV. The inner belt is dense enough to cause radiation damage to satellites that pass through it, gradually degrading their solar cells. Manned space flights stay below the belt.

The 1958 discovery of the inner belt was the first major achievement of Earth satellites. The belt seems to originate in the neutron albedo, in secondary neutrons from the collisions between cosmic ray ions and nuclei in the high atmosphere. The trapping of its particles seems very stable, enabling them to accumulate over many years. The planet Saturn also seems to have an inner belt, with the neutrons coming from cosmic ray collisions with the planet's rings.

inshore   The portion of a beach profile lying between the foreshore and offshore.

insolation   Shortwave solar radiation (UV, visible, near infrared) per unit area that is received in the Earth's atmosphere or at its surface, taking into account the angle of incidence to the horizontal.

Institute of Space and Astronautical Studies (ISAS)   Japanese space agency responsible for a number of successful scientific spacecraft, including the solar missions Hinotori and Yohkoh.

integrated Sachs–Wolfe effect   In cosmology, in the linear regime where fluctuations can be considered small perturbations on an expanding homogeneous isotopic cosmology, the contribution to temperature anisotropies from varying gravitational potentials along the line of sight. For a flat universe with no cosmological constant, the gravitational potentials do not evolve with time and this effect is zero. In open and flat models with cosmological constant (*see* Friedmann models) it can dominate over the gravitational potential variations across the last scattering surface.

intensity   The radiant power in a given direction per unit solid angle per unit wavelength interval [W sr^{-1} nm^{-1}].

interaction of galaxies   The gravitational attraction between two or more galaxies, which can induce notable modifications in their morphology, as well as in their photometric and spectroscopic properties. Interacting galaxies are often classified as peculiar, since their morphology does not fit the criteria of any of the main classification schemes for galaxies. The effect of interaction among galaxies depends strongly on their mutual distance from a companion galaxy, depending on the inverse cube of the distance. Extensive observation, as well as computational simulation, suggests that for disk galaxies, the effects of interaction on morphology encompass the formation or enhancement of a spiral pattern, the formation of a bar, and, in more extreme cases, the formation of tidal tails, or of a prominent outer ring, as in ring

galaxies or, ultimately, the production of a remnant (after a merger) which resembles an elliptical galaxy. Collisions involving only elliptical galaxies may lead to the production of ripples, extended halos, and asymmetries in the photometric profiles, but they do not produce such spectacular features as tidal tails. *Interaction of galaxies* has been linked to an enhancement of star formation in the host galaxies and, more speculatively, to the occurrence of quasar-type nuclear activity. *See* elliptical galaxies, galaxy, spiral galaxy, starburst galaxy.

intercloud medium   The warm (≈ 3000 K) low density ($\approx 10^{-2}$/cm^3) gas in rough pressure equilibrium with interstellar clouds.

intercommutation   In the physics of cosmic strings, after a cosmic phase transition with the generation of cosmic strings, the resulting network is free to evolve and multiple interactions between single strings will take place. For the simplest (Abelian–Higgs) model, numerical simulations show that two strings will, in general, interchange extremes when crossing. This exchange of partners is *intercommutation* (also sometimes referred to as reconnection). In strings that might be generated from more complicated fields (non-Abelian $\pi_1(\mathcal{M})$) string entanglement may take place, due to the fact that exchange of partners is topologically forbidden. *See* cosmic string, cosmic topological defect, cusp (cosmic string), kink (cosmic string).

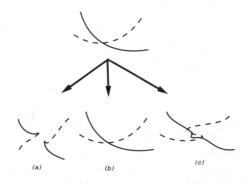

String interactions: (a) intercommutation, with the possible appearance of kinks and cusps, (b) no intercommutation, i.e., the strings cross each other and are left essentially unaffected, and (c) entanglement.

interconnection field   A magnetic field component, derived from a scalar potential (*see* harmonic model), added to an empirical model of the closed magnetosphere to turn it into a model of the open magnetosphere. It may be added either to obtain a better fit to data, or to create a testbed model of the open magnetosphere for testing various theoretical notions. Often a constant, southward-directed field is used to obtain this effect.

interface-controlled creep   In rheology, a deformation mechanism for diffusion creep. In an *interface-controlled creep,* macroscopic strain is caused by diffusion mass transport involving creation and/or deposition of vacancies at grain-boundary and their diffusion, through some rearrangement of atomic structures at grain-boundaries. Interface-controlled creep assumes that the interface reaction is fast compared to diffusion, so the overall rate of deformation is controlled by grain boundary reaction. One example is the case where the rate of generation or absorption of vacancies is controlled by the mobility of grain-boundary dislocations. The constitutive equation for interface-controlled creep can be expressed as

$$\varepsilon = C \frac{\sigma^2 \Omega D}{\kappa T \mu b d}$$

where D is the diffusion coefficient, b the thickness of grain boundary, σ the differential stress, d the grain size, Ω the atomic volume, κ Boltzmann's constant, C a numerical coefficient depending on the grain shape and the boundary conditions for σ, T the temperature, μ the shear modulus, and ε the strain.

interference fading   Results from interference between signals arriving at a receiver by two or more different paths. The multiple paths may arise from sky wave and ground wave paths, two polarization modes, and various combinations of multi-mode sky-wave paths, scattering from ionization irregularities or combinations of any of these options. *Interference fading* is a collective term for the observed fading statistics whereas multipath fading specifically describes the path structure of the received radio signal. In a more restricted sense, interference fading

(sometimes called diffraction fading) may describe multiple paths formed within the region of the ionosphere where the radio wave for a single mode is reflected and the resulting fading is attributed to local ionization irregularities. Interference fading occurs with an associated period of 1 sec or larger. *See* fading.

interglacial   For the last 4 million years covering the Holocene (recent) and the Pleistocene epochs, the Earth has been primarily in ice ages. An *interglacial* is a period of relative warming. The last interglacial occurred about 120,000 years ago; the current interglacial began about 20,000 years ago. The record indicates that interglacials have a length of about 20,000 years.

interior Schwarzschild space-time   In general relativity, the spacetime described by the metric

$$ds^2 = -\left(3\sqrt{1 - \frac{r_1^2}{R^2}} - \sqrt{1 - \frac{r^2}{R^2}} \right)^2 dt^2$$
$$+ \frac{dr^2}{1 - \frac{r^2}{R^2}}$$
$$+ r^2 \left(d\theta^2 + \sin^2 \theta d\phi^2 \right)$$

describing the gravitational field interior to a static, incompressible perfect fluid sphere of radius r_1, density $\mu = \left(\frac{3}{8\pi R^2} \right)$, and pressure

$$p = \mu \frac{\sqrt{1 - \frac{r^2}{R^2}} - \sqrt{1 - \frac{r_1^2}{R^2}}}{3\sqrt{1 - \frac{r_1^2}{R^2}} - \sqrt{1 - \frac{r^2}{R^2}}} .$$

The total mass of the sphere is $m = \frac{4\pi}{3} \mu r_1^3$. The pressure is regular at the center provided $r_1^2 < (8/9)R^2$. This condition sets an upper limit on the mass of the sphere.

intermediate shock   Results from the steepening of an intermediate MHD wave that is a purely traverse wave with velocity perturbations perpendicular to both the wave vector $\mathbf{k_o}$ and the average magnetic field $\mathbf{B_o}$. *Intermediate shocks* can exist only in anisotropic media. In an isotropic plasma, such as the solar wind, a steepened intermediate wave does not form a shock but a rotational discontinuity. The propagation speed parallel to the magnetic field equals the Alfvén speed; thus, the intermediate shock is sometimes also called Alfvén shock.

intermittency   Even in non-stratified, homogeneous turbulence, variations of the dissipation of turbulent kinetic energy (ε) and scalar variance (χ) are large and follow lognormal probability distributions. In natural waters, *intermittency* is larger because, in addition to the inherent intermittency, the production and transport of turbulent kinetic energy is usually variable in time and patchy in space. Due to the variations of the physical processes and intensities leading to turbulence, turbulence parameters, such as ε and χ, are usually not lognormal distributed. The standard deviation of $\ln(\varepsilon_i)$ is sometimes called intermittency factor.

internal energy   The total energy in a thermodynamic system not associated with bulk motion; an extensive thermodynamic potential U given by
$$U = ST + PV \, ,$$
where S is the entropy, T is the temperature, P is the pressure, and V is the volume of the system. The change of the *internal energy* is the maximum work that can be extracted from a reversible closed system at constant V. For a reversible process at constant S and V, work stored as internal energy can be recovered completely.

internal friction angle   A term used in the study of soil mechanics; denotes the maximum sustainable angle for a pile of the soil. Also referred to as angle of repose.

internal wave   In oceanography, a wave that occurs within seawater whose density changes with depth either gradually or abruptly at an interface discontinuity.

International Atomic Time (TAI)   A weighted average of atomic clocks in many countries, each having its rate adjusted by a very small amount to bring its rate, as transmitted to sea level (the geoid) by telemetry, to that which it would exhibit if situated on the geoid. Because the combination of the results from many clocks, weighted according to their estimated accuracy, requires hours or days to calculate, TAI is not available in real time, but signals from several sources, such as NIST and the USNO, are very close to TAI and corrections are available after date. First, "Free-running Atomic Time" (EAL) is produced and then TAI is computed from it after frequency adjustments are applied to correct for instrumental errors in the length of the second, as determined from measurements of primary frequency standards (atomic clocks) at timing laboratories. Like Coordinated Universal Time (UTC), TAI is referenced to the Greenwich meridian time zone. Because it increases continuously, when UTC is held back by the introduction of leap seconds, TAI may be thought of as being equal to UTC + (cumulated leap seconds). Thus, TAI noon precedes Greenwich (UTC) solar noon by that number of seconds. *See* geoid.

international dateline   A broken line running from the north to south pole on the Earth roughly along the meridian at 180° East longitude = 180° West longitude, 180° away from the prime meridian through Greenwich. The line along which a discontinuity in the current date occurs. Legal dates are arranged across this line so that the Eastern side of the line has a date 24 hours earlier than the Western side. Thus, on traveling eastward across this line, the legal date becomes one day earlier, e.g., Sunday August 10, 1997 became Saturday August 9, 1997. For political reasons the legal date line does not follow exactly the 180° longitude.

International Geomagnetic Reference Field (IGRF)   *See* harmonic model.

International System of Units (SI)   The unit system adopted by the General Conference on Weights and Measures in 1960. It consists of several base units (meter, kilogram, second, ampere, kelvin, mole, candela), plus derived units and prefixes.

International Temperature Scale (ITS-90) The official international temperature scale adopted in 1990. It consists of a set of fixed points and equations that enable the thermodynamic temperature to be determined from operational measurements.

interplanetary dust particles (IDPs)   The small debris that exists between the planets. Most of this material comes from comets and from collisions between asteroids. Study of *interplanetary dust particles* provides information about the primitive material out of which the solar system formed, material which has not been subsequently processed by heat, pressure, and chemical reactions as has happened on the larger bodies of the solar system. Sunlight reflecting off the IDPs gives rise to the zodiacal light and gegenschein.

interplanetary magnetic field   *See* heliospheric magnetic field.

interplanetary magnetic sector   A contiguous region observed in interplanetary space, usually near Earth, where for locally observed magnetic field lines, the ends anchored on the sun have all the same polarity. If those ends are directed towards the sun, the region is a "towards sector"; if away from it, it is an "away sector". Sector structure is caused by the waviness of the heliospheric current sheet (HCS) and the regions above the sun's poles and away from the HCS do not experience sector variations as the sun rotates.

Since the Earth's orbit is inclined to the solar magnetic equator, near sunspot minimum, when the waviness of the HCS is small, the Earth generally samples two different sectors. At times of greater solar activity, when the solar magnetic field is more complex, the HCS becomes rather wavy, and the Earth can cross 4, 6, or even 8 sectors per solar rotation.

interplanetary propagation   The propagation of energetic charged particles through the interplanetary medium is influenced by the following processes:

(a) pitch-angle scattering of particles at the solar wind turbulence is a stochastic process, described by the pitch-angle diffusion coefficient $\kappa(\mu)$ which depends on pitch-angle μ, particle rigidity, level of interplanetary turbulence, and location. In addition, pitch-angle diffusion coefficients might be different for electrons and nuclei of the same rigidity. Pitch angle scattering is a stochastic process. *See* pitch angle diffusion, quasi-linear theory, slab model.

(b) focusing in the diverging interplanetary magnetic field is a systematic process decreasing the particle's pitch angle as it propagates outward. *See* focusing.

(c) field-parallel propagation of the particle. *See* focused transport equation.

(d) convection of particles with the solar wind leads, in particular for particles with speeds in the order of the solar wind speed, to a more efficient propagation outwards rather than inwards. *See* convection.

(e) adiabatic deceleration leads to a cooling of the cosmic ray gas as it expands with the expanding solar wind that is a transport in momentum. *See* adiabatic deceleration. The transport equation then reads

$$\frac{\partial F(\mu, s, p, t)}{\partial t}$$

streaming:

$$+ \frac{\partial}{\partial z} \mu v F$$

convection:

$$+ \frac{\partial}{\partial z} \left(1 - \mu^2 \frac{v^2}{c^2} \right) v_{\text{sowi}} \sec \psi \, F$$

focusing:

$$+ \frac{\partial}{\partial \mu} \frac{v}{2\zeta} \left(\frac{E'}{E} + \mu \frac{v_{\text{sowi}}}{v} \sec \psi \right) \cdot (1 - \mu^2) F$$

differential convection:

$$- \frac{\partial}{\partial \mu} v_{\text{sowi}} \left(\cos \psi \frac{\mathrm{d}}{\mathrm{d}} r \sec \psi \right) \cdot \mu (1 - \mu^2) F$$

scattering:

$$- \frac{\partial}{\partial \mu} \frac{\kappa}{2} \frac{\partial}{\partial \mu} \frac{E'}{E} F$$

deceleration:

$$- \frac{\partial}{\partial p} p v_{\text{sowi}} \cdot$$

$$\cdot \left(\frac{\sec \psi}{2\zeta} (1 - \mu^2) + \cos \psi \frac{\mathrm{d}}{\mathrm{d}r} \sec \psi \mu^2 \right) F$$

source:

$$= Q(\mu, s, t, s, v)$$

with F being the density of particles in a given magnetic flux tube, t the time, v the particle speed, s the distance along the magnetic field spiral, p the momentum in the solar wind frame, v_{sowi} the solar wind speed, ψ the garden hose angle between the magnetic field line and the radial direction, ζ the focusing length, and $E'/E = 1 - \mu v_{\text{sowi}} v \sec\psi/c^2$ the ratio of the total energy in the solar wind frame to that in a fixed frame.

This transport equation can be solved only numerically. Different approximations exist: the focused transport equation can be applied to MeV particles in the inner heliosphere (*see* focused transport equation); the diffusion convection equation, on the other hand, is applied to particles with energies in the keV range, to observations beyond 1 AU, and to the modulation of galactic cosmic rays (*see* diffusion-convection equation).

interplanetary scintillation (IPS) Solar wind induced scintillation of cosmic radio signals.

interplanetary scintillation (IPS) observations Remote-sensing observations of the inner solar wind, based on analysis of how a radio beam from a distant source (natural or from a spacecraft) is disturbed as it passes near the sun.

interplanetary sector structure *See* heliospheric magnetic field.

interplanetary shock wave *See* hydromagnetic shock wave.

interplanetary stream structure *See* heliospheric stream structure.

interplate earthquake An earthquake that occurs on the boundary fault dividing two lithospheric plates, as opposed to intraplate earthquakes that occur within a plate.

interstellar clouds Dark material along the spiral arms of our galaxy, consisting of clouds of neutral hydrogen, or of interstellar molecules.

interstellar dust Small dust grains of size 5×10^{-9} m to 2×10^{-7} m, apparently consisting of refractory materials (silicates and graphite) ($\approx 10^8$ atoms/grain) with a power law distribution $n(r) \propto r^{-3.5}$ where r is the radius of the particle. Additional components consisting of polyaromatic hydrocarbons, and of water ice, are also present. *Interstellar dust* is responsible for extinction (especially visible in dark nebulae), and for reddening (a general phenomenon) of starlight.

Dust production presumably occurs in cool red giants, which are old stars with enhanced metals and substantial stellar winds, and also in supernova explosions, which can transport large amounts of metals into the interstellar environment.

interstellar gas The gas lying between stars in the plane of the galaxy, composed principally ($\approx 90\%$) of atomic hydrogen, H, approximately half of which is neutral (HI), observable by its 21 cm radiation, and half is ionized (HII, ionized by hot O and B stars). The number density varies from 10^{-2} to 10^6 cm^3, averaging about 1 cm^3. There is a small admixture of "metals".

interstellar medium The matter composed of dust and gas in diffuse form that fills the space between the stars of a galaxy. Interstellar matter varies widely in temperature, density, and chemical composition. The *interstellar medium* reaches to the outer layers of the atmosphere of stars where the density and temperature of the gas increase steeply, but far from the stars its density drops to less than 1 particle cm^{-3}. However, in a galaxy there are many places where the density of interstellar matter increases significantly. These concentrations or regions of the interstellar medium, sometimes called clouds, or shells because of their appearance, have different physical and chemical characteristics that depend on the amount of light and energy that they receive.

interstellar molecules Inorganic (silicates); and especially organic molecules identified via infrared spectroscopy in cool dense dark interstellar clouds (of typically solar system size) in the nucleus and along the spiral arms of the galaxy; about 10^{-4} the interstellar mass. At present (AD 2000) over 100 organic molecules have been fairly securely identified, involving H,

C, N, O, S, ranging from the simplest CO (carbon dioxide), NO, OH, to H_2O (water), C_2S, and NH_3, to molecules with up to eight carbons, and unusual (or surprising) small molecules like CN, CP, CSi, HCl, KCl, HF, MgCN, and SiH_4.

intertropical convergence zone (ITCZ)   A fairly narrow band where surface southeasterly and northeasterly trade winds meet and converge. The convergence forces moist surface air to rise and as a result, water vapor condenses, forming high clouds and releasing condensational heat. Thus, the *ITCZ* also appears as a longitudinally oriented, latitudinally narrow band of high clouds, making itself one of the most visible features of the Earth from space. Over continents, the ITCZ moves back and forth seasonally across the equator following the sun, while the oceanic ITCZ's position is determined by the underlying sea surface temperature. In particular, the ITCZ over the eastern Pacific and Atlantic Oceans stays north of the equator for most times of a year (5° to 10° N) because sea surface temperatures are generally higher north of the equator than south. The development of equatorial asymmetry in Earth's climate despite the nearly symmetric distribution of annual-mean solar radiation is a result of ocean-atmosphere interactions and land-sea distributions.

intraplate earthquake   Any earthquake that cannot be associated with plate margin processes. Three very large *intraplate earthquakes,* with magnitudes of about eight, occurred near New Madrid, Missouri in the winter of 1811–1812.

intrinsic permeability (k)   The ability of a porous medium to transmit fluid, independent of the fluid properties themselves, increases with increasing size and shape of the pore space, $k = Nd^2$, where N is a dimensionless constant that accounts for the shape of the pore space and d is the mean grain diameter. In practice, the *intrinsic permeability* is measured with a device called a permeameter, ranging in natural systems from 10^{-21} to $10^{-9} m^2$.

intrusion   In geophysics, a magma body that has solidified at depth.

invariant   *See* adiabatic invariant, longitudinal invariant.

inverse Compton effect   Compton scattering of a photon by a particle, typically an electron, whose kinetic energy is comparable to, or larger than, the energy of the photon. In contrast to the usual Compton scattering, where the photon loses energy, in *inverse Compton scattering* part of the kinetic energy of the particle is transferred to the photon. The emerging photon has its energy increased by up to a factor γ^2 where $\gamma^2 = 1/(1 - (v/c)^2)$, where c is the speed of light. If electrons are relativistic $v/c \sim 1$, then $\gamma \gg 1$. In this case inverse Compton scattering can dramatically increase the energy of initially low-energy photons. Inverse Compton scattering has been proposed as the mechanism producing hard X-rays and gamma rays in active galactic nuclei. *See* Compton scattering.

inverse Compton radiation   The radiation produced when a high energy (relativistic) electron interacts with a much slower moving photon thereby transferring energy from the electron to the photon. Potentially important in large solar flares where the ambient population of low energy thermal photons interact with the high energy non-thermal electrons accelerated in the flare.

inverse problems   A general classification for problems in which (ir)radiance measurements are used to infer inherent optical properties of the medium and sources.

inverse theory   A model of the world can be tested against measurements from the real world by evaluating the difference between the measurements and predictions from the model for the measurements. Predicting measurements from models is termed "forward modeling", while the reverse — using data to create models, by using the data to infer values for parameters in the model — is termed "inverse modeling". A very simple case is least squares fitting of a straight line to data. A slightly more complex case would be the least squares fit of a model with many parameters to data, where the function that predicts the data **d** from the model parameters **p** is linear, and can be written as a matrix **G** so that:

$$\mathbf{d} = \mathbf{Gp}$$

in which case a least-squares solution $\mathbf{p}'$ can be obtained via:

$$\mathbf{p}' = \mathbf{G}^T G^{-1} \mathbf{G}^T \mathbf{d} \,.$$

However, it may be that other solutions fit the data equally well, in which case the inversion is said to be nonunique. Complications may be introduced by a nonlinear relationship between **p** and **d**, by constraints on the parameters, or by biases concerning the model (e.g., a preference for simple models). *See* nonuniqueness.

inverted barometer effect (inverse barometer response)   The inverse response of sea level to changes in atmospheric pressure. A static reduction of 1.005 mb in atmospheric pressure will cause a stationary rise of 1 cm in sea level. The dynamic response, especially at periods shorter than a few days is not constant.

Io   Moon of Jupiter, also designated JI. Discovered by Galileo in 1610, it is one of the four Galilean satellites. Its orbit has an eccentricity of 0.004, an inclination of $0.04°$, a precession of $48.6°$ yr^{-1}, and a semimajor axis of 4.22×10^5 km. Its radius is 1815 km, its mass 8.89×10^{22} kg, and its density 3.55 g cm^{-3}. It has a geometric albedo of 0.61, and orbits Jupiter once every 1.769 Earth days. Io is tidally heated by Jupiter, and has active volcanoes as a result.

ion acoustic waves   *Ion acoustic waves* are among the simplest wave motions supported by a plasma. They behave in a fashion similar to normal sound waves in neutral gas, with one principal difference: the ion acoustic wave speed is determined by the ion's inertia and the combined pressure of both ions and electrons in the plasma. In this wave, the ions and electrons oscillate synchronously, coupled by the electric field generated.

ionization   The process by which ions are produced, typically occurring by collisions with atoms or electrons ("collisional ionization"), or by interaction with electromagnetic radiation ("photoionization").

ionization equilibrium   A plasma at a given temperature is in *ionization equilibrium* if the ratio of the number of ions in a given ionization state for a given element, n_{ion}, to the total number of ions of that element, n_{el}, is governed by the equations of thermodynamic equilibrium. Departures from ionization equilibrium imply changes in the temperature of the peak abundance of the various ions, so that the corresponding lines would be formed at temperatures significantly different from the formation temperature deduced from equilibrium.

ionizing radiation   Electromagnetic radiation with energy exceeding the typical binding energy of electrons in molecules; electromagnetic radiation with wavelength shorter than that of ultraviolet light; also applied to particulate radiation of similarly high energy.

ionogram   The conventional display obtained from an ionosonde which contains information about the ionosphere. An *ionogram* is constructed by displaying the returned signal as a function of frequency (on the horizontal axis) and time delay (on the vertical axis). The vertical axis, called the virtual height, is the height that the returned signal would have reached had it traveled in free space and been reflected from a perfect reflector above the ionosonde. The E and F layers of the ionosphere can be recognized easily on most middle and low latitude ionograms, recognition being more complex at high latitudes due to the presence of particle effects. The layers of the ionosphere can be parameterized in terms of their peak electron density (measured in megahertz, the units used on an ionogram for measuring electron density) and base height (measured in virtual height units, usually kilometers), both of which can be readily measured from ionograms. Conventionally, these are only recorded for the ordinary ray polarization, information from the extraordinary ray, when present on the ionogram, being used to improve these estimates. The most familiar internationally agreed parameters are hE, foE (the E region), hF, foF1 (the daytime F1 region), and hF2, foF2 (the daytime F2 region). At night these become just hF and foF2 (the nighttime F region). When the F1 region is not visible, the daytime F region parameters are hF and foF2.

Each sporadic E layer can be described by three parameters: hEs, foEs, and fbEs. Finally, two further important parameters are f_{min}, the minimum frequency returned from the ionosphere and recorded on the ionogram and M(3000)F2. These parameters are tabulated hourly, together with other information, and exchanged between World Data Centers. There is now a serious attempt to collect these data in real time and construct global maps of the ionosphere. To obtain the electron density profile in real height units, sometimes called true height, it is necessary to invert the ionogram, and correct for the time delay caused by the ionosphere. *See* critical frequency, ionosonde, ionospheric sounding, plasma frequency.

ionosonde A swept frequency pulsed radio transmitter on the ground cosited with a receiver forming a vertically directed radar system. A single frequency pulse is transmitted upwards and may be reflected from the ionosphere, the reflected signal being observed with the receiver. This process is repeated at a sequence of frequencies usually starting at a frequency of 1.0 MHz and finishing at around 20 MHz. A picture of the ionization above the *ionosonde* is then built up into an ionogram. This is constructed by displaying the returned signal as a function of frequency (on the horizontal axis) and time delay (on the vertical axis). *See* ionogram.

ionosphere The layer of the Earth's atmosphere extending from the top of the mesosphere (typically 80 to 90 km above the surface) to about 500 km. The thermosphere, or *ionosphere,* is the uppermost layer of a planetary atmosphere. It is also often called the upper atmosphere. It is characterized by low density ($\leqslant 3 \times 10^{-9}$gm/cm^3; number density $\leqslant 7 \times 10^{-11}$/cm^3) and pressure ($\leqslant 0.1$ Pa), and substantial ionization of the atoms, resulting in significant free electron density, which is a function of solar activity. The temperature increases rapidly with increasing altitude up to about 200 km, followed by a leveling off in the 300 to 500 km region. This heating is the result of photoionization and photodissociation of molecules into atoms and ions by direct absorption of solar photons.

ionospheric absorption Radio waves passing through the ionosphere experience interactions with the free electrons and may lose energy by these interactions if the electrons experience collisions with the neutral atmosphere. This process is called absorption, or sometimes non-deviative absorption, because the radiowave path through the absorbing region is not affected by the absorption. Collisions between electrons and the neutral atmosphere increase with decreasing altitude, so the D region is particularly important for absorption processes. Absorption varies as the inverse square of the frequency, so the losses are larger at low HF frequencies. In normal circumstances, during the daytime, *ionospheric absorption* prevents radio frequencies greater than about 2 or 3 MHz propagating via the ionosphere unless very large powers are transmitted. During nighttime, when the D-region ionization drops to very low levels, these same transmissions can propagate by the ionosphere to large distances. During a solar flare, X-ray radiation from the sun can increase significantly, resulting in greater ionization of the D region and subsequently larger absorption of radio waves. These increases may be sufficiently large to prevent all HF signals propagating by the ionosphere. *See* D region, short wave fadeout, solar flare.

ionospheric index An index usually based on observations of the F region critical frequency (foF2), which gives a measure of the solar cycle effects on the ionosphere. There are a variety of *ionospheric indices* of which the best known are IF2 and T. Both these indices are calculated in similar ways. First, past data are used to estimate calibration curves then, using current data with the calibration curves, an index is calculated that best represents the solar cycle effects for this epoch. Ionospheric indices are effective because the F region is strongly dependent on changes in solar activity. The correlation between the monthly median hourly foF2 and the smoothed sunspot number, for a location, can be higher than 0.95. Ionospheric indices are often used with HF propagation models for predicting propagation conditions. *See* ionospheric radio propagation path.

ionospheric polarization   In an ionized medium, such as the ionosphere, in the presence of a magnetic field, a plane-polarized radio wave is split into two characteristic waves. In the ionosphere these are circularly polarized waves, each propagating independently. The wave that most closely approximates wave propagation in the absence of the magnetic field is called the ordinary (o) wave and the other is the extraordinary (x) wave. For the ordinary (extraordinary) wave propagating in the direction of the magnetic field the electric vector rotates in the opposite (same) direction to that of an electron gyrating in the field and for propagation perpendicular to the magnetic field, the electric vector oscillates parallel (perpendicular) to the magnetic field. The partitioning of the energy between the two components depends on the angle the radio wave makes with the Earth's magnetic field. At low frequencies, the extraordinary wave is heavily attenuated relative to the ordinary wave.

ionospheric propagation mode   A single ray path between the transmitter and receiver. An ionospheric propagation path may comprise several different individual paths, all of which can propagate the signal from the transmitter to the receiver. The mode is labeled in terms of the number of reflections it experiences for each of the layers from which it is reflected. The simplest ionospheric mode is a single hop path having one reflection in the ionosphere (e.g., 1E, 1F). More complex modes can have several reflections from a layer (e.g., 2F) and may also have reflections from more than one layer (e.g., 2Es1F). *See* ionospheric radio propagation path.

ionospheric radio propagation path   The radio propagation path may describe any path traversed between two or more terminals by a radio wave carrying a signal from one terminal to the other(s). The terminals of a propagation path are the points from which a signal is transmitted or received. The terms "radio link", "radio circuit", and "circuit" are used interchangeably to describe the path along which information from a transmitter is passed to a receiver. Thus, a broadcast to an area may comprise many individual circuits. When the *radio propagation path* includes the ionosphere, then the properties of the ionosphere can often dom-

inate the design and management of the link. A simple ionospheric link may have a single reflection from the ionosphere and the ionospheric properties at the mid-point of the link are important. Likewise, for an Earth-satellite path it is the location of the sub-ionospheric point, or pierce point, where the Earth-to-satellite path passes through the ionosphere, that is important. Radio waves propagated by the ionosphere are called HF sky waves, or just sky waves. Because of the nature of the antennas used, and the ionosphere, several paths may be possible simultaneously. Each path is called a propagation mode. A mode may contain one or more ionospheric reflections. Thus, a one-hop mode will have one ionospheric reflection. One level of link management uses prediction programs to decide on the state of the ionosphere to determine optimum operational conditions for an HF system. Many of these programs treat the propagation mode in terms of control points and the ionospheric properties at these points are used to determine the basic MUF of the layer. The control point is the point in the ionosphere where the radio path is reflected, or refracted, back to the Earth's surface. A one-hop mode will have one control point, at the center of the great circle path joining the transmitter and receiver sites. Each mode, and the whole path, may be described in terms of the maximum and minimum frequencies that can be used. Several terms are used to describe this. Among them are the LUF, MUF, OWF, and MOF. *See* atmospheric noise, basic MUF, diversity reception, frequency of optimum traffic, galactic noise, ionospheric propagation mode, Lowest Useable Frequency, Maximum Observable Frequency, operational MUF, Optimum Working Frequency, radio frequency interference, radio frequency spectrum, skip fading.

ionospheric refraction   When an obliquely propagating radio wave enters the ionosphere it encounters a medium where the refractive index decreases as the electron density increases. This causes the wave to bend away from the vertical, refracting it back towards the direction from which it has come. This process continues as the wave penetrates further into the ionosphere until the refracted angle is 90° and the wave has penetrated to its deepest point. It is subsequently refracted towards the vertical as it leaves the ionosphere. The amount of refraction depends on the electron density of the ionosphere and the operating frequency. Although oblique radio waves are actually refracted in the ionosphere, it is often referred to as being reflected from the ionosphere. The latter is common in connection with simple mirror models of the ionosphere used for estimating HF system parameters (e.g., ASAPS, IONCAP) or when referring to ionospheric propagation modes. *See* ionospheric radio propagation path.

ionospheric regions   The ionized part of the Earth's upper atmosphere above about 70 km is known as the ionosphere consisting of overlapping ionized layers. These layers are produced by the action of solar electromagnetic radiation (ultraviolet and X-rays) and cosmic rays on the neutral particles. As these regions are ionized, the electrons affect the radio frequency propagation between any points on Earth or from outer space to the Earth. The number density of positive ions equals the electron density to keep charge neutrality at all heights. The layers have an altitude of maximum, above and below which the ionization drops off. The altitude profiles of ionization are functions of solar activity, time of day, latitude, season, and extra-terrestrial events such as magnetospheric storms, solar wind, interplanetary magnetic field and energetic particles, and solar flares to name a few.

The D region is situated normally around 85 km and present only during the day. The E region peak is situated around 110 km. The D and E region consist mainly of NO^+, O_2^+ ions and electrons. The F region consists of two regions, F1 and F2. The F1 layer is situated at a height of about 180 to 200 km and is absent at night. The F2 region peaks around 250 to 300 km. The F region consists mainly of O^+, NO^+, N^+ and electrons. The region above F2 is called the topside ionosphere. As we go higher in the topside ionosphere, we encounter the region called "heliosphere" above about 400 km, where helium ions are predominant. The region above about 600 km is called as "protonsphere", where atomic hydrogen ions are predominant. At higher latitudes, H^+ and He^+ may escape along field lines as "polar wind" into the magnetosphere. At low and middle latitudes, the

ions are trapped along inclined magnetic lines of force and do not escape. This region is called the "plasmasphere". The outer edge of the plasmasphere where the magnetic line of force is open to the magnetosphere is called "plasmapause" and is located along the line of force mapping down to about 60° magnetic latitude.

ionospheric sounder   Usually, a radio transmitter on the ground probing the electron density distribution of ionosphere below the density maximum. It is based on the tendency of a plasma to reflect radio signals below its plasma frequency. The sounder sends out a series of signals of increasing frequency, and measures the delay at which their echoes return: each delay gives the lowest altitude at which the electron density sufficed to reflect the signal.

Based on ionospheric soundings, the so-called D, E, and F layers of the ionosphere were identified long ago, the latter resolved into the F1 and F2 layers. Above the F-layer the ion density slowly decreases again, and its density profile there was first probed in 1962 by the orbiting *ionospheric sounder* aboard the Canadian satellite Alouette 1. That plasma ends at the plasmapause.

ionospheric sounding   The process of obtaining ionospheric information using a radio sounding technique. This usually refers to operating an ionosonde.

ionospheric storm   A global disturbance in the F region of the ionosphere, which occurs in connection with geomagnetic activity and especially a geomagnetic storm. The high latitude atmosphere is heated during geomagnetic activity, causing neutral winds to blow towards the equator. These winds substantially alter the chemistry of the atmosphere and thereby alter the global ionosphere. During the first few hours of a storm, the positive phase, the daytime F-region ionization can increase, sometimes significantly. It is generally attributed to the action of winds forcing ionization up the magnetic field lines and is most apparent in the middle latitudes. Generally, this phase ceases after sunset and is often followed by the negative phase, when a decrease in F-region ionization occurs and sometimes lasts a few days. Individual storms can vary, and their behavior depends on geomagnetic latitude, season, and local time. Storms affect higher geomagnetic latitudes to a greater extent and are more prevalent in the summer hemisphere. Severe storms may also affect the lower F region. Occasionally, the upper F-region ionization (foF2) can drop below the lower F-region ionization (foF1). In these circumstances, the lower F1 region will support HF propagation instead of the F2 region. *See* geomagnetic storm, ionosphere.

ionospheric variability   The average ionospheric behavior is estimated by monthly medians of the hourly observations of the key ionospheric layer parameters (e.g., foF2). Diurnal, seasonal, and solar cycle dependencies for locations around the globe have been tabulated and modeled empirically and, possibly with less success, physically. However, there is also a large residual ionospheric variability in the hourly observations about the medians. *Ionospheric variability* is described by an accumulative statistic, often upper and lower quartiles, and is summed over many different physical processes. As an example, the F-region peak electron density, measured by foF2, may vary because of changes in several processes, some of which are solar ionizing radiation, geomagnetic activity including geomagnetic storms and the resultant ionospheric storms, and the effects of traveling ionospheric disturbances. Each of these sources of variability has different latitudinal and temporal properties. Ionospheric variability can also be described by correlation coefficients that depend on separation, as well as geographic latitude, longitude, season, and time of day. Generally, estimates of variability at a remote location can be improved by 50% using the locally observed ionospheric information, provided separations are less than typically 250 to 500 km in latitude and 500 to 1000 km in longitude. Since these are gross statistics, they should be interpreted with caution. *See* ionosphere.

ion torus   (**1.**) A thick accretion disk, whose existence has been suggested to explain the collimation of radio jets, and the low bolometric luminosity (compared to the mechanical energy needed to produce radio lobes) of powerful radio sources. An *ion torus* is a structure supported by the pressure of the ions of very hot plasma, with kinetic energies as high as $\sim$ 100 MeV. It is expected to surround a supermassive black hole with steep walls creating a narrow funnel which collimates the highest energy radiation produced from inside the funnel and, if a magnetic field is present, is expected to collimate the radio jet itself.

(**2.**) Donut-shaped cloud of neutral and ionized gases (plasma) along the orbit of the Jovian satellite Io, maintained by volcanic eruptions on Io.

ion trap   *See* Faraday Cup.

IRAF (Image Reduction and Analysis Facility)   A freely available general-use suite of software that is extensively used in many areas of study in observational astronomy. It is distributed by the IRAF project based at the National Optical Astronomical Observatories (NOAO) in Tucson, Arizona. The software and information can be accessed through their web page at: http://iraf.noao.edu/.

IRAF is organized into "packages", which contain collections of "tasks" (procedures). The packages tend to be organized according to projects (such as Hubble Space Telescope data) and/or types of analyses. Various packages include tasks for CCD reductions, astrometry, photometry, plotting, data management, etc.

Iris   Seventh asteroid to be discovered, in 1847. Orbit: semimajor axis 2.3854 AU, eccentricity 0.2304, inclination to the ecliptic 5°.52412, period 3.68 years.

IRIS (Incorporated Research Institutions for Seismology)   A name of a research project propelled by *Incorporated Research Institutions for Seismology*. The project started in the 1980s, and globally deployed highly efficient long-period seismographs such as STS. The purpose of the project is to propel research on Earth's internal structure and source processes of large earthquakes, aiming at building up global seismic networks of next generation seismological observations. For more information, refer to http://www.iris.washington.edu/.

iron K_α line   A spectral line at energy of approximately 6.4 keV from the transition between the L and K shells (i.e., the second innermost and the innermost atomic shells, corresponding to quantum number $n = 2$ and $n = 1$, respectively) of an iron atom. The iron fluorescence K_α line is a strong emission feature in the X-ray spectra of active galactic nuclei and of cataclysmic variables. It can be produced by recombination following photoionization in gas irradiated by an intense X-ray source, as in systems powered by accretion. An electron of the K-shell may also be removed by Compton scattering due to an X-ray photon. The line is excited collisionally in the hot gas of stellar flares, supernova remnants and in the intra-cluster medium in clusters of galaxies. Observation of the line has been made possible by X-ray space-borne observatories. The instruments on board the orbiting Japanese observatory ASCA have revealed, in the K_α line profile of several Seyfert galaxies, characteristic effects predicted for radiation coming from a gaseous, rotating disk at a few gravitational radii from a black hole. *See* accretion disk, Seyfert galaxies.

iron meteorite   A meteorite consisting of solid nickel-iron alloy. Such an object must have formed in the core of a substantial planet which was then disrupted, presumably in a collision or close encounter with another object. Iron (or iron-nickel) meteorites are classified as Hexahedrites (less than 6.5% nickel); Octahedrites (6.5% to 13% nickel); and Ataxites with greater amounts of nickel.

irradiance   The radiant power per unit area per unit wavelength interval [W m^{-2} nm^{-1}].

1. [downward (upward) plane irradiance] The downward (upward) directed radiant power per unit area onto an upward (downward) facing horizontal surface [W m^{-2} nm^{-1}].

2. [downward (upward) scalar irradiance] The downward (upward) directed radiant power per unit area onto a spherical collecting surface [W m^{-2} nm^{-1}].

3. [net plane irradiance] The downward plane irradiance minus the upward plane irradiance [W m^{-2} nm^{-1}].

4. [irradiance ratio] The ratio of the upward plane irradiance to the downward plane irradiance.

5. [reflectance] irradiance ratio.

6. [scalar irradiance] The power per unit area incident from all directions onto a spherical collecting surface [W m^{-2} nm^{-1}]; it equals the downward scalar irradiance plus the upward scalar irradiance.

irregular galaxies   Galaxies that lack a central bulge, azimuthal symmetry, and that most often show a rather patchy appearance. They are of blue color, have high neutral gas content, and show evidence of ongoing star formation. *Irregular galaxies* are often dwarf galaxies, on average smaller and less massive than spiral galaxies, with typical masses $\sim 10^8$ solar masses. They also have lower rotational velocity, if any, and lower luminosity than spirals. Although a small fraction in large catalogs of galaxies, such as the Revised New General Catalog, irregular galaxies are thought to account for 1/2 to 1/3 of all galaxies in the local universe. They have been further subdivided into the Magellanic type (from the prototype galaxy, the Large Magellanic Cloud), and the amorphous type (or M82-type) galaxies. Magellanic irregulars have a patchy appearance due to clusters of hot stars in star-forming regions spread over the whole galaxy, while amorphous irregulars are smoother in appearance, with a single supergiant star-forming region at the center of the galaxy. This subdivision is now considered mainly of historical importance.

irregular waves   Waves without a single, clearly defined period. Also referred to as random waves.

irribarren number   A dimensionless parameter used to predict wave breaker type. Equal to $S/\sqrt{H_b/L_o}$, where S is beach slope, H_b is a breaking wave height, and L_o is deepwater wavelength.

irrotational   The property of a vector field $\mathbf{v}$ (e.g., in fluid dynamics a velocity field) that its curl vanishes: $\nabla \times \mathbf{v} = 0$. Hence, $\mathbf{v}$ is the gradient of a scalar function: $\mathbf{v} = \nabla\phi$.

ISC (International Seismological Centre)
ISC was formed in Edinburgh, Scotland in 1964 and moved to Newbury in southern England, and subsequently to Thatcham, near Newbury. ISC collects and compiles seismic observation data throughout the world, and determines hypocentral parameters (location, origin time, magnitude, fault plane solution), issuing monthly reports (Bulletin of the International Seismological Centre) with observation data. ISC also issues printed regional catalogs of earthquakes, extracting hypocentral parameters from the monthly reports, and catalogs of research papers in relation to seismology. These have become important fundamental data concerning earthquakes. For more information, refer to http://www.isc.ac.uk/.

isentropic coordinate When horizontal scales are large compared with vertical scales, it is sometimes advantageous to replace the z-coordinate with potential temperature θ to use with x and y as independent coordinates, and then the variables (x, y, θ) are known as *isentropic coordinates.*

island arc Ocean trenches (subduction zones) often have an arcuate structure; they form part of a circle or an arc. The line of volcanic islands associated with the subduction zone has a similar distribution; thus they form an *island arc.* The Aleutian islands in Alaska are an example.

isobar A line connecting points of equal pressure on a graphical representation of a physical system, or the actual physical surface of equal pressure.

isobaric coordinate When horizontal scales are large compared with vertical scales, it is sometimes advantageous to replace the z-coordinate with pressure to use with x and y as independent coordinates, and then the variables (x,y,p) are known as isobaric coordinates, or pressure coordinates. This set of coordinates is widely used in atmospheric science.

isogonic lines Lines of constant declination. *See* declination.

isoline A line along which some parameter is constant. For example, isobars are lines of constant pressure.

isometry A transformation of coordinates $x^\mu \mapsto x'^\mu = x'^\mu(x)$ is said to be an *isometry* of space if the form of the metric **g** of the space does not change under the transformation

$$\mathbf{g}'(x') = \mathbf{g}(x) \, .$$

For infinitesimal transformations, one can introduce Killing vectors which represent infinitesimal isometries. *See* Killing vector.

isopycnal processes Processes taking place along surfaces of equal density.

isostacy The process of compensation or hydrostatic equilibrium by which the densities of elevated mountain ranges are compensated by the low density of depressed crustal roots.

An area that is isostatically compensated will show no free air gravity anomaly. There are two types of *isostacy:* Airy isostacy, in which variations in crustal thickness ("roots") explain the compensation, and Pratt isostacy, which explains isostatic compensation in terms of variations in the density of materials.

isotherm A line connecting points of equal temperature on a graphical representation of a physical system, or the actual physical surface of equal temperature.

isothermal atmosphere An atmosphere in hydrostatic equilibrium in which the temperature is constant with height. In such an atmosphere the pressure and density in an ideal gas are proportional, and therefore the pressure decreases exponentially upward:

$$p = p_0 \exp\left(-gz/R_d T_v\right),$$

where R_d is the gas constant for dry air; T_v is the virtual temperature; g is the acceleration of gravity. Also called exponential atmosphere.

isotope A particular species of an element identified by the number of neutrons in the nucleus $A - Z$. (The number of protons in the nucleus, the atomic number Z, determines which element is under discussion. Here A is the atomic mass number.) For instance, hydrogen (atomic number $Z = 1$) has two stable isotopes. In each hydrogen nucleus there is one proton (equal to the atomic number). Normal hydrogen, ^{1}H ($Z = 1$, $A = 1$), is the isotope with no neutrons; deuterium, ^{2}H ($Z = 1$, $A = 2$), is the isotope with one neutron. There is also an unstable isotope, tritium, ^{3}H ($Z = 1$, $A = 3$), with two neutrons, which decays by beta decay in 12.26 years.

isotope delta value (δ)   Stable isotope compositions of low-mass (light) elements such as oxygen, hydrogen, and carbon are normally reported as δ values. Delta values are reported in units of parts per thousand (per mil or $°/_{\circ\circ}$) relative to a standard of known composition: $\delta(°/_{\circ\circ}) = (R_x/R_s - 1) \times 1000$, where R represents the ratio of the heavy to light isotope (e.g., $^{18}O/^{16}O$), R_x is the ratio in the sample, and R_s is the ratio in the standard, which can be rewritten, for instance, for the oxygen isotopes of water as

$$\delta^{18}O = \frac{\left(^{18}O/^{16}O\right)_{\text{sample}} - \left(^{18}O/^{16}O\right)_{\text{standard}}}{\left(^{18}O/^{16}O\right)_{\text{standard}}} \times 1000$$

isotope fractionation factor (α)   The fractionation associated with the equilibrium exchange reaction between two substances A and B (e.g., liquid and vapor phases of water) is: $\alpha_{A-B} = R_A/R_B$, where R is the ratio of the heavy isotope to the lighter isotope in compounds A and B. For the liquid-vapor system of water at 20°C:

$$\alpha_{\text{liquid-vapor}} = \frac{\left(^{18}O/^{16}O\right)_{\text{liquid}}}{\left(^{18}O/^{16}O\right)_{\text{vapor}}} = 1.0098$$

isotropic   A material whose properties are independent of direction. For instance, in hydrodynamics hydraulic properties (such as intrinsic permeability) do not depend on the direction of flow and are equal in all directions.

isotropic turbulence   Very large eddies tend to be anisotropic because they contain structure that is related to the energy input. Succes-

sive interaction between the three spatial components redistributes turbulent kinetic energy successively more equally among the components, leading to increased isotropy at smaller scales. However, because of stratification, complete isotropy is seldom attained, not even at the smallest scales.

isotropy   In geometry, independence of direction; a system is isotropic if there exists no preferred spatial direction at any point in the system; a space is said to be isotropic with respect to a point of view P if it looks the same in all directions when seen from P.

In relativity and cosmology, the metric of an isotropic D-dimensional space is form invariant with respect to all rotations of center P. The latter are thus generated by $N_I = D(D-1)/2$ Killing vectors.

If the space is isotropic with respect to all points, then it is also homogeneous and admits $N_O = D$ further Killing vectors. When the number of Killing vectors equals $N_I + N_O = D(D+1)/2$, then the space is maximally symmetric.

In cosmology, the assumption may be made in modeling the universe that it is at some level approximately isotropic. However, after anisotropies connected with the motion of the Earth around the sun, of the sun in our galaxy and of the galaxy itself towards the Great Attractor are subtracted, the temperature of the microwave background radiation (presumably emitted at redshift $z \sim 1,000$) is not quite isotropic around us showing a residual anisotropy: $\Delta T/T \approx 10^{-5}$, where $T = 2.73$ K is the average temperature of the radiation and ΔT are differences between the values of T in different directions. This is taken as an indication of early inhomogeneity leading to inhomogeneities in matter, which grew to the observed large-scale structure. *See* homogeneity, Killing vector, maximally symmetric space.

Israel–Wilson–Perjés space-times (1971)
Stationary space-times describing the external fields of an arbitrary array of spinning and electrically charged bodies held in equilibrium by the balance of the gravitational and electromagnetic forces. *See* equilibrium space-times.

J

J_2   A dimensionless coefficient in the $P_2(\cos\theta)$ term of the Legendre polynomial expression of the gravitational potential of a planet, where θ is the colatitude (90°-latitude). Assuming symmetry of mass distribution with respect to the rotational axis and to the equatorial plane and without considering the effect of rotation, the gravitational potential of a planet as a function of radial coordinate r and colatitude θ is

$$V = -\frac{GM}{r} + J_2\frac{GMa^2}{r^3}\left(\frac{3}{2}\cos^2\theta - \frac{1}{2}\right)$$
$$+ \text{ higher order terms of } \frac{1}{r}$$

where $G = 6.6725985 \times 10^{-11}$ m^3 kg^{-1} s^{-2} is the gravitational constant, M is the total mass of the planet, and a is the equatorial radius. J_2 is an important parameter in the gravitational and rotational dynamics of a planet because of its relation with the moments of inertia:

$$J_2 = \frac{C - A}{Ma^2}$$

where C and A are the polar and equatorial moments of inertia, respectively, and a is the mean radius of the Earth. The value of J_2 for the Earth is about 1.082626×10^{-3}. Redistribution of mass in the Earth causes changes to the gravitational potential and hence J_2.

Jacobian   In considering the transformation between a given coordinate $n-$dimensional system $\{x^i\}$ and the new one $\{y^i\}$, the elementary volume element undergoes a transformation:

$$dy^n = J dx^n ,$$

where J is the *Jacobian*, given by $det|\partial y/\partial x|$, where the matrix $J(y,x) = \{\partial y/\partial x\}$ is constructed with elements of the form

$$\partial y^i/\partial x^j$$

for all i and j. $J(y,x)$ is sometimes called the Jacobian matrix, and $J = det J(y,x)$ the Jacobian determinant.

Jacobi ellipsoid   A triaxial stationary configuration of rigidly rotating fluid with meridional ellipticity exceeding 0.91267.

Jansky (Jy)   A unit of spectral flux used in radio astronomy: $1 Jy = 10^{-26} W\ m^{-2} Hz^{-1}$.

Janus   Moon of Saturn, also designated SX. Discovered by A. Dollfus in 1966, it is now uncertain if Dollfus saw *Janus* or its co-orbital partner, Epimetheus. Its orbit has an eccentricity of 0.007, an inclination of 0.14°, and a semimajor axis of 1.51×10^5 km. Its size is $98 \times 96 \times 75$ km, its mass 2.01×10^{18} kg, and its density 0.68 g cm^{-3}. It has a geometric albedo of 0.8 and orbits Saturn once every 0.695 Earth days.

Japan current   An ocean current flowing northeast from the Philippines along the eastern coast of Japan.

JD   *See* Julian date.

Jeans escape   The process by which gases undergo thermal escape from an atmosphere. Although the temperature of the atmosphere may be too low for the average molecule to have escape velocity, there will be molecules at the tail of the Maxwell distribution that will have the required speed. If those molecules are at the base of the exosphere, they will be able to escape to space. This mechanism is called *Jeans escape*, and the resultant flux is the Jeans flux.

Jeans instability   James Jeans discovered that perturbations in a self-gravitating gas are unstable. Density fluctuations grow as time passes. Matter collapses in some regions (leading to the growth of structures) and becomes rarified elsewhere. Similar results hold in an expanding cosmology, with modifications for the overall expansion. On very large scales and always outside the cosmic horizon, the evolution of the fluctuations is essentially free fall, so the structure at any time reflects the initial pattern or fluctuations. *See* Jeans length.

Jeans length   Minimum length scale of a density perturbation that is gravitationally unstable; the critical size above which a smooth medium inevitably degenerates into a clumpy one because of gravitational attraction; the fluid collapses under its own weight. For scales smaller than this critical length, the fluid can build enough hydrostatic pressure to halt the collapse and perturbations oscillate as sound waves. For a perfect Newtonian fluid the *Jeans length* is given by:

$$\lambda_J = \left(\frac{\pi v_s^2}{G\rho} \right)^{1/2}$$

where v_s is the sound speed of the fluid, G the gravitational constant, and ρ the mean density. For a perturbation longer than the Jeans length collapsing in a non-expanding fluid, the density grows exponentially on a dynamical time scale $\tau_{dyn} = (4\pi G\rho)^{-1/2}$. If the fluid is expanding, the background perturbation growth is like a power law with an exponent close to unity, the exact value depending on the equation of state of the fluid. The Jeans length during the conditions believed to be present at the time after the Big Bang when decoupling of ordinary matter and radiation occurred was 6×10^{17} m, or about 60 light years.

Jerlov water type   A water clarity classification scheme based on the downward diffuse attenuation coefficient just below the sea surface.

jet   In fluid mechanics, a compact, coherent, collimated stream of relatively fast moving fluid impinging into still fluid.

In meteorology, a shorthand for *jet stream*.

In astrophysics, usually bipolar streams of high velocity gas emitted from an active region. Examples include jets from star formation regions (collimated by magnetic fields) and from black hole accretion in stellar mass or in supermassive black holes, in which the collimation is accomplished by the rotation of the black hole.

jet stream   Relatively strong west to east winds concentrated in a narrow stream up to 500 km across, located in a thin vertical range 10 to 15 km in altitude, with winds exceeding 90 km/h. Winds, especially in the winter, can approach 500 km/h. In North America, major cold outbreaks are associated with movement of the jet stream south. In summertime, and during

mild winter weather, the jet stream stays to the north. Its actual fluctuations range from northern Canada to over the Gulf of Mexico.

jetty   A structure at a tidal inlet. Commonly built as a rubble mound structure, to keep waves and sediment out of a harbor entrance.

jog   In geology, segments of dislocation line which have a component of their sense vector normal to the glide plane. *Jogs* are abrupt changes in direction taking the dislocation out of the glide plane. When the normal component extends over only a single interplanar spacing d, as for A and B, the jogs are *unit jogs,* or for brevity just *jogs.* If the normal component extends over more than one interplanar spacing, the jog is called *superjog.*

joints   Fractures in the Earth's crust across which there has been no relative motion. If relative motion has occurred, the fracture is a fault.

JONSWAP   Acronym for Joint North Sea Wave Project, an experiment involving measurement of ocean wave energy spectra yielding an empirical spectrum of the same name.

Joule (J)   The SI unit of energy, equal to Nm.

Joule heating   The dissipation of current in a plasma in the form of heat due to the presence of a finite electrical conductivity. The efficient dissipation of currents in the corona requires the currents to be confined to small volumes because of the extremely high conductivity. The heating is given by $W_J = \mathbf{j} \cdot \mathbf{E} = j^2/\sigma$ where $\mathbf{E}$ is the electric field, j is the current density, and σ is the electrical conductivity of the plasma. *Joule heating* is also known as Ohmic heating.

Jovian planet   A planet that has characteristics similar to Jupiter, the largest planet in our solar system. There are four *Jovian planets* in our solar system: Jupiter, Saturn, Uranus, and Neptune. The primary characteristics of Jovian planets are large size ($>$48,000 km in diameter), large distance from the sun ($>$5 astronomical units), low densities ($<$ 1600 kg/m^3, indicating gaseous compositions of primarily hydrogen and helium), many moons, and ring systems.

The characteristics of the Jovian planets contrast with those of the terrestrial planets, which are smaller, rocky, and close to the sun. The Jovian planets are sometimes also called the outer planets since they are located in the outer region of our solar system.

Julian calendar   The calendar introduced by Julius Caesar in 46 B.C. which divided the year into 365 days with a leap year of 366 days every fourth year. Thus, a century has exactly 36,525 days.

Julian Date and Modified Julian Date
The number of days elapsed since noon Greenwich Mean Time on January 1, 4713 B.C. Thus January 1, 2000, 0h (midnight) GUT was JD 2,451,543.5. This dating system was introduced by Joseph Scaliger in 1582. *Julian Dates* provide a method of identifying any time since 4713 B.C. as a real number independent of time zone. The Julian Date (JD) is counted in whole days and fractions, measured in days from noon, 4713 B.C., on the Julian proleptic calendar. The Julian proleptic calendar is a backwards extension of the Julian calendar.

Sometimes the Julian Date is defined by parsing the number just described into years, months, days, and a fractional day. In other cases, the Julian Day Number is taken to be the integer part of the Julian Date. For dates on or after Nov 17, 1858, the Modified Julian Date (MJD) is defined as the Julian Date less 2,400,000.5; it is zero at midnight on the date just mentioned (Gregorian calendar). Note that the fraction 0.5 moves the start of an MJD to midnight, while the JD starts at noon.

Julian Date Systems Prior to 1972

For dates before January 1, 1972, Julian Dates generally refer to Greenwich Mean Solar Time (GMT) unless otherwise specified. For example, the Julian Date can instead be used to represent Ephemeris Time (ET), used for planetary and other ephemerides. Ephemeris Time, which came into usage in 1956, was replaced by Dynamical Time in 1984. Dynamical Time was later replaced by the two kinds of dynamical time mentioned in the next paragraph.

Julian Date Systems after December 31, 1971
Starting in 1972, GMT was split into Coordinated Universal Time (UTC), the basis of civil timekeeping, and UT1, which is a measure of Earth rotation. Furthermore, in 1977, Dynamical Time was split into two forms, Terrestrial Dynamical Time (TDT) and Barycentric Dynamical Time (TDB). Julian Dates can represent any of these time systems, as well as International Atomic Time (TAI). Nevertheless, the Astronomical Almanac specifies UT1 as the system to be assumed in the absence of any other specification. This convention must be followed in any application whose result is to be a measure of Earth rotation, for example in finding sidereal time. When high accuracy time determinations are needed, one of the other systems, such as TAI, TDT, or TDB should be used, according to the situation. The differences among all these systems, now reaching to over a minute difference between TDT and UTC, can influence the whole day number near a day boundary. When the whole Julian Date is used, the problem is present near UTC noon. When Modified Julian Date is used, the problem is at midnight. When the time stream being used is clearly specified, and the decimals carried are sufficient to represent seconds, there is no real problem, but if the JD or MJD is truncated to whole days, the integer values can differ for different time streams when representing identical times.

Transformations among JD, MJD, and Gregorian calendar date
The Astronomical Almanac (published annually by the U.S. Naval Observatory) provides tables for the translation between Gregorian date and Julian Day number. Software for translating between Gregorian calendar date, Julian Date, and Modified Julian Date is available from the U.S. Naval Observatory, as follows: On a UNIX system, with e-mail, one can enter:

Mail -s cdecm adsmail@tycho.usno.navy.mil < /dev/null

A few explicit translations are given in the table.

See also *Explanatory Supplement to the Astronomical Almanac,* P.K. Seidelmann, Ed., (University Science Books, Mill Valley, CA, 1992), which provides algorithms for the translation.

Gregorian Date (midnight)	Julian Date (UTC)	Modified Julian Date (UTC)
Jan 1, 1900	2415020.5	15020
Jan 1, 1980	2444239.5	44239
Jan 1, 2000	2451544.5	51544

For more information, see:
http://www.capecod.net/ pbaum/date/back.htm
(and http://tycho.usno.navy.mil/systime.html)

Julian year   *See* year.

Juliet   Moon of Uranus also designated UXI. Discovered by Voyager 2 in 1986, it is a small, irregular body, approximately 42 km in radius. Its orbit has an eccentricity of 0.0005, an inclination of $0°$, a precession of $223°$ y^{-1}, and a semimajor axis of 6.44×10^4 km. Its surface is very dark, with a geometric albedo of less than 0.1. Its mass has not been measured. It orbits Uranus once every 0.493 Earth days.

Junge particle distribution   A power-law distribution function [particles m^{-3} μm^{-1}] often used to describe the particle concentration [particles m^{-3}] per unit size interval [μm] vs. equivalent spherical diameter of particles; the number density for particles of equivalent diameter x is proportional to x^{-k}. In natural waters k is typically 3 to 5.

Juno   Third asteroid to be discovered, in 1804. Dimensions 230 by 288 km. Orbit: semimajor axis 2.6679 AU, eccentricity 0.25848, inclination to the ecliptic $12°.96756$, period 4.358 years.

Jupiter   A gas giant, the fifth planet from the sun with orbital semimajor axis 5.20AU (778,330,000 km), orbital eccentricity 0.0483; its orbital period is 11.8623 years. Jupiter is the largest planet, with equatorial radius 71,492 km, and mass 1.9×10^{27} kg; its mean density is 1.33 gm/cm^3. It has a surface rotational period of 9.841 hours. Jupiter has an orbital inclination of $1.308°$ and an axial tilt of $3.12°$. Jupiter has an absolute magnitude of -2.7, making it usually the fourth brightest object in the sky, after the sun, moon, and Venus; occasionally Mars exceeds Jupiter in brightness.

The model structure of Jupiter envisions a rocky core of order 10 to 15 $M_\oplus$, at a temperature $\approx 2 \times 10^5$ K, maintained by the gravitational compression of the planet. Because of this heating, Jupiter radiates considerably more energy than it receives from the sun. Above this core is a layer of liquid metallic hydrogen (at pressure depths exceeding 4 million atmospheres). Liquid metallic hydrogen is an electrical conductor and large scale currents in this material produce Jupiter's magnetic field, via the dynamo effect. Jupiter's field is 14 times stronger at Jupiter's "surface" equator (4.3 Gauss), than is Earth's at its equator (0.3 Gauss). Jupiter's magnetic field is roughly dipolar, with its axis offset by 10,000 km from the center of the planet and tipped $11°$ degrees from Jupiter's rotation axis. Jupiter's magnetosphere extends more than 650 million kilometers outward, but only a few million kilometers sunward.

Above the metallic hydrogen is a region composed primarily of ordinary molecular hydrogen (90%) and helium (10%) which is liquid in the interior and gaseous further out.

Three distinct cloud layers exist in the atmosphere, consisting of ammonia ice, ammonium hydrosulfide, and a mixture of ice and water. Jupiter's atmosphere has high velocity winds (exceeding 600 km/h), confined in latitudinal bands, with opposite directions in adjacent bands. Slight chemical and temperature differences between these bands are responsible for the colored bands that dominate the planet's appearance. The reddish colors seem to be associated with the highest clouds, the bluish colors with the deepest. Intermediate clouds have brown-cream-white colors.

Jupiter posseses a very long-lived atmospheric phenomenon, the Great Red Spot, which has been observed for over 300 years. This is a reddish colored elliptical shaped storm, about 12,000 by 25,000 km.

Jupiter has three faint and small, low (about 0.05) albedo rings (called Halo, Main, and Gossamer) extending from 100,000 km to as much as 850,000 km from the center of the planet. They are comparable to Saturn's, but much less prominent.

Jupiter has 16 known satellites, the four large Galilean moons Io, Europa, Ganymede, and Callisto, all discovered by Galileo in 1610, and 12 small ones.

K

Kaiser–Stebbins effect (1984) Anisotropies in the cosmic microwave background generated by cosmic strings present after the time of the decoupling of the cosmic radiation. For a string moving transversely to the line of sight, light rays passing on opposite sides of the string will suffer different Doppler shifts. The result is then the existence of step-like blackbody effective temperature T discontinuities, with relative magnitude change on different sides of the string given by

$$\delta T / T = 8\pi G U \gamma_s \hat{n} \cdot (\vec{v}_s \times \hat{s})$$

where γ_s is the relativistic Lorentz factor, v_s is the string velocity with respect to the observer, $\hat{s}$ is the string segment orientation and $\hat{n}$ points in the direction of the line of sight. G is Newton's constant, units are chosen so that $\hbar = c = 1$ and U is the (effective) mass per unit length of the (wiggly) cosmic string. *See* cosmic string, cosmic topological defect, deficit angle (cosmic string).

katabatic wind A wind that is created by air flowing downhill; wind flowing down slope at night as the valley floor cools. Examples of drainage winds are the Mistral in France and the Santa Ana in California. Another example is winds from the Greenland plateau where a pool of cold dense air forms and if forced off the plateau seaward accelerates downslope.

K corona The K (or *kontinuerlich*) *corona* is generated by the scattering of photospheric radiation from coronal electrons. The high speeds of the scattering electrons smear out the Fraunhofer lines except the H- and K-lines. The K corona dominates below $\sim 2R_\odot$ where its intensity ranges from 10^{-6} to 10^{-8} of the disk-center intensity. The K corona is polarized by the electron scattering with the electric vector parallel to the limb.

k-correction A necessary correction in observing redshifted objects. Since the visual or blue magnitudes in such sources correspond to absolute luminosities at higher frequencies than is the case for nearer sources, the *k-correction* for this effect depends on modeling the spectrum of the source.

Kelvin–Helmholtz instability Also called "shear instability". An instability of an unbounded parallel shear flow to the growth of waves with phase speed in the flow direction approximately equal to the speed of the inflection point of the shear. The instability occurs due to a resonant coupling between wave-like disturbances on either flank of the shear flow where the gradient of the shear is non-zero.

Kelvin material Also called a Kelvin solid. A viscoelastic material that, in response to a suddenly imposed constant loading, deforms as an incompressible, linearly viscous fluid over a short time but as a linearly elastic solid over a long time. The stress for the *Kelvin material* is the combination of the elastic stress according to Hooke's law and the viscous stress according to the Newtonian flow law. The relation between the volumetric strain $\theta = \varepsilon_{ii}$ and pressure $p = \sigma_{ii}/3$ is elastic: $p = K\theta$, where K is the bulk modulus. The constitutive relation for the deviatoric strain $\varepsilon'_{ij} = \varepsilon_{ij} - \theta\delta_{ij}/3$ and deviatoric stress $\sigma'_{ij} = \sigma_{ij} - p\delta_{ij}$ is

$$\sigma'_{ij} = 2G\varepsilon'_{ij} + 2\mu \frac{\partial \varepsilon'_{ij}}{\partial t}$$

where G is shear modulus, and μ is viscosity. Quantity $\tau_K = \mu/G$ is called the Kelvin relaxation time, a time roughly defining the transition from predominantly viscous to predominantly elastic behavior after a suddenly imposed constant loading.

Kelvin temperature scale Absolute temperature scale. At 0 K all thermal motion of matter would be at a standstill; thus, in the Kelvin scale no negative temperatures can exist. 0 K correpsonds to $-273.15°C$; therefore, absolute temperatures T in the Kelvin scale are related to temperatures θ in centigrades by $T = \theta + 273.15$ K.

0-8493-2891-8/01/$0.00+$.50
© 2001 by CRC Press LLC

Kelvin (Thompson) circulation theorem   In a perfect fluid, the circulation of the velocity of a fluid element is constant as the element moves along flow lines.

Keplerian map   In planetary dynamics, an approximate but fast method to solve for the orbital evolution of a small body such as a comet in the solar system under the influence of the planetary perturbations. One assumes that planetary perturbations become significant only during the short duration around perihelion passage, and the orbit changes either by systematic, or in some cases approximately random, small jumps from one Keplerian orbit to another, as the small body passes perihelion.

Kepler shear   The velocity difference between particles in adjacent Keplerian orbits about a central massive body.

Kepler's laws   Observations of the motion of planets due to Johannes Kepler.

1. All planets move in elliptical orbits, with the sun at one focus of the ellipse.

2. The line connecting the sun and the planet sweeps out equal areas in equal times.

3. The square of the period T of orbit of a planet is proportional to the cube semimajor axis a of its orbit:

$$T^2 = ka^3$$

where the proportionality factor k is the same for all planets.

Kepler's supernova (SN1604, 3C358)   A supernova that occurred in 1604 and was first observed by Kepler on October 17 of that year. Kepler reported that the star was initially as bright as Mars, then brightened and surpassed Jupiter within a few days, suggesting a peak brightness of magnitude -2.25. It plateaued at this brightness as it was lost in twilight of November 1604. It reappeared in January 1605, and Kepler found it still brighter than Antares ($m = 1$). It remained visible until March 1606, a naked-eye visibility of 18 months. From its light curve, it was almost certainly type I supernova. A remnant is now found at Right Ascension: $17^h 27' 42''$ and Declination: $-21°27'$. It is now observable as a remnant of about 3 arcmin diameter, consisting of faint filaments in the optical, but as a shell in the radio and in X-ray. The distance is approximately 4.4 kpc.

Kerr black hole   (1963) A rotating black hole, i.e., a black hole with angular momentum associated to its spinning motion. The spin axis of the *Kerr black hole* breaks the spherical symmetry of a nonrotating (Schwarzschild) black hole, and identifies a preferential orientation in the space-time. In the vicinity of the hole, below a limiting distance called the static radius, the rotation of the hole forces every observer to orbit the black hole in the same direction as the black hole rotates. Inside the static radius is the event horizon (the true surface of the black hole). These two surfaces delimit the ergosphere of the Kerr black hole, a region from which a particle can in principle escape, extracting some of the rotational kinetic energy of the black hole. On theoretical grounds it is expected that gravitational collapse of massive stars or star systems will create spinning Kerr black holes, and the escape of particles from the ergosphere, may play an important role as the power source and collimation mechanism of jets observed in radio galaxies and quasars. *See* quasar.

Kerr metric   (1963) The metric

$$ds^2 =$$
$$-\left(1 - \frac{2Mr}{r^2 + a^2\cos^2\theta}\right)\left(du + a\sin^2\theta d\phi\right)^2$$
$$+ 2\left(du + a\sin^2\theta d\phi\right)\left(dr + a\sin^2\theta d\phi\right)$$
$$+ \left(r^2 + a^2\cos^2\theta\right)\left(d\theta^2 + \sin^2\theta d\phi^2\right)$$

discovered by R.P. Kerr and representing the gravitational field of a rotating Kerr black hole of mass M and angular momentum aM, when the condition $a^2 \leq M^2$ holds. When the condition is not met, the space-time singularity at $r = 0$ and $\theta = \frac{\pi}{2}$ is "naked", giving rise to global causality violation. Uniqueness theorems have been proven that (at least with a topology of a two-sphere of the event horizon) there are no other stationary vacuum black hole metrics, assuming general relativity is the correct theory of gravity. *See* cosmic censorship.

Kerr–Newman metric   The unique asymptotically flat general relativistic metric describing the gravitational field outside a rotating axisymmetric, electrically charged source. When the electric charge is zero, the metric becomes the Kerr metric.

The line element in Boyer–Lindquist coordinates (t, r, θ, ϕ) is

$$ds^2 = -\rho^2 \frac{\Delta}{\Sigma^2} dt^2$$
$$+ \frac{\Sigma^2}{\rho^2} \left(d\phi^2 - \frac{2\,a\,M\,r}{\Sigma^2} dt \right)^2$$
$$+ \frac{\rho^2}{\Delta} dr^2 + \rho^2 d\theta^2 \,,$$

where $\rho^2 = r^2 + a^2 \cos^2\theta$, $\Delta = (r - r_+)(r - r_-)$, and $\Sigma^2 = (r^2 + a^2)^2 - a^2 \Delta \sin^2\theta$. The quantity M is the ADM mass of the source, $a \equiv J/M$ its angular momentum per unit mass, and Q its electric charge. When $Q \neq 0$ one also has an electromagnetic field given by

$$
\begin{aligned}
F_{rt} &= Q \frac{r^2 - a^2 \cos^2\theta}{\rho^4} \\
F_{\phi r} &= a \sin^2\theta \, F_{rt} \\
F_{t\theta} &= Q \frac{a^2 r \sin 2\theta}{\rho^4} \\
F_{\theta\phi} &= \frac{r^2 + a^2}{a} F_{t\theta} \,.
\end{aligned}
$$

For $M^2 < a^2 + Q^2$ the metric describes a naked singularity at $r = 0$ and $\theta = \pi/2$. For $M^2 \geq a^2 + Q^2$ the metric describes a charged, spinning black hole. *See also* Reissner–Nordström metric, black hole, naked singularity.

Kerr–Schild space-times The collection of metrics

$$ds^2 = \lambda \ell_a \ell_b dx^a dx^b + ds_0^2$$

where λ is the parameter and ds_0^2 is the "seed" metric, often assumed to be the Minkowski metric. The vector ℓ is null with respect to all metrics of the collection. Many space-times, e.g., the Kerr metric and the plane-fronted gravitational waves, are in the Kerr–Schild class.

Keulegan–Carpenter Number A dimensionless parameter used in the study of wave-induced forces on structures. The *Keulegan–Carpenter Number* is defined as $\bar{u}_{\max} T/D$,

where $\bar{u}_{\max}$ is the maximum wave-induced velocity, averaged over the water depth, T is wave period, and D is structure diameter.

K-function Diffuse attenuation coefficient.

Kibble mechanism Process by which defects (strings, monopoles, domain walls) are produced in phase transitions occurring in the matter. In the simplest cases there is a field ϕ (called the Higgs field) whose lowest-energy state evolves smoothly from an expected value of zero at high temperatures to some nonzero ϕ at low temperatures which gives minimum energies, but there is more than one such minimum energy (vacuum) configuration. Both thermal and quantum fluctuations influence the new value taken by ϕ, leading to the existence of domains wherein ϕ is coherent and regions where it is not. The coherent regions have typical size ξ, the coherence length. Thus, points separated by $r \gg \xi$ will belong to domains with, in principle, arbitrarily different orientations of the field. It is the interfaces of these different regions (sheets, strings, points) which become the topological defects. In cosmology, this is viewed as occurring in the early universe. Because of the finite speed of light, ξ is bounded by the distance light can travel in one Hubble time: $\xi \overset{\sim}{<} H^{-1}$ where H is the Hubble constant expressed as an inverse length. In cosmology, the defects (cosmic domain walls, cosmic strings, etc.) can have an important effect on later formation of cosmic inhomogeneities.

Kibel, I.A. (1904–1970) Regarded by many as the founder of the Soviet school of modern dynamical meteorology. He made important contributions to the theory of gas dynamics, nonhomogeneous turbulence in a compressible fluid, the atmospheric boundary layer, cloud dynamics, global climate, and mesoscale wind systems.

Killing horizon The set of points where a non-vanishing Killing vector becomes null, i.e., lightlike. Examples are given by the horizons in the Schwarzschild, Reissner–Nordström, and Kerr–Newman metric.

Killing tensor   A totally symmetric n-index tensor field $K_{a_1...a_n}$ which satisfies the equation

$$\nabla_{(b} K_{a_1...a_n)} = 0 .$$

(the round brackets denote symmetrization). A trivial Killing tensor is a product of lower-rank Killing tensors. *See* Killing vector.

Killing vector   The vector generator $\boldsymbol{\xi}$ associated with an isometry. When the *Killing vector* exists it may be viewed as generating an infinitesimal displacement of coordinates $x^\mu \mapsto x^\mu + \epsilon\, \xi^\mu$ ($|\epsilon| \ll 1$), with the essential feature that this motion is an isometry of the metric and of any auxiliary geometrical objects (e.g., matter fields). Such a vector satisfies the Killing condition

$$\mathcal{L}_{\boldsymbol{\xi}}\, , g = 0 .$$

the vanishing of the Lie derivative of the metric tensor. In coordinates this is equivalent to $\xi_{\mu;\nu} + \xi_{\nu;\mu} = 0$ where ; denotes a covariant derivative with respect to the metric. Similarly, other geometrical objects, T, satisfy $\mathcal{L}_{\xi} T = 0$.

In a coordinate system adapted to the Killing vector, $\boldsymbol{\xi} = \frac{\partial}{\partial x^a}$, the metric and other fields do not depend on the coordinate x^a. The Killing vectors of a given D-dimensional metric space form a vector space whose maximum possible dimension is $D\,(D + 1)/2$. Wilhelm Killing (1888). *See* isometry.

K index   The K index is a 3-hourly quasi-logarithmic local index of geomagnetic activity based on the range of variation in the observed magnetic field measured relative to an assumed quiet-day curve for the recording site. The range of the index is from 0 to 9. The K index measures the deviation of the most disturbed of the two horizontal components of the Earth's magnetic field. *See* geomagnetic indices.

kinematical invariants   The quantities associated with the flow of a continuous medium: acceleration, expansion, rotation, and shear. In a generic flow, they are nonzero simultaneously.

kinematics   The study of motion of noninteracting objects, and relations among force, position, velocity, and acceleration.

kinematic viscosity   Molecular viscosity divided by fluid density, $v = \mu/\rho$, where v = kinematic viscosity, μ = molecular (or dynamic or absolute) viscosity, and ρ = fluid density. *See* eddy viscosity, dynamic viscosity.

kinetic energy   The energy associated with moving mass; since it is an energy, it has the units of ergs or Joules in metric systems. In nonrelativistic systems, for a point mass, in which position is described in rectangular coordinates $\{x^i i = 1 \cdots N$, where N = the dimension of the space, the kinetic energy T is

$$
\begin{aligned}
T &= \frac{1}{2}m \sum \left(v^{i2} \right) \\
&= \frac{1}{2} m \delta_{ij} v^i v^j
\end{aligned}
$$

$\delta_{ij} = 1$ if $i = j$; $\delta_{ij} = 0$ otherwise. Since δ_{ij} are the components of the metric tensor in rectangular coordinates, it can be seen that T is $\frac{m}{2}$ times the square of velocity $\mathbf{v}$, computed as a vector $T = \frac{m}{2}(\mathbf{v} \cdot \mathbf{v})$. This can be computed in any frame, and in terms of components involves the components g_{ij} of the metric tensor as

$$T = \frac{m}{2} g_{ij} v^i v^j ,$$

where v^i are now the components of velocity expressed in the general non-rectangular frame.

In systems involving fluids, one can assign a kinetic energy density. Thus, if ρ is the mass velocity, the *kinetic energy* density (Joules/m^3) is $t = \rho \mathbf{v}\mathbf{v}$.

For relativistic motion, the kinetic energy is the increase in the relativistic mass with motion. We use $E = \gamma m c^2$, where m is the rest mass of the object, c is the speed of light, $c \cong 3 \times 10^{10}$ cm/sec, and $\gamma = \frac{1}{\sqrt{1 - \frac{v^2}{c^2}}}$ is the relativistic dilation factor and $T = E - mc^2$. If v is small compared to c, then Taylor expansion of γ around $v = 0$ gives

$$T = m^2 c \left[1 + \frac{1}{2}\frac{v^2}{c^2} + \frac{3}{8}\frac{v^4}{c^4} \cdots \right] - mc^2$$

which agrees with the nonrelativistic definition of kinetic energy, and also exhibits the first relativistic correction. Since $v \ll c$ in everyday experience, relativistic corrections are not usually observed. However, these corrections are extremely important in many atomic and nuclear processes, and in large-scale astrophysical processes. *See* summation convention.

kinetic temperature   A measure of the random kinetic energy associated with a velocity distribution of particles. *Kinetic temperatures are often given for gases that are far from thermal equilibrium, and the kinetic temperature should not be confused with thermodynamic temperature, even though the two can be expressed in the same units. Kinetic temperatures are most often given in Kelvin, though they are sometimes expressed in electron volts or other energy units. If the velocity distribution is isotropic, the kinetic temperature T is related to the mean square velocity $\langle v^2 \rangle$ by

$$3kT = m \left\langle v^2 \right\rangle ,$$

where m is the particle mass, k is Boltzmann's constant, and the velocity is reckoned in a reference frame in which the mean velocity $\langle \mathbf{v} \rangle = 0$. In a gyrotropic plasma it is common to speak of kinetic temperatures transverse and parallel to the magnetic field,

$$2kT_\perp = m \left\langle v_\perp^2 \right\rangle ,$$

and

$$kT_\parallel = m \left\langle v_\parallel^2 \right\rangle .$$

Note that electrons and the various ion species comprising a plasma will in general have differing kinetic temperatures. *See* plasma stress tensor.

kink   In materials, abrupt changes in the dislocation line direction which lie in the glide plane. In cosmic strings, the discontinuity in cosmic string motion occurring when two cosmic strings intersect and intercommute because segments of different strings previously evolving independently (both in velocity and direction) suddenly become connected. Similar behavior is seen in flux lines in superconductors and in vortex lines in superfluids. *See* cosmic string, intercommutation (cosmic string).

Kippenhahn–Schlüter configuration   *See* filament.

Kirchoff's law   In thermodynamical equilibrium, the ratio between emission coefficient ϵ_ν and absorption coefficient κ_ν is a universal function $B_\nu(T)$ which depends on frequency ν and temperature T:

$$\epsilon_\nu = \kappa_\nu \cdot B_\nu(T) .$$

B_ν is also the source function in the equation of radiative transport. For a black-body, $B_\nu(T)$ is given by Planck's black-body formula. *See* black-body radiation.

Kirkwood gaps   Features (gaps) in the distribution of the asteroids discovered by Kirkwood in 1867. They are located at positions corresponding to orbital periods of the asteroid 4, 3, 5/2, 7/3, and 2 times the period of Jupiter. Because of the resonance with Jupiter such orbits are very rare, and this results in "gaps" in the asteroid distribution.

knoll   In oceanographic geophysics, a cone-shaped isolated swell with a height difference of less than 1000 m from its ambient ocean bottom. Believed to be of volcanic origin.

knot   A nautical mile per hour, 1.1508 statute miles per hour, 1.852 km per hour.

Knudsen flow   The flow of fluids under conditions in which the fluid mean free path is significant, and the behavior has features of both molecular flow and laminar viscous flow.

Knudsen number   In a fluid, the ratio of the mean free path to a typical length scale in the problem, e.g., surface roughness scale. A large *Knudsen number* indicates a free-molecule regime; a small Knudsen number indicates fluid flow.

Kolmogorov scale (Kolmogorov microscale)   The length scale, at which viscous and inertial forces are of the same order of magnitude; the length scale at which turbulent velocity gradients in a fluid are damped out by molecular viscous effects. Kolmogorov suggested that this length scale depends only on parameters that are relevant to the smallest turbulent eddies. Hence, from dimensional considerations, the length scale is expressed by

$$L_K = \eta = \left(\frac{\nu^3}{\epsilon}\right)^{1/4} \left[\text{rad m}^{-1}\right]$$

where ν is the viscosity of the fluid and ϵ is the rate at which energy is dissipated by the smallest turbulent eddies (*see* dissipation rate). Values for η are of the order of millimeters in the ocean and in the atmosphere. This implies that for scales shorter than the Kolmogorov scale L_K, the *turbulent kinetic energy* is converted into heat by molecular viscosity ν. L_K is often referred to as the size of the smallest possible eddies to exist in a fluid for a given level of dissipation and viscosity.

Kolmogorov spectrum   The spectrum of the turbulent kinetic energy follows in the inertial subrange (scales smaller than the energy-containing range of eddies; larger than the Kolmogorov scale) the characteristic so-called $-5/3$-law: $\phi(\text{k}) = 1.56\varepsilon^{2/3}\text{k}^{-5/3}$ (*see* inertial subrange). The name of the spectrum is due to Kolmogorov (1941).

Kolmogorov wavenumber   Defines the wave numbers at which the turbulent kinetic energy is converted into heat by molecular viscosity ν, i.e., $k_K = (\varepsilon/\nu^3)^{1/4}$ [rad m^{-1}] or $k_K = (2\pi)^{-1}(\varepsilon/\nu^3)^{1/4}$ [cycle m^{-1}]. *See* Kolmogorov scale.

Kp index   A 3-hourly planetary geomagnetic index of activity generated in Göttingen, Germany, based on the K indices from 12 Magnetic Observatories distributed around the world. The *Kp index* is calculated by combining these indices using local weightings. The Kp index is often presented as a Bartels musical diagram, a presentation that emphasises the 27-day recurrent nature of much geomagnetic activity. *See* geomagnetic indices.

KREEP   Unusual lunar (and martian) rocks, with unusual amounts of "incompatible elements": K-potassium, Rare Earth Elements, and P-phosphorous, representing the chemical remnant of a magma ocean (caused by impact or volcanism).

Kruskal extension (1960)   The maximal analytic extension of the Schwarzschild space-time by the introduction of a coordinate system in which the coordinate velocity of light is constant. *See* maximal extension of a space-time.

krypton (Kr)   From the Greek kryptos or "hidden." A gaseous element, atomic number 36, one of the noble gasses, discovered by William Ramsay and M.W. Travers in 1898. Its naturally occurring atomic weight is 83.80. Natural krypton is found in the atmosphere at about 1 ppm. Its naturally occurring isotopes are ^{78}Kr, ^{80}Kr, ^{82}Kr, ^{83}Kr, ^{84}Kr, ^{86}Kr. ^{84}Kr is naturally most abundant.

K star   Star of spectral type K. Arcturus and Aldebaran are *K stars*.

Kuiper belt   Region beyond about 35 AU and extending to roughly 100 AU, in the ecliptic, that is the source of most short-period comets. Originally suspected on theoretical grounds in the early 1950s, it was not until 1992 that the first *Kuiper belt* objects were observed. The belt is roughly planar, and it is believed that interactions between Kuiper belt objects and the giant planets cause belt objects to occasionally cross the orbit of Neptune. Objects that have a gravitational encounter with that planet will either be ejected from the solar system, or perturbed into the region of the planets. The Centaur class of asteroids, which orbit the sun in the region between Jupiter and Neptune, as well as the planet Pluto/Charon, are believed to have originated in the Kuiper belt. Unlike the Oort cloud comets, Kuiper belt objects are thought to have been formed *in situ*.

Kuiper belt object, trans-Neptunian object, Edgeworth–Kuiper object   A minor body that resides in the Kuiper belt. Sizes can range up to a few hundred kilometers.

Kuroshio   *See* Japan current.

L

lagoon   A shallow, sheltered bay that lies between a reef and an island, or between a barrier island and the mainland.

Lagrange points   Five locations within a three-body system where a small object will always maintain a fixed orientation with respect to the two larger masses though the entire system rotates about the center of mass. If the largest mass in the system is indicated by M1 and the second largest mass is M2, the five *Lagrange points* are as follows: in a straight line with M1 and M2 and just outside the orbit of M2 (usually called the L1 point); in a straight line with M1 and M2 and just inside the orbit of M2 (L2); in a straight line with M1 and M2 and located in M2's orbit 180° away from M2 (i.e., on the other side of M1) (L3); and 60° ahead and behind M2 within M2's orbit (L4 and L5). In the sun-Jupiter system, the Trojan asteroids are found at the L4 and L5 locations. In the Earth-sun system, the Lagrangian points L1 and L2 are both on the sun-Earth line, about 236 R_E (≈ 0.01 AU) sunward and anti-sunward of Earth, respectively. The other points are far from Earth and therefore too much affected by other planets to be of much use, e.g., L3 on the Earth-sun line but on the far side of the sun. However, L1 (or its vicinity) is a prime choice for observing the solar wind before it reaches Earth, and L2 is similarly useful for studying the distant tail of the magnetosphere. Spacecraft have visited both regions — ISEE-3, WIND, SOHO and ACE that of L1, ISEE-3 and GEOTAIL that of L2. Neither equilibrium is stable, and for this and other reasons spacecraft using those locations require on-board propulsion.

Of the Lagrangian points of the Earth-moon system, the two points L4 and L5, on the moon's orbit but 60° on either side of the moon, have received some attention as possible sites of space colonies in the far future. Their equilibria are stable.

Lagrangian   In particle mechanics, a function $L = L(x^i, dx^j/dt, t)$ of the coordinate(s) of the particle x^i, the associated velocity(ies) dx^j/dt, and the parameter t, typically time, such that the equations of motion can be written:

$$\frac{d}{dt}\frac{\partial L}{\partial\left(\frac{\partial x^i}{\partial t}\right)} - \frac{\partial L}{\partial x^i} = 0 .$$

In this equation (Lagrange's equation) the partial derivatives are taken as if the coordinates x^i and the velocities dx^j/dt were independent. The explicit d/dt acting on $\frac{\partial L}{\partial(\frac{\partial x^i}{\partial t})}$ differentiates x^i and dx^i/dt. For a simple *Lagrangian* with conservative potential V,

$$L = T - V = \frac{1}{2}m\sum\left(dx^i/dt\right)^2 - V\left(x^j\right) ,$$

one obtains the usual Newtonian equation

$$\frac{d}{dt}m\frac{dx^i}{dt} = -\frac{\partial V}{\partial x^i} .$$

Importantly, if the kinetic energy term T and the potential V in the Lagrangian are rewritten in terms of new coordinates (e.g., spherical), the equations applied to this new form are again the correct equations, expressed in the new coordinate system.

A Lagrangian of the form $L = L(x^i, dx^j/dt, t)$ will produce a resulting equation that is second order in time, as in Newton's equations. If the Lagrangian contains higher derivatives of the coordinates, then Lagrange's equation must be modified. For instance, if L contains the acceleration,

$$a^j = \frac{d^2 x^j}{dt^2} ,$$

so that

$$L = L\left(x^i, dx^j/dt, d^2 x^k/dt^2, t\right) ,$$

the equation of motion becomes

$$-\frac{d^2}{dt^2}\frac{\partial L}{\partial\left(\frac{d^2 x^i}{dt^2}\right)} + \frac{d}{dt}\frac{\partial L}{\partial\left(\frac{\partial x^i}{\partial t}\right)} - \frac{\partial L}{\partial x^i} = 0 ,$$

which will in general produce an equation of motion containing third time derivatives. The presence of higher derivatives in the Lagrangian produces higher order derivatives in the equation of

0-8493-2891-8/01/$0.00+$.50
© 2001 by CRC Press LLC

motion, which generalize the terms given above and appear with alternating sign.

The Lagrangian arises in consideration of extremizing the action of a system, and a development from this point of view clarifies many of the properties of the Lagrangian. *See* action, variational principle.

Lagrangian coordinates   In hydrodynamics, physical parameters such as pressure, fluid velocity, and density can be expressed as functions of individual flowing particles and time. In this case the physical parameters are said to be represented in Lagrangian Coordinates (*see also* Eulerian Coordinates). Named after Joseph Louis Lagrange (1736–1813).

Lagrangian coordinates   In fluid mechanics, a coordinate system fixed to the fluid, so that the coordinates of a particular packet of fluid are unchanged in time. In such a frame some of the fluid behavior is easier to compute. However, transforming back to a lab frame may become difficult to impossible, particularly in complicated flows. *See* Eulerian coordinates.

Lagrangian representation   Description of a phenomenon relative to the moving water parcel. Floats, neutral buoys, and deliberately introduced tracers are typical applications to measure currents in the Lagrangian frame. *See* Eulerian representation.

Lagrangian velocity   That velocity that would be measured by tracking a dyed particle in a fluid. *See also* Eulerian velocity.

Laing–Garrington effect   (1988) The higher degree of polarization of the radio lobe associated to the jet, with respect to that associated to the counter-jet, observed in quasars, and to a lower level in radio galaxies. The *Laing–Garrington effect* is straightforwardly explained assuming that the there is no strong intrinsic difference between jet and counter-jet, and that the different surface brightness of the jet and counter-jet is due to relativistic beaming. Then the radio emission coming from the counter-jet is more distant from the observer. The source is expected to be embedded in a tenuous hot thermal medium which depolarizes intrinsically polarized radiation because of Faraday rotation. Radiation from the counter-jet then travels a longer path through the plasma and emerges less polarized.

Lambertian surface   A surface whose radiance, reflectance, or emittance is proportional to the cosine of the polar angle such that the reflected or emitted radiance is equal in all directions over the hemisphere.

Lambert's law   The radiant intensity (flux per unit solid angle) emitted in any direction from a unit radiating surface varies as the cosine of the angle between the normal to the surface and the direction of the radiation.

Lamé constants   Two moduli of elasticity, λ and G, that appear in the following form of Hooke's law:

$$\sigma_{ij} = \lambda \varepsilon_{kk} \delta_{ij} + 2G\varepsilon_{ij}$$

where σ and ε are stress and strain, respectively. Parameter G is also called the shear modulus or rigidity. The *Lamé constants* are related to Young's modulus E and Poisson's ratio v as

$$\lambda = \frac{vE}{(1+v)(1-2v)} \quad \text{and}$$
$$G = \frac{E}{2(1+v)}.$$

laminar flow   A smooth, regular flow in which fluid particles follow straight paths that are parallel to channel or pipe walls. In *laminar flow,* disturbances or turbulent motion are damped by viscous forces. Laminar flow is empirically defined as flow with a low Reynolds number.

Landau damping and instability   In a collisionless plasma, damping or instability associated with the $n = 0$ resonance; the damping of a space charge wave by electrons which move at the wave phase speed and are accelerated by the wave. Landau damping is of importance in space physics and astrophysics as a process for the dissipation of magnetoacoustic waves. *See* magnetoacoustic wave, resonant damping and instability.

Langacker–Pi mechanism   In cosmology, a mechanism that can reduce the number of magnetic monopoles arising from an early phase transition, through a later phase transition that creates cosmic strings linking monopole-antimonopole pairs. Because the cosmic strings are under tension, they pull together the pairs, enhancing the monopole-antimonopole annihilation probability and reducing the monopole density to consistency with cosmological observations.

Langmuir circulation   Wind-induced sets of horizontal helical vortices in the surface waters of oceans and lakes. The counter-rotating vortex pairs appear in series of parallel sets, which form tube-like structures. These structures are aligned within a few degrees of the wind direction, and they are visible at the surface as streaks or lines. These streaks form at the convergence zones of the counter-rotating vortices, where debris or foam floating on the water surface is collected into long, narrow bands. Appearance of the phenomenon requires a certain threshold speed of about 3 ms^{-1}. The horizontal spacing between vortex pairs can range from 1 m up to hundreds of meters, and smaller, more irregular structures can coexist among larger, widely spaced structures. The vortex cells penetrate vertically down to the first significant density gradient (seasonal pycnocline), and their aspect ratio is generally assumed to be about $L/2D$, where L is the horizontal spacing of the vortex pairs and D is their penetration depth. Typical spacing is of order tens of meters.

The generation of Langmuir cells is explained by the widely excepted CL2 model, developed by Craik and Leibovich. This model assumes a horizontally uniform current $U(z)$ and a cross-stream irregularity $u(x, y, z)$, where a right-handed coordinated system is considered with the x-direction pointing down-stream. The irregularity produces vertical vorticity $\omega_z = -\partial u/\partial y$ and a horizontal vortex-force component $-U_s \omega_z \mathbf{e_y}$ directed towards the plane of maximum u, where $U_s = U_s(z)$ is the Stokes drift and $\mathbf{e_y}$ is the unit vector in the y-direction. The vortex-force causes an acceleration towards the plane of maximum u, where, in order to satisfy continuity, the water must sink. Hence, surface water is transported downward at the streaks and upwelling occurs in between. If U decreases with depth and if shear stresses are ignored, conservation of x-momentum along the convergence plane requires that, as the water sinks, u must increase. Thus, the initial current irregularity is amplified, which then further amplifies the convergence. The vertical vorticity is rotated towards the horizontal by the Stokes drift, which results in even increased convergence and amplification of the velocity anomaly. Eventually, the vorticity is rotated completely into the horizontal and forms a set of helical vortices.

Langmuir waves   Fundamental electromechanical plasma oscillations at the plasma frequency. Also called plasma oscillations or space-charge waves. *Langmuir waves* are dispersionless in a cold plasma and do not propagate in a stationary plasma. They are important for solar physics since the Langmuir oscillations are readily converted into electromagnetic radiation.

La Niña   The 1 to 3 year part of the Southern Oscillation when there are anomalously cold sea surface temperatures in the central and eastern Pacific. *See* El Niño, Southern Oscillation Index.

Laplace equation   The equation

$$\nabla^2 \phi = 0 .$$

This equation describes the Newtonian gravitational potential, the electrostatic potential, the temperature field and a number of other phenomena (all in the absence of sources). For instance, in groundwater flow through homogeneous regions under steady-state conditions there is no change in the hydraulic head h with time, and groundwater flow can be described by combining Darcy's law with conservation of mass to obtain the Laplace equation:

$$\frac{\partial^2 h}{\partial x^2} + \frac{\partial^2 h}{\partial y^2} + \frac{\partial^2 h}{\partial z^2} = 0 .$$

Flow nets are graphical solutions to the *Laplace equation*.

lapse rate Vertical change rate of the air temperature. *See* adiabatic lapse rate, moist lapse rate.

Large Angle and Spectrometric Coronagraph (LASCO) A three-part coronagraph on board the SOHO spacecraft designed to provide white light images of the sun's outer corona. LASCO consists of three successively larger occulting disks providing coronal images at 1.1 – 3 $R_\odot$, 1.5 – 6 $R_\odot$, and 3.5 – 30 $R_\odot$, respectively.

Large Magellanic Cloud (LMC) An irregular galaxy in the southern constellation Dorado at Right Ascension $5^h 20^m$, declination $-6°$, at 55 kpc distance. The LMC has angular dimension of $650' \times 550'$, about 10 kpc. It has a positive radial velocity (away from us) of $+ 13$ km/s. Both the *Large Magellanic Cloud* and the Small Magellanic Cloud orbit the Milky Way. On February 24, 1987 supernova 1987A, a peculiar type II supernova, occurred in the Large Magellanic Cloud, the nearest observed supernova since Kepler's supernova. *See* Small Magellanic Cloud, Kepler's supernova.

large-scale Structures in any physical system obtained by ignoring or averaging over small scale features; in oceanography, structures of water properties (such as CTD profiles) with scales larger than tens of meters (finestructure) representing the overall stratification of the natural water body.

Larissa Moon of Neptune, also designated NIV. It was first seen on Voyager photos in 1989. Its orbit has an eccentricity of 0.00138, an inclination of 0.20°, a precession of $143°$ yr^{-1}, and a semimajor axis of 7.36×10^4 km. Its size is 104×89 km, but its mass is not known. It has a geometric albedo of 0.056, and orbits Neptune once every 0.555 Earth days.

Larmor frequency The angular frequency of gyration of a charged particle in a magnetic field; in cgs units, $qB/m\gamma c$, where q and m are the charge and rest mass of the particle, B is the magnetic field strength, c is the vacuum speed of light, and γ is the Lorentz factor. Equivalent terms are gyrofrequency and (nonrelativistically) cyclotron frequency. Named after Joseph Larmor (1857–1942).

Larmor radius The radius of gyration of a charged particle in a magnetic field, equal to the Larmor frequency multiplied by the component of velocity transverse to the magnetic field. Equivalent terms are gyroradius and (nonrelativistically) cyclotron radius. *See* Larmor frequency.

last scattering In physical cosmology, the universe is apparently now transparent to microwave photons, which are observed as the cosmic microwave background, but was not transparent earlier. Thus, those photons underwent a *last scattering*. In cosmological descriptions, the temperature is approximately uniform at any given instant of cosmic time, and is given by $T = T_{now} a_{now}/a$ where a is the cosmological length scale. In a Big Bang universe such as we inhabit, $a_{now} > a$ for any previous epoch. T_{now} is observed to be about 2.7 K. An identical expression holds for the redshift z of previously emitted radiation (ν is the frequency $= c/\lambda$, λ is the wavelength)

$$\nu = \nu \left(\frac{a_{now}}{a} \right) = \nu_{now}(1 + z) .$$

At a redshift of $z \sim 1000$, the temperature was around 3000 K, hot enough for thermal photons to ionize hydrogen to a plasma of protons and electrons. The free electrons strongly scattered photons (Thomson scattering). As the universal temperature dropped through ~ 3000 K, the proton-electron plasma combined (perversely called recombination in the literature) to form neutral hydrogen, which is transparent to thermal photons with $T < 3000$ K. In simple physical descriptions of the universe, this was the last scattering of those photons, which are now observed as a microwave background. Although the physics is unclear, it is possible that sources of ionizing radiation (quasars, supernovae) may have been prolific enough to reionize the universe during the epoch between $z \sim 30$ and $z \sim 5$. In that case the last scattering occurred at the much more recent epoch of $z \sim 5$. The collection of points at which the last scattering of currently observed radiation occurred is called the last scattering surface. *See* quasar, supernova, Thomson scattering.

latent heat *Latent heat* of vaporization is the amount of heat per unit mass which is required to vaporize the liquid; latent heat of fusion is the equivalent definition for melting.

latitude Angular distance from the Earth's equator, counted positive northward and negative southward. It is zero on the equator, 90° at the North Pole, and −90° at the South Pole. Geodetic latitude ϕ, defined as the angle of the local normal to the Earth ellipsoid with the Earth's equatorial plane, is used for most mapping. Geocentric latitude ϕ' for any point P is the angle between the line $\overline{OP}$ from Earth's center O to the point, and Earth's equatorial plane, counted positive northward and negative southward. It is useful for describing the orbits of spacecraft and other bodies near the Earth. Astronomical latitude is the angle between the local vertical (the normal to the geoid) and the Earth's equatorial plane, and is generally within 10" arc of geodetic latitude in value. Geodetic latitude is derived from astronomical latitude by correcting for local gravity anomalies. The maximum difference between geometric and geodetic latitude is about 10' arc, and occurs at mid-latitudes.

If f is the Earth ellipsoid flattening factor $(A - C)/A$ (*see* Earth ellipsoid), then for points on the ellipsoid, the geodetic and geocentric latitudes are related by:

$$\tan(\phi') = (1 - f)^2 \tan(\phi) .$$

For points at a height h, this equation must be replaced by the following: The cylindrical coordinates (ρ, Z) of such a point can be found from:

$$\rho = \left[h + A / \sqrt{1 - (2f - f^2) \sin^2 \phi} \right] \cos \phi$$

$$Z = \left[h + A (1 - f)^2 / \sqrt{1 - (sf - f^2) \sin^2 \phi} \right]$$

$$\sin \phi .$$

The geocentric latitude is then

$$\phi' = \tan^{-1} (Z/\rho) .$$

The inverse of the transformation can be obtained analytically by solving a quartic equation, but the inversion is usually done iteratively.

lattice preferred orientation (LPO) The geometric and spatial relationship of the crystals making up polycrystalline aggregates, in which substantial portions of the crystals are oriented dominantly with one particular crystallographic orientation. *Lattice preferred orientation* in general results from plastic deformation, controlled by dislocation glide.

Laurasia Prior to the opening of the North Atlantic Ocean, North America and Europe were attached and formed the continent Laurasia.

lava Material that is molten on the surface of a planetary body. (When the molten material is still underground, it is usually called magma.) The composition of the *lava* depends on the body — terrestrial lava flows are usually composed of silicate minerals, while those on Jupiter's moon of Io are primarily composed of sulfur and lava flows on icy bodies are composed of volatile elements such as water and ammonia. The features formed by lava depend on various characteristics of the volcanism which produces the lava, including eruption rates, viscosity of the material, and gas content. Low viscosity lavas produced with high eruption rates often give rise to flood lavas (also often called flood basalts on the terrestrial planets) while slightly lower eruption rates give rise to shield volcanos. More viscous lavas will produce shorter flows and more explosive features, such as the composite volcanos and pyroclastic deposits.

law of equal areas Kepler's second law of planetary motion: The line joining the sun and a planet sweeps out equal areas in equal times (even though the speed of the planet and its distance from the sun both vary). This is now recognized as a consequence of the conservation of the angular momentum of the planetary motion.

law-of-the-wall layer Boundary layer zone where the logarithmic velocity profile applies; i.e., $\partial u/\partial z = (u^*/\kappa) \ln(z/z_0)$, where u* is friction velocity, κ is von Kàrmàn constant (0.41 ± 0.01), z is depth, and z_0 is roughness length. In natural waters *law-of-the-wall layer* is typically found in the bottom boundary layer, as well as in the surface layer below the wave-affected surface layer. *See* law-of-the-wall scaling.

law-of-the-wall scaling   In a law-of-the-wall boundary layer zone, where the logarithmic velocity profile applies, a layer's turbulence depends on the surface stress τ_0 (or equivalent $u_* = (\tau_0/\rho)^{1/2}$) and on depth z. Dimensional analysis provides the *law-of-the-wall scaling* relations (often simply called wall-layer scaling) given by the table.

laws of black hole physics   Classical black hole solutions of general relativity (*see* Schwarzschild metric, Reissner–Nordström metric, Kerr–Newman metric) obey four laws which are analogous to the laws of thermodynamics:

1. *Zeroth law:* The surface gravity κ is constant on the horizon.

2. *First law:* The variation of the ADM mass M is given by the Smarr formula

$$\delta M = \frac{\kappa}{8\pi}\,\delta\mathcal{A} + \Omega\,\delta J + \Phi\,\delta Q\,,$$

where $\mathcal{A}$ is the area of the horizon, J is the angular momentum of the black hole, Ω is the angular velocity at the horizon, Q is the charge of the black hole, and Φ is the electric potential at the horizon.

3. *Second law:* No physical process can decrease the area $\mathcal{A}$ of the horizon,

$$\delta\mathcal{A} \geq 0\,.$$

4. *Third law:* The state corresponding to vanishing surface gravity $\kappa = 0$ cannot be reached in a finite time.

From the zeroth and third law, one notices a similarity between the surface gravity $\kappa/2\pi$ and the temperature of a classical thermodynamical system. The ADM mass M behaves like the total energy (first law) and the area $\mathcal{A}/4$ as an entropy (second law). The analogy was further substantiated by the discovery due to S. Hawking that a black hole in a vacuum emits radiation via quantum processes with a Planckian spectrum at the temperature $\kappa/2\pi$. *See* ADM mass, black hole, Kerr–Newman metric, Reissner–Nordström metric, Schwarzschild metric, surface gravity.

layering   *See* double diffusion.

leader spot   A sunspot or sunspot group lying on the western side of an active region complex, i.e., leading in the direction of solar rotation. The largest spots tend to form in the side of the bipolar group that is preceding in the direction of solar rotation. The growth of the longitudinal extent of an active region as it develops is achieved primarily through the rapid forward motion of the preceding spot.

leap second   A second inserted in Universal Time at the end of June 30 or December 31 when, in the judgment of the Bureau International des Poids et Mesures, its addition is necessary to resynchronize International Atomic Time with Universal Time, specifically UT1. The latter is determined by the Earth's rotation, which is generally slowed down by oceanic and atmospheric tidal friction. Leap seconds have been added since June 30, 1972.

leap year   A year in the Julian and Gregorian calendars in which an extra day is inserted (February 29) to resynchronize the calendar with sidereal time.

Leda   Moon of Jupiter, also designated JXIII. Discovered by C. Kowal in 1974, its orbit has an eccentricity of 0.148, an inclination of 26.07°, and a semimajor axis of 1.109×10^7 km. Its radius is roughly 8 km, its mass 5.7×10^{15} kg, and its density 2.7 g cm^{-3}. Its geometric albedo has not been well determined. It orbits Jupiter once every 238.7 Earth days.

lee wave   The atmospheric wave that is generated in the lee of isolated hills. Typical values of the wavelengths observed in the atmosphere are 10 to 20 km.

left-lateral strike-slip fault   The horizontal motion on this strike-slip fault is such that an observer on one side of the fault sees the other side moving to the left.

Lemaître, Georges   (July 17, 1894 – June 20, 1966). Belgian priest, mathematician, physicist, and astrophysicist. He is most famous for his rediscovery of one of the cosmological models

Law-of-the-Wall Scaling Relations

Length	$L_{Ls} \sim z$
Time	$\tau_{Ls} \sim z u_*^{-1}$
Velocity	$w_{Ls} \sim u_*$
Dissipation of turbulent kinetic energy	$\varepsilon_{Ls} \sim u_*^3 (\kappa z)^{-1}$
Temperature (fluctuations)	$\Theta_{Ls} \sim F_{th}/(\rho c_p u_*)$

found by Friedmann (*see* Friedmann–Lemaître cosmological models) and several contributions to relativistic cosmology based on these models. His other achievements include the theory of "primaeval atom" — an overall vision of the origin and evolution of the universe (now replaced by the more detailed and better confirmed observationally Big Bang theory), several papers on physics of cosmic rays (which he interpreted as remnants of what is now called the Big Bang), mathematical physics, celestial mechanics, and automated computing (even before electronic computers were invented).

Lemaître–Tolman cosmological model   Inhomogeneous cosmological model first described in 1933 by G. Lemaître, but now commonly called the Tolman model. *See* Tolman model.

length of day   The *length of day* has decreased with geological time. The rotational period of the Earth has slowed due to tidal dissipation.

lens   A transparent solid through which light can pass, and which has engineered-in properties to deflect or focus the light. *See* gravitational lens.

Lense–Thirring precession   The dragging of space and time by a rotating mass, most evident in cases of rapidly rotating compact objects, such as Kerr black holes. Predicted using the equations of general relativity by J. Lense and H. Thirring in 1918, the effect has been presumably detected by the extremely tiny effects on satellites orbiting Earth, and around distant, rotating objects with very intense gravitational field, such as neutron stars and black holes. The Lense–Thirring effect gives rise to a precessional motion if an object is not orbiting in the equatorial plane of the massive body.

In the vicinity of a rotating black hole, within the ergosphere, the dragging is strong enough to force all matter to orbit in the equatorial plane of the black hole. *See* accretion disk, Kerr black hole.

lepton   A fundamental spin 1/2 fermion that does not participate in strong interactions. The electrically charged *leptons* are the electron, the muon, the tau, and their antiparticles. Electrically neutral leptons are called neutrinos and have very small (or zero) mass. The neutrinos are observed to have only one helicity state (left-handed). Their antiparticles have positive helicity.

leveling (survey)   A geodetic measurement to obtain height difference between two points. The height difference between the two points can be obtained by erecting leveling rods at two distant points from several to tens of several meters, and then by reading scales of the leveling rods using a level that is placed horizontally at the intermediate distance between the two points. Repeating this operation, height difference between substantially distant two points can be measured. Height above the sea at a point can be obtained from control points whose height above the sea has already been determined (bench marks) or from height difference between the point and tide gage stations. This kind of leveling is referred to as direct leveling, whereas trigonometric leveling and barometric leveling are called indirect leveling.

Lewis number   The non-dimensional ratio $L_C = D_C/D_T$, where D_C is molecular diffusivity of substance C and D_T is molecular diffusivity of heat, expresses the ratio of the rate of transfer of molecules of C to that of heat. L_C in water is strongly temperature-dependent and of the order of 0.01. The turbulent *Lewis num-*

ber is expected to approach 1, and the turbulent diffusivities should become independent of the considered substance C, if turbulent transport is due to intermingling neighboring water parcels. For double-diffusion, in water the so-called "apparent" diffusivities of salt and heat are identical only in the asymptotic case of stability ratio $R_\rho \to 1$.

Lg wave When we observe a shallow earthquake at a point with epicentral distance of several hundred to several thousand kilometers, there are some cases that predominant and substantially short-period (2 ∼ 10 s) seismic waves compared to usual surface waves appear. The wave was discovered by F. Press in 1952. The apparent velocity of the *Lg wave* is about 3.5 km/s, and the wave is considered to be generated by multi-reflection of body waves within a continental crustal structure, namely, the granitic layer. Characteristics of Lg wave are similar to Love wave, and the name Lg was taken from the first letter of Love wave, followed by that of granite. It is known that Lg wave does not propagate in oceanic crustal structures. This is thought to be because the oceanic crust is thin and the trapped waves leak into oceanic water.

libration The behavior of an angular variable in a dynamical system which exhibits a small oscillation around an equilibrium configuration. The moon is known to be in synchronous rotation: It completes one orbit around the Earth in 29.5 days while it rotates once on its own axis so that in general only 50% of the lunar surface should be visible from the Earth. However, it exhibits three kinds of *libration,* the libration in longitude about the selenographic origin over $\pm 7°45'$ due to its orbital eccentricity, the libration in latitude about the selenographic origin over $\pm 6°44'$ due to its rotational axis being tilted with respect to the normal to its orbital plane, and the diurnal libration about the selenographic origin over $\pm 1°$ simply due to the rotation of the Earth causing parallax to the earthbound observer. In the orbital motion of an artificial satellite around the Earth, it is known that its argument of perigee librates around 90° when its orbital inclination stays in the neighborhood of 63.4° as the specific shape of the Earth causes gravitational perturbations. An-

other example is a Trojan asteroid the average ecliptic longitude of which librates around that of Jupiter $\pm 60°$ (Lagrange points L_4 and L_5) due to Jovian gravitational perturbations.

Lie derivative Derivative operator associated with a one-parameter group of diffeomorphisms. Let ξ^a be the generator of the group. The *Lie derivative* of a field $\Phi(x)$ of geometric objects at point x of a differentiable manifold is defined by the limit

$$\mathcal{L}_\xi \Phi = \lim_{t \to 0} \frac{\Phi(x) - \Phi^*(x)}{t}$$

where $\Phi^*(x)$ is the image of the field at $x + \xi t$ under the diffeomorphism map. Expressed in components, the Lie derivative of a tensor is

$$\mathcal{L}_\xi T^{ab...p}{}_{qr...t} = \xi^u \frac{\partial}{\partial x^u} T^{ab...p}{}_{qr...t}$$
$$- T^{ub...p}{}_{qr...t} \frac{\partial}{\partial x^u} \xi^a$$
$$- \cdots - T^{ab...u}{}_{qr...t} \frac{\partial}{\partial x^u} \xi^p$$
$$+ T^{ab...p}{}_{ur...t} \frac{\partial}{\partial x^q} \xi^u$$
$$+ \cdots + T^{ab...p}{}_{qr...u} \frac{\partial}{\partial x^t} \xi^u .$$

In theoretical physics the Lie derivative is important in establishing classifications of spacetimes in general relativity (isometry groups, or groups of motions, and the Bianchi types in spatially homogeneous cosmologies); in formulation of the Noether theorem in field theory; in the theory of reference frames; in some methods of generation of exact solutions of Einstein's equations; in some approaches to the second quantization postulate in quantum field theory, and in physical applications of the theory of Lie groups. *See* tensor, vector.

lift coefficient A dimensionless coefficient that appears in the equation for lift force induced by a flow around an object. The lift force is typically vertical, but in general need only be perpendicular to the flow direction.

lifting condensation level The level at which a parcel of moist air lifted adiabatically becomes saturated.

light bridge   Observed in white light, a bright tongue or streak penetrating or crossing a sunspot umbra. The appearance of a *light bridge* is frequently a sign of impending region division or dissolution.

light cone   The 3-dimensional hypersurface in spacetime which is generated by light rays (i.e., light-like geodesics) originating at a single event p (future light cone) and meeting at this event (past light cone); the event p is referred to as the vertex of the *light cone*. In the flat spacetime of special relativity such hypersurfaces with vertices at different events of the flat spacetime are isometric (i.e., are identical copies of each other). In general spacetimes, the hypersurfaces may be curved and may end at folds, or self-intersections. In curved spacetimes, light cones with different vertices are in general not isometric.

The notion of a light cone is important in observational astronomy: the image of the sky seen by an observer at the moment $t = t_0$ is a projection of the observer's past light cone at $t = t_0$ onto the celestial sphere. All the events in our actual spacetime that could have been observed by a given observer up to the moment $t = t_0$ on his clock lie on or inside the past light cone of the event $t = t_0$ of the observer's history, a restatement of the fact that an event q can be observed by an observer O only by the time when the light-ray emitted at q reaches O.

light-curve   A plot of light received from a star vs. time. A variable star is one that has a *light-curve* with peaks and troughs. Often these features are periodic (in the case of eclipsing binaries) or nearly (or quasi-) periodic in the case of variations in an isolated star, such as sunspots (starspots) or stellar oscillations.

For planetary bodies, the variation in brightness due to an object's rotation. A light-curve can be due to shape effects (seeing different cross-sectional areas of an object as it rotates), albedo changes on the surface (seeing different bright and dark patches as the object rotates), or some combination of the two.

light deflection   The curving of light rays by the gravitational field. In the absence of all gravitational fields, i.e., in the flat spacetime of special relativity, light rays are straight. However, in curved spacetime containing a massive body, a light ray passing by the massive body curves toward the body. If a body of mass M is spherical, and the closest distance between the ray and the center of the body is r, then the angle of deflection $\Delta\phi$ is given by the approximate Einstein formula $\Delta\phi = 4GM/(c^2 r)$, where G is the gravitational constant and c is the velocity of light. The formula applies only when $r \gg 2GM/c^2$ (i.e., when the angle $\Delta\phi$ is small). For a ray grazing the surface of the sun, $\Delta\phi = 1.75$ sec of arc. This effect was measured by Eddington in 1919, and more recently confirmed with greater precision for radio-waves from radio-sources; *deflection of light* by the sun has been verified to parts in 1000 by observations of the direction to extragalactic radio sources, as the sun passes near their position in the sky. An observer placed far enough behind the deflecting mass can see light rays that passed on opposite sides of the deflector. For such an observer, the deflector would be a gravitational lens. In the case of distant sources, deflection by intervening galaxies or clusters of galaxies causes lensing, leading to the appearance of multiple images, and of rings or arcs of distorted images. Light rays approaching the very strong gravitational field of a black hole can describe very complicated orbits.

light-harvesting complex   A pigment-protein complex containing chlorophyll that supplies additional energy in photosynthesis.

lightlike current (cosmic string)   *See* electric regime (cosmic string).

lightning   Electric discharge in the atmosphere that produces a lightning flash. Two fundamental types can be distinguished: ground flashes where the discharge occurs between the ground and the cloud with the flash either propagating upward or downward, and cloud flashes or intracloud flashes, where the discharge occurs between clouds and the flash does not strike the ground.

light pollution   Deleterious effects of outdoor lighting on astronomical observations (which require dark sky and which are severely affected by specific spectral lines arising from outdoor lighting using ionized gas lamps).

light year (l.y.)   A unit of distance used in astronomy, defined as the distance light travels in one year in a vacuum. Its approximate value is 9.46073×10^{15}m.

LIGO (Laser Interferometric Gravitational Observatory)   A laboratory for detecting gravitational waves by using laser interferometry. At each of two sites, one at Livingston, Louisiana, the other at Hanford, Washington there is an L-shaped Fabri–Perot light resonator (a version of a Michelson interferometer). The length of the arms is 4 km. The Washington site also contains a shorter ($\ell = 2$ km) interferometer inside the same vacuum pipe. Laser interferometer gravitational wave detection occurs by observing an alteration of the relative length of the two arms of the interferometer caused by the passage of the wave. Operation is expected to start in 2001. *See* GEO, Virgo.

limb   The edge of the visible disk of an astronomical object. For instance, one refers to the *limb* of the sun, or to the limb of the moon.

limb brightening   The apparent brightness increase observed in long wavelength (1 mm) photospheric observations towards the limb of the sun. The transition from limb darkening to *limb brightening* provides a diagnostic for determining the location of the temperature minimum in the photosphere.

limb darkening   The reflectivity of a body with an atmosphere is determined in part by the absorption and reflectivity properites of the atmosphere. For a spherical body, the amount of atmosphere encountered by a light ray will increase towards the limb of the body. The reflectivity of the body will therefore vary from the center of the disk to its limb. For most atmospheres this results in a lowering of the reflectivity, hence the term *limb darkening*. In the sun, the fall off in photospheric brightness towards the solar limb due to the effects of the decreasing temperature of the higher photospheric layers. Predominantly observed in white light

and at wavelengths, $\lambda < 1$ mm, i.e., visible light and infrared.

Lindblad resonance   A resonance in the orbital angular speed developed in non-axisymmetric gravitational potentials, such as the potential in a weakly barred galaxy, or of a planet slightly deviating from spherical symmetry, named after B. Lindblad (1965). In the case of a barred galaxy, the bar supposedly rotates like a rigid body with a "pattern speed" Ω_P and thus gives rise to a non-axisymmetric gravitational potential. In non-axisymmetric potentials, orbits are not generally closed; if the deviation from axisymmetry is small, their orbital motion can be thought of as due to the rotation associated to circular motion plus small radial oscillations. *Lindblad resonances* occur for stars orbiting at angular speed $\Omega = \Omega_P \pm \frac{\kappa}{m}$, where κ is the resonant angular frequency for radial oscillation, usually in the range 1 to $2\,\Omega_P$, and m is an integer. The plus sign and the minus sign define inner and outer Lindblad resonances, respectively. Radii at which such resonances occur are called Lindblad radii. Rings or ring-like features are expected to form at and close to the Lindblad radii.

linearized gravitation   A treatment of weak gravitational fields. The space-time metric is written in the form $g_{ab} = \eta_{ab} + h_{ab}$ where η_{ab} is the Minkowski (i.e., the flat space, zero gravity) metric and the quantities h_{ab}, together with their derivatives, are treated as infinitesimal quantities. This form of the metric is preserved by the infinitesimal coordinate transformation $x^{a'} = x^a + \xi^a$ where the functions ξ^a are infinitesimal. *Linearized gravitation* is widely used in various perturbation problems and is the basis of the post-Newtonian approximation scheme. It also predicts the gravitational radiation from motions accurately described in the Newtonian regime.

linear momentum   Momentum $\mathbf{p} = \gamma m \mathbf{v}$, where

$$\gamma = 1/\sqrt{1 - v^2/c^2}\,,$$

where c = speed of light.

For nonrelativistic motion, $\gamma \sim 1$, and momentum $\mathbf{p} = m\mathbf{v}$.

linear wave theory   Also referred to as Airy wave theory, or Stokes first-order theory. A theory for description of water waves on a free surface, such as waves on the ocean. Referred to as linear because of the fact that higher order terms in the boundary conditions are neglected during derivation of the solution.

line of apsides   In celestial mechanics, the line connecting the two extrema of an elliptical (or nearly elliptical) orbit.

line of nodes   The line connecting the center of mass of the solar system with the point on a planet's orbit where the planet crosses the ecliptic in a northward direction. Can be similarly defined for Earth centered orbits, and for artificial satellites.

LINER   Acronym for Low Ionization Nuclear Emitting Region. *LINERs* are narrow emission line galaxies that show optical and UV spectra with notable differences from classical Seyfert-1 and Seyfert-2 galaxies, namely lower nuclear luminosity and stronger forbidden lines from low-ionization species such as neutral oxygen, singly ionized sulfur, and nitrogen. The ionization mechanism of the line-emitting gas is unclear. An appealing explanation sees LINERs as the lowest luminosity active galactic nuclei, photoionized by a non-stellar continuum weaker than that of more powerful active nuclei. Alternatively, gas may be heated mechanically by shocks, or may be photoionized in dense clouds embedding hot stars of early spectral types. LINERs are frequently observed in the nuclei of both spiral and elliptical galaxies, and might be detectable in nearly half of all spiral galaxies. *See* Seyfert galaxies.

line squall   A squall or series of squalls that occur along a line and advance on a wide front, caused by the replacement of a warmer by colder body of air. *Line squalls* are often associated with the passage of a cold front; it defines the line of the cloud and wind structure. Warmer air is overrun by cold air to produce the line squall. A line squall may extend for hundreds of miles, with a sudden rapid change of wind direction, generally from the southeast or south to west, northwest and north, a rapid rise in barometric pressure and relative humidity, a rapid fall in temperature, and violent changes in weather — heavy cloud, thunderstorms, heavy rain, snow, or hail.

line wing   That portion of a line profile (either absorption or emission) on either side of the central line core. The shape of a line profile is related to the Maxwellian velocity distribution of gas particles (and therefore to the temperature of the gas), the density of the gas, and the natural line width (*See* curve of growth). The larger the spread in velocities, the broader the *line wings;* and the denser the gas, the deeper the central core of the line profile. Even if the gas has a temperature of absolute zero (0 K), the line profile would still have a non-zero width due to the quantum mechanical uncertainty in the energy (Heisenberg's uncertainty principle); this is called the natural line width and has a Lorentzian shape. In realistic circumstances, i.e., gases with T > 0 K, a Maxwellian distribution in velocity broadens the line profile, and is the dominant contributor to the line wings. Other phenomena also cause a broadening of the line profile: pressure broadening, macro- and micro-turbulence in the gas, and bulk rotation of the gas, as in a rotating star. Rotational broadening can be dramatic, depending on the rotation velocity of the star. In early type stars, (O and B) rotation velocities can be as high as 400 km/s. At this speed, the power in the line core is so smeared out into the wings that photospheric lines are completely washed out.

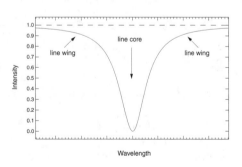

Liouville's theorem (the equation of continuity)   The substrate of points in phase space behaves like an incompressible fluid, that is for an observer moving with the flow, the phase space density is constant. Liouville's theorem is a consequence of the equation of continuity (*see* equation of continuity) in phase space

$$\frac{\partial f}{\partial t} + \nabla_6 \left(f \vec{C} \right) = 0$$

with f being the phase space density, C the speed in phase space, and ∇_6 the divergence in phase space. If there are collisions,

$$df/dt = \left(f_{,t} + F^i f_{,p^i} + p^i f_{,x^i} \right) = Coll,$$

where *Coll* represents the collision of particles located at the same position in physical space, which diminish or replenish particles of momentum p_i, and the subscript commas denote partial differentiation.

liquefaction   When loosely accumulated alluvial sandy ground is shaken by strong earthquake motion, friction acting among sand particles decreases due to increase of pore water pressure, and the ground loses shear resistance, causing loss of the supporting force. This phenomenon is referred to as *liquefaction*. To generate liquefaction, necessary conditions are: the ground is saturated with water; the sand is loosely accumulated; the average diameter of sand particles ranges from about 0.02 to 2 mm. At the time of liquefaction, a large amount of underground water sometimes springs up with sand.

LISA (Laser Interferometric Space Antenna) A proposal to put an array of several laser devices forming an interferometric gravitational wave detector on a heliocentric orbit. The sensitivity of gravitational wave amplitude $h \approx 10^{-23}$ will be achieved in the frequency range 10^{-2} to 10^{-3} Hz. *See* GEO, LIGO, Virgo.

lithophile   Elements that display a strong affinity to combine with oxygen. Such elements are concentrated in the crusts of the terrestrial planets and in stony meteorites.

lithosphere   The outermost rigid layer of a terrestrial planet is called the *lithosphere*. The Earth's lithosphere is comprised of the crust and

the upper part of the mantle. It is the region of a planet where stress and strain result in fracturing of the brittle rocks. The lithosphere is the portion of the Earth that is broken up into plates (*see* plate tectonics) and the motions of these plates give rise to most of the earthquake and volcanic activity on our planet. On Earth the lower boundary of the lithosphere is defined by the temperature below which rocks behave rigidly, typically 1400 K. The typical thickness of the lithosphere is 100 km although it will be much thinner in zones of active volcanism such as ocean ridges. On other planets, such as Venus and Mars, the lithosphere is not apparently broken up into plates, but extensive fracturing indicates that these surfaces also experience tectonic processes.

lithostatic pressure   The *lithostatic pressure* $p_l(y)$ is the pressure within the earth as a function of depth y taking into account the variation of local gravitational acceleration and density ρ as a function of depth.

little ice age   A period between about 1550 and 1860 during which the climate of the middle latitudes became generally colder and there was a world-wide expansion of glaciers. However, it was not always colder during this period. There were three cold subperiods. At 55°N Europe, they are 1541–1680, 1741–1770, 1801–1890. In China, they are 1470–1520, 1620–1720, and 1840–1890. An explanation for the *little ice age* is very low solar activity during the Maunder Minimum period. Little ice age effects have been recorded in the Alps, Norway, and Iceland, where farm land and buildings were destroyed. There were times of especial severity during the early 1600s when glaciers were particularly active in Chamonix valley, in the French Alps.

littoral   A term used to refer to the coastal region that is affected by wind and wave-driven sediment transport.

littoral barrier   A physical barrier to longshore sediment transport. A groin, placed to try to stabilize a shoreline, represents such a barrier.

littoral cell   A littoral region that does not receive or lose sediment to adjoining upcoast and downcoast areas. Sediment is transported within the *littoral cell,* and may enter it via a source such as a river, or leave via a submarine canyon or other sink.

littoral current   A mean flow of water within the littoral zone, in the longshore direction. *See* longshore current.

littoral drift   A term used to denote the sediment transport along a coast by wave action. *See* longshore sediment transport. Often measured in units of yd^3/y or m^3/y.

littoral drift rose   A polar graph that illustrates the potential longshore sediment transport rate at a site as a function of incident wave energy.

littoral transport   Transport of sediment within the littoral zone. Generally used to refer to sediment transport in the longshore direction. *See* longshore sediment transport.

lobe   In general, a roundish division or projection of an object. *See* lobe dominated quasars, lobes, high latitude.

lobe dominated quasars   High luminosity, radio-loud active galactic nuclei whose radio emission is dominated by extended lobe emission. *Lobe dominated quasars* have radio power and morphologies similar to those of Fanaroff–Riley class II radio galaxies. In quasars, the jet appears one-sided, and both core and jet have higher luminosity than in Fanaroff–Riley II galaxies. *See* Fanaroff–Riley radio galaxies.

lobes, high latitude   (also known as tail lobes) Two regions extending into the tail of the magnetosphere (the magnetotail), located north and south of the plasma sheet. Each tail lobe contains a bundle of magnetic field lines connected to one of the Earth's polar caps: the lines of the northern lobe lead into the region around the north magnetic pole, while those of the southern lobe come out of the region near the southern pole.

The magnetic field of the tail lobes is relatively strong and nearly uniform — about 20 to 30 nano Tesla (nT) near Earth, diminishing to 9 to 10 nT in the distant tail. Their ion density is very low, around 0.01 ion/cm^3, and they seem to stretch well past 100 to 200 R_E, at which distance they become infiltrated by plasma sharing the flow of the solar wind (possibly originating in the plasma mantle). The magnetic energy stored in the lobes is considerable and is widely believed to be the main energy source of substorms.

local acceleration of gravity   The force per unit mass of gravitational attraction toward the Earth or other large local concentration of matter, $\mathbf{F}_{grav}/m$. For distance r from a spherical "Earth" of mass $M_\oplus$ the local acceleration of gravity is

$$\mathbf{g} = -\hat{z}GM_\oplus/r^2 \,,$$

where $\hat{z}$ is a vertical unit vector.

Local Group   A group of approximately 30 galaxies, which includes the galaxy and its closest neighbor galaxies within a distance ≈ 1 Mpc. Most members are thought to form a gravitationally bound system; the *Local Group* is therefore the closest example of a cluster of galaxies. The brightest members are our galaxy, the spiral galaxy Messier 31 (the Andromeda galaxy), and the Sc spiral M33, although the majority of galaxies belonging to the Local Group are dwarf galaxies, either dwarf spheroidal or irregular galaxies. With present-day instruments, several galaxies of the Local Group can be resolved into stars.

local thermodynamic equilibrium (LTE)
The assumption that a localized volume of emitting gas is in thermodynamic equilibrium such that the source function for the emitted radiation is given by the Planck function at the local temperature. *LTE* can be assumed to apply to plasma which is sufficiently dense that most of the photons are absorbed and thermalized before they travel a distance over which the temperature changes considerably.

local topological defect   In cosmic topological defects formed as the result of the breakdown of a gauge (or local) symmetry, an important role is played by the gauge fields. In fact, these help to compensate the growing energy gradients of the Higgs field far away from the defect. In this way they achieve the confinement of the energy of the configuration to a small radius given by the Compton wavelengths of the Higgs field ($\propto m_H^{-1}$) and the gauge field ($\propto m_{gauge}^{-1}$), which form the core of the defect. *See* Abelian Higgs model, Nielsen–Olesen vortex, t'Hooft–Polyakov monopole.

loess   An unstratified windblown deposit composed of loosely arranged silt-sized particles (1/256 to 1/16 mm in diameter). *Loess* deposits occur primarily in semiarid and temperate zones, never being found in tropical regions or areas covered by ice in the last ice age. Loess deposits provide excellent soil for agriculture and some of the most productive grain-raising areas (such as the Great Plains of the U.S.) are located in loess deposits. Most experts consider loess to be windblown sediment associated with glacial activity — glacial and accompanying fluvial activity produced the small particles, which are then moved by wind to their current location. Loess usually occurs as plains, although some is also found as mantling deposits on mountains and hills. Loess is found on Earth primarily in the $24° - 55°N$ and $30° - 40°S$ latitude ranges, and has been suggested to occur on Mars as well.

Long Duration Event (LDE)   A solar flare with a long gradual phase mostly associated with coronal mass ejections. When seen in soft X-rays, such flares are composed of a small number of prominent loops of order 10^5 km in length.

longitude, terrestrial   The *longitude of a terrestrial* point P is an angle between a meridian through P and the Greenwich meridian, measured positive eastward and negative westward. Longitudes meet at the $\pm180°$ meridian, where the longitude is discontinuous. The International Date Line runs close to this meridian, but does not always coincide, zigzagging a bit in the interest of keeping civil time zones continuous within certain political domains. Longitude is defined on other planetary bodies by choosing a prime meridian.

longitudinal invariant   An adiabatic invariant associated with the bounce motion of trapped particles, equal to $I = \int p_{||} ds$, where $p_{||}$ is the momentum component parallel to the field line, ds is the element of length along the field line, and the integration is usually between mirror points.

The value of I determines onto which of the neighboring field lines a particle trapped along a given line will drift, averaged over its motion along the given line: namely, to that of the neighboring lines, for which the integral I between the particle's mirror points has the same value as before (the mirror points are determined by the magnetic moment, which is also an adiabatic invariant).

longitudinal wave   In mechanical systems, a propagation of a signal in which the microscopic motion is in the direction of the wave motion. Sound propagating in ordinary isotropic fluids (e.g., air, water) constitutes a *longitudinal wave*.

long period variables   *Long period variable* stars (LPVs) are either in the red giant stage or the asymptotic giant branch stage of stellar evolution. The variation of light, with typical periods of 400 days or longer, is caused by pulsations of the star's atmosphere due to a pumping mechanism just below the photosphere, where partial ionization of H or He causes changes in opacity, trapping radiation and then releasing it, like a steam engine. LPVs are classified as SR (semi-regular, with periods of typically 200 days and light variations less than 2.5 magnitudes in the visible band), Miras (regular pulsators with generally longer periods than the SRs and deep amplitudes of pulsation causing light variations of greater than 2.5 magnitudes and often as much as 4 or 5 magnitudes), and Lb (irregular variables). The conditions for this type of pulsation require low surface gravity (low density in the atmosphere) and cool temperatures, thus limiting the LPVs to cold red giants or asymptotic giant branch stars. Most are of spectral type M, N, R, or S.

longshore bar   A bump or rise in the seafloor in the vicinity of a wave breakpoint. A beach profile may have several *longshore bars*. They run generally parallel to the coastline.

longshore current   A mean flow of water, in or near the vicinity of the surf zone, in a direction parallel to the coast.

longshore sediment transport Transport of sediment in a direction parallel to the trend of the coast. *See* littoral transport.

long slit spectroscopy A technique employed to obtain spectra of extended objects, such as galaxies or planetary nebulae. The spectrograph aperture on the focal plane of the telescope is limited by a slit, whose width is typically a few seconds of arcs or less, and whose height may cover an angular size of several minutes of arc. Only light coming from the narrow strip defined by the slit is allowed to enter the spectrograph to avoid contamination by adjacent strips: nearby sources could produce spectra that would overlap spatially on the detector. *Long slit spectroscopy* has been employed in the measurement of continuum, absorption, and emission lines from every extended object. An example is the construction of radial velocity and rotation curves of galaxies. *See* velocity curve.

look-back time The finite speed of light means that objects are seen as they were at some time prior to the observer's time of observation. Thus T, the *look-back time*, is $T = d/c$, where d is the distance to the observed object, and c is the speed of light.

Lorentz boost Lorentz transformation of space-time coordinates from one system of reference to another moving at a constant velocity with respect to each other. These transformations distinguish themselves from general Lorentz transformations in that they do not include rotation of spatial coordinates. Named after Hendrik Lorentz (1853–1928). *See* Lorentz transformation.

Lorentz factor, γ The quantity

$$\gamma = \frac{1}{\sqrt{1 - \frac{v^2}{c^2}}}$$

where v is the speed and c is the speed of light. γ is an indicator of special relativistic effects and enters into length contraction and time dilitation, for instance.

Lorentz–Fitzgerald contraction The decrease of the length of a physical body when measured in a uniformly moving reference system rather than in the reference system of the body, as calculated by the Lorentz Transformations in the Special Theory of Relativity. Named after Hendrik Lorentz and George Francis FitzGerald. *See* coordinate transformation in special relativity.

Lorentz force equation Equation describing the force on a charged particle moving in specified electric (**E**) and magnetic (**B**) field, with velocity **v**:

$$\mathbf{F} = q(\mathbf{E} + \mathbf{v} \times \mathbf{B}),$$

where this is a vector equation; the charge is in coulombs, the electric field has units of Volt-meters, and the magnetic field is measured in Tesla. The presence of the cross product $\times$ indicates that the magnetic force is orthogonal to both the direction of the magnetic field, and to the direction of motion of the charged particle.

Lorentzian metric The metric of a four-dimensional manifold with one negative and three positive eigenvalues, thus of signature $(-, +, +, +)$ (or one negative and three positive). An example is Minkowski space-time, having the metric $ds^2 = -c^2 dt^2 + dx^2 + dy^2 + dz^2$ written in Cartesian coordinates, where c is the speed of light. The space-time model of general relativity is also one with a *Lorentzian metric*.

Lorentz invariance The invariance of physical expressions under Lorentz transformations from one coordinate system to another coordinate system moving uniformly with respect to the first. In special relativity all physical laws must be *Lorentz invariant*. Named after Hendrik Lorentz (1853–1928). *See* coordinate transformation in special relativity.

Lorentz transformation In special relativity, the coordinate transformations relating distance and time measurements in two reference systems ("frames") in relative motion. If written in rectangular coordinates $\{t, x, y, z\} = \{x^0, x^1, x^2, x^3\}$ the relation between two frames moving relatively along the x-axis is:

$$\begin{pmatrix} t' \\ x' \\ y' \\ z' \end{pmatrix} = \begin{pmatrix} \gamma & \frac{-\gamma v}{c^2} & 0 & 0 \\ -\gamma v & \gamma & 0 & 0 \\ 0 & 0 & 1 & 0 \\ 0 & 0 & 0 & 1 \end{pmatrix} \begin{pmatrix} t \\ x \\ y \\ z \end{pmatrix}$$

where $\gamma = \dfrac{1}{\sqrt{1-\frac{v^2}{c^2}}}$, the unprimed coordinates relate to the first frame, and the origin $0'(x' = 0, y' = 0, z' = 0)$ in the primed frame is moving at speed v in the $+x$ direction, as measured in the unprimed frame. This is typically written in a more compact way in matrix notation as

$$\mathbf{x}' = \Lambda(v)\mathbf{x} .$$

Here x' stands for the column vector, and Λ is the Lorentz matrix. The inverse to Λ is the transformation with opposite velocity $-v$:

$$\mathbf{x} = \Lambda(-v)\mathbf{x}' .$$

By writing out the transformations we obtain:

$$\begin{aligned} x' &= \gamma(x - vt) \\ t' &= \gamma\left(t - \frac{v}{c^2}x\right) . \end{aligned}$$

The first of these is like the Newtonian expression except for the factor $\gamma = \dfrac{1}{\sqrt{1-\frac{v^2}{c^2}}}$, which is very near unity for most terrestrial motions ($v \ll c$). The very surprising feature of the second is that time as measured in relatively moving frames has an offset that differs at different positions. Further, there is a rate offset proportional to γ. Again, these effects are noticed only when dealing with relative motions near the speed of light $v \sim c$.

For general motion not along an axis:

$$\begin{aligned} t' &= \gamma\left(t - \frac{\mathbf{v} \cdot \mathbf{x}}{c^2}\right) \\ \mathbf{x}' &= \mathbf{x} + \mathbf{v}\left[\frac{(\gamma - 1)}{v^2}(\mathbf{v} \cdot \mathbf{x}) - t\right] . \end{aligned}$$

Here the $\mathbf{x}$ are 3-vectors, and the complicated form of the expression for $\mathbf{x}'$ is just a separation and separate transformation of components parallel and perpendicular to the motion.

If one chooses differential displacements dx^i and dt corresponding to motion at the speed of light, then

$$0 = -c^2 dt^2 + \delta_{ij} dx' dx^i \qquad i, j = 1, 2, 3 ;$$

summed on i, j; ($\delta_{ij} = 1$ if $i = j$; 0 otherwise). The *Lorentz transformations* ensure that this statement remains true, and of exactly the same form (and c has the same value) when expressed in the "primed" frame. Thus, the speed of light is

$$c^2 = \delta_{ij} \left(\frac{dx^i}{dt}\right)\left(\frac{dx^j}{dt}\right) .$$

In this situation, the numerical value is unchanged when carrying out such transformations:

$$c = 29979245620 \times 10^{10} \text{ cm/sec}$$

in every observation. *See* coordinate transformation in special relativity.

Lorentz transformation [electric and magnetic fields]   *See* electromagnetism, Lorentz transformation and.

Lorentz transformation [implications]   An important implication is that time is not an absolute variable; that is, its value varies for different reference systems. This leads towards a four-dimensional space-time coordinate system. That is, when transforming observations from one reference system to another, not only must one transform the three spatial coordinates but also time. Direct consequences of the Lorentz transformation are Time Dilation (*see* time dilation) and Length Contraction (*see* Lorentz–Fitzgerald contraction). Another important aspect of Lorentz transformations is that the equations that describe the relationship between electric and magnetic field, their causes and effects, are invariant under them. Special relativity postulates that physical laws must remain unchanged with respect to two uniformly moving reference systems, and thus they must be invariant under Lorentz transformations. This postulate leads to Relativistic Mechanics, which on one hand reduce to Newton's Mechanics at low velocities, and on the other hand are invariant under Lorentz transformations. Thus, one finds that special relativity presents a consistent theoretical groundwork for both mechanics and electromagnetism.

Love numbers   These are numbers that relate the elastic deformation of the Earth to applied deforming forces. If a potential V_n is applied to the Earth in the form of a spherical harmonic of degree n (for example, the tidal potential due to the moon, which is a zonal harmonic of degree 2), then if the Earth were a fairly inviscid fluid it would quickly adopt the shape of the new equipotential: locally, the Earth's surface would move vertically and sideways. The new potential is not the simple sum of the old potential and the deforming potential: There is an additional potential due to the deformation of the Earth by the applied potential. h_n is the ratio of the height of the resultant elastic deformation to that of the deforming potential, l_n is the equivalent ratio for the horizontal displacement, and k_n is the ratio of the additional potential associated with the Earth's deformation to the deforming potential. There are additional *Love numbers* h'_n, k'_n, and l'_n associated with surface loading (e.g., by an ice sheet), and also h''_n, k''_n, and l''_n that represent the effect of shear forces on the surface of the Earth (such as winds). The fact that the Earth rotates and is not spherically symmetric introduces complications.

Love wave   A horizontally polarized (SH) elastic surface seismic wave. *Love waves* propagating near the free surface of a two-layer medium, in which the shear wave speed in the lower layer is greater, are a result of the multiple reflection of the horizontally polarized shear wave from the layer interface beyond the critical angle of refraction and from the free surface. Love waves are dispersive, with the phase velocities ranging between the shear wave speeds of the two layers. Love waves also occur as guided waves in an embedded layer of low shear wave speed.

lower hybrid waves   Electrostatic ion oscillations at a frequency intermediate to the electron extraordinary wave (high frequency) and the magnetosonic wave (low frequency). Nonlinear wave processes at the lower hybrid resonance frequency are extremely important in transferring energy between different particle populations and fields in astrophysical plasmas.

lower mantle   The region of rock in the Earth's interior reaching roughly 3500 to 5700 km in radius.

Lowes power spectrum   The power of the geomagnetic field, as a function of spherical harmonic degree. The magnetic field is commonly written as the gradient of a potential $\mathbf{B} = -\nabla V$ which is then expanded in Schmidt quasi-normalized spherical harmonics:

$$V = r_e \sum_{n=1}^{\infty} \sum_{m=0}^{n} \left(\frac{r_e}{r}\right)^{n+1}$$
$$\left[g_n^m \cos m\phi + h_n^m \sin m\phi\right] P_n^m (\cos\theta)$$
$$+ r_e \sum_{n=1}^{\infty} \sum_{m=0}^{n} \left(\frac{r}{r_e}\right)^{n}$$
$$\left[q_n^m \cos m\phi + s_n^m \sin m\phi\right] P_n^m (\cos\theta)$$

where r_e is the radius of the Earth and the coordinate system is the spherical system (r, θ, ϕ). Terms in g_n^m and h_n^m represent fields generated by internal sources, and terms in q_n^m and s_n^m represent fields generated by external sources. The radial portion of the field is $B_r = -\partial V / \partial r$. The Lowes power spectrum R_n is usually defined separately for the internal and external fields in terms of the mean square value of the field at the Earth's surface. For the internal field:

$$R_n = (n + 1) \sum_{m=0}^{n} \left(g_n^{m\,2} + h_n^{m\,2}\right)$$

and for the external field:

$$R_n = n \sum_{m=0}^{n} \left(q_n^{m\,2} + s_n^{m\,2}\right) .$$

The *Lowes power spectrum* for the internal field is consistent with roughly equal power for each degree between degrees 2 and 13 when the power spectrum is extrapolated down to the core-mantle boundary, and roughly equal power for degrees greater than around 15 at the Earth's surface.

Lowest Useable Frequency (LUF)   Defined as the minimum operating frequency that permits acceptable performance of a radio circuit between given terminals at a given time under specified working conditions. The LUF is determined by the absorption, the radio noise background, the radio frequency interference, and system parameters such as the transmitted power and antenna gains. *See* ionospheric radio propagation path.

Low frequency radio emission from planets

1. Mercury: No radio emissions have been detected because there is only a very weak magnetic field.

2. Venus: No radio emissions have been detected coming from Venus due to a lack of a magnetic field. Venus does have an ionosphere, however, and orbiters have detected "whistlers" there.

3. Earth: In addition to all the manmade radio emissions, Earth emits a natural radio emission called Auroral Kilometric Radiation (AKR). It is triggered by the interaction of Earth's magnetic field with the solar wind and ranges in frequency from about 30 to 800 kHz (wavelength 10 to 0.4 km).

4. Mars: Mars does not have a global magnetic field, only a patchy remnant field at places near the surface. Mars does not have radiation belts; therefore, it does not emit any *low frequency radio emission.*

5. Jupiter: Four well-established bands of planetary radio emissions have been established which correspond to spectral peaks in the Jovian emission: kilometer wavelength radiation (KOM), hectometer wavelength (HOM), decameter wavelength (DAM), and decimeter wavelength (DIM). The KOM frequencies range from 10 to 1000 kHz corresponding to wavelengths of 30 to 0.3 km, respectively. The DAM frequencies range from about 3 to 40 MHz (100 to 7.5 m); the HOM ranges from 300 to 3000 kHz (1 to 0.1 km); the DIM ranges from 100 MHz to 300 GHz (3 m to 1 mm).

The spectral peak intensity of the radiation occurs at about 8 MHz in the powerful and bursty DAM radio emissions. These emissions occur in the plasmasphere surrounding Jupiter from interactions with the satellite Io and with the sun. They are the only extraterrestrial planetary radio sources capable of being observed with ground-based radio telescopes. All other sources must be observed from space due to the absorptive properties of the Earth's ionosphere. The HOM is more continuous and is triggered by particle-magnetic field interactions in the plasmasphere surrounding Jupiter. The KOM emission is found to be coming from plasma interactions within Jupiter's magnetosphere and from the relatively dense torus of plasma that surrounds Jupiter at the orbit of the satellite Io. The torus arises from volcanic eruptions from the surface of Io and the ejected particles get ionized by the solar UV radiation and interact with Jupiter's magnetic field triggering various radio emissions. The DIM radio emission is caused by relativistic electrons in Jupiter's inner magnetosphere. These particles are trapped in belts that are similar to Earth's Van Allen radiation belts. The high magnetic field strengths cause the particles to be accelerated to high speeds and to emit radio waves at high frequencies.

6. Saturn: Saturn emits radio waves at kilometric wavelengths and the emission is called Saturnian Kilometric Radiation (SKR). SKR is believed to be similar to Earth's AKR triggered by interactions of the planet's magnetic field with the solar wind. The emission occurs over a frequency range of about 20 to 1200 kHz.

7. Uranus: Uranus also has a magnetic field and interactions with the solar wind cause Uranian kilometric radiation (UKR) over a frequency range of about 60 to 850 kHz.

8. Neptune: The Neptunian magnetic field has a surface field strength similar to that of Uranus and the Earth (approx. 0.1 to 1.0 Gauss). Because of the presence of the magnetic field, and thus a magnetosphere, interactions with the solar wind trigger Neptunian Kilometric Radiation (NKR) from about 20 to 865 kHz.

9. Pluto: No spacecraft have flown near Pluto; therefore, no magnetic field has been directly measured. No radio emissions have been detected nor are any expected.

low-velocity zone   The region of the Earth beneath the lithosphere where seismic velocities are low. The asthenosphere is a *low-velocity zone.*

luminosity class   The classification of a star based on the appearance of its spectrum, usually the relative strengths of emission and absorption lines, compared to the spectra of standard stars of the classification system. The classical Morgan and Keenan (MK) system is a two-dimensional system: spectral type and luminosity class. In the MK system, classification was defined at moderate (3 Å) resolution in the blue region (4000 to 5000 Å).

The primary luminosity classes range from I (supergiant) to III (giant) to V (dwarf or main-sequence), with classes II and IV as intermediate cases. The spectral criteria that define *luminosity class* are primarily a function of surface gravity (reflecting atmospheric density and envelope size), with the larger, supergiant stars having a lower surface gravity and less presure broadening of the lines.

luminosity distance Distance to an astronomical object obtained from the measured flux once its intrinsic luminosity is known: if L is the luminosity of the object and F its measured flux, then $d_L = \sqrt{L/4\pi F}$. Other definitions of distances exist. They all coincide for a non-expanding Euclidean universe, but they differ in the real expanding universe at redshifts close to unity and larger. Luminosity distance can be easily related to d_A, the angular diameter distance: $d_L = d_A(1 + z)^2$. *See* magnitude.

luminosity function of galaxies A function specifying the number density of galaxies per unit luminosity (or, equivalently, per unit magnitude). From counts and measurements of the integrated magnitude of galaxies in rich clusters, P. Schechter derived the following law:

$$\Phi(L) = \text{const} \times \left(L/L^*\right)^{\alpha} \exp\left(-L/L^*\right) ,$$

where L is the galaxy luminosity, $L^* \approx 3 \times 10^{10} L_{\odot}$ is a turnover luminosity in units of solar luminosity, and α is found to be in the range ≈ -1.0 to -1.5. This law suggests that the most luminous galaxies are the rarest, and that the number of galaxies increases with decreasing luminosity. According to Schechter's law, a galaxy population in a magnitude limited sample — where galaxies are counted down to a fixed limit of brightness — is dominated by galaxies of luminosity near to L^*. On the contrary, in a volume limited sample — where ideally all galaxies are identified up to a fixed distance — the faintest galaxies would be by far the most numerous, and would contribute to the vast majority of light. Recent results suggest that Schechter's law predicts even fewer faint galaxies than observed.

luminous blue variables The brightest known single stars, near absolute magnitude -10. They have already evolved slightly off the main sequence and typically have vigorous stellar winds that will gradually reduce their masses. The extreme luminosities and winds result in erratic variability, including unpredictable outbursts. A well-known example is Eta Carinae, which, in 1843, brightened to become the second brightest star in the sky. It has been fading ever since (though with occasional recoveries) and is no longer a naked-eye star. *Luminous blue variables* probably evolve to Wolf Rayet stars.

luminous efficiency Commission de l'Eclairage (CIE), 1924. A roughly Gaussian curve centered at 555 nm with value unity there, decreasing to zero at 425 nm and at 700 nm, meant to represent the response of human vision to the same physical flux at different wavelengths.

Luminous power (lumens) For monochromatic radiation, 683 times radiant power(watts) times luminous efficiency. For a mixture of wavelengths, the sum of the luminous powers for the individual wavelengths. *See* Abney's law of additivity, luminous efficiency.

lunar eclipse A darkening of the full moon because the Earth is directly between the sun and the moon; the shadow of the Earth darkens the moon.

lunar highlands The lighter-color areas on the moon, classically called the "Terrae" (plural of the Latin, terra, "land"), which is usually translated as "highlands" or uplands. Portions of the highlands were removed in massive meteor impacts and were subsequently filled with younger, low lying volcanic flows to form the lunar maria. Compared to the maria, the highlands are an older surface, not altered since the heavy cratering era of planetary formation, rough and broken on a large scale. Highlands rise higher than the maria. (When the moon is in a phase where sunlight hits it at an angle, such as First Quarter phase, it is possible to see, using binoculars, the highlands casting shadows on the lower maria, and inside craters.) Chemically, the *lunar highlands* differ from the lunar lowlands mainly in their concentrations of the metal oxides, which are richer in calcium (Ca) and aluminum (Al) than are the lowlands.

Highlands geology consists of craters 1 m to more than 1000 km across, and overlapping layers of material ejected from craters. Rocks brought back from the highlands vary in age (the time since it last solidified from molten liquid) between 3.84 and 4.48 billion years old. The highlands are the oldest parts of the moon, not having been resurfaced by mare lava flows.

lunar libration   The apparent rocking of the orientation of the moon due to geometrical and orbital relations between the Earth and the moon. These arise from (a) the ellipticity of the moon's orbit ($e = 0.055$), (b) the slight non-alignments of the lunar orbit: the lunar equator is tilted from the ecliptic by $1°\,32'$ from the ecliptic and the lunar orbital plane is tilted $5°\,9'$ from ecliptic; thus the lunar equator is tilted from its orbital plane by $6°\,41'$, (c) parallax from the Earth (different viewpoints between moonrise and moonset). This leads to longitudinal librations of about $7.7°$ due to the ellipticity, latitudinal librations of about $6.7°$ due to the tilt of the lunar equator from the ecliptic, and about $2°$ total due to parallax from the Earth. As a result, about $15°$ additional surface of the moon is visible at various times from the Earth.

Lunar Maria   Darker low lying areas on the moon which are large basaltic lava flows. The maria (latin plural of *mare,* "sea") are younger (they have fewer impact craters), and are richer in magnesium (Mg) and iron (Fe) than are the highlands. The maria were apparently formed by very large impacts and clearly represent flow into low lying regions. They solidified near or after the end of the cratering epoch because little subsequent cratering occurred. Although the impacts forming the maria may have occurred about 4 billion years ago, the maria were apparently kept molten by heating from heavier radioactive elements which flowed up to the surface following the impacts that formed them. Mare rocks have been measured to be between 3.15 and 3.77 years old, which provides an estimate of the time of the solidification of the mare. The maria occur essentially only on the side of

the moon facing the Earth; the far side is essentially all highlands.

lunar mascons   The Lunar Maria are generally associated with strong positive gravity anomalies. Because the maria are low-lying, the gravity anomalies are attributed to buried positive density anomalies or loads. These loads are known as mascons.

lunar meteorites   Currently there are 13 meteorites believed to be from the moon. After the discovery of young meteorites believed to be from Mars (*see* martian meteorites), scientists began searching for *lunar meteorites.* Since Martian meteorites are believed to be ejected from Mars by meteorite impact, it was argued that the lower escape velocity from the moon should allow lunar meteorites to be common as well. The first meteorite recognized to be from the moon was discovered in 1981–1982. The lunar meteorites are similar to lunar rocks returned by the Apollo and Luna missions, reflecting both basaltic (from the Lunar Maria) and anorthositic (from the lunar highlands) compositions. However, none of the lunar meteorites is exactly identical to the chemical composition of the returned lunar samples, which indicates the lunar meteorites are from locations on the moon not visited by the human and robotic explorers.

Lunar Prospector   A NASA Discovery Mission launched to the moon on January 6, 1998, that marked the first NASA Moon mission in 25 years. It uses a complement of five instruments that address questions concerning the moon's resources, its structure, and its origins. During its primary 1 year polar orbiting mission, data returned suggesting that the craters at the north and south poles contain up to 300 million metric tons of frozen water, as implied by the detection of hydrogen. The ice is probably in the form of frost mixed with lunar soil. The general consensus is that the water was supplied over billions of years via cometary bombardment. Because the sun makes a shallow angle at the moon's poles, the bottoms of the polar craters never see sunlight and so trap the frozen water. The amount of ice may indicate the frequency of cometary hits and how long the poles have been in their present orientation.

Prospector then dropped to an altitude of 30 km above the surface and took additional data at significantly higher resolutions that pertain to hydrogen concentrations at the north and south poles, and the moon's magnetic and gravity fields. Global maps of the moon's elements will also benefit from these high-resolution data.

Lunar Rille   Trenchlike or cracklike valley, up to several hundred kilometers long and up to a few kilometers wide, on the moon's surface. Some rilles may be relatively straight. However, many rilles may be extremely irregular with meandering courses ("sinuous rilles"). They are lava channels and collapsed lava tubes which were probably active during the maria formation (many have a sinuous appearance resembling river meanders, so were once thought to be dry river beds). These sinuous rilles typically begin at craters, and end by fading into the mare surface. Channels are U-shaped or V-shaped. Scales are much larger than equivalent terrestrial lava tubes, presumably because of differences in physical conditions and/or outflow rates.

Lux   An illuminance equal to one lumen per square meter.

Lyapunov exponent   The time for the distance between two chaotic trajectories to increase by a factor e when the initial conditions are altered infinitesimally. It indicates how fast nearby trajectories diverge and how unpredictable such trajectories become. A dynamical system is said to be chaotic if it possesses sensitive dependence on the initial conditions. Consider a one-dimensional mapping $x_{n+1} = f(x_n)$. That the mapping is chaotic means that randomly chosen very close initial values x_0 and $x_0' = x_0 + \delta x_0$ generate totally different trajectories after a long time. Let x_n evolve from x_0 while x_n' evolves from x_0'. For a chaotic map an exponential increase of the difference $|x_n - x_n'| \sim |\delta x_0| \exp(\lambda n)$ is observed for an infinitesimal $|\delta x_0|$. The long-time average of the separation rate λ is called the *Lyapunov exponent* of the map. Chaotic trajectories correspond to a positive Lyapunov exponent while stable trajectories correspond to a negative exponent. In a continuous n-dimensional dynamical

system $dx/dt = F(x)$ a more general definition of the Lyapunov exponent of this system is given by $\lambda = \lim_{t \to \infty} (1/2t) \ln\{\mathrm{Tr}[L^\dagger(t)L(t)]\}$ where L is a square matrix of dimension n, $L^\dagger$ is its hermitian conjugate, and the infinitesimal variation in the solution $\delta x(t)$ satisfies a linearized equation $\delta x(t) = L(t)\delta x(0)$. If L is independent of time, λ is the greatest real part of eigenvalues of L.

Lyman α forest   A large number of narrow (width ~ 10 km s^{-1}) absorption lines observed in quasars shortward of the wavelength of the hydrogen line Lyman α. Spectra of many moderate and high redshift quasars show a characteristic "eroded" appearance due to the high number of absorptions per unit wavelength. It is very difficult to explain the Ly α forest as due to matter associated to the quasar; the current view is that the narrow lines are produced by relatively low density, cold hydrogen in shreds or clouds between the quasar and the observer. The absence of strong absorption lines of heavy elements suggests that the chemical composition is very different from the chemical composition of the sun, with heavy elements 10 to 100 times less abundant in the absorbing clouds than in solar gas.

Lyman alpha (Ly α)   A strong ultraviolet emission line of hydrogen, at 1216 Å (121.6 nm). Ly α is a major component of the geocoronal glow, observed in space from the region around Earth.

Lyman limit   In spectroscopy of hydrogen, the Lyman series (in the ultraviolet) corresponds to transitions between the ground state and higher states. The emitted or absorbed wavelength approaches from above the limit given by

$$\lambda^{-1} = R_\infty ,$$

called the *Lyman limit* (where R_∞ is the Rydberg constant). Because of the state structure of the hydrogen atom, there are an infinite number of consecutive lines of the Lyman series between the longest wavelength $(4/(3R_\infty))$ and the limit, and the distance between the lines decreases and approaches a continuum as the limit is approached. For wavelengths shorter than this limit (912 Å), photons are energetic enough to ionize hydrogen from the ground state. Thus, hydrogen is opaque to radiation of wavelength shorter than the Lyman limit. There are corresponding limits associated with transitions from the second, third, fourth, fifth ... excited states, called the Balmer, Paschen, Brackett, Pfund ... limits.

Lyman series   The set of spectral lines in the far ultraviolet region of the hydrogen spectrum with frequency obeying

$$\nu = cR_\infty \left(1/n_f^2 - 1/n_i^2 \right) ,$$

where c is the speed of light, R_∞ is the Rydberg constant, and n_f and n_i are the final and initial quantum numbers of the electron orbits, with $n_f = 1$ defining the frequencies of the spectral lines in the *Lyman series*. This frequency is associated with the energy differences of states in the hydrogen atom with different quantum numbers via $\nu = \Delta E / h$, where h is Planck's constant, and where the energy levels of the hydrogen atom are:

$$E_n = hcR_\infty/n^2 .$$

Lysithea   Moon of Jupiter, also designated JX. Discovered by S. Nicholson in 1938, its orbit has an eccentricity of 0.107, an inclination of 29.02°, and a semimajor axis of 1.172×10^7 km. Its radius is approximately 18 km, its mass, 7.77×10^{16} kg, and its density 3.2 g cm^{-3}. Its geometric albedo has not been well determined, and it orbits Jupiter once every 259.2 Earth days.

M

M51   Object 51 in the Messier list, a bright Sc spiral galaxy notable for the grand design of its spiral arms (and sometimes referred to as the "Whirlpool" galaxy). M 51 is perturbed by a smaller companion galaxy which appears to be in touch with and distorts one of its spiral arms. Optical spectra of the M 51 nucleus show emission lines suggestive of non-thermal nuclear activity, and whose intensity ratios are intermediate between those of LINERs and Seyfert 2 nuclei.

M87   (NGC 4486) Giant elliptical galaxy, type E1, in Virgo at RA $12^h30^m.8$, dec $+12°24'$, $m_V = 8.6$, angular diameter $7'$. Distance 18 Mpc, giving a linear diameter of about 40 kpc at the center of the Virgo cluster. Also a radio source, Virgo A; also a strong source of X-rays. Total mass about $3 \times 10^{12} M_\odot$. M87 has two very notable features. One is a strong jet (~ 2 kpc in length), which is also a source in the radio and connects the central core of M87 with the more diffuse emission of its northwestern lobe. The other remarkable feature is the large number (up to 15,000) of globular clusters associated with M87.

M87 has also been identified as containing a probable supermassive black hole, of mass about $3 \times 10^9 M_\odot$. *See* cD galaxies.

M101   (NGC 5457). The "Pinwheel" galaxy, type Sc, in Ursa Major at RA $14^h03^m.2$, Dec $+54°21'$, $m_V = 7.9$, angular diameter $22'.0$. Distance, by recent measurements using the period-magnitude relation of Cepheid variable stars detected by the Hubble Space Telescope, is 8.3 Mpc, giving a physical diameter of about 50 kpc.

Mach, Ernst (1838–1916)   An outstanding experimental physicist whose seminal works (essentially, his criticisms of the Newtonian world picture) paved the way for the major revision of physics in the beginning of the 20th century. He is sometimes considered more as a philosopher of science, though he persistently denied this title.

Born on February 18, 1838, in 1855 he enrolled in Vienna University, and received from it his Ph.D. degree in 1860. He continued as a Privat–Dozent delivering lectures on mathematics, psychophysics, and Helmholtz's studies in the perception of sound. From 1864, he held the chair of mathematics in Graz, then (from 1867) for 28 years he taught at the experimental physics department of the Charles University (its German branch). In 1895, Mach moved to Vienna as the philosophy professor of the university (the theory and history of inductive sciences) where he worked up to his retirement in 1901. He died on February 19, 1916.

Mach's scientific interests were primarily focused on acoustic phenomena and processes in continuous media (for example, the unit of velocity in terms of the velocity of sound in a medium, now generally used in aerospace techniques, bears his name — mach), but he was in general a deep thinker, and his achievements in experimental studies were intimately connected with his fundamental methodological quest as well as with his research in the organs of perception physiology and, finally, the epistemology (he influenced these areas profoundly). His sharp critical analysis of the basic concepts of Newtonian mechanics is widely known, and it influenced the development of both the relativities by Einstein. Especially popular among gravitational physicists is Mach's principle (the term introduced by Einstein) which relates local inertial properties of matter with its global distribution (and, finally, motion). Mach was the first who clearly formulated the idea of Gedankenexperiment (thought experiment) and pointed out its great importance for physics.

Mach's ideas were greatly appreciated by Einstein and L. Boltzmann, and disdained by Planck (who was shocked by Mach's negative attitude toward religion) and Lenin (who accused Mach of fideism, i.e., a point-blank acceptance of religion).

Mach number   Ratio of the speed of an object to the sound (or by extension, the light) speed in the medium:

$$M = v/v_p \, ,$$

where v_p is the propagation speed.

Mach's principle   The idea that the local inertial rest frame (non-accelerating, non-rotating) is determined by the influence of all the matter in the universe. Einstein's general relativity includes some of the features of Mach's principle. *See* general relativity.

Maclaurin Spheroid   A uniform density ellipsoid used to describe self-gravitating rotating objects. *Maclaurin Spheroids* are secularly unstable when the rotational kinetic energy of the object exceeds roughly 14% of its potential energy and, given sufficient time, will evolve into Jacobi or Dedekind ellipsoids. If the rotational kinetic energy exceeds 27% of the potential energy, the object is dynamically unstable and this evolution process will occur on a dynamical timescale.

macroscopic description (cosmic string)
See equation of state (cosmic string).

Madden–Julian oscillation (MJO)   A dominant mode of variability in the tropical troposphere with typical time scales of 30 to 60 days, thus also called intraseasonal oscillation. It was first discovered by R.A. Madden and P.R. Julian in 1972. Its signals are observed in pressure, temperature, and wind velocity, predominantly in the zonal wavenumber $= 1$ component. It is generally believed to arise from interaction of deep convection and large-scale flow field. The associated variability in precipitation is largest in the Indian and western Pacific Oceans, where the sea surface temperatures are among the highest in the world ocean. A typical MJO starts with enhanced convective activity in the Indian Ocean, and then moves eastward through the maritime continents into the western equatorial Pacific. This is accompanied by eastward shifts in the rising branch of the Walker circulation on intraseasonal time scales.

Magellanic Clouds   Satellite galaxies of the Milky Way galaxy that are visible to the naked eye in the southern hemisphere. The Large Magellanic Cloud (LMC) is at a distance of about 50 kpc, and subtends about 8° centered at RA $= 5.3$ h and Dec $= -68.5°$. The Small Magellanic Cloud (SMC) is at a distance of about 58 kpc, and subtends about 4° centered at RA $= 0.8$ h and Dec $= -72.5°$. Both the LMC and SMC were originally classified as irregular galaxies, though they exhibit a bar-like spiral structure that also supports SB(s)m classification.

Magellan Mission   The first planetary spacecraft to be launched by a space shuttle (the shuttle Atlantis). On May 4, 1989, Atlantis took Magellan into low Earth orbit, where it was released from the shuttle's cargo bay, and then a transfer engine took it to Venusian orbit.

Magellan used an imaging radar (in order to image through Venus' dense atmosphere) to make a highly detailed map of 98% of the Venusian surface during its four years in orbit between 1990 to 1994. Each orbital cycle lasted eight months.

In Magellan's fourth orbital cycle (September 1992 to May 1993), the spacecraft collected data on the planet's gravity field by transmitting a constant radio signal to Earth. At the end of this cycle flight controllers lowered the spacecraft's orbit using a then-untried technique called aerobraking. This maneuver produced a new, more circularized orbit that allowed Magellan to collect better gravity data in the higher northern and southern latitudes near Venus' poles. Gravity data was obtained for 95% of the planet's surface.

On October 11, 1994, Magellan's orbit was lowered a final time to plunge the spacecraft to the surface by catching it in the atmosphere. Although much of the spacecraft would have vaporized, some sections will have hit the planet's surface intact.

magma   Term given to molten rock when it is still underground. Once the *magma* has been erupted onto the surface of a body by volcanism, it is usually called lava.

magma chamber   Beneath many active volcanos there is a *magma chamber* at a depth of 2 or more km where the magma resides before an eruption occurs and it ascends to the surface.

magmatic water Water that forms in phase separation from magma. This water carries large amounts of dissolved minerals which it deposits in veins as it travels to the surface of the Earth. Also called juvenile water.

magnetic anomaly The local deviation of the geomagnetic field from the global field generated by the dynamo in the core and the external field, due to the magnetization of nearby crust. This is usually obtained by measuring the local magnetic field and then subtracting the prediction of a global field model such as an IGRF. Crustal rock may be magnetized *in situ* after a heating event by cooling down through its Curie temperature in the ambient magnetic field of the Earth, or else magnetized particles may be oriented on or after deposition in a sedimentary environment by the Earth's field so that the rock is left with net magnetization. Magnetic anomalies in oceanic plate show reversals in magnetization as one travels perpendicular to the mid-ocean ridge that produced them, due to the occasional reversal of the Earth's magnetic field so that rock that cools at different times may have oppositely oriented magnetization. This provided evidence for the hypothesis of sea floor spreading. Continental anomalies may be associated with geological structures such as subduction or collision zones or volcanos, where the crust has been heated. Many anomalies, including one at Bangui in Africa, are large enough to be detected through satellite magnetic measurements.

magnetic bay A dip in the trace of a high-latitude magnetogram, resembling a bay in a shoreline. *Magnetic bays* observed this way in the 1950s and earlier were later identified as magnetic substorms.

magnetic carpet The distribution of magnetic field in small scales over the quiet sun which is observed to be recycled over a period of 40 h. A discovery of the Michelson Doppler Imager on the SOHO spacecraft.

magnetic cloud The interplanetary counterparts of coronal mass ejections (CMEs) that exhibit the topology of helical magnetic flux ropes. *Magnetic clouds* make up about one third of all solar wind streams that are identified as interplanetary consequences of CMEs.

magnetic crochet Accompanying a solar flare, dayside magnetic fields may show a sharp change, called a crochet, starting with the flare, but shorter-lived, due to electric currents set up in the lower ionosphere as a result of the flare induced ionization and conductivity changes. *See* short wave fadeout.

magnetic declination The azimuth, measured from geographic north, of the geomagnetic field vector at a given location.

magnetic diffusivity A measure of the ability of a magnetic field to diffuse out of a volume of plasma. A perfectly conducting plasma has zero *magnetic diffusivity*. The solar atmosphere has a magnetic diffusivity comparable to that of copper wire. However, because of the large scales involved on the sun, compared to the typical length of a copper wire, the diffusion times are extremely long.

magnetic dipole The simplest source of a magnetic field envisaged to consist of two magnetic poles of equal strengths but opposite signs a small distance apart. The first-order geomagnetic field is a dipole field.

magnetic field In general, the conceptualization due to Faraday, and rigorized by Maxwell of the laws relating the magnetic forces between current loops (or fictitious magnetic monopoles) as embodied in a space-filling collection of vectors (a field) which collectively define magnetic flux tubes. In geophysics, the magnetic of the Earth or of other planets. On Earth, the field is closely approximated by that of a dipole. The north and south magnetic poles are relatively close to the geographic poles. The *magnetic field* is generated by dynamo processes in the Earth's liquid, iron-rich core. The rotation of the Earth plays an important role in the generation of the magnetic field and this is the reason for the close proximity of the magnetic and geographic poles. The Earth's magnetic field is subject to near random reversals on time scales of hundreds of thousands of years.

In solar physics, the sun's magnetic field is presumably also generated by a dynamo process, but the resulting field is much more complex than that of a simple dipole, with alternating polarities in small regions on the solar surface. Magnetic fields capture energetic charged particles into radiation belts, like the Earth's Van Allen belts. These magnetic fields thus help to protect a planet's atmosphere and/or surface from the bombardment and erosion caused by these charged particles. As charged particles spiral in towards the magnetic poles, they can interact with molecules in the planet's atmosphere, creating vivid aurora displays.

magnetic field spiral   *See* Archimedian spiral.

magnetic helicity   One of the quadratic invariants occurring in the theory of hydromagnetic turbulence. The cross helicity within a volume $\mathcal{V}$ is defined as

$$H_M = \int \mathbf{A} \cdot \mathbf{B} d^3 x \ ,$$

where $\mathbf{A}$ is the magnetic vector potential, $\mathbf{B} = $ curl $\mathbf{A}$ is the magnetic field, and the integral is taken over $\mathcal{V}$. In an incompressible dissipation-free fluid, for suitable boundary conditions, H_M is conserved and is gauge invariant. In fully developed three-dimensional dissipative hydromagnetic turbulence, the magnetic helicity exhibits self-organization in the sense that it cascades from small eddies up to large-scale eddies. This inverse cascade process may be contrasted to the direct cascade and small-eddy decay exhibited by the energy. *See* cross helicity, helicity, hydromagnetic turbulence.

magnetic inclination   The angle of the geomagnetic field vector with the horizontal at a given location. It is defined to be positive when the field vector dips downward.

magnetic latitude   The latitude of a point on Earth in a system of spherical coordinates centered on the magnetic poles defined by the

dipole component of the main magnetic field of the Earth.

magnetic local time (MLT)   The magnetic longitude (converted to hours) in a system whose zero meridian is shifted to local midnight. Thus MLT = 0 is the magnetic meridian through midnight, MLT = 12 is the one through noon, and MLT = (6,18) are the ones at dawn and dusk, respectively.

magnetic longitude   The longitude of a point on Earth in a system of spherical coordinates centered on the magnetic poles defined by the dipole component of the main magnetic field of the Earth. Zero *magnetic longitude* marks the magnetic meridian which overlaps the geographic meridian on which the north magnetic pole (off northern Canada) is located.

magnetic mirror   Charged particles spiral around magnetic field lines. Since the magnetic field can do no work on the particle, the kinetic energy of the motion is conserved. As a particle moves toward a stronger magnetic field, the orbiting kinetic energy increases, slowing the translational motion, and eventually reversing the drift. The location of this reversal, which depends on the strength and gradient of the magnetic field, and on the kinetic energy of the particle, is a *magnetic mirror*.

magnetic pole   Loosely speaking, if the Earth's field is approximated by a dipole, the two points on the surface of the Earth corresponding to the poles of that dipole.

Several more precise definitions exist, and should be carefully distinguished. The *magnetic poles* of the main dipole are the ones obtained by neglecting all non-dipole terms of the harmonic model of the geomagnetic potential: the positions of the resulting poles on the surface have north-south symmetry. The poles of the eccentric dipole are similarly based on the eccentric dipole model but are not symmetric. The dip poles (favored by early explorers) are the surface points where the magnetic dip angle is 90°.

magnetic pressure   Basic concept in magnetohydrodynamics. Graphically, *magnetic pressure* can be described as the tendency of neighboring magnetic field lines to repulse each other. Thus, an inhomogeneity in the magnetic field $\mathbf{B}$ gives rise to a force density $\mathbf{f}$ pushing field lines back from regions of high magnetic density into low density areas:

$$\mathbf{f} = \frac{1}{4\pi} (\nabla \times \mathbf{B}) \times \mathbf{B} .$$

In contrast to the gas-dynamic pressure, the magnetic pressure is not isotropic but always perpendicular to the magnetic field and can be defined as $p_\mathrm{M} = B^2/(8\pi)$.

magnetic reconnection   The dissipation of magnetic energy via magnetic diffusion between closely separated regions of oppositely directed magnetic field. The result of the dissipation is that the oppositely directed field lines form a continuous connection pattern across the diffusion region resulting in a change in the magnetic configuration.

magnetic regime (cosmic string)   *See* electric regime (cosmic string).

magnetic reversal   In geophysics, the Earth's magnetic field is subject to near random reversals on time scales of hundreds of thousands of years. These reversals are attributed to the chaotic behavior of the Earth's dynamo. In solar physics, similar but much more rapid and local effects occur.

magnetic Reynolds number   In ordinary hydrodynamics, the Reynolds number gives the ratio between inertial and viscous forces. If in a flow the Reynolds number exceeds a critical value, the flow becomes turbulent. A similar definition can be used in magnetohydrodynamics, only here the viscous forces do not depend on the viscosity of the fluid but on the conductivity of the plasma:

$$R_\mathrm{M} = \frac{UL}{\eta} = \frac{4\pi\sigma UL}{c^2}$$

with U being the bulk speed, L the length scale, σ the conductivity, and $\eta = c^2/(4\pi\sigma)$ the magnetic viscosity. The coupling between the particles therefore does not arise from collisions as in ordinary fluids but due to the combined effects of fields and particle motion.

magnetic secular variation   Time variations of the Earth's magnetic field, usually taken to imply the time variation of the part of the field generated by the dynamo in the Earth's core. This generally includes most variation in the magnetic field on periods of decades or longer: external variations in the field, and their associated induced internal counterparts, tend to be on diurnal timescales, although some time averaging of the field may cause measurements to exhibit power on longer timescales such as that of the solar cycle. On the other hand, the highest frequency on which the core field varies is not well known: geomagnetic jerks, for example, appear to have timescales of around a year or two.

magnetic shear   The degree to which the direction of a magnetic field deviates from the normal to the magnetic neutral line, defined by the loci of points on which the longitudinal field component is zero. A sheared magnetic field indicates the presence of currents since $\nabla \times B \neq 0$.

magnetic tension   Basic concept in magnetohydrodynamics. Graphically, *magnetic tension* can be interpreted as a tendency of magnetic field lines to shorten: if a magnetic field line is distorted, for instance by a velocity field in the plasma, a restoring force, the magnetic tension, acts parallel but in opposite direction to the distorting flow. Thus, magnetic tension can be interpreted as a restoring force like the tension in a string. This concept can be used, for instance, to derive Alfvén waves in a simple way.

magnetism   Magnetic fields and interactions. In solar system physics, solar and planetary magnetic fields are usually produced by moving electric currents in an interior conducting layer. In the sun, this is internal motion of the ionized gas. In the case of the Earth, the conducting layer is the liquid iron outer core. In the case of Jupiter and Saturn, the magnetic fields are produced by motions within a metallic hydrogen layer, and for Uranus and Neptune an interior slushy-ice layer is thought to give rise to the magnetic fields. Mercury's magnetic field is produced by its very large iron core, but debate continues as to whether this is an active (i.e., currently produced by interior currents) or a remnant field. The moon and Mars currently do not have active magnetic fields, but rocks from both bodies indicate that magnetic fields were present in the past. The large Jovian moons also appear to have weak magnetic fields. The Earth's magnetic field has undergone polarity reversals throughout history, the discovery of which helped advance the hypothesis of sea floor spreading, which eventually led to the theory of plate tectonics. The area of space where a planetary body's magnetic field interacts with the solar wind is called the magnetosphere.

magnetoacoustic wave   Any compressive hydromagnetic wave; frequently called magnetosonic wave. *Magnetoacoustic waves* are characterized by fluctuations of density and magnetic field strength, and by linear polarization of fluctuations in the velocity and magnetic field vectors. Their phase and group velocities are anisotropic in the sense that they depend on the direction of propagation with respect to the mean magnetic field.

The theory of magnetoacoustic waves can be developed in the framework of either magnetohydrodynamics or kinetic theory. The magnetohydrodynamic approach gives two different kinds of magnetoacoustic wave, conventionally called the fast and slow magnetoacoustic modes. For propagation parallel to the mean magnetic field, the fast (slow) mode propagates at the greater (lesser) of the sound speed C_S and Alfvén speed C_A; for propagation transverse to the mean magnetic field, the fast mode propagates at the speed

$$C_\perp = \left(C_A^2 + C_S^2 \right)^{\frac{1}{2}},$$

and the slow mode has zero propagation speed. The other qualitative difference between the two modes is that the fluctuations in density and magnetic field strength are correlated positively for the fast mode, negatively for the slow mode. In the limit of strong magnetic field, the slow mode disappears, and the fast mode propagates at C_A for all directions; in the limit of weak magnetic field, the slow mode disappears once again, and the fast mode is simply a sound wave. The phase speed of a magnetoacoustic wave, unlike the Alfvén wave, depends on amplitude, so

that magnetoacoustic waves can steepen to form shock waves.

The kinetic-theory approach yields analogs of the fast and slow modes, but additionally gives other, very strongly damped modes. Moreover, in a hot plasma like the solar wind, even the fast and slow modes are subject to strong Landau damping. Indeed, although solar wind fluctuations have often been identified with Alfvén waves, they can rarely be associated with magnetoacoustic waves. Landau damping of magnetoacoustic waves may be an important mechanism of plasma heating in some circumstances. In particular, it may severely restrict the formation of shock waves from magnetoacoustic disturbances. *See* Alfvén wave, hydromagnetic wave.

magneto-fluid mechanics   *See* magnetohydrodynamics.

magnetogram   Graphic representation of solar magnetic field strength and polarity. *Magnetograms* show a hierarchy of spatial scales in the photospheric magnetic field from sunspots to the general magnetic network.

magnetohydrodynamics (MHD)   The study of the dynamics of a conducting fluid in the presence of a magnetic field, under the assumption of perfect, or partial (resistive MHD) locking of the plasma to the field lines. Important in many branches of physics, but in geophysics it is primarily important in the study of planetary magnetospheres, ionospheres, and cores. In these non-relativistic cases, Maxwell's equations for electrodynamics and the Lorentz field transformation yield the induction equation (Faraday's law):

$$\frac{\partial \mathbf{B}}{\partial t} = \nabla \times (\mathbf{V} \times \mathbf{B}) - \nabla \times (\eta \nabla \times \mathbf{B})$$

where $\mathbf{B}$ is the magnetic field, $\mathbf{V}$ is the velocity of the conducting fluid, and η is the magnetic diffusivity of the fluid. The first term on the right-hand side represents the effects of fluid flow on the magnetic field, and the second term on the right-hand side represents diffusion of the magnetic field, and in a region where η is constant may be written $+\eta \nabla^2 \mathbf{B}$. In the case of a simple conducting fluid in the presence of a magnetic field, the force due to the magnetic field (the Lorentz force) per unit volume is:

$$\mathbf{f}_B = \frac{1}{\mu} \left(\nabla \times \mathbf{B} \right) \times \mathbf{B}$$

where μ is the magnetic permeability.

In a medium characterized by a scalar electrical conductivity σ, the electric current density $\mathbf{J}$ in cgs units is given by Ohm's law

$$\mathbf{J} = \sigma \left(\mathbf{E} + \mathbf{V} \times \mathbf{B}/c \right) \ ,$$

where $\mathbf{E}$ is the electric field, $\mathbf{B}$ the magnetic field, and c the (vacuum) speed of light. Many fluids of importance for space physics and astrophysics have essentially infinite electrical conductivity. For these systems to have finite current density,

$$\mathbf{E} + \mathbf{V} \times \mathbf{B}/c = 0 \ .$$

Thus, the electric field can be eliminated from the *magnetohydrodynamic* equations via Faraday's law. Also, the current density has disappeared from Ohm's law, and has to be calculated from Ampère's law (the displacement current is negligible in nonrelativistic magnetohydrodynamics). Thus, for a nonrelativistic inviscid fluid that is a perfect electrical conductor, the Euler equations for magnetohydrodynamics are the equation of continuity

$$\frac{\partial \rho}{\partial t} + \nabla \cdot \left(\rho \mathbf{V} \right) = 0 \ ,$$

where ρ is the mass density, the momentum equation

$$\rho \left(\frac{\partial}{\partial t} + \mathbf{V} \cdot \nabla \right) \mathbf{V} = -\nabla P + f_B$$

where P is the pressure, and $\mathcal{F}$ is the body force per unit volume (due, for example, to gravity), plus some additional condition (such as an energy equation, incompressibility condition, or adiabatic pressure-density relation) required for closure of the fluid-mechanical equations. Faraday's law completes the magnetohydrodynamic equations.

Magnetohydrodynamics has been applied broadly to magnetized systems of space physics and astrophysics. However, many of these systems are of such low density that the mean-free

paths and/or times for Coulomb collisions are longer than typical macroscopic length and/or time scales for the system as a whole. In such cases, at least in principle, a kinetic-theory rather than fluid-mechanical description ought to be used. But plasma kinetic theories introduce immense complexity, and have not proved tractable for most situations. Fortunately, experience in investigations of the solar wind and its interaction with planetary and cometary obstacles has shown that MHD models can give remarkably good (though not perfect) agreement with observations. This may be so partly because, absent collisions, charged particles tend to be tied to magnetic field lines, resulting in a fluid-like behavior at least for motion transverse to the magnetic field.

Moreover, moments of the kinetic equation give continuity equation and momentum equation identical to their fluid counterparts except for the fact that the scalar pressure is replaced by a stress tensor. Turbulent fluctuations can also influence transport properties and enforce fluid-like behavior. On the other hand, predictions from kinetic theory can and do depart from the MHD description in important ways, particularly for transport phenomena, instability, and dissipation.

magnetometer   An instrument measuring magnetic fields. Until the middle of the 20th century, most such instruments depended on pivoted magnetized needles or needles suspended from torsion heads. Compass needles measured the direction of the horizontal component (the magnetic declination, the angle between it and true north), dip needles measured the magnetic inclination or dip angle between the vector and the horizontal component, and the frequency of oscillation of a compass needle allowed the field strength to be determined. Alternatively, induction coils, rotated rapidly around a horizontal or vertical axis, produced an induced e.m.f. which also gave field components.

Induction coils have also been used aboard spinning Earth satellites, but most spaceborne instruments nowadays use fluxgate *magnetometers,* also widely used for geophysical prospecting from airplanes and for scientific observations on the ground. These are not absolute but require calibration, and for precision work therefore alkali vapor instruments are often used, generally in conjunction with fluxgates. These are absolute instruments, observing the intensity of the magnetic field by the frequency at which a swept-frequency radio signal causes enhanced absorption of a light beam in a glass cell filled with alkali vapor. Rubidium or strontium vapor is generally used, and the beam is emitted from the same element, in a narrow frequency range. The mechanism is based on optical pumping of energy sub-states of these atoms.

Other instruments (used mainly on the ground) include the proton precession magnetometer, on which the widespread technique of nuclear magnetic resonance in medicine is based. The Overhauser effect magnetometer is related to this but it is more precise, and it can be used as an alternative to the alkali vapor instrument in geomagnetic survey satellites.

magnetopause   The boundary of the magnetosphere, separating plasma attached to the Earth from that flowing with the solar wind. The *magnetopause* is defined by the surface on which the pressure of the solar wind is balanced by that of the Earth's magnetic field. The "nose" of the magnetopause, on the sunward side of the Earth is $\sim$ 15 Earth radii away, on average. As the pressure of the solar wind changes, the magnetopause shrinks or expands accordingly. The idea of a magnetopause has been around since about 1930 when Chapman and Ferraro proposed a theory that explained geomagnetic storms as interactions of the Earth's magnetosphere with plasma clouds ejected from the sun. The magnetopause was first discovered in 1961 by NASA's Explorer 12 spacecraft.

magnetosonic wave   *See* magnetoacoustic wave.

magnetosphere   The planetary region of space where a body's magnetic field can be detected. As the solar wind plasma embedded with interplanetary magnetic field (IMF) flows around the planet, it interacts with the Earth's magnetic field and confines it to a cavity called the *magnetosphere*. Since the solar wind is supersonic, a shock known as Bow shock is formed on the sunward side of the magnetosphere. The solar wind then flows across the bow shock and

its speed is reduced from supersonic to subsonic. The solar wind ahead is deflected at a boundary between the magnetosphere and the solar wind known as the magnetopause. The subsolar point on the magnetopause is about 10 Earth radii from the center of the Earth. The bow shock is about 3 Earth radii from the sunward side of the magnetopause at the subsolar point. The region between the bow shock and the magnetopause is called the magnetosheath.

At the magnetopause, the solar wind pressure outside is balanced by the magnetic field pressure inside the magnetosphere. As the solar wind sweeps past the Earth, the magnetic field lines from the polar cap are pulled toward the nightside to form a geomagnetic tail. This tail can be observed as far as 1000 Earth radii, as the combined pressure of the field and plasma prevent its closing on the night side. Moreover, the polar cap magnetic field lines do not close resulting in a thin sheet in the equatorial plane known as the neutral sheet across which an abrupt field reversal occurs. The region where magnetic field lines from sub-auroral latitudes tend to close is known as plasma sheet of thickness about 1 Earth radii. The plasma sheet consists of energetic particles in the energy range of about 1 to 10 kev and is the source of radiation belt particles.

magnetospheric substorms   Apart from quiet variations in the Earth's magnetic field like solar quiet (S q) and lunar variations resulting from the generation of ionospheric currents by solar and lunar tides, other observed variations are far from constant and without any periodicity. Such variations result from the interaction of solar wind with the geomagnetic field of the Earth and are denoted by D. These are called magnetospheric storms and have characteristic times of minutes to days. The magnetospheric storm consists of frequently occurring small storms called *magnetospheric substorms*. The intensity of the storm is given in terms of the AE (auroral electrojet) index and of the substorm by the Dst index. Each substorm is associated with injection of energetic protons of few tens of kev in the Van Allen belts. When frequent injection occurs, a proton belt known as ring current is formed whose strength is given in terms of Dst. At first there is a sudden commencement (SC) in the geomagnetic field, an impulsive increase in the H component with a risetime of a few minutes with an amplitude of several nano teslas observed over the entire Earth with a spread in arrival time of less than 1 min. The increase in H during SC is maintained for a few hours known as the initial phase until a large H decrease of the main phase begins, resulting from the ring current buildup. The final recovery phase, of typically a day or more, is faster at first and then slower, and results from a decrease in ring current to prestorm values. The effects of the magnetospheric storms and substorms in the ionosphere are called ionospheric storms.

magnetostratigraphy   The use of measurements of the magnetization of strata for absolute or relative dating purposes. Sedimentary rock containing magnetized particles that tend to orient with the Earth's field upon or soon after deposition will thus make a record of the geomagnetic field: similarly, hot rock containing minerals with a high magnetic susceptibility will record the geomagnetic field at the time at which they are cooled to below their Curie temperature. Hence, strata that was deposited or cooled at a particular site at different times may exhibit differently oriented magnetizations, as the Earth's field changes with time and also the location of the strata may have moved or rotated with respect to the Earth's rotation axis. Since the pattern of reversals of the Earth's field is fairly well known for several epochs, a good record of the variation of the magnetization at a particular site may sometimes be used to date the rocks.

magnetostrophic   A dynamic state of a rotating conducting fluid which is threaded by a magnetic field. The leading order force balance is between the Lorentz (magnetic) force, the Coriolis force, the pressure gradient, and buoyancy. Other terms, such as that associated with inertia or viscosity, are deemed much smaller. This is the force balance commonly assumed to hold for the Earth's core, as the viscosity of the core fluid is thought to be very low. There are several types of *magnetostrophic* waves including MAC (magnetic-archimedean-coriolis) waves which involve the archimedean force (i.e., buoyancy)

and MC (magnetic-coriolis) waves that do not. A magnetostrophic flow should generally satisfy Taylor's condition that the azimuthal component of the Lorentz force integrates to zero on axial cylinders, except in certain circumstances where the gravitational force has an azimuthal component. *See* Taylor state.

magnetotail   The long stretched out nightside of the magnetosphere, the region in which substorms begin. It ranges from about 8 Earth radii nightwards of the Earth and has been observed out to 220 Earth radii. The *magnetotail* is created by the interaction of the solar wind with the Earth's magnetosphere. The solar wind particles are deflected around the Earth by the bow shock, dragging the nightside magnetosphere out into a long tail. It is estimated that the magnetotail may extend to over 1000 Earth radii from the Earth.

magnitude   In astronomy, the system of logarithmic measurement of brightness of stars was invented by the ancient Greeks. The Greeks had six equal divisions from the brightest star to the faintest observable by eye, the brightest being first *magnitude* and the faintest sixth magnitude. The human eye is sensitive to logarithmic changes in brightness, and so this scheme was logarithmic in nature. The modern definition of magnitude takes the difference in one magnitude of brightness being equal to $^5\sqrt{100}$. The magnitude system is related to the brightness of an object, such that

$$m = -2.5 \log(F) + K$$

where F is the observed flux in units of erg s^{-1} cm^{-2}. The quantity m is the apparent magnitude. Flux can be written as a function of the intrinsic luminosity of the star such that

$$F = \frac{L}{4\pi R^2}$$

where L is the intrinsic luminosity and R is the distance to the star. The difference in two magnitudes is then

$$m_1 - m_2 = 2.5 \log\left(\frac{F_2}{F_1}\right).$$

The constant K is defined by assigning magnitudes to a number of standard stars, which are then used as calibration standards in measurements. Such calibration is typically accurate about $0^m.01$.

One may also assign an absolute magnitude M to any star, defined as the apparent magnitude the star would have if it were 10 parsecs from the observer on the Earth.

Maia   Magnitude 4.0 type B9 star at RA 03h 45m, dec $+24°21'$; one of the "seven sisters" of the Pleiades.

main field   Main magnetic field of the Earth, the magnetic field originating inside the Earth.

main sequence star   A star whose primary energy source is hydrogen burning in a core of about 10% of its total mass. Stars begin this phase of evolution when fusion begins in the core, terminating the proto-star era of stellar formation. The name arises because such stars make up a very large fraction of all the ones we see (in a volume-limited sample), and they occupy a diagonal stripe across a Hertzsprung–Russell diagram. The physics of main sequence stars are relatively simple, and qualitative relationships can be derived among their masses, luminosities, radii, temperatures, and lifetimes. The range of masses is 0.085 to 100 solar masses (for a composition like the sun). Lower masses are brown dwarfs and configurations with larger masses are quite unstable and quickly blow off enough surface material to come down to 100 solar masses or less. The relationships are:

$$L \propto M^4$$

(steeper at low mass, flatter a high mass)

$$R \propto M^{\frac{1}{2}} \quad T \propto M^{\frac{1}{2}}$$

lifetime $\propto M^{-2}$ to M^{-3} .

For masses less than about 1.5 solar masses, the main energy source is the proton-proton chain (whose rate depends on central temperature as about $T^{4.5}$, so that the stars have radiative cores). These lower mass stars have atmospheres in which the hydrogen is not totally ionized, so that their envelopes are convective (*see* convection). Conversely, stars of larger masses derive their energy from the CNO cycle and have convective cores, but the hydrogen

in their atmospheres is already ionized, so that they have radiative envelopes. No star of more than about 0.3 solar masses is fully convective after its formation stages, and so a core of helium gradually builds up at the center as a star ages from the the zero age main sequence toward the Hertzsprung gap. The main sequence phase takes up roughly 90% of a star's (pre-white-dwarf or pre-neutron-star) life, and so 90% of the stars in a given volume will be *main sequence stars*. *See* brown dwarf, CNO cycle, convection, Hertzsprung Gap, Hertzsprung–Russell diagram, hydrogen burning, proton-proton chain.

major axis   In an ellipse, the distance corresponding to the maximum length measured along one of the two symmetry axes, or this axis itself.

Majumdar–Papapetrou space-times   (Majumdar, 1946; Papapetrou, 1947) In relativity, static space-times containing only electric fields in which the level surfaces of the gravitational potential V and electric potential ϕ coincide. Solution of the gravitational equations yields $V = A + B\phi + (1/2)G\phi^2$ where A and B are arbitrary constants and G is the gravitational constant.

Manning coefficient   A measure of roughness of an open channel. Denoted by the symbol n and appears in the Manning equation for calculation of velocity or flowrate in a channel with a free surface, such as a river, canal, or partially full pipe. The *Manning coefficient* increases as roughness increases.

Manning's equation   An equation that can be used to compute the average velocity of flow in an open channel, based on channel geometry and roughness:

$$U = \frac{\kappa}{n} R_H{}^{2/3} S^{1/2}$$

where U is the average velocity, n is the dimensionless Manning coefficient with typical values of 0.02 for smooth-bottom streams to 0.075 for coarse, overgrown beds, R_H is the hydraulic radius and is defined as the ratio of the channel cross-sectional area to the wetted perimeter, S is the slope or energy gradient or the channel, and the constant κ is equal to $1 \text{ m}^{1/3}\text{s}^{-1}$ in SI units.

manometer   A device for measuring pressure or pressure differences. It consists of a u-shaped tube, preferably of glass, partially filled with a liquid (mercury, oil, or water are commonly used). The difference in height, h, of the liquid in the two arms of the tube gives the pressure difference as $\Delta p = \rho g h$, where ρ is the fluid density, and g is the local acceleration of gravity. If the meter-kilogram-second units are used, the pressure has units of Pascal (Pa). Pressure difference may also be indicated by giving h and the fluid being used, as in "$mm\ Hg$", "inches of Mercury", "inches of water".

mantle   The terrestrial planets are differentiated into layers, generally consisting of an outer crust, interior mantle, and central core. The *mantle* is usually the largest (by volume) of these layers and is composed of minerals that are intermediate in density between the iron comprising the core and the lighter materials (such as feldspars) which make up the crusts. The Earth's mantle is composed primarily of magnesium, iron, silicon, and oxygen; the mantles of the other terrestrial planets are believed to be similar. Seismology indicates that there are several transition zones within the mantle, the result of changes in the densities and other properties of the component minerals with depth. Seismic studies of the Earth's mantle indicate it is composed of solid rock, but due to the high temperatures and pressures, the mantle rocks can deform and flow over thousands to millions of years. This ductile flow of the mantle rocks is called mantle convection and is an efficient process by which heat is transported from the hot interior of the Earth to the cooler surface. This convection is believed to be the driving mechanism for the plate tectonics operating on the Earth. Plumes of hot material, not necessarily associated with the mantle convection cells, also occur and produce the hot spots that are responsible for intra-plate volcanism such as the Hawaiian Islands.

mantle convection   On geologic time scales the Earth's mantle behaves as a fluid due to solid-state creep processes. Because of the heat loss from the interior of the Earth to its surface, the deep rocks are hot and the shallow rocks are cool. Because of thermal contraction the cool near surface rocks are more dense than the hot deep rocks. This leads to a gravitational instability with the near surface dense rocks sinking and the deeper light rocks rising. Plate tectonics, with the subduction of the cool, dense lithosphere, is one consequence.

mantle transition zone   A zone that is located between depths of 410 and 660 km where seismic velocity and density increase markedly with increasing depth. Sometimes the mantle transition zone is taken to include the uppermost portion of the lower mantle down to depths of about 750 to 800 km. Depth distributions of seismic velocity and density, and their discontinuities at depths of 410 and 660 km can be explained by pressure and temperature dependency, and phase transformations of major composite minerals (olivine) in pyrolite composition.

March 5th event   A gamma ray transient, observed with 12 spacecraft in 1979, that caused great controversy. Its properties demonstrated that it could be a distinct, new class of high-energy transient. The *March 5th event* was clearly identified with the supernova remnant N49 in the Large Magellanic Cloud (from 55 kiloparsecs away). Its intensity, however, made it the brightest "gamma ray burst" to date, and led many to conclude that its apparent N49 source location must be accidental and that it came instead from a few parsecs away in the nearby interstellar region, then thought typical of gamma ray bursts. A few much smaller events (*see* soft gamma repeaters) were seen from the same direction over the years, contrasting to the apparent lack of repetition of usual gamma ray bursts, adding to the confusion. Now it is known, ironically, that gamma ray bursts originate from cosmological sources, considerably more distant than even the neighboring galaxies, and that the March 5th event and the four known soft gamma repeaters do originate in "magnetar" galactic and LMC supernova-remnant neutron stars, including N49. Only one other transient similar to the March 5th event has been detected in three decades of space-age monitoring, seen on August 27, 1998, confirming all aspects of this interpretation.

mare   Name for a flat area on the moon. Maria are now known to be dust-covered cooled melt flows from meteorite impacts.

Marianas Trench   Undersea trench running roughly south from Japan at about 140°E. Formed by the subduction of the Pacific tectonic plate below the Philippine plate. The deepest ocean waters in the world occur in the Challenger deep, part of the *Marianas Trench.*

marine snow   In oceanography, particles of organic detritus and living forms whose downward drift, in a dense concentration, appears similar to snowfall.

Markarian galaxies   Galaxies showing an excess of blue and near UV emission, identified by B.E. Markarian through an objective prism survey with the 1-m Schmidt telescope of the Byurakan observatory. The lists Markarian published in the 1970s include approximately 1500 objects, of which $\approx 10\%$ are Seyfert galaxies, $\approx 2\%$ are quasars, $\approx 2\%$ are galactic stars, and the wide majority are galaxies with enhanced star formation, such as star-forming dwarf galaxies and starburst galaxies. *See* dwarf galaxy, Seyfert galaxies, starburst galaxy.

Mars   The fourth planet from the sun. Named after the Roman god of war, *Mars* has a mass of $M = 6.4191 \times 10^{26}$ g, and a radius $R = 3394$ km, giving it a mean density of $3.94\,\mathrm{g\,cm^{-3}}$ and a surface gravity of 0.38 that of Earth. The rotational period is $24^h 37^m 22.6^s$ around an axis that has an obliquity of 23° 59'. This rotation is, in part, responsible for the planet's oblateness of 0.0092. Mars' orbit around the sun is characterized by a mean distance of 1.5237 AU, 2.28×10^8 km, an eccentricity of $e = 0.0934$, and an orbital inclination of $i = 1.85°$. Its sidereal period is 687 days, and its synodic period is 779.9 days. An average albedo of 0.16 gives it an average surface temperature of around 250 K, varying from 150 to 300 K. Its atmosphere is more than 90% CO_2, with traces of O_2, CO, and H_2O. The atmospheric pressure at the surface is 3.5 mbars. Mars continues to be well studied in part because there is evidence of past liquid water on the surface, and thus Mars may have once harbored life. Mars has a highly varied terrain of mountains, canyons, and craters that are kilometers in height and depth. Mars has a silicate mantle and core which is probably a mixture of Fe and S. Its moment of inertia has recently been measured to be $I = 0.365 M R^2$. Mars has two satellites (Phobos and Deimos), which orbit the planet in synchronous rotation.

Mars Climate Orbiter (MCO)   An orbiting spacecraft that belongs to the Mars Surveyor '98 program along with the Mars Polar Lander (MPL). MCO was launched on December 11, 1998, to reach Mars 9 1/2 months later. The spacecraft was destroyed on September 23, 1999 when a navigational error pushed it too far into the Martian atmosphere. The error was attributed to a confusion between metric and English units of force.

Mars Global Surveyor (MGS)   A spacecraft launched on November 11, 1996, that signifies America's successful return to Mars after a 20-year hiatus. Surveyor took 309 days to reach Mars, and entered into an elliptical orbit on September 12, 1997. During its first year at Mars, the Orbiter Camera observed evidence of liquid water in the Martian past, extensive layered rock, boulder-strewn surfaces, volcanism and new volcanic features, the Martian fretted terrain, the polar layered deposits, and the work of wind on the Martian surface. The orbit has since been circularized through aerobraking. During this period, the Global Surveyor was able to acquire some "bonus" science data, such as a profile of the planet's northern polar cap.

The primary mapping phase of the mission began on March 15, 1998. It ended 687 Earth days later on January 31, 2000. In addition to making a photographic map of the entire planet, *Mars Global Surveyor* studied the planet's topography, magnetic field, mineral composition, and atmosphere. Surveyor is also used as a communications satellite to relay data to Earth from landers on the surface of Mars. This phase of the mission is scheduled to last until January 1, 2003, or until the spacecraft's maneuvering propellant runs out.

Mars Microphone   A microphone developed for the Planetary Society by the University of

California, Berkeley Space Sciences Laboratory, for the Mars Polar Lander (launched on January 3, 1999, and destroyed on landing in Mars on December 3, 1999).

Mars Microprobe　The Mars Microprobe Mission, also known as Deep Space 2 (DS2) was launched aboard the Mars Polar Lander on January 3, 1999.

The microprobes were two basketball-sized aeroshells designed to crash onto the Martian surface at a velocity of about 200 m per second and release a miniature two-piece science probe into the soil to a depth of up to 2 m. The microprobes were to have separated from the Mars Polar Lander on December 3, 1999 prior to entry into the Martian atmosphere, and return signals relayed through the Mars Global Surveyor. However, no signal was received from them. Subsequent studies indicated severe design defects in the probes. The Mars Polar Lander also crashed on Mars on December 3, 1999. *See* Deep Space 2, Mars Polar Lander.

Mars Pathfinder Mission　The second of NASA's low-cost planetary Discovery missions, originally called Mars Environmental Survey, or MESUR, Pathfinder. It was launched on December 4, 1996, arrived at Mars on July 4, 1997, and continued to operate for 4 months. The mission consisted of a stationary lander and a surface rover (Sojourner), which was controlled by an Earth-based operator. The rover survived for 3 months, significantly longer than its designated lifetime of 1 week. After landing, Mars Pathfinder returned 2.6 billion bits of information, including more than 16,000 images from the lander and 550 images from the rover (which infer that some of the rocks may be sedimentary, thus implying a significant fluvial history), as well as more than 15 chemical analyses of rocks and extensive data on winds and other weather phenomena.

The spacecraft entered the Martian atmosphere without going into orbit and landed on Mars with the aid of parachutes, rockets and airbags, taking atmospheric measurements on the way down. It impacted the surface at a velocity of about 18 m/s (40 mph) and bounced about 15 m (50 ft) into the air, bouncing and rolling another 15 times before coming to rest

approximately 2.5 minutes after impact, about 1 km from the initial impact site. The landing site, in the Ares Vallis region, is at 19.33°N, 33.55°W and has been named the Sagan Memorial Station.

Mars Polar Lander (MPL)　A spacecraft launched on January 3, 1999, that crashed on landing on Mars on December 3, 1999. The lander was designed to directly enter the atmosphere, deploy a parachute, then fire rockets and soft land on the surface. Subsequent studies have indicated that a coupled software/hardware error led to premature shutdown of the landing rocket. Instruments on the Mars Polar Lander included cameras, a robotic arm, soil composition instruments, a light detection and ranging instrument (LIDAR), and two microprobes which were presumably deployed before the lander entered the atmosphere but were also lost.

Martian geophysical epochs　Similar to assignments on Earth, names have been assigned to geophysical epochs on the planet Mars. The names and their approximate age correspondences are: Amazonian, 0 to 1.8 Gy BP; Hesperian, 1.8 to 3.5 Gy BP; Noachian, more than 3.5 Gy BP.

Martian meteorites　Currently 13 meteorites display characteristics that indicate they are probably from Mars. Twelve of these meteorites have very young formation ages (ranging from 1.8×10^8 yr to 1.3×10^9 yr) and have compositions indicative of formation in basaltic lava flows. These young meteorites are called the Shergottites, Nakhlites, and Chassigny (or SNC meteorites), after the three meteorites which display the major chemical characteristics of each group. The young, basaltic compositions suggested a body that was volcanically active very recently. This process of elimination resulted in the Earth, Venus, and Mars as the likely parent bodies for these meteorites. Other chemical evidence (primarily from oxygen isotopes) eliminated the Earth from consideration and orbital dynamic considerations eliminated Venus as a possibility. The Martian origin of these meteorites was largely confirmed by the discovery of trapped gas within some of the meteorites; the isotopic composition of this gas is statistically identical to the composition of the Martian atmosphere. The 13th meteorite is much older than the other 12 (dating from 4.5×10^9 yrs ago) and may contain evidence of fossilized Martian life. These meteorites are believed to have been ejected from the planet's surface during the formation of impact craters on Mars. The subsequent discovery of lunar meteorites indicates that this is a common process, at least for the smaller solar system bodies.

mascon "Mass concentration"; concentration of density beneath the surface of the moon, detectable because of its effect on satellite orbits. *Mascons* associated with lunar craters typically lie under the center of the craters; there is usually a mass deficit in a ring associated with the crater wall.

maser Acronym for microwave amplification by stimulated emission of radiation: amplification of radiation coming from excited states of molecules. While a *maser* can be a laboratory instrument, especially interesting are the natural astrophysical masers. Amplified lines have been observed at frequencies in the range of 1 to 100 Ghz, in association with dense molecular clouds associated to star forming regions, or in circumstellar envelopes of cold late-type stars, such as giant and supergiant M stars, carbon and S stars, where diatomic or more complex molecules are not dissociated by the radiation from the star. Masers are produced by stimulated emission: a photon of frequency matching the frequency of a particular transition between two states induces the emission of a second photon, whose frequency and phase are identical to the first. Net amplification of the line radiation is achieved by population inversion, a condition realized when the higher energy level is more populated than the lower by a "pumping" source of energy, like a background radiation field. Astrophysical masers have been observed at several frequencies corresponding to rotational transition of di- and tri-atomic molecules. The most luminous masers have been observed in external galaxies at the frequency of 22.235 Ghz corresponding to a rotational transition of the water vapor molecule; they are known as water mega-masers.

mass coefficient *See* inertia coefficient.

mass-defect The binding energy E_b of a system expressed in terms of Einstein's famous formula $E_b = M_b c^2$. (M_b is taken positive if the system is bound.) In Newtonian gravity, and in Einstein's general relativity,

$$M_{\text{grav}} = M_0 - M_b$$

where M_0 is the sum of the masses of the system's constituents if it were infinitely dispersed, and M_{grav} is the active gravitational mass that determines, e.g., planetary orbits. In other relativistic theories of gravity this simple relation does not hold, and different kinds of binding energy make different contributions to the active gravitational mass.

mass extinctions Extinctions of a large percentage of flora and/or fauna have occurred throughout Earth's history. The definition of a *mass extinction* is that more than 25% of the existing species disappear from the Earth within a short time period (generally $< 10^5$ yrs). Since the boundaries between terrestrial geologic epochs are defined by changes in the fossil record, many of these boundaries are coincident with mass extinctions. Mass extinctions result from major environmental changes that occur rapidly enough that many species do not have time to adapt. Several mass extinctions are now associated with large impact events on the Earth, the most famous being the extinction of the dinosaurs and other species at the end of the Cretaceous Period (65×10^6 yrs ago) by the creation of the approximately 200-km-diameter Chicxulub impact crater in Mexico. Debris tossed into the atmosphere by these large impacts decreases the amount of sunlight reaching the surface, causing a collapse of the food chain and accompanying starvation, which leads to the mass extinctions. Mass extinctions are also suspected to occur when plate motions cause changes in ocean currents, which leads to climate change, and perhaps by periods of enhanced volcanic activity.

massive ghost A theoretical massive particle with unphysical properties which spoils the physical content of a quantum particle theory.

For instance, in a model higher derivative quantum gravity, the massive spin-2 ghosts have negative kinetic energy and they interact with physical particles such as gravitons. As a result they lead to the instability of the classical solutions and, on a quantum level, to the loss of conservation probably in interaction (loss of unitarity of the physical S matrix). *See* higher derivative theories.

mass-luminosity relation   The relation between luminosity and mass for main sequence stars. As the mass of a star increases, the pressure in its core, and hence its temperature, also increases. Hence, as the mass of a star increases, it burns hydrogen more rapidly in its core and is consequently brighter. The *mass-luminosity relation* for main sequence stars is: $L \propto M^x$ where x is roughly 3. This relation only holds when comparing stars lying on the main-sequence (hydrogen burning stars). A low mass giant star can be brighter than a higher mass main sequence star.

mass transfer   Mass transfer occurs in binary star systems when one star overfills its Roche-lobe and accretes onto its companion. Binary mass transfer is separated into three cases depending upon the evolutionary phase of the mass-losing star (Case A, Case B, Case C). Case A denotes the mass transfer that occurs during hydrogen burning. Case B mass transfer occurs after hydrogen burning, but before helium core ignition. Case C mass transfer denotes any mass transfer which occurs after helium core ignition. If the orbital angular momentum of the system is conserved during the mass transfer phase, the mass transfer is denoted "conservative". The orbital separation of the binary is uniquely determined:

$$\frac{a}{a_0} = \left(\frac{M^1}{M_0^1}\right)^{-2} \left(\frac{M^2}{M_0^2}\right)^{-2},$$

where the subscript 0 denotes the initial conditions. Note that if the mass-losing star is less massive than the accreting star, the orbit widens, and if it is more massive, the orbit decreases. In nature, however, orbital angular momentum is lost from the system either as mass escapes the

binary or as mass forms an accretion disk and spins up the accreting star.

mass wasting   Also called mass movement, *mass wasting* is the downhill movement of soil or fractured rock under the influence of gravity. Common features associated with mass wasting include landslides (including rockslides, mudflows, Earthflows, and debris avalanches), rock falls (also called talus slopes), soil creep (the very slow (usually imperceptible) downhill flow of soil under the influence of gravity), and solifluction (movement of frozen soil).

matter density perturbations   Inhomogeneities in the matter distribution in the universe. The density field $\delta(\vec{x}) = (\rho(x) - \rho_o)/\rho_o$ is usually characterized by its Fourier transform $\delta(\vec{k})$. It is often assumed that the Fourier modes are Gaussian random variables. In this case, the power spectrum $P(k) = <|\delta(k)|^2>$, suffices to characterize all the statistical properties of the density field. Two theories compete to explain the origin of density perturbations: (1) they were generated during inflation, favored by the data on cosmic microwave background temperature anisotropies, or (2) they are the result of distortions produced by topological defects (cosmic strings or domain walls) on a homogeneous background. Inflation generically predicts that today $P(k) \propto k^n$ with $n \sim 1$ on scales close to the horizon size.

The exact shape of the power spectrum is one of the most challenging problems of observational cosmology. Given an initial density field, density perturbations can be evolved in time and the power spectrum at present can be computed. The final processed spectrum depends on the geometry of the universe, the baryon fraction, and the amount and nature of the dark matter. In the figure on page 308 we plot a compilation of the power spectrum obtained from the distribution of galaxies and clusters in different catalogs. The largest scale sampled by the data is $600h^{-1}$ Mpc. The amplitude is different since clusters and galaxies are biased tracers of the matter distribution. *See* biasing parameter.

Maunder Minimum   The period from 1645 to 1715 during which the number of sunspots on the solar disk was severely depressed and,

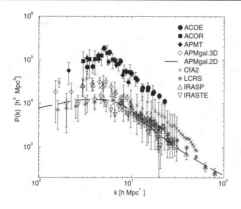

Observed power spectrum of galaxies and clusters of galaxies: ACO-E and ACO-R are spectra for Abell-ACO clusters as derived with two different techniques; APM-T is the spectrum of APM clusters; APM-gal.3D and APM-gal.2D are spectra of APM galaxies found from 3-D and 2-D data; CfA2 is the spectrum of the SSRS+CfA2 130 Mpc/h sample; LCRS is the spectrum of the Las Campanas Redshift Survey; IRAS-P and IRAS-TE are spectra of IRAS galaxies found by two different groups. Notice the difference in amplitude when the power spectrum corresponds to galaxies and clusters. The difference is due to bias.

despite frequent and systematic attempts to observe them, sunspots were detected only occasionally. During this period, when sunspots were detected, they were observed at relatively low latitudes. The solar rotation was slower than today, in addition, and the differential rotation was more pronounced. This period of unusual low solar activity is also documented indirectly by an unusually small number of aurora observations and unusual high counts in cosmogenic nuclides such as ^{14}C and ^{10}Be. The ^{10}Be concentrations in ice cores suggest that a solar cycle still was existent but its duration was reduced to 9 years. The combination of these observations with models of solar activity suggests that during the *Maunder Minimum* the solar constant might have been reduced by 0.2 to 0.5%. The Maunder Minimum coincides with an unusual cold time period, the Little Ice Age.

maximal extension of a space-time   A space-time is called extendible if it has an extension. A *maximal extension of a space-time* is an extension that is not extendible. *See* extensions of space-times.

maximally rotating black hole   A Kerr black hole for which the angular momentum per unit mass (which in geometrized units has dimensions of [mass]) reaches a maximum value equal to the mass of the black hole, in these units. Thus, the maximum angular momentum expected from the solution of Einstein's field equation for a rotating black hole is equal to GM^2/c, and corresponds to a black hole rotating at the speed of light at a radius equal to GM/c^2, i.e., at the event horizon for this maximally rotating case. (In the nonrotating case the event horizon is at $r = 2GM/c^2$.) As shown by Thorne, a rotating black hole interacting with its environment can come close to, but cannot reach, this limit.

maximally symmetric space   A metric space which admits the maximum number N of linearly independent Killing vectors. If D is the dimension of space, then $N = D(D+1)/2$ and the space is isotropic with respect to each point; therefore, it is also homogeneous and has constant curvature.

For $D = 3$, one has $N = 6$ and the space-time is described by one of the Robertson–Walker metrics. *See* homogeneity, isotropy, Killing vector, Robertson–Walker cosmological models.

Maximum Observable Frequency (MOF)
Defined as the highest frequency at which a reflection from the ionosphere occurs during oblique incidence sounding between two points. The MOF is a measurement made on a real propagation path using an oblique sounder capable of exciting all possible paths for the circuit. It may also refer to an identified mode, in which case the mode would be shown with the MOF. *See* ionospheric radio propagation path, oblique ionogram.

maximum probable wave   (or probable maximum wave) A description of the largest wave that is expected to exist in a particular sea condition. Used for design purposes.

Maxwell–Boltzmann distribution   An expression for the fraction of molecules $f(v)$ in a gas that have velocity v within a small specified interval, divided by that interval:

$$f(v) = 4\pi \left(\frac{M}{2}\pi RT\right)^{\frac{3}{2}} v^2 e^{\frac{-Mv^2}{2RT}}$$

where M is the molar mass, R is the molar gas constant, and T is the temperature.

The *Maxwell–Boltzmann distribution* is a classical distribution; it does not take account of quantum indistinguishability, which are described by Bose–Einstein distribution for integer spin quantum particles, and by the Fermi–Dirac distribution for half-integer spin quantum particles.

Maxwell material   Also called a Maxwell fluid or Maxwell solid. A viscoelastic material that behaves as a linearly elastic solid in response to high frequency loading but as an incompressible, linearly viscous fluid in response to low frequency loading. The strain rate for the *Maxwell material* is a combination of the rate of elastic strain according to Hooke's law and the rate of viscous strain according to Newtonian flow law. The relation between volumetric strain $\theta = \varepsilon_{ii}$ and pressure $p = \sigma_{ii}/3$ is elastic: $\theta = p/K$, where K is the bulk modulus. The constitutive relation for the deviatoric strain $\varepsilon'_{ij} = \varepsilon_{ij} - \theta\delta_{ij}/3$ and deviatoric stress $\sigma'_{ij} = \sigma_{ij} - p\delta_{ij}$ is

$$\frac{\partial \varepsilon'_{ij}}{\partial t} = \frac{1}{2G}\frac{\partial \sigma'_{ij}}{\partial t} + \frac{1}{2\mu}\sigma'_{ij}$$

where G is shear modulus, and μ is viscosity. The quantity $\tau_M = \mu/G$ is called the Maxwell relaxation time, a time that roughly defines the transition from predominantly elastic to predominantly viscous behavior after a suddenly imposed constant loading. In modeling the rheology of the Earth or other planets, the above constitutive relation is often modified by replacing the Newtonian flow law with a nonlinear flow law (such as the Power law). In geophysics, such nonlinearly viscoelastic materials are usually still referred to as the Maxwell materials.

mean anomaly, M   Measured in degrees, the ratio M/360 is equal to the ratio of the time elapsed since last periapse to the orbital period. Thus, at periapse, M = 0 , and at apoapse, M =

180. For a circular orbit the *mean anomaly* and the true anomaly are the same.

mean celestial equator   The great circle on the celestial sphere perpendicular to the mean celestial pole. Its intersections with the ecliptic define the mean vernal equinox, and the mean autumnal equinox.

mean celestial pole   The mean direction of Earth's north rotation pole when the short timescale (days to decades) variations, called nutation, are averaged out. This hypothetical pole executes a circle of radius approximately 23.44° about the North Ecliptic Pole, in about 25,800 y. (There is a corresponding mean south celestial pole, opposite to the north one.) The precise derivation of the motion of this pole is given by J.H. Lieske, T. Lederle, W. Fricke and B. Morando in *Astron. and Astrophys.,* **58**, 1 (1977).

mean cosine of scattering angle   The integral over all directions of the volume scattering function multiplied by the cosine of the scattering angle, divided by the integral over all directions of the volume scattering function; also called the single-scattering asymmetry factor.

mean diameter, mean grain size   The mean particle size in a soil sample. Not as frequently used as median grain size, d_{50}.

mean field dynamo   A mathematical simplification of the physics governing the behavior of magnetic field in a dynamo such as that in the Earth's core so that the evolution of the magnetic field can be simulated. Essentially, the system is considered to be axisymmetric; that is, variations along lines of latitude are neglected. However, it turns out that a purely axisymmetric dynamo cannot exist (Cowling's theorem), as a vital part of the axisymmetric portion of the magnetic field can only be generated from longitudinal variations in the magnetic field. In the *mean field dynamo,* such effects (which are commonly assumed to be associated with fluid turbulence) are parameterized. This simplification of the equations allows dynamo solutions to be found that satisfy both the equation governing the evolution of the magnetic field and the force balance equation of the conducting fluid. In historical terms, this was in some ways an improvement on "kinematic dynamo" models which only solve for the magnetic field and not for the forces within the fluid, although the kinematic models did not have the axisymmetric simplification. The advent of high performance computing has allowed fully three-dimensional simulation of the dynamo process, which alleviates the need for mean field parameterization.

mean free path   In normal spatial diffusion the average distance a particle travels between two subsequent collisions with particles from its surroundings. The specified type of interaction may be some particular types, such as collision, scattering, capture, or ionization. The total *mean free path* is the definition that specifies for all types of interaction. If we consider scattering processes where the direction of motion is changed only slowly by successive interactions, such as in pitch-angle scattering, the mean free path can also be interpreted as the average distance traveled by a particle before its direction of motion is reversed. In one-dimensional diffusion, the relation between the diffusion coefficient D and the mean free path λ is $D = v\lambda/2$, in three-dimensional diffusion it is $D = v\lambda/3$, (where v is the mean particle speed). In photon-matter interactions, the mean free path equals the inverse of the beam attenuation coefficient [m].

mean motion   For an object in motion, $n = 2\pi/P$, where P is the period of the orbit, an average of the angular velocity of the orbit.

mean motion (short-period) resonances
The relationship between motions of two bodies orbiting a common central object where the mean motions n, that is, the average angular speeds of these bodies are in the ratio of whole numbers. For instance, the Neptune-Pluto 3:2 mean motion resonance means $n_N : n_P = 3 : 2 = 1/167$ yr : $1/250$ yr. Thus, while Pluto makes two revolutions around the sun, Neptune makes three revolutions and the two planets pass one another once, which results in a perturbative force with a short period of the order of the synodic period $1/(n_N - n_P) = 500$ years. The

$(i + N) : i$ *mean motion resonances* are called those of order N.

mean sea level   The average level of sea by averaging height of the sea surface for all stages of the tide. It is obtained by using long period tidal height observational data. The longer the tidal data is, the more reliable and stable the value of the *mean sea level* will be. It is a widely used standard datum level in geodesy.

mean solar day   The average length of the day over one year. Because the Earth's orbit is an ellipse with the sun at one focus, noon-to-noon time varies over the year and is shortest in Northern hemisphere winter. This variation is described by the equation of time.

mean solar time   Time based conceptually on the diurnal motion of a fictitious mean sun, under the assumption that the Earth's rotational rate is constant, as opposed to apparent solar time, to which it is related by the equation of time.

mean water level   The average water level at a site, with the averaging period taken to be long enough to remove wind waves. It may, however, include the influence of wave setup on water level. As reported in tide tables, the *mean water level* typically represents a 19-year average.

mechanics   The study of forces and the resulting motions that arise from contact between bodies, or from interaction with conservative fields (e.g., gravitational or electromagnetic fields).

mechanics, Newtonian   The fundamental laws on which classical dynamics is based. (1) A physical body will maintain a state of rest or uniform motion in a straight line unless forces act upon it. (2) The rate of change of momentum is proportional to the applied force, and it is in the same direction as the applied force. (3) For every force of one body acting on a second, there is an equal force of the second body acting on the first of the same magnitude and in opposite direction. Named after Isaac Newton (1642–1727).

median diameter, median grain size   The median particle size in a soil sample. Generally reported in millimeters or ϕ-units.

Meiyu front   A persistent surface front that separates cold air to the north and warm, humid air to the south over East Asia and northwestern Pacific in early summer. It slants slightly northeastward along which precipitating low pressure systems develop and travel eastward. It first forms in May in South China, progresses northward and covers the Yangtze (Changjiang) River, Korea and Japan in mid-June. In late July, it starts to move further northward and weakens substantially in intensity. Called Changma in Korea and Baiu in Japan, the Meiyu (Plum Rain in Chinese) is the single most important climate feature that controls the floods and droughts in East Asia.

membrane paradigm   A viewpoint in black hole physics which regards the event horizon of a black hole to be a two-dimensional membrane made of a viscous charged electrically conducting fluid. The physics of a black hole are then governed by standard equations of fluid mechanics, electrodynamics, and thermodynamics. Although the *membrane paradigm* is not equivalent to the full mathematical theory of black holes, it is useful for intuitive understanding of complex astrophysical processes involving black holes.

Mercalli intensity scale   An arbitrary scale, devised by Italian geologist Giuseppi Mercalli in 1902, to quantify the intensity of earthquake shaking at a given location, regardless of the magnitude of the earthquake. The scale ranges from I (detectable only by sensitive instruments) to XII (almost total destruction). It has been modified by different countries according to local conditions and has been largely replaced by the Richter scale.

Mercury   Named after the Roman messenger god, *Mercury* is the planet closest to the sun, having a mass of $M = 3.3022 \times 10^{26}$ g, and a radius of $R = 2444$ km, giving it a mean density of $5.43 \, \mathrm{g \, cm^{-3}}$ and a surface gravity of 0.38 that of Earth. Its rotational period of 58.65 days is exactly 2/3 of its orbital period around the sun, 87.97 days. The rotation axis has an obliquity of 0°. The slow rotation means that the planet's oblateness is very nearly 0. Mercury's orbit around the sun is characterized by a mean distance of 0.39 AU, an eccentricity of $e = 0.206$, and an orbital inclination of $i = 7.0°$. Its synodic period is 116 days. An average albedo of 0.11 gives it an average surface temperature of 620 K on the day side, and around 100 K on the night side. It does not have any significant atmosphere. Mercury is probably composed of a large Fe-rich core surrounded by a silicate mantle, but because its moment of inertia is not known, the relative sizes of the core and mantle can only be estimated from the mean density of the planet. Mercury has no known satellites. Less than half the surface of Mercury was mapped during the Mariner 10 flybys in the 1970s so relatively little is known about this planet.

merger   The remnant of a collision between two galaxies, in which the stars of the two galaxies have been mixed together to form a single galaxy. In the process of merging, two galaxies behave like highly inelastic bodies that interpenetrate, and transfer part of their orbital energy to their stars. Collisions among galaxies of similar mass with orbital velocity comparable to their internal velocity dispersion (typically a few hundred km s^{-1}) can produce *mergers*. If at least one of the merging galaxies is gas-rich, the compression of the gas in the disk leads to extensive star formation which, in turn, temporarily enhances the galactic luminosity. The morphology of a late-stage merger is expected to be nearly indistinguishable from that of an elliptical galaxy. It is, however, unclear which fraction of elliptical galaxies has been formed by merging of disk galaxies. A prototypical merger is the galaxy NGC 7252, also known as the "Atoms for Peace" galaxy. NGC 7252 appears still highly disturbed in a deep image taken with a long exposure time, where tidal tails are visible, but resembles a regular elliptical galaxy in its central parts, with radial surface brightness profile following a de Vaucouleurs' law, in an image taken with a shorter exposure. *See* de Vaucouleurs' law, starburst galaxy.

merging, magnetic   *See* reconnection.

Merian's equation   An equation for determination of the natural periods of a basin of water, such as a lake or harbor. The natural period is also referred to as the seiching period.

meridian   A great circle passing through the poles of a sphere; in astronomy the *meridian* of an observer is the meridian passing through the zenith (i.e., directly overhead).

meridional   In the direction of the meridians, the lines of constant longitude (i.e., north-south). *See also* zonal.

meromixis   Refers to a natural water body, which does not experience deep convection (such as due to cooling) on a regular basis (e.g., annual). *Meromixis* is mostly linked to permanent salinity gradients at greater depth than occurs in isolated ocean basins (such as in the Canada Basin) as well as in many deep lakes (such as Baikal, Tanganyika, Malawi, etc.), as well as in eutrophic lakes.

Merope   Magnitude 4.2 type B5 star at RA 03^h46^m, dec $+23°56'$; one of the "seven sisters" of the Pleiades.

mesogranulation   Intermediate scale of convection on the sun. Mesogranules are typically 5,000 to 10,000 km in size with speeds (vertical or horizontal) of $\sim$60 m s^{-1}.

meson   An unstable bosonic massive particle which undergoes strong nuclear interactions.

mesoscale   A motion scale in the atmosphere smaller than a synoptic scale like the scale of a weather system, but larger than a microscale like the scale of an individual fair weather cumulus clouds. *Mesoscale* phenomena include the weather fronts, tornadoes, and storm and maintain waves. In the ocean, the scale of the Gulf Stream eddies or rings is considered a mesoscale.

mesosphere   In atmospheric physics, the atmospheric layer which lies between the stratosphere and the thermosphere. It is bounded on the bottom by the stratopause and at the top by the mesopause. On Earth, the *mesosphere* lies between 50 and 85 km altitude. The mesosphere is characterized by a decrease in temperature with altitude. The transport of radiation within the terrestrial mesosphere is primarily due to the absorption and emission of heat by carbon dioxide. The mesosphere and the stratosphere combined are often referred to as the middle atmosphere.

In solid earth geophysics, the lower portion of the upper mantle to the lower mantle underlying the asthenosphere. The Earth's crust and mantle can be divided into three different portions, namely, lithosphere, asthenosphere, and mesosphere, in terms of their depth range and dynamic properties. The mesosphere has difficulty flowing compared to the asthenosphere. Since motion of the mesosphere is considered to be much slower than plate motions near the Earth's surface, absolute plate motion is understood as relative motion to the mesosphere.

mesotrophic water   Water with moderate concentrations of phytoplankton biomass; chlorophyll *a* concentration ranges from between approximately 0.5 to 10 mg m^{-3}.

Messier   The Messier catalog is a list of nebulous objects in the sky compiled by Charles Joseph Messier (1730–1817). *Messier* was a comet hunter and kept a list of fuzzy objects in the sky which were not comets, starting with the Crab Nebula. There are 103 objects in the catalog and each is indicated by the letter "M" followed by its number in the list. The catalog includes galaxies (e.g., M31, the Andromeda Galaxy), planetary nebulae (e.g., M42, the Great Nebula in Orion), and star clusters (e.g., M13, the Great Cluster in Hercules).

metallic hydrogen   When hydrogen molecules are placed under sufficiently high pressure, the molecules dissociate and the resultant atomic phase has electrons that are free to move about the lattice, giving the solid metallic properties. The exact pressure of the transition depends on the temperature, but shock experiments indicate a transition pressure of 140 GPa at 3000 K.

metallicity   The abundance, in stars or galaxies, of all elements heavier than hydrogen and helium (for which the early tracers were metals like Ca, Fe, and Na). The solar value is about 1.7%, or $Z = 0.017$. Observed values range from 10^{-4} of solar metallicity (for the oldest stars in the halos of our own and other galaxies) to perhaps 2 to 3 times solar (for stars in the centers of large elliptical galaxies and the ionized gas near the centers of quasars). Frequently *metallicity* is given as the $\log_{10}$ of the ratio of iron/hydrogen of the star divided by iron/hydrogen in the sun. This is abbreviated [Fe/H]. *See* elliptical galaxies, halo, quasar.

metamorphism   If sedimentary rocks are buried in a sedimentary basin, the temperature can become sufficiently high so that chemical reactions can occur. These reactions can transform a sedimentary rock into a metamorphic rock.

meteor   Any phenomenon in the air (a cloud is a humid meteor), but especially the glowing track of a meteoroid as it travels through the air.

Such *meteors* also are often called shooting stars or falling stars. Most of the debris creating meteors is extremely small (dust to sand-sized) and burns up during the atmospheric passage. If the material survives the passage through the atmosphere and lands on the surface, it is called a meteorite. A meteor shower occurs when large numbers of meteors are seen radiating from the same area of the sky.

meteorite   Meteoroids that survive to impact onto a body's surface. *Meteorites* are classified as stones, meteorites that are composed primarily of silicates and are similar to terrestrial rocks; stony-iron meteorites which have approximately equal amounts of iron and silicates; iron meteorites, composed almost entirely of nickel-iron; and further subdivided according to compositional details. Stony-irons are the least common. Most meteorites show thermal processing, and are believed to be fragments of larger parent bodies, probably asteroids, although Martian meteorites and lunar meteorites have also been identified. Meteorites are also often characterized as falls (seen to fall and recovered shortly thereafter) and finds, which are found long after they have landed. Stony meteorites are the most common types of falls and are thus believed to represent the majority of meteoritic material. Irons are the most common types of finds simply because they do not erode as quickly, nor are they easily confused with terrestrial rocks. *See* meteorite parent body, SNC meteorites, and under the different classes of meteorites.

meteorite parent body   Different classes of meteorites have been differentiated to various degrees, and have been subjected to elevated temperatures and pressures. These processes presumably occurred while the meteorite was still part of a parent body. The breakup of such parent bodies resulted in the various meteortite classes observed. Particular asteroid classes have been identified with parent bodies of various meteorite types based on spectral similarities.

meteoroid   A metallic or rocky small body orbiting the sun in space; small particles that may derive from comets; larger (1 cm or larger) particles that probably derive from the asteroid belt. *Meteoroids* seen passing through an atmosphere are called meteors, and if they survive to land on the surface, they are called meteorites. Small meteoroids are also often called interplanetary dust particles (IDPs). Sunlight reflecting off this material gives rise to the zodiacal light and the gegenschein.

meteor shower   A large number of meteors radiating from the same area of the sky. *Meteor showers* are caused when the Earth passes through an area of space with an enhanced amount of small debris. Most meteor showers are associated with cometary debris left behind in the comet's orbit. Meteors in a meteor shower appear to radiate from the same area of the sky since the material causing the shower is located in a relatively small region of space. The constellation encompassing the area of the sky from which the meteors appear to radiate provides the name of the meteor shower (e.g., the Perseids appear to radiate from the constellation Perseus, the Leonids come from the constellation Leo, etc.). Debris large enough to survive passage through the atmosphere and land on the surface as a meteorite is very rare during meteor showers.

methane   A flammable, explosive, colorless, odorless, tasteless gas, CH_4. Concentration in the air approximately 1.8 ppm. A greenhouse gas with 25 times the effect of CO_2. Its current contribution to the greenhouse effect is estimated at 13%. The chief constituent of natural gas. Produced in all sorts of biological decay. Boiling point $-161.6°C$, freezing point $-182.5°C$. Probably also present in clathrates in abyssal ocean floors.

Metis   Moon of Jupiter, also designated JXVI. Discovered by S. Synnott in 1979, its orbit lies very close to that of Adrastea, with an eccentricity and inclination that are very nearly 0, and a semimajor axis of 1.28×10^5 km. Its radius is 20 km, its mass 9.49×10^{16} kg, and its density 2.8 g cm^{-3}. It has a geometric albedo of 0.05, and orbits Jupiter once every 0.295 Earth days. Also, an asteroid, ninth asteroid to be discovered, in 1848. Orbit: semimajor axis 2.3865 AU, eccentricity 0.1217, inclination to the ecliptic 5°.579, period 3.69 years.

metric   The array of coefficients (components) which, in principle, depend on position and are needed to calculate the length of a curve segment when the coordinates of the ends of the segment are given, or the abstract operation (tensor) which is computed in a particular reference frame using these components. The notion can be applied in a space with an arbitrary number of dimensions and with an arbitrary curvature. The metric is the tensor that acts on vectors to return their length; it can also produce the scalar product of two vectors. The simplest example is that of a Euclidean space in rectangular Cartesian coordinates. Suppose the space is four-dimensional, and the end-points of the straight segment (vector) have coordinates (x^1, x^2, x^3, x^4) and $(x^1 + \Delta x^1, x^2 + \Delta x^2, x^3 + \Delta x^3, x^4 + \Delta x^4)$. The length of the segment, Δs, is then given by the Pythagorean theorem:

$$(\Delta s)^2 = \left(\Delta x^1\right)^2 + \left(\Delta x^2\right)^2 + \left(\Delta x^3\right)^2 + \left(\Delta x^4\right)^2,$$

and components of the metric form the array

$$g = \begin{pmatrix} 1 & 0 & 0 & 0 \\ 0 & 1 & 0 & 0 \\ 0 & 0 & 1 & 0 \\ 0 & 0 & 0 & 1 \end{pmatrix}.$$

The form of the metric expressed in components depends on the reference frame, and the zeros here stand for the absent terms $\Delta x^1 \Delta x^2$, $\Delta x^1 \Delta x^3$, etc. that would be present in nonrectangular coordinates.

In 3-dimensional Euclidean space, in rectangular coordinates (x, y, z) the metric is diag(1, 1, 1) similar to the above, but if the distance is expressed spherical coordinates (r, θ, φ) where

$$x = r \sin \vartheta \cos \varphi, \quad y = r \sin \vartheta \sin \varphi, \quad z = r \cos \vartheta,$$

then the components of the metric are

$$g = \begin{pmatrix} 1 & 0 & 0 \\ 0 & r^2 & 0 \\ 0 & 0 & (r \sin \vartheta)^2 \end{pmatrix}.$$

From here, the metric of the surface of a sphere $r = a = $ const can be read out:

$$g = \begin{pmatrix} a^2 & 0 \\ 0 & (a \sin \vartheta)^2 \end{pmatrix}.$$

This metric on the 2-d surface of a sphere has nonconstant components no matter what reference frame is used for the 2-d surface, showing that the metric contains implicit information about the curvature, in fact, all geometrical properties of the space. In general relativity, the metric describes the geometry of the spacetime. Components of the metric are the unknown functions in Einstein's equations. If the curvature implied by a given metric is nonzero, then the corresponding spacetime is a model of a system with a gravitational field (e.g., the interior of a star, the neighborhood of a star, the whole universe). In this case, the components of the metric are not globally constant in any coordinate system. If the curvature is zero, then the gravitational field is absent, and the metric describes the flat space that is the background of the special relativity theory. In rectangular Cartesian coordinates, it has the components diag($-1, 1, 1, 1$). The coordinate that is distinguished from the others by this "-1" is the time.

metricity of covariant derivative   The requirement that the covariant derivative have no effect on the metric:

$$\nabla g = 0,$$

or in components,

$$g_{\alpha\beta;\gamma} = 0 \, ,$$

where ; denotes the covariant derivative. This form can be used to define the affine connection coefficients in terms of derivatives of the metric, the torsion, and the structure coefficients associated with the basis as $\Gamma^{\rho}_{\beta\gamma} = g^{\rho\mu}\Gamma_{\mu\beta\gamma}$, where

$$\begin{aligned}\Gamma_{\mu\beta\gamma} &= \tfrac{1}{2}\left(g_{\mu\beta,\gamma} + g_{\mu\gamma,\beta} - g_{\beta\gamma,\mu}\right.\\ &+ C_{\mu\beta\gamma} + C_{\mu\gamma\beta} - C_{\beta\gamma\mu}\\ &\left. - T_{\mu\beta\sigma} + T_{\beta\mu\sigma} + T_{\sigma\mu\beta}\right) \, ,\end{aligned}$$

where $C_{\beta\gamma\mu} = g_{\lambda\mu}C_{\beta\gamma}{}^{\lambda}$ where $C^{\beta}{}_{\mu\gamma}$ are the structure coefficients and $T^{\mu}{}_{\gamma\beta} = g^{\mu\rho}T_{\rho\gamma\beta}$ is the torsion tensor. *See* affine connection, structure coefficients, torsion.

metric radio burst   Solar radio emission in the metric range that is at frequencies between some MHz and about 500 MHz. Metric radio bursts are characterized by a frequency drift from high to low frequencies. The radio emission is caused by streams of energetic electrons that excite Langmuir oscillations as they travel through the solar corona. The frequency of such Langmuir oscillations depends on the electron density n_e according to

$$\omega_{\text{pe}} = \sqrt{\frac{4\pi n_e e^2}{m_e}} \, .$$

Thus with a density model of the corona, the height of the radio source at a certain time and thus also its propagation speed can be inferred from the frequency drift. *Metric radio bursts are classified into five different types: the continuous radio noise from the sun (see type I radio burst), the slow (see type II radio burst) and fast (see type III radio burst) drifting bursts giving evidence for shocks and streams of energetic electrons, and the continuum emission following these bursts (see type IV radio burst, type V radio burst).*

MHD   *See* magnetohydrodynamics.

MHD condition   The mathematical condition $\mathbf{E} = -\mathbf{v} \times \mathbf{B}$, obeyed to a good approximation in highly conducting fluids or plasmas and in collision-free plasmas. The *MHD condition* prescribes the components of the electric field $\mathbf{E}$ which are perpendicular to the local magnetic field $\mathbf{B}$, in a fluid obeying it which is moving with bulk velocity $\mathbf{v}$. It signifies the vanishing of the perpendicular (non-relativistic) electric field in a frame of reference moving with the fluid velocity $\mathbf{v}$. In many situations, the component parallel to $\mathbf{B}$ can be assumed to vanish as well.

The MHD condition also assures the freezing-in of magnetic flux, into the fluid or plasma moving with bulk velocity $\mathbf{v}$.

MHD simulation   The numerical simulation of the behavior and motion of a plasma assumed to obey the MHD equations, using a fast computer. *MHD simulations* of the global magnetosphere, its tail region, shocks and reconnection in plasmas, comet behavior, the expansion of the solar wind and other phenomena in space plasma physics are widely used to study situations that cannot be duplicated in the lab and are not easy to observe in nature with sufficient resolution.

Some results of MHD simulations, e.g., for collision-free shocks, are very encouraging. However, MHD equations do not cover all details of plasma behavior, some of the boundary conditions (e.g., in the atmosphere) must be approximated, and results may be hard to check against observations. Still, the use of this method is growing.

Michelson Doppler Imager/Solar Oscillations Investigation (MDI/SOI)   Helioseismology instrument aboard SOHO spacecraft which analyzes the vibrational modes of the sun and also measures the sun's magnetic field in the photosphere.

Michelson–Morley experiment   In 1887, in Cleveland, Ohio, Albert A. Michelson and Edward Morley attempted to detect a difference in the speed of light in two different directions: parallel to and perpendicular to the motion of the Earth around the sun. Such an effect would be expected if there was a fixed ether through which the Earth moves. The experiment, which was a repeat of one done in 1881 by Michelson, used an interferometer with right angle arms of 11.0 m optical path length. If light travels with constant speed c with regard to a fixed ether (the name given to this hypothetical substance), and the Earth moves through this ether with speed v, then the travel time for light along the arms should be aligned along and across the motion should differ by $.5(v/c)^2$; the "cross stream" time is shorter. Note that this difference is second order in the ratio of the velocity to the speed of light. In the interferometer, this path difference will lead to a phase difference and an interference between the two beams of light in the interferometer. Furthermore, the interferometer can be turned (Michelson and Morley mounted theirs on granite and floated it in mercury to facilitate this) and the interference pattern should shift as the device is turned. To their amazement, Michelson and Morley found no such effect, with an experimental accuracy of about 2 to 3%. Multiple repeat experiments arrived at the same result. This result was debated for a very long time. It is now regarded as one of the fundamental experiments supporting special relativity.

Miche–Rundgren theory   A theory for description of wave-induced forces on a wall (non-breaking waves).

micrometeorite   A meteorite less than 1 mm in diameter. *Micrometeorite* strikes are a major source of erosion on the moon, and of the production of the lunar surface (the regolith).

microstructure   Fluctuations on scales at which entropy is generated by the smoothing effect of molecular viscosity and diffusivity are referred to as *microstructure*. This scale lies typically below 1 m in the ocean and lakes; often structures in CTD profiles ($<$ some dm) resolving the Kolmogorov or Batchelor scale are generally called microstructure; *see* Kolmogorov scale, Batchelor scale.

microwave background radiation   In observational cosmology, the radiation field at microwave frequencies with a black body spectrum corresponding to heat radiation at about 3 K; the radiation left over from the early period when the universe was dense and hot, 10^5 years (10^{13} sec.) after the Big Bang. Initially the temperature was so high that ordinary elementary particles could not exist. Matter emerged from the Big Bang in the form of a mixture of its sim-

plest components: protons, neutrons, electrons, photons, and neutrinos. (Cosmology also considers still earlier epochs, e.g., inflation; and the epoch when even protons and neutrons had not existed because the cosmic matter was a mixture of quarks.) At first, matter was so hot that no stable atoms could form and the particles remained in thermodynamical equilibrium with photons. However, the universe was cooled because of expansion, and later, 10^5 years after the Big Bang and at the temperature $\approx$ 3000 K, the atomic nuclei that had come into existence in the meantime (these were hydrogen [protons], deuterium, tritium, helium, lithium, beryllium, and boron, formed in reactions of the protons and neutrons that were there from the beginning) could capture the electrons. At this moment, the radiation was emitted for the last time (this moment is called last scattering) and it has evolved without significant contact with matter until now. It has kept the spectrum of a black body radiation, but its temperature is constantly decreasing. Exactly this kind of radiation was detected in 1965, with the temperature at 2.73 K. Its black-body spectrum has been verified with a very high precision, and it comes to us from all directions in space, with the relative fluctuations of temperature ($\Delta T / T$) not exceeding 10^{-5} (this result is obtained after the anisotropies in T caused by the motion of the Earth on its orbit, of the sun in the galaxy and of the whole galaxy in the local group have been subtracted). This discovery eliminated the steady-state models and is still the strongest confirmation that the idea of a Big Bang is correct. The radiation was in fact detected in 1935 by A. McKellar, through excitations in the CN-molecules in interstellar space, but the results were not understood at that time. The existence of the background radiation was predicted by George Gamow and co-workers in 1946–1948 on the basis of theoretical speculations. Robert Henry Dicke and James Peebles with co-workers were preparing an experiment to detect the radiation in 1965, when it was actually (and accidentally) discovered by Arno Penzias and Robert W. Wilson in the course of a quite different experiment, as an irremovable noise in a microwave antenna.

microwave burst   A transient enhancement of solar radio emission in the mm–cm wavelength range, normally associated with optical and/or X-ray flares. *Microwave bursts* provide a powerful diagnostic of energetic electrons in the solar atmosphere.

mid-ocean ridge A location of seafloor spreading (divergence zone). The more or less continuous line centered in the oceans where new material reaches the surface of the Earth, driven by convection in the Earth's mantle. The Mid-Atlantic ridge begins in the North polar region, wanders south between W15° and W45°, essentially staying in the middle of the Atlantic Ocean. It connects to the Southwest Indian Ridge at about latitude S50°, longitude 0°. The Southwest Indian Ridge connects with the Indian Ridge at about S30°, E70°. From this junction, the Indian Ridge extends north as the Central Indian Ridge through the Red Sea; it extends South and East as the Southeast Indian Ridge along about S45°, becoming the Pacific Antarctic Ridge at S60°, E150°. This becomes the East Pacific Rise at about S60°, W120°. The East Pacific Rise runs north along approximately W100°, up the west coast of North America, ending around N50° as the Juan de Fuca ridge.

Mie scattering Scattering of light by a spherical particle. Given the complex index of refraction of the particle, and the ratio of its radius to the wavelength of the light, it is possible to exactly solve Maxwell's equations to find the fraction of light absorbed, and the fraction scattered, as well as the phase function of the scattered light and its polarization. The detailed solution of this problem is called *Mie scattering.*

Mie size parameter For scattering by spheres, the ratio of a sphere's circumference to the wavelength.

migration (seismic) This is a technique which, when implemented on data from seismic reflection surveys, can help to elucidate the structure of the underlying rock. In a reflection survey, seismic waves are generated at one point on the Earth's surface and recorded at geophones (i.e., seismic wave detectors) distributed nearby (for example, in a linear fashion behind a ship). Each time the seismic signal encounters a "reflector", i.e., a rock layer that causes part of the seismic energy to be reflected back to the surface, a seismic pulse is returned to the geophones. The time delay between the generation of the seismic signal and its detection at a geophone depends on the speed of seismic waves in the rock, the depth of the reflector, and the horizontal distance between seismic source and geophone, but also on the geometry of the reflector. For example, if the reflector is inclined upwards toward the geophone from the direction of the source, then the part of the reflection that is recorded at the geophone will have been reflected closer to the geophone, rather than midway between the geophone and the source if the reflector had been horizontal. *Migration* corrects this effect, and also removes other geometrical artifacts such as diffractions associated with scattering centers in the rock.

Milankovich cycle Cyclic variations in climate driven by periodic changes in orbital and Earth orientation parameters. Climate in a particular area depends on the solar flux, and therefore on both the distance from the Earth to the sun and the angle between the surface of the area in question and the sun's rays over the course of a day (overhead sunlight leading to a greater flux than tangential sunlight). These both cause the seasonal variations in weather. The importance of the former depends on the eccentricity of the Earth's orbit, which varies on a 96,000-year cycle, while the importance of the latter depends on the obliquity, i.e., the angle between the planes defined by the Earth's equator and by the Earth's orbit around the sun (the ecliptic plane), which varies between 21° and 24° (it is currently at 23.5°) and which varies on a 41,000-year cycle. The effect of the eccentricity may be either to augment the seasonal variations in either the northern or southern hemisphere while reducing the variations in the other hemisphere (if the closest approach of the Earth to the sun occurs near a solstice, i.e., northern or southern winter), or to have a relatively neutral effect (if it occurs near an equinox, i.e., northern or southern spring). The orientation of the orbit is also cyclic, so that these effects vary on a 22,000-year timescale. As long, hard winters are thought to be important for growing ice sheets, these cycles may therefore have significant impact on global climate, although they do not directly cause global warming or cooling. Such astronomical cycles occur for other planets as well, particularly Mars.

Miles–Phillips–Hasselmann theory   A theory for description of the growth of wind wave energy in the sea.

Milky Way   The band of light in the night sky resulting from the stars in the galactic plane. The term is also used to denote the galaxy in which the sun is located. *See* galaxy.

millibar   A measure of pressure. One one-thousandth of a bar. One bar is equivalent to 750 mm (29.53 in.) of mercury, or 10^5 Pascals (Pa), and represents 98.7% of normal atmospheric pressure.

Mimas   Moon of Saturn, also designated SI. Discovered by Herschel in 1789. Its surface is dominated by the crater Herschel which has a radius of 65 km, nearly 1/3 the radius of the entire moon. This is probably the largest crater the body could have sustained without disruption. Its orbit has an eccentricity of 0.02, an inclination of 1.53°, a precession of 365° yr^{-1}, and a semimajor axis of 1.86×10^5 km. Its radius is 196 km, its mass 3.80×10^{19} kg, and its density 1.17 g cm^{-3}. It has a geometric albedo of 0.5, and orbits Saturn once every 0.942 Earth days.

minimal coupling   The expression of the behavior of some field ϕ, which may be a scalar, spinor, vector, or tensor in a second field, written by simply replacing the partial derivatives in the description of the free motion of ϕ, by covariant derivatives involving the connection from the second field. An example is the equation for a scalar wave ϕ in a gravitational background, where the equation of motion is

$$g^{\alpha\beta} \nabla_\alpha \nabla_\beta \phi = 0 \, ,$$

where ∇_α is the covariant derivative along α. If there is no gravitational field (flat space), then in rectangular coordinates $g^{\alpha\beta}$ is the simple form diag$(-1, 1, 1, 1)$, and ∇ is the ordinary partial derivative, but in a general spacetime additional terms in the equation arise beyond second partial derivatives. Nonetheless, this is in a geometrical sense the simplest way to express an influence of the gravitational field or ϕ. *See* affine connection.

minimum-B surface   Another name for the equatorial surface of the magnetosphere. Because of the weak field of the cusps, this surface splits into two sheets on the dayside of the Earth, in the region near the magnetopause. The sheets that lead to the cusps are associated with butterfly distributions.

minisuperspace   In general relativity, the space of geometries obtained by restricting the form of the spatial metric in superspace so as to have a finite number of degrees of freedom (functions of time only). For such choices of the metric tensor, Einstein's equations greatly simplify, and it is possible, for instance, to carry out quantization of the gravitational field, a problem which has no known unambiguous solution in the general case. *See* geometrodynamics, superspace, Robertson–Walker cosmological models.

minor axis   In an ellipse, the distance corresponding to the minimum length measured along one of the two symmetry axes, or this axis itself.

Mira   Refers to a class of cool variable stars called long period variables, or LPV. The star *Mira* (*o* Ceti) is the proto-type, first observed in 1596. Mira is a red giant classified as spectral type M5e – M9e and luminosity class III. Mira variables typically have long, regular periods of variation, usually 300 days or longer. The light curve is typically asymmetric with the rise from minimum to maximum light being faster than the decline. The defining characteristic of Mira variables is the large change in brightness from minimum to maximum light, being at least 2.5 visible magnitudes and often as much as 10 magnitudes for some Miras. LPV stars (*see also* semi-regular, irregular variables) are thought to be past the He core-burning red giant stage and in a very brief end stage of evolution called the asymptotic giant branch, where the inert C, O core is surrounded by concentric shells of He-burning and H-burning material. About 90% of LPVs are oxygen-rich and 10% are carbon-rich (known as carbon stars).

mirage   Appearance of the image of some object, usually in distorted form, arising from refraction in air with strong index of refraction variations, caused by temperature variations. The refraction generically bends light rays toward air with greater density (lower temperature). Hence, mirages near the hot sand surface in a desert produce a false pond (the light paths are concave upward, so objects above the ground appear to be reflected in a surface near the ground); and those at sea over cold ocean produce looming (the light paths curve downward and follow the curve of the Earth, allowing far distant objects to be seen, and stretched out in the vertical direction).

Miranda   Moon of Uranus, also designated UV. It was discovered by Kuiper in 1948. Its orbit has an eccentricity of 0.0027, an inclination of $4.7°$, a semimajor axis of 1.30×10^5 km, and a precession of $19.8°$ yr^{-1}. Its radius is 236 km, its mass is 6.3×10^{19} kg, and its density is 1.14 g cm^{-3}. Its geometric albedo is 0.27, and it orbits Uranus once every 1.413 Earth days. *Miranda* is known for its highly irregular terrain.

mirror instability   Instability of a thermally anisotropic collisionless plasma, usually associated with an excess of pressure transverse to the magnetic field. The general instability criterion is complicated; it takes a simple form for a bi-Maxwellian hydrogen plasma with equal proton and electron anisotropy:

$$\frac{8\pi}{B^2}\left(P_\perp - P_\parallel\right) > \frac{P_\parallel}{P_\perp}$$

where B is the mean magnetic field strength, and $P_\perp$ and $P_\parallel$ are, respectively, the pressures transverse and parallel to the mean magnetic field.

mixed layer   In atmospheric physics, over land, a layer of 1 to 2 km thick where the vertical transport of sensible heat, moisture, and momentum is so efficient that potential temperature, mixing ratio, and wind show relatively little change with height. Also called planetary boundary layer. In oceanography near-surface waters subject to mixing by wind and waves; there is little variation in salinity or temperature within the mixed layer. Typically a layer of 100 m or less thick, it is separated from the colder waters of the deep ocean by the thermocline.

mixing efficiency   Parameter commonly used in oceanic turbulence studies, which relates the buoyancy production of turbulent kinetic energy B to the dissipation rate ϵ: $\gamma = B/\epsilon$. The mixing efficiency is related to the flux Richardson number Ri$_f$ by

$$\gamma = \frac{\text{Ri}_f}{1 - \text{Ri}_f}.$$

Most estimates for shear-induced turbulence in natural stratified water fall within the range of $\gamma_{\text{mix}} = 0.15 \pm 0.05$ (Ivey and Imberger, 1991), but oceanic observations of γ range from 0.05, obtained from measurements at open ocean locations, to 0.7 in highly energetic flows in tidal channels. *See also* Richardson number.

mixing length   In the theory of the turbulent flow of fluid, the average distance through which each turbulent cell moves by turbulent motion before meeting other turbulent cells. Symbol is l'. Defined by Planck, who assumed that moving within the mixing length, the properties of turbulent cell will not change. In the parameterization of turbulent shear stress, the *mixing length* is expressed as

$$-\rho <u'w'> = p <l'^2> \left|\frac{\partial <\vec{V}>}{\partial z}\right| \frac{\partial <u>}{\partial z}$$

where l' is mixing length, $\vec{V}$ and u are horizontal wind speed and its component on the x direction, $< >$ means average value, and u' and w' are turbulent fluctuating velocity. A turbulent coefficient K_m can be defined as

$$K_m = <l'^2> \left|\frac{\partial <\vec{V}>}{\partial z}\right|.$$

Different effective mixing lengths may be observed for different fluid properties. Thus, usually the turbulent coefficient K_m is used instead of using mixing length l'. However, mixing length theory is still widely used in astrophysics.

mixing ratio   The ratio of the mass of vapor to the mass of dry air.

Mixmaster universe   A dynamically anisotropic but homogeneous theoretical cosmology (Misner, 1966) which expands from a Big Bang to a maximum size and recollapses to a final "big crunch". The constant-time 3-spaces are distorted 3-spheres. A nonlinear oscillation occurs because the spatial curvature generates a "force term" in the motion of the anisotropy, bounding the anistropy by an amount which is finite at any one time, though the bound itself diverges at the Big Bang and big crunch. The universe behaves like an expanding, then contracting, ball that shrinks to a flattened spindle and then puffs out first in one direction and then in another, changing into a pancake configuration, and so on. Any shape can eventually arise. Further, cosmological horizons can be eliminated in some situations, allowing mixing of matter and providing, in principle, a homogenization (though this has been shown to be of very small probability); thus the name, which also stems from the analogy with a particular kitchen appliance.

MJD   *See* Modified Julian Date.

MK system of stellar classification   In the 1890s Harvard Observatory began to classify stars according to hydrogen features, such as the strength of Balmer lines. In the 1920s this system of classification based on temperature was revised and compiled by E.C. Pickering, A.J. Cannon, and others, leaving us with O, B, A, F, G, K, M. The types were divided up into 10 subtypes, from 0 to 9, and special types R, N, and S were added for carbon-rich stars. However, often stars of the same temperature type differed greatly in luminosity, and so several luminosity classification schemes were used. In the 1930s W.W. Morgan and P.C. Keenan of Yerkes Observatory documented a luminosity classification in combination with a temperature type that is still used today. The system used Roman numerals along with the subdivided temperature classification. Class V stars are on the main sequence, class Ia, b are the supergiants, class II are the bright giants, class III are the normal giants, class IV are the sub-giants, class VI are the dwarfs, and class VII are the white dwarfs. Each of these luminosity classes occupy distinct positions on the Hertzprung–Russell diagram. Stars with the same temperature yet different luminosity must differ in surface area (*see* effective temperature). This difference produces measurable effects in some spectral features, such as pressure broadening, and can thus be measured. Later, Keenan developed, along with others, a dual type of classification for the difficult late-type carbon-rich stars, the carbon stars. Types N and R have been replaced by a C classification followed by two numbers, one indicating the temperature and the other the strength of the carbon abundance.

M magnitude   Stellar magnitude derived from observations in the infrared at a wavelength of 5 μm.

model atmosphere   A computationally constructed atmosphere model, created according to physics laws and representing the atmospheric status by discrete values on horizontal grid mesh and vertical levels. By inputting initial atmospheric state conditions, numerical atmospheric models can start time integration to simulate the real atmospheric processes approximately. Their accuracy depends on the model's resolution, the number and quality of contained physics processes, and parameterization packages to describe the sub-grid processes in the models. By simulating real atmospheric processes, numerical atmospheric models can provide products for weather forecasts, research on different atmospheric science issues, and on climate status in ancient times.

mode water   An ocean water mass characterized by small vertical gradient in temperature and density. For this, it is also called thermostad or pycnostad. Occupying a great depth range, this water mass stands out as a distinct mode in a volumetric census taken against temperature or density. Subtropical mode waters, found in northwestern parts of the North Pacific and North Atlantic subtropical gyres, are the most famous ones, with a characteristic temperature centered on 18°C, they are sometimes called 18°-waters. They form in the deep mixed layer under intense cooling in winter and are advected southwestward by the wind-driven subtropical gyres. Other *mode waters* with different temperature characteristics are found in other parts of the oceans.

Modified Julian Date (MJD)   The Julian date minus 2400000.5 days. MJD is used in atomic time work.

modulation of galactic cosmic rays   The intensity of galactic cosmic rays (GCRs) in the inner heliosphere varies in anticoincidence with the 11-year cycle of solar activity: during times of high solar activity, the intensity of GCRs is diminished while it is maximal during times of low solar activity. This temporal variation is called modulation; it is most pronounced at energies between about 100 MeV/nucl, decreases to about 15 to 20% at energies around 4 GeV/nucl, and vanishes at energies higher than about 10 GeV/nucl.

Modulation of GCRs results from the variation of the structure of the heliosphere during the solar cycle, allowing cosmic rays to propagate faster and/or in larger numbers from the outer borders of the heliosphere (such as the termination shock) into the inner heliosphere during solar minimum, when the heliospheric structure is rather simple with a small tilt angle and only very few transient disturbances such as magnetic clouds and interplanetary shocks. Since different effects influence the propagation of energetic particles in the interplanetary magnetic field, modulation is determined by these processes as well. Of all the processes influencing interplanetary propagation (*see* interplanetary propagation) for modulation only those processes that vary during the solar cycle are important:

1. Drift in the heliospheric current sheet. The heliospheric current sheet separates magnetic field lines of opposite polarity. Thus, gradient drift allows particles to propagate along this neutral line from the outer skirts of the heliosphere toward the orbit of Earth. During solar minimum conditions, when the tilt angle is small, the heliospheric current sheet is rather flat. During maximum conditions, however, the tilt angle becomes much larger, leading to a wavier heliospheric current sheet and consequently a much longer drift path for particles traveling into the heliosphere, thus diminishing intensities of galactic cosmic rays at times of solar maximum conditions.

2. Drift from polar to equatorial regions and vice versa. This drift should lead to a charge separation of electrons and positively charged particles and therefore in the equatorial plane different intensity-time profiles for both species should be observed depending on the polarity of the solar magnetic field. The influence of this effect is still debated because observations are inconclusive.

3. Transient disturbances. In contrast to the drift, transient disturbances cause step-like intensity decreases, called Forbush decreases (*see* Forbush decrease). These transient disturbances are shocks and magnetic clouds; thus, their number increases with increasing solar activity. Therefore, part of the decrease in cosmic ray intensity during solar maximum can be understood as a fast sequence of decreases related to transient disturbances.

4. Merged and grand merged interaction regions (MIRs and GMIRS). Not only traveling shocks but also shocks at corotating interaction regions cause a reduction in cosmic ray intensity. This reduction becomes even more pronounced when at larger radial distances corotating interaction regions and transient disturbances merge, eventually forming a closed torus around the inner heliosphere. Galactic cosmic radiation then can be blocked efficiently and globally by this torus.

Mohorovičić discontinuity   A discontinuity in the speed of seismic waves which marks the transition from the Earth's crust to its mantle. Also called the *moho* for short, it is found at depths of between 25 and 40 km under continents, and about 5 km under the ocean floor. Its location is relatively easily determined because seismic waves reflect off of it.

Mohr's circle   A circle whose diameter is the difference between maximum and minimum principal stresses which are on an axis of normal stress as the abscissa, and an axis of shear stress as the ordinate. Using Mohr's circle, we can obtain the relation between normal stress and shear stress exerted on a plane with arbitrary direction within a plane containing the axes of the two principal stresses. All the *Mohr's circles* at the time of fracture drawn at a variety of state of stresses are tangent to a straight line represented by Coulomb's formula. An area of stresses for the upper side of the line becomes that of fracture.

moist lapse rate   The lapse rate calculated on assumption that the air remains saturated and that all the liquid water formed by condensation is removed as precipitation (without affecting the buoyancy of the parcel). This is also called the pseudoadiabatic lapse rate. The *moist lapse rate* is smaller than the adiabatic lapse rate. *See* adiabatic lapse rate.

molar gas constant (R)   Constant of proportionality of the equation of state of an ideal gas when the amount of gas is expressed in number of moles n, i.e., $PV = nRT$. A mole of a substance is the amount of a substance whose total mass, when expressed in grams, numerically equals the dimensionless molecular weight of the substance. The 1998 CODATA (Committee on Data for Science and Technology) recommended value for the Molar Gas Constant is $R = 8.314472(15)$J mol^{-1}K^{-1}.

molecular cloud   A part of the interstellar medium where the gas is composed mostly of molecules and contains substantial dust. Multiatom molecules are observed in these clouds, with hydrogen and carbon monoxide being the most abundant. The mass of these clouds can be up to 10^6 solar masses and their size from 10 to 100 parsecs. The temperature varies from 10 to 80 K, and the density is 100 to 10^6 particles cm^{-3}.

molecular torus   A thick structure within the innermost few parsecs of a galaxy, made up of dense molecular clouds, which obscures the innermost part of active galactic nuclei if the torus axis is seen at large angles. Concentration of molecular gas in the regions surrounding the nuclei is observed in starburst and several active galaxies, for example in the case of the prototype Seyfert-2 galaxy NGC 1068. However, properties such as the geometry, the extension, the thickening mechanism, and the frequency of occurrence of molecular tori in active galactic nuclei are largely hypothetical. *See* Seyfert galaxies, starburst galaxy.

moment   (torque) The cross product of a force vector and a position vector. Has units of force times distance (ft-lb, N-m) and is used to measure twisting forces.

moment of inertia   In Newtonian dynamics, a 3-tensor whose components are quantities describing the angular response of a body under an applied torque, computed by the volume integral of the density of the body weighted by functions of the displacement from an assumed origin. In general, the moment of inertia is written as a 3×3 matrix, but in a specific Cartesian frame centered on the center of the mass of the body, aligned with the principal axes of the body (which coincide with the symmetry axis of the body, if the latter exist) the moment of inertia is diagonal: (I_x, I_y, I_z) where

$$I_x = \int \rho dV (y^2 + z^2) dV$$

and I_y, I_z are defined by cyclically rearranging indices. By relabeling axes, we may assume $I_z \geq I_y \geq I_x$. With this ordering, rotation about the axis corresponding to I_x or I_z is stable, rotation about the y-axis is not.

Monin–Obukhov length   Parameter to compare the relative importance of mechanical and convective turbulence having the dimension of a length scale; the depth z, at which law-of-the-wall scaling and convective scaling provide the same value for the dissipation of turbulent kinetic energy. Defined by

$$L_M \equiv -\frac{u_*^3}{\kappa \alpha g \overline{wT'}}$$

where u_* is the friction velocity, κ is the Karman constant, α is the thermal expansivity, g is the constant of gravity, and $\overline{wT'}$ is the heat flux. The *Monin–Obukhov length* is mainly conceived for atmospheric turbulence studies, but it can also be applied to oceanic applications. L_M is the distance from the boundary at which shear production and the buoyant destruction of turbulence are of the same order. The boundary can be either the bottom boundary (land or ocean bottom), or the sea-surface where L_M is measured downward. For both stable and unstable conditions, at a distance $z \ll |L_M|$ the effects of stratification are slight and turbulence is dominated by mechanical processes (quantified by u_*). For $z >> |L_M|$ the effects of stratification dominate. For unstable conditions, at distances $z >> -L_M$ turbulence is generated mainly by buoyancy forces in the form of free convection, and at these distances shear production of turbulence is negligible.

monopole   Pointlike topological defects as described in some field theories which may carry magnetic charge, possibly produced in the very early universe. *Monopoles* arise in symmetry breaking schemes in Grand Unification theories. Monopoles are massive, and their predicted abundant production may conflict with observation unless other processes intervene to dilute their density after production. *See* cosmic topological defect, homotopy group, Langacker–Pi mechanism, monopole excess problem, spontaneous symmetry breaking, t'Hooft–Polyakov monopole.

monopole annihilation   In cosmology, the potential destruction of (perhaps magnetic) monopoles by combining a monopole and an anti-monopole, destroying both and releasing energy. Unfortunately, such a pair will first form a bound state before decaying, and these bound states can survive for a very long time. Thus, the annihilation rate cannot keep pace with the expansion rate and the monopole excess persists from the early universe to the present day. *See* cosmic topological defect, Langacker–Pi mechanism, monopole excess problem.

monopole excess problem   Grand unified theories assume the existence of a simple compact symmetry group G which is subsequently broken to H during a cosmic phase transition. A generic prediction of such symmetry breaking theories is the occurrence of monopoles.

Applied to cosmology, their remnant energy density Ω_M (in units of the critical density) in the universe is found to be very large. In fact, calling h the Hubble constant (in units of 100 km·sec^{-1}·Mpc^{-1}), we have

$$\Omega_M h^2 \simeq 10^{11} \left(\frac{T_C}{10^{14}\text{Gev}} \right)^3 \left(\frac{m_M}{10^{16}\text{Gev}} \right),$$

which for grand unified monopoles ($T_C \geq 10^{14}$ GeV, $m_M \geq 10^{16}$ GeV) is some 11 orders of magnitude above any acceptable value today. The inflationary paradigm and particular scenarios such as the Langacker–Pi mechanism were proposed to solve this problem.

A more direct understanding of the *monopole excess problem* arises at the various experimental constraints derived from monopole nondetection. With the density given above, the expected flux, calculable as their number density times their characteristic velocity, namely on average

$$\langle F_M \rangle = n_M \langle v_M \rangle ,$$

is numerically found to be of order (in units of cm^{-2} · sr^{-1} · sec^{-1})

$$\langle F_M \rangle \simeq 10^{-3} \left(\frac{T_C}{10^{14}\text{Gev}} \right)^3 \left(\frac{v_M}{10^{-3}\text{c}} \right) .$$

Such a large flux is well within the present day experimental possibilities and should therefore be easy to detect (in contrast to what happens in reality). *See* cosmic topological defect, critical density, Kibble mechanism, Langacker–Pi mechanism, monopole, monopole annihilation, Parker limit.

monsoon   A seasonal wind which blows with regularity and constancy during one part of the year, and which is absent or blows from another direction during the remainder of the year. In English, this word comes from Arabic in which it means "season". In the late 17th century, E. Hadley first suggested that *monsoons* result from the temperature difference between continent and ocean. As the direction of the monsoon changes, weather is changed correspondingly. The climate of the Indian subcontinent is especially characterized by the monsoon, where a distinct rainy season (from June to September) occurs in the southwesterly monsoon. Since the path of monsoon is long, the direction of the monsoon is greatly influenced by the Coriolis effect. After World War II, as high level observation data became more abundant, the monsoon theory became more complete and detailed. Currently, it is believed that monsoons depend on three factors: sea-continent distribution, planetary circulation, and orographic effects. Monsoons can be divided into four types:

(1) sea-continent monsoon, (2) high level monsoon, (3) plateau monsoon, and (4) planetary monsoon. Every monsoon phenomenon is the result of mixing these four types. According to the latitude distribution, monsoons can be divided as tropical monsoons, sub-tropical monsoons, and temperate monsoons. The Indian monsoon is caused by the seasonal change of planetary wind belts which is enhanced by the influence of sea-continent distribution effects, and the eastern Asian monsoon is caused mainly by sea-continent distribution factor. The Tibetan Plateau is a major plateau monsoon region which is due to its huge plateau's orographic effects. Other major areas of monsoons are southeastern Asia, the west African coast (latitude 5°N - 15°N), and northern Australia.

monsoon climate   A climate associated with the tropical regions of the Indian Ocean, Southwestern Brazil, and northwest South America. Characterized by high annual temperatures ($\geq$ 27° C), heavy high sun (summer) rain, and a short low sun (winter) dry period. Total annual rainfall is approximately 3 m.

month   The period of one revolution of the moon about the Earth or the calendar unit approximating this. A "tropical month" is the period between successive passages by the moon of the vernal equinox (27.32158214 days of mean solar time). A "sidereal month" is the period between two successive transits through the hour circle of a fixed star (27.32166140 mean solar days). An "anomalistic month" is the period between two successive transits of the moon through its perigee (27.5545505 mean solar days). A "draconitic" or "nodal month" is the period between successive passages through the ascending node of the moon's orbit (27.212220 mean solar days). A "synodical month" is the period between two identical phases of the moon (29.5305882 mean solar days).

moody diagram   A diagram for determination of a dimensionless friction factor which appears in the Darcy–Weisbach equation for description of head loss in a full conduit carrying a fluid. The friction factor is generally determined graphically once the Reynolds Number

for the flow and the relative roughness of the pipe are known.

moon   (**1.**) Any natural body which orbits a planet. The term "moon" is often used synonymously with "natural satellite". A moon can actually be larger than some planets (as is the case with Ganymede, Titan, and Callisto all larger than Mercury and Pluto).

(**2.**)   The twin planet of the Earth. The airless satellite of Earth of radius 1740 km, mass 7.35×10^{22} kg, orbital semimajor axis 3.844×10^5 km, eccentricity 0.0549, albedo 0.07. The moon is tidally locked to synchronous rotation to the Earth.

Dating of returned lunar rocks gives an age of 2.5×10^9 years ago for the oldest lunar surface. The surface shows evidence for heavy meteor bombardment ending about 3.5×10^9 years ago.

The moon is responsible for the majority of tidal effects on the Earth. Phases of the moon occur because its orbit is close to the ecliptic and carries it from positions between the Earth and the sun, when the side visible from Earth is unilluminated, to positions beyond the Earth from the sun, where the side visible from Earth is fully illuminated. Because the moon's angular size on the sky is closely the same as the sun's, it may cause solar eclipses — the shadow of the Moon on the Earth (when its orbit crosses the ecliptic at new moon). Lunar eclipses occur when the moon moves through the Earth's shadow in space; they occur when its orbit crosses the ecliptic at full moon.

moonbow   A rainbow that arises from the refraction and reflection of moonlight on rain drops or mist. Because of the faintness of the source (moonlight), its colors are usually very difficult to detect.

moraines   Glacial deposits of unconsolidated, unsorted material left at the margin of an ice sheet. Several types of *moraines* are recognized. Ground moraines are deposits of material with no marked relief. This material is believed to have been transported at the base of the glacier. A lateral moraine is a moraine formed along the side of a valley (alpine) glacier and composed of rock scraped off or fallen from the valley sides. A media moraine is a long strip of rock debris carried on or within a glacier. It results from the convergence of lateral moraines where two glaciers meet. A terminal moraine is a sinuous ridge of unsorted material pushed ahead of an advancing glacier and deposited when the glacier begins to recede.

Moreton wave   The chromospheric component, seen in H_α radiation, of a solar flare-induced wave which propagates away from the flare site. A *Moreton wave* travels with a roughly constant velocity of ~ 1000 km s^{-1} and is attributed to MHD fast mode shocks generated in the impulsive phase of the flare.

morning cloud (Mars)   *Morning clouds of Mars* appear in the early morning and disappear in midday. They reappear the next morning at the same place. They tend to appear in the equatorial region and mid-latitudes of the northern hemisphere in spring to summer of the northern hemisphere. In late spring to mid-summer of the northern hemisphere morning clouds are seen everywhere in low latitudes. They do not disappear even in midday and shift to evening clouds, so that the equatorial region is surrounded with a cloud belt. A notable morning cloud is the one that appears in an area centered at 10 N and 120 W, surrounded by gigantic volcanos of Olympus Mons and Tharsis Montes. The altitude of the cloud is lower than the tops of volcanos. The vertical optical depth of the cloud is about 2 in the early morning. A large morning cloud extends from low latitudes of the southern hemisphere to mid-latitudes of the northern hemisphere in the most active period. *See* evening cloud, afternoon cloud.

morphodynamics   A term used to describe the changing morphology, or form and structure, of a coastal area.

morphology-density relationship   The observational fact that more elliptical galaxies are found near the center of a cluster of galaxies where the number density of galaxies is higher while more spiral galaxies are observed in the outskirts of the cluster where the number density is lower (Dressler, 1980). The relative abundance of ellipticals, lenticulars, and spirals is now known to vary smoothly over the range

from 10^4 galaxies per Mpc^3 near the cluster center to 10^{-2} galaxies per Mpc^3 near the cluster boundary. It is not completely understood if the segregation of galaxy types is primordial, but is widely believed to be the result of evolutionary effects subsequent to the formation of the galaxies.

mountain climate   Climate of relatively high elevations. Typically populated by northern (or arctic) biome. Conditions include short growing season (in nontropical mountains), decreased nighttime temperature, reduced pressure and oxygen availability, substantial rainfall, and increased ultraviolet solar radiation.

moving magnetic features   Small-scale regions of magnetic flux which migrate across the solar surface. When associated with the growth of sunspots these flux regions emerge and approach sunspots with speeds of 0.25 - 1 km s^{-1}.

M star   Star of spectral type M. Betelgeuse and Antares are *M stars*.

MUF fading   *See* skip fading.

multicell storm   Storm with multicells. *Multicell storms* come in a wide variety of shapes, sizes, and intensities and most of them move systematically toward the right of the environmental winds in the middle troposphere. However, the storm propagates toward the right in discrete jumps as individual cells form and dissipate. *See* supercell storm.

multifingered time   The concept that, in general relativity, a time function may be defined which is very general, and in particular does not correspond to uniform proper time advance into the future. *See* proper time.

multipath fading   Multipath, for any propagation link, is the condition where more than one ray can pass between the transmitter and the receiver. Multipath, on an ionospheric propagation path, is a common situation arising from the different paths a ray can follow through the ionosphere. The most obvious source of multipath occurs when several different modes are excited (e.g., the 2E-mode, 1F-mode, and 2F-mode may all be possible). This can lead to severe amplitude fading, and hence signal distortion, because of the small time delays (up to 1 msec and more for several modes) in radio signals traveling by different modes. The effects of this form of multipath may be minimized by choosing a working frequency that is supported on fewer modes, and by selecting antennas that favor one mode over others. A single ionospheric mode (e.g., 1-F mode) may encompass four separate paths: the ordinary and extraordinary paths and a high and low ray for each of these polarizations. This situation is common near the maximum useable frequency for the circuit and time delays of the order of 0.5 msec are possible. *See* fading.

multiple ring basins   Very large impact basins often display a pattern of three or more complete or partial rings of mountainous materials. These *multiple ring basins* provide information on the thickness of the crust at the time the basin formed. Multiple ring basins are created by impact into a rigid surface layer (lithosphere) which overlies a more fluid layer (asthenosphere). If the crater resulting from the impact is smaller than the thickness of the lithosphere, a normal impact crater is produced. However, if the impact crater is greater than the thickness of the lithosphere, the asthenosphere can flow inward beneath the portion of the lithosphere which is pushed down by the impact. This flow exerts a drag force on the descending segment of the lithosphere, which can cause fracturing in the lithosphere surrounding the crater. The number of rings in a multi-ring basin is related to the thickness of the crust at the time of basin formation — fewer rings are produced in a thicker crust, while many rings indicate a relatively thin crust.

multiple shock   An earthquake composed of several spatially and temporally discontinuous subevents. Most large earthquakes are *multiple shocks*. From detailed analysis of seismograms, we can find that an earthquake source fault does not spread continuously and smoothly, but expands intermittently and heterogeneously. This kind of rupture pattern is considered to reflect the inhomogeneous structure of a source region.

mushy zone   As a melt containing more than one component (e.g., a rock magma with diverse chemistry, or the fluid of the Earth's outer core, which is thought to be composed of iron, nickel, and some lighter elements) is cooled, the composition of the first crystals to solidify are generally different from the bulk composition of the melt as a whole. As cooling continues, the chemical makeup of the material being solidified evolves, both because temperature of the melt is changing and because the chemical makeup of the remaining melt is itself changing (becoming more enriched in the material excluded from the first crystals). These factors also come into play at the boundary between the Earth's inner core and outer core: Between the depths where a parcel of outer core fluid would start to crystallize and finish crystallizing, the boundary region may be a mixture of liquid and solid material, commonly referred to as a *mushy zone*. As neither the composition of the core nor the thermodynamic conditions at the inner core boundary are particularly well constrained, the existence and depth of the mushy zone are not known, but it may be a factor both for the seismic properties of the top of the inner core and the dynamic boundary conditions for the geodynamo at the base of the outer core. The crystallization of aqueous analogs reveals some of the nature of mushy zones: liquid circulates through the zone, creating organized structures such as channels called "chimneys".

N

Nabarro–Herring creep   When material is subjected to differential stress, the differential stress sets up a vacancy concentration gradient at the scale of grain-size, which causes spatial distribution of vacancy. Therefore, vacancy flux will occur, producing creep. Creep caused by motions of vacancies through a self-diffusion mechanism is called *Nabarro–Herring creep*, which was first proposed by Nabarro in 1948 and improved by Herring in 1950. *Nabarro–Herring creep* is described in the form of

$$\varepsilon = \alpha \frac{D_{SD}}{d^2} \frac{\sigma \Omega}{kT}$$

where T is temperature, D_{SD} is the coefficient of self-diffusion, σ is the differential stress, d is the grain size, Ω is the atomic volume, k is Boltzmann's constant, and α is a numerical coefficient depending on the grain shape and the boundary conditions for σ. It is a linear rheology ($n = 1$, n is the stress sensitivity of creep rate at steady-state stage) with a high sensitivity to grain-size (the strain rate depends on the grain size as $\varepsilon \propto 1/d^2$).

nadir   The direction directly downward from the observer, as defined by the local horizontal. A stone dropped from a bridge, for example, falls toward the *nadir* point. An Earth-orbiting spacecraft has two commonly used nadir directions: *geocentric nadir,* the direction towards the Earth center, and *geodetic nadir,* the direction of the normal dropped to the Earth ellipsoid. *See also* altitude, topocentric system, zenith.

nadir angle   The angle between a given direction and the downward vertical direction.

Naiad   Moon of Neptune also designated NVIII. Discovered by Voyager 2 in 1989, it is a small, irregularly shaped body approximately 29 km in radius. Its orbit has an eccentricity of 0.00033, an inclination of 4.74°, a precession of 626° yr^{-1}, and a semimajor axis of 4.82 × 10^4 km. Its mass has not been measured. It orbits Neptune once every 0.294 Earth days.

naked singularity   In general relativity, a singularity that is not hidden inside a black hole. A singularity is a place where currently known laws of physics break down. Hence, it can, in principle, emit matter and radiation at unpredictable rates and in unpredictable amounts. Einstein's equations do not automatically forbid the creation of any *naked singularities* (*see* cosmic censorship); examples of exact solutions of Einstein's equations are known in which naked singularities exist. Whether naked singularities can be created during collapse of massive bodies is a question of active theoretical interest. There have been no observations suggesting astrophysical naked singularities. The only singularity of relevance to observational astronomy has thus far been the one that created our universe, i.e., the Big Bang. *See* Cauchy singularity, future/past event horizon, real singularity.

nakhlite   One of a small class (only three are known) of meteorites, composed principally of pyroxene, a silicate containing calcium, iron, and magnesium. *Nakhlites* appear to be of Martian origin. Named for Nakhla, Egypt, the location of the first reported fall in 1911.

nano-flares   Proposed small-scale impulsive energy releases in the solar atmosphere due to reconnection: the photosphere, as the top of the solar convection zone, is in continuous motion with bubbles rising and falling and plasma flowing in and out. Frozen-in into this plasma, the magnetic field is shuffled around as well. Thus, on small scales, magnetic field configurations suitable for reconnection are likely to form, thereby converting magnetic field energy into thermal energy. Observational evidence for such small-scale impulsive energy releases is found in the so-called bright X-ray points and in small-scale exploding EUV events. *Nano-flares* are suggested to play an important role in coronal heating.

nanotesla (nT)   A unit of magnetism often used in geomagnetism. 1nT = 1.0×10^{-9} tesla, and is equivalent to 1.0×10^{-5} gauss. The older name for this magnetic field strength was the gamma, where $1\gamma = 1$ nanotesla $= 10^{-5}$ gauss.

narrow emission line galaxies A broad, heterogeneous class of galaxies showing permitted and forbidden emission lines in their optical nuclear spectra. The term "narrow" is used in juxtaposition to Seyfert-1 galaxies, which show much broader permitted lines. The line profiles of *narrow emission line galaxies,* whose width is several hundred kilometers and may reach 1000 km, are broad for normal galaxies. The term narrow is used to distinguish these lines and lines whose full width at half maximum is several thousands kilometers, emitted in a region of active nuclei (the Broad Line Region) distinct from the Narrow Line Region. The strongest lines in the optical spectrum of the Narrow Line Region are the Balmer lines of hydrogen, the forbidden lines of oxygen twice ionized at 500.7 nm and 495.9 nm, and the lines from ionized nitrogen at 654.8 and 658.3 nm. From the presence of strong forbidden lines, it is inferred that the density of the Narrow Line Region gas must be $\sim 10^4$ to 10^5 particles per cubic centimeter. The absence of variability and the observation of nearby active galaxies, where the Narrow Line Region is partly resolved, suggest that the Narrow Line Region spans distances in the range from a few parsecs to a few hundred parsecs from the galactic center. The excitation mechanism of the line emitting gas is probably photoionization by the radiation from the active nucleus; it has been suggested that mechanical heating might also play a role. In several nearby AGN, emission lines whose intensity ratios are similar to those observed for the Narrow Line Region extend far out from the nucleus, to a distance that can be a significant fraction of the size of a galaxy, typically several kiloparsecs, in an exceptional case even 20 kpc. Since these regions appear resolved in long slit spectra, they are often referred to as the Extended Narrow Line Region.

Narrow emission line galaxies are also collectively referred to as type-2 active galactic nuclei, again in juxtaposition to Seyfert-1 galaxies. Narrow emission line galaxies include Seyfert-2 galaxies or LINERs, and all other type-2 active galactic nuclei; among narrow emission line galaxies several authors also include galaxies whose nuclei show spectra of HI regions, which are not type-2 AGN.

Narrow Line Region (NLR) A region of active galaxies, where narrow permitted and forbidden emission lines are produced. The term narrow is used to distinguish lines whose width is typically $\gtrsim 300$ km s^{-1}, hence already unusually broad for non-active galaxies, and lines whose full width at half maximum is several thousand km s^{-1}, emitted in a region of active nuclei (the Broad Line Region) distinct from the NLR. The excitation mechanism of the line emitting gas is probably photoionization by the radiation from the active nucleus; it has been suggested that mechanical heating might also play a role. Images of several Seyfert-2 galaxies obtained with HST suggest that the morphology of the line emitting gas is closely related to radio plasma ejected from the nucleus. *See* narrow emission line galaxies.

Nasmyth universal spectrum Empirical energy spectrum of ocean turbulence in the inertial subrange. The spectrum is based on ocean turbulence data measured by Nasmyth in an energetic tidal flow. This spectral form is considered characteristic for oceanic turbulence in various flow regimes, and most oceanic turbulence measurements are routinely compared to Nasmyth's universal form.

According to Kolmogorov's hypothesis, the spectrum, Φ, of turbulent kinetic energy in the inertial subrange depends only on the wavenumber k, viscosity ν, and the rate of dissipation of turbulent kinetic energy ϵ. Dimensional arguments then require that

$$\Phi(k) = \epsilon^{2/3} k^{-5/3} F\left(k/k_\eta\right)$$

where $F(k/k_\eta)$ is a universal function. The function $F(k/k_\eta)$ was experimentally determined by Nasmyth and it is shown in the figure on page 329.

natural line broadening Minimum width $\Delta\lambda$ of a spectral line profile which arises from the finite Δt lifetime of the state via Heisenberg's uncertainty principle:

$$\frac{\Delta\lambda}{\lambda} = \left(2\pi\frac{c\Delta t}{\lambda}\right)^{-1}$$

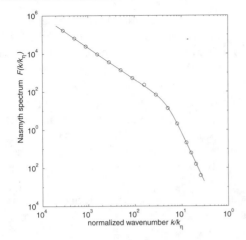

Universal spectrum of oceanic turbulence according to Nasmyth. The circles are the measured points and they are plotted against the cyclic wavenumber normalized by the Kolmogorov wavenumber k_η.

and has a Lorentzian shape given by:

$$\phi(\Delta v) = \frac{\gamma/4\pi^2}{\gamma/4\pi^2 + (\Delta v)^2}$$

where γ is the radiation damping constant. In a description in terms of classical radiation theory,

$$\gamma = \frac{8\pi^2 e^2}{3m_e c \lambda^2},$$

where e and m_e are the charge and mass of the electron, and c is the speed of light.

natural remanent magnetism   Many volcanic rocks contain magnetic minerals. When these rocks cool from the molten state, they are permanently magnetized by the Earth's magnetic field at that time. This *natural remanent magnetism* provides information on the time evolution of the Earth's magnetic field and continental drift.

nautical mile   A unit of length equal to exactly 1852 m.

nautical twilight   *See* twilight.

Navier–Stokes equations   The complicated set of partial differential equations for the motion of a viscous fluid subject to external forces.

neap tide   The tide produced when the gravitational pull of the sun is in quadrature, i.e., at right angles, to that of the moon. They occur twice a month at about the times of the lunar first and last quarters. In these situations the difference between high and low tides is unusually small, with both the high tide lower and the low tide higher than usual.

NEAR   A Near Earth Asteroid Rendezvous (NEAR) spacecraft that was the first Discovery mission launched on February 17, 1996. NEAR took photographs of Comet Hyakutake in March 1996, flew by C-asteroid 253 Mathilde on June 27, 1997, and S-asteroid 433 Eros on December 23, 1998, and also past the Main Belt asteroid Illya. It began orbiting S-asteroid 433 Eros in February 2000.

Its primary mission was to explore 433 Eros in detail for approximately 1 year. The mission will help answer basic questions about the nature and origin of near-Earth objects, as well as provide clues about how the Earth itself was formed. Studies were made of the asteroid's size, shape, mass, magnetic field, composition, and surface and internal structure. It achieved this using the spacecraft's X-ray/gamma ray spectrometer, near infrared imaging spectrograph, multispectral camera fitted with a CCD imaging detector, laser altimeter, and magnetometer. A radio science experiment was performed using the NEAR tracking system to estimate the gravity field of the asteroid. Periapsis of the orbit was as low as 24 km above the surface of the asteroid. Although NEAR has no landing legs, on February 12, 2001 it was landed on the surface of Eros at a speed of about 1.9 m/s. It photographed on the way down, the last image from 120 m altitude. Remarkably it continued to function, after impact sending back γ-ray data.

nearly diurnal free wobble   The retrograde motion of the Free Core Nutation (FCN) as viewed from points fixed on the Earth. The wobble appears from the Earth to have frequencies of $\omega_{\text{wobble}} = \omega_{\text{nutation}} - \Omega$, where Ω is the frequency of the Earth's rotation. Very long baseline interferometry observations, which measure the nutation in space relative to fixed stars, place the FCN frequency $\omega_{\text{nutation}} \approx 433.2$ cycles/sidereal day.

nebular hypothesis   The hypothesis that the solar system, and typical systems around other stars, formed by condensation of material in a molecular cloud in which the star first formed. Because there will be some initial angular momentum in such a cloud, cooling will lead to the formation of a disk. Planetary formation in the environment of a flattened disk gives an explanation for the general uniformity of rotational behavior among the planets, and for the fact that the planetary orbits all lie close to the same plane, the ecliptic. The composition of the planets, with rocky planets near the sun, and gas giants at larger distances can be explained by early strong radiation from the forming sun. However, several extrasolar planetary systems are now known, where this arrangement of planetary bodies does not seem to hold.

nebular lines   Forbidden emission lines observed in optical spectra of gaseous nebulae, typically planetary nebulae, HI regions, and external galaxies. The strongest *nebular lines* are observed at 495.9, and 500.7 nm, at 654.8 and 658.4 nm, and at 372.7 nm. Nebular lines are forbidden lines, not observed in laboratory spectra. Nebular lines were shown (Bowen, 1928) to be due to forbidden transitions between the lowest terms of singly and doubly ionized abundant atomic species, such as oxygen and nitrogen. *See* forbidden lines.

needle-probe method   A method for measuring thermal diffusivities of unconsolidated sediments or other soft materials. A needle-like probe containing a calibrated electric heat source and temperature sensors is inserted into the specimen. The diffusivity of the material is determined by analyzing the rate of the temperature rise after the heat source is turned on but within the time window when the line source of heat can be assumed to be infinitely long and the medium to be infinitely large.

neon   (From the Greek neos, new.) Noble gas, Ne, discovered in 1898 by Ramsay and Travers. Atomic number 10, natural atomic weight 20.179. Colorless, odorless, inert gas. Present in the atmosphere as 15 ppm. Melting point 24.56 K, boiling point 27.07 K.

The dominant natural isotope is ^{20}Ne, which constitutes more than 90% of the abundance. The other two stable isotopes are ^{21}Ne (0.25%) and ^{22}Ne (9.25%).

neon burning   The set of nuclear reactions that convert neon to heavier elements, including magnesium and silicon. It occurs in hydrostatic equilibrium in the evolution of stars of more than about 10 solar masses (between carbon burning and oxygen burning). It is not a major energy source even for them because neon is never very abundant.

Neptune   Giant gas planet, eighth planet from the sun. (Occasionally, as at present, Pluto is actually inside the orbit of Neptune.) Outermost gas planet. Mass 1.02×10^{26} kg, diameter 49528 km. Average distance from sun 30.06 AU, orbital eccentricity 0.009. Rotation period 19.1 hours, orbital revolution period 164.8 years. Inclination to orbit 29.6°. The visible surface has a composition closely matching primordial amounts: 74% hydrogen, 25% Helium. Its surface mean temperature is 48 K. Its surface albedo is of order 0.5.

Nereid   Moon of Neptune, also designated NII. It was discovered by G. Kuiper in 1949. Its orbit has an eccentricity of 0.751, an inclination of 27.6°, a precession of 0.39° yr^{-1}, and a semimajor axis of 5.51×10^6 km. The orbit, which has the highest eccentricity of any satellite in the solar system, indicates that *Nereid* is a captured Kuiper Belt object. Its radius is 170 km, and its mass is not known, but has been estimated to be about 2×10^{19} g based on an assumed density of 1 g cm^{-3}. It has a geometric albedo of 0.155, and orbits Neptune once every 360.1 Earth days.

net photosynthetic rate   The total rate of photosynthetic CO_2 fixation minus the rate of loss of CO_2 in respiration.

network   In solar physics, general name given to the distribution of photospheric magnetic field outside of sunspots and faculae. The network pattern consists of magnetic fields with characteristic cell dimensions of 20,000 to 40,000 km, covering the quiet photosphere. There is a corresponding chromospheric network defined by bright emission in Ca K spectroheliograms.

Neupert effect   Statement that the temporal derivative of the observed soft X-ray emission during a solar flare reproduces the observed time development of the hard X-ray emission. This is found in many flares and indicates that a single energization process is responsible for the production of both the non-thermal and thermal radiation.

neutral equilibrium   In mechanics, a configuration in which the system experiences no net force (equilibrium), and if the system is moved slightly from this state, no forces arise, so a nearby configuration is also an equilibrium. The paradigm is a ball on a smooth horizontal surface. *See* stable equilibrium, unstable equilibrium.

neutral point   In electromagnetism, a point in a magnetic configuration in which the field intensity B drops to zero, and where therefore the direction of the magnetic field is undefined. Expanding the field $\mathbf{B}$ around an isolated *neutral point* gives $\mathbf{B} = \mathbf{B}_0 + \mathbf{r} \cdot \nabla \mathbf{B}_0$ in the immediate neighborhood of the point, with subscript zero marking values at the point. Since $\mathbf{B}_0 = 0$, the constant dyadic $\nabla \mathbf{B}_0$ determines the character of the field. In an "x-type neutral point," $\nabla \mathbf{B}_0$ has three real characteristic roots and three field lines cross at the point, giving a configuration like the letter "x" but usually an intricate small-scale geometry. In an "o-type neutral point" $\nabla \mathbf{B}_0$ has just one real root and the point is the limiting point of nested closed field lines.

Neutral lines of x-type and o-type form in two-dimensional geometries (extending unchanged in the third dimension), where the number of real roots is 2 or 0. The nearby field lines either form an x-shaped pattern outlined by the eigenvectors of $\nabla \mathbf{B}_0$, or form nested o-shaped loops.

Neutral points are of great interest in space plasma physics because theory associates them with magnetic reconnection. The polar cusps of the Earth magnetosphere are also evolved (x-type) neutral points.

In atmospheric physics, the points where the degree of polarization of sky diffuse radiation

equals zero. Due to aerosol scattering, multiple scattering and reflecting from surfaces, polarization of sky light is not consistent with the ideal status; there will be some abnormal neutral points. According to Rayleigh scattering theory, under ideal conditions the solar point (the direction of solar incident ray) and anti-solar point are neutral points. In the real atmosphere, in general, there are three neutral points: Arago neutral point, which locates about 15° to 25° above the anti-solar point; Babinet neutral point, which locates about 12° to 25° above the sun; and Brewster neutral point, which locates about 15° to 25° below the sun. All these angles and positions may change due to the variations of the position of sun, the atmospheric turbidity and the reflective properties of the Earth surface.

neutral stability   In meteorology, the stratification status when the lapse rate of air temperature (γ) equals its adiabatic lapse rate (γ_d).

neutrino   A stable elementary particle which carries zero electric charge, angular momentum of $\frac{1}{2}\hbar$, little or no mass, but finite momentum and energy. Invented by Pauli in the 1930s to make energy, momentum, and angular momentum conservation hold in weak nuclear decays, the electron *neutrino* was identified in the laboratory by Cowan and Reines in 1953. Two additional types, called the μ and τ neutrinos (for the other particles co-produced with them), have been found since. Each has an anti-particle, leading to a total of six types. In astrophysics, electron neutrinos are produced in stellar nuclear reactions, including the proton-proton chain and CNO cycle. All three types are produced by Type II Supernovae, and there should be a thermal sea of all types left from the early universe, corresponding to the cosmic microwave background radiation. Solar neutrinos and those from one supernova, 1987A, have been observed in laboratory detectors.

neutrino annihilation   In core-collapse supernovae and gamma-ray burst models, the number density of neutrinos and anti-neutrinos can become large, and the rate of their annihilation and production of electron/positron pairs and photons can be an important part of the explosion energy. However, the annihilation cross-section depends sensitively upon the incident angle of the neutrino and anti-neutrino, peaking for head-on collisions. For supernovae and spherically symmetric gamma-ray bursts, this angular dependence drastically diminishes the energy produced by neutrino annihilation. However, for asymmetric gamma-ray burst models, such as the class of black hole accretion disk models, neutrino annihilation can be quite high along the disk axis, producing a strongly beamed jet.

neutrino viscosity Stress transport and entropy generation carried out by neutrinos whose mean free time t_c is comparable to the time scale τ of some system. In cosmology, *neutrino viscosity* could have been significant when the age of the universe $\tau_u \sim H^{-1}$ was about 1 sec. Strictly "viscosity" applies when $Ht_c \ll 1$, but extended simulations show that the entropy generation peaks when $Ht_c \simeq 1$. (Here H is the Hubble parameter at the time in question.)

neutron albedo Neutrons emitted outwards from the Earth's atmosphere, following nuclear collisions by incoming cosmic ray ions, with typical energies of 10 to 50 MeV. Free neutrons decay with a lifetime of about 10 min, and such neutrons have a small but non-zero probability to decay in the region close to Earth where the magnetic field can trap the resulting proton. The inner radiation belt is believed to arise in this fashion.

neutron star A remnant after a core collapse (type II) supernova explosion. Chandrasekhar showed that a cold star of mass greater than $\sim 1.4 M_\odot$ cannot support itself by electron degeneracy pressure, but must undergo a collapse. A *neutron star* is a possible (much smaller) equilibrium configuration. In a neutron star, the gravitational compression has raised the Fermi level of electrons so high that inverse β-decay has occurred, and protons have been converted to neutrons, and the star is predominantly neutrons with relatively small amounts of protons, electrons, and other particles. Thus, the star is supported by neutron degeneracy pressure. Neutron stars which can be observed in binary pulsars have mass near $1.4 M_\odot$, and hence (from theoretical models) radii around 20 km. Be-

cause the exact constituents of the neutron star, and the exact behavior of the equation of state, are unknown, the maximum mass of a neutron star can only be estimated, but a wide range of models suggest $\sim 2.5 M_\odot$. Above such a mass the neutron star has no support and will collapse into a black hole. Known neutron stars are pulsars and typically rotate rapidly (observed periods of multiseconds to seconds) and possess high magnetic fields (up to 10^{12} Gauss equal to 10^8 Tesla). Radiation processes in magnetospheres of neutron stars — regions of high-energy plasma around the star — are responsible for radio-emission that makes them observable as radio pulsars. A neutron star forming a close binary system with a normal star can accrete matter from the normal companion due to its strong gravitational field; this may lead to strong emission in X-ray range. Such systems are observed as X-ray pulsars and bursters. Some theoretical models also link γ-ray bursts to neutron stars. *See* black hole.

neutron star kicks Observations of the pulsar velocity distribution and specific neutron star associations with supernova remnants require that at least some neutron stars receive large velocities (~ 450 km s^{-1}), or "kicks", at or near the time of formation. A wide variety of mechanisms to produce these kicks exist, most of which assume some asymmetry in the ejecta or the neutrino emission in supernova explosions. Many proposed causes of these asymmetries exist: convection in the supernova, rotation, magnetic fields, asymmetric supernova collapse, asymmetric fallback, etc.

Newman–Penrose formalism (1962) In general relativity, a method that combines two-component $SL(2, C)$ spinor calculus with complex null tetrads for the treatment of partial differential equation systems. Applications are mainly directed to the field equations of general relativity. Major results: perturbative treatment of black holes, asymptotic behavior of the gravitational field, exact solutions, new conserved quantities of isolated systems. *See* spin coefficients.

New Millennium Program (NMP) A National Aeronautics and Space Administration

(NASA) initiative to test and prove new technologies by flying them on missions similar to science missions of the future; thus, the missions are technologically driven. Included in them are Deep Space 1 (propelled by xenon ions), launched October 24, 1998; Deep Space 2 (which was to probe beneath the surface of Mars), launched January 3, 1999 and lost on Mars on December 3, 1999; Deep Space 3 (with telescopes flying in formation), scheduled for launch in 2005; Deep Space 4 (a voyage to the heart of a comet); Earth Orbiter 1 (with an advanced land imager); and Earth Orbiter 2 (with a laser wind instrument on the Space Shuttle). These missions will return results promptly to future users about whether or not the technologies work in space. The missions are high risk because they incorporate unproven technologies, probably without backup, that require flight validation.

Newtonian   A type of reflecting telescope invented by Isaac Newton, with a small flat secondary mirror mounted in front of the primary mirror, to deflect rays approaching a focus out one side of the support tube, where they are viewed using a magnifying lens (eyepiece).

Newtonian cosmology   A collection of models of the universe constructed by the rules of Newtonian mechanics and hydrodynamics. A Newtonian model describes the universes via the evolution of a portion of a fluid, assuming the fluid moves in 3-dimensional Euclidean space. The historically first such model (Milne and Mc-Crea, 1934) demonstrated that some basic predictions of the Friedmann–Lemaître cosmological models could have been deduced much earlier from Newtonian physics. However, Newtonian models are of limited value only. Descriptions constructed in general relativity theory automatically imply the laws of propagation of light in the universe, and the geometry of space. The Newtonian models say nothing about geometry and cannot describe the influence of the gravitational field on the light rays. Moreover, they are unable to describe gravitational waves (which simply do not exist in Newtonian physics).

Newtonian fluid   A viscous fluid in which shear stress depends linearly on the rate of shear strain. In tensor notation, the flow law is $\sigma'_{ij} = 2\mu\dot{\varepsilon}_{ij}$, where σ'_{ij} is deviatoric stress, $\dot{\varepsilon}_{ij}$ is the rate of strain, and constant μ is the viscosity. A solid such as a rock deforms like a Newtonian fluid as a result of diffusion creep at elevated temperatures.

Newtonian gravitational constant   The constant $G = 6.673 \times 10^{-11} m^3 kg^{-1} s^{-2}$, which enters the Newtonian force law:

$$\mathbf{F} = Gm_1 m_2 \hat{r} r^{-2}$$

where $\mathbf{F}$ is the attractive force, m_1 and m_2 are the masses, r is their separation, and $\hat{r}$ is a unit vector between the masses.

Newtonian gravitational fields   The region in which one massive body exerts a force of attraction over another massive body according to Newton's law of gravity.

Newtonian gravity   The description of gravity due to Newton: The attractive gravitational force of point mass m_1 on point mass m_2 separated by displacement $\mathbf{r}$

$$\mathbf{F} = -\hat{\mathbf{r}} Gm_1 m_2 / r^2 \,,$$

where G is Newton's gravitational constant

$$
\begin{aligned}
G &= 6.67 \times 10^{-11} m^3 / \left(kg\ sec^2\right) \\
&= 6.67 \times 10^{-8} cm^3 / \left(g\ sec^2\right) \,.
\end{aligned}
$$

In *Newtonian gravity* forces superpose linearly, so the force between two extended bodies is found by summing (integrating) the vector forces between infinitesimally small masses in the two bodies. Newton proved that the force on a point mass due to a spherical extended mass is as if all the extended mass were concentrated at its center.

One can also show that the force on a point mass m_1 due to a second mass is given by m_1 times the gradient of the potential, defined in general by integration over the extended mass:

$$\mathbf{F} = -m_1 \nabla\phi$$

with

$$\phi = \int G\rho dV' / r' \,,$$

where ρ is the mass density of the object at the point $\mathbf{r}'$ and dV' is the volume element at that position. From this latter expression one also has the partial differential equation

$$\nabla^2 \phi = 4\pi\rho$$

where ∇^2 is the Laplacian. *See* density, gradient.

Newtonian invariance   *See* Galilean invariance.

Newtonian relativity   *See* Galilean relativity.

Newtonian simultaneity   In Newtonian Mechanics, since time is absolute, the simultaneity of two events is independent of the reference system. That is if one observer finds that two given events are simultaneous, then all other observers regardless of their respective reference systems will also observe that the two events are simultaneous. This does not in general hold in Relativistic Mechanics.

Newton (N)   Unit of force equal to 1 m kg s^{-2}.

Newton's laws of motion   **1.** A body at rest remains at rest, and a body in uniform motion remains in uniform motion, unless acted on by an external force.
2. The acceleration of a body is equal to the net applied force, divided by the mass of the body:

$$\mathbf{a} = \mathbf{F}/m .$$

3. For every force there is an equal and opposite reaction force.

The third law is exemplified by a person attracted by gravity toward a planet. There is simultaneously an equal opposite force attracting the planet toward the person.

NGC 5195   Irregular Galaxy, Canes Venatici at RA $13^h 30^m.0$, dec $+47°16'$, $m_V = 9.6$. Companion of M51, lying about $4'35''$ North of M101.

Nielsen–Olesen vortex   (1973) In condensed matter physics, and in cosmology, a defect (a string) that arises from a breaking of a high temperature symmetry, in a model with a complex

(i.e., 2-component) scalar (Higgs) field ϕ, and a massless gauge vector field A_μ. The phase transition requires a potential for ϕ which has a minimum at a nonzero value σ of $|\phi|$ and which is indifferent to the complex phase of ϕ. One finds a physical "magnetic" flux proportional to the circulation of the gauge field $\oint \vec{A} \cdot \vec{d\ell}$ (A_μ is the gauge field associated with the local $U(1)$ symmetry). Here, $\vec{d\ell}$ is taken along an arbitrary closed loop encircling the defect in physical space. Moreover, this flux is quantized, since solutions of the Higgs field ϕ can have a complex phase that increases by multiples of 2π as one moves around the vortex.

Kinetic terms of the Higgs field contribute to the overall energy of the configuration in the form of a logarithmically divergent energy per unit length. The presence of the gauge field A_μ cancels this divergence and hence the energy of the configuration is confined to a cylindrical core of radius roughly given by the Compton wavelengths of the Higgs and gauge bosons. Such vortices exist in superconductors, and represent the typical model for the description of cosmic strings. *See* Abelian Higgs model, cosmic string, local topological defect, winding number.

nimbus   *See* cloud classification.

nitrogen   Colorless, odorless gas, N, atomic number 7. Natural atomic weight 14.0067. Melting point 63.05 K, boiling point 77.36 K. Naturally occurring isotopes ^{14}N (99.6%) and ^{15}N (0.4%). Relatively inert at low temperatures, but its compounds are vital for biological processes. Nitrogen constitutes about 78% (volume) of the Earth's atmosphere.

Noachian   Geophysical epoch on the planet Mars, more than 3.5 Gy BP, exemplified by the oldest and heavily cratered regions lying in the southern hemisphere of Mars.

nocturnal   Pertaining to occurrences during the night, as opposed to diurnal occurrences.

nodal month   *See* month.

node   A point of zero displacement. Standing waves in a water basin have *nodes* and antinodes.

no-hair theorems   (Also known as uniqueness theorems.) A set of propositions proved in classical general relativity which restricts possible black hole geometries. All black hole metrics (including dilatonic black holes arising in some string theories) belong to a small number of families. Each family is identified by the set of Killing vectors on the event horizon (*see* Killing horizon) or, equivalently, by the isometry of the asymptotically flat space-time outside the event horizon.

For instance, if the black hole is spherically symmetric then it is described by the Schwarzschild metric if electrically neutral (*see* Birkhoff's theorem) and by the Reissner–Nordström metric if the net electric charge is not zero. Rotating cases are given by the (electrically neutral) Kerr metric and the (charged) Kerr–Newman metric. Thus, general relativistic black holes have no properties that make them individually distinguishable when their mass, charge, and angular momentum are identical. A few other quantities can contribute to define classes of black holes, such as Yang–Mills charges. *See* black hole, dilatonic black hole, domain of outer communication, future/past event horizon, isometry, Kerr–Newman metric, Reissner–Nordström metric, Schwarzschild metric.

non-Abelian string   In condensed matter or in cosmology, a string defect in which the symmetry describing the string has non-commuting components: the order of applying different symmetry actions leads to different results. For instance, the isotropic-nematic transitions in liquid crystals, where molecules have the form of rods, and are end-to-end symmetric. In the high temperature phase, the rods can point in any direction, but at low temperatures they line up and are symmetric under rotations around the reflections reversing their direction. Since a left-handed rotation turns into a right-handed one under reflection, the rotation symmetry does not commute with the reflection. Non-Abelian strings are principally characterized by their different behavior under collision; in particular, intercommutation is not then mandatory in string collisions.

nonderivative absorption   *See* ionospheric absorption.

nonminimal coupling   The expression of the behavior of some field in a second field, which includes additional interaction terms beyond those of minimal coupling. *See* minimal coupling.

nonsimultaneous Big Bang   The Big Bang in inhomogeneous models that occurs at different times in different places. In the Robertson–Walker cosmological models, the Big Bang is a single event in spacetime (i.e., all matter explodes out of a single point in the same instant). In generic inhomogeneous models, it is a process extended in time, which could in principle be determined by observing that different regions of the universe are of different ages.

non-thermal line broadening   Excess width seen in solar soft X-ray emission lines. The width of these lines often exceeds that of the width expected from a thermal (Maxwell–Boltzmann) velocity distribution and is thought to be due to a combination of effects including small-scale flows, turbulent motions, and wave-particle interactions.

non-thermal particles   A distribution of particles which does not conform to that of the Maxwell–Boltzmann distribution characteristic of particles subject to purely thermal motions. *See* Maxwell–Boltzmann distribution.

non-thermal radiation   Radiation emitted by the interaction of charged particles where the distribution of the incident particles is non-thermal (e.g., bremsstrahlung from flare-accelerated particles).

non-thermal spectral energy distribution
A spectral energy distribution of light produced if electrons bound to atoms are not distributed among atomic levels according to the Maxwell–Boltzmann law, or if the velocity distribution of free, radiating particles is not a velocity distribution that follows Maxwell's law. A non-thermal

spectrum differs substantially from black-body (following Planck's law) or thermal free-free emission, in the case when the source is optically thick or optically thin, respectively. Stars are regarded as thermal sources *par excellence;* hence, non-thermal spectrum is often a synonym for non-stellar spectrum. A typical non-thermal spectrum is the spectrum of synchrotron radiation. In this case, radiating electrons are accelerated to relativistic speed, and do not follow a Maxwellian distribution.

non-topological soliton   Solitons are non-trivial solutions of a nonlinear field theory, usually associated with some topological conservation law, i.e., for fields having particular boundary conditions imposed, leading to a conserved number. In some cases, however, these boundary conditions yield no conserved quantity, even though the solution is still non-trivial. These configurations, usually unstable, are called non-topological solitons. *See* Derrick theorem, soliton.

nonuniqueness   When models are compared with (or "inverted" from) measurements by examining the error between the measurements and those predicted by the models, it may sometimes be found that the best possible fit to the data can be achieved by more than one model. This means that the data cannot uniquely determine a best-fit model. It may be due to the sparse or restricted nature of the observations, but may also be related to fundamental physical difficulties. For example, the gravitational potential ϕ outside of source regions (i.e., where the density ρ is non-zero) obeys Laplace's equation, $\nabla^2 \phi = 0$, and hence a uniqueness theorem tells us that if the gravitational acceleration on a surface external to the Earth is known then the gravitational field is known down to the Earth's surface. However, sources of gravitational anomalies lying within the Earth cannot be uniquely determined using this knowledge alone because many different distributions of ρ within the Earth will allow the gravitational Poisson's equation $\nabla^2 \phi = 4\pi G \rho$ to match the field at the Earth's surface, and therefore external to the Earth. Nevertheless, the range of possibilities is restricted by the available measurements. *Nonuniqueness* may be alleviated either

through adding other data or by making other assumptions that help to pick between the models — for example, by penalizing some measure of model complexity in order to find the simplest of the best fit models. *See* inverse theory.

norite   A granular crystalline igneous mafic rock found in the lunar highlands composed of plagioclase and pyroxene; an orthopyroxene gabbro.

normal fault   In geophysics, a type of dip-slip fault, where displacement occurs up or down the dip (i.e., the angle that the stratum makes with the horizontal) of the fault plane. A *normal fault* is a dip-slip fault upon which the deformation is extensional. The direction of slip motion in a normal fault is opposite that of a reverse fault. Parallel normal faults can give rise to downdropped valleys called *graben* or elevated ridges called *horsts.*

normal incidence frame   Frame of reference in which the shock is at rest. The upstream plasma flow is normal to the shock but oblique to the magnetic field, the downstream flow is oblique to both the shock normal and the magnetic field. *See* de Hoffman–Teller frame.

normal incidence optics   At normal incidence light strikes the surface more or less at right angles, as with an ordinary wall mirror; X-rays and extreme ultraviolet light (EUV) are mostly absorbed at these angles.

normal modes   In seismology, the free oscillations of the Earth, i.e., the set of elastic standing waves that are natural to the Earth. There are different types of standing waves that the Earth can sustain, such as "spheroidal" modes that involve radial motions such as a simple, spherically symmetric compression and expansion of the Earth, or the oscillation of the Earth between oblate and prolate spheroidal shapes, and "torsional" modes that involve only toroidal motions (i.e., no radial component) and which involve twisting motions of the Earth. For each of these types of motion, there is a discrete spectrum of permitted frequencies and wavenumbers. Nevertheless, any elastic deformation of the Earth can be represented as a sum of these normal modes. *Normal modes* are excited by earthquakes, and there have also been claims of excitation by atmospheric processes. The frequency and spatial structure of each normal mode depends on Earth properties such as density, bulk modulus, and shear modulus, so since different normal modes involve differing amounts of motion in the various parts of the Earth, observations of normal modes can be used to constrain Earth structure.

North Atlantic current   A warm ocean current, an extension of the Gulf Stream, flowing northeastward from south of Greenland at about 45°N extending north of the British Isles and along the coast of Scandinavia.

The current then joins the North Equatorial Current in the eastern tropical Atlantic Ocean and turns west. The Northern Equatorial Current mixes with water from the South Equatorial Current and the Guyana Current and moves past the Windward Islands into the Caribbean Basin.

North Atlantic Oscillation (NAO)   A sea level pressure (SLP) seesaw between the Icelandic low and Azores high pressure systems, most pronounced in winter months. A NAO index is often defined as the SLP difference between Azores and Iceland. At the positive phase of the NAO when the SLP rises on the Azores but falls on Iceland, western Greenland and eastern Canada tend to be anomalously cold, while northern Europe experiences warm and humid winters as the strengthened westerly winds enhance inland advection of the warm Atlantic air. Recent studies indicate that the NAO may be part of the Arctic Oscillation.

North Equatorial current   A westward flowing ocean current fed on the East from the southern end of the Canary current, and extending westward into the Caribbean basin.

Northern Plains (Mars)   The bulk of plains material on Mars is found in the northern lowlands. They cover an area equivalent to half the size of the U.S. and are flatter than the Sahara Desert. Crater studies show they formed over a protracted period of time from the early Hesperian until the Amazonian. Two alternative hypotheses exist for their formation. The first is

sedimentary and the second is volcanic. A sedimentary origin has been proposed on the basis that Mars may have had a significant climatic history. Examples cited of sedimentary deposits include alluvial fans and polygonal terrain. The general consensus, however, is that the *Northern Plains* are principally of volcanic origin as indicated by the combined presence of lobate flow fronts, wrinkle ridges, and lava rilles. Additionally, marginal locations of the plains show no erosion but are either lobate or thin gradually. Images taken by the Mars Global Surveyor spacecraft also support a volcanic origin having revealed huge plates of solidified volcanic lava.

nova   The subset of cataclysmic variables in which hydrogen accreted onto a white dwarf has ignited while degenerate and so burned explosively. The system survives and the explosions repeat in periods from about 30 years (recurrent novae) up to 10^{4-5} years (classical novae). Most or all of the accreted hydrogen and its burning products are expelled at speeds of about 1000 km/sec, leading to a visible nebula called a nova remnant. These remain detectable for decades until the gas (only about 10^{-5} solar masses) dissipates. A few novae are found in the Milky Way each year, and the total number is probably 20 to 50, many of which are hidden behind galactic dust. The ejecta make some contribution to the galactic inventory of carbon and of rare isotopes like nitrogen-15 and (probably) aluminum-26.

nuclear fusion   The exothermic process by which nuclei combine to form heavier nuclei releasing vast amounts of energy. The process that generates the emitted energy of the sun via the "burning" of hydrogen into helium.

nuclear reactions   Reactions that rearrange the protons and neutrons in the nuclei of atoms (in contrast to chemical reactions, which merely rearrange the outer electrons of the atoms). Typically, two particles come together, and two different ones go out. The nucleus of an atom is represented by its chemical symbol, with a superscript indicating the total number of neutrons + protons in the nucleus, for instance ^{12}C or ^{16}O (formerly written C^{12} or O^{16}, and properly still spoken in the latter order). The incoming and outgoing particles are abbreviated:

$$p = \text{proton}$$
$$n = \text{neutron}$$
$$\alpha = \text{alpha particle}$$
$$\gamma = \text{photon (gamma ray)}$$
$$\bar{\nu}_e = \text{electron anti-neutrino}$$
$$e^- = \text{electron}$$
$$\nu_e = \text{electron neutrino}$$
$$d = \text{deuteron } (^2H)$$
$$e^+ = \text{positron}$$

Typical examples are discussed under s process, pp chain, CNO cycle, and elsewhere.

Nuclear reactions will occur spontaneously if the products have a total mass less than the sum of the incoming particles (and if certain quantum numbers have the right values). The energy corresponding to the extra mass is liberated, and nuclear reactions are thus the primary energy source for all stars. The general format is target nucleus (incoming particles, outgoing particle) product nucleus, as, for instance, $^{13}C(p, \gamma)^{14}N$.

nuclear time scale   The time scale on which a star changes when its luminosity is derived from nuclear reactions. In the case of hydrogen burning, this is about 10^{10} years for the sun, and proportional to $(M/M_o)^{-2}$ for stars of other masses. Main sequence evolution occurs on the *nuclear time scale,* as does the evolution of a binary system in the Algol state, when the mass donor is the less massive star.

nuclear winter   A strong surface cooling phenomenon due to nuclear war. From numerical simulation results, nuclear war can cause a serious change of global weather and climate. Nuclear explosions on the surface can carry a large amount of dust into atmosphere, and nuclear explosions in the air can leave heavy smoke and fog in the atmosphere. They will strongly reduce the short-wave solar radiation received by Earth's surface, cause the outgoing long-wave radiation from the surface to be larger than the reaching solar radiation to the surface. Thus, the surface temperature will decrease rapidly below the freezing point, and can even reach to $-20°$ to $-25°C$.

nucleation   Condensation or aggregation process onto a "seed" particle. The seed may be of the same material as that condensing, or different, but with a similar crystal structure. Rain and snow condense onto airborne particles, in one example of this process.

nucleus   In nuclear physics, the central positively charged massive component of an atom, composed of a number of protons and neutrons held together via the nuclear force, even in the presence of the repulsive force associated with the accumulated positive charge. In astronomy, the actual physical body of a comet, a few kilometers in size, composed of ices and silicates. The *nucleus* is embedded in the head of the comet, which may be hundreds of thousands of kilometers across. Outgassing and loss of dust from the cometary nucleus is the source of the head and the tail(s) of the comment.

In astronomy, the roughly spherical central region of many spiral galaxies. The popuation of stars in the nucleus is much more evolved than that of the spiral arms.

nucleus (of a comet)   The solid "body" of the comet. The *nucleus* contains water-ice, organic and silicate dust grains, and other frozen volatiles such as CO and CO_2. These volatiles may also be trapped in the water–ice matrix as clathrates. The density of the nucleus of comet Halley has been found to be roughly $0.5 \, g \, cm^{-3}$. This implies that the nucleus is very porous. Comet nuclei range in size from several kilometers to several hundred kilometers. They can be highly non-spherical and can have low albedos, indicating that the surface is covered by a dust layer. Rotation rates of comet nuclei have been estimated from light curves, and vary from ~ 10 hours to days.

null infinity   In general relativity, a domain $\mathcal{I}$ of a (weakly) asymptotically simple spacetime which is isometric with a neighborhood of conformal infinity of Minkowski space-time. In other words, it encodes the ultimate radiation behavior of regions of matter and gravitation-free spaces far from any source. *See* asymptotically simple space-time, conformal infinity.

null vector   In special relativity (and in general relativity), the tangent to a light ray. Because of the indefinite signature $(- + ++)$ in special relativity, such a vector has zero length even though its components are non-zero. In psuedo-Reimannian geometry, an element ℓ of a linear space with metric g such that $g(\ell, \ell) \equiv g_{ab}\ell^a\ell^b = 0$.

numerical cosmology (computational cosmology) The technique, and a collection of models of the universe obtained by solving Einstein's equations numerically. Because they proceed in terms of discretization, these models are approximate, but well-formulated models show a convergence as the discretization is refined. For reasonable resolution in 3-dimensional simulations, large computers are required.

numerical model A set of discretized mathematical equations, solved on a computer, which represent the behavior of a physical system.

numerical relativity The study and solutions of the differential equations of general relativity by discretization of the equations and the computational solution of the resulting finite algebraic problem. In many cases without overall symmetry, analytic methods fail and *numerical relativity* is the only method to elucidate and understand generic solutions for spacetimes in general relativity. Computational simulation has become a real research tool, for instance, the critical behavior and naked singularity found by Choptuik. Similarly, the naked singularity that may exist in some circumstances at the center of the Lemaître–Tolman cosmological model was first identified in numerical calculations. Numerical investigations have amassed considerable knowledge about problems too complicated for analytical treatment: collisions of rotating black holes, emission of gravitational waves by rotating nonsymmetric bodies or by binary systems, collapse of a binary system to a single body, etc. *See* binary black holes, critical

phenomena in gravitational collapse, gravitational wave.

Nusselt number A dimensionless number quantifying the efficiency of heat transfer in a convecting system. It is the ratio of total heat flux to the conductive heat flux. For example, the conductive heat flux across a uniform layer of thickness h with differential temperature ΔT is $\lambda \Delta T / h$. If the total heat flux including convective heat transfer is q, the Nusselt number in this case is

$$Nu = \frac{q}{\lambda \Delta T / h} \, .$$

nutation This is a variation of the orientation (the obliquity) of the Earth's axis of rotation associated with the Earth's forced precession. In the theory of the rapid rotation of a rigid body, it is used to imply an analog of the free (i.e., unforced) precession of a body in the case where there is also a forced precession, which causes the body's axis with the largest principal moment of inertia to "nod" backwards and forwards even as it is precessing. On Earth, this effect is known as the Chandler wobble, and the term *nutation* may also refer to motions due to the fact that the Earth is deformable and its layers imperfectly coupled, as well as motions caused by time variations in the torque acting on the Earth. This happens because the couple is produced by the tidal interactions with the moon and sun, which vary because the relative orientations of the Earth, moon and sun change with time in various different ways. There are nutations at various periods, the largest nutation being at 18.6 years. Nutation is one of the contributors to the Milankovich cycles which cause climatic variations. In the case of the Earth, nutation causes the obliquity to vary from $21°$ to $24°$ (the current value of Earth's obliquity is $23.5°$) over a period of 41,000 years. *See* precession.

O

obduction In tectonic activity, the process in which part of a subducted plate is pushed up onto the overriding plate. The process responsible for the emplacement of ophiolites.

Oberon Moon of Uranus, also designated UIV. It was discovered by Herschel in 1787. Its orbit has an eccentricity of 0.0008, an inclination of 0.10°, a semimajor axis of 5.83×10^5 km, and a precession of $1.4°$ yr^{-1}. Its radius is 762 km, its mass is 3.03×10^{21} kg, and its density is 1.54 g cm^{-3}. Its geometric albedo is 0.24, and it orbits Uranus once every 13.46 Earth days.

objective grating A course grating, often consisting of a parallel array of wires, which is placed in front of the first element of a telescope (the objective) which produces a low dispersion spectrogram of each object in the field; used for rapid survey of the properties of sources in a relatively wide region of sky.

objective prism A prism of narrow apex angle (a few degrees) located at the objective of a telescope, most often of a Schmidt telescope. The *objective prism,* acting as a disperser, provides the spectrum of each object in the field of the telescope. A large number of low resolution spectra (up to $\sim 10^5$) can be recorded on a plate or electronic detector. Surveys based on objective prism spectroscopy have been very efficient in finding objects with peculiar spectral energy distribution, such as galaxies with UV excess, or objects with very strong and broad emission lines, such as quasars. Spectral resolving power($\lambda/\Delta\lambda$) achieved in common usage are ~ 100. Higher spectral resolving power can be achieved with wider apex angle prism.

oblateness For an oblate spheroid, the ratio of the difference between equatorial and polar radii to the equatorial radius. In general, the *oblateness* ϵ of a rotating inhomogeneous spheroid with north-south symmetry depends on the rotation parameter $m = \Omega^2 R^3 / GM$ and the zonal harmonic coefficients $J_l (l = 2, 4, 6, ...)$ in a complicated manner. Ω is the rotational angular speed, R is the equatorial radius, G is the gravitational constant, M is the total mass of the spheroid, and $J_l = -(1/M) \int d\mathbf{r}(r/R)^l \rho(r, \theta)$ $P_l(\cos \theta)$ where the mass density of the spheroid ρ depends only on the polar radius r and the polar angle θ measured from the north pole, and P_l is the l-th order Legendre polynomial. In planetary physics, analyses of velocity variations during the spacecraft fly-bys at major gaseous giant planets have permitted us to determine their first few zonal harmonic coefficients to a relatively high accuracy, which has in turn provided us with their dynamical oblatenesses. To the first order in J_2 (quadrupole moment, i.e., anisotropy in moments of inertia), there is a simple relation oblateness $\epsilon = (3J_2 + m)/2$, which shows that the oblateness is a critical indicator of how fast a body is rotating. For a model planet consisting of a small dense core of mass M_c clad with a homogeneous spheroidal envelope of mass $M_e = M - M_c$ and equatorial radius R, $J_2/m = (1 - \delta_c)/(2 + 3\delta_c)$ and hence we find $\epsilon/m = 5/(4 + 6\delta_c)$ where $\delta_c = M_c/M$ is the fraction of core mass. Thus, with the total mass M, the equatorial radius R, and the rotation rate Ω all being equal, a homogeneous spheroid without a core is more oblate than an inhomogeneous spheroid containing a core.

oblique ionogram The conventional display obtained from an oblique ionosonde, containing information about the ionosphere. The synchronized transmitter and receiver for an oblique ionosonde are separated each at the terminal of an HF link. An *oblique ionogram* is constructed by displaying the received signal as a function of frequency (on the horizontal axis) and time delay (on the vertical axis). The structure of the oblique ionogram is highly dependent on the path length between the receiver and transmitter. For short paths, less than 100 km, the oblique ionogram will look similar to a vertical ionogram. However, as the path length increases, the F modes will gradually become more obvious. For frequencies below the maximum usable frequency (MUF), for a single mode, the shape of an oblique ionogram shows that two paths are possible. These two paths are for a high and low elevation angle. The high elevation path is called the Pederson ray, or the high ray. The high ray penetrates deeper into the ionosphere and suffers greater retardation than for the low ray, which is therefore the preferred propagation path because time dispersion is lower. However, for very long paths the low ray elevation angle may be low enough for it to be obscured, leaving only the high ray to use. At the MUF, the high and low rays are coincident. Fewer characteristic parameters are measured from oblique ionograms than from vertical ionograms. The most important is the maximum operating frequency (MOF) for the link although ideally, an MOF should be measured for each identifiable mode. On longer paths, the MOF will usually be an F mode. Oblique ionograms contain much useful information about a single HF link and this information may contribute to effective link management. *See* ionospheric radio propagation path.

oblique-slip fault   A fault that combines the motions of a dip-slip fault (such as a normal or reverse fault) and a strike-slip fault. Motion thus occurs in both vertical and horizontal directions along an *oblique-slip fault*.

obliquity   The measure of the tilt of a planet's rotation axis. It is measured from the perpendicular of the planet's orbital plane. *Obliquity* can change due to the gravitational effects of other bodies (nutation), which in turn can affect the body's climate (*see* Milankovitch Cycles). Planetary obliquities range from 0° (for Mercury) to 177° (for Venus). The term inclination is sometimes used as a synonym.

obliquity factor   *See* basic MUF.

Occam's razor   A maxim attributed to William of Occam, a 14th century English logician, which, literally translated, states that "Plurality should not be posited without necessity". However, in modern science this maxim has been strengthened to a form that states that for any two competing interpretations (or theories) the one with the least assumptions is the better one. This corrupted form has been, more appropriately, called a "law of parsimony" or "rule of simplicity".

occluded front   A complex weather frontal system occurring when a cold front overtakes a warm front. Because of the greater density of cold air, the warm air is lifted. Such a situation is likely to produce rainfall as water vapor suspended in the warm air condenses out at the interface between the two. Also known as an occlusion.

occultation   The temporary diminution of the light from a celestial body O_1 (e.g., a star) caused by its passage behind another object O_2 (e.g., the moon) located between the observer and O_1. This is called the *occultation* of O_1 by O_2. On August 26, 1981 during the encounter of Voyager 2 with Saturn, its ultraviolet spectrometer and photopolarimeter recorded the brightness variation of the star δ Scorpii (O_1), at 1300 Å and 2640 Å, respectively, as it was occulted by the C, B, and then A rings (O_2) located within Saturn's shadow. As shown in this example, the observer or receiver is not necessarily Earth bound. Occultation is one of the most powerful techniques to probe the structure and composition of planetary rings and atmospheres. Stellar occultation by the moon is useful in determining the lunar position with high precision. Occultations of radio sources of small angular sizes lead us to determine their precise positions and spatial distributions of radio intensity. In 1977, Uranian rings were serendipitously discovered during measurements of stellar occultation by the central planet, which spurred the subsequent discovery of Neptunian arc rings by similar but ground-based stellar occultations in 1984. Stellar occultations by asteroids have also helped determine the asteroids' sizes and shapes.

ocean   The great interconnected mass of salt water (average salinity about 3.5% by weight), covering about 71% of the surface of the Earth, or any of the five main subdivisions of this great *ocean:* Pacific, Atlantic, Indian, and Arctic; the southern portions of the first three, which converge around Antarctica, are known collectively as the Antarctic Ocean. Area of about 361,000,000 km^2, average depth about 3,730 m, and a total volume of about 1,347,000,000 km^3. Current theory holds that ocean water originated through release at seafloor spreading sites.

Ocean water is a vast source or sink of heat, dissolved gases, and dissolved minerals. Its currents transport of these quantities worldwide and the total terrestrial climate and biosphere depend on these ocean processes. The tremendous heat capacity of the oceans contributes strongly to the stability of the Earth's temperature.

ocean color   A generic term referring to the spectral dependence of the radiance leaving a water body.

ocean color sensor   Any instrument for sensing of ocean color, usually from aircraft or satellites.

oceanic optics   *See* optical oceanography.

Ockham's razor   *See* Occam's razor.

offshore   A region extending from the seaward edge of the surf zone to the edge of the continental shelf. Also used to denote a direction or relative position.

Olbers paradox   A paradox stated in 1826 by the German astronomer Heinrich Wilhelm Matthäus Olbers that shows the inconsistency of Newtonian cosmology. According to this paradox, in an infinite unchanging Newtonian universe filled with stars at a constant number density, the combined radiant energy from all stars should make the night sky bright. Indeed, if the absolute luminosity of a typical star is L and the number density of stars in the universe is a constant n, then the total radiant energy at any point is given by

$$E = \int_0^\infty \frac{L}{4\pi r^2} 4\pi r^2 n dr = Ln \int_0^\infty dr \ .$$

The integral diverges and gives an infinite quantity.

Olbers attempted to solve the paradox by assuming that light from distant stars gets absorbed by interstellar matter. This does not in fact solve the problem, since in an unchanging universe a state of thermal equilibrium will eventually be achieved, and the absorbing matter will radiate exactly the same amount of energy as it absorbs.

Solution of *Olbers paradox* is possible in the framework of relativistic cosmology, according to which the universe is neither infinite in size nor unchanging. It is now known that the universe undergoes a cosmological expansion, due to which the light from distant stars and galaxies is shifted towards smaller frequencies, thus greatly reducing the radiated energy reaching a distant observer.

oligotrophic water   Water with low phytoplankton biomass, typical in many open ocean regions; chlorophyll *a* concentration is below 0.5 mg m^{-3}.

Olympus Mons   A young, shield volcano in the Tharsis Province of Mars that stands 25 km above the plains, and measures over 700 km across, and is considered to have formed by successive eruptions of low-viscosity lavas. The volcano has a volume 50 to 100 times that of the largest volcano on Earth, Mauna Loa. A basal scarp exists around *Olympus Mons,* which is up to 6 km high in places. The shield has a complex summit caldera 80 km in diameter, which consists of several coalesced collapse craters with wrinkle ridges on their floors.

The enormous height of the shield volcano has been attributed to the absence of plate tectonics on Mars. In this way, Olympus Mons may have developed by continuous and prolonged eruption from a point source vent over a thermal anomaly in the mantle, to result in one volcano of enormous height, compared with comparatively small volcanoes (e.g., the Hawaiian Emperor chain) strung along segments of moving crust on Earth.

omega bands   Ribbons of aurora in a meandering pattern across the sky, which gives their segments a shape like the Greek capital letter omega (Ω).

ω effect   This a mechanism by which some simple fluid motions — those of differential rotation — can generate magnetic field, and hence contribute to the maintenance of a dynamo. Differential rotation occurs in a fluid rotating axisymmetrically about an axis, but where the rotation rate varies within the fluid. Magnetic flux in a highly conducting fluid behaves as if attached to the fluid itself (apart from the effects of magnetic diffusion, which are related to the resistivity of the fluid). Therefore, the magnetic field is stretched parallel to the shear in the fluid associated with the differential rotation. For nearly spherical situations, such as the dynamo in the Earth's core, it is useful to consider a poloidal/toroidal decomposition of the magnetic field, in which case it can be seen that the ω *effect* is particularly efficient at generating toroidal field from poloidal field, although not the reverse. To maintain a dynamo, poloidal field must also be generated (from either poloidal or toroidal field), which can be accomplished through the α effect, and Earth-like dynamos are often described as $\alpha\omega$ dynamos (although it should be noted that this term may be used to imply a type of mean-field dynamo simulation). Some astrophysical dynamos are thought not to rely upon the ω effect, and are often called α^2 dynamos. *See* α effect.

One-form, 1-form   A covariant vector. The paradigm is the gradient df of a function f, which defines a geometrical structure, namely the collection of level surfaces of the function, which show how rapidly the function varies. In a coordinate system $\{x^\alpha\}$, the gradients of the coordinate functions are 1-forms dx^β, and the collection of these gradients $\{dx^\beta\}$ is a basis for 1-forms.

Oort cloud   Roughly spherical region between about 3×10^4 and 5×10^4 AU that is the source of long period comets. *Oort cloud* comets are thought to have been formed in the region of the outer planets, and been ejected to the Oort cloud by means of gravitational interactions, mostly with Jupiter and Saturn. Since the Oort cloud comets are very weakly bound to the sun, interactions with passing stars or interstellar clouds, or even the gravitational gradient as the sun passes through the galactic mid-plane, will be sufficient to change significantly the orbits of these comets. They will then either be ejected completely from the solar system, or fall into the inner solar system, where they will appear as moving on highly eccentric orbits. The number of comets in the Oort cloud is not known, but it may be as high at 10^{12}, and contain a substantial fraction of the mass of the solar system.

open cluster   A group of several hundred to several thousand stars lying in a small region (of order 10 pc across) in the plane of the galaxy. Typically the site of active star formation, and of many young stars of closely similar ages.

open field region   A region of the sun with predominantly unipolar magnetic field and where the field is regarded as "open", i.e., not returning to an opposite polarity region on the solar surface. In *open field regions,* the solar corona is not in hydrostatic equilibrium but is continuously expanding outwards as the solar wind.

open magnetosphere   The observed behavior of the Earth's magnetosphere in which reconnection at the dayside magnetopause, caused by dissipation and ionic drift, allows field lines to move backward around the Earth in the solar wind.

open ocean   Water seaward of the edges of the continental shelves; often, but not always, case 1 waters. *See* case 1 water.

open universe   A model of the universe in which the space has infinite extent. If the cosmological constant is zero, such a universe will, according to the description of general relativity, expand forever.

operational MUF   The operational MUF is defined as the highest frequency that would permit acceptable performance of a radio circuit between given terminals at a given time under specified conditions. The operational MUF is an anticipated performance parameter derived, possibly, from an HF prediction program. It is a frequency that could be used, and may be observed on a circuit. The operational MUF differs from the basic MUF in that it includes system parameters in its calculation. *See* ionospheric radio propagation path, oblique ionogram, basic MUF.

Ophelia   Moon of Uranus also designated UVII. Discovered by Voyager 2 in 1986, it is a small, irregular body approximately 16 km in radius. Its orbit has an eccentricity of 0.01, an inclination of $0°$, a precession of $419°$ yr^{-1}, and a semimajor axis of 5.38×10^4 km. It is the outer shepherding satellite for Uranus' epsilon ring. Its surface is very dark, with a geometric albedo of less than 0.1. Its mass has not been measured. It orbits Uranus once every 0.376 Earth days.

ophiolites   In general, the mantle of the Earth lies beneath the crust. In some cases the more dense and basic mantle rocks have been brought to the surface by mountain building processes. These are *ophiolites*. Important ophiolites occur in Cyprus, Yemen, and other places.

Oppenheimer–Snyder (OS) model   The simplest general relativistic model of gravitational collapse. It consists of a sphere of dust (perfect fluid without pressure) in vacuum which collapses under its own weight to a point-like singularity (*see* Schwarzschild metric). Since there are no forces other than the self-gravity of dust, the surface of the sphere moves along a geodesic on the external Schwarzschild metric with an ADM mass given by the matching conditions with the internal Robertson–Walker comoving metric. Such mass equals the proper energy of dust only if the surface of the sphere starts collapsing from rest with infinite radius, otherwise it is either bigger or smaller than the rest mass depending on the velocity of the surface at a given radius.

The absence of internal pressure makes the OS model unrealistic for modeling the collapse of a star, but does show the qualitatively expected features. *See* ADM mass, black hole, comoving frame, Robertson–Walker cosmological models.

opposition   Orientation of planets so that the angle planet-Earth-sun equals $180°$. For outer planets, the planets are as close to the Earth as possible in their orbits. *Opposition* is geometrically impossible for inner planets.

optical closure in measurement   Making consistent measurements; examples include independently verifying that the sum of the absorption coefficient plus the scattering coefficient equals the beam attenuation coefficient and that the integral over all directions of the volume

scattering function equals the scattering coefficient.

optical closure in models   Development of consistent analytical and/or numerical models that make predictions consistent with observations.

optical closure in scale   Making the transition from small scale (e.g., single particle) data to bulk scale data in a manner consistent with observations.

optical constants   The nonnegative real and imaginary parts of the refractive index; the constants often strongly depend on frequency.

optical density   *See* absorbance.

optical depth   *See* optical distance.

optical distance   The integral of the dimensionless product of the beam attenuation coefficient $[m^{-1}]$ multiplied by an infinitesimal unit of distance [m] along the direction of travel.

optical double star   A pair of stars very close to each other in the sky but at different distances, so that they do not really make up a binary star system. *See* binary star system.

optical oceanography   The subdiscipline of oceanography concerned with the propagation and interaction of radiation, typically at wavelengths between ~ 350 and ~ 750 nm with seawater.

Optimum Working Frequency (OWF)   The frequency that is exceeded by the operational MUF, at a given time, between a specified pair of terminals via any available propagation mode, during 90% of a specified period, usually a month. The operational OWF is an anticipated performance parameter derived, possibly, from an HF prediction program. Unlike the operational MUF, the OWF is based on known statistical properties of the ionosphere and offers a safe estimate of a reliable frequency for a propagation path. When planning HF services, the OWF is often used. However, by its nature, 90% of the time it is possible to find frequencies higher than the OWF that will be better to use. Conventionally, the OWF has probably been the most important ionospheric parameter deduced for HF propagation. The OWF is also sometimes called the Frequency of Optimum Traffic (FOT). *See* ionospheric radio propagation path.

orbital elements   Seven quantities that must be specified to specify the position of a point mass in a Newtonian orbit: the longitude of the ascending node (Ω), the inclination i of the object's orbital plane to the ecliptic (or to the plane of the sky, for extrasolar objects), the argument of the perihelion (ω), the semimajor radius of the orbit, a, the eccentricity of the orbit, ϵ, and the mean anomaly, M, which increases linearly with time. *See* separate entries for these quantities.

orogeny   The creation of a mountain range.

orographic cloud   A cloud that forms because of forcing of the air motion by the Earth's topography, as cap clouds, lee wave clouds, etc.

Orowan's equation   The main relationship for the physics of plastic deformation, expressing the strain rate (ε) caused by the motion (glide or climb) of dislocations of Burgers vector **b**, with an average velocity ($\bar{v}$) and the density of mobile dislocations being ρ_m,

$$\varepsilon = \rho_m \mathbf{b} \bar{v} .$$

It is the master equation, from which many constitutive equations expressing microscopic deformation models can be derived.

orthogonal   Perpendicular; lines drawn perpendicular to the crests of water waves are referred to as wave *orthogonals*.

Osborn–Cox method   *See* Cox number, dissipation of temperature variance.

Osborn model   Method to infer turbulent density fluxes from the measurement of the dissipation rate of turbulent kinetic energy ϵ. The vertical turbulent density flux is normally expressed via the eddy diffusivity K_ρ

$$\overline{w'\rho'} = -K_\rho \frac{\partial \rho}{\partial z}$$

where w is the vertical velocity in the z direction, the primes denote turbulent fluctuations from the mean (*see* Reynolds decomposition), and the over-bar denotes a suitable spatial average that is longer than the typical length scales of the turbulence, but shorter than any energy containing scales. Under the assumption that the turbulence is steady in time and homogeneous in space, the equation describing the turbulent kinetic energy (TKE) reduces to the so-called TKE-balance equation

$$P = B + \epsilon$$

where $P = -\overline{u'w'}\,(\partial U/\partial z)$ represents the mechanical production of turbulent kinetic energy, $B = g/\rho^{-1}\overline{w'\rho'}$ the destruction of TKE by buoyancy, and ϵ is the rate of dissipation of TKE. The buoyancy term can be parameterized by the non-dimensional flux Richardson number $\mathrm{Ri}_f \equiv B/P$. Substitution of Ri_f into the TKE-balance equation expresses the vertical density flux in terms of the dissipation rate ϵ

$$g\rho^{-1}\overline{w'\rho'} = \frac{\mathrm{Ri}_f}{1 - \mathrm{Ri}_f}\,\epsilon = \Gamma\epsilon .$$

The quantity $\Gamma = \mathrm{Ri}_f / (1 - \mathrm{Ri}_f)$ is often referred to as the mixing efficiency.

oscillating universe   If the universe is closed, i.e., it has enough mass or equivalently expands slowly enough for the gravitational pull of matter to cause it to contract back to zero size in the far future, this may be repeated as a cycle. The temperature and density would increase without limit as the contraction intensifies, would go formally to infinity, and then decrease again as the universe expands. This model of cosmic evolution parallels the ancient idea of the cyclic universe — one that undergoes a never-ending sequence of rebirths, each time rising phoenix-like from the ashes of its previous demise. In the modern viewpoint, the bounce may occur in a very high temperature quantum gravity epoch.

O star   Star of spectral type O.

Ostriker–Thompson–Witten scenario (1986) Closed loops of superconducting cosmic strings formed during string network interactions will oscillate and radiate gravitationally. This energy release will make the loop shrink, and therefore its current will augment in a way inversely proportional to its size. By the end of the loop life, the energy output due to this increased electromagnetic power may supersede that in gravity waves. In such a case, plasma in the vicinity of the string loop will be subject to a strong radiation pressure, and thus thin expanding plasma shells will be generated. These dense gas shells would then eventually be the loci for the formation of galaxies and other structures.

While in standard scenarios cosmic strings attract surrounding matter via their gravitational fields, here instead string loops repel shells of matter via electromagnetic pressure, forming large-scale structures on the shell boundaries, and voids in the shell interiors. This model is importantly constrained by present measurements of distortions in the 3K cosmic microwave background radiation, notably by the detectors in the COBE satellite. *See* conducting string, cosmic string, Witten conducting string.

Ottawa 10.7 cm flux   The solar radio flux observed at a wavelength of 10.7 cm (or 2800 MHz) by the Ottawa Radio Observatory at 1700 UT (local noon) has been accepted as a standard measurement of solar fluxes. Normally, the flux is reported in solar flux units (1 sfu $= 1 \times 10^{-22}$ webers m^{-2} Hz^{-1}) The variations of the 10.7 cm flux, of 2800 MHz flux as it is sometimes called, is closely associated with enhanced thermal radiation from solar active regions, and thus to the overall level of solar activity.

outer core   In geophysics, the Earth's *outer core* has an inner radius of 1215 km and an outer radius of 3480 km. It is primarily composed of molten iron with other components dissolved in it. Convection in the outer core produces the geodynamo that generates the Earth's magnetic field.

outfall   A pipeline placed along the seafloor to disperse wastewater into the sea.

outflow channel   Very large dry river channels seen on Mars. They typically begin in a region of jumbled terrain (called chaotic terrain) and cross terrain of various ages, often termi-

nating in the lower-lying northern plains. The channels have widths between 10 and 100 km and are 400 to over 2000 km in length. Features associated with *outflow channels* suggest formation by catastrophic floods, probably caused by groundwater breaking out of the source region. Outflow channels are young, typically with formation ages less than 3×10^9 years.

outgoing longwave radiation (OLR)   Infrared radiation emitted by the Earth toward the space. The OLR displays strong temporal and spatial variations, affected by surface temperature, moisture, and clouds. The maximum annual-mean OLR occurs over the Saharan desert where the surface temperature is high and the atmosphere is dry. Satellite-measured OLR is used to determine the temperature and hence the approximate height of clouds. Over the tropics where heavy precipitation is caused mostly by deep convection, OLR can be used as a proxy of convective activity and precipitation rate.

oval equation   The partial differential equation

$$\chi(u)\frac{\partial^2}{\partial\sigma^2}\left[(1+\sigma^2)\alpha + \frac{1+u^2}{\alpha}\right] + \lambda(u)\frac{\partial^2\alpha}{\partial u^2} = 0$$

for the unknown functions $\alpha = \alpha(\sigma, u)$, $\chi = \chi(u)$ and $\lambda = \lambda(\sigma)$. The field equation describing stationary and axially symmetric spacetimes in general relativity containing only electric fields. Additionally, the cylindrical radius function ρ given by

$$\rho^2 = \left[(1+\sigma^2)\alpha - \frac{1+u^2}{\alpha}\right]^3 \alpha\lambda\chi$$

is required to be a harmonic function in these solutions.

overtopping   A term used to denote the process of waves pushing water over the top of a structure such as a breakwater or jetty.

overturning scale   *See* Ozmidov scale.

overwash   A term used to describe the flow of water across a barrier island and into the bay behind. An overwash fan of deposited sediments generally results.

OVV quasar   *Optically violently variable quasar.* Their optical spectrum closely resembles that of BL Lac objects, with a "blue bump" in some cases. They exhibit strong variability high polarization, and a flat radio spectrum.

oxygen   Colorless, odorless gas, O, atomic number 8, appears in the form of molecules O_2 and the unstable O_3, called ozone. O_2 is a strongly reactive gas, comprising about 21% of the sea level atmosphere. Naturally occurring atomic weight 15.9994. Naturally occurring isotopes: ^{16}O (99.76%), ^{17}O, ^{18}O melting point 54.8 K, boiling point 90.2 K. Constituent of water (H_2O). Produced by photosynthesis from CO_2 by plants, essential for respiration of most animals.

Ozmidov scale   The vertical scale, at which the buoyancy force is of the same magnitude as the inertial forces, is defined as $L_O = (\varepsilon/N^3)^{1/2}$, where ε is the dissipation rate of turbulent kinetic energy and N is the stability frequency. The *Ozmidov scale* quantifies the maximum size of overturning eddies for a given level of turbulence (characterized by ε) and stratification (characterized by N). Experimental evidence indicates that the Ozmidov scale (L_O) and the Thorpe scale (L_T) are strongly related and about equal (Dillon, 1982). *See* buoyancy scale.

ozone   A form of molecular oxygen, each molecule consisting of three oxygen atoms. It is colorless but has a characteristic odor. Its molecular formula is O_3, and molecular weight is 47.9982. *Ozone* is a common atmospheric constituent when oxygen is abundant, as is the case for the Earth. Ozone is important for the development and survival of life on Earth since it absorbs the deadly ultraviolet radiation produced by the sun. Most of the Earth's ozone is found within the stratosphere, although ozone is also a primary component of the smog produced by photochemical reactions with industry and car exhaust in cities. Within the stratosphere, the ozone layer continuously absorbs the incoming ultraviolet radiation, which causes the ozone molecule to split apart into an O_2 molecule and oxygen (O) atom. Normally the O_2 molecule and oxygen atom combine to reform ozone and the process repeats. However, the introduction of chlorofluorocarbons (CFCs) into the atmosphere by humans alters this ozone production process since chlorine atoms will combine with the oxygen atoms. The result is a thinning of the ozone layer (and production of the Antarctic ozone hole), which allows increased amounts of ultraviolet radiation to reach the surface.

ozone layer   The region in the stratosphere between altitudes of approximately 15 and 50 km (approximately coinciding with the stratosphere) where most of the atmosphere's ozone (O_3) is found, with the maximum concentrated ozone at about 20 to 25 km. At the maximum concentrated *ozone layer,* ozone is 1 to 10 parts per million. Ozone is an important constituent of the atmosphere as it acts to absorb most of the ultraviolet radiation from the sun in the wavelength range 230 to 320 nm, which is damaging to the complex molecules found in the cells of living organisms. In addition, heating by ozone absorption plays a profound role in determining the structure of the atmosphere. The heat absorption affects the lapse rate of the stratosphere in a way to make it much more stable, inhibiting vertical air movements, and stratospheric circulation. Ozone is formed by the breakdown of molecular oxygen (O_2) caused by short-wavelength solar radiation to form atoms of oxygen (O). These atoms collide with molecular oxygen (O_2) to form ozone (O_3), which in turn absorbs solar radiation for future dissociation to O and O_2. The distribution of atmospheric ozone layer varies at the different latitudes and seasons. The minimum concentration of ozone appears near the equator and the maximum locates at 60°N and 60°S. Ozone has its maximum concentration in spring and minimum in fall. Additionally, the ozone concentration has clear diurnal variation with higher concentration during night. This is because the dissociated oxygen atoms can continuously combine with oxygen molecules at night, yet the ozone decomposing process does not exist without solar radiation. The vertical distribution of ozone concentration has discontinuous or jump phenomena. The distribution of ozone is additionally influenced by a complex set of chemical destruction mechanisms involving atmospheric trace gases, as well as by transport by stratospheric winds.

Although significant decreases in ozone have been observed over much of the globe since the 1970s, the mechanisms are presently only partly understood. The appearance of the Antarctic ozone hole in the early 1980s demonstrated that polar ozone can be drastically reduced by man-made substances, namely chlorofluorocarbons (CFCs). These CFCs provide a source of chlorine in the atmosphere. Atomic chlorine is produced from Cl_2 or from CFCs by photodissociation in sunlight. Hydrogen chloride and chlorine nitrate are the most important two stratospheric chlorine reservoirs. Especially in polar regions, stratospheric clouds can serve as the reaction sites to produce molecular chlorine and nitric acid from the reservoirs. Chlorine is extremely damaging to ozone, because it acts catalytically:

$$Cl^* + O_3 \rightarrow ClO^* + O_2$$
$$O^3 + h\nu \rightarrow O^* + O_2$$
$$ClO^* + O^* \rightarrow Cl^* + O_2 \; .$$

The net result is to reform the chlorine radical, and to convert two ozone molecules to three O_2 molecules. The Montreal Protocol, signed in 1987, and its subsequent amendments, provide an international framework for the phasing-out of CFCs, although the extent and timescale of the recovery in the ozone layer are yet to be determined.

ozonosphere The atmospheric layer with higher concentrated ozone, which lies about 10 to 50 km altitude and thus is essentially synonymous with the stratosphere. The maximum concentration of ozone occurs at about 20 to 25 km.

P

package effect In oceanography and other areas, discrepancy between the spectral absorption coefficient of a particle suspension, consisting of a spatially nonuniform distribution of pigment molecules in cells, and the corresponding coefficient of a homogeneous solution containing the same amount of pigment.

Palatini (first order) formalism In general relativity, in the framework of the first order formalism the metric $g_{\mu\nu}$ and the affine connection $\Gamma^{\alpha}_{\beta\gamma}$ are considered as independent variables for the purpose of obtaining the Einstein equations from a variational principle. Taking variational derivatives of the otherwise standard Hilbert form of the variational action for general relativity with respect to those variables, one arrives at the dynamical Einstein equations in first order formalism. In the action

$$S_{EH} = \int d^n x \sqrt{-g} g^{\mu\nu} R_{\mu\nu}(\Gamma)$$

where $R_{\mu\nu}(\Gamma)$ depends on the connection only, but not on the metric, the equation of motion for $\Gamma^{\alpha}_{\beta\gamma}$ gives the same result as the metricity condition, that is, it fixes connection $\Gamma^{\alpha}_{\beta\gamma}$ to be the Cristoffel symbol $\left\{ \begin{matrix} \alpha \\ \beta\gamma \end{matrix} \right\}$. Taking this constraint into account, the equations for the metric become usual Einstein equations. One may apply similar concepts in gauge theories or in other variations of gravity theories. *See* christoffel symbol, curvature tensor, metricity of covariant derivative.

paleoclimate The science that treats the global climates throughout the geological ages. Its data are the distribution of glacier deposits, nature of plant and animal fossils, topography and geography of former periods, and character of sedimentary rocks. Based on the information extracted from these data, the ice period, the distribution of temperature and moisture can be identified.

palimpsest In planetary physics, circular features found on icy moons (particularly Ganymede) that appear to be the imprints of impact craters without any topographic relief. They typically display a bright interior zone, thought to represent the excavated crater, surrounded by a rough region representing part of the ejecta blanket. Although it may be argued that these are ancient craters which have relaxed over time due to the low strength of ice, *palimpsests* are more likely pristine forms that resulted from impact into ice which then melted and deformed as part of the cratering process. They probably represent features formed during the time when Ganymede's ice crust was solidifying.

Pallas One of the larger main belt asteroids. Second asteroid to be discovered, in 1802. Its diameter is estimated at 540 km, and a crude mass estimate, based on close encounters with Vesta and Ceres, gives $M = 2.16(\pm0.44) \times 10^{24}$ g. The best estimate of its density (good to about 35%) is 2.6 g cm^{-3}. It has a visual albedo of 0.66 and the spectral reflection characteristics of an S-type asteroid. Its composition has not been well determined. Orbit: semimajor axis 2.772 AU, eccentricity 0.22965, inclination to the ecliptic 34°.84603, period 4.62 years.

Pan The innermost of Saturn's known moons, also designated SXVIII. It was discovered by M. Showalter in 1990 in Voyager photos, and orbits within the Encke Gap in Saturn's A ring. It has a radius of approximately 10 km. Its orbit has a semimajor axis of 1.34×10^5 km. Its mass has not been measured.

pancake distribution In plasma physics, a pitch angle distribution in which the highest intensities are orthogonal to the magnetic field so that a contour of the flux density against polar angle θ is an oblate spheriod (pancake). *See* cigar distribution.

pancake dome A volcanic feature on the surface of Venus composed from viscous lava. The lava was probably quartz-rich, granitic, or a frothy, gas-rich basaltic lava. *Pancake domes* are widely scattered, often forming small groups or clusters. Dome clusters comprise 5 or 6 domes. Most occur in the lowland plains far from the largest shields, and often near the lowland coronas. "Tick-like" features also show a pattern similar to the pancake domes, both in their spatial distribution and morphological similarities. Thus, it is thought that many ticks are modified dome features.

Pandora   Moon of Saturn, also designated SXVII. It was discovered on Voyager photos in 1980. Its orbit has an eccentricity of 0.004, an inclination of essentially 0, and a semimajor axis of 1.42×10^5 km. Its size is $57 \times 42 \times 31$ km, its mass is 2.2×10^{17} kg, and its density 0.71 g cm^{-3}. It has a geometric albedo of 0.9, and orbits Saturn once every 0.629 Earth days. *Pandora* is the outer shepherd of Saturn's F ring.

Pangaea   In geophysics, the single supercontinent in which all the continental areas were connected, and which broke up under continental drift about 180 million years ago.

Papapetrou spacetimes (1953)   Stationary empty space-times for which the norm and curl scalars of the Killing vector are in a functional relationship. Their asymptotic expansion lacks the mass monopole term, so they are in a sense zero-mass gravitational fields.

PAR   *See* photosynthetically available radiation.

paraboloid   A figure of revolution based on a parabola. The preferred shape for reflecting telescope primaries because it brings all rays parallel to the axis to a focus at the same unique point.

parallax   Motion of nearby stars against the more distant background stars, as a result of the different vantage points from the Earth's orbit at different times of the year. Essentially a baseline of 2 AU (astronomical units) is used to give a perspective on the distance to the star. *See* astronomical unit, parsec.

parallel electric field   The component E$_{||}$ of the electric field in a plasma parallel to the magnetic field vector. Because charges can easily flow along magnetic field lines, in many plasma configurations E$_{||}$ is expected to be virtually zero. However, observations on auroral field lines (e.g., that of beams of ions) suggest that a non-zero E$_{||}$ may exist there, balanced either by the mirror force (*see* quasi neutral equilibrium) or by turbulent plasma processes. The ambipolar electric field, in which a relatively small E$_{||}$ is balanced by gravity, is another example.

Parker limit   A limit on the cosmological flux of magnetic monopoles:

$$F_M \leq 10^{-15} \, \text{cm}^{-2} \cdot \text{sec}^{-1} \, .$$

Acceleration of magnetic monopoles in the presence of a magnetic field B (like that of a galaxy, for instance), makes the magnetic field energy density decrease with time. The characteristic decay time is

$$\tau_d = \frac{eB}{4\pi n_M \langle v_M \rangle} \, ,$$

where e is the electric charge, n_M the magnetic monopole number density, and $\langle v_M \rangle$ their average velocity. As the time necessary to regenerate the magnetic field in a galaxy is comparable to the galactic rotation period $\tau_r \sim 10^8$ years, the limit results from the requirement that the decay time τ_d be larger than τ_r (so that magnetic monopoles do not affect galactic physics noticeably). The maximum allowed flux is much less than what is expected for monopoles formed at the grand unified energy scale. *See* monopole, monopole excess problem.

parsec   The distance corresponding to a *par*allax of one *sec*ond of arc. A unit of distance defined as the distance at which 1 astronomical unit (AU) subtends an angle of 1 sec of arc. It is equal to 206264.806 AU $= 3.08568 \times 10^{16}$ m $= 3.26166$ light years.

partial derivative   For a function f of multiple independent variables x^a, the derivative with respect to one variable, say x^i, computed by treating the other independent variables x^a, $a \neq i$ constant. Notation:

$$\frac{\partial f}{\partial x^i}$$

partial pressure   In any mixture of gases, of each component gas, the pressure that the gas would exert alone, i.e., if all the other gases had been removed. The total pressure equals to the summation of all the partial pressures.

particle acceleration   In astrophysics, any process by which ambient particles gain energy. Typically applied to high energy radiation signatures. Planetary atmospheres, the sun, galactic nuclei, and accretion disks are all sites of *particle acceleration.*

particle horizon   The causal horizon determined by the maximum distance from which a particle can receive physical signals. Since no signals can travel faster than light, the *particle horizon* is given at a fixed time by the past light-cone originating from the space-time position of the particle.

For example, in the Big Bang model of the universe, two particles that emerged from the surface of last scattering (when matter decoupled from radiation) could see each other by now only if enough time has passed such that a beam of light could have traveled the distance that separates the two particles. The proper distance between two particles locally at rest in some space-time models (e.g., De Sitter and generally inflationary models) grows faster than light speed (although no signal is transmitted and causality is not violated), in which case the horizons of the particles actually shrink in time.

pascal (Pa)   A unit of pressure, equal to 1 Newton/m^2. Atmospheric pressure at sea level is slightly greater than 10^5 Pa.

Paschen series   The series of lines in the spectrum of the hydrogen atom which corresponds to transitions between the state with principal quantum number $n = 3$ and successive higher states. The wavelengths are given by $1/\lambda = R_H(1/9 - 1/n^2)$, where $n = 4, 5, 6, \cdots$ and R_H is the Rydberg constant. The first member of the series ($n = 3 \leftrightarrow 4$), which is often called the P_α line, falls in the infrared at a wavelength of 1.875 μ m. *See* Rydberg constant.

Pasiphae   Moon of Jupiter, also designated JVIII. Discovered by P. Melotte in 1908, its orbit has an eccentricity of 0.378, an inclination of 145°, and a semimajor axis of 2.35×10^7 km.

Its radius is approximately 25 km, its mass 1.9×10^{17} kg, and its density 2.9 g cm^{-3}. Its geometric albedo has not been well determined, and it orbits Jupiter (retrograde) once every 735 Earth days.

passive continental margin   A continental margin that is not a plate boundary. A *passive continental margin* is the result of a process of continental rift, seafloor spreading, and creation of the oceanic crust.

passive earth force   The force exerted (either laterally or vertically) by stationary soil against a surface.

passive margins   The boundaries between the oceans and the continents are of two types, active and passive. *Passive margins* are not plate boundaries and have little or no seismic activity or volcanism. The boundaries of the Atlantic Ocean are almost entirely passive margins.

patchiness and intermittence   Oceanic turbulence exhibits strong spatial and temporal variability. Spatially, the turbulence is distributed in layers or "patches" having vertical scales from about 0.5 to 25 m and horizontal scales ranging from several meters to several hundreds of meters. Thinner layers occur more frequently and they are generally less active than thicker layers. Temporally, turbulence is intermittent and occurs in "bursts" caused by local dynamic instabilities in high-gradient regions, such as the breaking of internal waves.

patera   A large, low-sloped ($< 1°$) volcano found on Mars. *Paterae* tend to have very highly dissected flanks, indicating they are covered with an easily erodible material such as ash. Most of the Martian paterae have high crater densities, indicating an old age. No terrestrial analogs are known, but many planetary scientists believe patera formed by the interaction of hot magma with surface or near-surface water. When hot magma encounters cold liquid water, a steam explosion causes the creation of fine-grained ash rather than lava flows. It is believed that early in Mars' history there was much more water on or near the surface than today and that the rates of volcanic activity were also higher, which could have led to the formation of the patera.

pathline   A line drawn through a region of fluid that indicates the path followed by a particular particle within the fluid.

P-Cygni profile   An emission/absorption line profile which is the result of an expanding shell of gas around a star; the star P Cyg is the prototype. Its visible spectrum shows lines that have a wide emission component and an absorption component, with the absorption blue shifted and the emission line approximately at the star's rest velocity. A simple explanation is that the absorption arises in material in front of the star along our line of sight toward the continuum source. Since this material is expanding directly at us, we see the absorption line blue shifted with respect to the star's rest frame. The part of the expanding gas that is moving at right angles to our line of sight, on either side of the star, without the stellar continuum in the background, produces emission lines that are roughly centered at the star's rest velocity. The degree in which the absorption portion is shifted depends on the velocity of the gas that is producing the lines, and thus these lines allow the measurement of wind velocities in astronomical objects. The winds in OB stars and in cataclysmic variables were originally detected through the observation of these line profiles in their spectrum.

Peclet number   A dimensionless number quantifying the importance of advective heat transfer. It is the ratio of heat flux by advection to heat flux by conduction. For example, if the characteristic flow velocity in the direction of heat conduction is v over a characteristic length L, the *Peclet number* is defined as $Pe = vL/\kappa$, where κ is the thermal diffusivity. If $Pe \gg 1$, heat transfer is mainly advective. If $Pe \ll 1$, heat transfer is mainly conductive.

peculiar motion   In astronomy, the physical or angular velocity of a source with reference to a fixed frame of interest. For stars, the transverse *peculiar motion* is typically given in seconds of arc per century. For extragalactic objects, transverse motion is not measurable, but the radial

peculiar motion is measured by means of red shifts, and reported in terms of km/sec.

peculiar velocity   *See* peculiar motion.

peeling property   The Weyl tensor (a form of the curvature tensor) of an asymptotically flat space-time, when expanded in power series of the affine parameter r of outgoing null geodesics, has the form

$$C_{ijkl} = \frac{N_{ijkl}}{r} + \frac{III_{ijkl}}{r^2} + \frac{II_{ijkl}}{r^3} + \frac{I_{ijkl}}{r^4} + \mathcal{O}\left(\frac{1}{r^5}\right)$$

where the coefficients N_{ijkl}, III_{ijkl}, etc. are of the Petrov type indicated by the kernel letter. In particular, gravitational radiation encoded in N_{ijkl} has an amplitude falloff of r^{-1}. *See* Petrov types.

penetrative convection   Sinking plumes in a convectively mixing surface layer (or rising plumes in a convectively mixing bottom layer) impinge with their characteristic velocity scale onto the adjacent stratified fluid. As the plumes have momentum, they reach a limited penetration of the depth scale before the plumes stop due to entrainment and buoyancy. For instance, substantial, convective motions are present in the upper layer of the sun. At the base of the convection zone there is a gradual transition to the stable, radiative interior, the *convective penetration* zone. Below it, energy is transported outwards entirely by radiative diffusion. At the penetrative convection depth, the gas becomes suddenly less opaque, becomes unstable, and convective transport becomes dominant.

Penman equation (or combination equation) Evaporation (E) can be estimated by combining mass-transfer and energy-balance approaches in a theoretically sound and dimensionally homogeneous relation:

$$E = \frac{s(T_a)(K + L) + \gamma K_E \rho_w \lambda_v u_a \left[e_{sat}(T_a)\right](1 - RH)}{\rho_w \lambda_v \left[s(T_a) + \gamma\right]}$$

where $s(T_a)$ is the slope of the saturation vapor pressure vs. air temperature (T_a), K is net shortwave radiation, L is net longwave radiation, γ is the psychrometric constant, K_E is a coefficient that reflects the efficiency of vertical transport of water vapor by turbulent eddies of the wind, ρ_w is the density of liquid water, λ_v is the latent heat of vaporization, u_a is the wind speed, e_{sat} is the saturation vapor pressure at air temperature T_a, and RH is the relative humidity. The *Penman equation* assumes no heat exchange with the ground, no water-advected energy, and no change in heat storage, and can be expressed in simplified form as

$$E \propto \frac{s\,(T_a) \times \text{net radiation} + \gamma \times \text{``mass transfer''}}{s\,(T_a) + \gamma}.$$

Penman–Monteith equation   Monteith showed how the Penman equation can be modified to represent the evapotranspiration rate (ET) from a vegetated surface by incorporating canopy conductance

$$ET = \frac{s\,(T_a)\,(K + L) + \gamma \rho_a c_a C_{at}\,[e_{sat}\,(T_a)]\,(1 - RH)}{\rho_w \lambda_v\,[s\,(T_a) + \gamma\,(1 + C_{at}/C_{can})]}$$

where variables from the Penman equation have been supplemented with the density of air ρ_a, the heat capacity of air c_a, the atmospheric conductance C_{at}, and the canopy conductance C_{can}.

Penrose diagram   Space-times $\mathcal{M}$ in general relativity often have infinite extent, which makes their pictorial representation difficult. However, it is always possible to apply a suitable conformal mapping to the metric,

$$\mathbf{g'(x)} = \Phi(\mathbf{x})\,\mathbf{g(x)},$$

such that $\mathcal{M}$ is mapped into a new manifold $\mathcal{M'}$ which has finite extension and thus (sections of it) can be drawn on a sheet of paper. This implies certain behavior of the conformal factor $\Phi(x)$.

The transformation above leaves the causal structure of space-time untouched. Further, it is possible to choose coordinates and representations so that light travel is along 45° lines when plotted in $\mathcal{M'}$.

Penrose process   A process in which energy can be extracted from the rotation of a black hole, by exploding particles in free fall in the ergosphere near the black hole; with correct choice

of the orbits, one of the two fragments reaches infinity with a total energy exceeding the total rest mass-energy of the pieces.

penumbra   (**1.**)   The lighter region of a sunspot surrounding the umbra. The *penumbra* consists of light and dark radial filaments that are typically 5000 to 7000 km long and 300 to 400 km wide. Individual penumbral filaments endure 0.5 to 6 hours compared to the days or months for the sunspot as a whole.

(**2.**)   During a solar eclipse, the part of the shadow from which part of the sun is still visible.

percolation   The process of a fluid passing through pores in a medium such as soil.

perfect fluid   A fluid in which energy transport occurs only by means of matter flow (there is no viscosity or heat flow). A single-component *perfect fluid* (a fluid composed of atoms or molecules of the same chemical element or compound) is characterized by two functions of state, e.g., pressure and mass-density, from which all other functions (e.g., the temperature and entropy) can be calculated by means of the equation of state and other thermodynamical relations. A multicomponent fluid is not chemically homogeneous and can be perfect only if either there are no chemical reactions or all chemical reactions are reversible (i.e., with no emission of heat). In either case, densities of the chemical components must be specified separately, and when chemical reactions take place, the chemical potentials are also necessary to describe the state of a fluid. In cosmology, it is often assumed that the matter in the universe behaves like a single-component perfect fluid or a dust; the latter is pressureless and so obeys an even simpler thermodynamic. For stellar interiors, a perfect fluid is only a rather crude first approximation to a description of thermodynamical properties.

perfect fluid space-times   Space-times containing a perfect fluid with the stress-energy tensor

$$T^{ab} = (\mu + p)u^a u^b + p g^{ab}.$$

Here the functions μ and p are the energy density and the pressure of the fluid. The timelike four-velocity u^a is normalized using the metric, $u^a u^b g_{ab} = -1$. *See* tensor.

perfect gas   A gas which at the molecular level can be taken to have pointlike noninteracting molecules; and which is thus described by the *perfect gas* law: $pV = nkT$, where p is the pressure, V is the volume, T is the (absolute) temperature, k is Boltzmann's constant, and n is the number of gram-moles of gas involved. Real gases always have some intermolecular interaction, but perfect gas behavior can be approached by dilution. Among other properties, an ideal gas has internal energy which is a function of temperature alone, and constant values for its specific heats (e.g., C_V at constant volume and C_p at constant pressure).

periapsis   The point in an elliptical orbit where the orbiting body is the closest distance to the body being orbited. (The apoapsis is the point of the farthest distance.) When the sun is the central body, the point of *periapsis* is called the perihelion.

periastron   In planetary motion, the closest distance achieved to the gravitating central star. Generically one says *periapse*. Specific applications are *perihelion* when referring to the motion of planets in our solar system; *perigee,* when referring to orbits around the Earth. Similar constructions are sometimes invented for orbits about the moon or about other planets.

perihelion   The point in an elliptical orbit around the sun that is closest to the sun. (The aphelion is the point farthest from the sun.) The time of *perihelion* passage for the Earth is typically around January 3.

perihelion shift   The parameter of an orbit of an astronomical object (usually a planet) that tells, by what angle the planet's *perihelion* is displaced during a unit of time (the preferred unit is different for different situations, for Mercury it is seconds of arc per century, for the binary pulsar PSR 1913+16 it is degrees per year). If the central body (e.g., the sun) were perfectly spherical and if it had only one companion (planet) on an orbit around it, then Newtonian

mechanics and gravitation theory predict that the orbit would be an ellipse, with the central body placed in one of the foci of the ellipse. In reality, the central body is not exactly spherical, and there are several planets and moons in the planetary system that perturb each other's orbit. Because of those perturbations, the real orbits are not ellipses even according to Newtonian mechanics. The orbit is then a curve that is close to an ellipse during one revolution of the planet around the star, but the axes of the ellipse rotate during each revolution by a small angle in the same sense as the planet's motion. The angle by which the perihelion is rotated in a unit of time is the *perihelion shift.* The nonsphericity of the sun gives a negligible effect, but the perturbations from other planets sum up to a perihelion shift of 530 sec of arc per century in the case of Mercury. On top of that comes the additional perihelion shift predicted by Einstein's general relativity description of gravity. According to this description, even with a perfectly spherical central star and a single planet orbiting it, the orbit would not be an ellipse, but a rotating ellipse, like a perturbed Newtonian orbit. For Mercury, relativity predicts this additional perihelion shift to be 42.95 sec of arc per century. This additional component in Mercury's orbital motion was detected in 1859 by LeVerrier, and called an "anomalous perihelion motion of Mercury". It was a major problem for 19th-century astronomy that several astronomers tried (unsuccessfully) to solve by methods of Newtonian mechanics, e.g., by postulating that the additional shift is caused by interplanetary dust, the solar wind, the nonsphericity of the sun, or even by a thus far unobserved additional planet. Explanation of this anomaly was the first, and quite unexpected, triumph of Einstein's general relativity theory. For other planets, the relativistic perihelion motion is small and became measurable only with modern technology; it is 8.62 for Venus, 3.84 for the Earth, and 1.35 for Mars. (All numbers are in seconds of arc per century. Values observed differ from these by 0.01 to 0.03.) The fastest perihelion motion observed thus far is the periapse shift of the binary pulsar PSR1913+16; it amounts to $4.2°$ per year. Its shift has been observed via the phase of pulses from this binary pulsar.

period   The amount of time for some motion to return to its original state and to repeat its motion.

period-luminosity relation   The relation between the luminosity of Cepheid variable stars and the variability period of these stars. The basic trend is that those stars with longer periods are brighter. By measuring the period of Cepheid variables, astronomers can use this relationship to deduce the intrinsic luminosity of these stars. Combined with a measurement of the apparent luminosity, the distance of Cepheid variables can be estimated. The Hubble Space Telescope has been used to detect Cepheid variables out to the Virgo cluster, measuring the distance of the Virgo cluster to roughly a 10% accuracy. *See* Cepheid variable.

permafrost   Soil or subsoil that is permanently frozen for two or more years, typical of arctic regions, or in other arctic climates (such as high altitude) where the temperature remains below 0°C for two or more years.

permeability   In electromagnetism, the relation between the microscopic magnetic field **B** and the macroscopic field (counting material polarizability) **H**. In fluid flow, *permeability,* also called the intrinsic permeability, characterizes the ease with which fluids flow through a porous medium. Theoretically, permeability is the intrinsic property of a porous medium, independent of the fluids involved. In reality, the permeabilities of some rocks or soils are affected by the fluid. Permeability in fluid flow has the dimension of area. For an anisotropic porous medium, the permeability is a tensor. *See* Darcy's law.

permeability coefficient   In electromagnetism, the coefficient μ in the relation $\mathbf{B} = \mu\mathbf{H}$ between the microscopic magnetic field **B** and the macroscopic field **H**. In general, μ may be a 3×3 linear function of position, with time hysteresis, though it is often taken to be a scalar constant. In fluid flow, the coefficient measuring how easily water molecules can cross the surface film. For clear water (Davies and Rideal, 1963), it is equal to 5 ms^{-1}.

persistent current   *See* current generation (cosmic string).

perturbation theory   A tool that can be applied to questions ranging from classical mechanics to quantum theory. Independent of the actual question, the basic idea is to find an approximate solution of the equations for a complex system by first solving the equations of a physically similar system chosen so that its solution is relatively easy. Then the effects of small changes or perturbations on this solution are studied. In classical mechanics, for instance, the motion of a planet around the sun is studied first and the influences of the other planets are added later.

Formally, a complex system is described by a set of coupled non-linear partial differential equations. If only small-amplitude disturbances are considered, such a system can be simplified by linearization of the equations: whenever two oscillating quantities are multiplied, since both are small, their product is a higher order term and can be ignored. If these results are to be applied to a real situation, however, one always has to take one step back and justify whether the amplitudes calculated in the real situation are small enough so that the non-linear terms actually are negligible compared with the linear ones.

perturbative solution   An approximate solution, usually found to a problem that is too difficult to be solved by an exact calculation. In a *perturbative solution* there exists at least one small dimensionless parameter (ε) that must be distinctly smaller than 1. Then, $\varepsilon^2 \ll \varepsilon$. The equations to be solved are expanded in power series with respect to ε and in the first step all terms proportional to ε^n where $n > 1$ are assumed to be equal to zero. In the next step, the already found first approximation is substituted for the terms proportional to ε, the terms proportional to ε^n where $n > 2$ are assumed equal to zero both in the equation and in the solution, and the coefficient of ε^2 is determined. The procedure can, in principle, go to an arbitrarily high degree of approximation, but in practice the calculations often become prohibitively complicated in the second step (this happens, e.g., with Einstein's equations in almost every case). Perturbative solutions are ubiquitous in astrophysics. Planetary orbits and the positions of planets on the orbits are calculated by such perturbative methods where the small parameter is the eccentricity of the orbit. In cosmology, the most commonly used description of the formation of galaxies is by solving Einstein's equations perturbatively, with the small parameter being the size of the matter deviation from the spatially homogeneous background. Perturbative solutions are often possible where an exact solution seems prohibitively difficult, but they have their own difficulties. In applications in general relativity care must be taken to distinguish between genuine perturbations and spurious perturbations.

Peru current   A cold ocean current running up the west coast of the South American continent, feeding into the South Equatorial current.

Petchek reconnection   Fast reconnection at an X-point or in a small localized region. About three-fifths of the inflowing magnetic energy is converted into kinetic energy behind the shock waves, the remaining two-fifths heat the plasma. With $v_{A,in}$ being the Alfvén speed in the inflowing plasma, L being the length scale of the reconnection region, σ being the conductivity, and c being the speed of light, the reconnection rate R_p can be written as

$$R_p = \frac{\pi}{8} \ln\left(\sqrt{\frac{c}{L\sigma v_{A,in} 4\pi}}\right) .$$

Petchek reconnection varies only weakly with the conductivity σ and is very efficient in mixing fields and plasmas. It probably does not play a role in magnetospheric plasmas but might be important in solar flares.

Petchek reconnection is stationary reconnection: the onset of reconnection does not destroy the general field configuration. Stationary reconnection results in an equilibrium between inflowing mass and magnetic flux, magnetic diffusion, and outflowing mass and magnetic flux.

Petrov types   Types in a classification of gravitational fields according to the algebraic structure of the Weyl tensor (a tensor algebraically constructed from the Riemann tensor to be completely traceless), introduced by Petrov (1966).

In general relativity the traces of the Riemann tensor are pointwise determined by the stress-energy tensor. The trace-free part, the Weyl tensor C_{abcd} is symmetric in the skew pairs ab and cd of indices. There are six independent components of each skew pair; hence the Weyl tensor can be viewed as a symmetric 6×6 matrix. The eigenvalues of this matrix sum to zero. An alternate (calculationally simpler) development can be given in terms of spinors. The eigenvectors determine null vectors; for the general type I there are the four different principal spinors α_A, β_A, γ_A and δ_A which determine different principal null directions. For types II, III, and N, respectively, two, three, and all four of the principal directions coincide. Type D arises when there are two pairs of coincident principal null directions. When the Weyl curvature vanishes, the type is O. *See* principal spinor, Weyl tensor.

Petzold data   A widely used data set containing volume scattering functions for various waters ranging from very clear to very turbid; the scattering phase function is highly peaked at small scattering angles and is somewhat independent of water type.

Pfund series   The set of spectral lines in the extreme infrared region of the hydrogen spectrum with frequency obeying

$$\nu = cR\left(1/n_f^2 - 1/n_i^2\right) ,$$

where c is the speed of light, R is the Rydberg constant, and n_f and n_i are the final and initial quantum numbers of the electron orbits, with $n_f = 5$ defining the frequencies of the spectral lines in the *Pfund series.* This frequency is associated with the energy differences of states in the hydrogen atom with different quantum numbers via $\nu = \Delta E/h$, where h is Planck's constant, and where the energy levels of the hydrogen atom are:

$$E_n = hcR/n^2 .$$

phase   In planetary astronomy; variance in the total amount of the visible disk of the moon, or of the inner planets, which appears illuminated at any one time. *Phases* arise because our vantage point sometimes allows the unlit side of the object to be viewed, as with the new or old moon, close to the sun in the sky. Phases can also be observed for Venus and Mercury, though they are behind the sun when at full phase as viewed from the Earth. For the moon, the phases are new (rising very close to the sun after dawn in the sky), waxing crescent (rising after dawn, becoming noticeable just at sunset), first quarter (rising at noon), waxing gibbous, full (rising at sunset; near zenith at midnight), waning gibbous, last quarter (rising at midnight), waning crescent.

In physical chemistry, a phase is a homogeneous state of matter that is (conceivably) physically separable from any other phase in a system. A phase is also used loosely to describe a mineral species, that has different solid forms of different structure and density, in a metamorphic mineral assemblage.

In signal physics, in periodic or nearly periodic systems, the point in the waveform measured in degrees or radians from a fiducial event such as a positive-going zero crossing. The phase of $360° = 2\pi$ is assigned to the interval between two such fiducial events (one period).

phase angle   The angle from the sun to an object to an observer. An object's *phase angle* is equal to the angular separation (elongation) between the sun and the observer as viewed from the object.

phase frequency threshold   In cosmology, the current trapped in cosmic strings can be timelike as well as spacelike. In the former case, the energy it contains is derivable essentially from the energy of a bound current carrying particle, which, in order for a bound state to exist, must have energy less than the free energy equivalent of the mass of the particle.

One therefore cannot increase the energy indefinitely. In fact, as it increases towards mc^2, the state it describes becomes closer and closer to that of a free particle, namely, it starts spreading. Also, as the energy tends to mc^2 the integrated current increases indefinitely and it is seen to diverge. This limit is referred to as the *phase frequency threshold*. It simply reflects the fact that no bound state can be obtained with energy greater than the rest mass energy equivalent of the free particle. Spacelike currents are not

subject to this constraint, since they can be set arbitrarily far from a free particle state; however, spacelike currents are limited also by current saturation.

At this phase frequency threshold, the energy per unit length diverges while the string tension decreases also without limit. The interesting consequence is that at some stage the tension vanishes (before eventually becoming infinitely negative) so that springs could be formed and the corresponding strings become violently unstable. *See* Carter–Peter model, cosmic spring, current saturation, duality in elastic string models, Witten conducting string.

phase function   In optics, the change in the brightness of an object as a function of the phase angle. In general, an object gets brighter as the phase angle approaches 180° or 0°. The function is usually fairly smooth except for at small phase angles where there may be a "spike" of increased brightness. The *phase function* is usually described as the change in magnitude (brightness) per degree of phase angle. In descriptions of wave propagation in scattering media, the ratio of the volume scattering function to the scattering coefficient [sr^{-1}]; the integral of the phase function over all directions is unity.

phase space   For Hamiltonian systems in classical mechanics, the even-dimensional space coordinatized by the configuration coordinates q^i and their conjugate momenta p_i. Here $i = 1 \cdots N$; N is the physical dimension of the system under study, so the dimension of *phase space* is $2N$. The Hamilton equations describe evolution in phase space that proceeds in a way that preserves an antisymmetric differential form (covariant antisymmetric 2-tensor)

$$\Omega \equiv \sum_{i=1}^{N} \left(dp_i \wedge dq^i \right)$$

where the "$\wedge$" indicates an antisymmetric tensor product. Thus, in these canonical coordinates the components of the tensor Ω are

$$\Omega_{ij} = \begin{pmatrix} 0 & 1 & & & & \\ -1 & 0 & & & & \\ & & 0 & 1 & & \\ & & -1 & 0 & & \\ & & & & 0 & 1 \\ & & & & -1 & 0 \end{pmatrix}$$

where the order of the indices is taken as:

$$q^1, p_1, q^2, p_2 \cdots \cdots .$$

Canonical transformations can be viewed as passive motions of the coordinates past a system point in phase space, or as active motion (dragging) of points in phase space. The flow under Hamilton's equations can be viewed as such an active dragging. This invariance of Ω implies the conservation of elementary volume in phase space, Liouville's theorem, since the volume element is proportional to

$$\underbrace{\Omega_\wedge \Omega_\wedge \cdots _\wedge \Omega}_{N \text{ terms}}$$

and each of the terms is separately conserved. Thus, one says the motion in phase is incompressible. Depending on the specific dynamical system (particularly if the system undergoes chaos) this elementary volume can become distorted into an arbitrarily long, twisted filament, and "mixed" throughout phase space during the evolutions.

A point in phase space represents the instantaneous state of the system. Hamilton equations are first order equations for the time derivative of this state:

$$\dot{q}^i = \frac{\partial H}{\partial p_i}$$

$$\dot{p}_i = -\frac{\partial H}{\partial q^i}.$$

Since the Hamiltonian is a function $H = H(p_i, q^i, t)$, an immediate consequence is that at a particular time, the direction of motion through a particular point in phase space is single valued. Orbits in phase space cannot cross.

phase speed   The speed deduced from tracking individual wave crests rather than the group profile in a wave:

$$v_{ph} = \omega / k ,$$

where ω is the angular frequency [sec^{-1}] and k is the wavenumber [m^{-1}] of the wave. Also referred to as celerity.

phase transition   Chemical elements or compounds are often capable of being ordered in different crystal structures, or "phases". The stable phase for fixed temperature and pressure is then the phase which minimizes the Gibbs free energy. In geophysics, because the Earth has gradients in both temperature and pressure, the same rock material may be found to contain different types of crystal at different depths. Lines dividing the areas of stability of different phases may be drawn on a temperature-pressure graph, in which case the slope of a line dividing two phases is given by the Clasius–Clapeyron equation:

$$\frac{dT_t}{dP} = \frac{\Delta V}{\Delta S}$$

where T_t is the temperature of the phase transition at some pressure P, ΔV is the difference in specific volume between the phases, and ΔS is the difference in specific entropy. ΔS may be either positive or negative for Earth materials, so the slope of the line may be either positive or negative. In the Earth, a positive slope would imply that a *phase transition* would be found higher where it is locally cool, and lower where hot, which would aid convection (as the phase at higher pressure is more dense). Similarly, a negative Clapeyron slope would impede convection. The 400 km deep transition in the mantle is often identified with the olivine/spinel transition, and would be an example of the former case, while the 670 km deep transition is identified with the spinel/perovskite transition, and would be an example of the latter case. The inner core/outer core boundary is an example of a transition between a liquid and a solid phase.

In cosmology, in the early universe, there may have been several phase transitions as the elementary particles filling the universe cooled and settled into particular low temperature configurations. Remnants of these transitions may still exist in exotic forms such as cosmic strings or magnetic monopoles.

phase velocity   *See* phase speed.

pheophytin   A phytoplankton pigment that is inert and does not contribute to photosynthesis.

phi unit   A measure of sediment size commonly used in geology. Sediment diameter and ϕ-size are related by $D = 2^{-\phi}$, so that a larger particle has a smaller ϕ value.

Phobos   One of the two moons of Mars, also designated MI. *Phobos* (meaning fear) was discovered by A. Hall in 1877. Its orbit has an eccentricity of 0.015, an inclination of 1.0°, a precession of 158.8° yr^{-1}, and a semimajor axis of 9378 km. It is one of the three satellites in our solar system whose period (7h 39m) is less than the rotational period of the primary planet (24h 37m for Mars); rises in the west and sets in the east, often twice a day. It is losing orbital energy to the surface tides it raises on Mars; its orbit is decaying at a rate of 1.8 m/century and in about 4×10^7 yr it will either be disrupted by tidal forces (in this way, it may become a ring plane about Mars within the next 50 million years) or it will crash into the surface of Mars. Its size is $13.5 \times 10.8 \times 9.4$ km, its mass is 1.08×10^{16} kg, and its density is 1.9 g cm^{-3}. Its geometric albedo is 0.06 and its surface is similar in reflectivity to C-type asteroids. It may be a member of that group that was captured in the past. New temperature data and close-up images of Phobos gathered by NASA's Mars Global Surveyor indicate the surface has been pounded into powder, at least 1 m (3 ft) thick, by eons of meteoroid impacts. Meteorite impacts also initiated landslides along the slopes of some craters. High temperatures were measured at -4°C (25°F) and lows at -112°C (-170°F). (Phobos does not have an atmosphere to hold heat in during the night.)

Phoebe   Moon of Saturn, also designated SIX. It was discovered by Pickering in 1898. Its orbit has an eccentricity of 0.163, an inclination of 177°, and a semimajor axis of 1.30×10^7 km. Its radius is 110 km, its mass is 4.0×10^{18} kg, and its density is 0.72 g cm^{-3}. Its geometric albedo is 0.05, much lower than the other Saturnian satellites (except the dark part of Iapetus). It orbits Saturn once every 550.5 Earth days. *Phoebe's* retrograde orbit indicates that it is probably a captured satellite.

photodissociation region   A part of the interstellar medium where the molecules have been destroyed into atoms by ultraviolet light, and the chemical composition and temperature are determined by the intensity of the light. *See* interstellar medium.

photoelectrons   Electrons emitted by the action of light, in particular (in magnetospheric research) of short-wave sunlight impinging on either the upper atmosphere or on the surface of a satellite. *Photoelectron* emission from the upper atmosphere is the major source of the ionosphere. Photoelectron emission from satellite surfaces causes spacecraft charging and can affect the operation of sensitive instruments in space. *See also* differential charging.

photoinhibition   The decrease in photosynthetic rate with increasing irradiance, caused by surpassing the photosynthetic capacity.

photoionization   Ionization of atoms and ions by photons. A photon with energy greater than a threshold energy, the ionization potential, is absorbed, and an electron previously bound to an atom is freed. The energy gained by the electron is equal to the difference between the photon energy $h\nu_0$ and the ionization potential ϵ_0, according to the formula $h\nu = h\nu_0 - \epsilon_0$. Spectral lines are emitted following *photoionization* when the electron recombines. In astronomy, such photoionized nebulae include HI regions, planetary nebulae, extended envelopes surrounding hot stars, and the line-emitting regions of active galactic nuclei.

photometric binary   An orbiting pair of stars, neither of which eclipses the other, but whose gravitational fields distort each others' shapes enough so that the surface area we see (hence the brightness of the system) changes through the orbit period. Also ellipsoidal variable or ellipsoidal binary.

photometry   The accurate measurement of the flux of light from a source.

photon   An elementary particle of zero mass and spin one. The quantum of electromagnetic radiation; the smallest "bundle of energy" of light of a particular frequency, with energy $E = h\nu = hc/\lambda$ where h is Planck's constant, c is the speed of light, ν is the frequency, and λ is the wavelength of the light.

photosphere   The layer of the sun or another star that we see in visible light. The *photosphere* is a quite thin layer and at a temperature between about 2,000 and 150,000 K (5,600 K for the sun). A stellar photosphere produces a continuous spectrum crossed by absorption lines due to the atoms and molecules of the elements and compounds in it.

photosynthesis   The manufacture of carbohydrates from carbon dioxide and water in the presence of chlorophyll, by utilizing radiant energy and releasing oxygen; the chemical change induced in chlorophyll by the absorption of a quantum of radiant energy.

photosynthetically available radiation   The integral over visible wavelengths (350 to 700 nm or sometimes 400 to 700 nm) of the number of photons available for photosynthesis [photons s^{-1} m^{-2}]; computed by integrating the spectrally dependent scalar irradiance divided by the photon energy at each wavelength.

photosynthetic capacity   The maximum photosynthetic rate per unit of biomass.

photosynthetic pigment   Molecules whose structures efficiently absorb light within the 400 to 700 nm range.

phytoplankton   Plant forms of plankton, generally with sizes from less than 1 to several hundred μm.

piezometer   The elevation of the water table or potentiometric surface as measured in a nonpumping well generally constructed from a small-diameter pipe with screened openings through which water can enter.

piezometric head   *See* hydraulic head.

Pileus cloud   A smooth cloud forming at the peak of a mountain, or at the top of a thundercloud.

pillow basalt   The form of an eruption that has occurred under water, leading to characteristic rock formations.

pitch angle   The angle between the velocity vector of a charged particle relative to the guiding magnetic field lines along which the particle is moving. It is given by $tan\theta = v_{perp}/v_{par}$ where subscripts *perp* and *par* refer to perpendicular and parallel to the direction of the magnetic field.

pitch angle diffusion   *See* diffusion, in pitch angle.

Pitot tube   (After Henri Pitot, 1695–1771) An open-ended tube used to measure the stagnation pressure of the fluid for subsonic flow; or the stagnation pressure behind the tube's normal shock wave for supersonic flow. In application, it is immersed in a moving fluid with its mouth pointed upstream. When combined with a "static" measurement, the pressure difference can be used to determine the speed of the tube through the fluid. The *Pitot tube* is in wide use to determine airspeed in aircraft, where the static pressure is obtained through an opening in the instrument case within the aircraft.

plage   An extended chromospheric emission feature of an active region overlying photospheric faculae.

plagioclase (triclinic feldspar)   Common mineral, essential constituents of most igneous rocks; composed of mixture of $NaAlSi_3O_8$, $Ca Al_2 Si_2O_8$, and occasionally barium silicates. Also called oligoclase.

Planck constant (h)   The elementary quantum of action, which relates energy to frequency through the equation $E = h\nu$. In metric units $h = 6.6261971 \times 10^{-27}$ gm cm^2/sec. Also, Planck's reduced constant, $\hbar = 1.05459 \times 10^{-27}$ gm cm^2/sec.

Planck length   The unique combination of the Planck constant $\hbar$, the gravitational constant G, and the speed of light c of dimension length:

$$\ell_P = \left(G\hbar/c^3\right)^{1/2} \sim 1.616 \times 10^{-33}\text{cm}.$$

The *Planck length* is the approximate scale at which quantum effects become strong in gravitating systems.

For a black hole, the typical length is set by the mass (half the gravitational radius), $R_H/2 = Gm/c^2$. The typical associated quantum length is the Compton wavelength $r_c = \hbar/mc$. The Planck length can be seen to coincide with R_H and r_c when the two are equal. Smaller masses (smaller gravitational radii) are strongly quantum because the quantum effect (their Compton wavelength) extends outside the classical black hole. Thus, the Planck length is the scale at which a quantum gravity description (not yet perfected) becomes necessary.

Planck mass   The mass of an object whose Compton wavelength equals the Planck length:

$$m = \left(\frac{\hbar c}{G}\right)^{1/2} \simeq 2.177 \times 10^{-5} \text{ grams} .$$

Planck's black body radiation   Radiation at all frequencies, such as would be emitted by a "black body". A black body is one that absorbs all radiation incident upon it. The emissivity of a black body is unity so the radiation that it emits is a function of temperature only. The peak of a black body spectrum is given by $\lambda_{\text{peak}} = (2.9 \times 10^6)/\text{T}$ nm, where T is measured in degrees Kelvin.

Planck's Law   That the energy of light is directly proportional to its frequency is known as *Planck's Law*: $E = h\nu$, where E is the energy of a photon, ν the photon's frequency, and $h = 6.626076 \times 10^{-34}$ J s, known as Planck's constant. This is a result of Planck's attempts to explain the observed black body radiation curves and is regarded as the beginning of the quantum theory of radiation and ultimately led to the foundation of quantum mechanics. Planck's law for the energy density of black body or cavity radiation is

$$\Psi(\lambda)\, d\lambda = 8\pi ch\lambda^{-5}[\exp(ch/\lambda kT) - 1]\, d\lambda$$

where c is the speed of light, h is Planck's constant, k is the Boltzmann constant, T is the temperature, and λ is the wavelength of the radiation. This law contains the Wien displacement law and the Stefan–Boltzmann law and

avoids the "ultraviolet catastrophe" inherent to Rayleigh's derivation of black body radiation.

Planck's radiation law   Electromagnetic radiation propagates in discrete quanta, each with an energy equal to the product of Planck's constant times the frequency of the radiation.

Planck time   The shortest length of time for which the classical theory of gravitation (general relativity) is useful, $\sim 10^{-43}$ sec. Below this scale general relativity must be improved to include quantum theory. It can be expressed as

$$t_P = \ell_P/c = \left(G\hbar/c^5\right)^{1/2} \simeq 5.4 \times 10^{-44} \text{ sec}$$

where G is Newton's gravitational constant, $\hbar$ is Planck's constant, and c is the velocity of light in vacuum.

plane-fronted waves   Gravitational waves in which normal of the wave fronts is a covariantly constant null vector. *See* pp-waves.

planet   A large body (generally larger than 2000 km in diameter) which orbits around a star. In our solar system, nine planets orbit around the sun.

planetary boundary layer   The turbulent boundary layer of the atmosphere, about 1 to 1.5 km altitude from ground in which the flow field is strongly influenced directly by interaction with the Earth's surface. In the planetary boundary layer, the frictional effects of the underlying surface generate turbulence; the air movement is completely turbulent, and the turbulent friction force is equal or larger than the orders of geostrophic force and pressure gradient force. In theoretical analysis one usually assumes that the *planetary boundary layer* can be represented by stationary homogeneous turbulence. The planetary boundary layer is the heat and moist source and the momentum sink of whole atmosphere. The free atmosphere interacts with the surface through the planetary boundary layer. Based on the different features of turbulence, the planetary boundary layer can be divided into three layers: the laminar sublayer, which is just above the ground less than 2 m thick; the surface layer above the laminar sub-layer, tens of meters thick; and the Ekman layer which can reach to about 1 to 1.5 km altitude.

planetary circulation   The circulation systems and circulation cells with 8,000 to 10,000 km horizontal scale. *Planetary circulation* includes the planetary wave, high and low pressure centers, mean meridional circulations, and mean zonal circulations as well as planetary scale monsoon circulations. The variation in planetary circulations is very slow with more than 1 to 2 week time scales. Thus the planetary circulations are the basic factors that determine medium- and long-term weather evolutions. They are also the background circulation fields for short-term weather evolutions. In general, planetary circulations are generated by forcing effects from large-scale orography, thermal effects of heat sources and sinks.

planetary magnetic fields   Magnetic fields believed to be generated within the deep interior (inner or outer core) by the rotation of the planet creating currents in molten material that conducts electricity. These electrical currents cause a magnetic field to form and the boundaries of that field extend far beyond the surface of the planet. A simple model of the field can be represented by a dipole (having North and South magnetic poles like a bar magnet) and is represented by field lines that originate at the North magnetic pole and terminate at the South magnetic pole. *Planetary magnetic fields* vary in size and shape because of the different properties within the planets generating the fields, and because of external influences like the sun's magnetic field and the solar wind. Earth's field is generally dipole-like until about 70,000 km from the planet. There the solar wind causes the Earth's field to be swept away from the sun, creating a teardrop shaped cavity where the influence of Earth's field dominates on the inside of the teardrop and the sun dominates on the outside. The long tail of the teardrop results from Earth's magnetic field lines being pushed downwind of the sun similar to the wake behind a rock in a stream.

planetary magnetosphere   The region of space surrounding a planet where the influence of the planetary magnetic field dominates over any external fields. The magnetic field originates within the body of the planet and the field extends outward until the magnetopause boundary. This boundary is the point where the intensity of the planetary magnetic field equals that of some other external field (usually the sun). The upper boundary can also be explained as the point where the most distant field lines are closed and are connected back to the planet. The lower boundary is the point where the neutral atmosphere dominates over the ionized (charged) region in space. For Earth, the magnetosphere in the direction toward the sun begins at a distance of about 50,000 km above the surface of the planet and ends at approximately 70,000 km. For the midnight side (pointing away from the sun), the magnetosphere can stretch millions of kilometers from the Earth due to the long magnetotail formed by the solar wind.

planetary nebula   The ejected outer envelope of a 0.8 to about 8 solar mass star that has completed hydrogen and helium burning, leaving a carbon-oxygen core. The envelope is ejected over a period of about 10^4 years, while the star is near the tip of the asymptotic giant branch. *Planetary nebulae* often show spectroscopic evidence of the helium, carbon, nitrogen, and other elements produced by nuclear reactions in the parent stars. They expand at a few $\times 10$ km/sec. A few tens of thousands of years after the gas begins to glow, a "fast wind" from the remaining material is blown out at high velocity (300 km/s or more), and snowplows through the slower moving gas. It is the interaction of the "slow" and "fast" moving material that contributes to the extravagant shapes of planetary nebulae. The nebulae then fade as the gaseous envelope becomes diffuse and the central star cools and no longer provides much ionizing radiation. About one planetary nebula is born in the Milky Way per year, leading to an inventory of 10^4 nebulae, of which about 1000 have been cataloged. Their complex shapes are due to some combination of the effects of companions, rotation, and the collisions of the primary wind with material ejected earlier and later. The nuclei of planetary nebulae fade to white dwarfs. The term "planetary nebula" is a misnomer that originated with the discovery of these faint, fuzzy objects. At first, astronomers thought they were planets like Uranus.

planetary radio astronomy   The study of radio wavelength emissions (long wavelength, low energy) from the planets within our solar system. The radio astronomy spectrum consists of the wavelength range from approximately 1 mm to 300 km corresponding to a frequency range of 300 GHz to about 1 kHz. Planetary radio emissions result from plasma interactions within the planetary magnetosphere and can be driven by the solar wind or some other particle source such as a satellite of a planet (e.g., Io in Jupiter's system). Due to the absorptive properties of Earth's ionosphere and the relative strengths of the radio emissions from the planets, only certain frequencies can be monitored from ground-based radio antennas (Jupiter is the only planet with intense enough radio emission so that it is observable from ground-based antennas). The other frequency bands must be monitored via spacecraft.

planetary rotation periods   The rotation periods of planets can be determined by using several different techniques: optical observations, RADAR measurements, and radio observations. Optical observations are appropriate for terrestrial planets when the surface can be seen (Mercury, Earth, Mars), but cloud penetrating radar is needed for Venus. Rotation period determinations are more complex for the gas giant planets since they do not have a "true" surface that is observable. Rotation period measurements can be made from cloudtop observations, but those were quickly shown to be variable. Observations of the radio emissions have proven to be the most reliable since the radio emission source regions are directly linked to the planetary magnetic field. The magnetic field is believed to be generated in the planetary interior; therefore, it rotates at the same rate as the interior (inner or outer core) of the planet. Radio rotation periods are the most accurate measurements of the rotation rate for Jupiter, Saturn, Uranus, and Neptune. Pluto's rotation period has been determined from changes in albedo markings on the surface and also from its tiny moon Charon which is locked in a synchronous orbit. The current *rotation periods of the planets* are:

Mercury — 58.646 days
Venus — 243.019 days (retrograde)
Earth — 23h 56.11m
Mars — 24h 37.44m
Jupiter — 9h 55.5m
Saturn — 10h 39.4m
Uranus — 17h 14.4m (retrograde)
Neptune — 16h 6.5m
Pluto — 6.3872 days.

planetesimal   A small planetary solar system object of size 1 to 100 km.

plankton   Passively drifting or weakly swimming organisms.

plankton bloom   An unusually high concentration of phytoplankton, usually producing a discoloration of the water body.

plasma   An ionized gas containing ions and electrons whose behavior is controlled by electromagnetic forces among the constituent ions and electrons.

plasma frequency   Characteristic frequency of electron plasma oscillations, fluctuations associated with deviations from charge neutrality. The *plasma frequency* in cgs units is $\omega_p = \sqrt{\left(4\pi N_e e^2/m_e\right)}$, where N_e is the electron number density, e is the electron charge, and m_e is the electron mass. An analogous frequency for ions is sometimes defined. Transverse electromagnetic waves cannot propagate through a plasma at frequencies lower than the local plasma frequency; this fact can be used to infer an upper limit on plasma density. *See* critical frequency.

plasma mantle   The boundary region separating the high latitude lobes of the Earth's magnetic tail from the magnetosheath region outside it. It is believed to contain plasma which has passed through the reconnection region or the polar cusp. Many details of the plasma mantle are uncertain, mainly because of the scarcity of observational data, but it is widely believed that it widens gradually with increasing distance down the tail, until it fills the entire tail, contributing the flowing plasma observed in the tail 100 R_E or more past Earth.

plasma sheet boundary layer (PSBL) In the Earth's magnetotail, the region between the lobe and the central plasma sheet. According to some theories, this region is linked to the "distant neutral line" at which lobe field lines reconnect and begin to return sunward, which might explain rapidly flowing streams of electrons and ions observed in it.

plasmasphere The region from about 250 km, the F-layer peak of the ionosphere or the upper atmospheric thermosphere, to about 1500 km is known as "topside" ionosphere. Above the topside ionosphere and roughly below about 4 to 6 earth radii is the region known as the *plasmasphere.* It is populated by thermal ions and electrons of energy around 1 ev. The plasmasphere like the ring current overlaps parts of the inner and outer radiation belts. The outer surface of the plasmasphere is called the plasmapause, where the electron density drops by a factor of 10 to 100 a few electrons within a distance of fraction of the Earth radius. Magnetic tubes inside the plasmasphere are significantly depleted during geomagnetic disturbances and are refilled during recovery periods. The plasmasphere does not corotate with the Earth, and the plasmapause moves closer to the Earth during geomagnetic disturbances. Beyond the plasmapause, the magnetic field lines are less dipole like and solar wind – magnetospheric interaction dominates the particles and the fields.

plasma stress tensor In plasma kinetic theory, the second moment $\mathcal{P}$ of the velocity distribution,

$$\mathcal{P} = m \int (\mathbf{v} - \mathbf{V})(\mathbf{v} - \mathbf{V}) f d^3 v ,$$

where $f(\mathbf{v}, \mathbf{x}, t)$ is the velocity distribution, m is the mass of the particles whose statistical properties are described by the kinetic theory, $\mathbf{v}$ and $\mathbf{x}$ are the velocity and spatial coordinates, and t is the time. Also, V is the mean velocity. $\mathcal{P}$ is also referred to as the pressure tensor; in the literature the term "stress tensor" is sometimes used for $-\mathcal{P}$. In the momentum equation for each charge species, the tensor divergence $\nabla \cdot \mathcal{P}$ replaces the pressure gradient of ordinary fluid mechanics.

For spatial and temporal variations of hydromagnetic scale in a collisionless plasma, the velocity distributions are gyrotropic, and the stress tensor is of the form

$$\mathcal{P} = P_\perp \mathbf{1} + \left(P_\parallel - P_\perp \right) \mathbf{BB}/B^2 ,$$

where $\mathbf{1}$ is the unit tensor, and the two scalars $P_\perp$ and $P_\parallel$ are the pressures transverse and parallel to the magnetic field $\mathbf{B}$. The pressures may also be written in terms of the transverse and parallel kinetic temperatures $T_\perp$ and $T_\parallel$, defined by $P_\perp = NkT_\perp$ and $P_\parallel = NkT_\parallel$, where N is the particle number density and k is Boltzmann's constant. Note that in general the electrons and the various ion species that comprise a plasma have differing number densities and kinetic temperatures. *See* Vlasov equation.

plastic anisotropy A plastic property for a given crystal; when a crystal is subject to a plastic deformation its plastic strength (or plasticity) is predominantly controlled by crystallographic orientation, and changes substantially with respect to the orientation from which the crystal is deformed, this property is called *plastic anisotropy.*

plastic (permanent) deformation Deformation of a body that is not recovered upon unloading, as opposed to elastic (temporary) deformation. During plastic deformation, work done by the loading force is dissipated into heat; while during elastic deformation, the work is converted into strain energy.

plateau material A thin mantling of material over the southern lunar highland plains, the surface of which is often etched, and degraded by runoff channels. It contains a morphologically distinct crater type, which lacks a raised rim, evidence for an ejecta blanket, and in which the floors are leveled by the plateau material. The origin of the plateau material is uncertain. It may be volcanic or produced by erosional stripping, for example, by transient ice-rich deposits.

plate tectonics The theory and study of how the segments of the Earth's lithosphere form, move, interact, and are destroyed. The concept was introduced by A. L. Wegener in 1912, and has now been validated by matching fossil types across (now separated) continental coast-lines, by similarities of geology in such situations, and by direct measurement via laser satellite geodesy, very long baseline interferometry, and the Global Positioning System. The theory states that the Earth's lithosphere is broken up into segments, or plates, which move atop the more fluid asthenosphere. Currently the Earth's surface is composed of six major plates and nine smaller plates. The interaction of the plates along their boundaries produces most of the volcanic and tectonic activity seen on Earth. New crust is created and plates move away from each other at divergent boundaries (mid-ocean ridges). Plates collide and are destroyed or severely deformed at convergent boundaries, where one plate is subducted under the other (deep sea trenches) or the two plates are uplifted to form a mountain range. Thus, the Atlantic Ocean is growing, and the Pacific Ocean is shrinking. Plates slide past one another at transform boundaries. The driving mechanism for *plate tectonics* is believed to be convection cells operating in the Earth's asthenosphere. It is understood that a giant hot plume ascends from the core-mantle boundary, whereas cold slabs drop through the mantle as a cold plume, producing descending flow. The surface expression of hot plumes are the volcanism of Hawaii, Iceland, and other hot spots not directly associated with plate tectonics.

The plates move with velocities of between 1 and 8 cm/yr. Earth appears to be the only body in the solar system where plate tectonics currently operates; remnant magnetism suggests plate tectonics may have been effective in the past on Mars.

platonic year *See* year.

Pleiades Young star cluster, Messier number M45, which is a bright star forming region, and has numerous young hot (B-type) stars, of which seven are prominent; the *Pleiades* are the seven sisters, daughters of Atlas.

Pleione Variable type B8 star at RA 03^h49^m, dec +24°07'; "Mother" of the "seven sisters" of the Pleiades.

Plimsoll's mark After Samuel Plimsoll; the name given to the load marks painted on the sides of merchant ships indicating the legal limit of submergence. The British began the system around 1899, and the U.S. adopted a similar system in 1930. Load lines include FW (fresh water), S (summer), W (winter) WNA (winter in the North Atlantic), and IS (Indian Summer) for the relatively calm period October to April in the Indian Ocean.

Pluto The ninth planet from the sun. Named after the Roman god of the underworld, *Pluto* has a mass of $M = 1.29 \times 10^{25}$ g, and a radius of $R = 1150$ km, giving it a mean density of 2.03 g cm^{-3} and a surface gravity of 0.05 that of Earth. Its rotational period of 6.3867 days, and its rotation axis has an obliquity of 122.5°. The slow rotation means that the planet's oblateness should be very nearly 0, but its small size may allow significant deviations from sphericity. Pluto's orbit around the sun is characterized by a mean distance of 39.44 AU, an eccentricity of $e = 0.250$, and an orbital inclination of $i = 17.2°$. Its large eccentricity means that for part of its orbit it is closer to the sun than Neptune, most recently during the 1980s and 1990s. Its sidereal period is 247.69 years, and its synodic period is 367 days. An average albedo of 0.54 gives it an average surface temperature of about 50 K. Its atmosphere is very thin, and mostly CH$_4$. Pluto is probably composed of a mixture of ices, organic material, and silicates. Because of its small size and icy composition, Pluto is more properly an inner member of the Kuiper belt, rather than a full fledged planet. Pluto has one satellite, Charon, which is nearly as massive as it is.

pluton A magma body (intrusion) that has solidified at depth in the Earth's crust.

plutonic Indicative of igneous rock consisting of large crystals. This suggests slow crystallization, as would happen at depths beneath the Earth.

pocket beach A beach comprised of sediments that are essentially trapped in the longshore direction by headlands which block longshore sediment transport.

point object   Astronomical source that has an angular diameter smaller than the resolving power of the instrument used to observe it.

point spread function (PSF)   In an optical system, the apparent radiance due to an unresolved Lambertian (cosine-emitting) point source, as normalized to the source intensity in the direction of maximum emission [m^{-2}]; numerically equal to the beam spread function. The PSF is the intensity distribution of a point source as it appears on the detector. The PSF is the result of any distorting factors induced by the full optical path including that of the instrument and environmental effects (e.g., atmospheric seeing).

Poiseuille's law   The mean velocity (U) for steady laminar flow through a uniform pipe can be described as $U = -(dp/dx)(D^2/32\mu)$, where D is the diameter of the pipe, μ is the viscosity of the fluid, and dp/dx is the pressure decrease in the direction of flow. *Poiseuille's law* is derived from the Bernoulli equation, assuming that elevation along the pipe is constant. We can generalize the equation by allowing the pipe to be inclined at an angle and adding a term related to the elevation difference, giving

$$U = -\frac{d}{ds}\left[\frac{p}{\rho g} + z\right]\frac{D^2 \rho g}{32\mu}$$

where ds is the length of a short section of pipe, ρg is the fluid weight, and z is the elevation difference. Note that the term in brackets is the hydraulic head h, a term used frequently in hydrology.

Poisson's ratio   When an isotropic elastic object is subject to elongation or compressive stress T_z a strain ε_z results in the direction of the stress. At the same time a strain arises in the transverse directions, ε_x, ε_y. If ε_z is an elongation, then $\varepsilon_x = \varepsilon_y$ are compressive. The ratio

$$\nu = -\frac{\varepsilon_x}{\varepsilon_z} = -\frac{\varepsilon_y}{\varepsilon_z}$$

is called *Poisson's ratio*.

polar cap   The region surrounding one of the poles of the Earth. In magnetospheric physics, the region surrounding the magnetic pole, in general the one inside the auroral oval. That region is believed to be connected to "open" field lines which extend to great distances from Earth and which to all intents and purposes can be considered to be linked to Earth at one end only. The existence of *polar cap precipitation* (infalling charged particles) suggests the other end is linked to the interplanetary magnetic field.

polar cap absorption (PCA)   At the polar cusps, even solar energetic particles with relatively low energies (in general protons with energies between about 1 MeV and 100 MeV) from a solar flare or coronal mass ejection can follow magnetic field lines to penetrate down to the ionosphere and stratosphere into heights between 30 and 90 km. These particles lead to an increased ionization, which in turn leads to the absorption of radio waves that may last up to days. Since this absorption is limited to the polar regions, this phenomenon is called *polar cap absorption (PCA)*. Owing to the propagation time of particles between the sun and the Earth, a polar cap absorption generally starts a few hours after a flare. In contrast, a similar effect on radio waves on the dayside atmosphere, the sudden ionospheric disturbance (*see* sudden ionospheric disturbance), starts immediately after the flare because it is caused by hard electromagnetic radiation. Polar cap absorptions are limited to a small latitudinal ring around the geomagnetic pole. Part of the particles' energy is transferred to electromagnetic radiation in the visible range, the polar glow aurora, a diffuse reddish glow of the entire sky. Since the incoming protons also influence the upper atmospheric chemistry, in particular the NO production, strong polar cap absorptions lead to a decrease in the total ozone column at high latitudes.

polar cap arc (or sunward arc)   A type of aurora most often observed by satellite-borne imagers. Polar cap arcs extend into the polar cap, usually starting near midnight and stretching sunward. Polar cap arcs can extend many hundreds of kilometers or even all the way across the polar cap, in which case they are known as the theta aurora because the auroral configuration seen by orbiting imagers — the auroral oval plus the arc stretching across it — resembles the Greek letter theta (Θ). *Polar cap arcs* are associated with northward interplanetary magnetic fields and their origin is not well understood.

polar cap (Mars)   Snow fields covering polar regions. Their main composition is CO_2 ice (dry ice). In winter the atmospheric temperature of the polar region is below the freezing point of the CO_2 gas which accounts for 95% of the Martian atmosphere. The polar region in winter does not receive sunlight (the inclination of Mars to its orbit is quite similar to that of Earth). Moreover, the Martian polar regions are covered by the polar hood cloud. Because it grows in darkness, observations have not been made of a growing polar cap. In late winter or early spring, the polar cap appears to us. The north polar cap extends from around 60°N. It recedes slowly during early spring and then quickly until a permanent cap is exposed in late spring. The north permanent cap extends from about 75°N. The south polar cap extends from 55°S in early spring and recedes with constant speed. Mars is near aphelion when the southern hemisphere is in winter, so that the south seasonal cap grows larger than the north seasonal cap. In the south, the permanent polar cap extends from about 85°S in summer. According to observations of the surface temperature by Viking spacecraft, the north permanent cap consists of H_2O ice and the south cap consists of CO_2 ice. There is no current explanation of the fact that the south permanent cap consists of dry ice. Martian atmospheric pressure varies with waxing and waning of seasonal polar caps. It reaches primary minimum in late winter of the southern hemisphere. The amplitude of the atmospheric pressure suggests 8.5×10^{15} kg for the mass of the south seasonal cap in late winter. *See* polar hood.

polar cap precipitation   The polar cap region is roughly the position of the so-called "inner" auroral zone and is a circular region around the geomagnetic pole of radius of about 10°. There are three types of low energy electron precipitation in the polar cap ionosphere: polar rain, polar showers, and polar squalls.

The polar rain particles seem to be of magnetosheath origin and fill the entire cap with thermal electrons of about 100 ev and a flux of about 10^2 ergs/cm^2/s.

The polar showers are embedded in the polar rain and consist of enhanced fluxes of precipitating electrons of mean energy around 1 kev. The showers are probably responsible for "sun aligned" arcs. Polar squalls are also localized intense fluxes of electrons of several kev during geomagnetic storms, which occur as a result of field aligned accelerations.

polar crown   Region around poles of sun at about latitude 70° of filaments oriented nearly parallel to the equator. The polar crown is occupied by prominences and large arcades which can erupt and result in CMEs.

polar cusp   *See* cusp, polar.

polar dunes   Dune-like features evident on Mars. The similarity in size and form between dunes on Earth and Mars indicates that surface materials have responded to wind action in the same way on both planets, despite differences in atmospheric density, and wind speeds. However, the global distribution of dunes differs greatly on the two planets. On Earth the most extensive sand dunes are in the mid- to low-latitude deserts, whereas on Mars most dunes are in high latitudes. In the Martian North polar region, an almost continuous expanse of dunes forms a collar, in places 500 km across, around the layered terrain, while in the south dune fields form discrete deposits within craters. The dunes imaged by Mars Global Surveyor Orbiter Camera are classic forms known as barchan and transverse dunes. These two varieties form from winds that persistently come from a single direction (in this case, from the southwest).

The source of material involved in the formation of the Martian dunes is unclear. As an alternative to quartz (silicic rocks are thought to be lacking on Mars) garnet has been proposed. (It is sufficiently hard to withstand the erosive action of wind.) Alternatively, the sand-sized particles could be produced by electrostatic aggregation, or frost cementation in the polar regions, of smaller particles.

polar glow   Reddish proton aurora at heights between 300 and 500 km. The *polar glow* covers large areas and is less structured than the more common electron aurora and, in contrast to the electron aurora, it is observed in the dayside instead of the nightside magnetosphere. The polar glow is formed when solar wind protons penetrate along the cusps into the dayside magnetosphere. In this case, the excitation of the neutral atmosphere is due to charge exchange: the penetrating proton is decelerated and becomes an excited hydrogen atom, emitting either the Lα line in UV or the Hα line in the red.

polar hood (Mars)   A steady cloud entirely covering polar regions of Mars from early autumn through early spring. *Polar hoods* are easily visible in blue, but not in red. The north polar hood appears in late summer and becomes stable in early autumn. It extends from about 40 N in the early morning and recedes to about 60 N in the afternoon. Bright spots of cloud appear frequently near the edge of the north polar hood in autumn. They drift east and disappear in several days. The south polar hood has not been investigated as much as the north polar hood, for observations are difficult owing to the tilt of the polar axis. It appears to be less stable than the north polar hood; its brightness varies from place to place and day by day, and its boundaries fluctuate frequently. Large basins in south midlatitudes are often covered with thick clouds in late autumn to early winter, which seem to be extensions of the polar hood. They recede poleward in late winter. The cloud on Hellas is especially bright and stable.

Polaris   A second magnitude F-star, located at RA 02^h 31.5^m, dec 89°16$'$. *Polaris* is about 200 pc distant. This is the North Pole star because it is within a degree of the pole (at present).

Polaris is a spectroscopic binary with at least two additional 12th magnitude components, and the principal star is a Cepheid variable with a period of 3.97 days and a variability of 0.15 magnitude.

polarization   The orientation of the electric vector of an electromagnetic wave along a preferential direction. *Polarization* can be linear or circular. In the first case, the direction of the electric vector is fixed in space; in the plane perpendicular to the direction of propagation of the wave. In the case of circular polarization, the electric vector rotates in the plane with constant angular frequency equal to the frequency of the light. Light coming from thermal sources such as stars is typically unpolarized, i.e., the electric vector oscillates randomly in all directions in the plane. Scattering by interstellar dust grains and charged particles can polarize previously non-polarized light analogously to nonpolarized sunlight becoming polarized after being reflected by the sea. Non-thermal radiation can be intrinsically highly polarized. For example, synchrotron radiation, which is an important astrophysical emission mechanism at radio frequencies can have a high degree of linear polarization ($\lesssim 70$ %).

polarization brightness   In solar physics, a measure of the polarized light emanating from the white light corona of the sun defined as the degree of polarization multiplied by the brightness. The emission is due to Thomson scattering of photospheric light by coronal free electrons and as such can be directly related to the electron density integrated over the line-of-sight.

polarization fading   The polarization of signals propagated by the ionosphere depends on the electron density along the propagation path and changes in this will result in changes in the polarization. When two polarization components are present, each responding to electron density changes, then the two modes will interfere resulting in a fading signal. This is *polarization fading* and lasts for fractions of a second to a few seconds (or the order of 1 Hz or less). *See* fading.

polarization state   The description of a light field using four components. *See* Stokes parameters.

polar motion   The motion of the rotation axis of the Earth relative to the geographical reference frame. The motion is described in coordinates fixed in the Earth, with x measured along the Greenwich meridian, and y, 90° west. The zero point is the CTRS pole. The motion has three major components: A free oscillation with period about 435 days (Chandler wobble) and an annual oscillation forced by the seasonal displacement of air and water masses cause an approximately circular motion about the mean rotation pole. The mean pole itself has an irregular drift in a direction $\sim 80°$ west. Thus, the wobble is centered at about 3.4" west of the CTRS pole as of this writing. Worldwide observations of *polar motion* of the CTRS pole are reduced and provided to the public by the International Earth Rotation Service.

polar plume   Bright ray-like solar structure of out-flowing gas which occurs along magnetic field lines in coronal holes. Plumes deviate noticeably from the radial direction and tend to angle towards lower latitudes as expected if they follow the global solar magnetic field. Most prominent at times near solar minimum.

polar wander   *See* polar motion.

polar wind   In the polar regions of the Earth where the geomagnetic lines of force are open, the continuous expansion of the ionospheric plasma consisting of O^+, H^+, He^+ leads to supersonic flow into the magnetospheric tail. This is known as *polar wind* and is analogous to the solar wind from the sun. This supersonic flow is reached beyond about 1500 km for lighter ions depending upon the plasma temperature.

pole-on magnetosphere   When the solar wind directly hits one pole of a planet's magnetic field, a *pole-on magnetosphere* results. Such a magnetosphere is cylinder-symmetric with an axial-symmetric neutral and plasma sheet around the axis of the magneto-tail. Neptune's magnetosphere oscillates between such a pole-on magnetosphere and an earth-like magnetosphere: its magnetic field axis is tilted by 47° with respect to its axis of rotation, which itself is inclined by 28.8° with respect to the plane of ecliptic. Thus during one rotation of the planet (0.67 days), the inclination of the magnetic dipole axis with respect to the plane of the ecliptic, and thus the solar wind flow, varies between $90° - 28.8° - 47° = 14.2°$, which leads to a nearly pole-on magnetosphere, and $90° + 28.8° - 47° = 71.8°$, which gives an earth-like magnetosphere.

poles of Mars   Regions with distinctly different physiography from the rest of Mars. First, layered deposits are unique to the polar regions and extend outwards for a little over 10°. They are arranged in broad swirls such that individual layers can be traced for hundreds of kilometers. The layer deposits are almost entirely devoid of craters, indicating they are among the youngest features of the planet. In the north the deposits lie on plains and are surrounded by a vast array of sand dunes, which form an almost complete collar around the polar region. Dune fields of comparable magnitude do not occur in the south. Elevations from the Mars Orbiter Laser Altimeter show the northern ice cap has a maximum elevation of 3 km above its surroundings, but lies within a 5-km deep hemispheric depression that adjoins the area into which the outflow channels emptied.

The polar caps are at their minimum size at the start of fall, and are at their maximum size in the spring. Clouds form a polar hood over the north polar cap in the winter months. In the south, on the other hand, the Viking Orbiter showed only discrete clouds existed in the polar regions during the fall and winter.

Pole star   North pole star, Polaris. There is no bright star near the southern celestial pole.

Pollux   1.14 magnitude star of spectral type K0 at RA07^{h}45^{m}18.9^s, dec +28°01'34".

poloidal/toroidal decomposition   It can be shown that a sufficiently differentiable vector field **v** that vanishes at infinity may be written in the form:

$$\mathbf{v} = \nabla\phi + \nabla \times (T\mathbf{r}) + \nabla \times \nabla \times (P\mathbf{r})$$

where **r** is the radius vector from the origin of the coordinate system. The three potentials ϕ, T, and P are, respectively, the scaloidal, toroidal, and poloidal potentials. If **v** represents a solenoidal vector such as magnetic field or incompressible flow, then the term in ϕ does not contribute ($\nabla\phi = 0$), leaving the other two terms as the *poloidal/toroidal decomposition* of **v**. This is a convenient representation in many geophysical systems because of the near spherical symmetry of the Earth. The toroidal portion of **v** has no radial component, i.e., $\nabla \times (T\mathbf{r})$ lies entirely on spherical surfaces, while the poloidal potential P can only be zero on a spherical surface if the radial component of **v** is also zero on that surface. It can further be shown that the curl of the toroidal term is poloidal, and the curl of the poloidal term is toroidal.

polycyclic aromatic hydrocarbons (PAHs)

Flat molecules made primarily of carbon atoms arranged in the graphitic form; in hexagons so that their skeleton looks like chicken wire. Pollutants on Earth, they are commonly observed in meteorites, and are widely distributed throughout the galaxy, accounting for 15 to 20% of the galactic carbon. The UV absorption of these stable molecules may be the source of the diffuse interstellar bands, and their emission in the IR is seen towards H II and star forming regions.

polynya An area of the ocean that persists in being either partially or totally free of sea ice under surface conditions where the sea would be expected to be ice covered. *Polynyas* appear in winter when air temperatures are well below the freezing point of sea water and are bordered by water that is covered with ice. They are typically rectangular or elliptical in shape and tend to recur in the same regions of the Arctic and Southern Ocean. The size of polynyas can range from a few hundred meters to hundreds of kilometers.

Polynyas form via two mechanisms, which often operate simultaneously. In the first, ice is continually formed and removed by winds, ocean currents, or both. The latent heat of fusion of the ice provides the energy necessary to maintain open water. The second mechanism requires oceanic heat to enter a region in quantities sufficient to prevent local ice formation.

Polynyas are sites for active brine formation, which may affect the local water density structure and current field, which in turn modifies large-scale water masses. They are also an interface for gas exchange between the ocean and atmosphere in polar regions. The large sensible heat fluxes (along with fluxes due to evaporation and longwave radiation) tend to dominate regional heat budgets. They are also of biological interest because their regular occurrence makes them important habitats.

polytropic process In thermodynamics, any process that can be described by the statement $pV^n = $ constant (for some fixed n). Here p is the pressure in the experiment, and V is the volume of the sample. Thus $n = 0$ yields $p = $ constant (constant pressure process), $n = \infty$ yields $V = $ constant (constant volume process). For a process at constant temperature in an ideal gas, $n = 1$, and for a constant entropy process in an ideal gas, n is the ratio of specific heat at constant pressure to that at constant volume.

poorly graded sediment A sediment with grain sizes that have a small standard deviation. Most grains are very close to the median grain size.

Pop I *See* Population I.

Pop II *See* Population II.

Population I Stars having a composition similar to the sun. *See also* main sequence star, metallicity.

Population II Stars significantly deficient, relative to the sun, in elements beyond helium. *See also* metallicity, main sequence star.

pore fluid pressure ratio The ratio of pore fluid pressure to lithostatic pressure, defined as

$$\lambda = \frac{p_f - p_s}{p_l - p_s}$$

where p_f is the pore fluid pressure, p_s is pressure at the surface, and p_l is the lithostatic pressure, the weight of the rock column above containing the pore fluid.

poroelastic medium A porous medium that deforms elastically under loading. A *poroelastic medium* can be envisaged to consist of an elastic matrix frame hosting an interstitial fluid. The elastic moduli of the matrix frame are defined under the drained condition, a situation in which the pore fluid freely enters and exits the medium such that the pore fluid pressure does not affect matrix deformation. In a poroelastic medium, strain and effective stress of the matrix frame are governed by its constitutive law (such as Hooke's law for infinitesimal strain), and fluid flow obeys Darcy's law. *See* effective stress, Darcy's law.

porosity (φ)   The ratio of the volume of void spaces in a soil or rock to the total volume of the rock or soil: $\phi = (V_a + V_w)/V_s$, where V_a is the volume of air in the sample, V_w is the volume of water in the sample, and V_s is the total volume of soil or rock in the sample.

Portia   Moon of Uranus also designated UXII. Discovered by Voyager 2 in 1986, it is a small, irregular body, approximately 55 km in radius. Its orbit has an eccentricity of 0, an inclination of 0°, and a semimajor axis of 6.61×10^4 km. Its surface is very dark, with a geometric albedo of less than 0.1. Its mass has not been measured. It orbits Uranus once every 0.513 Earth days.

position vector   In a Euclidean space, the vector pointing from the origin of the reference frame to the location of a particle, thus specifying the coordinates of the position of the particle.

positron   The antiparticle of the electron. It has the same mass and spin as an electron, and an equal but opposite charge.

POSS   Acronym for Palomar Observatory Sky Survey, a survey that produced a collection of several hundreds of wide field, deep blue and red photographic plates originally covering the northern sky down to declination $\delta \approx -24°$ obtained with the Oshkin telescope at Palomar observatory, in the years from 1950 to 1955. With a scale of 67"/mm, and a limiting magnitude of ≈ 20 in the red, the POSS plates have been instrumental to any source identification, including faint galaxies and quasars, for which later observations were being planned. The POSS plates have been digitized and supplemented with observations obtained in the southern hemisphere of similar scale and limiting magnitude to cover the entire sky. The Digitized Sky Survey (DSS) is stored on a set of 102 commercially available compact disks, covers the entire sky and includes an astrometric solution for each Schmidt plate, to readily obtain equatorial coordinates. In more recent years a second generation survey has been carried out at Palomar employing plates with finer emulsions. This second generation survey, known as POSS-II, was almost completed and in large part digitized as of 1999.

post-flare loops   A loop prominence system often seen after a major two-ribbon flare, which bridges the ribbons. *Post-flare loops* are frequently observed to emit brightly in soft X-rays, EUV, and H_α.

postglacial rebound   Because the solid Earth's mantle has a fluid behavior, a glacial load depresses the Earth's surface. The center of Greenland is depressed below sea level due to the load of the Greenland ice sheet. During the last ice age, from 20,000 to 10,000 years ago, great ice sheets covered much of North America and Europe and depressed these regions. After the melting of these ice sheets the regions rebounded. *Postglacial rebound* continues in northern Canada and Scandinavia at rates of approximately 1 cm per year. Evidence for this rebound comes from dated, elevated wave cut terraces. Postglacial rebound indicates that the interior of the Earth deforms as a viscous or viscoelastic material and the rate of postglacial rebound provides a quantitative measure of the viscosity of the Earth's mantle.

post-seismic relaxation   Gradual decrease of earthquake-induced deviatoric stress as a result of the viscoelastic behavior of the Earth material.

potassium-argon age   A naturally occurring isotope, ^{40}K (about 1% of natural abundance), is radioactive and decays with a half life of 1.277×10^9 years to two different daughter products, ^{40}Ca (β decay) and argon-40 (branching ratio 10.72%, electron capture). Argon is a gas; whenever rock is melted to become magma or lava, the argon tends to escape. Once the molten material hardens, it again begins to trap the argon produced from its potassium. In this way the potassium-argon clock is clearly reset when an igneous rock is formed. The main complication to this simple determination is contamination from included air containing argon that was in radioactive decay since the last melting.

Potassium-argon ages are reliable to hundreds of millions of years.

potential   In nonrelativistic conservative mechanical systems, a scalar function ϕ from which the mechanical forces are obtained as proportional to $\nabla\phi$. Examples are the electric potential and the gravitational potential.

In fluid mechanics, in situations where the fluid velocity vector $\mathbf{v}$ is irrotational, $\mathbf{v}$ may be obtained as the gradient of a velocity potential.

potential density   The density that a parcel of fixed composition would acquire if moved adiabatically to a given pressure level (called reference pressure).

potential energy   The ability to do a specific amount of work. In conservative mechanical systems, work (energy) can be done on the system and stored there. This stored work is called *potential energy* (units of Joules or ergs in metric systems). A prototypical example is a system consisting of a stone moved to the top of a mountain. Work is required to move it to the top; once there, the system contains potential energy, which can be recovered (as heat or as thermal energy) by allowing the stone to roll down the mountain. A related gravitational example is in the storage of water behind a dam; potential energy is converted to work by allowing the water to flow through a turbine. Electrostatics provides another example, in which work must be done to bring a positive charge from infinity to add to a collection of positive charge. Such a collection has potential energy which can be recovered by allowing the charges to freely accelerate away from one another; the energy can then be collected as kinetic energy, as the charges move off to infinity.

Potential energy may also be mechanical (a compressed spring) or chemical, such as the energy stored in unstable compounds (e.g., nitroglycerine) or in flammable substances (e.g., H_2 and O_2 gases mixed). It also exists in extractable form in the nuclei of heavy atoms (e.g., U^{235}), which release it as kinetic energy when undergoing fission; and in a system of neutrons and protons, which will, at appropriate densities and pressures, release some potential energy as kinetic, in fusion to form helium and heavier nuclei.

potential height   Dynamic height.

potential instability   Also called "convective instability". Stratification instability caused by convective activities, i.e., the lower layer has higher moisture and becomes saturated first when being lifted, and hence cools thereafter at a slower rate than does the upper, drier portion, until the lapse rate of the whole layer becomes equal to the saturation adiabatic and any further lifting results in instability. In general, use $\frac{\partial\theta_{sw}}{\partial Z} < 0$ or $\frac{\partial\theta_{se}}{\partial Z} < 0$ as the criterion of convective instability. (θ_{sw} and θ_{se} are pseudo-wet-bulb potential temperature and pseudo-equivalent potential temperature, respectively.)

potential temperature   The temperature Θ that a parcel of fixed composition would acquire if moved adiabatically to a given pressure level (called reference pressure). For instance in saltwater, Θ is the temperature that would be measured after a water parcel, which shows the measured in situ temperature T, has been moved isentropically (without exchanging heat or salt) through the ambient water masses to a reference depth, usually taken at the surface: $\frac{d\theta}{dz} = \frac{dT}{dz} - \left(\frac{dT}{dz}\right)_{ad}$. The difference between Θ and T is given by $\theta(z) = T(z) - \int_z^{z_0} \left(\frac{dT}{dz}\right)_{ad} dz'$.

potential theory   In geophysics, the area that utilizes gravitational potentials to determine the interior structure of a body. Study of a body's gravitational potential (obtained by perturbations on an orbiting body) provides information about the distribution of interior mass, how much the primary body varies from hydrostatic equilibrium, the response of a body to tidal forces, and how much isostatic equilibrium surface features have undergone.

potential vorticity   (Rossby, 1940) In a shallow homogeneous layer, the *potential vorticity* is a conserved dynamic quantity Q as the ratio between the absolute vorticity and the thickness of the layer, defined by $Q = (f + \zeta)/H$, where f is the planetary vorticity (i.e., Coriolis parameter) and ζ is the vertical component of relative vorticity; and H is the layer thickness.

In a stratified fluid, the potential vorticity is defined by

$$\mathbf{Q} = \frac{\omega + 2\mathbf{\Omega}}{\rho} \cdot \nabla\Theta$$

where ω is relative vorticity, $\mathbf{\Omega}$ is the Earth's rotation, ρ is density, and Θ is a conserved quantity. This form is called Ertel potential vorticity (Ertel, 1942). If Θ is potential temperature, this is also called isentropic potential vorticity.

potentiometric surface (or piezometric surface) Water will rise to the *potentiometric surface* in a well or piezometer penetrating a confined aquifer. The elevation of the potentiometric surface above an arbitrary datum is the sum of the pressure head and the elevation head, or the hydraulic head h in Darcy's law. The water table is the potentiometric surface in an unconfined aquifer.

power The rate of doing work, energy per unit time. In mechanics, power $P = \mathbf{F} \cdot \mathbf{v}$. The units of *power* are erg/sec, or Joule/sec. Note that 1 Joule/sec $\equiv$ 1 Watt.

power-law fluid A viscous fluid in which the rate of shear strain is proportional to the power of shear stress. The flow law is often generalized to be $\dot{\varepsilon}_{ij} = C\sigma^{n-1}\sigma'_{ij}$, where $\dot{\varepsilon}_{ij}$ is the rate of strain, σ'_{ij} is deviatoric stress, $\sigma = (\sigma'_{ij}\sigma'_{ij}/2)^{1/2}$ is an invariant of σ'_{ij}, and C is a quantity independent of strain rate and stress. An effective viscosity can be defined as $\mu_e = (2C\sigma^{n-1})^{-1}$, such that the flow law takes the form of the Newtonian fluid $\sigma'_{ij} = 2\mu_e\dot{\varepsilon}_{ij}$. A solid such as a rock deforms like a power law fluid as a result of dislocation creep at high temperatures and stresses.

Poynting–Robertson effect A drag force arising on particles orbiting the sun, because solar radiation striking the leading surface is blueshifted compared to that striking the following surface. Thus, the particles receive a component of momentum from the radiation pressure which is opposite to the direction of motion. This drag force causes interplanetary dust particles to spiral inward towards the sun, removing such parti-cles from the solar system. For 10 micron-sized dust particles at the Earth's orbit, the *Poynting–Robertson effect* will cause these particles to spiral into the sun on a time scale of only 1 million years. The presence of interplanetary dust particles even today thus indicates that such material is continuously being resupplied by comets and collisions among asteroids. In very accurately tracked satellites (e.g., LAGEOS) the Poynting–Robertson effect is important in long-term accurate modeling of the orbit.

p process The set of nuclear reactions responsible for producing the rare nuclides of the heavy elements with proton to neutron ratio higher than the ratio in the most tightly bound nuclides of each element. It acts on products of the s process and r process either by adding protons or (more likely) removing neutrons, and probably occurs in supernovae and their environs. No element has a p-process nuclide as its most abundant isotope, and so we know nothing about the abundance of *p-process* products outside the solar system.

pp-waves (1923) In general relativity, a particular description of gravitational waves in a matter-free space, described by the metric

$$ds^2 = 2\,du\,dv - 2\,d\zeta d\bar{\zeta} + 2H(u, \zeta, \bar{\zeta})\,du^2$$

discovered by Brinkmann. The Weyl conformal curvature is Petrov type N, and the principal null direction is covariantly constant.

$$H = f(\zeta, u) + \bar{f}\left(\bar{\zeta}, u\right) \ .$$

Here ζ is a complex stereographic spherical coordinate. In another representation

$$ds^2 = 2H(x, y, z)\,dt^2 - dx^2 - dy^2 - 2\,dz\,dt$$

where the function $H(x, y, z)$ satisfies the elliptic equation

$$\left(\frac{\partial^2}{\partial x^2} + \frac{\partial^2}{\partial y^2}\right) H = 0 \ .$$

The normals of the wave fronts are covariantly constant vectors, hence the waves are planefronted. *See* Petrov types, Weyl tensor.

prairie An extensive level or rolling grassland consisting of rich soil and a variety of grasses with few trees except along riverbanks. *Prairies* covered much of central United States and Canada prior to the development of farming and fire control methods in these areas in the late 19th century.

Prandtl number   A dimensionless number (P_r) given by the ratio of the kinematic viscosity to the diffusivity. It expresses the ratio of the diffusivity of momentum to that of temperature through a fluid.

$$P_r = \frac{\nu}{\chi}$$

where ν is kinematic viscosity, χ is heat diffusivity coefficient, $\chi = \frac{K}{\rho C_p}$, in which, K, ρ, and C_p are thermal conductivity, density, and specific heat at constant pressure, respectively. For air, $P_r \approx 1.4$. P_r is strongly temperature-dependent. The turbulent Prandtl number is defined analogously by $Pr_t = \nu_t/\chi_t$ where ν_t is turbulent viscosity and χ_t is turbulent diffusivity. For very intense turbulence $Pr_t \rightarrow 1$, whereas for very weak turbulence (laminar), $Pr_t \rightarrow Pr$.

Prasad–Sommerfield limit   A limiting case of parameters describing (magnetic) monopoles so that they have mass equal to a minimum value called the Bogomol'nyi bound. For this set of parameters, for some theoretical models also including a scalar field (called a Higgs field) it becomes possible analytically to solve the field equations, to find that the force between two equal (magnetically repelling) monopoles exactly vanishes because this force gets compensated by the long-range Higgs field attractive interaction. *See* Bogomol'nyi bound, cosmic topological defect, monopole, t'Hooft–Polyakov monopole.

Pratt compensation   In *Pratt compensation,* the density ρ_P varies above a depth of compensation W in order to balance the mass of elevated topography.

Pratt isostasy   An idealized mechanism of isostatic equilibrium proposed by J.H. Pratt in 1854, in which the crust consists of vertical rock columns of different densities with a common compensation depth independently floating on a fluid mantle. A column of lower density has a higher surface elevation. *See* Airy isostasy.

precession   The slow, conical movement of the rotation axis of a rotating body. In the case of solar system objects (planets), precession is caused by the gravitational torques of other nearby objects. The Earth's orbit around the sun defines a plane (the plane of the ecliptic) and the Earth's axis of rotation is at an angle of $66.5°$ to this plane. The Earth's "obliquity" is the angle between the rotation axis and a normal to this plane, i.e., $23.5°$. The Earth's rotation creates a bulge around the equator, which itself defines a plane at $23.5°$ to the ecliptic. Both the sun and the moon exert a tidal couple on the Earth's bulge, which results in the forced *precession* both of the Earth's rotational axis and of the material orientation of the Earth itself. This is analogous to the precession of a rapidly spinning gyroscope pivoted at one end in a gravitational field: The material orientation of the gyroscope's symmetry axis rotates so as to form a cone, while the axis of rotation of the gyroscope (which lies very close to the symmetry axis) follows. Similarly, the Earth's pole swivels around the normal to the ecliptic so as to describe cones. The period of this precession is 25,730 years. Superimposed on the simple rotation of the Earth's pole are irregularities termed nutations, as well as the Chandler wobble and a longer term variation in the obliquity to the ecliptic. A notable effect of precession is the changing of the North Star (today the north celestial pole is near the star Polaris, but around 2000 BC it was near the star Thuban) as the Earth's rotation axis points toward different locations in space.

precession of the equinoxes   Slow shift in the celestial coordinates of astronomical objects because of the precession of the Earth's pole direction. Celestial positions in published ephemerides are typically correct at the epoch 2000.0 (i.e., correct for noon UT, January 1, 2000), and correction tables and formulae correct the coordinates to any particular epoch. (Older tables give coordinates correct at the epoch 1950.) *See* precession.

precipitation   Any form of liquid or solid water particles that fall from the atmosphere and reach the Earth's surface. Liquid precipitation includes drizzle and rain. Solid precipitation can include snow and agglomerated snow flakes, frozen rain (sleet), or hailstones.

precursor   In solar physics, a *precursor,* sometimes also called pre-flare phase, can be observed prior to very large flares as a weak brightening of the flare site in soft X-rays and Hα, indicating a heating of the flare site. The precursor can last for some minutes to even 10 minutes. *See* flare electromagnetic radiation.

PREM (Preliminary Reference Earth Model) A standard but preliminary model concerning interior of the Earth proposed by A.M. Dziewonski et al. (1981). On the basis that the Earth is anisotropic to a depth of 220 km, depth distributions for depths of seismic discontinuities, seismic velocity, density, physical properties, and pressure of the Earth's interior are given, using many observation data concerning earthquakes such as Earth's free oscillation and travel times.

pressure   A scalar quantity indicating force per unit area.

pressure altitude   The altitude calculated from the barometric height formula, under the assumption of standard atmospheric conditions and setting 1013.25 Pa as the pressure at zero altitude. This is the basis of the pressure altimeter.

pressure anisotropy   *See* plasma stress tensor.

pressure coordinate   *See* isobaric coordinate.

pressure head ($p/\rho g$)   The *pressure head* has units of length and is a component of the hydraulic head that may be thought of as the "flow work" or the work due to pressure per unit fluid weight, $p/\rho g$, where p is the fluid pressure, ρ is the fluid density, and g is the acceleration of gravity. Pressure in hydrology is conventionally measured as gage pressure, which is defined as zero at the piezometric surface ($p_{surface} = 0$),

thus $p \geq 0$ for groundwater (saturated) systems. In unsaturated flow, the pressure head is negative and is equal in magnitude to the tension head ψ.

primary production   The amount of organic matter produced from inorganic matter by photosynthesis, e.g., in [g C (Carbon) m^{-3}] or, for a water column, in [g C (Carbon) m^{-2}].

primary productivity   The rate of production of organic matter from inorganic matter by photosynthesis, e.g., in [g C (Carbon) $m^{-3} h^{-1}$] or, for a water column, in [g C (Carbon) $m^{-2} h^{-1}$].

primary rainbow   The bright rainbow produced when light undergoes one internal reflection in water droplets. The colors range from blue on the inner part of the arc at about 40° from the center of the arc, to red on the outside of the arc at about 42° from the center. If a secondary rainbow is seen, it lies outside the *primary rainbow.*

prime focus   The location where rays reflected from the primary mirror of a reflecting telescope meet, with no intervening secondary reflections. In very large telescopes, detectors are placed at *prime focus.*

primordial black holes   Black holes produced at the early stage of the cosmological expansion of the universe. Study of the gravitational collapse of stars indicates that only sufficiently heavy stars can produce a black hole, and black holes of masses much less than the solar mass cannot form at the present stage of the evolution of the universe. However, the density of matter a short time after the Big Bang was enormous, and inhomogeneities in expansion could have produced small black holes. When the quantity $L = cT$, where c is the speed of light and T is the time elapsed since the Big Bang, became of the order of the size of a typical gravitational inhomogeneity, formation of *primordial black holes* became possible. Primordial black holes are a candidate constituent of cosmological dark matter, though it is not clear if they could be a major constituent.

principal null directions   In pseudo-Riemannian geometry and general relativity, one of the null (i.e., lightlike) eigenvectors obtained from the Weyl (conformal) tensor, or from an equivalent spinor representation. The behavior (and possible coincidence of one or more of these directions) is used in the Petrov classification. *See* Petrov types.

principal spinors   The spinors α_A, β_B, $\cdots$, determined up to a complex multiplying factor, in the symmetrized decomposition

$$\psi_{AB...P} = \alpha_{(A}\beta_B, \ldots \pi_{P)}$$

of a symmetric p-index $SL(2, C)$ spinor ψ.

principle of equivalence   *See* equivalence principle.

probable maximum hurricane   A parameterized storm used for design purposes or to predict hurricane damage. *See also* standard project hurricane.

progressive vector diagram   A diagram that shows the displacement a particle would have if it had the velocity observed at the fixed position of the current meter at all times. For infinitesimal motion, this coincides with the particle trajectory.

progressive wave   A (water) wave that exhibits a net horizontal translation, such as waves in the open ocean, as opposed to a standing wave.

Prometheus   Moon of Saturn, also designated SXVI. It was discovered by S. Collins and others in Voyager photos in 1980. Its orbit has an eccentricity of 0.003, an inclination of 0°, and a semimajor axis of 1.39×10^5 km. Its size is $73 \times 43 \times 31$ km, its mass is estimated at 2.7×10^{17} kg, and its density at $0.7 \, \text{g cm}^{-3}$. This low density indicates that *Prometheus* is probably a porous body. It has a geometric albedo of 0.6, and orbits Saturn once every 0.613 Earth days. It is the inner shepherd satellite of Saturn's F ring.

prominence   A structure in the solar corona consisting of cool dense plasma supported by magnetic fields. *Prominences* are bright structures when seen over the solar limb, but appear dark when seen against the bright solar disk (*see* filament). Prominences are characterized as quiescent or active with the former being very stable, lasting for months, and the latter being mostly associated with solar flares and having lifetimes of minutes or hours.

proper time   The time measured by an observer with the clock that the observer carries with him/herself, i.e., time measured in a reference frame in which the clock or process is at rest. One of the effects predicted by special relativity is that if an observer moves with respect to another, with a velocity comparable to the velocity of light, then the clock of the moving observer goes more slowly than the observer at rest measures. The effect is relative: each of the observers sees that the other one's clock slows down. However, if one of the observers moves noninertially (e.g., going to a distant star, reversing and returning), then on comparison at return, the moving clock shows less elapsed time. (*See* time dilatation.) The increment $d\tau$ of the proper time τ is related to the coordinate time and the velocity $\frac{dx^i}{dt}$, $i = 1, 2, 3$ by

$$c^2 d\tau^2 = c^2 d^2 t - \left(dx^2 + dy^2 + dz^2\right) ,$$

where $\{dx^\alpha\} = \{dt, dx, dy, dz\}$ are the change in the event coordinates during the interval $d\tau$.

General relativity is locally equivalent to a flat space (because of the equivalence principle). In general, for any coordinate frame, even in general relativity, one can compute: $c^2 d\tau^2 = -g_{\alpha\beta} \, dx^\alpha dx^\beta$, where $g_{\alpha\beta}$ are components of the metric assumed to have signature $(-+++)$. *See* summation convention.

proplyd   A proto-star or young stellar object whose residual gas reservoir is being ionized away by ultraviolet radiation from nearby more massive stars.

Proteus   Moon of Neptune, also designated NIII. It was discovered in 1989 on Voyager photos. Its orbit has an eccentricity of 0.00044, an inclination of 0.04°, and a semimajor axis of 1.18×10^5 km. Its size is $218 \times 208 \times 201$ km, but its mass is unknown. It has a geometric albedo of 0.064, and orbits Neptune once every 1.122 Earth days.

proton   A stable elementary particle (a baryon) of unit positive charge and spin $\hbar/2$. *Protons* and neutrons, which are collectively called nucleons, are the constituents of the nucleus. *See* Planck constant.

proton flare   Any solar flare producing significant fluxes of 10 MeV protons in the vicinity of the Earth.

proton-proton chain   The set of nuclear reactions for hydrogen burning that begins with two protons interacting directly via the weak interaction to form a deuteron. This occurs at somewhat lower temperatures than the CNO cycle and so dominates energy production in stars of less than about 1.5 solar masses (including the sun). It must also be the mechanism of hydrogen fusion in stars of the first generation, before any heavy elements are present. The main chain of reactions (*see* nuclear reactions for an explanation of the notation) is

$$^1H\left(p, e^+\nu_e\right){}^2H(p,\gamma){}^3He\left(^3He, 2p\right){}^4He .$$

The chain can be completed in other ways that are much more strongly dependent on temperature. These other closures are also the main sources of high energy neutrinos (the ones to which the first experiment was sensitive; *see* solar neutrinos). The additional reactions are

$$^3He(\alpha,\gamma){}^7Be\left(e^-, \nu_e\right){}^7Li(p,\alpha){}^4He$$
$$^7Be(p,\gamma){}^8B\left(, e^+\nu_e\right){}^8Be(, 2\alpha){}^4He .$$

See CNO cycle, deuterium, hydrogen burning, nuclear reactions, solar neutrinos, weak interaction.

protostar   A condensation of interstellar gas and dust that will become a star after accreting enough matter to start nuclear reactions at its center; the stage of stellar evolution immediately preceding the main sequence. The time scale is the Kelvin–Helmholtz one. Many *protostars* are surrounded by accretion disks, in which planets can or may form. They also typically display mass loss along their rotation axes, in a bipolar pattern. The most massive protostars are completely shielded from optical study; less massive ones can be studied at infrared and optical wavelengths and evolve into T Tauri stars. The bipolar outflow often ionizes gas clouds at some distance from the proto-star. These are called Herbig–Haro objects.

protosun   The early sun, forming from a molecular cloud, before or just after it began to generate internal energy by nuclear reactions.

protuberance   Bright arc extending high above the photosphere on the solar limb, covering angular distances up to some 10°. When viewed against the photosphere, the *protuberance* becomes visible as a dark strip and is called a filament. *See* filament.

Proxima Centauri   The nearest star from the sun at 1.31 pc (4.2 light-years) and a member of the ternary α Centauri system. Proxima Cen is a cool M5.5 main sequence dwarf, about 20,000 times intrinsically fainter than our sun, of apparent magnitude 11.2, and undergoes irregular flares when it brightens by as much as one magnitude but for only several minutes. Its mass is 0.1 solar with a radius 0.093 times that of our sun. Its orbital period within the α Centauri system is estimated to be about one million years.

pseudobreakup   Rapid motion and brightening of aurora in the midnight sector, resembling the ones that characterize the breakup phase of substorms. *Pseudobreakups* often precede the breakup phase and differ primarily in failing to develop into a full substorm.

pseudovector   A 3-dimensional quantity that behaves like a vector, except that its definition is such that it does *not* change sign under an inversion $\{x, y, z\} \longrightarrow \{-x, -y, -z\}$ of the coordinate system. An example is the angular momentum $\mathbf{L} = \mathbf{r} \times \mathbf{p}$ where both $\mathbf{r}$ and $\mathbf{p}$ *do* change sign under an inversion; hence $\mathbf{L}$ does not; $\mathbf{L}$ is a pseudovector. *See* vector cross product.

PSF   *See* point spread function.

psychrometer   A wet-bulb/dry-bulb hygrometer. Depending on the ambient humidity, the wet bulb cools by evaporation to a lower temperature than the dry bulb. The difference in temperature is calibrated to indicate relative humidity.

psychrometric constant   The *psychrometric constant* is often used in the calculation of evapotranspiration:

$$\gamma = c_a P / 0.622 \lambda_v = -0.066 \text{ kPa/K}$$

at 293.2 K (20°C) and 100 kPa, where c_a is the heat capacity of the atmosphere, P is the atmospheric pressure, and λ_v is the latent heat of vaporization.

Puck   Moon of Uranus also designated UXV. Discovered by Voyager 2 in 1986, it is a small, roughly spherical body, approximately 77 km in radius. Its orbit has an eccentricity of 0, an inclination of 0.3°, a precession of 81° yr^{-1}, and a semimajor axis of 8.60×10^4 km. Its surface is very dark, with a geometric albedo of less than 0.1. Its mass has not been measured. It orbits Uranus once every 0.762 Earth days.

pulsar   A neutron star in which a combination of rapid rotation and strong magnetic field result in radio waves or other forms of electromagnetic radiation being emitted primarily along narrow beams or cones, so that an observer sees pulses of radiation at the rate of one or two per rotation period. Young, single *pulsars* are sometimes found in supernova remnants, for instance PSR 0531+21 in the Crab Nebula (formed in 1054 CE, rotation period 0.033 sec) and a 0.016 sec period pulsar in the Large Magellanic Cloud: SNR N157B. These fade in about 10^7 years as their rotation slows. Neutron stars in close binaries can appear as X-ray sources (X-ray binaries), radiating energy that comes from material transferred from their companions. These are sometimes called accretion powered pulsars. The mass transfer can spin them up to rotation periods as short as 0.0015 sec and, when accretion stops, they will appear as millisecond pulsars, either in binary systems or single if they are liberated by the explosion of the second star.

pulsar velocities   Measurements of the transverse velocities of radio pulsars, most reliably made by proper motion measurements, indicate that these pulsars are moving through the galaxy at a mean velocity of 450 km s^{-1}. These high velocities require that neutron stars, which become pulsars, receive a 450 km s^{-1} kick at, or near, the time of formation in a supernova explosion (for kick mechanisms, *see* neutron star kicks). The actual velocity depends sensitively upon the model of electron density distribution in the galaxy, from which the distance to the pulsars, and hence velocities, are calculated.

pycnocline   The region of large gradient of potential density in oceans.

pycnostad   A layer of the ocean in which density varies little with depth. Such regions are usually well-mixed parts of the ocean with little variation in temperature.

pyconuclear reaction   Nuclear reaction at zero temperature in condensed matter arising because zero point fluctuations of nearby ions result in the penetration of the Coulomb barrier. The rate of *pyconuclear reactions* increases steeply with density, and this is presumably the limiting factor in white dwarf masses; pyconuclear reactions induce explosive burning if the central density exceeds about 10^{10} gm/cm^3, though the exact value is uncertain.

pyroclastic   Involving explosive volcanic activity. A *pyroclastic* rock is one that is formed by the accumulation of volcanic rock fragments that were scattered by explosive activity. Pyroclastic deposits are usually composed of ash and cinders that resulted from the eruption of gas-rich viscous magma; such deposits are also called tuffs and ignimbrites.

pyroxene   A silicate mineral common in igneous and high-temperature metamorphic rocks such as basalt and gabbro; a mixture among $FeSiO_3$, $MgSiO_3$, and $CaSiO_3$. Abundant in lunar rocks.

Q

Q "Quality"; a measure of attenuation in an oscillating system. It is conventionally defined in terms of the energy in the oscillation:

$$Q = 2\pi \frac{E}{\Delta E}$$

where E is the energy stored in the oscillation, and ΔE is the energy dissipated over the period of one cycle. If there is no other process at work, then Q can be expressed in terms of the real and imaginary parts of the complex frequency:

$$Q = \frac{\omega_r}{2\omega_i}$$

where the amplitude of the oscillation is governed by

$$A = A_0 e^{i\omega t} = A_0 e^{i(\omega_r + i\omega_i)t} = A_0 e^{i\omega_r t} e^{-\omega_i t} .$$

Q may also be written in terms of (or measured from) the width of the peak associated with the oscillation in a power spectrum (that might be obtained from data through a fourier transform):

$$Q = \frac{\omega}{\Delta\omega}$$

ω in this case is real and is the frequency of the center of the peak, while $\Delta\omega$ is the width between the half power points on each side of the peak. Q may be a function of frequency. There are many different geophysical oscillations for which Q may be defined, such as the free seismic oscillations of the Earth, torsional oscillations of the core, and gravity waves in the atmosphere.

QCD Quantum Chromo–Dynamics, the quantum theory of the strong interactions.

Q factor Quality factor, Q.

QSO Acronym for QuasiStellarObject. QSOs are high luminosity active galactic nuclei which show optical appearance almost undistinguished from stars, and a spectrum with strong and broad emission lines invariably shifted to the red. The optical emission line spectrum closely resembles that of Seyfert-1 galaxies. Searches for quasars have been carried out in several regions of the electromagnetic spectrum, from the infrared to the X-ray. The final identification of a quasar is, however, made on the identification of redshifted spectral emission lines. Many *QSOs* are strong radio sources, but a large class is radio-quiet. The terms QSO and radio-quiet QSO are a synonym of radio-quiet high luminosity active galactic nucleus. While in typical Seyfert galaxies the luminosity of the nucleus is comparable to the luminosity of the host galaxy, in QSOs the nucleus can be hundreds of times more luminous. It is customary to define as QSOs all radio-quiet active galactic nuclii (AGN) above the luminosity of $10^{11} L_\odot$. This subdivision is somewhat arbitrary, since at this limit there is no break in continuity of the luminosity distribution.

Other defining properties of QSOs are large UV flux, broad emission lines, large redshift, and time-variable continuum flux. Some of the brightest radio sources are associated with QSOs whose host galaxy is not clearly visible. The most distant QSOs are now being observed at redshift $\lesssim 5$. The 8th Edition of *A Catalogue of Quasars and Active Nuclei* by M.- P. Vé-Cétty and P. Véron lists more than 11,000 QSOs (both radio-quiet and radio-loud) known as of early 1998. *See* Seyfert galaxies.

quadrupole formula In general relativity the radiated power of a slowly varying source of gravitational waves is given by the *quadrupole formula*

$$P = \frac{G^2}{45c^5} \sum_{\mu,\nu=1}^{3} \left(\frac{d^3 Q_{\mu\nu}}{dt^3} \right)^2 ,$$

where the quadrupole moment of a point mass m at x_μ is

$$Q_{\mu\nu} = 3m \left[x_\mu x_\nu - (1/3)\delta_{\mu\nu} \left(x_\rho \right)^2 \right] .$$

quanta meter An instrument to measure the number of photons (e.g., photosynthetically available radiation), as opposed to energy.

quantization of redshift   A controversial systemic trend in the radial velocity difference between pairs of galaxies, deduced from redshift between pair members, preferentially an integer multiple of 72 km/s, i.e., $cz = n72$ km/s.

quantum efficiency   The ratio of the number of photons incident on a detector to the number of electrons produced in the first stage of the detector; the inverse of the number of photons needed on average to produce one such photoelectron.

quantum field theory in curved spacetime
Quantum field theory which is extended from its original definitions in flat (gravity-free) spacetimes, to situations where the curvature of spacetime (i.e., the gravitational field) affects the quantum field, but gravity itself is completely classical and is not quantized.

quantum gravity   A yet-to-be-defined description of gravity as a quantum theory. Because of the tensor nature of general relativity, it is not renormalizable as a field theory in perturbation from flat space. Various attempts to quantize general relativity have thus far been unsuccessful. String theory contains general relativity in some limit, and gravity may eventually be quantized in the context of the quantum theory of strings.

quantum yield   The number of CO_2 molecules fixed in biomass per quantum of light absorbed by a plant; it is linearly related to energy conversion efficiency.

quark   An elementary entity that has not been directly observed but is considered a constituent of protons, neutrons, and other hadrons.

quasar   Acronym of quasi stellar radio source, also QSO; an active galactic nucleus that is of high luminosity, radio loud, and that does not show, in visual images, any clear evidence of an underlying galaxy. It is customary to define as *quasars* all radio-loud active galactic nuclei above the luminosity $\approx 10^{11} L_{\odot}$, although this subdivision is somewhat arbitrary. Other defining properties of quasars are large UV flux, broad emission lines, large redshift, and time-variable continuum flux. Some of the brightest radio sources are associated with quasars. The term "quasars" designates often, albeit improperly, the broader class of both distant radio quiet and radio loud active galactic nuclei whose image is nearly stellar, i.e., whose host galaxy is not clearly visible.

quasar-galaxy association   The observation of an unusually large number of quasars surrounding bright, nearby galaxies. For example, more than 40 quasars have been found in a field by $\approx 3 \times 3$ square degrees centered on the spiral galaxy NGC 1097. This excess of quasars has been explained as due to gravitational lensing of the light of very far, background quasars by the matter associated with a nearby galaxy. H. Arp and co-workers, on the contrary, have argued that the quasars surrounding the nearby galaxy must be physically associated with it, and that they are, therefore, not very far background objects as implied by Hubble's law and by their redshift. This is a distinctly non-standard interpretation, at variance with the opinion of most of the astronomical community. It is complementary to the usual interpretation that quasars are in fact the extremely active nuclei of very distant galaxies, and in this sense there is an intimate *quasar-galaxy association.*

quasi-biennial oscillation   The alternation of easterly and westerly winds in the equatorial stratosphere with an interval between successive corresponding maxima of 20 to 36 months. The oscillation begins at altitudes of about 30 km and propagates downward at a rate of about 1 km per month.

quasi-hydrostatic approximation   An approximate use of hydrostatic equilibrium for large scale weather systems. Hydrostatic equilibrium is such that in an ideal atmosphere, for an air parcel, the upward vertical pressure gradient force is equal to the gravitation force:

$$\frac{\partial p}{\partial z} = -\rho g$$

where p is pressure, z is altitude, ρ is density, and g is gravitational acceleration. The negative sign shows pressure decreases as altitude increases. Thus, for an atmosphere in hydrostatic equilibrium, air parcels have zero vertical acceleration. There is no absolutely zero vertical acceleration in real atmosphere, especially in small scale weather systems. However, for large scale weather systems, comparing to the gravitational acceleration, the vertical acceleration is very small and can be neglected, and thus the *quasi-hydrostatic approximation* is appropriate.

quasi-linear theory   In plasma physics, a theory based on perturbation theory; interactions between waves and particles are considered to be first order only. As in perturbation theory, all terms of second order in the fluctuating quantities should be small enough to be ignored. Only weakly turbulent wave-particle interactions can be treated this way: The particle distribution is only weakly affected by the self-excited waves in a random-phase uncorrelated way. This requirement not only corresponds to small disturbances in perturbation theory but even directly results from it as the waves are described in the framework of perturbation theory. The waves generated by the particles will affect the particles in a way that will tend to reduce the waves. Thus, the plasma is assumed to be a self-stabilizing system: Neither indefinite wave growth happens nor can the particles be trapped in a wave well.

The basic equation of *quasi-linear theory* is the Vlasov equation. All quantities then are split into a slowly evolving part and a fluctuating part. The long-term averages of the fluctuating part vanish. In quasi-linear theory, the slowly evolving quantities give the evolution of the system under study, while in case of the plasma waves the fluctuating quantities are of interest because they describe the wave.

With the index 'o' describing the slowly evolving quantities and the index '1' giving the fluctuating quantities, the average Vlasov equation as the basic equation of quasi-linear theory can be written as

$$\frac{\partial f_0}{\partial t + \mathbf{v} \cdot \nabla f_0} + \frac{q}{m} \frac{\mathbf{v} \times \mathbf{B}_0}{c} \cdot \frac{\partial f_0}{\partial \mathbf{v}}$$
$$= -\frac{q}{m} \left\langle \left(\mathbf{E}_1 + \frac{\mathbf{v} \times \mathbf{B}_1}{c} \right) \cdot \frac{\partial f_1}{\partial \mathbf{v}} \right\rangle .$$

with f being the phase space density, $\mathbf{v}$ the plasma speed, q the charge, m the mass of the charged particles, $\mathbf{E}$ the electric, and $\mathbf{B}$ the magnetic field. The term on the right-hand side contains the non-vanishing averages of the fluctuations and describes the interaction between the fluctuating fields and the fluctuating part of the particle distribution. These interactions, combined with the slowly evolving fields on the left-hand side lead to the evolution of the slowly evolving part of the phase space density. Note that no assumption about the smallness of the fluctuations enters: The only limitation is a clear separation between the fluctuating part and the average behavior of the plasma. *See* perturbation theory.

quasi-neutral equilibrium   A plasma condition apparently occurring on magnetic field lines in the auroral oval, in which a parallel electric field is balanced by the mirror force on precipitating electrons. The parallel voltage may originate in the circuit that powers Birkeland currents linking the auroral ionosphere with the distant magnetosphere. It accelerates the auroral electrons and at the same time expands the loss cone, allowing an increase in the current intensity.

quasi-separatrix layer   Generalizations of magnetic separatrices to magnetic configurations with a non-zero magnetic field strength everywhere in a given region. They are the locations where drastic changes in field-line linkages occur in the solar atmosphere.

quasi-single-scattering approximation   A radiative transfer model that accounts for only single scattering of photons with the assumption that forward-scattered light is treated as unscattered.

quasi stellar object   *See* QSO.

quasi-viscous force   A force transferring momentum through the magnetopause, from the solar wind flowing past Earth to the plasma inside the geomagnetic tail. A *quasi-viscous force* was proposed in 1961 by Ian Axford and Colin Hines as a possible mechanism for causing global convection in the Earth's magnetosphere. Magnetic reconnection, proposed the same year by James Dungey, is an alternative mechanism, and many researchers believe both mechanisms are active, with the quasi-viscous force being the weaker but steadier factor.

quay   A wharf used for docking, loading, and unloading ships.

quenching   *See* current saturation (cosmic string).

quicksand   *See* liquefaction.

quiet sun   The state of the sun when no sunspots, solar flares, or solar prominences are taking place. Also refers to the regions of the sun outside the active regions containing no filaments or prominences. The *quiet sun* typically consists of regions of mixed-polarity magnetic field ("salt-and-pepper"), which are continuously being reprocessed in a series of flux cancellations and flux emergence. Quiet sun coincides with minimal magnetospheric storms and minimal solar wind flux on the Earth. The last quiet sun minimum occurred in 1989.

R

R 136   *See* 30 Doradus.

Raadu–Kuperus configuration   *See* filament.

rad   In describing the interaction of nuclear particles with matter, an amount of radiation equivalent to the deposition of an absorbed energy of 100 ergs per gram of the absorbing material.

radar   An acronym for **RA**dio **D**etection **A**nd **R**anging. Although early radar operated at a frequency of tens of Megahertz, current radars operate in the microwave portion of the electromagnetic spectrum and work by "shining" a pulsed or continuous beam of microwave energy at an object and measuring the time delay for energy reflected back. *Radar* is often referred to as L-Band (1–2 GHz), S-Band (2–4 GHz), C-Band (4–8 GHz), X-Band (8–12 GHz), and K-band (12–18, 27–40 GHz). These designations were assigned by the military during the early development of radar in WWII. Each designation has a specific application. Imaging radars are generally considered to include wavelengths from 1 mm to 1 m. Navigation radars make use of the longer wavelengths (lower frequencies).

radian   A (dimensionless) measure of angle defined by the ratio of the length of the arc s on a circle contained within the angle θ, to the radius r of the circle:

$$\theta = s/r \, .$$

Because a full circle has circumference of $2\pi r$, the radian measure of a circle (360°) is 2π. Hence,

$$1° = 2\pi/360 \text{ radians } \sim 0.01745 \text{ radians} \, ,$$

and

$$1 \text{ radian} = 360/(2\pi)° \sim 57.296° \, .$$

radiance   The radiant power in a beam per unit solid angle per unit area perpendicular to the beam per unit wavelength interval [W m^{-2} sr^{-1} nm^{-1}].

radiation   Any sort of propagation of particles or energy from a source. Examples include electromagnetic radiation (visible light, X-rays, gamma rays, radio, infrared, ultraviolet), neutrino radiation, gravitational radiation, and particle streaming (alpha radiation, beta radiation, neutron flux, heavy ion radiation).

radiation belts   Regions containing substantial densities of energetic particles, typically electrons or protons, confined to the near magnetic equatorial regions of a planet, due to the planet's magnetic field. Particle injection into the *radiation belts* is from solar wind and solar-magnetosphere interactions.

The Van Allen belts are the terrestrial example. Trapping occurs as a result of the Earth's near-dipolar shape of the magnetic field characterized by magnetic field lines that converge at high latitudes resulting in an increased field strength compared with the equatorial region. The radiation belts were one of the first major discoveries made by satellites and are often called the "Van Allen radiation belts" after J. A. Van Allen. The Van Allen belts are divided into two regions, an inner region centered at 1.5 Earth radii at the equator, and an outer region centered at 3.5 Earth radii. Particle populations in the inner belt are very stable and long-lived, while the outer zone particle populations vary with solar and geomagnetic activity. Inner belt particles mostly have high energies (MeV range) and originate from the decay of secondary neutrons created during collisions between cosmic rays and upper atmospheric particles. Outer belt protons are lower in energy (about 200 eV to 1 MeV) and come from solar wind interactions. The outer belt is also characterized by highly variable fluxes of energetic electrons sensitive to the geomagnetic storms. During storm periods the particles may undergo radial and pitch angle diffusion propelling them toward Earth where they collide with exospheric or geocorona particles. The disturbed electrons can be precipitated into the middle ionosphere enhancing ionization and thus affecting radio propagation. The giant planets (Jupiter, Saturn, Uranus, and Neptune) also carry such radiation belts, as does Mercury, but Venus and Mars have very weak magnetic fields and do not.

radiation cooling   The transfer of heat energy away from an object by the net outward transport of radiation. In geophysics, at night, there is no incoming solar radiation, and the outgoing radiation from the Earth's surface exceeds the incoming radiation. The loss of radiant energy lowers the temperature of the surface, and the *radiational cooling* in surface layers of the atmosphere often causes a strong inversion of lapse rate. In atmosphere radiational processes, the vertical variations are much larger than their horizontal variations. Thus, the radiation cooling rate C_R can be expressed approximately by its vertical component. That is

$$C_R = \frac{1}{\rho C_p} \frac{\partial F_N}{\partial h}$$

where ρ is air density, C_p is specific heat at constant pressure, F_N is vertical component of net radiation flux, and h is height.

radiation gauge   In linearized (weak field) descriptions of gravity, one may use the restricted coordinate transformations to achieve $h^\alpha_\alpha = 0$, $h_{0\mu} = 0$, and $h^{\alpha\mu}{}_{,\mu} = 0$ (where $\mu = 0, 1, 2$ or 3) for the linearized gravitational potential h_{ab} in a source-free region. (The comma denotes partial derivative.)   Gravitational potentials meeting these requirements are said to be in the *radiation gauge*. *See* linearized gravitation, summation convention.

radiation pressure   The transfer of momentum to an object by radiation through the scattering, absorption, and emission of the radiation. The momentum transported by a single photon is given by $p = h\nu/c$ where h is Planck's constant, ν is the frequency of the photon, and c is the speed of light. The change in the momentum of a flux of photons as it is reflected (or absorbed) results in a net pressure being applied to the object. If the radiation is reflected, then the momentum transferred by the photon is twice the normal incident momentum.

radiation tide   Periodic variations in sea level primarily related to meteorological changes, such as the twice daily cycle in barometric pressure, daily land and sea breezes, and seasonal changes in temperature. Other changes in sea level due to meteorological changes that are random in phase are not considered to be part of the *radiation tides*.

radiation zone   An interior layer of the sun, lying between the core and the convection zone, where energy transport is governed by radiation. In the *radiation zone* of the sun, the temperature is a little cooler than the core, and as a result some atoms are able to remain intact. Their opacity influences the flow of radiation through this zone.

radiative-convective equilibrium   An equilibrium state where the outgoing radiation would be equal to the absorbed radiation at all latitudes.

radiative transfer equation   The linear integrodifferential equation that describes the rate of change with distance of the radiance in a collimated beam at a specified location, direction, and wavelength; the equation accounts for all losses (e.g., due to absorption and scattering out of the beam) and gains (e.g., by emission or scattering into the beam).

radio absorption   Absorption occurring to radio waves as a result of interaction between an electromagnetic wave and free electrons in the ionosphere. *See* ionospheric absorption.

radioactive   Containing isotopes that undergo radioactive decay to another isotope, or another element, by emitting a subatomic particle: typically electron (beta particle) or helium nucleus (alpha particle). In some cases, the radioactive nucleus undergoes an internal rearrangement and emits an energetic gamma ray (a photon), without changing its atomic number or atomic weight.

radioactive decay   The process by which unstable atomic nuclei (radioisotopes) spontaneously break down into one or more nuclei of other elements. During this process, energy and subatomic particles are released. The radioactive element is called the parent element, while the element produced by the decay is called the daughter element. There are three major processes by which *radioactive decay* occurs: alpha decay (loss of a helium nucleus, hence the atomic number decreases by two and the atomic weight decreases by four), beta decay (a neutron decays to a proton and electron, leading to an increase in the atomic number by one), and electron capture (where an electron is captured and combines with a proton to form a neutron, thus decreasing the atomic number by one). The half-life is the amount of time it takes for one-half of the parent to decay into the daughter. The radioactive decay of isotopes of uranium, thorium, and potassium is responsible for the heating of the Earth's interior.

radio burst   Short, pulse-like radio emission.

1. *Radio bursts* or pulses can be emitted from astrophysical objects (radio pulsars) where the radiation is created as synchrotron radiation of streaming electrons in the pulsar's magnetic field.

2. In the heliosphere, radio bursts can be observed on the sun, on Jupiter, and in interplanetary space. On Jupiter the source of the radio emission is the gyro-synchrotron radiation of electrons in the Jovian magnetosphere, and on the sun and in interplanetary space radio bursts result from the excitation of Langmuir waves by streams of electrons. *See* metric radio burst, type i radio burst, $i = I \ldots V$.

radio core   Compact radio emitting region at the center of radio galaxies and quasars. *Radio cores* are usually unresolved at angular resolution $\gtrsim 0.1$ sec of arc. At resolution of the order of 1 milliarcsecond, achieved with very long baseline interferometers, cores are resolved into one-sided jets, whose size is typically of the order of a parsec in nearby radio-galaxies. There is always a continuity with larger scale jets connecting the core to the radio lobes, although the jets often bend significantly passing to larger scale. The spectral energy distribution of a core is nearly constant over frequency in the radio domain, so that radio cores are said to be flat spectrum sources, with spectral index $\approx$ 0. Core-dominated (or equivalently, flat spec-

trum) radio sources do not show evident radio jets extended on scales of kiloparsecs, nor lobes.

radio emission: types I-IV   Emissions of the sun in radio wavelengths from centimeters to dekameters. *Radio emission* is classified by four types:

1. Type I: A noise storm composed of many short, narrow-band bursts in the range of 50 to 300 MHz.

2. Type II: Narrow-band emission that begins in the meter range (300 MHz) and sweeps slowly toward dekameter wavelengths (10 MHz). Type II emissions occur in loose association with major flares and are indicative of a shock wave moving through the solar atmosphere.

3. Type III: Narrow-band bursts that sweep rapidly from decimeter to dekameter wavelengths (500 to 0.5 MHz).

4. Type IV: A smooth continuum of broadband bursts primarily in the meter range (30 to 300 MHz). These bursts are associated with some major flare events beginning 10 to 20 min after the flare maximum and can last for hours.

radio frequency interference (RFI)   The degradation of a wanted radio signal due to the effect of an unwanted or interfering signal. Interference, unlike natural noise sources, is assumed to be manmade in origin and may be minimized by frequency management or, if the source can be identified, by negotiation. At night, RFI rather than natural noise is often the limiting factor for HF propagation.

radio frequency spectrum   The electromagnetic spectrum extends from DC to gamma rays. The *radio frequency spectrum* is part of this spectrum. It is further split into a series of frequency bands. A frequency band is a continuous set of frequencies lying between two specified limiting frequencies and generally contains many channels. The radio frequency bands are listed in the table on page 388.

radio galaxy   An active galaxy that radiates strongly in the radio. Instead of a point source, the emission comes from large radio lobes extending from jets that are located on either side of the parent galaxy, up to megaparsecs in scale.

Radio Frequency Bands

Designation	Band	Frequency	Wavelength
Extremely Low Frequency	ELF	0.003 – 3 kHz	100,000 – 100 km
Very Low Frequency	VLF	3 – 30 kHz	100 – 10 km
Low Frequency	LF	30 – 300 kHz	10 – 1 km
Medium Frequency	MF	300 kHz – 3 MHz	1 km – 100 m
High Frequency	HF	3 – 30 MHz	100 – 10 m
Very High Frequency	VHF	30 – 300 MHz	10 – 1 m
Ultra High Frequency	UHF	300 MHz – 3 GHz	1 m – 10 cm
Super High Frequency	SHF	3 – 30 GHz	10 – 1 cm
Extra High Frequency	EHF	30 – 300 GHz	1 cm – 1 mm

radioisotopes   The isotopes of elements which are unstable and change into other elements via radioactive decay.

radio lobe   Extended, often irregular and filamentary region of radio emission observed at the end of radio jets on opposite sides of the nucleus, in powerful radio galaxies and quasars. *Radio lobes* often extend beyond the optical image of a galaxy and typically reach end-to-end sizes of 50 kpc to 1 Mpc (the largest radio lobes known span 9 Mpc). The radio spectral energy distribution of lobes decreases sharply with increasing frequency; lobes are said to be steep spectrum sources in opposition to radio cores which show a flat spectral specific flux distribution. Emission is due to synchrotron processes, as in radio cores, but produced by electrons with lower energy. A prototypical lobe-dominated radio source is the elliptical galaxy Centaurus A (the brightest radio source in the Centaurus constellation), with two lobes extending far beyond the optical image of the galaxy, each $\sim$ 300 kpc in projected linear size.

radiometer   An instrument used to measure radiant energy, as opposed to the number of photons.

radiometric dating   A method of obtaining absolute ages of geological specimens by comparing relative concentrations of parents and daughter elements for a particular radioactive decay.

radiometry   The science of the measurement of radiant energy.

radio stars   Stars whose radio emission is strong enough to be detected from Earth. This is rather rare and generally associated with extreme youth (*see* stellar activity) or presence of a companion (*see* RS Canum Venaticorum stars). *Radio stars* are particularly important because they permit coordinate systems on the sky determined separately from radio and optical data to be combined.

radius of deformation   A horizontal length scale defined as $r = c/f$, where c is the phase speed of long gravity waves in the ocean or atmosphere, and $f = 2\Omega \sin\theta$ is the Coriolis parameter with Ω being the angular velocity of Earth's rotation and θ latitude. For motion of horizontal scales larger than the deformation radius, rotation effects are dominant, whereas for motion of smaller horizontal scales, rotation effects are less important. On the equator, f vanishes and the equatorial radius of deformation is defined as $r_E = \sqrt{c/\beta}$ where $\beta = 2\Omega\cos\theta$ is the meridional gradient of the Coriolis parameter.

radon   A radioactive colorless noble gas, *Rn*. Atomic number 86, natural atomic number 222. Melting temperature 202 K; boiling temperature 211.3 K. *Radon* exhibits an orange to yellow fluorescense in the solid state. Radon is produced as a daughter element in $\alpha-$ decay of radium. The longest-lived isotope is ^{222}Rn which has a half-life of 3.825 days, much longer than that of the other known isotopes (^{210}Rn to ^{221}Rn). All isotopes decay by α emission; ^{210}Rn and ^{211}Rn also have an electron capture branch, and ^{221}Rn also has a $\beta-$ decay branch.

Recently radon buildup in buildings has been a concern because inhaling the element directly irradiates the lungs, potentiating lung cancer, especially in conjunction with smoking. U.S. standards recommend action if the air radon load exceeds four picoCuries per liter.

rain   Liquid water precipitation that reaches the ground. Rain drops (water droplets) accumulate on condensation centers in clouds until they are too heavy to remain suspended and then fall earthward.

rainbow   An arc that displays a spectrum of colors and appears opposite the sun when solar rays are refracted and reflected in raindrops, spray, or mist. Bright, primary rainbows have a color pattern ranging from blue on the inside at about 40° from the center of the arc, to red on the outside of the arc at about 42° from the center.

Raman scattering   Inelastic photon scattering in which the energy of the scattered photon equals the energy of the incident photon plus or minus energy determined by the vibrational and rotational frequencies of the molecule; characterized by a volume scattering function that has forward-backward directional symmetry. *See* scattering cross-section.

random waves   In oceanography, waves that include a variety of wave periods, as opposed to a single period, which would define a monochromatic sea. If the range in wave periods is small, a narrow-banded energy spectrum results. A wide range of periods will yield a broad-banded energy spectrum.

Rankine–Hugoniot relations   Relations describing the conservation of mass, momentum, energy, and the electromagnetic field across the shock front. With the abbreviation $[X] = X_u - X_d$ describing the change of a property X from the upstream to the downstream medium, the *Rankine–Hugoniot relations* are:
Conservation of mass:

$$[m\mathbf{vn}] = [\varrho\mathbf{vn}] = 0 .$$

Momentum balance, with the last two terms describing the magnetic pressure perpendicular

and normal to the shock front:

$$\left[\varrho\mathbf{u}(\mathbf{un}) + \left(p + \frac{\mathbf{B}^2}{8\pi} \right) \mathbf{n} - \frac{(\mathbf{Bn})\mathbf{B}}{4\pi} \right] = 0 .$$

Energy balance, containing flow, internal and electromagnetic energy:

$$\left[\mathbf{un} \left(\frac{\varrho\mathbf{u}}{2} + \frac{\gamma}{\gamma - 1} p + \frac{\mathbf{B}^2}{4\pi} \right) \right]$$
$$\left[-\frac{(\mathbf{Bn})(\mathbf{Bu})}{4\pi} \right] = 0 .$$

Continuity of the normal component of the magnetic field (this is a direct consequence of Gauss' law of the magnetic field, $\nabla \cdot \mathbf{B} = 0$):

$$[\mathbf{B} \cdot \mathbf{n}] = 0 .$$

Continuity of the tangential component of the electric field:

$$[\mathbf{n} \times (\mathbf{v} \times \mathbf{B})] = 0 .$$

In the above relations $\mathbf{v}$ is the plasma flow speed, $\mathbf{B}$ the magnetic field, $\mathbf{n}$ the shock normal, ϱ the density, p gas-dynamic pressure, and γ the specific heat ratio.

In purely gas-dynamic shocks, all terms containing the electromagnetic field vanish, the Rankine–Hugoniot relations then are limited to the conservation of mass, momentum, and energy.

rare earth elements   The elements Sc, Y, and the lanthanides (La, Ce, Pr, Nd, Pm, Sm, Eu, Gd, Tb, Hy, Ho, Er, Tm, Yb, Lu).

Raychaudhuri equation   An equation relating the local expansion, rotation, and shear of a set of nearby trajectories of either material particles, or photons representing speed-of-light matter:

$$\dot{\theta} = -\frac{1}{3}\theta^2 - \sigma_{ij}\sigma^{ij} + \omega_{ij}\omega^{ij} - R_{\alpha\beta}u^\alpha u^\beta + a^\alpha_{;\alpha} .$$

(We sum over repeated indices.)

The dot indicates a time derivative associated with proper time for a central particle or with affine parameter for a photon. Here θ is the expansion, defined as the time rate of change of the local volume containing the matter (hence

$\theta = -\dot{\rho}/\rho$, where ρ is the density of the points being followed.) θ can also be defined in terms of the local velocity field u^{μ} : $\theta = u^{\mu}_{;\mu}$, the divergence is taken using the 4-velocity and 4-metric. The ";" is the metric covariant derivative.

Similarly the shear, σ_{ij} is the 3-tensor defining non-spherical motion

$$\sigma_{ij} = \left(u_{\alpha;\beta} + u_{\beta;\alpha} - \frac{1}{3}g_{\alpha\beta}\theta\right) P_i^{\alpha} P_j^{\beta}$$

and P_i^{β} is a projection into the 3-space associated with the central particle. The quantity a^{α} is the acceleration of the velocity u^{α} : $a^{\alpha} = u^{\alpha}_{;\beta}u^{\beta}$.

Finally

$$\omega_{ij} = \left(u_{\alpha,\beta} - u_{\beta,\alpha}\right) P_i^{\alpha} P_j^{\beta}$$

is the rotation 3-tensor. The appearance of the Ricci tensor $R_{\alpha\beta}$ in the *Raychaudhuri equation* can be related via Einstein's equations to the gravitational focusing power of matter. It can be seen that θ and σ_{ij} act to decrease the expansion, ω_{ij} acts to increase it. *See* summation convention.

Rayleigh distribution   A probability density function. It has non-zero skewness and describes the distribution of certain non-negative parameters (such as wave height in deep water).

Rayleigh–Jeans approximation   An approximation to Planck's blackbody equation for monochromatic specific intensity at long wavelengths. For long wavelengths, we can make the approximation $h\nu << kT$, where h is Planck's constant, ν is the frequency of the radiation in Hertz, k is Boltzmann's constant, and T is the excitation temperature; then a Taylor expansion in temperature of the Planck equation gives:

$$I_{\nu} = \frac{2kT}{\nu^2 c^2}$$

where I_{ν} is the intensity of the radiation in units of erg s^{-1} cm^{-2} Hz^{-1} sr^{-1}. The Rayleigh–Jeans approximation is useful to describe blackbody radiation in the domain of submillimeter and radio astronomy. If the intensity is in units of wavelength, then the approximation becomes

$$I_{\lambda} = \frac{2ckT}{\lambda^4}.$$

See Planck's law.

Rayleigh number   Fluid, heavier by $\Delta\rho/\rho$ on top of lighter fluid may remain laminar despite the static instability (*see* stability), as long as the *Rayleigh number* Ra $\tilde{<}$ 1700$\pm$50. Ra is defined as the ratio of (inertial time scale)^2/ (thermal diffusive time scale $\times$ viscous diffusive time scale) or explicitly in terms of the relevant parameters Ra $=$ g$(\Delta\rho/\rho)$H$^3/(\nu D_T)$, where $\Delta\rho/\rho$ is density difference between top and bottom, g is gravitational acceleration, H is thickness of the unstable layer, ν is kinematic viscosity, and D$_T$ is thermal diffusivity. Critical Rayleigh numbers, allowing the fluid to become unstable can be achieved in laboratory applications. Natural convective systems (such as atmosphere, ocean, lakes) show very large Ra and always turn turbulent under unstable conditions (*see* convective turbulence).

Rayleigh scattering   The scattering of light by a body small compared to the wavelength of the light. For a given refractive index, the scattering cross-section varies like the wavelength to the inverse fourth power. The light is scattered symmetrically forward and backwards, with the percentage of light scattered to a cone at an angle θ from the forward direction being proportional to $cos^2\theta$.

Rayleigh wave   A seismic surface wave resulting from the coupling of the P wave and the vertically polarized shear wave (SV) near the free surface. The speed of the *Rayleigh wave* is less than that of the shear (S) wave. In Rayleigh wave propagation, the particle motion forms a retrograde ellipse with the major axis vertical and the minor axis in the direction of wave propagation. For this reason, the Rayleigh wave is also called "ground roll" in exploration seismology. Rayleigh waves along the surface of a layered crust are dispersive. *See* dispersion.

ray parameter   A geometrical quantity that is preserved by a seismic wave as it is refracted through a medium with a variable wave speed. In particular for a spherically symmetric stratified Earth, the ray parameter p is defined by:

$$p = \frac{r \sin \theta_i}{v}$$

where r is the distance from the center of the Earth, v is the local appropriate wave speed, and θ_i is the angle that the direction of phase propagation makes with a radial vector from the center of the Earth. It can be shown that

$$p = \frac{dT}{d\Delta}$$

where T is the travel time of the ray from source to receiver, and Δ is the angle between lines from the two to the center of the Earth. In a horizontally stratified situation (which is approximately true near the Earth's surface, where variations in r can be neglected to first order) the ray parameter simplifies to $p = R_\oplus \sin \theta_i / v$. If v is known as a function of radius or depth, then it is easy to use p to calculate the path taken by a ray emitted from a seismic source at a particular angle θ_i.

ray tracing   Determining the path of a wave propogation by the use of the laws of reflection and refraction, ignoring diffraction.

ray tracing: wave packet approximation
The method by which the dominant wave behavior of light or other waves is taken into account, in which the central motion of approximately described wave packets determines the ray paths.

R Coronae Borealis stars   Highly evolved, asymptotic giant branch stars that have shed (or burned) all of their hydrogen, leaving an atmosphere made mostly of helium and carbon. The carbon sporadically condenses into dust, veiling the visible light from the stars, so that they fade by many magnitudes in a few weeks. The dust is gradually blown away, and the stars brighten back to normal, only to fade again in a few years. Most are also pulsationally unstable (*see* instability strip) and show rather subtle periodic changes in size and brightness with periods of one to a few months.

R_E   *See* Earth radius.

real singularity   In general relativity a region in space-time where some invariant geometrical quantities diverge, or where extension of finite acceleration curves is impossible.

Examples are the centers of black holes and white holes (e.g., *see* Schwarzschild metric).

Singularities can be space-like (occurring everywhere "simultaneously") or time-like (occurring over a period of time). *See* black hole, white hole.

recombination (in atomic and molecular physics)   Any of a number of processes in which an atom or molecule with positive charge reacts with another species or particle of negative charge. The different types of *recombination* can be represented:

$X^+ + e^- \rightarrow X + h\nu$ (radiative)

$X^+ + e^- \rightarrow X^* \rightarrow X + h\nu$ (dielectronic)

$XY^+ + e^- \rightarrow X + Y^*$ (dissociative)

$X^+ + Y^- + M \rightarrow Z + M$ (three–body)

$X^+ + Y^- \rightarrow X + Y^*$ (mutual neutralization)

where $h\nu$ represents a photon released in the process, e^- an electron, Y^* represents an excited state of species Y, M is a third particle that can absorb the energy released, and Z can be X, Y, or a new resulting species.

recombination line   An emission line produced by the quantum transition between two states of an atom or molecule, in which the energetically higher state was produced by the recombination of an atom or molecule with an electron either directly or by other transitions following the recombination event. *See* emission line, recombination.

reconnection   A process in a magnetized plasma, such as the solar magnetosphere, wherby diffusion of the ions allows adjacent oppositely directed flux lines to merge and cancel, lowering the total energy of the field. *See also* intercommutation (cosmic string).

recurrent novae   Ordinary novae of the sort involving a nuclear explosion on the surface of a white dwarf, but where the hydrogen to be burned collects fast enough again that more than one event has been seen in historic times. Only about six examples are known, and particular systems are often added to or subtracted from the inventory as we learn more about them.

red giant   The evolutionary phase of a star of 0.8 to about 5 solar masses during which the primary energy source is hydrogen burning (by the CNO cycle), in a thin shell around an inert helium core (more massive analogs are called red supergiants). The star is about 10 times as bright as it was on the main sequence (*see* HR diagram), and as the core grows, the star leaves the main sequence track of the Hertzsprung–Russell diagram and ascends the red giant branch, gaining in luminosity (on the order of 10^3 $L_\odot$) yet decreasing in surface temperature (near 3800 K). The phase lasts about 10% of the main sequence lifetime and is terminated by the ignition of helium burning. As the star eventually depletes its core supply of He, hydrogen fusion begins again in a thin shell around the core. Eventually, the core becomes inert C and O with separate, concentric shells where He-fusion and H fusion continues. At this stage, the star is considered to be in the asymptotic giant branch stage of evolution and has achieved a luminosity of 10^4 $L_\odot$ or greater and a surface temperature of 1800 to 3500 K. The sun will become a *red giant* about 5 billion years in the future, increasing the surface temperature of the Earth to about 900 K, where some metals melt and sulfur boils.

red line   A coronal line observed at 6374 Å resulting from a forbidden transition in highly ionized iron atoms (Fe X). Important for study of coronal structures at temperatures of order 1 MK.

redshift   A displacement of a spectral line toward longer wavelengths. This can occur through the Doppler effect or, as described by general relativity, from the effects of a star's gravitational field or in the light from distant galaxies. The relative displacement toward longer wavelength (or equivalently, of lower energy) of light is measured from spectral features like emission or absorption of lines and is defined as

$$z = (\lambda - \lambda_0)/\lambda_0 \,,$$

where λ_0 is the rest wavelength of a spectral feature. The displacement toward a longer wavelength is conventionally termed a *redshift* even if the photons are not in the visible spectral range. Galaxies and quasars, with the only exceptions of several galaxies very close to the Milky Way, invariably show all their spectral lines shifted to the red, with redshift increasing with distance according to Hubble's Law. Redshifts up to $z \approx 5$ have been (in 1998) observed for quasars.

Red Spot   A feature in Jupiter's atmosphere, some 10,000 by 50,000 km in extant. It was probably first observed by Hooke in 1664, but a continuous record of observations only goes back to the end of 1831. Its position is roughly fixed, with some drift relative to the average position. Although a number of suggestions have been forwarded to explain this feature, the most likely is an atmospheric disturbance, similar to a hurricane. Scaling arguments indicate that such a disturbance would be much longer lived on Jupiter than on Earth.

reduced gravity   Reduced gravity g' is defined by $g' = g(\rho_2 - \rho_1)/\rho_2$ where ρ_1 and ρ_2 are the density of the upper and lower layer.

reduced gravity model   A model treating the ocean thermocline as an infinitesimally thin interface separating a warm upper layer and an infinitely deep abyssal. In such a model, the motion in the abyssal is negligible, and the upper layer motion is described by a set of shallow water equations with the gravitational acceleration g replaced by a reduced gravity $g' = g(\rho_1 - \rho_2)/\rho_2$, where ρ_1 and ρ_2 are water density in the upper layer and abyssal, respectively. The *reduced gravity model* is particularly useful in modeling the longest wavelength internal mode in the tropical oceans where the thermocline is tight and shallow.

reef   A ridge of coral or rock which lies at or near the surface of the sea.

reference frame   A standard of reference used for physical systems, which is in principle equivalent to observations by a real observer, via measurements using the observer's clock and a local set of Euclidean measurement axes with a length standard. The *reference frame* is expressed in terms of a set of basis vectors $\mathbf{e}_{(a)}$. In principle another set $e'_{(b')}$ of basis vectors can be defined by a nonsingular (not necessarily constant) linear combination

$$\mathbf{e}'_{(b')} = A_{b'}{}^a \mathbf{e}_{(a)} \,.$$

General vectors can be written as a linear combination of the basis vectors:

$$\mathbf{A} = A^a \mathbf{e}_{(a)} \ .$$

See summation convention.

In Newtonian mechanics, global inertial frames are always possible, but not necessary. From the viewpoint of a non-inertial (accelerated and rotating) observer, fictitious (inertial), e.g., the centrifugal and Coriolis, forces appear to act on bodies and particles. This description proved to be of great use in application to the study of mechanics. In general relativity (curved, thus inhomogeneous spacetimes), there exist no global inertial frames. Since general relativity embraces not only mechanics, but the theory of all physical fields as well, their field theoretical counterparts appear alongside the relativistic inertial forces as well. Even in general relativity, to a local freely falling observer inertial forces do not appear, only physically arising forces, e.g., via Coulomb's law.

reflecting telescope   A telescope that is constructed using a mirror as the primary optical element, and which forms a focus by reflection, in contrast to a refracting telescope, where the primary optical element is a lens.

reflection   Any wave or particle phenomenon in which an incoming flux strikes a layer and is returned obeying the law: $\Theta_n = \Theta_i$, where the incident and excident angles Θ_i, Θ_r are measured from the normal to the layer. In geophysics, seismic waves are reflected by discontinuities in seismic velocities within the Earth. Important *reflections* are the Earth's core, inner core, and Moho. Unconformities in the sedimentary column act as reflecting boundaries and provide data on sedimentary structures, which is important for geophysical exploration.

reflection coefficient   In any essentially 1-dimensional wave phenomenon, a typically complex dimensionless number giving the amplitude and phase of a reflected wave compared to the incident wave. In some cases, the term is used to mean the magnitude of this quantity (ignoring the phase). For instance, for water waves it is defined as the ratio of the reflected wave height to the incident wave height.

$K_{\text{refl}} = H_r / H_i$, where K_{refl} = reflection coefficient, H_r = height of reflected wave, and H_i = height of incident wave.

reflection nebula   A part of the interstellar medium that reflects the light of nearby stars. *Reflection nebulae* are associated with stars that are not hot enough to ionize the gas.

reflectivity   The fraction of incident energy in a particular bandpass reflected in one encounter with a reflecting surface. *See* emissivity.

refracting telescope   A telescope that is constructed entirely using lenses and forms focused images by refraction, in contrast to a reflecting telescope, where the primary optical element is a mirror.

refraction   Deflection of direction of propagation of a wave on passing from one region to another with different wave propagation speed. Applied, for instance, to lenses, where the phase speed of light is lower in the glass lens than in air (the glass has a higher refractive index), resulting in a deflection which can be arranged to bring parallel light rays to a focus. In geophysics, the bending of seismic body waves upwards to the Earth's surface as they pass through the mantle. This bending occurs because the velocity of seismic wave propagation increases with depth. *See* refractive index.

refraction coefficient   In oceanography, a non-dimensional coefficient that represents the increase or decrease in wave height due to wave refraction for water waves. $K_r = \sqrt{cos\theta_1/cos\theta_2}$, where K_r = refraction coefficient, θ_1 = the angle that the wave crest makes with the bathymetric contours at a particular point 1, and θ_2 = the angle between the wave crests and bathymetric contours at a point 2 that has a different depth.

refractive index   The complex number for which the real part governs refraction (change of direction) at interfaces, and the imaginary part governs absorption; the real part is the ratio of the velocity of light in vacuum to the phase velocity in the medium (e.g., about 1.34 for seawater).

refractory   Material that resists heat, or in geology, elements or compounds with high melting or dissociation temperatures that remain in formations even after heating. *See* volatile.

region of anomalous seismic intensity   A region where seismic intensity becomes markedly intense for the magnitude of an earthquake or for the epicentral distance. For instance, there are many cases showing that the Pacific coast from Hokkaido to Kanto districts in Japan is a region of anomalous seismic intensity for deep earthquakes beneath the Japan Sea and shallow earthquakes along the Pacific coast. This is thought to be because seismic waves, including short-period components, reach the region, passing through the subducted plate with low attenuation. In a more typical situation, seismic waves passing through a high attenuation region away from the subducted plate do not produce sensible ground motion. Recent research also suggests that the existence of a low velocity layer overlying the subducted plate plays an important role in producing a *region of anomalous seismic intensity*.

Regolith   The fine powdery surface of an atmosphereless planet (e.g., the moon) which is produced by radiation and micrometeoric impacts on the surface.

Regulus   1.38 magnitude star of spectral type B7 at RA10$^{\rm h}$ 08$^{\rm m}$ 22.2$^{\rm s}$, dec $+11°58'02''$.

Reissner–Nordström (RN) metric   In general relativity, the unique, asymptotically flat metric describing the spherically symmetric gravitational and electric fields of an isolated mass M with electrical charge.

In spherical coordinates (t, r, θ, ϕ), the line element takes the form (in geometric units with speed of light $c = 1$, and Newton's constant $G = 1$)

$$ds^2 = -\left(1 - \frac{2M}{r} + \frac{Q^2}{r^2}\right) dt^2$$
$$+ \left(1 - \frac{2M}{r} + \frac{Q^2}{r^2}\right)^{-1} dr^2$$
$$+ r^2 \left(d\theta^2 + \sin^2\theta \, d\phi^2\right),$$

The electric field is described by a potential $\phi_E = Q/r$.

For $Q^2 > M^2$ the *Reissner–Nordström metric* is regular everywhere except for the real (naked) singularity at $r = 0$. For $Q^2 \leq M^2$, the Reissner-Nordström metric describes a black hole very similar to the Schwarzschild metric; there is an inner and an outer horizon $r_\pm = M \pm \sqrt{M^2 - Q^2}$ instead of just one horizon. Outside the outer horizon r_+ (i.e., for $r > r_+$) the solution has the expected properties of a mass and charged spherical mass (Region I). For $r_- < r < r_+$ (Region II) the coordinates switch meaning so that trajectories of worldlines have a spatial separation labeled by t, and evolve with a time parameter related to r. (This latter behavior is found for $r < 2M$ in the uncharged, Schwarzschild, black hole.) The surface $r = r_+$ is a future event horizon. For $0 < r < r_-$, r takes on again the meaning of a spatial coordinate (and t of time) (Region III). The timelike line $r = 0$ is a real singularity, visible to observers in Region III.

The Reissner–Nordström metric can be analytically extended by transforming to Kruskal-like coordinates so as to cover a larger manifold which contains an infinite chain of asymptotically flat regions of type I connected by regions II and III, the latter being bounded by time-like real singularities. In this case time-like curves exist which thread from regions I ("our universe outside the black hole") through the hole to another region I. At face value it appears possible to fall into such a hole, and later emerge (elsewhere). However, the sensitivity of the solutions to small perturbations may prevent this possibility. The maximal analytic extension was obtained by Graves and Brill. *See* ADM mass, black hole, Cauchy singularity, domain of outer communication, future/past event horizon, Kruskal extension, real singularity, Schwarzschild metric.

relative depth   A non-dimensional measure of water depth, used in the study of water waves. Defined as the ratio of water depth to wavelength, h/L.

relative humidity   The ratio of the amount of moisture in a given volume of space to the amount which that volume would contain in a state of saturation, i.e., the ratio of the actual vapor pressure (e) to the saturation vapor pressure (E). The relative humidity f is (as a percentage)

$$f = \frac{e}{E} \times 100\% .$$

Relative humidity expresses the saturation status of air. Generally the relative humidity decreases during the daytime as temperature increases, and increases at night as the temperature falls.

relativistic jets   Collimated ejection of matter at a velocity close to the speed of light, giving rise to highly elongated, often knotted structures in radio-loud active galactic nuclei, for example, radio quasars and powerful radio galaxies. Radio jets usually originate from an unresolved core and physically connect, or point, to extended lobes. Linear sizes of jets mapped at radio frequencies in external galaxies range from several kpc or tens of kpc, down to the minimum size resolvable with very long baseline interferometers (~ 1 parsec). Parsec-size jets show several indications of relativistic motion, including apparent superluminal motion, and jet one-sidedness ascribed to relativistic beaming of radiation. It is less clear whether jets observed at scales of several kiloparsecs are still relativistic. It is thought that only the most powerful radio galaxies, class II according to Fanaroff and Riley, and quasars may sustain a relativistic flow along kiloparsec-sized jets. Jets emit radiation over a wide range of frequencies. This suggests that the radio emission is electron synchrotron radiation. Galactic objects, like the evolved binary system SS 433, and galactic superluminal sources are also believed to harbor relativistic jets. *See* superluminal source.

relativistic time delay   The elongation of the travel-time of an electromagnetic signal, caused by the signal's passing through a gravitational field. This typically occurs when a radar signal is sent from the Earth, reflected back from another planet to be received at the Earth, perhaps passing near the sun on the way. When the path of the signal does not come near the sun, the time-lapse between the emission of the signal and its reception on the Earth is almost the same as calculated from Newton's theory. However, if the signal passes near the sun, the distance covered by the signal becomes longer because the sun curves the spacetime in its neighborhood. When the planet and the Earth are on opposite sides of the sun, and the path of the signal just grazes the sun's surface, one finds for the relativistic time delay Δt:

$$\Delta t = \frac{GM}{c^3} \ln \left(4 R_E R_p / d^2 \right) ,$$

where M is the mass of the sun, R_E is the distance from the sun to the Earth, R_p is the distance from the sun to the planet, and d is the radius of the sun. Here G is Newton's gravitational constant, and c is the speed of light. The planets used in the measurement are Mercury, Venus, the artificial satellites Mariner 6 and 7, and Mars (where artificial satellites orbiting Mars or landed on Mars were used as reflectors). The value of Δt predicted by Einstein's relativity theory for the situation described above is 240 ms (Shapiro, 1964). This is now one of the standard experimental tests of the relativity theory (along with perihelion shift, light deflection, and the gravitational redshift).

relativity   *See* general relativity, special relativity.

relaxation time   Characteristic time to represent rate of relaxation phenomena. For instance, for a viscoelastic body which is characterized by Maxwell model, stress decreases exponentially with time when a deformation is applied at a time and is subsequently kept constant. The time that stress becomes $1/e$ (e is base of natural logarithms) of the initial value is called *relaxation time*. Each material has its own relaxation time.

rem   An obsolete unit of radiation dose equivalent, equal to 100 ergs per gram.

remote sensing reflectance   In oceanography, the ratio of the "water-leaving" radiance in air to the downward plane irradiance incident onto the sea surface [sr^{-1}].

residence time (replacement time or average transit time) (T_R)   The average length of time that a parcel of water spends in a hydrologic reservoir; the minimum period required for all of the water in a basin (e.g., the Mediterranean) or region to be replaced. For systems at steady state, *residence time* can be easily calculated for a reservoir S because the inflow (i) and outflow (q) rates are identical: $T_R = \frac{S}{q} = \frac{S}{i}$.

residual circulation   The time-averaged circulation in a water body, after averaging over a long period (greater than the wind wave period and tidal period).

resolution   A measure of the minimum angular separation between two sources at which they can unambiguously be distinguished as separate.

resonance   In planetary dynamics, an orbital condition in which one object is periodically subjected to the gravitational perturbations caused by another object. The two bodies are usually in orbit around a third, more massive object and their orbital periods are some whole number ratios of each other. For example, Io and Europa are in a *resonance* as they orbit around Jupiter: Io orbits twice for every orbit of Europa. Europa and Ganymede are also in a resonance, with Europa orbiting twice for every orbit of Ganymede. In the case of Io and Europa, the resonances cause tidal heating to be a major internal heat source for both objects. Resonances can also create gaps, such as the Kirkwood Gaps within the asteroid belt (when asteroids are in resonance with Jupiter) and several of the gaps within Saturn's rings (caused by resonances with some of Saturn's moons).

resonance scattering   A magnetohydrodynamical phenomenon: the scattering of particles by waves can be described as a random walk process if the individual interactions lead to small changes in pitch-angle only. Thus, a reversal of the particle's direction of propagation requires a large number of such small-angle scatters. If, however, the particle motion is in resonance with the wave, the scattering is more efficient because the small-angle scatterings all work together in one direction instead of mostly cancelling each other. Thus, pitch-angle scattering will mainly occur at the magnetic field fluctuations with wavelength in resonance with

the particle motion parallel to the field:

$$k_\| = \frac{\omega_c}{v_\|} = \frac{\omega_c}{\mu v}$$

with $k_\|$ being the wave number of the waves leading to the particle scattering, $v_\|$ the particle speed parallel to the magnetic field, and μ the cosine of the particle's pitch-angle.

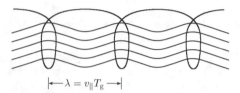

$$|\!\!\leftarrow \lambda = v_\| T_g \rightarrow\!\!|$$

Resonance scattering.

From the resonance condition we can see that for a given particle speed particles resonate with different waves, depending on their pitch-angle. Since the amount of scattering a particle experiences basically depends on the power density $f(k_\|)$ of the waves at the resonance frequency, scattering is different for particles with different pitch-angles although their energy might be the same. Therefore, the pitch-angle diffusion coefficient $\kappa(\mu)$ depends on pitch-angle though μ. *See* slab model.

resonant absorption   In solar physics, in a closed magnetic loop, resonant frequencies appear at multiples of $v_A/2L$, where $v_A = B/\sqrt{4\pi m_p n}$ is the Alfvén speed and L is the coronal length of the loop. The resonances occur because of reflections off the transition region parts of the loop. *Resonant absorption* occurs when the frequency of a loop oscillation matches the local Alfvén frequency. This creates a resonant layer in which there is a continuous accumulation of energy and, consequently, may result in the heating of the coronal plasma. *See* Alfvén speed.

resonant damping and instability   In a collisionless plasma, wave damping or growth associated with the interaction between a wave and particles moving with a velocity such that the wave frequency is (approximately) Doppler-shifted to zero, or to a multiple of the particle's Larmor frequency. Thus, in a magnetized plasma, for a wave of angular frequency ω and wave vector $\mathbf{k}$, the resonant particles have a component of velocity $v_\parallel$ along the magnetic field satisfying

$$\omega - k_\parallel v_\parallel \pm n\Omega \approx 0 \;,$$

where n is an integer and Ω is the particle Larmor frequency. The $n = 0$ resonance is called the Landau resonance, and the other resonances (especially $n = 1$) are called cyclotron resonances. These resonances are of importance in wave dissipation and the pitch-angle scattering of charged particles. *See* cyclotron damping and instability, Landau damping and instability.

resonant layer   In solar physics, a narrow, typically cylindrical, shell <250 km across in which the dissipation of Alfvén waves is thought to occur via the process of resonant absorption.

rest mass   The mass of an object as measured by an observer comoving with the object, so that the observer and object are at rest with respect to one another. In relativistic physics there is a mass increase for moving objects; the term *rest mass* is used to denote the intrinsic, unchanging, mass of the object.

retrograde motion   In observation of planets from Earth, westward motion of a planet against the background stars. Because of their motion in orbit, planets move in an average easterly direction against the background stars. However, when Earth is closest to another planet, the motion of the Earth gives a parallax which can lead to temporary *retrograde* (westward) *motion.*

retrograde orbit   An orbit of a satellite of a planet, in which the orbital angular momentum lies in the hemisphere opposite that of the planet's angular momentum.

retrograde rotation   Rotation of a celestial body such that the angular velocity associated with its rotation has a negative projection on the angular velocity associated with its orbital motion; equivalently, an inclination to the orbital plane greater than 90°. Venus and Uranus have *retrograde rotation* (though Uranus' inclination of 97°55′ means that its pole is essentially lying in its orbit, with a slight retrograde net rotation).

return current   Particle acceleration in solar flares is thought to result in a beam of electrons formed in the corona which propagates through the corona to dissipate in the chromosphere. The charge separation leads to the development of an equal and opposite *return current* which is set up in the ambient plasma in order to replenish the acceleration region.

return stroke   In lightning flashes, the main current flash in a lightning stroke, in which current flows from the ground up the channel opened by the stepped leader. *See* stepped leader, dart leader.

reverse fault   A fault where the rocks above the fault line move up relative to the rocks below the fault line. A *reverse fault* has the direction of motion opposite that of a normal fault. A reverse fault is a type of dip-slip fault, where displacement occurs up or down the dip (i.e., the angle that the stratum makes with the horizontal) of the fault plane. A reverse fault where the dip is so small that the overlying rock is pushed almost horizontally is called a thrust fault.

reversible process   In thermodynamics, a process that occurs so slowly that the system is very close to equilibrium throughout the process. In such a case no entropy is produced (neither the entropy of the system nor that of its surroundings is changed), and the system can be returned to its original state by infinitesimally changing the external conditions.

revetment   A man-made structure constructed to protect soil from erosion. Frequently made of natural stone or concrete.

Reynolds decomposition   Separation of a state variable into its mean and the deviation from the mean:

$$x = \overline{x} + x' \;.$$

The over-bar denotes the average or mean over a timescale τ and the prime denotes the fluctuations or deviations from the mean. Ideally, the time scale τ should be chosen such that $\tau = 2\pi/\omega_g$ where ω_g is the spectral gap frequency of the energy spectra $\phi(\omega)$.

Reynolds number (Re)   A dimensionless quantity used in fluid mechanics, defined by $Re = \rho v l / \eta$, where ρ is density, v is velocity, l is length, and η is viscosity. In each case "typical" values are used. The *Reynolds number* gives a measure of the relative importance of acceleration to viscosity in a given situation.

Flows that have the same Re are said to be dynamically similar. Above a certain critical value, Re_c, a transition from laminar flow to turbulent flow occurs. Its value serves as a criterion, low values being associated with high stability, for the stability of laminar flow. Depending on the geometry of flow, once Re becomes larger than 10^5, turbulence is likely to occur. In open channel flow, the hydraulic radius R_H is used for the diameter of the passageway and turbulent flow occurs when $R > 1000$. For subsurface flow, the mean grain diameter is substituted for the diameter of the passageway, with turbulent flow occurring when $R > 10$. In oceanographic applications, it is often useful to consider the buoyancy Reynolds number

$$\Re_b \equiv (L_N / K)^{4/3} = \epsilon \rho / \eta N^2$$

where L_N is the buoyancy scale, K is the Kolmogorov scale, ϵ is the dissipation rate of turbulent kinetic energy, and N^2 is the buoyancy frequency. In this formulation, the Reynolds number describes the turbulent state of the flow based on the stratification and intensity of the turbulence.

Reynolds stress   The stress, that occurs as a result of the turbulent exchange between neighboring fluid masses, defined by $\rho < u_i u_j >$ to form a 9-term tensor. If the fluid is turbulent, the *Reynolds stress* is much larger than the viscous stress.

Rhea   Moon of Saturn, also designated SV. It was discovered by Cassini in 1672. Its orbit has an eccentricity of 0.001, an inclination of $0.35°$, a semimajor axis of 5.27×10^5 km, and a precession of $10.16°$ yr^{-1}. Its radius is 765 km, its mass, 2.49×10^{21} kg, and its density 1.33 g cm^{-3}. It has a geometric albedo of 0.7, and orbits Saturn once every 4.518 Earth days.

rheological constitutive equation (creep law, flow law)   The relation,

$$\dot{\varepsilon} = F(T, P, c, X, \sigma),$$

relating creep rate (ε, in the steady-state stage of creep), temperature (T), pressure (P), chemical environment (c), microstructural parameter (X), and deviatoric stress (σ). A *rheological constitutive equation* can be determined through a series experiments performed under well-controlled conditions with a general form as

$$\dot{\varepsilon} = \dot{\varepsilon}_0 f_{O_2}^{m1} f_{a_i}^{m2} \sigma^n \exp \left(-\frac{\Delta H_0(\sigma) + P \Delta V^*}{RT} \right)$$

where ε_0 is a constant that often depends on the orientation of stress axis with respect to the crystal lattice, $f_{O_2}^{m1}$ and $f_{a_i}^{m2}$ reflect the contribution from chemical environments (oxygen fugacity and component activity), ΔH_0 and ΔV^* are the activation enthalpy and volume, respectively, of the controlling process, and P is the hydrostatic pressure. The term $(\Delta H_0(\sigma) + P \Delta V^*)$ defines the activation energy of the controlling process.

rheology   The study of the flow of liquids and deformation of solids. *Rheology* addresses such phenomena as creep, stress relaxation, anelasticity, nonlinear stress deformation, and viscosity. Rheology can be a function of time scales. For instance, in geophysics, on short time scales, say less than about 1000 years, the Earth's mantle and crust are elastic, the outer core is a fluid and the inner core is solid. On longer time scales, the Earth's mantle beneath the lithosphere behaves as a fluid, resulting in postglacial rebound and mantle convection.

Ricci rotation coefficients   In geometry, the components of a connection 1-form

$$\gamma_{mnp} = e_n^a e_p^b \nabla_b e_{am}$$

where e_m^a are the coordinate components of a complete collection of smooth orthogonal vectors e_m. Here $e_p^b \nabla_b e_{am}$ is the covariant derivative of the vector e_m along the vector e_p. (Bold indices **m, n, p** label the vector; a, b, c label components.) Without some other condition on

γ_{mnp}, this equation is definitional; the γ_{mnp} can be specified as arbitrary smooth functions. *See* metricity of covariant derivative, torsion.

Ricci tensor   In Riemannian and pseudo-Riemannian geometry, the symmetric two index tensor R_{ab} representing the trace of the Riemann curvature tensor: $R_{ac} = R_{abc}{}^{b}$. Some works (among others, L.P. Eisenhart's *Riemannian Geometry*) define the *Ricci tensor* by contraction in the first index, thus reversing the sign of the Ricci tensor. *See* Riemann tensor.

Richard's equation   Combining Darcy's law with the continuity equation provides an expression of mass conservation and water flow in the unsaturated zone, along with a history of the tension distribution in a vertical soil column:

$$\frac{d\theta}{\partial t} = \frac{\partial}{\partial z}\left[K\theta\left[\frac{\partial\psi}{\partial z}+1\right]\right]$$

where θ is the volumetric soil moisture, t is the time interval, z is depth in the unsaturated zone, ψ is the tension head, and $K(\theta)$ is the unsaturated hydraulic conductivity. For steady flow, the time derivative of the volumetric moisture content θ is zero and a single integration of the right-hand side of *Richard's equation* yields $q_z = -K(\theta)[d\theta/dz + 1]$. For saturated conditions $d\theta/dz$ is also zero, $K(\theta)$ becomes a constant K, the tension head becomes positive, and Richard's equation reduces to the Laplace equation for one-dimensional flow.

Richardson, Lewis F.   (1881–1953) English physicist and meteorologist. He made the first numerical weather prediction by using finite differences for solving the differential equations, but at that time (1922) the computations could not be performed quickly enough to be of practical use.

Richardson number   Dimensionless parameter that compares the relative importance of mechanical turbulence (quantified by velocity shear) and convective turbulence (quantified by the buoyancy frequency) and gives the stability criterion for the spontaneous growth of small-scale waves in a stably stratified flow with vertical shear of horizontal velocity, Ri, defined by

$$\text{Ri} = \frac{N^2}{(\partial\mathbf{V}/\partial z)^2}$$

where N is Brunt–Väisälä (buoyancy) frequency, $\mathbf{V}$ is horizontal velocity, and z is upward coordinate. The *Richardson number* is a measure of the ratio of the work done against gravity by the vertical motions in the waves to the kinetic energy available in the shear flow. In general, the smaller the value of Ri the less stable the flow with respect to shear instability. The most commonly accepted value for the onset of shear instability is Ri $= 0.25$.

For a continuously stratified fluid, the bulk Richardson number can also be defined by

$$\text{Ri} \equiv \frac{N^2 l^2}{U^2}$$

where N is the buoyancy frequency, l is the characteristic length scale, and U is the characteristic velocity scale of the flow.

The ratio of the buoyant destruction of turbulent kinetic energy, $g\alpha\overline{wT'}$, to the shear production of turbulent kinetic energy, $-\overline{uw}\,(dU/dz)$, is called the flux Richardson number, defined by

$$\text{Ri}_f = \frac{g\alpha\overline{wT'}}{-\overline{uw}\,(dU/dz)}$$

where $\overline{wT'}$ is the heat flux, $\overline{uw}$ is the Reynolds stress, g is the constant of gravity, and $\alpha = \rho^{-1}(\partial\rho/\partial T)$ is the thermal expansivity of sea water. For $\text{Ri}_f > 1$, buoyancy dampens turbulence faster than it can be produced by shear production. However, laboratory observations and theoretical considerations suggest that the critical value is less than unity and that a $\text{Ri}_f \gtrsim 0.25$ turbulence ceases to be self-supporting.

Richter magnitude scale   An earthquake magnitude scale originally devised by American seismologist C.F. Richter in 1935 for quantifying energy release in local California earthquakes. It was originally defined to be the base 10 logarithm of the maximum amplitude traced on a seismogram generated by a then standard seismograph at 100 km epicentral distance. It was later substantially generalized. Since the early 1980s, the most widely used scale is the moment magnitude scale.

ridge push   *See* ridge slide.

ridge slide (ridge push)   A mid-ocean ridge is elevated compared to the surrounding ocean basin. The oceanic plate is driven to move away from the ridge by the component of gravitational force parallel to the plate in the down slope direction. This plate driving force is called *ridge slide* or *ridge push*. *See* seafloor spreading.

Riemann-Christoffel tensor   *See* Riemann tensor.

Riemann tensor   In geometry, the curvature tensor of an affine space. In terms of a 1-form basis $\{\omega_a\}$ and its second covariant derivatives, the *Riemann tensor* $R_{abc}{}^d$ is defined as

$$\nabla_a \nabla_b \omega_c - \nabla_b \nabla_a \omega_c = -R_{abc}{}^d \omega_d .$$

In a coordinate basis, the components R^a_{bcd} may be written as

$$R^a_{bcd} = \Gamma^a_{bd,c} - \Gamma^a_{bc,d} + \Gamma^a_{fc}\Gamma^f_{bd} - \Gamma^a_{fd}\Gamma^f_{cb}$$

where Γ^a_{bc} is the connection, and the "," denotes a partial derivative. Note that when Γ^a_{bc} is symmetric (no torsion) $R_{abcd} = -R_{abdc}$ and $R_{abcd} = R_{cdab}$. *See* affine connection, covariant derivative, sign convention.

Rigel   (β Orionis), Spectral type B, RA 05^h $14^m 32^s.3$, dec $-08°12'06''$. The brightest star in the constellation Orion and the seventh brightest star in the sky (apparent magnitude +0.12). *Rigel* is a blue (very hot) supergiant (absolute magnitude -8.1), about 100 times bigger in diameter than the sun. It is approximately 430 pc distant.

right ascension   An angle coordinate on the celestial sphere corresponding to longitude, and measured in hours thus dividing the circle of 360° into 24 hours, measured eastward from the location of the vernal equinox (also called the first point of Aries). At the time of the vernal equinox the *right ascension* of the sun is 0 hours. The direction to the sun at this time is then $0^h RA$, 0° declination, and this direction on the celestial sphere is referred to as the Vernal Equinox.

right-hand coordinate system   One in which $\hat{z} = \hat{x} \times \hat{y}$ in which $\{\hat{x}, \hat{y}, \hat{z}\}$ are the unit vectors in the coordinate directions.

right-hand rule   A prescription to determine the direction in a right-hand coordinate system, of a vector cross product $\mathbf{A} \times \mathbf{B}$: right-hand fingers in the direction of $\mathbf{A}$, curl fingers towards $\mathbf{B}$; thumb points in the direction of product.

right-lateral strike-slip fault   In geophysics, a strike-slip fault such that an observer on one side of the fault sees the other side moving to the right. The San Andreas fault in California is a right-lateral strike-slip fault.

rigidity   *See* shear modulus.

rigid lid approximation   A boundary condition applied when the surface displacements are small compared with interface displacements. In the continuity equation of flow motion, one neglects the time change in the surface displacements, but retains the time change in the interface displacements.

rille   Channel formed by flowing lava. *Rilles* are old lava tubes where the roof has collapsed along part or all of the channel length. Lava tubes (hence rilles) are formed by very low viscosity lavas. On the moon, rilles have widths from a few tens of meters to a few kilometers and lengths of hundreds of kilometers. Longer channels are seen on Venus, although the composition of the flows that created these channels is still debated. Rilles typically are widest at the upstream end and narrow as they progress downstream. Rilles show no delta-like deposits at their termini — they typically just fade away into a lava plain.

Rindler observer   A uniformly accelerated observer in flat Minkowski spacetime. The observer's 4-dimensional trajectory is given by

$$\begin{cases} x = ca^{-1} e^{a\xi} \cosh(a\tau/c) \\ t = c^2 a^{-1} e^{a\xi} \sinh(a\tau/c) , \end{cases}$$

with y, z, constant. Here x, y, z are Cartesian coordinates (x along the direction of motion), t is the time of an inertial observer, τ is the observer's proper time, ξ is a parameter related to the position of the turning point, and c is the speed of light. The constant proper acceleration is given by $a\,e^{-a\xi}$.

Because of the form of the trajectory, which approaches the light cones $x \pm t$ for $t \to \pm\infty$, the *Rindler observer* is prevented from getting signals from events where $x < t$, and cannot send signals to events where $x < -t$. Thus $x = t$ is a future causal horizon for the observer and $x = -t$ is a past causal horizon. *See* future/past causal horizon.

ring   Rings of small icy debris orbit around each of the four Jovian planets. The *rings* around Jupiter, Uranus, and Neptune are composed of very small particles (generally only a few microns in diameter) while the major rings of Saturn contain ice particles ranging in size from dust to boulders a few meters in diameter. Only Saturn's rings are easily visible through ground-based telescopes. The origin of the different ring systems is still controversial. Saturn's thick rings (primarily the A and B rings) were probably formed either by tidal disruption of an icy moon which wandered within the planet's Roche Limit, or by the collision and destruction of two or more icy moons. The thinner rings around Saturn (primarily the E, F, and G rings) and the rings around Jupiter, Uranus, and Neptune, may result from material ejected off the small inner icy moons which orbit these bodies — this material may be ejected either from volcanism (as is proposed for Saturn's moon Enceladus) or by meteorite impact (suggested for Jupiter's ring). Resonances with moons may explain some of the gaps seen within rings, and shepherd moons are responsible for the maintenance of some rings, but the structure and narrowness of many rings are still not well understood.

ring current   The dipole-like topology and the inhomogeneity of the Earth's magnetic field cause the energetic particles to be trapped in the magnetosphere in the "inner" and "outer" radiation belts. The magnetic dipole field decreases with increasing distance from the Earth's center by the inverse cube law, hence the particles will not follow exactly circular orbits while gyrating about a line of force. While gyrating, the magnetic field will be stronger when the particle is closer to the dipole than when it is farther away from it. Therefore, the radius of curvature is smaller at lower altitudes than in the higher and hence after gyration the particle will have drifted longitudinally. Electrons will drift eastward and protons will drift westward, thus producing a net westward electric current known as *ring current*. It is situated normally in the 4 to 6 Earth radii region and overlaps parts of the outer radiation belt. During geomagnetically disturbed periods, particularly during the "main" phase, the ring current is intensified. The activity of the ring current during disturbed periods is measured by an hourly Dst index, which is the magnitude of the horizontal component of the disturbed magnetic field. High and low latitudes are avoided in the Dst determination to avoid the effects of the auroral and equatorial electrojets. The intensity of the ring current varies from about 1 to 2 nA/m^2 in quiet times to up to 10 nA/m^2 in disturbed periods.

ring galaxy   A galaxy showing a bright, prominent outer ring encircling the whole galactic body. *Ring galaxies* are thought to be produced by the head-on encounter, nearly perpendicular, between a disk galaxy and a second galaxy: After the collision, an outward-moving density wave within the disk gives rise to the ring, which appears to surround a central area of much lower surface brightness. Ring galaxies are of rare occurrence (only a few tens are presently known), consistent with their formation requiring a special orbit orientation for the encounter. A prototypical example of ring galaxies is the Cartwheel galaxy, believed to have been crossed by a smaller companion galaxy visible in its vicinity.

riometer   For "Relative Ionospheric Opacity Meter", an instrument measuring the amount of ionization below the usual height of the main ionospheric layers, caused by energetic particles that precipitate into the Earth's atmosphere. It detects a decrease in the radio noise observed on the ground and coming from the distant universe. *See* polar cap absorption.

rip current   A shore-normal current at a beach that carries water from a region landward of a sand bar through a gap in the sand bar and into the offshore region.

riprap   A term used to refer to the heavy material used to armor a shoreline. May be comprised of rock, concrete elements or debris, or some other heavy material.

Robertson–Walker cosmological models
A class of spacetimes describing homogeneous isotropic expanding universes. The cosmic matter is assumed to be a perfect fluid. The metric can be written in spherical coordinates (t, r, θ, ϕ) as

$$ds^2 = -dt^2 + R^2 \left[\frac{dr^2}{1 - kr^2} + r^2 \left(d\theta^2 + \sin^2 \theta \, d\phi^2 \right) \right].$$

An alternate equivalent formulation is:

$$ds^2 = R^2 \left[d\chi^2 + \text{sinn}^2\chi \left(d\theta^2 + \sin^{-2}\theta \, d\varphi^2 \right) \right]$$

where

$$
\begin{aligned}
\text{sinn}\chi \quad &= \quad \sin \chi, \, k = +1 \\
&= \quad \chi, \, k = 0 \\
&= \quad \sinh \chi, \, k = -1 \, .
\end{aligned}
$$

The curvature parameter $k = 0, \pm 1$ for, respectively, flat, spherical, and hyperbolic 3-space.

In all cases, $R(t)$ is an arbitrary function of the time-coordinate called the scale factor. The Einstein equations equate the Einstein tensor (calculated from derivatives of the metric coefficients) to the stress-energy tensor, which must have the same symmetry as the geometry, and so contains only spatially homogeneous matter-density ρ and pressure p. The time-evolution of the model is undetermined until an equation of state is imposed on ρ and p; the equation of state results in determining the function $R(t)$. With $p = 0$ ("dust"), the *Robertson–Walker models* become the Friedmann–Lemaître cosmological models; the dust solutions also apply to the interior of a sphere of homogeneous dust. *See* comoving frame, Friedmann–Lemaître cosmological models, maximally symmetric space, Oppenheimer–Snyder model, Tolman model.

Robinson–Trautman space-time (1962)   In general relativity, an algebraically matter-free

special space-time in which the repeated principal null direction diverges but has no shear or rotation. It is thought to describe the gravitational fields of radiative sources. The vacuum metric, discussed by I. Robinson and A. Trautman, has the form

$$ds^2 = 2 \, dr \, du - 2r^2 P^{-2} d\zeta \, d\bar{\zeta} + H du^2$$

where

$$H = 2P^2 \frac{\partial^2 \log P}{\partial \zeta \partial \bar{\zeta}} - 2r \frac{\partial \log P}{\partial u} - \frac{2m}{r} \, .$$

The function P satisfies

$$\frac{\partial \log P}{\partial u} = k \Delta \Delta \log P, \qquad \Delta = 2P^2 \frac{\partial^2}{\partial \zeta \partial \bar{\zeta}}$$

and ζ is a complex stereographic spherical coordinate. No physically well-behaved Robinson–Trautman solutions other than the (nonradiative) Schwarzschild space-time are known.

Roche limit   The closest orbital distance, R, at which an orbiting satellite can hold itself together gravitationally, in spite of tidal forces from its primary. Thus, the gravitational acceleration holding a particle on the surface of a body of mass, m_b of radius R_b is:

$$a_g = \frac{Gm_b}{R_b^2} \, .$$

On the other hand, the relative tidal acceleration between the center of the satellite and a point radially towards or away from the primary is:

$$a_t = \frac{GM}{R^2} - \frac{GM}{(R + R_b)^2} \approx \frac{2GM}{R^3} R_b \, .$$

If the ratio a_t/a_g exceeds unity, the satellite will be destroyed due to the tidal forces:

$$a_t/a_g = \frac{2M}{R^3} \Big/ \frac{m_b}{R_b^3} \, . \tag{1}$$

The R for which $a_t/a_g = 1$ is the Roche limit:

$$R_{\text{Roche}} = \left(\frac{2M}{m_b} R_b^3 \right)^{\frac{1}{3}} \, . \tag{2}$$

where M is the mass of the central body and m_b and R_b^3 refer to the satellite. Note that according to (1), the ratio of accelerations is twice the ratio of densities (in the case of the central body, as if it were spread out to the orbital radius).

Small solid bodies such as rocks or artificial satellites are held together by molecular bonds and hence can survive the tidal stresses, but bodies with little internal strength (such as "rubble pile" asteroids) are easily pulled apart. The destruction of bodies with low internal strength by movement within the *Roche limit* may explain the formation of some ring systems. The accretion of very small satellites into a larger moon is also inhibited in this region.

Roche lobe   (Also inner Lagrangian surface.) In a binary system, in a corotating frame, the lowest-lying equipotential surface which envelopes both members. If one of the stars fills its *Roche lobe,* then mass will flow from it towards the companion star. Stars that form or move so close together that both stars fill their Roche lobes are contact binaries or W Ursa Majoris stars. In the more common case, an evolved star which has already shed a good deal of matter, continues to transfer material from the smaller side of the lobe through a gas stream to the more massive companion, perhaps through an accretion disk. If the accretor is a main sequence star, we would see an Algol system. If it is a white dwarf, we would see a cataclysmic variable.

Rocket effect (string loop)   One way of diminishing the mass-energy density of a network of cosmic strings is by the continuous generation of loops. These will form through string interactions, like intercommutation.

Loops will oscillate with characteristic frequencies given by the inverse of their size (in appropriate units). This allows energy to be radiated away in the form of gravity waves. These waves carry not only energy but also momentum and therefore highly asymmetric loops might be propelled like rockets, reaching velocities as high as a tenth of the velocity of light. *See* cosmic string, intercommutation (cosmic string), scaling solution (cosmic string).

Rosalind   Moon of Uranus also designated UXIII. Discovered by Voyager 2 in 1986, it is a small, irregular body, approximately 27 km in radius. Its orbit has an eccentricity of 0, an inclination of $0.3°$, a precession of $167°$ yr^{-1}, and a semimajor axis of 6.99×10^4 km. Its surface is very dark, with a geometric albedo of less than 0.1. Its mass has not been measured. It orbits Uranus once every 0.559 Earth days.

Rossby, Carl-Gustav   (1898–1957) Swedish-American meteorologist. He founded the first meteorology department in the U.S. at the Massachusetts Institute of Technology in 1928. He was among the first to recognize the important role of transient mid-latitude wave disturbances in the atmospheric general circulation. He established many of the basic principles of modern meteorology and physical oceanography during the 1930s. Some of the theoretical ideas that he developed during that time were instrumental in the development of numerical prediction models during the following decades.

Rossby number   The *Rossby number* ϵ is a non-dimensional parameter defined as the ratio between the horizontal advection scale and the Coriolis scale in the horizontal momentum equation, that is, $\epsilon = \frac{U}{fL}$ where U and L are the velocity scale and length scale of motion and f is the Coriolis parameter. When this number is much less than unity, the flow is said to be geostrophic.

Rossby radius   *See* Rossby radius of deformation.

Rossby radius of deformation   (Rossby, 1938) This is the horizontal scale at which the Earth's rotation effects become as important as buoyancy effects and is defined by $L_R = C/|f|$ where C is the gravity wave speed and f is the Coriolis parameter. This has been called the *Rossby radius* or the *radius of deformation.* In baroclinic flow, C is the wave speed of each baroclinic mode in a non-rotating system. It describes the radius Ro $=$ u/f[m] of an inertial circle which a freely moving water parcel would follow with a speed u under the effect of the Coriolis parameter f.

Rossby wave   A wave in a rotating fluid body, involving horizontal motion, for which the sole restoring force is the Coriolis effect. In neutron stars the retrograde mode, which rotates slower than the body itself, can grow by emitting gravitational waves and can therefore radiate angular momentum. In meteorology, westward propagating atmospheric waves. Weather systems in the atmosphere are often associated with these *Rossby waves,* and they are important for weather patterns, including the formation of cyclones. The planetary Rossby waves play a fundamental role in large-scale ocean and atmosphere dynamics and long-term ocean climate. These waves take only a few months to cross an ocean basin in the tropics, but years to decades at higher latitudes in the ocean.

rotation   The mathematical quantity (an antisymmetric rank-2, 3-dimensional tensor) that determines the angular velocity with which neighboring particles of a continuous medium revolve around a given particle (*see also* acceleration, expansion, kinematical invariants, shear and *see* tidal forces). All astronomical objects for which *rotation* can be measured (like stars, planets, galaxies) do rotate. The origin of their rotation is easily understood: The net angular momentum of all matter that eventually entered the object, with respect to the current position of the axis of rotation, was nonzero. Only planetoids and moons with no fluid core rotate like rigid bodies. Gaseous bodies (like the sun or Jupiter) and planets with hot fluid cores (like the Earth) display complicated patterns of local rotation. *See* rotation of the universe.

rotational discontinuity   *See* hydromagnetic wave.

rotation curve   In astronomy, a curve that describes the rotational velocity as a function of radius, typically of gas or stars in a disk galaxy. The rotation curve is obtained from the radial velocity measured on absorption or emission lines detected along a galaxy's apparent major axis, and corrected for the inclination of the galaxy. A rotation curve, in a diagram where velocity in km s^{-1} is plotted against angular or linear distance, is usually made up of a linear segment, where velocity increases linearly from zero with radius; a turning point, where the curve flattens; and a long, nearly flat or slowly decreasing segment. The so-called Keplerian trace, or a part of it, in which the orbital velocity decreases with

distance according to Kepler's third law, after all the matter of a galaxy is left behind, is almost never observed. The maximum rotational velocity depends on the galaxy's morphological type; typical values are $\approx$ 200 to 300 km s^{-1}. *See* long slit spectroscopy.

rotation of the universe   The rotation tensor (*see* rotation) in our actual universe. In principle, rotation could be detected and even measured on the basis of observations of distant galaxies, but in practice the transverse motions necessary to detect rotation in any reasonable rotating model of the universes are far too small to measure.

r process   The process of rapid neutron capture that is responsible for production of the neutron-rich isotopes of elements beyond iron (roughly germanium to uranium, and probably on to at least a few of the transuranics that no longer exist in the solar system). The "r" means that successive neutron captures occur more rapidly than beta decays back to the most stable nuclide at a given mass number. Therefore, since the beta-decay process depends in part on the total mass of the nucleus, the *r-process* of nucleosynthesis produces elements different than the *s*-process where neutron capture is slow with sufficient time for beta-decay between successive bombardments. The process almost certainly occurs in Type II supernovae, where both neutrons and iron peak nuclides are copiously available, but some details of how the products are ejected remain to be worked out. *See* supernova.

RR Lyrae star   Stars on the horizontal branch whose brightness varies due to radial pulsation of the star. They fall within the instability strip in the H-R diagram, have periods of about half a day, brightnesses of about 40 times that of the sun, and are found both in globular clusters and in the halos of our own and other galaxies. Because they are all about the same brightness, they are useful distance indicators.

RS Canum Venaticorum stars   A class of close binary stars in which the orbital period is from 1 to 14 days, where both stars are of roughly solar type and mass, but slightly evolved, the more massive component is an F, G, or K main-sequence or sub-giant star, and there is a significant amount of chromospheric activity. Chromospheric activity refers to the degree of magnetic field related phenomena near the stellar surface, e.g., star spots and stellar flares. This combination, especially the interaction between the two winds, heats material to 10^7 K or more, so that RS CVn's are often observed to have powerful flares which have durations of a fraction of an hour up to a few days, radiate 10^{33} to 10^{38} ergs, and emit light from radio waves through X-rays. Thus, the energy for these flares is believed to arise from the reconnection of tangled magnetic fields into simpler geometries. This large-scale reorganization releases a fraction of the energy stored within the magnetic field.

runoff channels   One channel type considered to indicate fluvial activity on Mars. *Runoff channels* are pervasive throughout the plateau material of the southern hemisphere old cratered terrain, where they form immature drainage networks around large craters and hilly mountainous regions. They are mainly located in low to middle latitudes. The apparent "loss" at high latitudes has been explained as an observational artifact owing to lower resolution imagery; a consequence of terrain softening; or as a consequence of polar wandering.

Runoff channels are also present on the flanks of Ceraunius and Hecates Tholus, and Alba Patera, volcanoes in the northern hemisphere. On Alba Patera they are highly integrated, and have a greater drainage density than those in the plateau material. On Ceraunius and Hecates Tholus their morphology is intermediate between those found in the plateau material and those found on Alba Patera. The northern hemisphere runoff channels are younger than those in the plateau material.

runup   The vertical excursion of water that results when a wave strikes a beach or structure.

RV Tauri stars   Highly evolved, asymptotic giant branch (probably) stars of relatively low mass which are so extended and bright that successive cycles of their pulsation interfere with each other, so they exhibit brightness minimum of alternating large and small amplitudes.

Rydberg constant (R_∞)   The fundamental constant that appears in the equation for the energy levels of hydrogen-like atoms; i.e., $E_n = hcR_\infty Z^2 \mu/n^2$, where h is Planck's constant, c is the speed of light, Z is the atomic number, μ is the reduced mass of nucleus and electron, m is the mass of the electron, and n is the principal quantum number ($n = 1, 2, \cdots$).

$$R_\infty = me^4/4\pi c\hbar^3$$
$$= 109737.312 \text{ cm}^{-1}.$$

S

Sagittarius A* Strong radio source located in Sagittarius and apparently located at the center of the Galaxy at RA $17^h45^m.6$, dec $-28°56'$ (at galactic longitude 0, latitude 0). Measurements of stellar velocity profiles near the center of the object indicate the presence of an object of mass approximately $2.6 \times 10^6 M\odot$ in a sphere of radius 0.15 pc. This implies the existence of a massive black hole at the center of our galaxy, at about 8 kpc.

saline contraction coefficient Coefficient β expressing the relative change of density due to the change in salinity at constant potential temperature Θ and pressure p; i.e., $\beta = \rho^{-1}(\partial\rho/\partial S)_{\Theta,p}$. β does not vary strongly for fresh-water to marine saline to hypersaline water (~ 0.7 to 0.85). For actual values *see* Algorithms for computation of fundamental properties of seawater, *UNESCO Tech. Pap. Mar. Sci.,* 44, 53 pp.

salinity The total fractional amount of dissolved material in seawater defined as the total amount of solid materials in grams contained in 1 kg of seawater when all the carbonate has been converted to oxide, the bromine and iodine replaced by chlorine, and all organic matter completely oxidized. The average salinity of seawater varies from about 3.5 to 3.7% by weight. There is an observed constant ratio of chlorinity to salinity: salinity in parts per thousand = (1.80655)(chlorinity in parts per thousand). *See* chlorinity.

salinity inversion *See* density inversion.

saltation Eolian or fluvial transport of small material by the bouncing of that material along a surface. Saltation is particularly effective for the transport of sand-sized material (1/16 to 2 mm in diameter).

salt fingering *See* double diffusion.

salt wedge A horizontal tongue of saline water located beneath a layer of fresh water within an estuary.

sand budget A description of all of the sand within a littoral system and a description of where it is going.

sand bypassing A process, either natural or as a result of man's activities, whereby sand is transported past a tidal inlet or entrance channel and reaches the beach on the other side.

sand spit A long, thin deposit of sand, deposited by longshore sediment transport. Typically located at the downdrift end of an island.

sand trap A trench dredged to capture sand. Such a trap is often used to facilitate sand bypassing. The trap is dredged near the updrift side of a channel or harbor entrance, captures sand, and the sand is later dredged out and placed on the downdrift beach.

Sargasso Sea Area in the north Atlantic Ocean between the West Indies and the Azores, at about 30 N lat. This relatively still sea is bounded by the North Atlantic, Canary, North Equatorial, and Antilles Currents, and the Gulf Stream. Named for the large amount of free floating seaweed (Sargassum) accumulated there.

saros 223 synodic (lunar) months = 19 eclipse years (almost exactly), about 18 years 11 days, a period of time over which the relative orientation of the moon and the sun repeat, so that lunar eclipses recur in similar patterns.

satellite altimetry A space-based ranging technique, which usually analyzes laser or radar pulses reflected off of the surface of a planet. When combined with precise orbit determination of the satellite, these rangings can be used to determine ocean, ice, and land surface topography. *See also* TOPEX/POSEIDON.

Satellite Laser Ranging (SLR) Determination of the distance between an artificial satellite with reflecting prisms and an observation station on the Earth's surface, by radiating laser beam

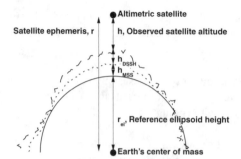

Satellite ephemeris, r

Altimetric satellite

h, Observed satellite altitude

h_{DSSH}

h_{MSS}

r_{el}, Reference ellipsoid height

Earth's center of mass

Fundamental measurements of satellite altimetry.

from the station, reflecting it from the prisms, and receiving the reflected laser beam. Since the location of the satellite can be determined independently, geodetic location of the station can be determined within an accuracy of several centimeters if the two-point distances are measured for at least three different locations of the satellite. Such observations are currently carried out at different stations in the world; one application is to measure the relative motion of tectonic plates. *Satellite laser ranging* is also used to provide information on Earth rotation and Earth orientation.

saturation   *See* current saturation.

saturation adiabat   A type of adiabatic process in movement of saturation moist air. Under the *saturation adiabatic* condition, when a saturated air parcel ascends, part of the vapor will condense to water or ice, and the air parcel will be still saturated. The latent heat released partly compensates the ascending adiabatic cooling, and the condensed water or ice remains in the saturated air parcel.

saturation adiabatic lapse rate   The adiabatic cooling rate of rising air parcel which is saturated, and in which condensation is taking place as it rises, so that the latent heat release moderates the adiabatic cooling, and causes the saturation adiabatic lapse rate to be less than the dry adiabatic lapse rate. The *saturation adiabatic lapse rate* is not a constant, and varies inversely with the temperature and somewhat with change of pressure. The difference between the saturation adiabatic lapse rate and dry adiabatic lapse rate is determined by the possibly released latent heat. Since the temperature at

higher level atmosphere is lower, the saturation adiabatic lapse rate has a smaller value at low level and larger value at high level which can approach or equal the dry adiabatic lapse rate.

saturation vapor pressure   The pressure exerted by the water vapor which is equilibrium with its condensed phase, which may be water or ice, the moist surface (or ice surface) saturation vapor pressure. In this state the same number of water molecules leave the water (ice) into the air as move from the air into the water (ice). *Saturation vapor pressure* is only a function of temperature. Its formula for water surface is

$$E = E_0 \ 10^{7.45t/(t+235)}$$

where E is saturation vapor pressure, $E_0 = 6.1078$ hPa, is the saturation vapor pressure at $0°C$, t is vapor temperature($°C$).

The ice surface saturation vapor pressure is

$$E = E_0 \ 10^{9.5t/(t+265)} \ .$$

Saturn   The sixth planet from the sun. Named after the Roman elder god of agriculture, *Saturn* has a mass $M = 5.6841 \times 10^{29}$ g, and an equatorial radius of $R = 60,330$ km, giving it a mean density of 0.71 g cm^{-3} and a surface gravity of 1.08 that of Earth. Its rotational period is 10 h 14 min. The rotation axis has an obliquity of $26.7°$. The rapid rotation gives the planet an oblateness of 0.09. Saturn's orbit around the sun is characterized by a mean distance of 9.54 AU, an eccentricity of $e = 0.056$, and an orbital inclination of $i = 2.5°$. Its sidereal period is 29.46 years and its synodic period is 378 days. Its average albedo is 0.47 but its atmospheric temperature is determined by an internal heat source. At one bar the temperature is about 140 K and increases inward. Saturn is a "gas giant" being composed primarily of hydrogen and helium and is the least dense, most oblate planet in the solar system. Saturn has prominent rings composed of large particles and numerous moons including Pan, Atlas, Prometheus, Pandora, Epimetheus, Janus, Mimas, Enceladus, Tethys, Telesto, Calypso, Dione, Helene, Rhea, Titan, Hyperion, Iapetus, and Phoebe.

scalar   A function or physical property that depends only on the position and time at which it is evaluated. Examples are the value of pressure beneath the ocean as a function of position, or the value of the temperature as a function of position and time in a room. *See* vector, tensor.

scale factor   The function of time in the homogeneous isotropic cosmological models that determines the changes of distance between any two points of space as time passes. In the simplest such model, the one with flat Euclidean spaces of constant time, the metric is

$$ds^2 = -dt^2 + R^2(t)\left[dx^2 + dy^2 + dz^2\right]$$

where $R(t)$ is the scale factor. The distance between points $P_1 = (x_1, y_1, z_1)$ and $P_2 = (x_2, y_2, z_2)$ at any time t is thus

$$l = R(t)\left[(x_2 - x_1)^2 + (y_2 - y_1)^2 + (z_1 - z_1)^2\right]^{1/2},$$

a function of time. The dependence of R on time is obtained by substituting the metric, and a matter source into the Einstein equations. For many sources, $R(t)$ vanishes at a particular time (conventionally $t = 0$), so the distance between any two points is zero at that time, the Big Bang. *See* Robertson–Walker cosmological models.

scaling solution (cosmic string and cosmic texture)   In theoretical cosmology, a state of evolution of a network of topological defects in which the energy in the structure (strings or textures) is radiated away as gravitational radiation or in other massless forms of energy. In such a situation, the energy in the defects maintains a constant ratio to the energy content in the radiation of the universe. Since energy density of the radiation eventually falls below that of ordinary matter, this guarantees that the defects do not dominate the universe at late times. *See* cosmic string, cosmic texture, cosmic topological defect, intercommutation (cosmic string), Goldstone boson.

scarps   A more or less straight or sinuous line of cliffs, seen on the Earth, the moon, or other terrestrial planets. The cliff may be the result of one or more processes including tectonic, volcanic, impact-related, or degradational processes.

scatterance   The fraction of the incident power at a prescribed wavelength that is scattered within a volume.

scattered disk, scattered disk object   The minor bodies in high eccentricity orbits in the ecliptic plane beyond Neptune. Objects in the scattered disk may be escapees from the Kuiper belt and/or may be scattered Uranus–Neptune planetesimals.

scatter-free shock acceleration   Dominant mechanism for particle acceleration at quasi-perpendicular magnetohydrodynamic shocks. The relative motion of the shock with respect to the plasma generates an electric induction field $E = -\mathbf{u}_u \times \mathbf{B}_u = -\mathbf{u}_d \times \mathbf{B}_d$ in the shock front, with $\mathbf{u}_i$ being the plasma flow speeds in the shock rest frame and $\mathbf{B}_i$ the magnetic fields in the upstream and downstream medium, respectively. The field is directed along the shock front, perpendicular to both the plasma flow and the shock normal. The shock-front is also associated with a jump in the magnetic field strength. Thus, gradient drift forces particles to move along the shock front. The drift direction depends on the charge of the particle and is always such that the particle gains energy. With increasing energy, the particle's velocity component perpendicular to the shock increases, too, eventually becoming larger than the shock speed. Then the particle escapes from the acceleration site. The details of the particle trajectory, in particular whether it traverses the shock or is reflected back into the downstream medium, strongly depend on the particle's initial energy and pitch-angle, leading to a characteristic angular distribution of the accelerated particles: an initially isotropic angular distribution of particles in both the upstream and downstream medium is converted to a very strong field-parallel beam in the upstream medium and to a smaller beam perpendicular to the field in the downstream medium. The energy gain depends on the time the particle can drift along the shock front. This in turn depends on the particle speed perpendicular to the shock: if this is small, the particle sticks to the shock; if it is large, the particle escapes. The energy gain can be estimated from the conservation of the magnetic moment across the shock. For an exactly perpendicular shock, the momentum perpendicular to both shock and magnetic field after the interaction between particle and shock is

$$\frac{p_{2\perp}}{p_{1\perp}} = \frac{B_d}{B_u} = r_B \ .$$

The normal component of the momentum remains unchanged. Thus, the change in momentum (and energy) is determined by the magnetic compression r_B, the ratio between the downstream and upstream magnetic field strength. For oblique shocks, the gain in momentum is smaller than in the above equation. A crude approximation for the energy gain here is $\Delta E \sim ou_u/\theta_{Bn}$ with u_u being the upstream plasma speed and θ_{Bn} as angle between magnetic field direction and shock normal. On average, the energy gain during one interaction between particle and shock is a factor between 1.5 and 5. Additional energy gain, for instance to accelerate MeV particles out of the solar wind plasma, requires repeated interactions between shock and particle and therefore sufficiently strong scattering in the upstream medium to reflect the particle back to the shock front. *See* diffusive shock acceleration.

scattering albedo   *See* albedo of single scattering.

scattering angle   The angle between the directions before and after scattering.

scattering coefficient   The limit of the ratio of the incident power at a prescribed wavelength that is scattered within a small volume to the distance of photon travel as that distance becomes vanishingly small [m^{-1}].

scattering cross-section   When a beam of radiation or particles with a uniform flux f of energy or particles (e.g., N particles per second per square centimeter) is incident on a scatterer, the amount scattered per unit time is: $f' = f A$, which defines the *scattering cross-section A*. Cross-section has units of area.

scattering efficiency factor   The ratio of the scattering cross-section to the geometrical cross-section of the particle.

Schmidt camera   Type of reflecting telescope invented by Bernhard Schmidt, which uses a thin correcting lens ("correcting plate") at the front of the telescope, which corrects the spherical aberration arising from the spherical primary mirror. This design leads to a large field of view (degrees across). Detection is usually made at prime focus in a *Schmidt camera*. *See* prime focus.

Schwarzschild black hole   In general relativity, a spherical, non-spinning black hole. *See* black hole, Schwarzschild metric.

Schwarzschild metric (Schwarzschild solution)   The unique metric of empty space-time outside an uncharged spherical source (*see* Birkhoff theorem), found by K. Schwarzschild in 1915.

In spherical coordinates (t, r, θ, ϕ), the line element takes the form (c is the speed of light and G is Newton's constant)

$$ds^2 = -\left(1 - \frac{2\,GM}{c^2\,r}\right) dt^2$$
$$+ \left(1 - \frac{2\,GM}{c^2\,r}\right)^{-1} dr^2$$
$$+ r^2 \left(d\theta^2 + \sin^2\theta \, d\phi^2\right)$$

where M is the mass parameter of the source. The radial coordinate r ranges from r_0 (corresponding to the surface of the source) to $+\infty$. Further, for $r_0 = $ constant to be a sensible (that is time-like) trajectory, one needs $r_0 > R_G$ (region I), where $R_G \equiv 2\,GM/c^2$ is the so-called gravitational radius, or Schwarzschild radius.

If one assumes that the space-time is vacuum everywhere, the above form is valid for any values of r from 0 to $+\infty$ and one encounters a singularity at $r = 0$, a real singularity where the curvature tensor diverges. The value R_G instead is a coordinate singularity (the curvature tensor is smooth and finite at $r = R_G$). The properties of the Schwarzschild solution led to the introduction of the notion of a black hole. The surface $r = R_G, t = +\infty$ is the future event horizon which screens the Schwarzschild black hole of mass M. The area of the horizon is

$$\mathcal{A} = \frac{16\pi\,G^2 M^2}{c^4} \ ,$$

and the surface gravity is

$$\kappa = \frac{c^2}{4GM}.$$

For $r < R_G$ (region of type II), r becomes a time coordinate. This means that inside the horizon everything evolves towards the central singularity, which occurs at time $r = 0$, and where tidal forces grow so strong so as to disintegrate any kind of matter.

The *Schwarzschild metric* can be analytically extended by transforming to Kruskal/Szekeres coordinates so as to cover a larger manifold which contains, besides the black hole with its future event horizon and future singularity, a white hole with its past event horizon and another copy of the asymptotically flat region. Early applications of the geodesics in the Schwarschild solution included the calculation of the perihelion shift of Mercury's orbit, which was the first observational test of general relativity. The geodesics were also used to calculate the angle of light deflection and the relativistic time-delay in the sun's vicinity. *See* ADM mass, asymptotic flatness, black hole, black hole horizon, future/past event horizon, Kruskal extension, real singularity, surface gravity, trapped surface, white hole.

Schwarzschild solution *See* Schwarzschild metric.

Schwarzschild spacetime *See* Schwarzschild metric.

scintillation Random fluctuations in the amplitude, phase, and direction of arrival of an electromagnetic signal. Scattering from small irregularities in electron density anywhere along the signal path causes *scintillation,* which tends to decrease with increasing frequency. The S_4 scintillation index, where $S_4 = [(< I^2 > - < I >^2)/ < I >^2]^{1/2}$ and I is the signal intensity with $<>$ indicating time-averaging, is the most commonly used parameter for characterizing the intensity fluctuations. Scintillation occurring in the Earth's ionosphere is usually observed on a transionospheric satellite communication link. There are two terrestrial regions of intense scintillations: at high and low latitudes. At high latitudes, scintillation frequency and intensity are greatest in the auroral oval, although it is also strong over the polar caps. Geomagnetic storms may cause scintillation to increase at any phase of the solar cycle. At low latitudes, nighttime scintillation shows its greatest range of intensity, with both the quietest and most severe of conditions being observed. The source of the scintillation is the irregularities on the walls of spread-F bubbles and is most prevalent during solar maximum. It may cause peak-to-peak fluctuations on 1.5 GHz and can be of the order of 56 dB above the geomagnetic equator, and as high as 20 dB for hours at a time in the equatorial anomaly region. At mid-latitudes scintillation is less severe and is positively correlated with spread-F and, to a lesser degree, with sporadic E. Interstellar scintillation, arising from similar mechanisms on a much larger scale, is ubiquitous in radio observations of pulsars. *See* spread F.

Scorpius X-1 The brightest X-ray source in the sky, identified with the spectroscopic binary V818 Scorpii at $16^h 19^m 55^s$, $-15°38'2''$. V818 Scorpii has a magnitude range of 11.80 to 13.20 with period 0.787313 days. The X-ray source has variations with the same period, and quasiperiodic oscillations of approximately 165 sec.

scour The process by which flowing water removes sediment from a seafloor, lake, or river bed. Typically most pronounced where engineering structures such as piles or jetties meet the seafloor.

scri ($\mathcal{I}$, Script I) Null infinity in an asymptotically simple or weakly asymptotically simple space-time. *See* asymptotically simple space-time, weakly asymptotically simple space-time.

sea The ocean, or a smaller body of salt water. Also used to describe the locally generated wind waves.

seafloor magnetic stripes Alternating bands of magnetic polarization found in the seabed floor. The Earth's magnetic field reverses with a period of hundreds of thousands of years. Seafloor spreading from the central seafloors occurs at a rate of centimeters per year, consistent with seafloor magnetic field alternations on the scale of 5 to 50 km. A similar banding was found on Mars via the magnetometer/electron reflectometer instrument on board NASA's Mars Global Surveyor spacecraft. An area in the southern highlands near the Terra Cimmeria and Terra Sirenum regions, centered around 180° longitude from the equator to the pole was surveyed. The bands are oriented approximately east-west and are about 150 km wide and 1000 to 2000 km long, suggesting past active tectonics on Mars as well.

seafloor spreading   The process by which new seafloor crust is created by the volcanism at divergent boundaries. As buoyant basaltic magma moves up from the interior and erupts at the rift zone of divergent boundary, new crust is attached to the plates on either side of the rift zone (mid-ocean ridge). The continual upwelling of new magma pushes the plates apart to make room for the new crust, causing the seafloor to spread at a half rate of up to several centimeters per year to either side. The new oceanic crust will retain the orientation of the terrestrial magnetic field polarity which existed at the time of its solidification, hence study of the polarity reversals (which are mirror images on either side of the rift zone) which results in alternating directions of magnetization of the oceanic crust and can provide information about the rate at which the plates are diverging. The discovery of *seafloor spreading* (which says the ocean floors are moving) in the 1960s, combined with the earlier theory of continental drift (which says that the continents are moving), led to the development of the theory of plate tectonics to explain the geologic activity of the Earth.

sea level   There have been major variations of *sea level* on geological time scales. Low stands in sea level (50 m or more) are associated with glaciations (on time scales of 10,000 years). Major high stands of sea level, for example in the Cretaceous some 80 million years ago, are associated with periods of rapid plate tectonics when the large number of ocean ridges displaced ocean water causing high sea levels of 200 m or more. Many variations in sea level are not understood. Increase in sea level of or-

der 10 cm/century is attributed to recent anthropogenic increases.

seamount   A cone-shaped isolated swell with height difference of more than 1000 m from its ambient ocean bottom. A swell with height difference of less than 1000 m is referred to as a knoll, and the one with relatively large, flat top is called a guyot. Most *seamounts* originate from submarine volcanoes. There are many seamounts from the mid to western Pacific ocean. Recently, many cobalt-rich crusts were discovered on slopes of seamounts, worthy of notice as valuable submarine resources in the future.

seasonal thermocline   *See* thermocline.

sea stack   Remnant small islands within or near the surf zone, composed of weather rock that has not yet been eroded away.

seawall   A man-made wall, typically vertical, built to resist wave impact and erosion of the sediments that are situated landward.

SeaWiFS   The Sea-viewing Wide Field-of-view Sensor satellite launched in 1998 that measures radiance in wavelengths centered at 412, 443, 490, 510, 555, and 670 nm (all with 20 nm bandwidth) and 765 and 865 nm (both with 40 nm bandwidth).

Secchi depth   The depth at which a Secchi disk disappears from view as it is lowered in water [m].

Secchi disk   A disk of diameter 20 to 30 cm, used to assess water clarity. The disk is painted with a high-contrast painting scheme and lowered into the water, and the depth at which it is no longer visible is noted. Repeated measurements are often used to assess changes in water clarity.

second   Unit of time equal to 1/86400 of a day whose length was based on the rotation of the Earth (*see* mean solar time), and now, because of the nonuniformities in the former, is based on atomic time. The *second* is the unit of International Atomic Time and is included in the "International System of Units" (SI). Specifically, since 1968, the second has been defined as the duration of 9192631770 cycles of the radiation produced by the transition between the two hyperfine levels of the ground state of the cesium-133 atom at a temperature of 0 K. A leap second is occasionally added to Universal Time.

secondary craters   Impact craters distinguishable from primary craters by their morphology. Primary craters are often circular with a distinct raised rim, of approximately uniform height. *Secondary craters,* however, are generally elongate, have low irregular rims, and are more shallow than primary craters of corresponding size. Additionally, they tend to occur in clusters or crater chains at a distance of one crater diameter or more from the primary crater. They form when crater ejecta from the primary crater has sufficient velocity to form a secondary crater upon landing (thus, secondary craters always exist circumferentially to a primary crater).

secondary rainbow   The fainter rainbow that occasionally forms outside a primary rainbow, with its colors in reverse order from those of the primary (i.e., red on the inside, at about 52° from the center of the arc, to blue on the outside of the arc at about 54°.5 from the center). The *secondary rainbow* arises from multiple reflection paths of light through the water droplets creating the rainbow.

second fundamental form   Let S be a p-dimensional submanifold of the n-dimensional Riemannian manifold X. Let u be the tangent vector to S at the point x and v a differentiable vector. The *second fundamental form* k_x is the mapping $k_x : T_x S \times T_x S \rightarrow (T_x S)^\perp$ defined by

$$(u_x, v_x) \rightarrow k_x \equiv (\nabla_x v)^\perp_x$$

where $(\nabla_x v)^\perp_x$ is the component of the covariant derivative of v that is normal to S.

In string theory, one views the history of the string as a 2-dimensional world sheet, so the extrinsic curvature extends into the remaining dimensions of the spacetime. In general relativity, one often treats 3-spaces of constant time embedded in the full 4-dimensional spacetime. Then the second fundamental form is a vector

valued tensor in the 3-space, which takes values proportional to the unit normal of the 3-space.

sector boundary   In the solar wind, the area of demarcation between sectors, which are large-scale features distinguished by the predominant direction of the interplanetary magnetic field, toward or away from the sun.

sector structure   Polarity pattern of the solar magnetic field in the plane of ecliptic. The sector boundaries between the different magnetic field sectors are the neutral line on the source surface (*see* source surface) as it is carried outward by the solar wind. Thus, during solar minimum, when the neutral line does not deviate much from the heliographic equator, a pattern of two or four magnetic field sectors can be observed. During solar maximum conditions, on the other hand, the extremely wavy neutral line shows excursions to high latitudes and often crosses the heliographic equator. In turn, the *sector structure* carried outwards into space is more complex with a larger number of magnetic field sectors.

secular (long-period) resonances   In a gravitationally interacting system of three or more bodies such as our solar system, neither the longitude of perihelion nor that of ascending node is constant as a consequence of their mutual perturbations. These two longitudes slowly change with secular precession rates g_i and s_i ($i = 1$ for Mercury through $i = 8$ for Neptune). *Secular resonances* occur when a linear combination of the average precession frequency of longitude of perihelion of a small body $< \dot{\varpi} >$ and that of its longitude of ascending node $< \dot{\Omega} >$ matches a linear combination of the natural frequencies of the gravitationally interacting system. In the asteroid belt, the mean motion resonances are unstable and manifest themselves as the Kirkwood gaps.

secular variation of the geomagnetic field
Slow and small changes (with a time scale of decades) in the main field of the Earth. For instance, the dipole component of the field diminishing between 1840 and 1970 by about 5% per century, a rate which may have increased since then to 7% per century.

sedimentary   Referring to rock that has been formed by deposition and consolidation of pre-existing material. Divided into types: clastic — consilidation (typically under water) of fine "sand" particles; carbonate — including oolitic limestones, which are consolidated diatom shells; organic — coal, which formed from compressed organic material; and evaporite — rock salt, gypsum. Only sedimentary stones contain fossils, since they form in ways that will preserve the structure of organic material.

sedimentary basin   In geophysics, a region in which the Earth's surface has subsided and has been covered by sediments. One mechanism for subsidence is thermal subsidence: the lithosphere cools and thickens, and due to the increased density of the cooler rocks the lithosphere sinks. A typical thickness of a *sedimentary basin* is a few kilometers but some basins have a thickness of 10 km or more. Subsidence is often impeded by the rigidity of the elastic lithosphere and the resulting flexure of the lithosphere results in near circular or linear basins with a radius or width of about 200 km.

sediment budget   *See* sand budget.

sediment load   The amount of sediment being transported by moving water. Consists of bed load, suspended load, and wash load.

sediment sink   A place or process that results in sediment removal from a beach or littoral system. A submarine canyon might serve as a *sediment sink,* and sand mining provides another example.

sediment source   The opposite of a sand sink; a source of sand for the littoral system or beach. Cliffs that back a beach, or a river might serve as *sediment sources.*

sediment transport   The movement of sediment due to the action of wind or water.

seeing   The phenomenon of time dependent refraction in the Earth's atmosphere which moves stellar images around, and often breaks them into multiple subimages. For ground-based telescope, *seeing* is the dominant factor limiting resolution. Seeing arises principally in the air within 50 m of the telescope. Seeing is reported in seconds of arc, as the size of the apparent disk of a point source (star). The best Earth-based sites have seeing of the order of 2 arcseconds. Active optics in telescopes can reduce the effects of seeing by factors of up to 10, thereby allowing ground-based observations closer to the diffraction limit.

seiche   A stationary water wave usually caused by strong winds, changes in barometric pressure, or seismic activity. It is found in lakes, semi-enclosed bodies of water, and in areas of the open ocean. The period, T, of a *seiche* in an enclosed rectangular body of water is usually represented by the formula $\frac{2L}{\sqrt{gd}}$, where L is the length, d is the average depth of the body of water, and g is the acceleration of gravity. Typical periods range from a few minutes in small lakes to a few hours in gulfs.

seismic coupling factor   A parameter used almost exclusively for subduction zone thrust faults. It is defined as the ratio of the long-term average rate of fault slip taking place during subduction earthquakes to the rate of plate convergence for a given fault area.

seismic gap   First, a region that has a potential to cause large earthquakes, but where large earthquakes have not occurred for a long time (*seismic gap* of a first kind). Second, a region where seismicity of usual small earthquakes decreases drastically for some time (seismic gap of a second kind). The latter sometimes precedes a forthcoming large earthquake. Seismic gaps of the first and second kinds could be important clues for long-term prediction of large earthquakes.

seismic intensity   Numerical values which represent the intensity of earthquake motion at a point, divided into several classes. Seismic intensity is determined mainly based on degree of shaking that the human body feels, also referring to shaking of surrounding materials, degree of damage, and phenomena associated with the earthquake. Seismic intensity scales such as MM, MSK, and JMA seismic intensity scales, which are appropriate to each country, are used.

Since seismic intensity is determined promptly without using instruments, it has been widely used for prompt reports of earthquakes, giving rough information on shaking at different regions and magnitude of earthquakes. Recently, the Japan Meteorological Agency (JMA) developed a seismic intensity meter, taking account of period, displacement, velocity, and duration time of earthquake motion, and deployed seismic intensity meters at different places to provide automatic reporting of seismic intensity.

seismicity   Earthquake activity in a particular region.

seismic moment   A quantity related to magnitude of an earthquake, based on one of the two torques composing double couple in earthquake faulting. According to dislocation theory, the *seismic moment* is represented by the product μDS of rigidity of rocks composing source region (μ), the amount of average slip on a fault plane (D), and a fault area (S). Recently, the seismic moment for relatively large earthquakes ($Ms \geq 5.5$) has been made available within a short time wherever they occur throughout the world by extracting this information from broadband seismographic records.

seismic parameter   The ratio of the bulk modulus to the density within a planetary body. For the Earth (and any other planet where the speeds of seismic waves can be determined as a function of depth), the *seismic parameter* can be used together with the assumption of hydrostatic equilibrium to compute the density as a function of depth.

seismic wave   Longitudinal, transverse, or mixed body or surface elastic wave which propagates through the Earth generated by earthquakes, atmospheric and oceanic disturbances, and lunar activity, for instance intentionally by explosions or "thumpers" for the purpose of geophysical subsurface visualization. *Seismic waves* propagating the Earth's interior are called body waves (P-wave, S-wave), while those propagating only along the Earth's surface are called surface waves (Rayleigh wave, Love wave). Typical velocities of propagation are 1 to 5 km/sec. Seismic waves arrive in the order

of P-wave, S-wave, and surface waves on a seismogram. Surface waves are the waves that have large amplitude near the Earth's surface, which are generated by interference of many reflected and refracted body waves. Velocity structure of the Earth's interior, attenuation structure, and materials and their state in the Earth can be estimated through investigation of the propagation of seismic waves.

seismogram   The output of a seismometer during an earthquake. A record of the ground motions caused by the compressional and shear body waves and the surface waves.

seismograph   A recording accelerometer that records earth motions (as especially during earthquakes) and from which the intensity and timing of the ground vibrations can be determined.

seismology   The study of the motions of the solid Earth (earthquakes and motions initiated by impacts or explosions), and attempts to explain these motions, and to determine the interior structure of a body. It is the source of the most detailed information about planetary interiors. The science of *seismology* is based on the study of stress and strain within planetary materials. When brittle material can no longer withstand the stress/strain that is imposed on it, the material will fracture and the released energy, felt as an earthquake, propagates through the surroundings. Seismic energy is dispersed in two types of waves: body waves (which travel through the interior of the body) and surface waves (which travel only along the surface). Body waves can be either primary (P) waves, which are longitudinal waves (i.e., direction of particle motion and the direction of energy propagation are parallel), or secondary (S) waves, which are transverse waves (i.e., direction of particle motion and direction of energy propagation are perpendicular). S-waves cannot travel through liquids, and both P and S waves will vary in velocity depending on the characteristics of the material through which they pass. Thus, by having several seismic stations on the surface pick up the seismic waves from a distant earthquake, the paths of each of those seismic waves can be reconstructed and the structure of the body's interior can be determined. Surface waves are subdivided into several types, but all are concentrated along the surface rather than propagating through the body. Surface waves can cause much damage to structures on the surface of the planet but give little information about the body's interior structure.

seismometer   An instrument that measures ground motions.

seismotectonics   A research field to investigate the tectonics by means of earthquakes. *Seismotectonics* pursues the mutual relation between regional characteristics of earthquake generation and observed various geoscientific structure and phenomena. To be more concrete, the configuration of plate boundaries, relative plate motion, interplate coupling, intraplate deformations, and stresses are investigated based on data such as seismicity, fault parameters like fault plane solutions and seismic moment, stress drop, and rupture process.

selective heating   Solar process which, in impulsive flares, leads to an enrichment of ^{3}He and heavier ions such as Fe compared to coronal or solar wind abundances. *See* impulsive flare.

self-generated turbulence   *See* diffusive shock acceleration.

semiclassical gravity   The approximation in which gravitational degrees of freedom are assumed to be sufficiently well defined by classical general relativity or another classical theory of gravity, but sources (e.g., matter) are taken in terms of a quantum description of matter.

semi-diurnal tides   Tidal components that have a cycle of approximately one-half day. The largest amplitude tidal constituent, M_2, is semi-diurnal, and therefore the tidal motion in most of the world is dominated by two high waters and two low waters each tidal day. The tidal current is said to be semi-diurnal when there are two flood and two ebb periods each day. All tidal constituents with a subscript 2 are *semi-diurnal tides*.

semimajor axis   Half the major axis of an ellipse.

separator   A magnetic field line that connects any two null points of a magnetic configuration. The *separator* defines the intersection of two separatrix surfaces.

separatrix   In space plasmas, the magnetic surface that divides topologically distinct magnetic flux regions, for instance separating bundles of magnetic field lines with different connections, e.g., open and closed field lines.

sequence stratigraphy   A subdiscipline of stratigraphy (the study of stratified rocks, either volcanic or sedimentary). The phrase *sequence stratigraphy* is not uniquely defined but is generally understood to involve the division of rocks in a sedimentary system into distinct packages, which are then related to each other in time through observation of their spatial and lithological relationships and the application of some stratigraphic rules. In practice, this has been associated with the development of algorithms for interpreting seismic profiles of sedimentary basins ("seismic stratigraphy"), as it became clear that packets of rocks are laid down in similar fashions in different basins. These packets are related to different stages of local transgression and regression of the sea, which occurs across a broad range of frequencies. By inferring that local sea level is related to global sea level, and correlating sequences between different basins, one may derive a curve of global sea level with different "orders" of cycles from timescales of around 100 Ma (first order cycles, associated with supercontinent assembly and break-up) through tens of thousands of years (fourth order or higher, perhaps due to changes in global glaciation); however, the global correlation of higher order cycles is disputed and local subsidence may often be important.

Sérsic–Pastoriza (1965) galaxies   Galaxies that exhibit an anomalous luminosity profile near their center, and often structures due to unresolved, compact emission regions ("hot spots"), which are sites of intense star formation. Several *Sérsic–Pastoriza galaxies* have a nucleus showing a Seyfert-type or LINER spectrum, while other compact emission regions close to the nucleus show the spectrum of H II regions. Studies of Sérsic–Pastoriza galaxies have been aimed at clarifying a possible link between intense, localized, star formation, as in the hot spots, and the presence of non-stellar, nuclear activity.

seston   The total animate and inanimate particulate matter in natural waters.

setdown   The radiation stress associated with breaking water waves results in a setup, or superelevation of the mean water level, in the vicinity of the mean water line (where the water meets the shore), and a setdown, or drop in mean water level, in the vicinity of the wave breakpoint.

settling speed (velocity)   The speed at which a particular type of particle will fall through a column of still fluid (generally water).

settling tube   An experimental apparatus for measurement of settling velocity.

setup   The radiation stress associated with breaking water waves results in a setup, or superelevation of the mean water level, in the vicinity of the mean water line (where the water meets the shore). *See also* setdown.

Seyfert-1 galaxies   Seyfert galaxies showing two systems of emission lines, i.e., broad Balmer lines (and, in general, broad recombination lines, which include strong emission from singly ionized iron), and narrower forbidden lines in the optical and UV spectrum of their nuclei. The width of the Balmer lines can reach $30{,}000 \, \mathrm{km \, s^{-1}}$ at the line base, and it is typically several thousand $\mathrm{km \, s^{-1}}$ at half maximum. The width of the forbidden lines is usually restricted to several hundred $\mathrm{km \, s^{-1}}$ at half maximum. The presence of broad lines is a defining feature of "type-1" active galactic nuclei, a more general class that includes quasars and QSOs since they show emission line properties very similar to that of Seyfert-1 nuclei. Broad and narrow lines are emitted in different regions, called the "Broad Line Region" and the "Narrow Line Region", respectively. The 8th Edition of *A Cat-*alogue of Quasars and Active Nuclei* by M.-P. Véron–Cétty and P. Véron lists more than 1100 *Seyfert-1 galaxies* known as of early 1998.

Seyfert-2 galaxies   Seyfert galaxies showing permitted and forbidden emission lines with similar width, typically several hundred $\mathrm{km \, s^{-1}}$ in the optical and UV spectrum of their nuclei. In other words, *Seyfert-2 galaxies* do not show broad permitted lines as observed in Seyfert-1 galaxies. Seyfert-2 galaxies are of lower luminosity than Seyfert-1, and they are believed to be roughly two to three times more frequent than Seyfert-1 in a volume-limited sample. The discovery of broad Balmer lines in polarized light in the prototype Seyfert-2 galaxy NGC 1068, and in other nearby Seyfert-2 nuclei, has led to a model in which Seyfert-2 galaxies are actually Seyfert-1, whose Broad Line Region is obscured from view by a thick torus of molecular gas. Free electrons slightly above the torus should scatter the broad line photons into the line of sight, letting them be detected only in polarized light. The applicability of this model to most Seyfert-2 galaxies is as yet a subject of debate. Several works have pointed out that Seyfert-2 galaxies are a heterogeneous class, and that several Seyfert-2 galaxies could be genuinely different from Seyfert-1 galaxies, with features in their spectral energy distributions that could be related to intense star formation. The 8th Edition of *A Catalogue of Quasars and Active Nuclei* by M.-P. Véron–Cétty and P. Véron lists about 560 Seyfert-2 galaxies known as of early 1998.

Seyfert galaxies   Galaxies showing a bright, star-like nucleus, whose optical and UV spectrum shows prominent emission lines. *Seyfert galaxies* are typically identified by their optical spectrum, which shows emission lines of the Balmer series, along with strong forbidden lines such as the nebular lines of the oxygen twice ionized, and of singly ionized nitrogen. Singled out in 1943 by C. Seyfert as an independent class of galaxies, the importance of Seyfert galaxies was largely unappreciated until the discovery of quasars. Quasars and Seyfert galaxies are now considered part of the broader class of active galactic nuclei. In this view, Seyfert galaxies are radio-quiet, low luminosity active galactic nuclei, distinct from quasars since the host galaxy is visible and since their luminosity is lower. It is customary, albeit arbitrary, as there is no discontinuity between nearby active galactic nuclei and quasars, to distinguish between Seyferts and quasars, defining Seyfert nuclei as having luminosity lower than 10^{11} solar luminosity, the luminosity of a large galaxy.

sferic   The abbreviated term for an occurrence of atmospheric radio noise. *See* atmospheric noise.

shadow bands   Bright and dark bands that are observed running along the ground just before the onset of totality in a solar eclipse. They are thought to arise from the refraction of the light from a thin crescent of the sun behind the moon, varying due to turbulence only tens or hundreds of meters above the ground. They can have wavelengths of centimeters to close to 1 m, and move at rates of meters per second (determined by the rate of motion of the atmospheric inhomogeneities). Their contrast is typically low, about 1%.

shadow zone   Seismic rays passing the core of the Earth are refracted due to the rapid decrease in seismic wave speed in the outer core, such that direct P waves cannot be detected in a zone of about 105° to 140° epicentral distance. This zone is called the *shadow zone*.

shallow water   A relative measure of water depth used in the study of water waves. *Shallow water* is (arbitrarily) defined as a ratio of water depth to wavelength, h/L, of 1/25 (sometimes 1/10).

shallow water wave   A wave in water more shallow than 1/25 of a wavelength; in this case, the speed of propogation is $c = \sqrt{gh}$, where g is the acceleration of gravity and h is the water depth.

shatter cone   In geology, a distinctively striated conical rock structure ranging from a few centimeters to a few meters in length, along which fracturing has occurred, believed to have been formed by the passage of a shock wave following meteorite impact. The mechanism is postulated to be a focusing of the shock wave behind obstacles, concentrating its effect in conical regions.

shear   The mathematical quantity (a symmetric rank-2, 3-dimensional tensor) that determines the rate of change of shape of a given portion of a continuous medium surrounding a given particle after factoring out the overall change in volume (*see also* acceleration, expansion, kinematical invariants, rotation, tidal forces). Consider a fluid in which acceleration, expansion, and rotation all vanish while *shear* is nonzero. An observer could examine a parallelepiped whose one vertex is at his position and the other vertices are at neighboring particles of the medium. With time, this parallelepiped would change into another one whose volume is the same, but its shape is different (sheared) from the initial one. In real astronomical bodies (and in all fluid mediums in everyday life) shearing motion is typical.

shear modulus   Also called rigidity. The proportionality constant G between shear stress τ_{ij} and shear strain ε_{ij} for a linearly elastic medium: $\tau_{ij} = 2G\varepsilon_{ij}$. G is expressed in terms of Young's modulus E and Poisson's ratio v as

$$G = \frac{E}{2(1+v)} .$$

shear probe   *See* airfoil probe.

shear stress   A stress (force per unit area) exerted along a surface, as opposed to a stress normal to a surface. In viscous fluids the *shear stress* τ is $\tau = \mu(du/dz)$, where du/dz is a velocity gradient, u is the velocity in the x-direction, and μ is the viscosity of the fluid.

shear velocity   A measure of shear stress used in the study of sediment transport. Defined as $u_* = \sqrt{\tau/\rho}$, where τ is the shear stress on the bed, and ρ is density.

shear wave   A seismic body wave in which the displacement is perpendicular to the direction of propagation. *Shear waves* can propagate through solids but not through liquids or gases.

shear wave splitting When shear waves propagate in an elastic medium that is anisotropic in the normal plane of the common ray path, waves with different particle vibration directions have different speeds. As a result, waves propagating in the same direction but vibrating in different directions are separated and arrive at the same receiver at different times. This phenomenon is called *shear wave splitting.*

sheet pile A sheet, generally made of steel, that is driven into the ground to hold soil back.

shelf wave *See* continental shelf waves.

shepherding moon *See* shepherd satellite.

shepherd satellite A satellite that constrains the extent of a planetary ring in the radial direction through gravitational forces. Saturn's F-ring and the rings of Uranus and Neptune are kept very narrow by shepherd moons which orbit just inside and outside the rings and by their gravitation influence on the small particles of the rings. Their presence produces "wakes" and waves in the rings which tend to keep the ring particles together.

shergottite A class of achondritic stony meteorites that are geologically young, and have solidified from molten rock in an extraterrestrial setting, whose prototype was (seen to fall) found in Shergotty, India in 1865. *Shergottites* have distinct textures and mineralogies indicative of igneous processes. These meteorites have a higher volatile content than other igneous meteorites, and their composition shows evidence of a basaltic origin. In addition, shergottite may contain maskelynite, a plagioclase glass formed by preterrestrial impact. This shock-melted glass has gas abundances and isotope ratios very similar to gases in Mars' atmosphere. Because of this, it is now widely accepted that shergottites and other SNC meteorites were formed on Mars, thrown off, and later captured by the Earth.

shields diagram A graph for assessment of critical motion of sediment particles subjected to flow of water. A dimensionless plot of nondimensional shear stress vs. boundary Reynolds number.

shield volcano A large, broad, low-sloped volcanic cone built up by successive flows of low-viscosity (i.e., fluid) lavas. The flanks of *shield volcanos* often have slopes of no more than 2° to 3°. Lava is erupted from both the central caldera as well as from fissures along the flanks, but the eruptions are generally nonexplosive in nature. Shield volcanos are common features on many planets — examples include the Hawaiian volcanos on Earth, Olympus Mons on Mars, and Sif Mons on Venus.

shingle Coarse, weathered gravel on a beach.

shoal A bump in the floor of a sea, lake, or river that results in a local reduction in water depth. Also used as a verb in the discussion of water wave transformation.

shoaling The process of a water wave changing height due to changes in water depth.

shoaling coefficient A non-dimensional coefficient used to calculate the change in height of a water wave due to a change in water depth. Defined as $K_s = \sqrt{C_{g1}/C_{g2}}$, where K_s is the *shoaling coefficient,* and C_{g1} and C_{g2} are the wave group velocities at the two points of interest.

shock normal Direction perpendicular to the shock front. If the *shock normal* is known, in magnetohydrodynamics the angle θ_{Bn} between the magnetic field direction and the shock normal can be calculated. In magnetohydrodynamics, this angle is crucial for shock formation and particle acceleration and can be used to classify shocks in quasi-parallel and quasi-perpendicular and to determine the shock speed (*see* quasi-parallel shock, quasi-perpendicular shock). The direction **n** of the shock normal can be calculated from the magnetic field measurements using the coplanarity theorem (*see* coplanarity theorem).

$$n = \frac{(\mathbf{B_u} \times \mathbf{B_d}) \times (\mathbf{B_u} - \mathbf{B_d})}{|(\mathbf{B_u} \times \mathbf{B_d}) \times (\mathbf{B_u} - \mathbf{B_d})|} .$$

Here $\mathbf{B_u}$ and $\mathbf{B_d}$ are the magnetic field vectors in the upstream and downstream medium.

The shock speed then can be calculated from the mass conservation across the shock (*see* Rankine–Hugoniot relations) as

$$v_s = \frac{\varrho_d \mathbf{u}_d - \varrho_u \mathbf{u}_u}{\varrho_d - \varrho_u} \cdot \mathbf{n}$$

with ϱ being the plasma density, $\mathbf{u}$ being the plasma speed, and the indices u and d indicating the upstream and downstream medium, respectively.

shock spike   In magnetospheric physics, a short intensity increase, lasting for some 10 min, around the time of shock passage. *Shock spikes* are observed at quasi-perpendicular shocks only; they are a typical feature of shock drift acceleration.

shock wave   A *shock wave* is associated with a sudden change in properties of a continuous medium, in particular a sudden increase in gas-dynamic pressure. A shock forms when a propagation speed exceeds the typical signal speed in the medium: a body might move through a medium faster than the signal speed (e.g., a supersonic jet, traveling shock) or a supersonic flow might be slowed down at an obstacle (e.g., the solar wind at a planetary magnetosphere, standing shock). Another shock wave is the blast wave following an explosion. General characteristics of shock waves are: (1) the disturbance propagates faster than the signal speed, (2) at the shock front, the properties of the medium change abruptly, and (3) behind the shock, a transition back to the state of the undisturbed medium must occur. Here a reverse shock might form. The jump of properties at the shock front is described by the Rankine–Hugoniot relations. Shocks transport information; however, since their speed exceeds the signal speed no information can propagate ahead of the shock. The passage of a shock changes the properties of the medium irreversibly.

Shoemaker–Levy   A comet that impacted on the planet Jupiter after breaking into at least 21 major pieces. The impacts lasted from July 16 through July 22, 1994. The impacts created fireballs and raised plumes of up to 3000 km above the Jovian cloudtops. Remnants left on Jupiter for days included circular and arcate dark smudges.

shoreline   A loosely used term indicating the line separating water from land. This line actually moves back and forth as wave and tide conditions change. More precise definitions involve description of the mean water line, still water line, mean high tide line, etc.

short wave fadeout (SWF)   A solar flare is accompanied by an increase in X-ray emissions. The increased X-ray flux ionizes the dayside D region of the ionosphere leading to an increase in absorption of radio waves passing through it. A fadeout occurs when a radio frequency becomes too weak to use as a result of this absorption. The event is called a short-wave fade because all short-wave frequencies passing through the daytime ionosphere will be affected. In the past, when communications depended more heavily on HF radio, this meant that all short-wave HF radio transmissions could simultaneously cease to be heard. Most M-class and all X-class flares will cause short-wave fades. The increased ionization leads to a number of other effects. *See* magnetic crochet, solar flare, sudden frequency deviation (SFD).

side-cast dredge   A dredge that discharges dredged material off to one side of the dredge centerline. Often used to dredge navigation channels.

sidereal day   The interval of time required for a given right ascension to return to a particular observer's meridian; $23^h 56^m 4^s$.

sidereal month   *See* month.

sidereal time   Time measured in terms of the rotation of the Earth with respect to the stars. Because the Earth's sense of rotation is in the same direction as its revolution around the sun, it must rotate through more than $360°$ from noon to noon. Hence, there is one more revolution with respect to the stars, than with respect to the sun, in a given year. *Sidereal time* equals the right ascension of objects on the observer's meridian. Equivalently it is the hour angle of the autumnal equinox. Sidereal time includes the same nonuniformities as are in apparent solar time, as well as a secular term due to precession of the equinoxes. Mean sidereal time is referred to the mean (not true) equinox and, hence, does not include the effects of nutation.

sidereal year   *See* year.

siderophile   Element that displays a strong affinity to associate with iron. Such elements are concentrated in iron meteorites and probably in the iron cores of the terrestrial planets.

sieve analysis   A method for determination of sediment grain size distributions. Sediment is dried in an oven, placed into a stack of increasingly finer sieves, and the entire assembly is shaken. Sediment falls through the stack of sieves, and is sorted by size. Each sieve is weighed before and after the test to determine percent finer by weight.

Sievert   The SI unit of radiation dose equivalent, equal to an absorbed energy of 1 Joule per kilogram.

sigma (σ) coordinate   In atmospheric physics, a nondimensional pressure coordinate. In atmospheric and oceanographic numerical studies, the Earth's land surface and ocean bottom create problems in specifying the lower boundary condition in the conventional vertical coordinate system. The vertical coordinate z is replaced by σ so the coordinates are (x, y, σ), where $\sigma = p/p_s$ (p_s being the surface pressure) is a nondimensional, scaled, pressure coordinate which ranges from a value of unity at the earth's surface (regardless of the terrain height) to zero at the top of the atmosphere in meteorology. In oceanography $\sigma = z - \eta/(H + \eta)$ is a nondimensional, scaled, vertical coordinate which ranges from zero at the sea surface to -1 at the ocean bottom, where H is depth of the ocean, η is the sea surface height, and z is the upward coordinate variable.

signature   The pattern of signs in the diagonal form of metric of a space or spacetime. Euclidean space has plus signs, and the *signature* of 3d Euclidean space is $(+, +, +)$. Four-dimensional Euclidean space has signature

$(+, +, +, +)$, while 4-dimensional Minkowski space has signature $(-, +, +, +)$. For the purposes of doing tensor analysis, the overall sign is irrelevant, so Minkowski space equally has signature $(+, -, -, -)$. In this dictionary we always use $(-, +, +, +)$. Analysis must be carried out with a consistent use of signature and sign convention. *See* sign convention.

sign convention   Geometrical tensors formed from the basis, the metric, and the connection require arbitrary choices of defining signs, involving at least the signature of the space (or spacetime), the sign of the connection coefficient, and the sign of the curvature tensor. A consistent set of such choices is a *sign convention*. This dictionary uses the sign convention of Misner, Thorne, and Wheeler, *Gravitation* (Freeman, 1973).

significant figures   A decimal number can have an arbitrarily long representation. In physics, the number of digits in a decimal number is adjusted to indicate the accuracy with which the number is believed known. For instance, 3 indicates a number known to be between 2.5 and 3.5, while 3.000 indicates a number between 2.995 and 3.005. For numbers not close to unity, the power-of-ten notation allows the indication of the number of *significant figures;* for instance, 3×10^4 indicates a number known to be between 2.5×10^4 and 3.5×10^4, while 3.000×10^4 indicates a number between 2.995×10^4 and 3.005×10^4.

significant wave height (characteristic wave height)   The average height of the one-third highest waves of a given wave group.

significant wave period   The average of the highest third of all measured wave periods. A collection of N waves are measured and sorted by height. Then the largest $N/3$ wave periods are averaged to determine the *significant wave period.*

silent earthquake   *See* slow earthquake.

silent universe   In relativity, any dust spacetime whose velocity vector field **u** is irrotational and such that the Weyl tensor is purely electric-type with respect to the velocity field **u**. Such a spacetime describes the evolution of an expanding or collapsing distribution of dust in which there is no exchange of information between different dust elements, either by sound waves or gravitational waves.

silicon burning   The set of nuclear reactions that converts silicon (and adjacent elements) to iron-peak elements (especially ^{56}Ni which later decays to ^{56}Fe). It occurs very near the end of the lives of stars of more than about 10 solar masses. The reaction chain begins with the photodisintegration of some silicon nuclei to alpha particles (helium nuclei), which are then captured by other silicon nuclei to build up from 28 to 32, 36, 40, 44, 48, 52, and 56-particle nuclei of various elements.

silt   Sediment in the size range between 0.0039 and 0.0626 mm.

Simon moments (1984)   Multipole moments of electrically charged, stationary, and axisymmetric isolated sources in the theory of general relativity. They form two infinite series, the gravitational and electromagnetic complex moments. The real parts of the two series begin with the mass and the electric charge and exist in the absence of currents. The imaginary parts are the current moments. *See* Geroch–Hansen moments.

simple harmonic motion   Motion that is described by a displacement proportional to a sinusoidal function of time:

$$x = x_0 \sin(\omega t + \delta) ,$$

where x is a displacement, x_0 is a constant parameter (the maximum of $|x|$), ω is an angular frequency, t is the time, and δ is a phase offset. The paradigm is 1-dimensional motion of a mass m on a perfect spring of force constant k (force per unit extension), in which case the frequency ω is

$$\omega = \sqrt{(k/m)} .$$

simple harmonic wave   A monochromatic, sinusoidal wave.

simple pendulum   A compact mass m on an essentially massless string of length L, in a gravitational field of acceleration g, constrained to move in a plane. For small oscillations, the motion is simple harmonic with angular frequency

$$\omega = \sqrt{(g/L)} .$$

See simple harmonic motion.

simultaneity, in Newtonian mechanics   *See* Newtonian simultaneity.

sine-Gordon soliton   The nonlinear equation $\partial_\mu \partial^\mu \alpha + m^2 \sin \alpha = 0$, $\mu = \{t, x, y, z\}$. (Summation on double indices assumed.) The sine-Gordon equation is mischievously named after the Klein–Gordon equation, which it resembles, with the mass term replaced with a sine term. This equation appears in many quantum field theories, the function α usually having the meaning of a phase, hence the invariance of the equation under transformations where

$$\alpha \to \alpha + 2\pi .$$

The sine-Gordon equation admits solutions where this phase rapidly varies from 0 to 2π in the form

$$\alpha = 4 \tan^{-1}[\exp(mx)] .$$

For these solutions, called solitons, the energy is very localized around $x = 0$, over a distance scale given by the Compton wavelength m^{-1}. Seen in three-dimensional space, such a solution would appear as a domain wall. *See* axionic string, domain wall, hybrid topological defect, soliton, summation convention.

single scattering albedo   For the scattering of light by a single particle, the ratio of scattered light to incident light.

single scattering phase function   For the scattering of light by a particle, the fraction of light scattered into a given solid angle.

singularities   In general relativity and in other field theories, subsets of the spacetime on which some quantities (often curvature and matter-density) become infinite. Apparent *singularities* may be caused by the choice of the coordinate system; these are removed by a coordinate transformation. Examples of such apparent singularities occur at the poles on the Earth. At every other point on the surface of the Earth, the geographical longitude can be determined. However, at each pole the meridians converge to a point and the distance between points on two different meridians tends to zero. A coordinate system nonsingular at the pole can be easily constructed, for example, by placing the origin of a 2-d rectangular map at the pole. Genuine singularities signal the breakdown of the theory used to describe the process; a physical system emerging from a singularity has the initial values of some of its parameters undetermined, and some aspects of evolution of physical systems will be unpredictable. Consequently, for instance, the existence of the Big Bang implies that general relativity cannot completely predict the state of matter at very high densities. It is expected that the adequate theory to describe the universe in the vicinity of the Big Bang will be quantum gravity, a theory that combines classical relativity and quantum mechanics. In quantum gravity, the expectation goes, the state of infinitely large density would be replaced with a very high, but finite density, through which the evolution of the universe can be calculated. Localized singularities are often discussed in the context of the cosmic censorship hypothesis, at present a vague conjecture that states that every generic singularity is hidden in a black hole and so no signal from it can propagate outside the surface of the black hole. *See* cosmic censorship.

singularity theorems   A set of precise mathematical arguments that prove that a universe will contain a singularity in the past or future if a number of specific assumptions about its structure are true.

sinistral fault   Another term for a left-lateral fault.

Sinope   Moon of Jupiter, also designated JIX. Discovered by S. Nicholson in 1914, its orbit has an eccentricity of 0.275, an inclination of 153°, and a semimajor axis of 2.37×10^7 km. Its radius is approximately 18 km, its mass 7.76×10^{16} kg, and its density 3.2 g cm^{-3}. Its geometric albedo has not been well determined,

and it orbits Jupiter (retrograde) once every 758 Earth days.

sinusoidal wave   A wave that has a profile defined by a simple sinusoid. Corresponds to linear or Airy wave theory.

Sirius   -1.46 magnitude star of spectral type A1 at RA$06^h04^m08.9^s$, dec $-16°42'58''$.

SI (Systeme International)   The system of units based on the meter, kilogram, and second.

skewness   A measure of slant or preference for one side. The third moment of a statistical distribution about the mean.

skip fading   When the operating frequency is close to the MUF and above the maximum observed overhead critical frequency, then *skip fading* (sometimes called MUF fading) may occur. Changes in the ionosphere may alter the MUF, taking it below the operating frequency resulting in a sharp drop in signal strength. It is possible that traveling ionospheric disturbances could cause skip fading, although the most common time it is observed is near dawn and dusk, when the MUF is changing more regularly. The distance from the receiver to the point where the transmitted frequency can first be observed is the skip distance. *See* skip zone.

skip zone   If the operating frequency of an HF transmitter is higher than the highest frequency that can be reflected from the overhead ionosphere, then there will be a region about the transmitter where signals cannot be received. This is called the *skip zone*. As the elevation angle for the radio waves drops from overhead, the obliquity factor increases until a point is reached when propagation is just possible. The distance from this point to the transmitter is called the skip distance. It is the minimum distance from the transmitter for which a sky wave will return to Earth when the operating frequency exceeds the vertical incidence critical frequency. The only way to reduce the skip distance is to lower the operating frequency. *See* ionospheric radio propagation path.

sky The apparent dome over the Earth seen by an observer on the ground. Psychologically the *sky* is not a hemisphere but appears flatter so that the overhead part is not perceived as being as far away as the horizon. The sky is blue in daytime because of scattered blue light from the sun; the sun and moon appear to lie in the surface of the sky. It is black at night, with bright stars, planets, occasional comets, and meteors visible. The clouds inhabit the sky and are sometimes said to obscure the sky.

Skylab The *Skylab* space station was launched May 14, 1973, from the NASA Kennedy Space Center. Skylab carried several solar experiments including a soft X-ray telescope which produced images of the sun in X-rays with wavelengths from 6 to 49 Å. Skylab reentered the atmosphere and crashed over Australia on July 11, 1979.

sky-wave propagation Describes radio waves reflected by the ionosphere in traveling between ground locations. While in the ionosphere, the radio wave experiences dispersion and changes in polarization that are controlled by the Earth's magnetic field, the ionospheric electron density and collisions (described by the collision frequency) between the electrons and the neutral atmosphere. *See* ionospheric radio propagation path.

slab model Model assumption about the properties of interplanetary magnetic field fluctuations and their importance for particle scattering. In the *slab model,* only waves with wave vectors parallel to the magnetic field and axially symmetric transverse fluctuating components are considered. Then the magnetic field power spectrum $f(k_\|)$ of these fluctuations can be described by a power law

$$f\left(k_\|\right) = C \cdot k_\|^{-q}$$

with $k_\|$ being the wave number parallel to the field, q the slope of the spectrum, and C the power at a certain frequency. The pitch-angle diffusion coefficient $\kappa(\mu)$ is then related to the spectrum by

$$\kappa(\mu) = A\left(1 - \mu^2\right)|\mu|^{q-1}$$

with μ being the pitch angle and A being a constant related to the level C of the turbulence.

From the above pitch-angle diffusion coefficient a particle mean free path can be obtained as

$$\lambda = \frac{3}{8}v \int\limits_{-1}^{+1} \frac{\left(1 - \mu^2\right)^2}{\kappa(\mu)} \, d\mu$$

with v being the particle speed. Note that here the mean free path is not the average distance between two successive pitch-angle scatterings but the average distance traveled by a particle before its pitch-angle has been changed by $90°$, i.e., the direction of motion has been reversed. Thus, for the overall motion, the λ obtained from the above relation has a meaning comparable to the mean free path in spatial diffusion.

The method is falling from favor because mean free paths obtained from the magnetic field fluctuations under the assumption of the slab model in general are markedly smaller than the mean free paths obtained from fits of a transport model on the observed intensity and anisotropy time profiles (discrepancy problem).

slab penetration In geophysics, there are some cases that an oceanic plate subducted from a trench (slab) penetrates into the lower mantle deeper than 660 km without becoming stagnant in the upper mantle. The phenomenon is referred to as *slab penetration.* The existence of subducting slabs can be identified from Wadati–Benioff zones in the upper mantle, whereas slab penetration into the lower mantle was previously not clear because deep earthquakes do not occur there. On and after the 1980s, it was shown that slabs penetrate into the lower mantle in some regions such as the Marianas subduction zone, through techniques of residual sphere and seismic tomography. On the other hand, there are some regions where slabs (like slabs beneath central Japan) are lying horizontally just above the 660-km seismic discontinuity. Such regionality of the fate of slabs (to penetrate or remain stagnant) is thought to be closely related to patterns of mantle convection.

slip system The combination of slip direction and slip plane for a given crystal system. Dislocation theory predicts that the slip direction and plane are associated with the Burgers vector and glide plane, respectively, of gliding dislocations. The correlated combination of the Burgers vector and the glide plane in a given crystal system is then defined as a *slip system.* In general, one would expect glide planes to be the closest-packed plane, or these planes have the lowest $\{hkl\}$ indices; the shorter of the possible perfect-dislocation Burgers vectors should correspond to the slip direction.

slough   A long, thin tidal estuary.

slow earthquake   An earthquake that emits seismic waves with much longer periods than usual earthquakes because rupture velocity on a fault plane is abnormally slow and rupture duration time is abnormally long. Rupture duration times for usual earthquakes are almost determined according to their magnitude. For instance, rupture duration time for a great earthquake with a magnitude of 8 class is tens of several seconds. Sometimes rupture duration time for *slow earthquakes* which occur below ocean bottoms becomes more than several minutes, leading to a tsunami (tidal wave). Furthermore, when a slow earthquake ruptures spending more than tens of several minutes, its seismic waves cannot be recorded even using long-period seismometers. In such a case, no tsunami is generated. Such an earthquake is called a silent earthquake, for which detection efforts are being made, using high-sensitivity extensometers.

slow magnetohydrodynamic shock   A *slow magnetohydrodynamic shock* results from the steepening of slow MHD waves. The magnetic field decreases from the upstream to the downstream medium and the field is bent toward the shock normal because the field's normal component stays constant. The upstream flow speed exceeds the sound speed but is smaller than the Alfvén speed. Thus far, in the interplanetary medium less than a handful of slow magnetohydrodynamic shocks have been observed. In the corona, slow MHD shocks might be more common because both the Alfvén speed and the sound speed are higher. *See* fast magnetohydrodynamic shock.

slow shock wave   *See* hydromagnetic shock wave.

slow solar wind   The properties of the *slow solar wind* are highly variable. Often large-scale structures such as magnetic clouds or shocks are embedded in the slow stream. Plasma speeds range from 250 km/s to 400 km/s, proton temperatures are about 3×10^4 K, electron temperatures are twice as much, and densities at 1 AU are about 8 ions/cm^3. The helium content is highly variable, averaging 2%. Despite the differences, momentum flux and total energy flux on average are similar in fast and slow solar wind streams. Two possible sources of the slow solar wind have been suggested: the slow wind might stem from the closed magnetic field regions in the streamer belt or it might be over-expanded, and therefore slowed down, solar wind from the outer skirts of the fast solar wind stemming from the coronal holes.

small amplitude wave   A sinusoidal water wave. Corresponds to linear or Airy wave theory. Linear wave theory derivation involves assumption of a small wave amplitude.

Small Magellanic Cloud (SMC,NGC 292)
An irregular galaxy in the southern constellation Tucana at right ascension $0^h 50^m$, declination $-73°$, at 65 kpc distance. The SMC has angular dimension of $280' \times 160'$, about 10 kpc. It has a positive radial velocity of -30 km/s (toward us). Both the Large Magellanic Cloud and the *Small Magellanic Cloud* orbit the Milky Way. *See* Large Magellanic Cloud.

small-scale turbulence   Nearly isotropic, eddy-like state of fluid random motions, where the inertial forces in the eddies are larger than the buoyancy and viscous forces (i.e., Reynolds and Froude number exceed critical values). The length scale of the small-scale turbulent motion is smaller than the Ozmidov-scale but at least an order of magnitude larger than the Kolmogorov scale.

SNC meteorites   A subclass of achondrites named after the first three examples found: *S*hergotty, *N*akhla, and *C*hassigny. The abundances of isotopes of trapped nitrogen and rare gases differ from those of other meteorites but are similar to abundances in the Martian atmosphere. As a result it is believed that these meteorites formed on Mars, and were launched into space by a large impact. *See* Martian meteorites.

Snell's law   The law that describes the refraction of light at an interface between two media that have different real indices of refraction: $n_1 \sin \theta_1 = n_2 \sin \theta_2$ where n_1, n_2 are the indices of refraction and θ_1 θ_2 are the angles to the normal of the interface.

snow   Frozen precipitation (small ice crystals, or "flakes") that formed in clouds by accretion onto accretion centers at temperatures below freezing, and fell to Earth without being melted. Hence, it preserves the hexagonal symmetry with which the ice crystal formed.

soft gamma repeaters   Fractional-second increases in the count rate of 20-keV photons from space that were first found with Russian gamma-ray burst instrumentation about 20 years ago. Unlike typical gamma ray bursts, these events repeat from the same sources in sporadic episodes, from several per year to several per day. The total number of known sources now totals only four, each localized at or near supernova remnants in the Large Magellanic Cloud or in our galaxy. Exhibiting many of the properties of hard X-ray burst events, such as the rate of change of a several-second period, the sources are clearly confirmed as neutron stars that are highly magnetized ("magnetars"). In addition, only twice in 30 years has a 1000-fold more intense soft gamma outburst been detected, originally confusing as to its possible gamma-ray burst identity (*see* March 5 event), but now thought to be a neutron star crustquake.

Soft X-ray Telescope (SXT)   A broad-band X-ray imager on board the Yohkoh spacecraft. The SXT provides images of the solar corona over the soft X-ray wavelength range, 3 to 60 Å, with a maximum spatial resolution of 4.9 arcsecs (pixel size = 2.45 arcsec) or 3500 km. The dynamic range of the SXT and the flexible exposure control enables it to image a wide range of solar phenomena from flares to quiescent coronal loops. SXT is sensitive to plasma at temperatures 1 to 2 million K.

solar abundance   The relative abundances of the elements in the early sun. When used in the context of lower temperature environments, where the atoms combine into molecules, *solar abundance* refers to the abundances of the atoms themselves, regardless of the molecules in which they are found.

solar activity   Solar activity can be described on several time scales. Two are particularly important: short time scales of days or less and long time scales of several years. The two time scales are linked because short-term *solar activity* varies over a period of 11 years, called the solar cycle. Short-term solar activity can encompass any change in the sun's appearance or behavior. The most commonly referred to change is the appearance, evolution, and disappearance of active regions and sunspots over periods of one to three solar rotations ($\sim$27 days when viewed from the Earth). Shorter time-scale transient perturbations of the solar photosphere, evolving in seconds and lasting for hours, are solar flares; dramatic visual forms of short-term solar activity. Changes in solar activity can directly affect the Earth. *See* solar cycle.

Solar and Heliospheric Observatory (SOHO)
A joint European Space Agency and NASA mission launched in December 1995 designed to study the internal structure of the sun, its extensive outer atmosphere and the origin of the solar wind, the stream of highly ionized gas that blows continuously outward through the solar system. SOHO consists of 12 separate instruments including helioseismology experiments, atmospheric imagers and spectrometers, white light and UV coronagraphs, and *in situ* particle detectors.

solar atmosphere   The *solar atmosphere* starts above the photosphere, the visible surface of the sun where most of its light is emitted, and consists of three layers: the chromosphere, the transition region, and the corona. The bottom layer of the solar atmosphere is the chromosphere, a thin (a few thousand kilometers thick) layer, which can be seen as a small reddish ring during a total eclipse. Its name, chromosphere, means colored sphere and stems from this reddish color, which results from emission in the $H\alpha$ line of hydrogen. The chromosphere shows highly variable structures, giving it the nickname "burning prairie". In the chromosphere, the density decreases by about three orders of magnitude while the temperature stays roughly constant. The next layer is the transition region, where the temperature increases by a factor of about 200, the density decreases by one order of magnitude and the collision time increases by more than 4 orders of magnitude — over a thin layer which is only a few hundred kilometers thick. The outer solar atmosphere, the corona, fills the entire heliosphere as solar wind. Aside from just above the transition region, temperature and density are roughly constant. The corona can be seen as a broad structured ring around the occulting disk of the moon during a total eclipse. These structures, in particular arcs, rays, or helmet streamers, reflect the magnetic field pattern in the corona, allowing the distinction between open and closed magnetic fields. Despite its high temperature, the corona does not radiate like a black body because its density is too low. Instead, the radiation is photospheric light scattered by the electrons. Thus, the light intensity also reflects the electron density in the corona. *See* chromosphere, corona, photosphere, transition region.

solar B-angle   The angle, relative to the center of the solar disk, between the north pole of the sun and the zenith, measured from north to west in degrees.

solar constant   The rate at which solar radiant energy at all wavelengths is received outside the atmosphere on a surface normal to the incident radiation, at the Earth's mean distance from the sun. The amount of variability is still a subject of debate, but is certainly very small (less than 1%) apart from the long-term development in the history of the sun. The symbol of solar constant is S. Unit of S is Wm^{-2} or $cal \cdot cm^{-2} \cdot min^{-1}$. Since 99.9% of the solar radiation energy is within the wave band 0.2 to 10.0 μm, the measurment of solar constant does not need to involve a wide wave band. In 1981, World

Meteorological Organization suggested the use of $S = 1367 \pm 7 \ W/m^2$ as the solar constant value, equivalent to $1.945 \ cal \cdot cm^{-2} \cdot min^{-1}$.

solar cycle   The *solar cycle,* the 11-year periodicity in the occurrence of sunspots, first was recognized in 1843 by H. Schwabe. Its features can best be seen in a butterfly diagram: at the beginning of a solar cycle a few sunspots start to appear at latitudes around 30°. These spots are relatively stable and can often be observed over a couple of solar rotations. The spots move toward the equator while at their original latitudes new spots appear. The number of sunspots increases until solar maximum. Afterwards, only a few new sunspots emerge while the spots at lower latitude dissolve. The total number of sunspots decreases until just after solar minimum when new sunspots begin to emerge at higher latitudes. The average duration of such a cycle is 11 years with variations between 7 and 15 years. Successive cycles have opposite magnetic polarity, so that the full cycle is actually 22 years. The maximum number of sunspots also differs from cycle to cycle by up to a factor of 4. Occasionally, the solar cycle, and with it the sunspots, can disappear nearly completely over time scales of some 10 years. The best documented case of such "missing sunspots" is the Maunder Minimum. Solar Cycle #23 began in October 1996. *See* butterfly diagram, Hale's polarity law, Maunder Minimum, Spoerer's law, sunspot, sunspot cycle.

solar day   The length of time from noon to noon; It varies from the mean solar day (24 h) as described by the equation of time.

solar disk   The visible surface of the sun projected against the sky.

solar dynamo   The process by which the interaction between magnetic field and plasma deep in the solar interior is thought to occur. The dynamo process results in the intensification of magnetic fields via the induction of plasmas trying to cross field lines. The action of the *solar dynamo* is thought to explain the existence of the solar cycle, the butterfly diagram, Spörer's Law, and the reversal of the sun's polar fields near sunspot maximum.

solar eclipse   The blocking of the solar disk due to the passage of the moon between the sun and the Earth. The moon casts a shadow on the Earth — or from the Earth, the sun is screened off by the moon. *Solar eclipses* can come in one of three forms: total, annular, or partial. A total solar eclipse occurs when the moon and sun are in perfect alignment and the relative distances are such that they have the same angular diameter in the sky. Because of the eccentricity of the Earth's orbit around the sun and in the orbit of the moon around the Earth, the angular diameter of the sun is sometimes less than, or sometimes greater than the angular diameter of the moon. The annular eclipse also occurs when the sun and moon are in perfect alignment but when the moon is more distant, so has a slightly smaller angular diameter than the sun creating a bright ring or annulus at the time of maximum eclipse. Partial eclipses are more common and refer to the occasions when the moon only partly covers the solar disk as seen from the viewpoint on Earth. Of these three, the total eclipse is by far the most spectacular, occurring on average once per year somewhere on Earth with durations ranging from 1 to 7 min. During a total solar eclipse, the corona and chromosphere become visible.

solar electromagnetic radiation   On average, the sun emits a total of 3.86×10^{23} kW integrated over its surface or 6.3×10^4 kW/m^2. This radiation can be divided into five frequency bands:

1. X-rays and extreme ultraviolet (EUV) with $\lambda < 1800$ Å contributes to about 10^{-3} of the total energy output and is emitted from the lower corona and the chromosphere. This hard radiation varies strongly with the solar cycle; during flares the emission can be enhanced by up to orders of magnitude.

2. Ultraviolet (1800 Å $< \lambda \leq 3500$ Å) contributes to about 9% of the total flux and is emitted from the photosphere and the corona. Variations with the solar cycle are similar to the ones in X-rays, although the amplitude is smaller.

3. Visible light (3500 Å $< \lambda < 7500$ Å) contributes to 40% of the total flux and is emitted from the photosphere. Variations with the solar cycle are small (less than 0.1%); only in ex-

tremely strong flares can a local brightening in visible light be observed (white-light flare).

4. Infrared (7400 Å $< \lambda < 10^7$ Å) contributes 51% to the energy flux. Like the visible light, it is emitted from the photosphere and shows only small variations with the solar cycle.

5. Radio-emission ($\lambda > 1$ mm) contributes only 10^{-10} to the total energy flux. It is emitted from the corona and can be enhanced significantly during solar flares.

solar flare   A violent re-organization of intense magnetic fields in and above the solar photosphere. The event is accompanied by an increase in X-ray emissions from the vicinity of the flare. The extent of the X-ray emission depends on the size of the flare. The magnitude, or class, of the flare can be indicated by the intensity of the X-ray emissions. The main classes of X-ray flare are: C class, between 10^{-6} and 10^{-5} watts meter^{-2}; M Class, between 10^{-5} and 10^{-4} watts meter^{-2}; and X class, greater than 10^{-4} watts meter^{-2}. *Solar flares* usually occur in active regions on the sun and are more common near solar maximum. *See* short wave fadeout.

solar flux unit   Unit of radio emission from the sun. 1 sfu $\equiv 10^{-22}$ Wm^{-2}.

solar granulation   Solar granules consisting of upwelling convective regions in the sun's photosphere, of order 1000 km in size, separated by dark, intergranular lanes, where cool material is descending. The typical lifetime of a granule is of order 5 to 10 min.

solar limb   The apparent edge of the sun as it is seen in the sky.

solar magnetic sectors   When solar plasma moves deep into the interplanetary space, the embedded magnetic field adopts a spiral structure due to the solar rotation. Magnetic fields in the low speed solar wind are more tightly wound than those in the high speed solar wind, which are more radially aligned. The angle that the magnetic field makes with the sun-Earth line at the Earth's orbit is about 45° in the solar equatorial plane. The strength of the magnetic field is of the order of a few nanotesla. The direction of the radial component of the magnetic field is either toward or away from the sun. The region in which such an orientation is maintained is called a sector. There are either away or toward sectors depending on the direction of the radial component of the magnetic field. At the transition from away to toward or vice versa, the magnitude of the magnetic field is fairly constant. Normally two or four sectors are observed during a solar rotation. The sectors are dependent on the solar latitude and disappear at higher latitudes where the Interplanetary Magnetic Field (IMF) has the same sign as the appropriate solar pole.

solar maximum   The time at which the solar cycle reaches its highest level as defined by the 12-month smoothed value of the sunspot number. During *solar maximum* there are more active regions and sunspots on the sun, as well as more solar flares leading to greater numbers of geomagnetic storms at the Earth. The most recent solar maximum occurred in July 1989. The next is expected to occur sometime in the year 2000. *See* solar cycle.

solar maximum mission (SMM)   The SMM spacecraft was launched on February 14, 1980 near the height of the solar cycle, to examine solar flares in more physically meaningful detail than ever before. SMM recorded its final data in November 1989.

solar minimum   The time at which the solar cycle reaches its lowest point as described by the 12-month smoothed value of the sunspot number. During *solar minimum,* there may be no sunspots or solar flares. The most recent minimum occurred around October 1996. *See* solar cycle.

solar nebula   The cloud of gas and dust out of which our solar system formed. About 4.6×10^9 years ago, a slowly rotating cloud of gas and dust began to collapse under its own gravitational influence. As the cloud collapsed, it began to form a flattened disk with a central bulge. The central bulge collapsed down to eventually form the sun. In the outer part of the disk, at least two other blobs of gas and dust collapsed down to form the planets Jupiter and Saturn. Elsewhere, small material condensed out of the cloud and began to accrete to form the other planets, most of the moons, and the small icy and rocky debris that became the comets and asteroids. This scenario helps to explain the counterclockwise orbital motion of all the planets, the counterclockwise orbital motion of most of the larger moons, and the counterclockwise rotation direction of most of the planets by proposing that the *solar nebula* was rotating in a counterclockwise direction.

Alternately, the disk of gas and dust that surrounded the newly formed sun at the time of solar system formation. The planets were formed from the material in the nebula. If we take the current masses of all the planets, and add enough hydrogen and helium to make the composition identical to that of the sun, the resulting mass is the mass of a minimum mass solar nebula.

solar neutrinos   Low-energy electron neutrinos released in nuclear reactions in the sun and detectable from Earth. Fewer are seen than predicted by standard physics and astrophysics. The reactions of the proton-proton chain produce neutrinos of energy less than 1 MeV at the deuterium-production stage and more energetic ones in connection with the reactions involving beryllium and boron. The first experiment, using Cl^{37} as a detector, was constructed by Raymond Davis in the Homestake Gold Mine in Lead, South Dakota beginning in 1968. By 1971 it was clear that only about a third of the expected neutrinos were being captured, but the experiment was sensitive only to the Be and B products, thus explanations focused on mechanisms that might cool the center of the sun and reduce production of these high energy particles without affecting the main reaction chain. A second experiment, at Kamioka, Japan, confirmed the deficiency of B^8 neutrinos starting in 1989, but also showed that the ones being seen were definitely coming from the direction of the sun. Two experiments of the 1990 (SAGE in the Caucasus Mountains and GALLEX under the Alps) look at the lower energy neutrinos from the main reaction chain. They are also deficient by a factor of about two. The full pattern is best accounted for if the solar model is the standard one, but electron neutrinos can change into mu or tau neutrinos en route to us. A possible mechanism, called MSW (for its inventors Mikheyev, Smirnov, and Wolfenstein), attributes the change to a catalytic effect of atomic nuclei, so that it happens only inside the sun and other stars, not in empty space.

solar P-angle   The heliographical latitude, in degrees, of the center of the solar disk as would be seen from the center of the Earth for a given date.

solar quiet current system   The daily variations in the geomagnetic field during quiet solar conditions of a few tens of nanoteslas are the result of the *solar quiet* (Sq) *current system* flowing in the E region of the ionosphere. This system is due to the electric field generated in the manner of a dynamo by tidal winds produced by the solar heating of the atmosphere at E region heights. A part of the electric field originating in the high latitude magnetosphere is also transferred to the E region by field aligned currents. The current system is concentrated in a band of a few hundred kilometers wide near the magnetic dip equator where it is called equatorial electrojet. The equatorial electrojet corresponds to an east-west electric field of a few tenths of mv/m and a vertical polarization field of 10 mv/m. The strength and pattern of the Sq current system depend upon longitude, season, year, and solar cycle.

solar radiation   The radiation from the sun covers the full electromagnetic spectrum. Approximately 40% of the sun's radiative output lies in the visible wavelengths (380 to 700 nm). Ultraviolet radiation (<380 nm) contributes an additional 7% to this ouput and infrared radiation accounts for more than half of the sun's power. While of great scientific interest, the X-ray radiation (below 0.1 nm) and millimeter radio emission are negligible in the total solar luminosity. The spectral distribution of the *solar radiation* peaks in the blue-green at 450 nm and is distributed approximately like a black body spectrum at a temperature of 5762 K. The black body approximation is particularly good at wavelengths greater than 400 nm but at shorter wavelengths the solar radiative output falls significantly below what the Planck spectrum would predict. This is due to the increasing number of Fraunhofer absorption lines at these wavelengths. At visible wavelengths, the smooth continuum spectrum is modified by a number of absorption lines due to the presence of ions in the solar atmosphere. In the ultraviolet and X-ray range, however, the spectrum changes to one dominated by emission lines. Both the Fraunhofer absorption lines and the optically thin emission lines provide important diagnostic information about the sun's atmosphere.

solar rotation   The sun rotates on its axis once in about 26 days. This rotation was first detected by observing the motion of sunspots. The sun's rotation axis is tilted by about 7.25° from the axis of the Earth's orbit so we see more of the sun's north pole in September of each year and more of its south pole in March. The sun does not rotate rigidly. The equatorial regions rotate faster (taking only about 24 days) than the polar regions (which rotate once in more than 30 days).

solar spectral irradiance   The spectral distribution of the sun's radiation as observed from the Earth's orbit. The integral of this over wavelength is known as the total solar irradiance or solar constant.

solar system   The collection of planets, moons, asteroids, comets, and smaller debris (such as meteoroids or interplanetary dust particles) which are gravitationally bound to the sun. Our solar system includes the nine major planets, the 65 currently known moons orbiting the planets, the particles comprising the ring systems around the Jovian planets, the myriad of asteroids and comets, and all the smaller debris resulting from these larger objects. The edge of our solar system is the heliopause, defined as the point where the galactic magnetic field counterbalances the sun's magnetic field.

solar system formation   *See* nebular hypothesis, tidal formation of solar system.

solar ultraviolet measurements of emitted radiation (SUMER)   An ultraviolet spectrometer aboard the satellite SOHO designed for the investigation of plasma flow characteristics, turbulence and wave motions, plasma densities and temperatures, and structures and events associated with solar magnetic activity in the chromosphere, the transition zone, and the corona.

solar wind The supersonically expanding outer atmosphere of the sun; the stream of fully ionized atomic particles, usually electrons, protons, and α particles (helium nuclei about 4% by number) with a trace of heavier nuclei emitted from the sun outward into space, with a density of only a few particles per cubic centimeter. The *solar wind* emanates from coronal holes in the sun's outer atmosphere and extends outward into the solar system along the field lines of the solar magnetic field. The existence of the solar wind was first predicted in 1958 by E.N. Parker, who argued on theoretical grounds that it is not possible for a stellar atmosphere as hot as the solar corona to be in complete hydrostatic equilibrium out to large distances in the absence of an unreasonably large interstellar confining pressure. The solar wind has two components distinguished by their average speeds. The fast solar wind has speeds of around 700 to 800 kms^{-1} while the typical speed of the slow solar wind is around 300 to 400 kms^{-1}.

Representative properties of the solar wind, as observed in the ecliptic plane at the heliocentric distance of one astronomical unit, are summarized in the table below. The solar wind velocity, density, etc. are highly variable; the quantities represented in the table can and do vary by factors of two or more, and the values presented should be regarded as illustrative only. The temperatures given are kinetic temperatures.

Representative Solar Wind Properties (Ecliptic Plane, 1 AU)

Property	Typical Value
velocity (V)	400 km/s
number density (N_p)	5 protons/cm^3
H$^+$ temperature (T_p)	10^5 K
e^- temperature (T_e)	1.5×10^5 K
He^{++} temperature (T_α)	$4 \times T_p$
magnetic field (B)	5×10^{-5} gauss
composition	96% H$^+$, 4% He^{++}

The flow is hypersonic; for example, in the ecliptic plane at 1 AU the flow velocity is many hundreds of kilometers per second, whereas typical sound and Alfvén speeds are of order 50 km/s. Because the flow is hypersonic, bow shock waves are formed when the solar wind is deflected by planetary obstacles. If the obstructing planet has a strong magnetic field, as is the case for the Earth and the outer planets, the solar wind is excluded from a sizable magnetospheric cavity that surrounds the planet.

As the solar wind expands, it carries with it a magnetic field whose field lines are rooted in the sun. Because the sun is rotating, the field lines are wrapped into a spiral pattern, so that field lines are nearly radial near the sun but nearly transverse to the radial direction at large heliocentric distances; near the orbit of Earth the field typically lies in the ecliptic plane at about 45° from the radial direction and the total intensity of the interplanetary magnetic field is nominally 5 nT.

Except for a brief period near the time of maximum sunspot activity, the general solar field has an underlying dipole character, with a single polarity at high northern latitudes and the opposite polarity at high southern latitudes; when this condition obtains, solar wind originating at polar latitudes typically has very high speed (of order 800 km/s) and its magnetic field is oriented inward or outward in accord with the predominant polarity of the region of origin. In contrast, at equatorial latitudes the solar magnetic field can have either polarity (the solar dipole is, in effect, tilted with respect to the rotation axis), so that as the sun rotates a fixed observer near the ecliptic sees alternating inward and outward magnetic field. A spatial "snapshot" of this alternating polarity structure is called the interplanetary magnetic sector pattern.

The solar wind is frequently disrupted by transient disturbances originating at the sun, especially at times of high sunspot activity. Such events may involve explosive ejection of mass from the corona, with associated shock waves, magnetic field disturbances, energetic particle enhancements, etc.

When transient disturbances in the solar wind impact planetary magnetic fields, they may cause such magnetospheric phenomena as magnetic storms and enhanced aurorae. Transient solar wind disturbances can also reduce the local galactic cosmic ray intensity by sweeping cosmic rays into the outer heliosphere, and can induce magnetic fields in objects containing conductive material. As the solar wind interacts with the comae of comets, it pushes the material back into the comet tails.

As the solar wind flows outward, its ram pressure declines as r^{-2}, and eventually becomes comparable to the local interstellar pressure due to interstellar matter or the galactic magnetic field or cosmic rays.

At such distances the interstellar medium is an effective obstacle to the solar wind, which must adjust by passing through a shock to become subsonic; the shocked wind is swept away by and eventually merged with the interstellar gas. Thus, the termination of the solar wind is thought to be characterized by two boundaries, the termination shock where the flow becomes subsonic, and the "heliopause," the boundary between the (shocked) solar material and exterior interstellar material. Spacecraft have not yet (June 1998) reached either of these boundaries, which are expected to be at distances of order 100 to 200 AU from the sun.

solar year The amount of time between successive returns to the Vernal equinox. For precision we refer to tropical year 1900 as the standard because the *solar year* is lengthening by about 1 millisecond per century.

solar zenith angle The zenith angle is the angle between the overhead point for an observer and an object such as the sun. The *solar zenith angle* is zero if the sun is directly overhead and 90° when the sun is on the horizon.

solid Earth tides Astronomical bodies such as the sun and moon deform the heights of gravitational equipotentials around the Earth. The disturbance in the height of an equipotential due to an external body such as the moon takes the form of a second order zonal spherical harmonic, which rotates around the Earth with a period slightly greater than one day (because as the Earth rotates, the moon itself moves in orbit). If the Earth's surface were inviscid and fluid on a diurnal timescale, then its surface would move so as to coincide with an equipotential (although the redistribution of mass at the surface would itself adjust the equipotentials). This is what

happens with the world's oceans, leading to high and low tides approximately twice per day (as there are two tidal bulges rotating around the Earth). For the viscoelastic solid Earth, the relaxation to an equipotential is incomplete: the proportion of actual readjustment to theoretical readjustment can be found from the appropriate Love number h_2, which (using the PREM Earth model) is 0.612. Due to irregularities such as the fact that the plane of the Earth's orbit around the sun, that of the moon's orbit around the Earth and the Earth's equatorial plane are not all parallel, and that the orbits are elliptical and have different periods, there are in fact many different frequencies in the solid Earth tides. *See* Love numbers.

solitary wave A wave with a single crest; wavelength is thus undefined.

soliton A spatially localized wave in a medium that can interact strongly with other solitons but will afterwards regain its original form. In hydrodynamical systems whose description involves non-linear equations, one can find solitary waves that are non-vanishing only in small regions of space, stable, and which can travel with constant velocity. They were first experimentally evidenced in 1842 by J. Scott Russel.

In non-linear field theories, equivalent stable bound state solutions called *solitons* can also exist both at the classical and quantum level. Their most remarkable property is that they do not disperse and thus conserve their form during propagation and collision.

Topological solitons, such as topological defects, require non-trivial boundary conditions and are produced by spontaneous symmetry breaking; non-topological solitons require only the existence of an additive conservation law. *See* cosmic topological defect, non-topological soliton, sine-Gordon soliton.

solstice Dates on which the day in one hemisphere (Northern or Southern) is of greatest length. Dates on which the sun is at one of two locations on the ecliptic which are most distant from the celestial equator. Because the Earth poles are inclined by 23°27′ to its orbital plane, northern and southern hemispheres typically receive different daily periods of sunlight. At the *solstices* the location of the planet in its orbit is such as to provide the maximum tilt of one hemisphere toward the sun, and this hemisphere receives more direct and longer duration light than does the opposite hemisphere. Thus, the northern hemisphere experiences long days at the summer solstice, the official beginning of northern summer, while the southern hemisphere experiences short days on this date, the official beginning of southern winter. Similarly the northern hemisphere experiences short days at the winter solstice, the official beginning of northern winter, while the southern hemisphere experiences long days on this date, the official beginning of southern summer. "Solstice" has the meaning "sun standing still" and marks the extreme of the northward or southward motion of the sun on the celestial sphere.

sound wave   A longitudinal wave propagating in a supporting medium, with a frequency audible to humans: about 20 to 20,000 Hz.

source functions   The terms in the radiative transfer equation that describe the inelastic scattering and true emission contribution to a beam of radiation; some authors also include elastic scattering in the source term.

source surface   Fictious surface in the solar corona at a height of about 2.5 solar radii. Graphically, the *source surface* separates the lower altitudes where open and closed magnetic field structures co-exist from the higher altitudes where all field-lines are open, extending into the interplanetary medium. The source surface can be determined from the photospheric magnetic field pattern using potential theory. Constraints on the field are then: (a) the magnetic field at the source surface is directed radially and (b) currents either vanish or are horizontal in the corona. The magnetic field pattern on the source surface then also gives the polarity pattern carried outward into with the solar wind. In particular, a neutral line separating the two opposite polarities can be identified. This neutral line is carried outwards as the heliospheric current sheet; the maximum excursion of the neutral line from the heliographic equator defines the tilt angle. *See* sector structure.

South Atlantic Anomaly   A region of strong radiation (principally protons) above the Atlantic Ocean off the Brazilian Coast, arising from the inner Van Allen belt. The Van Allen belts are energetic particles trapped by the Earth's magnetic field. The offset between the Earth's geographical and magnetic axes leads to an asymmetry in the belt position, and the *South Atlantic Anomaly* is the resulting region of minimum altitude (about 250 km). Some of the quasi-trapped charged particles mainly electrons in the kev range from the radiation belts may dip to very low heights at around 100 km in the Earth's atmosphere, where collisions with neutral particles can cause ionization and photochemical excitation. The anomaly extends as far as the coast of Africa. The South Atlantic Anomaly is thus a region of energetic particles through which low-orbiting satellites frequently pass, and the radiation density can be harmful to electronics and to human crew. *See* Van Allen Belts.

South Equatorial current   An ocean current flowing westward along and south of the equator.

southern oscillation   Quasi-periodic phenomenon in sea level pressure, surface wind, sea surface temperature, cloudiness, and rainfall over a wide area of the tropical Pacific Ocean and adjacent coastal areas, south of the equator with a period from two to four years, characterized by simultaneously opposite sea level pressure anomalies at Tahiti, in the eastern tropical Pacific and Darwin, on the northwest coast of Australia. *See* El Niño, La Niña.

Southern Oscillation Index (SOI)   An index that is calculated to monitor the atmospheric component of the El Niño-Southern Oscillation (ENSO) phenomenon. It is usually defined as the monthly averaged sea level pressure anomaly at Tahiti minus the pressure anomaly at Darwin, Australia. The SOI is often normalized by the standard deviation over the entire record. Anomalously high pressure at Darwin and low pressure at Tahiti usually indicate El Niño (warm) conditions. *See also* El Niño.

space-like infinity   The endpoint i^0 of all space-like geodesics of a space-time.

space-like vector   An element t of a linear space with a Lorentzian metric g of signature $(-, +, +, +)$ such that the norm is $g(t, t) = g_{ab}t^a t^b > 0$. The norm of a space-like four vector in the theory of relativity represents the spatial distance of two events that appear simultaneous to some observer.

spacetime   The collection of all places and all moments of time in the whole universe. For various problems simplified models of our physical *spacetime* are constructed. For example, cosmology often ignores details of the geometry and matter distribution in the universe that are smaller than a group of galaxies. In investigating planetary motions, typically all stars other than the sun are ignored, and it is assumed that the empty space around the sun extends to infinite distances. A mathematical model of spacetime is a four-dimensional space (one of the four coordinates is time) with a given metric; two spacetimes are in fact identical if their metrics can be transformed one into the other by a coordinate transformation. Points of the spacetime are called events. Two events p and q can be in a timelike, lightlike (null), or spacelike relation. The metric makes it possible to calculate the lapse of time between the events p and q when they are in a timelike relation or the distance between them when they are in a spacelike relation. In fact the metric contains all the relevant information about the geometry and physics in its underlying spacetime, although some of the information may be technically difficult to extract. In general relativity, Einstein's equations relate the matter distribution to the geometry of the spacetime. Geodesics can be calculated to determine the trajectories of particles, and of light rays, moving under the influence of gravitational forces. If the properties of the spacetime are found to correspond to part of the observed world, then the mathematical spacetime model is considered realistic in the appropriate range of phenomena. Very few such analytically exact realistic models exist. Among them are the flat Minkowski spacetime that is the geometrical arena of special relativity, the spacetime of the Schwarzschild solu-

tion that describes spherical black holes, and the sun's dominant, spherically symmetric gravitational field in the solar system (but without taking into account the planets' own gravitational fields or the rotation and rotational deformation of the star), the spacetimes of the Friedmann–Lemaître and Robertson–Walker cosmological models that are used to model the whole universe, the Kerr spacetime describing the gravitation field of a stationary rotating black hole, and a few more spacetimes corresponding to simple patterns of gravitational waves and isolated structures in the universe (the Lemaître–Tolman cosmological model is among the latter). However, computational modeling is beginning to provide a much longer list of accessible realistic spacetimes.

space weather   The conditions and processes occurring in space which have the potential to affect the near Earth environment and, in particular, technological systems. *Space weather* processes include the solar wind and interplanetary magnetic field, solar flares, coronal mass ejections from the sun, and the resulting disturbances in the Earth's geomagnetic field and atmosphere. The effects can range from the unexpected (e.g., disruption of power grids, damage to satellites) to the common (e.g., failure of HF systems). Although space weather effects have been recognized and studied for many years, it has only recently developed as a recognized field of unified activity which attempts to forecast solar flares, magnetic storms and other space-related phenomena.

spallation   A nuclear reaction in which an atomic nucleus is struck by an incident high energy particle. As a result, particles typically heavier than an α-particle are ejected from the nucleus. Astrophysical amounts of the lightly bound isotopes ^6Li, ^9Be, ^{10}B, and ^{11}B are believed to have been formed by *spallation* by energetic cosmic rays.

special relativity   A description of mechanical and electromagnetic phenomena involving sources and observers moving at velocities close to that of light, but in the absence of gravitational effects. Maxwell's theory describes the dynamics of electric and magnetic fields and predicts a propagation velocity, c, which is numerically equal to the observed speed of the light.

Minkowksi, Lorentz, and especially Einstein in his 1905 exposition, modified concepts of space and time in a way that accommodates the fact that every observer, regardless of his motion, will always measure exactly the same speed for light as experimentally observed by Michelson and Morley. To accomplish this required accepting that time is involved in Lorentz transformation, the change of coordinate frame between moving observers, just as spatial position is involved. The Lorentz transformation also predicts length contraction, time dilation, and relativistic mass increase, and led Einstein to his famous result $E = mc^2$.

A consequence of the universality of the speed of light is that velocities are not additive. If the system B (say, an airliner) moves with respect to the system A (say, the surface of the Earth) with a given velocity v_1 (say, 900 km/h) and an object C (say, a flight attendant) moves in the same direction with a velocity v_2 (suppose, 4 km/h), then the velocity of the object C with respect to the system A is not simply $v_1 + v_2$, but $V = \frac{v_1+v_2}{1+(v_1 v_2/c^2)}$, where c is the velocity of light (in the example given, the flight attendant moves with respect to the surface of the Earth with the velocity $904 \cdot (1 - 0.31 \cdot 10^{-12})$ km/h). As is seen from the example, at velocities encountered in everyday life the corrections provided by special relativity to Newton's mechanics are negligible. However, for velocities large compared to c, the difference is profound. In particular, if $v_2 = c$, then $V = c$. Also, if $v_1 \leq c$ and $v_2 \leq c$, then $V \leq c$, i.e., the velocity of light cannot be exceeded by adding velocities smaller than c. The geometric arena of special relativity is not the three-dimensional Euclidean space familiar from Newton's theory, but a four-dimensional Minkowski spacetime. The Euclidean space is contained in the Minkowski spacetime as the subspace of constant time, but it is not universally defined, i.e., every observer will see a different Euclidean space in his/her own reference system. *See* Lorentz transformation, Michelson–Morley experiment.

specific   The adjective used to express a quantity per unit mass.

specific absorption coefficient   The absorption coefficient $[m^{-1}]$ per unit mass of material, e.g., for unit chlorophyll a concentration (Units: [mg chl a m^{-3}]); one obtains a specific absorption coefficient with units [m^2 (mg chl a) $^{-1}$].

specific discharge   *See* Darcy velocity.

specific energy   A term used in the study of open channel flow to denote the energy of a fluid relative to the channel bottom. Specific energy is defined as $E = y + V^2/2g$, where y is flow depth, V is flow speed, and g is acceleration of gravity. *Specific energy* has units of length and corresponds to energy per unit weight of fluid.

specific gravity   Ratio of the density of a substance to that of water.

specific heat   *Specific heats,* also called specific heat capacity or heat capacity, are defined under constant volume (c_v, C_v, also called isochoric specific heat) and constant pressure (c_p, C_p, also called isobaric specific heat). The constant volume specific heat of a pure substance is the change of molecular internal energy u for a unit mass (or 1 mole) per degree change of temperature when the end states are equilibrium states of the same volume:

$$c_v \equiv \left(\frac{\partial u}{\partial T}\right)_v \quad \text{and} \quad C_v \equiv \left(\frac{\partial \bar{u}}{\partial T}\right)_v,$$

where c_v is for unit mass and C_v is called the molal specific heat, u is the specific internal energy, $\bar{u}$ is internal energy for 1 mole; $C_v = M c_v$. The constant pressure specific heat of a pure substance is the change of enthalpy for a unit mass (or 1 mole) between two equilibrium states at the same pressure per degree change of temperature:

$$c_p \equiv \left(\frac{\partial h}{\partial T}\right)_p \quad \text{and} \quad C_p \equiv \left(\frac{\partial \bar{h}}{\partial T}\right)_p$$

where $h = (u + pv)$ is the specific enthalpy, and $\bar{h}$ is the enthalpy for 1 mole; $C_p = M c_p$.

specific humidity   The mass of water vapor per unit mass of air.

specific photosynthetic rate   Photosynthetic rate, net or gross, per unit biomass or per unit volume, e.g., [μmoles CO_2 (or O_2) (mg chl)$^{-1}$ h^{-1}].

specific storage (S_s)   The amount of water per unit volume of a saturated formation that is stored or released from a saturated aquifer in response to a unit increase or decrease in hydraulic head, with units of meter^{-1} (L^{-1}): $S_s = \rho g (\alpha + n\beta)$, where ρg is the fluid weight, α is the compressibility of the aquifer, and β is the compressibility of the fluid.

specific volume   In a dipole-like magnetic configuration, like that of the not too distant magnetosphere of the Earth, the integral $V = \int ds/B$ along a closed magnetic field line. (The integral is usually assumed to be computed from one end of the line to the other, but the value obtained between the two intersections of the line with the ionosphere is not much different.) The variable V is a useful quantity in the simplified formulation of MHD in which the plasma in any convecting flux tube is assumed to stay inside that tube and to remain isotropic. The equations of this formulation (e.g., the Grad–Vasyliunas theorem) are often used in theories and simulations of global convection. In particular, an adiabatic law $pv^{5/3}$ is expected to hold then, with p the plasma pressure.

spectral   The adjective used to denote either wavelength dependence or a radiometric quantity per unit wavelength interval.

spectral energy distribution   The relative intensity of light (or other electromagnetic radiation) at different frequencies, measured over a range of frequency; usually expressed in units of power per unit frequency or per unit wavelength. In astronomy often referring to the continuum spectrum of an astronomical object, with the absorption or emission line spectrum not considered. The *spectral energy distribution* is very different for different astronomical sources and emission mechanisms. For example, stars emit radiation whose spectral energy distribution show minor deviations from the Planck function, as expected for high-temperature black bodies; synchrotron processes produce a spectral energy distribution that can be described, over a wide spectral range, by a power-law as a function of frequency or of wavelength.

spectral energy distribution of active galactic nuclei   A spectral energy distribution which is very different from the those of stars and non-active galaxies and which is characterized by significant, in a first approximation almost constant, energy emission over a very wide range of frequencies, from the far IR to the hard X-ray domain (power-law function of frequency ($f_\nu \propto \nu^{-\alpha}$), with a spectral index $\alpha \approx 1$; hence, the emitted energy νf_ν does not depend on ν and is constant). In radio-loud AGN, roughly constant emission extends to radio frequencies; in radio quiet AGN, emission at radio frequencies is typically ~ 100 times lower. Superimposed to the power-law emission there is a broad feature, the big blue bump, which extends from the visible to the soft X-ray domain, with maximum emission in the far UV. The big blue bump has been ascribed to thermal emission, possibly by the putative accretion disk. This interpretation of the AGN continuum is, however, the subject of current debate. There are also significant differences between AGN subclasses. *See* big blue bump, spectral energy distribution.

spectral gap   Refers to a substantial local minimum in the energy spectrum $\phi(\omega)$ at frequency ω_g, which divides the energy spectrum in hydrodynamics into a mean flow (large-scale, advective) ($\omega < \omega_g$) and turbulence ($\omega > \omega_g$). The presence of a *spectral gap* is an ideal prerequisite for the Reynolds decomposition.

spectral line   The specific wavelengths of light which are exceptionally bright or exceptionally faint compared to neighboring wavelengths in a spectrum. When passed through a spectrograph, these are seen as bright or dark transverse lines in the spectrum. Bright lines arise from emission from specific atomic states; dark lines are absorption on specific atomic states.

spectral line profiles   The shape of an absorption or emission line profile depends on many factors. Among them are thermal Doppler broadening, pressure broadening, collisional damping, radiation damping, rotational broadening, and natural broadening. If we consider only quantum mechanical effects, a spectral line profile must have a minimum width based on Heisenberg's uncertainty principle; this is the natural line profile and is Lorentzian, rather than Gaussian, in shape. Convolved with the natural profile are the effects of the gas rotating as a whole, the velocity distribution of the gas particles, and the ambient pressure.

spectral type One of several ways of classifying stars in terms of the lines (absorption usually, but sometimes emission) that are most conspicuous in their optical (more recently sometimes infrared or ultraviolet) spectra. The classical Morgan and Keenan (MK) system is a two-dimensional system: spectral type and luminosity class. In the MK system, classification was defined at moderate (3 Å) resolution in the blue region (4000 to 5000 Å). The most important physical parameter is the surface temperature of the star, and most stars can be put into a linear sequence covering the range 100,000 K down to 3,000 K or less with the types called OBAFGKML and dominant lines as shown below:

O - ionized helium
B - neutral helium
A - hydrogen
F - hydrogen, ionized metals
G - ionized and neutral metals
K - neutron metals and molecules
M - molecules
L - molecules and clouds

The OB (blue) types are referred to as "early" type stars, and the KM (red) types referred to as "late" type stars. The spectral sequence, originally based on the progression of line patterns, was a function of ionization stages of chemical elements identified in the spectra, and has proven to be ordered from the hottest and most massive stars (O and B types) to the coolest and least massive stars (K and M types). A second dimension makes use of ratios of lines that are sensitive to ambient electron pressure and so provides an indicator of stellar luminosity (via surface gravity) from type I (supergiants) on down to V (main sequence stars) and VI (subdwarfs). Yet additional refinements indicate variations in the surface composition of the stars and intermediate temperatures and spectral

appearances between the primary types. Another significant type of star is the Wolf–Rayet (W-R) star, which has unusually strong emission lines and peculiar chemical abundances. Our sun is a G2 V star. *See* HR (Hertzsprung–Russell) Diagram.

spectrograph An instrument to record the spectrum of some source, usually consisting of a slit through which light or other electromagnetic radiation passes, a grating to disperse the radiation, and a recording medium (photographic emulsion or electronic detector).

spectrometer An instrument for measuring the intensity of radiation as a function of wavelength.

spectrophotometry The measurements of spectral line and spectral continuum fluxes, and their comparison, made over a wide range of frequency or, equivalently, of wavelength. Instrument sensitivity strongly depends on the frequency of the incoming light. To recover the radiation flux and the intrinsic spectral energy distribution due to an astronomical source reaching the telescope, a wavelength-dependent correction must be applied to the recorded spectrum. A spectrophotometric calibration is customarily achieved in optical and UV spectroscopy by the repeated observations of standard stars for which the intrinsic spectral energy distribution is already known. Accurate *spectrophotometry,* even if restricted to the visible spectral range ($\approx$ 400 to 800 nm) is difficult to achieve from ground-based observations, and has been made possible only recently by the employment of linear detectors such as the CCDs.

spectropolarimetry The measurement of the degree of polarization of radiation, and of the polarized radiation flux, as a function of frequency or wavelength. Optical spectropolarimetric measurements are achieved by inserting a half-wave or quarter-wave plate and a polarizer in the optical path of the light allowed into the spectrograph. Astronomical *spectropolarimetry* can in principle achieve high precision ($\sim$ 0.01%). However, partly oblique reflections due to telescope and spectrograph design may polarize intrinsically unpolarized radiation and limit observations. Some emission mechanisms like synchrotron processes can produce a high degree of polarization, but real sources typically show optical spectra with a small degree of polarization, reducing the collected flux. As a consequence, spectropolarimetric measurements have been, until recently, rather sparse and limited to relatively bright objects.

spectroradiometer   A radiometer that measures radiant energy as a function of wavelength.

spectroscopic binary   A binary star system that reveals itself because the stars have a component of their orbit velocities along the line of sight to us, so that Doppler shifts move the absorption (or emission) lines in their spectra back and forth in wavelength. The period over which the pattern repeats is the orbit period; the amount of the shift is the orbit velocity projected along the line of sight. When the system is also an eclipsing binary, then the projection angle is known and the orbit velocities can be used to compute the semi-major axes of the orbit leading to a measurement of the masses of the stars, using Kepler's third law of planetary orbits in modified form: $(M_1 + M_2)P^2 = a^3 \frac{M_1}{M_2} = \frac{K_2}{K_1}$ where the M's are the masses of the stars in solar masses, P is the orbit period in years (most spectroscopic eclipsing binaries have periods less than about a year), and a is the semi-major axis of the relative orbit, found from the two stellar velocities, K_1 and K_2 by $a = 2\pi \frac{(K_1+K_2)}{P}$. a must be converted to astronomical units for use in the formula as given. Additional complications arise, but systems can still often be analyzed usefully when (a) the orbit is not circular, (b) the system does not eclipse, or (c) only one set of spectral lines can be seen (normally attributable to the brighter of the two stars); these are called single line spectroscopic binaries, as opposed to double line spectroscopic binaries where both sets of spectral lines are seen. *See* Doppler shift.

spectroscopy   The study of a source by dispersing light into a spectrum of different wavelengths. This is usually accomplished by focusing light from the source on a narrow slit, then passing the light from the slit through a diffrac-

tion grating. The resulting dispersion shifts different wavelengths of light transversely to the slit, so the relative intensities of energies at different wavelengths can be determined.

spectrum   The separation of electromagnetic radiation into its component colors or wavelengths. Spectra of visible light from the sun are often punctuated with emission or absorption lines, which can be examined to reveal the composition and motion of the radiating source. Particle spectra give information on the distribution of a given particle population with energy.

spectrum binary   An orbiting pair of stars that appears as a single point of light in the sky, but whose spectrum shows absorption and/or emission lines of both stars, recognizable as such generally because they are of very different spectral type. *See* spectral type.

speed of light   The speed of electromagnetic radiation propagation in vacuum: 299792458 m/sec.

spherical harmonic analysis   A harmonic analysis method applied on a sphere. In meteorology, it is often used on harmonic analysis of geopotential height fields. For any grid point of the geopotential height field, the height can be expressed as

$$Z(\lambda, \mu)$$
$$= \sum_{m=0}^{\infty} \sum_{n=0}^{\infty} \left(a_n^m \cos m\lambda + b_n^m \sin m\lambda\right) P_n^m(\mu)$$

where λ is longitude, μ is $\cos(90° - \phi)$, ϕ is latitude, m is zonal wave number, n is meridional wave number, $P_n^m(\mu)$ is a normalized associated Legendre function, $\cos m\lambda P_n^m(\mu)$ and $\sin m\lambda P_n^m(\mu)$ are spherical harmonics, and a_n^m and b_n^m are coefficients of spherical harmonic analysis, as

$$a_n^m = \frac{1}{\pi} \int_s \int Z(\lambda, \mu) \cos m\lambda P_n^m(\mu) d\mu d\lambda$$

$$b_n^m = \frac{1}{\pi} \int_s \int Z(\lambda, \mu) \sin m\lambda P_n^m(\mu) d\mu d\lambda \,.$$

spherical pendulum   Physically the same as a simple pendulum, but with motion allowed anywhere on the surface of a sphere of radius equal to the length of the suspension string. As with a simple pendulum, for small oscillations the motion is simple harmonic with angular frequency

$$\omega = \sqrt{(g/L)} \ .$$

See simple harmonic motion.

Spica   0.98 magnitude star of spectral type B1 at RA $13^h 25^m 11.5^s$, dec $-18°18'21''$.

spicule   Chromospheric jet of plasma seen at the solar limb. *Spicules* are ejected from the high chromospheric part of supergranule boundaries and reach speeds of 20 to 30 km s^{-1} and heights of about 11,000 km before fading.   Spicules have typical lifetimes of 5 to 10 min, diameters of 50 to 1200 km, maximum lengths of 10,000 to 20,000 km, temperatures of 10,000 to 20,000 K and electron densities of 3×10^{10} to 3×10^{11} cm^{-3}.

spin coefficients   The connection components $\Gamma_{abcd'} = (\nabla_{CD'}\zeta_{aA})\zeta_b^A \zeta_c^C \bar{\zeta}_{d'}^{D'}$ in the SL (2,C) spinor dyadic basis ζ_a^A introduced by Newman and Penrose. They are equal to the complex Ricci rotation coefficients in the null basis defined by the spinor dyad, and are individually denoted

$$\Gamma_{abcd'} =$$

ab			
cd'	00	01	11
$00'$	κ	ϵ	π
$10'$	ρ	α	λ
$01'$	σ	β	μ
$11'$	τ	γ	ν

See affine connection.

spinel   Olivine, $(Fe,Mg)_2 SiO_4$, is one of the major constituents of the Earth's mantle, and is expected to be an important component of the mantles of the other terrestrial planets as well.

Under high pressures ($\sim 10^{10}$ Pa) the olivine crystal structure undergoes a phase change to a more compact phase with a resulting density increase of several percent. This more compact polymorph of olivine is called *spinel*. In the Earth, the olivine-spinel phase change is believed to be responsible for the density discontinuity at a depth of 400 km (pressure of 13.4 GPa).

spinor   A complex 2-dimensional vector. *Spinor* representations are typically used in quantum mechanics, e.g., to represent the spin of an electron, and in some formulations in general relativity.

S–P interval   Time difference between P- and S-waves arriving at a point. P-wave velocity is about $\sqrt{3}$ times as fast as S-wave velocity. When underground velocity structure can be regarded as homogeneous, S–P interval (τ sec) is proportional to hypocentral distance (R km). Thus, for shallow earthquakes

$$R = k\tau \quad \text{(Omori's formula for}$$
$$\text{hypocentral distance)}$$

with

$$k = \frac{Vp\,Vs}{Vp - Vs}$$

where Vp and Vs are P- and S-wave velocities, respectively. The value of k is dependent on locations and depths of earthquakes, taking values of about 6 to 9 km/s.

spin up   The transient initial stage of certain numerical ocean simulations when the various modeled fields are not yet in equilibrium with the boundary and forcing functions.

spiral arm   The part of the spiral pattern of a galaxy which is more or less continuously traceable. *Spiral arms* are very evident in the blue and visual bands (and hence were revealed since the early days of extragalactic astronomy) or, with a more "knotty" appearance, in narrow band images centered on the hydrogen Balmer line Hα. This indicates that spiral arms are sites of recent star formation. Spiral arms accordingly become less and less prominent at longer wavelengths, and almost undetectable in the near infrared, where most of the light is produced by more evolved stars. (Although all masses of stars are formed in the star forming region, the massive blue stars are brighter, and they live a substantially short time, so they remain near the spiral wave that triggered their formation; they expire before they have a chance to move far from their birthplace.) Dust lanes, filamentary absorption features that appear dark on the emission background of the galaxy's disk, are often associated with spiral arms. The physical origin of the spiral pattern of galaxies is a subject of current debate; it is presumably a density wave in the rotating galactic disk, moving through the disk at approximately 20 km/sec; the regions of compression become regions of star formation. "Grand design" spirals have been linked to the gravitational perturbation by a nearby companion galaxy, as in the case of M51. *See also* M51.

spiral galaxy   A galaxy showing a bright spiral pattern, superimposed to smooth disk emission. *Spiral galaxies* are composed of a spheroidal bulge and a flattened system of gas and stars (the disk), over which the spiral pattern is seen. While the bulge of a spiral galaxy loosely resembles an elliptical galaxy, the disk shows an exponential decrease in surface brightness with increasing distance from the nucleus. Both the prominence of the bulge and the shape of the spiral pattern vary along the Hubble sequence from *S*0, the lenticular galaxies, to *Sa* (which show large nuclii and faint spiral structure), to *Sb*, to *Sc* galaxies which have small nuclei and strong spiral arms. Many galaxies show a bar across the nucleus, in which case a *B* is added to the classification, as in *SBc*. In total count, *S*0 amount to 22%, and *Sa, b, c* galaxies to 61% of all galaxies observed. Spiral galaxies are most common in the "field", i.e., not in clusters ("morphology-density relation"). *See also* Hubble sequence, spiral arm.

split-hull barge   A barge that is constructed so that its hull can be opened to drop the contents of its hold. The hinge runs along the long axis of the barge.

Spoerer's law   Describes the spatial distribution of sunspots during the solar cycle: sunspots always start at relatively high latitudes (about 30°) and move towards the equator. In addition, during the solar cycle the latitude of emergence of sunspots moves also toward the equator.

spontaneous symmetry breaking   In condensed matter physics, the zero magnetization of an (initially) isotropic ferromagnetic system takes on a nonvanishing value (and thus, the magnetization points in a particular direction) as the temperature decreases below the critical temperature T_c. Thus, the initial symmery (the isotropy) is spontaneously broken. The case for particle physics is described by a scalar field ϕ called the Higgs field, and now the order parameter characterizing the transition is the vacuum expectation value $\langle |\phi| \rangle$ of this field. The vacuum expectation value is the expected value of a field in its lowest possible energy configuration.

Research done in the 1970s in finite-temperature field theory led to the result that the temperature-dependent effective potential for the Higgs field can be written roughly as

$$V_T(|\phi|) = -\frac{1}{2}m^2(T)|\phi|^2 + \frac{\lambda}{4!}|\phi|^4$$

with $T_c^2 = 24m_0^2/\lambda$, $m^2(T) = m_0^2(1 - T^2/T_c^2)$, this potential has one minimum energy (the "false vacuum") at $|\phi| = 0$ when T is large, but a different value given by $\langle |\phi| \rangle^2 = 6m^2(T)/\lambda$ where m_0 and λ are two positive constants when T is small. Because this minimum depends only on the magnitude of $|\phi|$, not on its sign or phase, there are possibly more than one low temperature minima, a situation called degenerate true vacuua.

For energies much larger than the critical temperature (in appropriate units), the fields are in the highly symmetric state characterized by $\langle |\phi| \rangle = 0$. But, when energies decrease the symmetry of the system is spontaneously broken: the scalar field rolls down the potential and sits in one of the nonvanishing degenerate new minima. *See* cosmic phase transition.

sporadic E   *Sporadic E* (Es) layers are transient localized thin patches of relatively high electron density occurring at (95 to 140 km) E layer altitudes. While the peak sporadic E electron density can be several times larger than the normal E region, it is independent of the regular solar produced E layer. Although the layers are called "sporadic" because of their varied appearance and disappearance times, they are a common phenomenon, mid-latitude ionograms rarely being free of some signs of sporadic E. Mid-latitude sporadic E layers are thought to be formed by a wind shear, in the presence of the local magnetic field, acting on long-lived metallic ions. At high latitudes, Es is attributed to ionization by incoming eneregetic particles. The resulting layers may vary in spatial extent from several kilometers to 1,000 km and have typical thickness of 0.5 to 2 km. Much statistical information has been collected on sporadic E from measurements made with ionosondes. Three parameters have been collected: (1) h′Es (the sporadic E layer height), (2) foEs (the maximum electron density of the layer), and (3) fbEs (the minimum electron density observable above the layer). The last parameters gives a measure of the layer patchiness. Sporadic E is most common in the summer daytime and diurnally there can be a pre-noon peak followed by a second pre-dusk peak in occurrence. There are also marked global differences; for instance, summertime sporadic E in southeast Asia, especially near Japan, are more intense (higher foEs) than at any other location. Mid-latitude sporadic E can support HF propagation of both wanted and unwanted (interference) signals as well as screen the F region, thereby preventing propagation. At low latitudes, equatorial sporadic E reflections from intense irregularities in the equatorial electrojet appear on iongrams as highly transparent reflections, sometimes up to 10 MHz. Equatorial sporadic E is most common during the daytime, and shows little seasonal dependence. The daytime electrojet irregularities can support propagation at VHF frequencies. At high latitudes, the auroral E region displays several phenomena including Es layers formed by particle precipitation. Combinations of these processes can give rise to rapid fading and multipath on HF circuits. *See* E region.

spread E   Ionization irregularities can form in the E region at any time of the day and night. While not as severe, or as frequent in occurrence as spread F, it can have significant effects on HF radio propagation. The irregularities in the E region have been shown to impose phase front distortions on radio waves reflected from higher layers and are also responsible for significant distortion of radio waves reflected in the E region. *Spread E* possibly results from small-structured ionization irregularities caused by large electric fields. *See* E region.

spread F   Irregularities in the F layer ionization can result in multiple reflections from the ionosphere forming a complex F-region ionization structure recorded on ionograms. The name *spread F* derives from the spread appearance of the echo trace on the ionogram. On occasions, the spread F traces can contain many discrete traces, other times there may be no distinct structure, this being suggestive of different scale size ionization irregularities. Spreading may be found only near the F-region critical frequency, generally called frequency spreading, and on other occasions range spreading may form near the base of the F region and extend upward in virtual height. These phenomena are observed during mid-latitude nighttime, more frequently during winter, and during ionospheric storms. Although spread F is observed mainly at night, during ionospheric storms, daytime spread F is often observed and is a clear indicator of major storm activity. Spread F may be related to traveling ionospheric disturbances. The ionization irregularities responsible for spread F are also responsible for high flutter fading rates at HF and VHF and for scintillation on satellite propagation paths. At high latitudes, spread F is common, and related to changes in particle fluxes into the ionosphere. At low latitudes, equatorial spread F is a distinctive form that restructures the overhead ionosphere removing all impression of the F region critical frequency on ionograms. This form shows a strong solar cycle dependence, being most prevalent at high solar activity, when ionization levels are highest. Coherent radars have shown this form is a result of bubbles of ionization moving upwards through the nighttime ionosphere. These irregularities tend to be field-aligned, roughly 100 to 2000 km in extent and extend from the bottomside up into the topside ionosphere (an altitude extent of 600 to 1000 km). Radio waves (HF, VHF) can be ducted within these large structures, that are thought to be one of the principal sources of trans-equatorial propagation. The equatorial Spread F affects the trans-ionospheric propagation and produces fluctuations in signals coming from the outer ionosphere known as Scintillations. Spread F tend to occur in the night. *See* F region, ionosphere.

spring   Season of the year in the northern hemisphere between the vernal equinox, about March 21 and the summer solstice, about June 21; in the southern hemisphere between the autumnal solstice, about September 21, and the winter solstice, about December 21. *See also* cosmic spring.

spring tide   Tide which occurs when the Earth, sun, and moon are nearly co-linear. Under these conditions the gravitational field gradients of the sun and moon reinforce each other. The high tide is higher and low tide is lower than the average. *Spring tides* occur twice a month near the times of both new moon and full moon.

sprites   Short-lived luminosities observed at high altitudes above thunderstorms, apparently associated with upward discharges of thunderstorm electricity.   They appear as columnar diffuse reddish glows between 30 km and 80 km above ground, branching into the upper atmosphere lasting tens of milliseconds, following large positive cloud-to-ground lightning strokes. It is currently believed that such lightning strokes leave in the cloud a residual charge of 200 Coulombs or more, creating a significant voltage difference between the cloud top and the ionosphere. This induces a heating of the middle atmosphere, which produces the *sprites* when electrons collisionally excite atmospheric nitrogen, that emits red light in flashes with several milliseconds duration.

Studied from the ground, aircraft and from the space shuttle, sprites appear to be most common above large mesoscale convection features, such as the storm systems of the American Midwest. Ground observations indicate that an active system can generate as many as 100 sprites in the course of several hours. Discovered in 1990. *See* elves, blue jet.

squall   A sudden strong wind or brief storm that persists for only a few minutes, often associated with thunderstorm activity and sometimes accompanied by rain or snow.

s (slow) process   The capture of free neutrons by nuclei of iron and heavier elements on a time scale slower than the average for beta-decays of unstable nuclei (cf. r process). The *s process* is the primary source of those isotopes of elements from $Z = 30$ to 82 with the largest binding energies (the "valley of beta stability"). It occurs in asymptotic giant branch stars and in the corresponding evolutionary stage of more massive stars. The main sources of free neutrons (which require some mixing between zones dominated by different nuclear reactions) are

$$^{13}C(\alpha, n)^{16}O \text{ and}$$
$$^{14}N(\alpha, \gamma)^{18}O(\alpha, \gamma)^{22}Ne(\alpha, n)^{25}Mg$$

(the latter being more important in massive stars). Among the important products of the s process are barium (because it is dominated by s process nuclides and can be seen in the atmospheres of cool, evolved stars) and technitium (because of its short half life). s-process production of elements does not reach beyond lead and bismuth because of the set of unstable elements before reaching uranium and thorium. A typical s-process chain is

$$^{174}Yb(n, \gamma)^{175}Yb(e^-\bar{\nu}_e)^{175}Lu(n, \gamma)^{176}Lu(n, \gamma)$$
$$^{177}Lu(e^-\bar{\nu})^{177}Hf(n, \gamma)^{178}Hf(n, \gamma)^{179}Hf$$
$$^{179}Hf(n, \gamma)^{180}Hf(n, \gamma)^{181}Hf(e^-\bar{\nu}_e)$$
$$^{181}Ta(n, \gamma)^{182}Ta\,(e^-\bar{\nu}_e)$$
$$^{182}W(n, \gamma)^{184}W(n, \gamma)$$

and so forth, onward to Re, Os, etc. The time scales of the beta decays range from minutes to months.

The cross-section to capture the next neutron is smallest for nuclides with filled shells, so that products of the s process with $N = 26, 50, 82$, and 126 are particularly abundant. *See* asymptotic giant branch star, beta decay, r process.

S stars   Cool (3000 to 3600 K) giant stars which have greatly enhanced *s*-process elements (such as zirconium, barium, yttrium, and technetium) and have an atmospheric C/O number ratio very close to unity. Their optical spectra are characterized by strong absorption bands of ZrO, as well as TiO, and carbon-rich compounds such as CN and C_2. Most are believed to be on the asymptotic giant branch (AGB) of the H-R diagram. Although *S stars* might be an intermediate stage of evolution between oxygen-rich (M stars) and carbon rich (C stars) AGB stars, it is possible that they are an end stage and will not go through a carbon rich phase. It has been shown that some S stars have undergone mass transfer from a companion and have white dwarf companions.

stability frequency   *See* buoyancy frequency.

stability of the water column   The strength of the stratification of a water column is expressed in terms of the intrinsic buoyancy frequency N, that a parcel experiences. The stability is given by

$$N^2 = -g\rho^{-1}(\partial\rho/\partial z)$$
$$= g(\alpha\partial\Theta/\partial z - \beta\partial S/\partial z$$
$$+ \text{ additional terms})$$

where g is gravitational acceleration, ρ density, z vertical coordinate (positive upward), α thermal expansivity, β saline contraction coefficient, Θ potential temperature and S salinity. Additional terms may become relevant, when (1) stability is very low (e.g., gradient of gases, such as CO_2 or CH_4, or silica in deep lakes), or when (2) inflowing river water or bottom boundary water contain high particle concentrations. N^2 varies over 10 orders of magnitude in natural waters from 10^{-11} s^{-2} (in well-mixed bottom layers or double-diffusive layers) to 1 s^{-2} in extreme halocline (merging fresh and salt water).

stability ratio   The non-dimensional ratio of the stability due to the stabilizing component divided by the destabilizing component of the stratification: $R_\rho = \left(\frac{\alpha\partial T/\partial z}{\beta_s\partial S/\partial z}\right)$ or $R_\rho = \left(\frac{\alpha\partial T/\partial z}{\beta_s\partial S/\partial z}\right)^{-1}$ (*see* double diffusion). Here T is the temperature, S is the salinity, and z is the vertical coordinate. The thermal and saline expansion coefficients of sea water, α and β, respectively, are defined such that they are always positive. Hence, the value of R_ρ indicates the type of the stability of the water column. The density of sea water is a function of temperature and salinity, and their vertical gradients determine the stability of the water column. In the oceans, temperature generally decreases with depth and

salinity increases with depth (i.e., for z positive upward, $\partial T/\partial z > 0$ and $\partial S/\partial z < 0$). A value of R_ρ in the open interval $(-\infty, 0)$ indicates a stable stratification, while a $R_\rho > 1$ indicates an unstable water column. Vice versa, if $\partial T/\partial z < 0$, then $-\infty < R_\rho < 0$ means the water column is unstable, while $R_\rho > 1$ indicates stable stratification.

In a two-layer situation in which two water masses of different salt and temperature composition are stacked vertically, a value of R_ρ in the interval between 0 and approximately 3 indicates the possibility of a double diffusive instability. For $\partial T/\partial z > 0$ the double diffusion will be in the form of salt fingers, for $\partial T/\partial z < 0$ in the form of layering.

R_ρ	$\partial T/\partial z > 0$	$\partial T/\partial z < 0$
$-\infty < R_\rho \leq 0$	Stable	Unstable
$R_\rho = 1$	Statically unstable	Statically unstable
$R_\rho > 1$	Unstable	Stable
$0 < R_\rho \lesssim 3$	Fingering	Layering

Infrequently R_ρ is called density ratio or Turner number.

stable auroral red (SAR) arcs   These are very stable, globe encircling arcs of pure red line emission in OI (^{3}P $-$ 'D)($\lambda\lambda$6300 $-$ 6364 Å) at about 50° geomagnetic latitude region and span several hundred kilometers in north-south direction. They are situated in the upper F region with intensity from several hundreds to several thousands of rayleighs. They are observed during highly geomagnetic disturbed periods close to the plasmapause. The excitation mechnanism is thermal excitation by high electron temperature conducted down from the magnetosphere. The energy comes from the decay of the ring current in the recovery phase of the geomagnetic storm, when it is closest to the plasmapause.

stable causality   Theorem (Hawking): A space-time is stably causal, if and only if, it admits a foliation with a family of space-like hypersurfaces.

Stable causality holds in a space-time if no arbitrary but small variation of the metric gives closed causal curves. Stable causality implies strong causality.

stable equilibrium   In mechanics, a configuration in which the system experiences no net force (equilibrium) and slight displacements from this state cause forces returning the system to the equilibrium state. The paradigm is a pendulum at rest at the bottom of its arc. *See* neutral equilibrium, unstable equilibrium.

standard atmosphere   An idealized atmospheric structure in the middle latitudes up to 700 km level. It is defined in terms of temperature at certain fixed heights; between these levels temperatures are considered to vary linearly and other properties such as density, pressure, and speed of sound are derived from the relevant formulas. In it, at the sea level, the standard gravitational acceleration is 9.80665 m/s^2, the temperature is 15°C, the pressure is 1013.25 hPa, density is 1.225 kg/m^3; From ground to 11 km is troposphere with 0.65°C/100 m lapse rate; From 11 to 20 km is the stratosphere with constant temperature; From 20 to 32 km, the lapse rate is -0.1°C/100 m.

standard candle   An astrophysical object with a specific, known, or calibrated brightness. Such an object can be used to measure distance by comparing apparent to (known) absolute magnitude. Supernovae of type Ia are *standard candles* for cosmological distance determinations since all type Ia supernovae have roughly equal maximum brightness.

standard pressure   Adopted values of pressure used for specific purposes; the value of a *standard pressure* of one atmosphere (1013.25 hPa) is defined as the pressure produced by a 76.0 cm height column of pure mercury whose density is 1.35951×10^4 kg/m^3 at temperature of 0°C, under standard gravitational acceleration $g = 9.80665$ m/s^2.

standard project hurricane   A theoretical, parameterized storm used for design purposes, "intended to represent the most severe combination of hurricane parameters that is reasonably characteristic of a region excluding extremely rare combinations" (U.S. Army Corps of Engineers, Shore Protection Manual, 1984).

standard temperature and pressure   The values of temperature and pressure defined in the standard atmosphere. *See* standard atmosphere.

standard time   The time in one of 24 time zones generally used as the civil time of the nations therein, except when daylight saving adjustments are made.

standing shock   Shock building in front of an obstacle in a super-sonic flow. Examples are planetary bow shocks (super-sonic solar wind hits a planet's magnetosphere).

standing wave   A wave pattern produced by oscillation on a finite domain with specific (in general mixed: a linear combination of value and first derivative) boundary conditions on the boundary of the domain.

In one dimension this can be visualized as a superposition of waves of equal frequency and amplitude, propagating in opposite directions. Oscillations of a piano string, for instance, constitute a transverse *standing wave.*

stand-off distance   Distance of the sub-solar point of a planetary magnetopause, that is the intersection of the magnetopause with the sun-planet line, from the center of the planet; in general measured in units of the planetary radius. The *stand-off distance* can be used as a crude measure for the size of the magnetosphere.

Stanton number (*St*)   A dimensionless quantity used in fluid mechanics, defined by $St = h/\rho v C_p$, where h is coefficient of heat transfer, ρ is density, v is velocity, and C_p is specific heat capacity at constant pressure.

starburst galaxy   A galaxy undergoing a strong episode of star formation. *Starburst galaxies* can be more quantitatively defined as galaxies whose total star formation rate cannot be sustained over the age of the universe (the Hubble time). This criterion is very general, and includes spiral galaxies whose nuclei show emission lines typical of HII regions, as well as star-forming dwarf galaxies. Starburst nuclei exhibit, along with optical and UV nuclear spectra typical of star forming regions, an excess of mid and far infrared emission with respect to normal galaxies of the same morphological type. Associated features include often high ultraviolet emission, galactic superwinds, dominant photoionization lines in the far infrared spectra, evidence for active periods in the recent star formation history, etc. Spectral features of Wolf Rayet stars (which are very bright and have broad, easily-identified emission lines) imply a recent burst of massive star formation, since these last only 3 to 4 Myr before they explode. Red supergiants, that last $\approx$ 10 Myr, signify a somewhat older burst. Starburst galaxies often show evidence of some sort of interaction, harassment or merger in their recent past. Prototype Starburst galaxies are the edge-on spiral NGC 253 and the dwarf irregular galaxy Messier 82. FIR-strong galaxies, blue galaxies, UV excess galaxies, or H$\textsc{ii}$ galaxies are often systems where star formation is enhanced with respect to normal galaxies, and can be synonyms of Starburst galaxies. The different names reflect the different technique and the different spectral range of observation and discovery.

Stardust   A spacecraft launched on February 7, 1999, that is the first space mission that will fly close to a comet and bring cometary material back to Earth for analysis. It was developed under NASA's Discovery Program of low-cost solar system exploration missions. Its primary goal is to collect comet dust and volatile samples during a planned close encounter with comet Wild 2 in January of 2004. The spacecraft will also bring back samples of interstellar dust including the recently discovered dust streaming into the solar system from the direction of Sagittarius. A unique substance called aerogel is the medium that will be used to catch and preserve comet samples. When *Stardust* swings by Earth in January 2006, the samples encased in a reentry capsule will be jettisoned and parachuted to a pre-selected site in the Utah desert. Ground-based analysis of these samples should yield important insights into the evolution of the sun and planets, and possibly into the origin of life itself.

Stardust is a collaborative effort between NASA, university and industry partners.

star formation   The process by which diffuse gas collapses to form a star or system of stars. This follows the concentration of the atomic gas within galaxies into dense, opaque, molecular clouds, and the further condensation of regions within these clouds into dense cores that become unstable to collapse. During collapse, a star and its surrounding protoplanetary accretion disk grow from infalling matter.

Stars are born with vastly less magnetic flux and angular momentum than their parent clouds. The loss of magnetic flux is attributed to ambipolar diffusion, the motion of neutral material across magnetic field lines, that operates in molecular clouds. Magnetic flux may be also lost by reconnection. Angular momentum is lost by magnetic braking before collapse, and then by the emission of a powerful, collimated wind powered by accretion through the disk. Such winds strongly affect the star-forming cloud by adding turbulent motion and by casting away material.

star formation rate (SFR)   The rate at which new stars are being formed in a galaxy, or in any star forming region, often measured in units of solar masses per year. The *star formation rate* can be estimated from measurement of the luminosity of the Hydrogen H_α spectral line, or from the luminosity emitted in the far infrared, if the initial mass function of stars is assumed. Galaxies of late morphological types, on the contrary, have yet to exhaust their molecular gas, and are still forming stars. The star formation rate in galaxies varies widely along the Hubble sequence: For galaxies in the local universe, typical values are close to zero for elliptical galaxies and $\sim$ 10 solar masses per year for Sc spirals.

Stark effect   The sub-division of energy levels within an atom caused by the external application of an electric field. For hydrogenic atoms, the energy shift is directly proportional to the applied field strength; this is known as the linear *Stark effect*. For other atoms and ions, the displacement of the energy levels is proportional to the square of the applied electric field strength, giving rise to the quadratic Stark effect. The latter results from the additional condition that the induced dipole moment is also proportional to the intensity of the applied electric field. For hydrogen atoms, electric fields of intensity > 10 V/cm are required to produce an observable Stark effect, with greater fields being required to produce the quadratic Stark effect.

star spots   The analog of sunspots on other stars. They are usually darker (cooler) than their surroundings (though bright spots also occur) and are associated with concentrations of magnetic field in their vicinity, as are sunspots. The field apparently inhibits energy transport by convection in the atmospheres of the relatively cool stars that have spots, leading to cooler, darker regions. *Star spots* are particularly conspicuous (and sometimes found near the poles rather than near the equator as in the sun) in young, rapidly rotating stars, and in close binary pairs, where the rotation period is synchronized to the short orbital period. They show up as changes in the brightness and spectrum of the star that approximately repeats at the rotation period.

statcoulomb   Unit of electric charge, 3.3356 $\times 10^{-10}$ coulomb. Previously defined as equal to the charge that exerts a force of 1 dyne on an equal charge at a distance of 1 centimeter in vacuum.

state parameter   *See* electric regime (cosmic string).

steady-state model   The cosmological model of Bondi, Gold, Hoyle and Narlikar in which matter is continuously created to fill the voids left as the universe expands. Consequently, such a universe has no beginning and no end and always maintains the same density. This model of the universe is thus derived from a stronger version of the cosmological principle, that the universe is not only homogeneous and isotropic (*see* homogeneity, isotropy), but also unchanging in time. This stronger version, the "perfect cosmological principle," was postulated as a universal law of nature. The rate at which new matter would have to appear to offset the dilution caused by expansion is one hydrogen atom in one liter of volume once in 10^9 years. There is no experimental evidence against this assumption since the rate is undetectably small. The steady state models were motivated by the discrepancy between the age of the universe and the age of the Earth — the latter was apparently longer than the former; the *steady-state model*

removed this discrepancy by making the universe eternally-lived. The discrepancy was later resolved as an error in estimating the value of the Hubble parameter that led to a drastic underestimation of the age of the universe. The steady-state model was definitively proven false when the microwave background radiation was discovered in 1965. This radiation finds a simple explanation in models with a Big Bang, and no natural explanation in the steady-state model.

steepness (wave)   A term used in the study of water waves, defined as the ratio of wave height, H, to wavelength, L, or H/L. A common rule of thumb is that the wave steepness cannot exceed 1/7 in deep water before wave breaking will occur.

Stefan–Boltzmann constant (σ)   Constant in the equation for the radiant emittance M (radiant energy flux per unit area) from a black body at thermodynamic temperature T: $M = \sigma T^4$.

$$\sigma = \frac{2\pi^5 k^4}{15 c^2 h^3} \quad = \quad 5.66956 \times 10^{-5}$$
$$\text{erg cm}^{-2} \text{deg}^{-4} \text{sec}^{-1}$$

where k is Boltzmann's constant, h is Planck's constant, and c is the speed of light.

stellar activity   The complex of X-ray, visible light, and radio wave phenomena associated with rapid rotation, strong magnetic fields, and the presence of a chromosphere and corona on a star (generally one whose photospheric temperature is less than about 6,500 K). These phenomena include emission lines of hydrogen, calcium, and sodium emitted by the chromosphere, X-rays and occasionally radio waves emitted by the corona, starspots, with magnetic fields up to several thousand gauss (generally detectable only on the sun), and flares.

stellar classification   A classification of stars according to the observed surface temperature; see below.

There is a further subdivision within each range with the numbers from 0 to 9, eg G5, A0.

The *stellar classification* is often called the Henry Draper system. It was first published (1918–1924) by Annie Jump Cannon of Harvard

Type	Temperature	Color	Spectral Lines	Example Star
O	35000 K	blue	singly ionized helium	spiral arm blue stars
B	11000 to 28000 K	blue-white	hydrogen, oxygen, neutral Helium in absorption	Rigel, Spica
A	10000 K	white	hydrogen Balmer	Vega, Sirius
F	6000 to 7500 K	yellow white	Hydrogen, singly ionized Calcium	Canopus, Procyon
G	5000 to 6000 K	yellow	Calcium, neutral and ionized metals	sun, Capella
K	3500 to 5000 K	orange	Strong H and K lines of Calcium	Arcturus, Aldebaran
M	less than 3500 K	red	molecular	Betelgeuse and Antares

College Observatory with the title Henry Draper Catalog because Draper's widow sponsored the project in his honor.

stellar evolution   The set of processes that gradually transform a newly-formed, chemically homogeneous (main sequence) star into sequential phases like red giant or supergiant, horizontal branch star, Cepheid variable, and so forth onward to death as a white dwarf, neutron star, or black hole. The star changes its brightness and surface temperature (and so can be followed on an HR Diagram). It also burns a sequence of nuclear fuels from hydrogen and helium burning (for low mass stars) on up to silicon burning (for stars of more than about 10 solar masses). It is the changes of chemical composition resulting from these nuclear reactions that are primarily responsible for the changes in luminosity and temperature that we see. Detailed calculations of *stellar evolution* are done by solving a set of four coupled, non-linear differential equations and allowing for these composition changes.

stellar population   The observed color of stars depends on the surface temperature. More massive stars have higher surface temperatures and are therefore bluer than lower mass stars. The surface temperature also depends on the chemical composition of a star. Many stars have a composition similar to the sun. Such stars are about 23% Helium and have traces of elements more massive than He (*see* metallicity). However, the majority of the mass is Hydrogen. Such stars are called Population I stars. Population II stars are those with a greatly reduced concentration of elements with an atomic number greater than 2 (i.e., Helium). Population II stars are deficient in these elements by factors up to 10,000 relative to Population I stars. This difference in composition causes the star to be less opaque to light and results in a much bluer star, for a given mass.

stellar winds   Gas blowing off the surface of a star, driven by some combination of magnetic energy and acoustic waves. The solar wind is a mild one, carrying only about 10^{-19} solar masses per year. Stars that are very bright for their mass or very rapidly rotating can have winds up to 10^{-4} solar masses per year, for brief periods of time. Stellar winds are one source of mass transfer in cataclysmic variables and X-ray binaries and a short-lived superwind phase at the end of the life on an asymptotic giant branch star is probably the origin of planetary nebulae.

stepped leader   In lightning strokes, the initiating process is that charge is forced down a broken path from the cloud, in steps of about 30 m in length taking about 1 ms for each step, with about 50 ms of pause between steps. This is called the *stepped leader. See* return stroke, dart leader.

steric height   An oceanographic quantity describing the depth difference between two surfaces of constant pressure. The *steric height h* is defined by

$$h(z_1, z_2) = \int_{z_1}^{z_2} \delta(T, S, p)\rho_0 \, dz$$

where z_1 and z_2 are the depths of the pressure surfaces, δ is the specific volume anomaly, T is the temperature, S is the salinity, p is the pressure, and ρ_0 is a reference density. It has the dimension of height and is expressed in meters.

stick-slip   The behavior of a fault that generates earthquakes. When the tectonic stress is low a fault "sticks". When the tectonic stress exceeds the rupture strength of the fault, "slip" occurs on the fault generating an earthquake.

stiff fluid   A fluid in which the energy-density ρ equals p/c^2 at every point (where p is the pressure and c is the speed of light), so that the velocity of sound is equal to the velocity of light. This is a theoretical construct (no known physical system has such a large pressure). The equality between the velocity of sound and the velocity of light greatly simplifies the dynamical properties of such a medium, and several problems that would be difficult to solve in realistic matter become solvable in a *stiff fluid*. One example is the propagation of soliton waves through such a medium. Also, solutions of Einstein's equations in which a stiff fluid is a source are easier to find than solutions with more realistic sources.

still water level   The level at which the water in an ocean or lake would be if all waves were absent. Differs from the mean water level because of the influence of wave setup and setdown.

stishovite   A high pressure polymorph of quartz having a density of about 1.6 times that of ordinary quartz thought to be naturally stable only in the Earth's mantle. *Stishovite* was first synthesized in the laboratory by Sergei Stishov in 1960, but later small crystals were discovered in nature at the site of the Barringer Meteor Crater in Arizona. Stishovite has the rutile structure and is one of the very few known materials in which Si occurs in octahedral coordination with oxygen.

stochastic acceleration   The energization of a population of particles in which the spread in momenta of the accelerated particles increases at the same time as the average energy per particle. Physically, *stochastic acceleration* of a particle occurs via a series of random kicks from the interaction with ambient particles.

Stokes' law   The statement, valid when viscous forces dominate inertial effects, that the viscous force F experienced by a sphere of radius a moving at velocity v in a medium of viscosity η is given by $F = -6\pi\eta av$.

Stokes parameters   Four parameters used to describe polarized radiation; one component describes the radiance, two components describe

the states of linear polarization, and one component describes the state of circular polarization.

Stokes parameters   Four quantities obeying $s_0^2 = s_1^2 + s_2^2 + s_3^2$, describing the polarization state of a beam of light. s_0 is the total intensity, s_1 is the difference in intensity of the two polarizations. The other two parameters describe the relative phases of the two polarization states. *Stokes parameters* can be defined in terms of linear, circular polarizations. (The resulting parameters differ in the two bases, for the same beam of light.)

Stokes polarimetry   The means by which to measure the amount of polarization in light from the sun in order to yield information about the direction of the magnetic field in three-dimensions. An accurate determination of the vector magnetic field, using the Zeeman effect, needs a precise measurement of the four Stokes parameters.

Stokes' Theorem   In three dimensions, the surface integral over a topological disk, of the dot product of the curl of a vector with the normal to the disk ($\mathbf{n} \cdot \nabla \times \mathbf{A}$) equals the line integral around the boundary of the surface of $\mathbf{A} \cdot d\mathbf{l}$, where $d\mathbf{l}$ is the directed line element in the boundary. In general, the integral of an exact differential form $d\omega$ in a region equals the integral of its potential ω in the boundary of the region.

Stokes wave theory   A theory for description of surface water waves. Generally refers to first-order (linear) wave theory, but higher-order terms may be retained during the derivation to yield an nth-order *Stokes wave theory*.

stony-iron meteorite   A meteorite that shows a combination of stone and nickel-iron. This type of meteorite may have been formed from the core-mantle boundary of a parent body that broke up to produce the meteorite, or may have been formed in an energetic collision that partly melted the colliding objects.

storativity (S)   The volume of water produced from a confined aquifer per unit area per unit decline in the potentiometric surface is a dimensionless number that is analogous to specific yield and is the product of the specific storage S_s and the aquifer thickness b, or $S = bS_s$.

Störmer orbits Orbits (trajectories) of energetic particles in the terrestrial magnetic field. These orbits first were calculated for auroral particles by C. Störmer by integrating the equation of motion along the particle orbit. Such calculations are of particular importance for cosmic ray particles because their gyro-radii are comparable to the size of the magnetosphere, thus concepts such as the adiabatic invariants cannot be applied. The particle orbit depends on the location and direction of incidence and can appear to be very irregular with large excursions in longitude and latitude before the particle either hits the denser Earth's atmosphere and is lost due to interactions or before it escapes back into interplanetary space.

The most important result of Störmer's calculations is the definition of allowed and forbidden regions on the ground that can be reached by a charged particle traveling towards the Earth. For instance, to hit the Earth at a certain geomagnetic latitude Φ_c, the particle's rigidity must be at least the cut-off rigidity $P_c = 14.9 \, \text{GV} \cdot \cos^4 \Phi_c$. The extent of the forbidden region, that is a sphere around the center of the Earth that cannot be penetrated by particles with a rigidity P, is given by the Störmer unit $C_{St} = \sqrt{M_E/P}$ with M_E being the magnetic moment of the Earth.

Störmer theory A theory of the motion of energetic ions in the field of a magnetic dipole, derived around 1907 by Carl Störmer. Though the theory was intended to explain the aurora, its main application was to cosmic rays. It showed that if all cosmic rays came from great distances and the Earth's magnetic field could be approximated by a dipole, particles of low energy were prevented from reaching parts of the Earth. Each point on Earth has (for protons) a cut-off energy below which protons from infinity cannot reach it, and at energies slightly above that protons coming from the west can arrive but those from the east are cut off. *See* east-west effect.

Störmer unit A unit of distance r_{st} used in Störmer theory; when distances are expressed in it, the equations become dimensionless. For particles of mass m, charge Ze (e the electron charge) and momentum p, $r_{st}^2 = Zem/p$.

storm sudden commencement (SSC) Geomagnetic storms may commence abruptly, the commencement of the storm being observed at nearly the same time at magnetic observatories around the globe. The sharp onset of storm conditions is called a *storm sudden commencements,* or SSC. The storm sudden commencement can usually be identified with a shock, propagating in the solar wind, reaching the Earth's magnetosphere. Since the shock and the solar wind particles responsible for the longer-lived geomagnetic storm effects propagate independently, associating a particular shock with a particular storm can be ambiguous. Nevertheless, historically, sudden commencement storms were associated with flares on the sun (although the processes are now recognized to be more complex), and are most dramatic geophysical events, although they may only last for a few days. *See* geomagnetic storm, gradual commencement storm.

storm surge Superelevation of the mean water level due to the effects of the reduced atmospheric pressure that accompanies a major storm, and the strong wind shear stress that results. *Storm surge* due to hurricanes or typhoons sometimes exceeds 6 m.

storm track An east-westward oriented zone where subtropical storms prefer to develop and travel. It is often defined as a region of large temporal variance of geopotential height in the upper troposphere. The formation of the storm tracks is closely related to the distribution of the westerly wind velocity. In the Southern Hemisphere, both the westerly wind speed and storm track are nearly uniform in the longitudinal direction. In the Northern Hemisphere, in contrast, winter storminess displays large variations in the longitudinal direction, reaching maximum over the North Pacific and North Atlantic. The North Pacific (North Atlantic) *storm track* is located downstream of the westerly wind speed maximum over Japan (eastern United States). These longitudinally localized jet streams and storm tracks are caused by large-scale orographic features such as the Tibetan

Plateau and the Rockies, and by thermal forcing such as land-sea heat contrast and tropical convection.

strain   The deformation of a body that results from an applied stress.

strain hardening   A solid yields plastically at a certain state of stress and strain, called its elastic limit. For some solids, subsequent plastic deformation requires greater loading beyond the current elastic limit. This phenomenon is called *strain hardening* because the solid becomes stronger with increasing plastic strain. It is due to a variety of processes such as dislocation tangles that lead to situations where greater differential stress is required to maintain a given strain rate. If the solid becomes weaker with increasing plastic strain, the phenomenon is called strain softening.

strain partitioning   The heterogeneous nature of rocks and their deformation means that on all scales from macro- to micro-, strain is partitioned between areas of higher strain and lower strain. Estimates for bulk strain obtained (on whatever scale) represent the average of these partitioned strains.

strain softening   Softening, in mechanical terms, can be expressed as a reduction in differential stress to maintain constant strain rate or an increase in strain rate can occur by a number of processes including: change in deformation mechanism, geometric softening, continual recrystallization, reaction softening and chemical softening. *See* strain hardening.

stratosphere   The part of the Earth's atmosphere extending from the top of the troposphere (typically 10 to 15 km above the surface) to about 50 km. The stratopause, at about 50 km, separates it from the higher mesosphere. Within the *stratosphere,* the temperature slowly increases with altitude, due primarily to the absorption of ultraviolet radiation by the ozone layer, and reaches 270 to 290 K at stratopause. This temperature variation inhibits vertical air movements and makes the stratosphere stable. Thus, the stratosphere has clear horizontal structure. The heating arises from the absorption of high-energy solar ultraviolet radiation by ozone molecules. The moisture content in the stratosphere is very small, and in general, there are no complex weather phenomena in the stratosphere. The stratosphere and mesosphere combined are often called the middle atmosphere.

stratovolcano   *See* composite volcanos.

stratus   *See* cloud classification.

streakline   The line that results if dye is injected at a fixed point within a moving fluid. Used in the study of fluids.

streamer belt   Region around the solar equator consisting of closed loops and helmet streamers. The *streamer belt* is associated with the active regions on the sun, thus its latitudinal extent as well as its inclination with respect to the solar equator varies with the solar cycle. The solar wind originating in the streamer belt can be slow or fast, depending on the topology. Therefore, in interplanetary space, fast and slow solar wind streams can be observed which at larger radial distances eventually form corotating interaction regions. At higher latitudes, that is above the streamer belt, the solar wind always is fast.

stream function wave theory   A description of surface water waves that involves a numerical solution for stream function. Typically solved via a numerical technique that determines a set of coefficients that yields a best fit to a particular set of boundary conditions.

streamline   A line drawn such that it is everywhere tangent to a velocity vector within a flowfield. *Streamlines* are used to visualize a flowfield at a particular instant of interest.

stress   Force per unit area applied to a body. Tensile *stress* tends to stretch and compressional stress to compress the body in the direction of the applied force; shear stress tends to distort the body.

stress drop   The amount of decrease in average shear stress along the rupture surface of an earthquake as a result of the earthquake. Can also be used to indicate any stress decrease.

stress energy tensor   A 4-dimensional symmetric tensor $T_{\mu\nu}$ combining the 3-dimensional stress tensor of a substance with its energy density and energy flux:

$$T_{oo} = \text{energy density}$$

$$T_{oi} = \text{momentum density} = \text{energy flux}$$

$$T_{ij} = \text{3-dimensional stress tensor.}$$

This tensor is the source of the Einstein equation for the graviational field:

$$G_{\mu\nu} = 8\pi T_{\mu\nu}$$

where $G_{\mu\nu}$ is the Einstein tensor. (Here μ, $\nu = 0, 1, 2, 3$; i, $j = 1, 2, 3$.)

striations, auroral   A banded structure of auroral arc, believed to be the overlap of moving sections that overtake each other.

strike-slip fault   A fault where the movement is horizontal and parallel to the strike (i.e., the angle between true north and the direction of any horizontal planar features) of the fault plane. *Strike-slip faults* can be right-lateral or left-lateral, depending on the direction of displacement. As one faces a strike-slip fault, if the rocks across the fault move toward the right, the fault is then a right-lateral strike-slip fault. Similarly if the rocks across the fault move toward the left. Transform boundaries at plate margins are usually composed of strike-slip faults.

string model   *See* cosmic string, elastic string model.

string network   *See* scaling solution (cosmic string).

stromatolite   The oldest known fossils, dating to 3.5 billion years; the oldest are found in Australia. Colonial structures formed by photosynthesizing cyanobacteria (blue-green algae) and other microbes. These prokaryotes (lacking a cellular nucleus) thrived in warm shallow environments and built layered reefs of calcium carbonate and sediment precipitating from the water, which were trapped within the colony. The colony then grew upward through the sediment to form a new layer. In some cases, bacteria was also fossilized within the mats. This process still occurs today particularly in Shark Bay in western Australia.

strong anthropic principle   The concept that the universe must be such as to admit intelligent observers or, even stronger, that the universe must produce such observers. *See* anthropic principle.

strong scattering limit   In magnetospheric plasma physics, a type of MHD for a guiding center plasma which assumes plasma particles are always isotropic, i.e., they flow with equal intensity in all directions. That is equivalent to assuming that some process keeps scattering the particles, thoroughly mixing their directions of motion. The specific volume V is an important variable in this theory, and Euler potentials are its natural coordinates.

Strouhal number (Sr)   A dimensionless quantity used in fluid mechanics, defined by $Sr = lf/v$, where l is length, f is frequency, and v is velocity.

structure coefficients   The set of functions C_{bc}^a appearing in the equation

$$d\sigma^a = \frac{1}{2}C_{bc}^a\sigma^b \wedge \sigma^c .$$

A set of basis 1-forms $\{\sigma^a\}$ defines the covariant vector basis. 1-forms are acted on by an exterior derivative to give a 2-form (an antisymmetric rank-2 covariant tensor). Here $\wedge$ is the antisymmetric tensor product, i.e.,

$$\sigma^b \wedge \sigma^c = \frac{1}{2}\left(\sigma^b \otimes \sigma^c - \sigma^c \otimes \sigma^b\right) .$$

If $\{\mathbf{e}_a\}$ is a vector basis dual to the $\{\sigma^b\}$, then the *structure coefficients* also give the commutation rules for the vectors:

$$[\mathbf{e}_a , \mathbf{e}_b] = C_{ab}^d\mathbf{e}_d .$$

The set of antisymmetric tensor products of the basis 1-forms is itself a basis for the 2-forms. *See* 1-form, 2-form, tensor.

structure toe   The place where the bottom of a structure meets the seafloor. Often a location of severe scour.

St. Venant Equations The equations of mass conservation and momentum for two-dimensional flow in an open channel. They assume negligible vertical acceleration, and therefore hydrostatic pressure distribution.

subcritical flow Flow in an open channel with a Froude Number less than unity. The Froude Number is defined as $Fr = V/\sqrt{gD}$, where V is the flow speed, g is acceleration of gravity, and D is the hydraulic depth.

subduction The process at ocean trenches where the oceanic lithosphere bends and descends into the interior of the Earth. *See* convergent boundary.

subduction zone A zone where a cold, dense oceanic plate is subducting beneath a less dense continental plate, located at the convergent boundary of the two plates. Two kinds of systems are formed at a *subduction zone:* island arc-trench systems and arc of continental margin-trench systems. An example of the former is the Japanese islands where the Pacific plate is subducting, accompanying marginal sea, while an example of the latter is the Andes where the Nazca plate is subducting. Circum-Pacific *subduction zones* are the most active seismic regions on the Earth, and trench-type great earthquakes take place frequently there. Deep earthquakes occur along inclined planes that deepen from a trench toward the continental side. Volcanic chains are formed in the continental plate, being almost parallel to the trench axis.

subdwarf In astronomy, a star that is smaller, less luminous and of lower metallicity than a main sequence (dwarf) star of the same spectral type.

subgiant A star in which sufficient helium has accumulated in its core to inhibit hydrogen burning. The core begins to contract and hydrogen shell burning begins. The star is then a subdwarf. It then moves off the main sequence and becomes larger, brighter and cooler as the core collapses and shell burning increases, and tracks across the HR diagram along the *subgiant* branch toward the red giant branch.

subgrain size Sub-grains develop in deformed crystals. Slightly misoriented regions (by a few degrees relative to the parent grain) of a deformed grain, separated by dislocation walls, are defined as subgrains. These subgrains form at the beginning of the deformation and their size (*subgrain size*) remains constant once they are formed. The subgrain size does not practically depend on temperature but only on the applied stress with the empirical law best written under as:

$$d = K\frac{\mu b}{\sigma}$$

where d is the average size of the subgrains, μ is the average shear modulus, **b** is the Burgers vector of the active dislocation, K is a non-dimensional constant of proportionality, and σ is the stress.

submarine canyon A steep valley (the location of a former river valley during a time of lower sea level) found in nearshore waters. Often located near the mouth of modern rivers.

subsidence The sinking of the Earth's surface via a variety of processes. When *subsidence* occurs, the surface is often covered with sediments to form a sedimentary basin.

subsolar point The point on the Earth's magnetopause closest to the sun, also known informally as the "nose" of the magnetosphere. The distance from the center of the Earth to the *subsolar point* is known as the subsolar distance; a typical value is 10.5 R_E, but cases are known of it ranging from 6 to 14 R_E. The subsolar points of the Earth's bow shock, and of magnetopauses and bow shocks of other planets, are similarly defined.

subsonic string model *See* elastic string model.

substorm injection A sudden increase in the flux of energetic ions and electrons at a point in space, associated with the arrival of substorm-accelerated particles. For ions in particular, whether the increase at lower energies starts after the one at higher energies ("velocity dispersion") or is simultaneous ("dispersionless injection") is believed to indicate whether the ions were convected from a distant location or were accelerated locally.

substorm phases Stages in a typical magnetospheric substorm, identified by S.-I, Akasofu, Robert McPherron and others. They begin with the growth phase, in which field lines on the night side are stretched, typically lasting 40 min. That is followed by onset or breakup, when the midnight aurora brightens and rapidly expands poleward (as well as laterally), tail lines snap back to less stretched shape ("dipolarization"), the auroral electrojets and Birkeland currents of the substorm wedge intensify, and large numbers of ions and electrons in the geotail are convected earthward and energized, either by convection or by a local mechanism. This typically lasts 20 min and is followed by a gradual recovery phase in which the auroral oval returns to its initial state.

substorm triggering The event that initiates a magnetospheric substorm, still a subject of controversy. An interval in which the north-south component B_z of the interplanetary magnetic field is directed southward is necessary to prepare the magnetosphere for a substorm, but in some storms a sudden northward turning is thought to be the final trigger.

substorm wedge An electric current circuit that becomes active in magnetospheric substorms. It seems to be caused by the diversion of some of the cross-tail current flowing across the plasma sheet, causing part of it to flow in a wedge-shaped circuit—Earthward along magnetic field lines, in the ionosphere along the midnight auroral oval, and back out to the tail. The physical mechanism of the diversion of current ("tail current disruption") is still poorly understood.

subtropical high-pressure belt High-pressure belt that lies over both the Pacific and Atlantic near 30°N and separates the predominantly easterly flow in the subtropics from the southwesterly flow over the high latitude oceans in the Earth's atmosphere. These subtropical high-pressure belts are characterized by light winds and absence of storms.

sudden frequency deviation (SFD) When a solar flare occurs, abrupt changes in HF radiowave frequency on a link, called a *sudden frequency deviation (SFD),* occur due to rapid changes in the propagation path as ionization changes. SFD are thought to have their origins in the lower F region. *See* short wave fadeout.

sudden ionospheric disturbance (SID) Sudden increase in ionization at the dayside ionosphere due to the enhanced hard electromagnetic radiation emitted in a flare. The enhanced X-rays and ultraviolet (UV) radiations during solar flares cause increased ionization at several levels of the ionosphere and may last 1 h or so. A number of ionospheric effects kown as *sudden ionospheric disturbances (SID)* result. These are:

1. Short wave fadeout (SWF). A sharp drop in high frequency transmission in the 2 to 30 MHz range.

2. Sudden Cosmic Noise Absorption (SCNA). A decrease in the constant galactic radio noise intensity in the 15 to 60 MHz range.

3. Sudden Phase Anolaly (SPA). A change in phase in very low frequency (VLF) transmissions in the 10 to 150 kHz range relative to a frequency standard.

4. Sudden Enhancement of Signals (SES). The enhancement in the VLF transmission intensity in the 10 to 150 kHz range.

5. Sudden Enhancement of Atmospherics (SEA). An increase in the background of VLF noise from distant thunderstorms.

6. Sudden Frequency Deviation (SFD). A short lived increase in the high frequency (HF) signal from a distant transmitter.

7. Solar Flare Effect (SFE) or Geomagnetic Crochet. A sudden variation in the H component of the Earth's geomagnetic field.

1, 3, 4, 5 are due to increase in the D region ionization from soft X-rays in the 1 to 8 A wavelength range. 2 and 6 are due to increased ionization and collision frequency in the F region. The lowering of the D region also contributes to 3. 7 is produced by increased D and E region conductivities and ionospheric currents.

sudden phase anomaly (SPA) Coincident with a solar flare, abrupt phase shifts may occur on VLF and LF radio signals propagated by the ionosphere. When the D-region ionization increases, the reflection height is lowered correspondingly and the altered path-length is responsible for the SPA. *See* short wave fadeout.

sudden stratospheric warming   A sudden increase temperature in the stratosphere that occurs in winter and can be driven by upward-propagating planetary waves. In a major event of this type, the temperature at the 10 mb (about 31 km) level at the North Pole may increase by 40 to 60 K in less than 1 week.

sulfur dioxide SO_2, a colorless, toxic gas. Freezing point $-72.7°C$; boiling point $-10°C$. Volcanic eruptions provide a natural source of *sulfur dioxide* in the atmosphere. However, sulfur dioxide is a major anthropogenic primary atmospheric pollutant, originating from fossil fuel combustion. In the atmosphere, sulfur dioxide is oxidized to form sulfur trioxide, that is extremely soluble in water. In the presence of water, droplets of sulfuric acid (acid rain) are formed. Sulfuric acid droplets also appear to act as nucleation centers, increasing the amount of rainfall. High airborne concentrations of sulfur dioxide may aggravate existing respiratory and cardiovascular disease, forming sulfuric acid in the bronchia.

summation convention   In equations involving vectors or tensors, and in matrix equations, one often encounters sums of the form $\sum_\alpha A_\alpha B^\alpha$, where $\{A_\alpha\}$ and $\{B^\alpha\}$ are collections of perhaps different kinds of objects, but of equal number labeled by and summed over the full set of integers $\alpha \in [\alpha_{min}, \alpha_{max}]$.

Einstein noted that in a large number of situations, the expressions unambiguously require the sum. In such cases we can omit writing the $\sum_\alpha$. This is the Einstein *summation convention*. In the unusual circumstance where one wants to examine the symbol $A_\alpha B^\alpha$ for one fixed value of α only, one appends a "no sum" to the expression. Formulae in this dictionary use the summation convention. Similar notations have been extended to continuous sums (i.e., integration).

summer solstice   The point that lies on the ecliptic midway between the vernal and autumnal equinoxes and at which the sun, in its apparent annual motion, is at its greatest angular distance north of the celestial equator. On the day of the *summer solstice,* that occurs on about June 21, the length of daytime is at its maximum in the northern hemisphere. After summer solstice, the northern length of daytime will decrease until the winter solstice. Because of complicated interactions with atmosphere ocean and land heat reservoirs, northern surface temperatures continue to increase for a period of time after the summer solstice. In the southern hemisphere, the longest day, and the beginning of southern summer, occur on the winter solstice, about December 21.

sun   Our star; an incandescent, approximately spherical star around which there exists a system of nine planets (the "solar system"), rotating in elliptical orbits. The *sun* is a main sequence star of type G2 V with an absolute magnitude of 4.8. The sun has a mass of 1.99×10^{30} kg (roughly 99.9% of all the matter in the solar system) a radius of 696,000 km, a mean distance from the Earth of 150 million kilometers (denoted an Astronomical Unit), a surface gravity of 274 m/s^2, a radiation emission of 3.86×10^{26} W, an equatorial rotation rate of 26 days and an effective temperature of 5785 K. The sun has a radiative interior surrounded by a convective zone both of which participate in the transportation of energy from the nuclear burning core to the surface.

The solar atmosphere consists of a number of distinct temperature regimes: the photosphere, chromosphere, transition region, and corona.

sunlit aurora   Aurora occurring in the upper atmosphere around the polar cusps of the Earth, observed in the ultraviolet by satellite imagers.

sunspot   A generally irregular dark spot, with considerable internal detail on the surface of the sun. A well-developed *sunspot* consists of a central circular or elliptical umbra, of lower luminosity, surrounded by a brighter penumbra. Sunspots are violent eruptions of gases cooler than the surrounding surface areas. The temperature of sunspots is about 1000 K lower than the typical temperature of the photosphere of the sun. Sunspots tend to occur in groups, are relatively short-lived, usually less than a day, but some sunspots can live as long as a month, and some special sunspots can live even to half a year. Sunspots occur in cycles of 11 or 12 years. They are restricted to regions (solar active regions) of 5°N −40°N and 5°S −40°S, moving from higher to lower latitudes during a sunspot cycle. At the beginning of a sunspot cycle, sunspots appear near 40°N and 40°S and move to lower latitude. Their number increase and then decrease during the moving processes, and reach a minimum at about 5°N and 5°S regions at the end of the sunspot cycle. Sunspots are characterized by strong magnetic fields (up to several thousand tesla), being concentrations of magnetic flux, typically in bipolar clusters or groups and are associated with magnetic storms on the Earth. Sunspots are the most obvious manifestation of solar activity. *See* sunspot cycle, sunspot number.

sunspot classification   A - A small single unipolar sunspot or very small group of spots without penumbra. B - Bipolar sunspot group with no penumbra. C - An elongated bipolar sunspot group. One sunspot must have penumbra. D - An elongated bipolar sunspot group with penumbra on both ends of the group. E - An elongated bipolar sunspot group with penumbra on both ends. Longitudinal extent of penumbra exceeds 10° but not 15°. F - An elongated bipolar sunspot group with penumbra on both ends. Longitudinal extent of penumbra exceeds 15°. H - A unipolar sunspot group with penumbra.

sunspot cycle   An approximately 11-year quasi-periodic variation in the twelve-month smoothed sunspot number. Other solar phenomena, such as the 10.7-cm solar radio emission, show similar cyclical behavior. The polarities of solar magnetic fields are now known to reverse with each cycle leading to a roughly 22-year cycle. This is sometimes called the Hale cycle.

sunspot number   The number of sunspots apparent on the sun at different times. At the beginning of a sunspot cycle, sunspots appear near 40°N and 40°S and move to lower latitude. Their numbers increase and then decrease during the moving processes and reach to the min-

imum at about 5°N and 5°S regions at the end of the sunspot cycle.

sunspot number (daily)   A daily index of sunspot activity. The sunspot number is computed according to the Wolf (1849) sunspot number $R = k(10g + s)$, where g is the number of sunspot groups (regions), s is the total number of individual spots in all the groups, and k is a variable scaling factor (usually <1) that indicates the combined effects of observing conditions, telescope, and bias of the solar observers. k is equal to 1 for the Zurich Observatory and adjusted for all other observatories to obtain approximately the same R number.

sunspot number (smoothed)   An average of 13 monthly sunspot numbers, centered on the month of concern. The 1st and 13th months are given a weight of 0.5.

sunward arc   *See* polar cap arc.

Sunyaev–Zeldovich (1972) effect   A modification of the spectral energy distribution of the background microwave radiation at temperature $\approx 2.7°$K, due to inverse Compton scattering of the microwave radiation photons by free electrons in the hot gas filling the intergalactic space in clusters of galaxies (the intra-cluster medium). The background photons gain energy, causing a minimal yet measurable change in the background radiation temperature. If the distribution of electron temperature and electron density of hot matter inside the cluster is known, then the value of the Hubble constant (and in principle of the deceleration parameter q_0) can be derived from observations of the Sunyaev–Zeldovich effect. Such measurements of H_0 have the advantage of being independent of the distance ladder built on optical distance indicators but require detailed X-ray observations of massive and dense clusters of galaxies. They have yielded, as of early 1998, values of $H_0 \lesssim 50$ km s^{-1} Mpc^{-1}, which are somewhat lower than other estimates. *See* inverse Compton effect.

supercell storm   The storm that is so organized that the entire storm behaves as a single entity, rather than as a group of cells. These so-called *supercell storms* account for most tornadoes and damaging hail. Most supercell storms move continuously toward the right of the environmental winds. *See* multicell storm.

super cloud cluster   Deep convection in the Tropics is organized into a hierarchy of spatial structures. A typical convective cloud is 10 km in horizontal scale. Such clouds often gather into a cloud cluster with horizontal dimensions of 1 to a few 100 km. A super cloud cluster is a group of such cloud clusters and has a longitudinal dimension of a few 1000 km. A *super cloud cluster* forms in the rising branch of a Madden–Julian Oscillation and is exclusively found in the large warm water region from the equatorial Indian to the western Pacific Ocean.

supercluster   A very large, high density cluster of rich clusters of galaxies that is flattened or filamentary in shape, with sizes as large as 150 Mpc. Superclusters appear to surround large voids creating a cellular structure, with galaxies concentrated in sheets; in higher density at the intersection of the sheets (edges); and in very high density superclusters at the intersection of edges. *Superclusters* typically contain several (2 to 15) clusters and are defined by a number density of galaxies taken to exceed some threshold, typically a factor 20 above the average galaxy number density. The mean separation between superclusters is reported to be approximately 100 Mpc.

superconducting string   *See* Witten conducting string.

supercooling   The phenomenon in which a pure material may be cooled below the usual transition temperature but not undergo a phase change. For instance, pure water may be cooled below 0°C without freezing. The addition of a freezing center, or of a small ice crystal will immediately initiate freezing in such a case. One can also speak of *supercooling* with respect to the condensation temperature.

supercritical flow   Flow in an open channel with a Froude number greater than unity. *See* Froude number.

superflare   A stellar flare having 100 to 10 million times the energy of the most energetic Solar flare. *Superflares* have been observed on normal solar-like stars. These flares have durations of a fraction of 1 h up to a few days, radiate 10^{33} to 10^{38} ergs, and emit light from radio waves through X-rays. One theory of their origin is that they arise from the sudden release of stored magnetic energy. *See* RS Canum Venaticorum stars.

supergeostrophic wind   Real wind which is stronger than the geostrophic wind. Along the axes of low level jets, it is often found that the wind speed is larger than the geostrophic wind. Both *supergeostrophic wind* and subgeostrophic wind are the non-geostrophic wind fluctuations caused by inertia gravitational waves. In general, there will be severe weather, such as heavy storms over the supergeostrophic wind regions.

supergiant   The evolved phase of the life of a star of more than about 5 solar masses. The distinction between giants, bright giants, and *supergiants* is somewhat arbitrary (*see* HR diagram), but, in general, supergiants will be the biggest (more than 1,000 solar radio), brightest (more than 10^4 solar luminosities), and the shortest-lived (less than 10^6 years). Supergiants, at least those whose initial mass was more than about 10 solar masses, will end their lives as supernovae of Type II. They have vigorous stellar winds that can reduce the initial mass of the star by a factor of two or more during their lives.

supergradient wind   Real wind which is stronger than the gradient wind. It is similar to supergeostrophic wind. When the pressure gradient force cannot be balanced by the horizontal Coriolis force and the centrifugal force, the *supergradient wind* will appear. Supergradient winds often can be found in tornado systems. In general, supergradient winds are not easy to create and will evolve to normal gradient winds due to the Coriolis effect. In practice, it is hard to determine the radius of curvature of wind required to determine the gradient wind; thus, it is hard to determine the existence of the supergradient wind.

supergranulation   Large scale of convection on the sun comprising the tops of large convection cells. Supergranules are irregular in shape and have diameters ranging from 20,000 to 50,000 km in size with horizontal motions ~0.3 to 0.4 km s^{-1} and vertical motions, at the edges of the convective cells, ~0.1 to 0.2 km s^{-1}. Individual supergranules last from 1 to 2 days.

superior conjunction   *See* conjunction.

superior mirage   A spurious image of an object formed above its true position by abnormal atmospheric refraction, when the temperature increases with height. Then light rays are bent downward as they propagate horizontally through the layer, making the image appear above its true position. *See* inferior mirage.

superluminal source   A radio source showing plasma apparently flowing at transverse velocity larger than the speed of light. *Superluminal sources* are discovered by comparing two high resolution radio maps obtained at different epochs with Very Long Baseline Interferometers, and the difference is seen as an angular position charge $\alpha > \frac{c\Delta t}{d}$ where c is the speed of light, d is the distance to the object, and Δt is the time interval between observations. Thus, simple geometry suggests motion faster than c, the speed of light. Superluminal motion is found usually in core dominated radio galaxies and quasars, and is made possible by the presence of highly relativistic motion and by a favorable orientation. The apparent transverse velocity is

$$v_{\text{trans}} = v \sin\theta / (1 - (v/c)\cos\theta) ,$$

where v is the velocity of the radiating particles, and θ the angle between the jet and the line of sight. For example, the radio quasar 3C 273 shows blobs of gas moving out along the jet at an angular speed of ≈ 0.67 milli-arcsec per year. If the radio jet points a few degrees from the line of sight, then the observed apparent velocity is 6.2 times the speed of light. About 70 extragalactic superluminal sources were catalogued by late 1993. In 1995, a galactic superluminal source was identified by Mirabel and Rodriguez, and as of early 1998, two galactic superluminal sources are known.

supermassive black hole   A black hole of mass $\sim 10^6 - 10^9$ solar masses. *Supermassive black holes* are believed to be present in the nuclei of quasars, and possibly, of most normal galaxies. A black hole of mass as large as 10^8 solar masses would have a gravitational radius $\approx 3 \times 10^8$ km and thus, linear dimension comparable to distance of the Earth from the sun. Evidence supporting the existence of supermassive black holes rests mainly on the huge luminosity of quasars and on the perturbations observed in the motion of stars in the nuclei of nearby galaxies that are probably accelerated by the central black hole gravity. If a massive dark object is present in the nucleus of a galaxy, the dispersion in the velocity of stars $\sigma(r)$ is expected to rise toward the center according to a Keplerian law i.e., $\sigma(r) \propto r^{-1/2}$. Such a rise has been detected, for example, in Messier 87 and suggests a central dark mass there of $\sim 5 \times 10^9$ solar masses.

supernova   A stellar explosion, triggered either by the collapse of the core of a massive star (Type II) or degenerate ignition of carbon and oxygen burning in a white dwarf (Type I). The former leaves a neutron star or black hole remnant; the latter disrupts the star completely. The appearance of the spectrum of the event and of the changes of brightness with time are rather similar for the two types. *Supernovae* occur at a rate of a few per century in a moderately large galaxy like the Milky Way. Ones in our galaxy, visible to the naked eye, occurred in 1054 (Crab Nebula), 1572 and 1604 (Tycho's and Kepler's supernovae) and a few other times in the past two thousand years. Many others undoubtedly have been hidden by galactic dust and gas from our sight. *See* supernova rates, supernovae, classification.

supernovae, 1987A   SN 1987A is the most studied supernova of the 20th century. Due to its close proximity, in the Large Magellanic Clouds, and the recent developments of CCD and neutrino detectors, it has become the Rosetta stone of Type II supernovae. Observations of the Large Magellanic Clouds prior to the supernova explosion recorded the progenitor star before its death, affirming that the progenitors of Type II supernovae were indeed massive supergiant stars. The neutrino signal (the first supernova neutrino detection) agreed well with the predicted flux from core-collapse models.

However, SN 1987A brought with it many more puzzles than it did answers. SN 1987A peaked twice at much lower magnitudes than most Type II supernovae. The progenitor was indeed a supergiant, but it was a blue supergiant, not a red supergiant as was predicted by theorists. The neutron star which should have formed in the collapse mechanism has yet to be detected. In addition, images and spectra of SN 1987A revealed the presence of interstellar rings, likely to be caused by asymmetric mass loss from the progenitor star due to a binary companion. The progenitor of SN 1987A may have been a merged binary which provides an explanation for the rings and for the fact that it was a blue supergiant.

supernovae, 1991bg   A peculiar Type Ia supernova. Although most Type Ia supernovae exhibit very little scatter in their peak luminosity (0.3 to 0.5 magnitudes in V and B), the peak luminosity for SN 1991bg was extremely subluminous (1.6 magnitudes less in V, 2.5 magnitudes less in B with respect to normal Type Ia supernovae). The late-time light curve decay was consistent with an explosion ejecting only $0.1 M_\odot$ of nickel. In addition, its expansion velocity ($10,000$ km s^{-1}) was slightly lower than typical Type Ia supernovae. SN 1991bg, and a growing list of additional low-luminosity (low nickel masses) supernovae (e.g., 1992K, 1997cn), may make up a new class of supernovae which are better explained by alternative Type Ia mechanisms (sub-Chandrasekhar thermonuclear explosions or accretion induced collapse of white dwarfs).

supernovae, 1993J   Supernova 1993J's early-time spectra had hydrogen lines and hence, is officially a Type II supernova. However, its light curve peaked early, dipped, and increased again, marking it as a peculiar supernova. Its late-time spectra exhibited strong oxygen and calcium lines with little hydrogen, very similar to the late-time spectra of Type Ib supernovae. This transition of supernova 1993J (and the similar 1987K) from Type II to Type Ib spectra suggests that Type II and Type Ib supernovae

are caused by a common core-collapse mechanism. The progenitor of 1993J was probably a massive star that went into a common envelope phase with a binary companion, removing most of its hydrogen envelope.

supernovae, accretion induced collapse (AIC)   Rapidly accreting C/O white dwarfs and most OMgNe white dwarfs, which accrete sufficient material to exceed the Chandrasekhar limit, collapse into neutron stars. The collapse proceeds similarly to the core collapse mechanism of Type Ib/Ic and Type II supernovae. These collapses eject up to a few tenths of a solar mass and may explain the low-luminosity Type Ia supernovae such as Supernova 1991bg. Like the core-collapse of massive stars, AICs produce neutron star remnants, but at a rate $< 1\%$ that of core-collapse supernovae. However, in special cases, such as globular cluster where it is difficult to retain neutron stars from Type II supernovae, AICs may form most of the neutron star population.

supernovae, classification   Supernovae are classified in two major groups: those with hydrogen lines in their spectra (Type II) and those without hydrogen lines (Type I). Type I supernovae are further subclassified by their spectra: Type Ia supernovae have strong silicon lines (Si II) whereas Type Ib/c supernovae do not. Type Ib supernovae exhibit helium lines (He I) which are absent in Type Ic supernovae. Type II supernovae are differentiated by their light curves: the luminosity of Type II-Linear supernovae peak and then decay rapidly whereas Type II-Plateau supernovae peak, drop to a plateau where the luminosity remains constant for ~ 100 days and then resume the light curve decay. Type Ia are more luminous than Type II supernovae (~ 3 magnitudes) and occur in all galaxies. Type Ib/c and Type II supernovae do not occur in ellipticals. Type Ia are thought to be the thermonuclear explosions of white dwarfs, whereas Type Ib/c and Type II supernovae are caused by the collapse of massive stars.

This classification scheme collapses the 5 class Zwicky system (Zwicky 1965) to these two separate classes (the Zwicky III, IV, and V explosions are all placed in the category of peculiar Type II supernovae). In addition, supernovae 1987K and 1993J exhibited hydrogen features in their early spectra, but their final spectra more closely resembled Type Ib supernovae suggesting a common link between Type Ib and Type II supernovae.

supernovae, core collapse mechanism Any supernova mechanism which involves the collapse of a massive star's core into a neutron star. Massive stars ($\gtrsim 10 M_\odot$) ultimately produce iron cores with masses in excess of the Chandrasekhar limit. The density and temperature at the center of this core eventually become large enough to drive the dissociation of iron and the capture of electrons onto protons. These processes lower the pressure support in the core and initiate its collapse. As the core collapses, the density and temperature increase, accelerating the rate of electron capture and iron dissociation and, hence, the collapse itself. The collapse process runs away, and very quickly (on a free-fall timescale), the core reaches nuclear densities. Nuclear forces halt the collapse. Only a small fraction (1%) of the potential energy released in this collapse is required to drive off the outer portion of the star and power the observed explosion. Several mechanisms have been proposed to harness this energy (*See* supernovae, prompt shock mechanism, supernovae, delayed neutrino mechanism). Core collapse supernovae leave behind neutron star (or black hole) remnants. Type II and Type Ib/Ic supernovae are thought to be powered by the core collapse mechanism.

supernovae, delayed neutrino mechanism Neutrinos carry away most of the gravitational potential energy released during the core collapse of a massive star (*see* supernova, core collapse mechanism). Roughly 1% of the energy of these neutrinos must be converted into kinetic energy to drive a supernova explosion. This 1% efficiency is achieved by absorbing neutrinos just above the proto-neutron star surface where densities are high, and the optical depth to neutrinos is also high. The neutrino-heated material convects upward and expands (and cools) before it can lose too much energy by the re-emission of neutrinos. The convection takes place after the prompt shock stalls (*see* supernova, prompt shock mechanism), and is thus a "delayed" mechanism.

supernovae, ΔM_{15}/Phillips relation The scatter in peak magnitudes of Type Ia supernovae can be correlated to the decay of the light curve. The relation championed by Phillips (1993), ΔM_{15} is defined as the amount (in magnitudes) that the B light curve decays in the first 15 days after maximum. This value can then be used to determine the absolute magnitude of any Type Ia supernova. This correction is similar to a technique used to correct for composition effects in the Cepheid Variable period-luminosity relation. The scatter in peak magnitudes typically lowers from 0.3 – 0.5 magnitudes to less than 0.1 magnitudes, making Type Ia very effective standard candles. A similar method using synthetic light curves, the light curve shape (LCS) method, has been developed with comparable results (Riess 1995). Like most standard candles, Type Ia supernova must be calibrated at low redshifts (typically with Cepheid Variables).

supernovae, distance indicators The high peak luminosities of supernovae and the homogeneity of Type Ia supernovae in particular, make them ideal standard candles for cosmological studies at high redshift. In addition, the scatter of Type Ia supernovae can be further reduced by correcting the peak luminosities using the relation between peak magnitude and light-curve decay (*see* supernovae, ΔM_{15}/Phillips relation). The absolute magnitudes of Type II-Plateau supernovae can be theoretically calculated from the colors and expansion velocities of the supernovae, placing them as one of the few standard candles which do not need to be calibrated at low-redshifts (*see* supernovae, expanding photosphere method).

supernovae, expanding photosphere method The expanding photosphere method (EPM) provides a physical model to calculate the absolute magnitudes of Type II-Plateau supernovae without any calibration. The expanding photosphere method combines observations of the photospheric radii and temperature to get a supernova luminosity (also known as the Baade–Wesselink method). Assuming the supernova emits as a blackbody, this luminosity is simply: $L_{\text{photosphere}} = 4\pi R_{\text{photosphere}} \sigma T_{\text{photosphere}}^4$ where $R_{\text{photosphere}} = V_{\text{photosphere}} t$, and the temperature is determined by the observed colors.

In practice, the blackbody assumption is not valid, and the luminosity is calculated using models of Type II supernovae.

This mechanism does not require calibration at low redshifts and is not limited by the uncertainties of the calibrator. Unfortunately, Type II supernovae are dimmer than Type Ia supernovae, and the low observed numbers limit their use as cosmological candles.

supernovae, fallback   *Supernova fallback* is that material that, during a core collapse supernova explosion, does not receive enough energy to escape the gravitational potential of the neutron star and eventually "falls back" onto the neutron star. Many physical effects can produce fallback. For instance, when the supernova shock slows down as it moves through the shallow density gradients of the exploding star's hydrogen envelope, a reverse shock is produced which drives material back onto the neutron star. The decay of ^{56}Ni also can reduce the velocity of some material to below the escape velocity, and this material will eventually fall back onto the neutron star. Neutron rich material is formed near the proto-neutron star surface due to the high neutrino emission and absorption rates. This material is nearest to the proto-neutron star and hence, most likely to fall back.

supernovae, gravitational waves   In the delayed neutrino mechanism, neutrino heating drives vigorous convection and ultimately powers the supernova explosion. This convection causes oscillations in the mass distribution which lead to the production of gravitational waves. The gravitational wave amplitude can be estimated from the time variation in the mass distribution and suggests detectability in current detectors (LIGO, Virgo, Geo) at tens of Mpc. Simulations of rotating core-collapses can produce wave amplitudes nearly 2 orders of magnitude higher. However, the gravitational wave amplitudes from supernova are all over an order of magnitude smaller than that of merging compact objects which will be detectable at hundreds of Mpc in current detectors.

supernovae, light curves   The time evolution of the luminosities (light curves) of Type Ia supernovae are dominated by ^{56}Ni decay and are remarkably homogeneous (with some exceptions: *see* supernovae, 1991bg). Type II *supernovae light curves* are characterized by an early-time peak as the material becomes optically thin and photons escape and a late-time exponential decreased determined by the decay of ^{56}Co to ^{56}Fe. The light curves of Type II supernovae vary dramatically (peak luminosities range over 4 magnitudes). Type II supernovae are classified roughly into two groups based on the light curve: Type II-Plateau supernovae whose decay after peak flattens for roughly 30 to 100 days, and Type II-Linear supernovae which do not exhibit any flattening.

supernovae, neutrino detectors   Supernova 1987A initiated the first *supernova neutrino detectors*. Both the Kamiokande and Irvine–Michigan–Brookhaven proton decay detectors observed Cherenkov radiation from relativistic particles accelerated by neutrinos. Current detectors fit into 3 basic categories based on the instrument design: water detectors, heavy water detectors, and scintillation detectors. Kamiokande, its successor, Super Kamiokande, and the Irvine–Michigan–Brookhaven detectors are all water detectors. Water detectors rely primarily upon anti-neutrino absorption by protons or neutrino scattering which then produces relativistic particles and Cherenkov radiation. Heavy water detectors, in addition to the processes in water detectors, also detect the dissociation of neutrons by neutrino scattering. Scintillation detectors rely upon emission from the decay of atoms excited by neutrino scattering.

supernovae, neutrino-driven wind mechanism   After the launch of a core collapse supernova explosion, material continues to blow off the hot proto-neutron star. The cooling neutron star emits copious neutrinos ~ 20 s after the initial explosion. The ejecta of this wind is neutron rich and thought to be a source of r-process nucleosynthesis products. However, the neutron fraction of the ejecta depends sensitively upon the relative absorption of the electron neutrino and anti-neutrinos and the true nucleosynthetic yield is difficult to determine.

The *neutrino-driven wind mechanism* is often confused with the delayed-neutrino mechanism. In the delayed neutrino mechanism, a convective region is trapped near the proto-neutron star by the ram pressure of infalling material. The explosion occurs only when the energy in the convective region overcomes this ram pressure. The neutrino-driven wind implies a more steady state ejection of material, not a sudden burst as occurs in the delayed neutrino supernova mechanism. Unlike the neutrino-driven wind mechanism, the electron fraction of the ejecta from the delayed-neutrino mechanism depends both upon the relative absorption and emission rates of electron neutrinos and anti-neutrinos.

supernovae, neutrinos   The emission of neutrinos in core collapse supernovae (Type Ib/c and Type II) is dominated by electron capture onto protons and by electron/positron annihilation. As the core collapses, its pressure is sufficient to overcome nuclear forces and capture electrons onto protons, producing a neutron and an electron neutrino. These neutrinos dominate the initial neutrino burst. As the core collapses further, it becomes so dense that it is optically thick to neutrinos, and the neutrinos become trapped in the core. The temperatures rise sufficiently to produce positron pairs, and the annihilation of electrons and positrons form neutrino/antineutrino pairs. In the first 20 s after collapse, neutrinos release over 10^{53} ergs of energy.

supernovae, prompt shock mechanism
When the collapsing core of a massive star reaches nuclear densities (*see* supernovae, core collapse mechanism), nuclear forces halt the collapse causing the infalling material to rebound. A bounce shock then propogates out of the star. The *prompt shock mechanism* argues that this shock carries sufficient energy to power a supernova explosion (1% of the gravitational potential energy released). However, as the shock expands outward, it loses energy through neutrino emission and the dissociation of outer material it hits. In most simulations, the shock stalls at ∼100 to 500 km; thus, failing to produce and explosion.

supernovae, spectra   The spectra from supernovae are the dominant characteristic used to classify supernovae types (*see* supernovae, classification). The spectral lines are generally broad (>10,000km/s) and blue-shifted due

to the rapid expansion velocities with different lines having different characteristic velocities. For a review, *see* Filippenko (1997).

supernovae, thermonuclear explosions
Type Ia supernovae are powered by the thermonuclear explosion of a white dwarf. The current "favored" model is the thermonuclear explosion of an accreting white dwarf which reaches the Chandrasekhar mass, triggering nuclear burning near the core (*see* supernovae, white dwarf accretion). This mechanism is further subdivided into several models: detonation, delayed-detonation, deflagration, etc. These models differ on the type of burning that occurs in the core. For instance, the delayed-detonation model is initiated by a core burning with a deflagration flame front (flame speed less than the sound speed) which then evolves into a detonation (flame speed greater than the sound speed).

Sub-Chandrasekhar thermonuclear explosions occur in white dwarfs less massive than the Chandrasekhar mass. The accreting material builds a helium layer on the white dwarf and ignites. The detonation of the helium layer drives pressure waves into the core, ultimately causing the white dwarf core to detonate. Currently, this mechanism has difficulty explaining the spectra of Type Ias and, hence, is not the favored Type Ia supernova mechanism.

supernovae, Type Ia   The kind of stellar explosion that occurs in all kinds of galaxies (not just ones with young stars) and whose spectra show no evidence for the presence of hydrogen. The cause of the explosion is the ignition of carbon and oxygen burning within degenerate material in a white dwarf. The star is completely disrupted; most of the carbon and oxygen fuses to iron and nearby elements releasing about 10^{51} ergs of energy, about 1% of which appears as visible light, and the rest of which blows the gas out into space at speeds of about 10,000 km/sec. *Type Ia supernovae* have no hydrogen in their spectra but exhibit strong silicon lines (most notably Si II $\lambda\lambda$6347, 6371). At late times, iron and cobalt lines become prominent. The expansion velocities inferred from the spectra are roughly ∼11,000 to 13,000 km s^{-1}. The absolute peak magnitudes of Type Ia show little scatter $\langle M_V \rangle = -18.5 \pm 0.3$ mag. The differences in the peak magnitude have been correlated to the decline rate of the supernova luminosity (*see* supernovae, ΔM_{15}/Phillips relation). Using this correlation, the peak magnitude variations can be removed, and Type Ia can thus be used as standard candles for measuring the Hubble constant, and other cosmological parameters.

The luminosity is powered by the decay of ^{56}Ni produced in the explosion. Some of the low luminosity outbursts (e.g., 1991bg) may be explained by the accretion induced collapse of white dwarfs (*see* supernovae, accretion induced collapse). Type Ia supernovae do not occur in extremely young stellar populations but do occur in all types of galaxies at a rate of 0.005 yr^{-1} per Milky Way-sized galaxy, consistent with the assumption that the progenitors of these systems come from low-mass stars. At peak brightness, a Ia SN can be as bright as its entire host galaxy. The precise nature of the progenitors is not clear. One popular candidate is a pair of white dwarfs whose total mass exceeds the Chandrasekhar limit in a binary system with orbit period less than about one day. Such a pair will spiral together in less than the age of the universe and explode as required when they merge, but we have not yet actually seen any white dwarf binaries with the required properties. Tycho's and Kepler's supernovae were probably Type Ia events.

supernovae, Type Ib/Ic   *Type Ib/Ic supernovae* exhibit neither hydrogen lines nor silicon lines. Type Ib supernovae are characterized by the existence of helium lines, absent in Type Ic supernovae. These two types of supernovae are otherwise very similar (both have oxygen and calcium in their late-time spectra) and are generally lumped together. They occur at a rate of 0.002 yr^{-1} per Milky Way-sized galaxy.

Type Ib/Ic supernovae are similar to Type II supernovae in that they are caused by the collapse of massive stars ($> 10 M_\odot$). However, at the time of collapse, Type Ib/Ic supernovae have lost most/all of their hydrogen envelope either through stellar winds or during binary evolution through a common envelope phase. Type Ic supernovae have lost not only their hydrogen envelope but most of their helium envelope as well. The connection between Type II supernovae and Type Ib/Ic supernovae comes from supernovae 1987K and 1993J which both ex-

hibited hydrogen in their spectra but mimicked Type Ib supernovae at late times. The two progenitors of these supernovae appear to have lost most, but definitely not all, of their hydrogen before collapse, implying a smooth continuum between Type Ib/Ic and Type II supernovae.

supernovae, Type II   The kind of supernova that results when the core of a massive star collapses to a neutron star or black hole (*see* core collapse). They are the primary source of new heavy elements (those beyond hydrogen and helium) made by nuclear reactions over the life of the star and expelled when energy from the core collapse blows off the outer layers of the star. The spectrum is dominated by lines of hydrogen gas, from the envelope of the star, but oxygen and other heavy elements are also seen. Core collapse, when the star has already lost its hydrogen envelope, produces hydrogen-free supernovae of Types Ib and Ic. It is estimated that about one SNII occurs in our galaxy each century. This class of supernovae is subdivided into two major groups based on their light curves: Type II-Linear (II-L) supernovae peak and then decay quickly, and Type II-Plateau (II-P) which, after their peak, decay only ~ 1 magnitude and then reach a plateau for ~ 100 days before a late-time decay similar to type II-L supernovae. The expansion velocities inferred from the spectra are roughly ~ 7000 km s^{-1}. The absolute peak visual magnitudes of *Type II supernovae* have considerable scatter ($\langle M_V \rangle = -17 \pm 2$ mag.). However, models of Type II-P supernovae allow a physical calibration of these objects, allowing them to be used as "standard" candles without relying upon a local calibrator such as Cepheid Variables.

Type II supernovae are caused by the core collapse of massive stars ($> 10 M_\odot$) and are powered by the potential energy released during this collapse (*see* supernovae, core collapse mechanism). Because their progenitors are short-lived, they only occur in young stellar populations, and none have been observed in elliptical galaxies. Type II (and Type Ib/Ic) supernovae form the bulk of the neutron stars in the universe and occur at a rate of 0.0125 yr^{-1} per Milky Way-sized galaxy.

supernovae, white dwarf accretion   Type Ia supernovae are powered by the thermonuclear explosion of accreting white dwarfs (*see* supernovae, thermonuclear explosions). The favored Type Ia supernova model requires that the white dwarf accrete up to a Chandrasekhar mass. However, accreting white dwarfs tend to lose mass during nova eruptions. Only for accretion rates greater than $\sim 10^{-9} - 10^{-8} M_\odot$ yr^{-1} are white dwarfs thought to gain mass from accretion. Rapidly accreting ($\gtrsim 10^{-6} M_\odot$ yr^{-1}) C/O white dwarfs and most accreting OMgNe white dwarfs, which gain enough matter to exceed the Chandrasekhar mass, are thought to collapse to neutron stars (Nomoto & Kondo) before nuclear burning can drive a thermonuclear explosion. Neutrino Urca processes prevent these white dwarfs from getting hot enough to burn efficiently. For more common accretion rates, C/O white dwarfs are thought to ignite their cores, initiating a thermonuclear explosion.

supernova rates   The rate at which supernovae of different types occur remains uncertain by factors of ~ 2. These rates can be indirectly determined by observing metal abundances (metals are produced almost entirely in supernovae) of galaxies and using a theoretically derived production rate of metals per supernova. Alternatively, the rate of core-collapse supernovae (Type II + Type Ib/c) can be calculated by combining the theoretical estimates of the mass range of stars with observations of the initial mass function and the star formation rate. However, the most direct and most accurate technique to determine these rates is to simply observe the supernovae that occur in the universe and correct for the biases intrinsic to the observed sample. Unfortunately, a large supernova sample does not exist, and biases such as luminosity differences and obscuration, make it difficult to exactly determine the *supernova rate*. Recent estimates of supernova rates are printed in the following table. (Cappellaro et al., 1997.) The units are per 10^{10} solar luminosities (in the blue) per century.

supernova remnant (SNR)   The expanding gas blown out during any type of *supernova* SN explosion. Masses can range from a few tenths to a few tens of solar masses. Expan-

Supernova Rates

Galaxy Type	Supernova Type		
	Type Ia	Type Ib/c	Type II
E-S0	0.15 ± 0.06		
S0a-Sb	0.20 ± 0.07	0.11 ± 0.06	0.40 ± 0.19
Sbc-Sd	0.24 ± 0.09	0.16 ± 0.08	0.88 ± 0.37

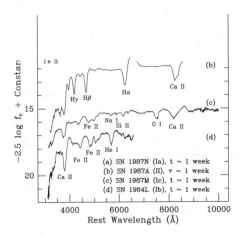

Early time spectra of supernovae. Type II supernovae have hydrogen lines. Type Ia supernovae have strong silicon lines whereas Type Ib/c supernovae do not. Type Ib supernovae exhibit helium lines which are absent in Type Ic supernovae. Figure courtesy of Alex Filippenko.

sion velocities range from 2,000 to about 20,000 km/sec. Young SNRs sometimes have pulsars in them (meaning that the SN was a core collapse event). A few hundred SNRs are known in our galaxy, from their emission line spectra, radio, and X-ray radiation which occurs by synchrotron and/or bremsstrahlung processes. Their spectrum sometimes shows emission lines characteristic of shocked and photoionized gas. SNRs add kinetic energy and heat to the interstellar medium and contribute to accelerating cosmic rays.

superposition principle   In linear systems any collection of solutions to a physical problem can be added to produce another solution. An example is adding Fourier harmonics to describe oscillations in the electromagnetic field in a cavity. In particular situations a limited form of superposition is possible in specific nonlinear systems.

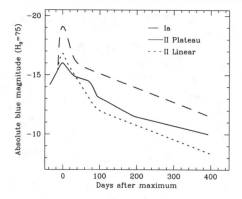

Typical supernovae light curves. Type II supernovae have a large amount of scatter (∼ 4 magnitudes in the peak luminosity). Type II supernovae are separated into two groups (Plateau and Linear) based on the light curve. Figure courtesy of Mario Hamuy.

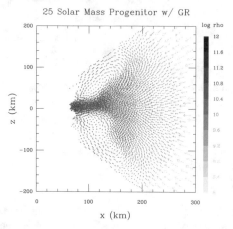

25$M_\odot$ core-collapse simulation 20 ms after bounce. In this simulation, the convective region stretches from roughly 50 km to 300 km.

supersonic   Involving speeds in excess of the local speed of sound.

supersonic string model   *See* elastic string model.

superspace   The space of inequivalent (under general coordinate transformations) spatial geometries and matter fields associated to the ADM decomposition of space-time (*see* geometrodynamics). Since the metric of space is a symmetric tensor of rank 2, there are apparently six gravitational degrees of freedom per space point. However, the latter are subjected to one Hamiltonian and three super-momentum

constraints which correspond to the transformations of the reference frame which leave the geometry unaltered. This leaves the equivalent of two degrees of freedom per space point or a total of $(\infty^3)^2$ gravitational degrees of freedom. A (generally time dependent) solution to Einstein's equations traces out a 1-parameter path through *superspace*.

One usually needs to impose some further restriction to the metric in order to handle the constraints and obtain manageable results. *See* minisuperspace, ADM form of the Einstein–Hilbert action.

surface boundary layer   The usually "weakly stratified" layer at the top of natural water, which is immediately and directly affected by either wind or convection.

surface gravity   The limiting value of the force applied at infinity to keep a unit mass at rest on the black hole horizon. For an explicit example, *see* Schwarzschild metric. The *surface gravity* is also related to the temperature of Hawking radiation. *See* black hole horizon.

surface tension (σ)   Molecules in the surface of liquid water are subjected to a net inward force due to hydrogen bonding with the water molecules below the surface. *Surface tension* is equal to the magnitude of that force divided by the distance over which it acts.

surface waves   Seismic waves that propagate along the surface of the Earth (as opposed to body waves). *Surface waves* are either Rayleigh (longitudinal, with a vertical component) or Love transverse waves. Surface waves are primarily responsible for the damage associated with earthquakes; they propagate more slowly than body waves so that the body waves provide a short precursory warning of the arrival of the destructive surface waves.

surf zone   The region at a coast where breaking waves are found.

surge   In solar physics, a relatively narrow active region jet of material in which plasma is accelerated outward at 50 to 200 km s^{-1} for a few tens of minutes. *Surges* reach coronal heights

($\sim 1.5 \times 10^5$ km) and then either fade or return into the chromosphere along the trajectory of ascent.

suspended load   The sediment that is transported in the water column in a river or ocean. *See also* bed load.

suspension   The transport of small material by mixing of that material into a fluid (air, water, etc.); or such a mixture itself. For *suspension* to occur, the gravitational force on the particle must be balanced by the uplifting force caused by turbulence in the fluid. Suspension is most effective for the transport of silt and clay sized material (less than 60 μm in diameter).

suture   The boundary between two colliding continents. When an ocean closes, the continents bounding the oceans collide. This is happening between India and China today resulting in the Himalayas.

Sv   Sverdrup, the volume transport unit, 1 Sv $= 10^6$ m³/s.

swash zone   The portion of a beach where the water rushes up and down due to wind wave action, alternately wet and dry.

Sweet–Parker reconnection   In plasma physics, the slow stationary reconnection in an extended X-line (the length L of the diffusion region is large compared to its width d). The outflow speed of the plasma is equal to the Alfvén speed $v_{A,in}$ in the incoming plasma flow, and the reconnection rate R_{SP} equals the Mach number of the incident flow:

$$R_{SP} = \sqrt{\frac{c}{L\sigma v_{A,in}4\pi}}$$

with σ being the conductivity. The reconnection rate then depends on the conductivity. In space plasmas, the conductivity is high and, therefore, a low rate of reconnection results. *Sweet–Parker reconnection* thus, is a slow process during which about half of the incoming magnetic energy is converted into the kinetic energy of the outflowing plasma, leading to two high-speed plasma streams flowing away from the reconnection site.

Sweet–Parker reconnection seems to play an important role at the magnetopause where the high-speed streams flowing away from the reconnection site can be detected *in situ*.

swell   A broad region of elevated topography generally associated with seismic hot spots. A typical example is the Hawaiian *swell* which is a near circular, domal structure with a diameter of about 1000 km and an elevation of about 1 kilometer.

"Swiss cheese" model   A model of the universe in which the background spacetime is the same as in the Friedmann–Lemaître dust-filled cosmological models, but there are holes in it, inside which the Schwarzschild solution applies. This model was used to discuss the influence of the expansion of the universe on the motion of planets; according to this model (Einstein and Straus, 1945) there is no such influence: since the planets remain in the region of the spacetime in which the Schwarzschild solution applies, they move as if the matter outside the hole does not exist. The mass-parameter in the Schwarzschild region and the active gravitational mass (counting binding) that was removed from the Friedmann background to make place for the planetary system are equal.

symbiotic star   The category of cataclysmic variable in which the donor star is a red giant. The spectrum shows both absorption features characteristic of the cool, bright red giant and emission features from the accretion disk around the white dwarf.

symmetric instability   A two-dimensional form of baroclinic instability in which the perturbations are independent of the coordinate parallel to the mean flow.

symmetry   An operation performed on a geometrical object after which the geometrical state of the object is indistinguishable from its initial state. Example: let the object be a sphere and the operation be any rotation around the center of the sphere. However, if the object is a cube, then only rotations by some definite angles around definite axes will be *symmetries*. (If the axis passes through the centers of opposite walls of the cube, then only rotations by multiples of 90° are symmetries. If the axis passes through the centers of diagonally opposite edges of the cube, then only rotations by multiples of 180° are symmetries.) Continuous symmetries of an n-dimensional space can be labeled by up to $\frac{1}{2}n(n+1)$ parameters. For example, among two-dimensional surfaces some have no symmetry (think about the surface of a loaf of bread), some have one-parametric symmetry (think about the surface of a perfectly smooth and round doughnut, which is symmetric with respect to rotations around an axis that passes through its center: the parameter is the angle of rotation), some have a two-parametric symmetry (think about the surface of an infinite cylinder, the parameters are the angle of rotation about the axis of the cylinder and the distance of displacement along the axis), and two have three-parametric symmetries. These latter are the plane and the surface of the sphere. For the plane, the symmetries are displacements along two perpendicular directions and rotations around an axis perpendicular to the plane. For the sphere, the symmetries are rotations around three mutually perpendicular axes. A characteristic property of each collection of symmetries is the difference between the following two operations: We take two of the basic symmetries (call them a and b), apply them to a point of the space in one order (first a, then b), and then apply them in the reversed order (first b, then a). The difference may be zero (like for the two translations on a plane) or nonzero (for any two rotations on a sphere or for a translation and a rotation on a plane). The fact that on a plane there exists a pair of symmetry transformations whose final result does not depend on the order in which they are executed, while no such pair exists on a sphere, shows that these two collections of symmetries are inequivalent. Real objects usually have no symmetry in the strict sense but may be approximately symmetric. (Example: the surface of the Earth is nearly a sphere. The departures from spherical shape caused by rotational flattening and by the mountains and other irregularities of the surface can be neglected for many purposes.) Symmetries are often assumed in physics in order to solve problems that would be too difficult to consider by exact methods without such an assumption.

Small departures from symmetries can then be taken into account by approximate calculations. A typical example is the calculation of orbits of the planets around the sun. Both in Newtonian mechanics and in general relativity a first approximation to the realistic solution was found under the assumption of spherical symmetry, which implies that the mass distribution inside the sun is perfectly spherical, and that the orbit of a given planet is not influenced by other planets. Then, small perturbations away from spherical symmetry caused by the other planets can be considered and corrections to the orbits calculated. Calculations without any symmetries assumed have been attempted only by numerical methods. A pronounced time dependent asymmetry in a gravitating system typically results in the emission of gravitational waves.

synchronous orbit   For a small body orbiting a more massive one, the orbital radius for which the orbital period is equal to the rotation period of the more massive body. A satellite in a zero inclination *synchronous orbit* above the equator will always appear to be above a particular point on the more massive body.

synchronous rotation   Rotation of a planet or other secondary body so that the same side always faces the primary; for instance, the moon *rotates synchronously* around the Earth. In the case of the moon, this is ensured by tidal locking; for commercial communications satellites, it is accomplished by active means, such as reaction wheels or thrusters.

synchrotron radiation   The radiation produced by a charged particle as it gyrates around magnetic field lines. The radiation is emitted at the gyrofrequency, $\omega = qB/\gamma mc$, where q is the particle charge, B is the magnetic field strength, m is the particle mass, γ is the Lorentz factor, and c is the speed of light. For an electron we have $\omega = 1.76 \times 10^7 B\gamma$. *Synchrotron radiation* from energetic electrons is an important production mechanism for microwave emission in the solar corona.

synchrotron self-Compton mechanism   A mechanism suggested for the production of high energy photons in radio loud active galactic nuclei. In the *synchrotron self-Compton scheme,* low-energy photons are produced in the radio domain by synchrotron emission by relativistic elections. If the source is very compact, the same relativistic electrons then turn the radio photons into higher energy photons by inverse Compton scattering. The synchrotron self-Compton mechanism suggests that the spectral shape of the seed synchrotron photons is maintained in the scattered spectrum. This prediction has been apparently confirmed by observations of the blazar 3C 279; however, the general validity of the synchrotron self-Compton mechanism for active galactic nuclei is as yet controversial.

synodical month   Synodic month.

synodic month   The time from one full moon to the next (about 29.53 days).

synodic period   In astronomy, the amount of time taken for a celestial body (usually the moon, but the concept can apply to other objects) to rotate once with respect to the Earth-sun line; the time between successive conjunctions of an orbiting body. Thus, the period of time between new moons (about 29.53 days) is the synodic month.

synoptic scale   In the atmosphere, the *synoptic scale* is referred to as the scale of moving weather systems associated with cyclones or anticyclones. This scale is larger than the mesoscale and less than the planetary scale.

synthetic aperture radar   A radar imaging technique in which the angular resolution is improved by combining the returns from an image as seen when illuminated by the radar transmitter at different points as it moves past the target. The essential point allowing this combination is precise geometric reconstruction based on the Doppler shift associated with specific target features. These signals are reduced to zero Doppler shift (maintaining phase coherence) and combined to produce the effect of a large aperture radar (an aperture, d, as large as the maximum distance between the source transmitter locations illuminating a particular target). This process thus, gives enhanced angular resolution, θ, according to the standard formula:

$$\theta \sim \frac{\lambda}{d}.$$

syzygy   An alignment of three celestial bodies in a straight line; in particular the position of the moon when it is new or full is a syzygy between the Earth, moon, and sun.

T

tachyon   A hypothetical subatomic particle that travels faster than the speed of light. There is no experimental or observational evidence of the existence of tachyons.

TAI   *See* International Atomic Time.

tail current   Current system in the tail of the magnetosphere, consisting of the cross-tail current separating the northern and southern lobe and the Chapman–Ferraro currents in the magnetopause. Thus, one closed current encircles each lobe, with both currents running together in the cross-tail portion.

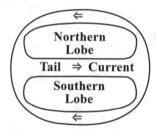

Tail current.

Talwani's method   An algorithm concerning analyses of magnetic anomaly and gravity anomaly proposed by M. Talwani (1965). For instance, the magnetic anomaly produced by a three-dimensional magnetic substance, such as a seamount, can be calculated, approximating the configuration of the substance by a stack of polygonal thin plates with homogeneous magnetic susceptibility. Similarly, the free air anomaly produced by a two-dimensional substance with a density contrast can be calculated, making an approximation for a cross-section of the substance by a polygon.

tangential discontinuity   *See* hydromagnetic wave.

tangential geostrophy   A force balance believed to be appropriate for fluid near the surface of the Earth's core and which is useful as an assumption for constructing models of the flow at the surface of the core from models of the magnetic field at the core-mantle boundary, because it reduces the ambiguity inherent to such flow modeling. Geostrophy is a balance between pressure and coriolis forces:

$$2\rho \boldsymbol{\Omega} \times \mathbf{u} = -\nabla p$$

where ρ is the density, $\boldsymbol{\Omega}$ is the rotation vector, $\mathbf{u}$ is the velocity, and p the pressure. *Tangential geostrophy* assumes that this balance holds in the horizontal direction (but not necessarily in the radial direction), i.e., that the Lorentz and viscous forces, buoyancy, and inertia are only important in the radial direction, if at all. The justification for this is that the Lorentz force may be reduced near the core-mantle boundary, since if the mantle is insulating, then the toroidal part of the magnetic field should drop to zero at the boundary, that gravity is predominantly radial, and that viscosity and inertia are small. These assumptions are arguable. However, the assumptions lead to a constraint on the flow:

$$\nabla_H \cdot (\mathbf{u}\cos\theta) = 0$$

where θ is the co-latitude, and ∇_H is the horizontal portion of the divergence operator. As the inverse problem for the flow is undetermined, this constraint can be useful for reducing flow ambiguity. Also, any part of the flow that represents torsional oscillations in the core will obey this constraint. *See* core flow.

Taygeta   Magnitude 4.4 type B7 star at RA 03^h45^m, dec +24.27'; one of the "seven sisters" of the Pleiades.

Taylor instability   Formation of rolls generated in a column of fluid bounded by differentially rotating cylindrical walls; governed by the dimensionless Taylor number: $T_a = r_i\omega^2h^3/\nu^2$, where r_i is the radius of the inner cylinder, ω is the rate of rotation, h is the space between cylinders, and ν is the kinematic viscosity.

Taylor number   A dimensionless number measuring the influence of rotation on a convecting system. It is also called rotational Reynolds number. Its value depends on the length scale of the convective system, the rotation rate, and kinematic viscosity. The Taylor number T_a is

$$T_a = (2\omega U)^2 / \left(v\frac{U}{H^2} \right)^2 = 4\omega^2 H^4 / v^2$$

where H is depth of fluid, Ω is rotational angular velocity, v is kinematic viscosity, and U is a typical velocity. If T_a is equal or greater than one, the rotational effects are significant. *See* Taylor instability for a slightly different form for T_a.

Taylor's hypothesis   *See* frozen field approximation.

Taylor state   A configuration of magnetic field in a conducting, rotating fluid so as to obey the constraint that the azimuthal component of the Lorentz force integrates to zero over the surface of cylinders coaxial with the rotation axis. This can be true for fluid in a magnetostrophic balance, i.e., balance of magnetic, pressure, buoyant, and coriolis forces, which may hold for the Earth's outer core if viscosity and inertia are both small. Taylor's constraint holds for a magnetostrophic flow if it is anelastic (i.e., $\nabla \cdot (\rho \mathbf{u}) = 0$ where ρ is the density and $\mathbf{u}$ the flow velocity) and if the gravitational force has no azimuthal (ϕ) component (or, at minimum, ρg_ϕ also integrates to zero on the cylinder). The constraint may be written as:

$$T = s \int [(\nabla \times \mathbf{B}) \times \mathbf{B}]_\phi \, d\Sigma = 0$$

where s is the distance from the rotation axis, $\mathbf{B}$ is the magnetic field, and Σ denotes the axial cylinder. Neither inertia nor viscosity are identically zero in the Earth's outer core, and it has been suggested that the Taylor torque T may be balanced by viscous forces acting on the cylinder at the core-mantle boundary, such as in Braginski's Model-Z dynamo. However, even in this case in the limit of vanishing viscosity, the above constraint is asymptotically satisfied. The effect of non-zero inertia is that a non-zero Taylor torque is balanced by a term representing acceleration, which leads to torsional oscillations. The net result is a "basic state" satisfying the above constraint with superimposed torsional oscillations.

TCB   *See* coordinate time.

TCG   *See* coordinate time.

TDB   *See* dynamical time.

TDT   *See* dynamical time.

tectonics   The study of the large scale movements and deformation of a planetary crust. Planetary crusts are affected by extensional and compressional forces caused by regional or global processes. The lunar maria display compressional features in their centers and extensional features along their edges, caused by subsidence of the dense lava flows which comprise the maria. Mercury is shrinking due to the cooling and solidification of its large iron core, which is creating compressional tectonics on its surface. The icy moons of the outer solar system display extensional tectonics due to the cooling and solidification of the ice in their interiors. Earth's crust is in a constant state of flux due to the influence of *plate tectonics,* which causes large segments of the crust to move. There is evidence from satellite magnetometer measurements of former surface tectonic activity on Mars.

tectosphere   A layer of rock up to several hundred kilometers thick at the top of the mantle underlying some of the continents, which has been observed by seismology to be distinct (perhaps colder) than the rest of the upper part of the mantle, i.e., as "continental roots". The *tectosphere* is thought to be relatively viscous and tightly coupled to the overlying continent, which implies that it is important for determining how the motions of the continent are coupled to those in the underlying mantle. There are different ideas as to how tectosphere may be formed, including simple cooling of relatively immobile mantle underlying a continental plate and the buildup of buoyant by-products of slab subduction.

Telesto   Moon of Saturn, also designated SXIII. Discovered by Smith, Reitsema, Larson, and Fountain in 1980, it orbits Saturn in the same orbit as Tethys, but leading it by 60°. This is one of the two stable Lagrange points in the Saturn–

Tethys system. Calypso orbits at the other. Its orbit has a semimajor axis of 2.95×10^5 km. Its size is $17 \times 14 \times 13$ km, but its mass has not been measured. It has a geometric albedo of 0.5 and orbits Saturn once every 1.888 Earth days.

temperature (**1.**) In thermodynamics, the integrating factor in the first law of thermodynamics:

$$dE = dU + TdS$$

where the energy E, the internal energy U, and the entropy S are functions of the state of the equilibrium system, and dE, dU, dS are their functional differentials; a function proportional to the pressure of a hypothetical perfect gas held at constant volume.

(**2.**) In statistical mechanics, a measure of the translational molecular kinetic energy.

(**3.**) The degree of hotness measured on a conventional temperature scale.

(Definitions (**1.**) and (**2.**) agree for equilibrium systems, as does (**3.**) for some appropriate range of temperature specific to the device.)

(**4.**) In equilibrium photon dynamics, the inverse multiplier of E/k (where k = Boltzmann's constant) in the Plack distribution $f = f(E/kT)$; the temperature of a black body radiation. In non-equilibrium situations, e.g., astronomical observations, a derived measure which agrees with one of the above definitions for some range of applicability.

If the object radiates like a true black body, all of these temperatures are equivalent; however, a perfect black body in nature is rare. A perfect black body is thermalized, that is, the atoms or molecules of the object are in perfect thermodynamic equilibrium. Most stars have a radiation distribution with wavelength that closely but not exactly approximates that of a black body. The sun, for example, has a radiation temperature that approximates a black body with $T_B = 6300$ K; however, its effective temperature, which compares the total output power to that of a black body, is better approximated by a $T_{\text{eff}} = 5800$ K, and the color temperature of the sun which compares the energy output ratio over two wavelength intervals is $T_c = 6500$ K. *See* excitation temperature, effective temperature, blackbody temperature, brightness temperature, color temperature.

temperature inversion An increase of atmospheric temperature with altitude. Under this condition the typical lapse rate is reversed, and great stability is created, which strongly damps vertical motions and vertical turbulent transport. Wind shear will exist between the top and the bottom of the temperature inversion layer. A *temperature inversion* acts as a ceiling, preventing further upward convection, and is generally the limit for cloud development. Marked and persistent inversions occur at lower levels, with subsiding air in major anti-cyclonic cells, such as the Azores high-pressure zone and cold anticyclones over continents. Temperature inversion arises for different reasons, such as frontal inversion, subsidence inversion, trade-wind inversion, radiation inversion, advective inversion, turbulence inversion, and stratospheric inversion. *See* density inversion.

temperature variance dissipation rate Measure of the rate at which gradient fluctuations of temperature are smoothed out by turbulence. It is defined by

$$\chi_T = 2\kappa_T \overline{(\nabla T)^2}$$

where κ_T is the thermal diffusivity, T is the temperature, z is the vertical coordinate, and ∇ denotes the gradient operator. The *temperature variance dissipation rate* can be regarded as the thermal equivalent of the dissipation rate of velocity fluctuations ϵ.

In oceanic or atmospheric studies, generally only one component of the temperature gradient is measured, and χ_T is estimated from

$$\chi_T \simeq 2\lambda\kappa_T \overline{\left(\frac{\partial T}{\partial z}\right)^2}.$$

The factor λ is a scaling parameter that reflects the level isotropy. For a completely isotropic turbulence field $\lambda = 3$.

tension (cosmic string) For a stringlike topological defect, the quantity

$$T = -2\pi \int r \, dr \, T^{zz},$$

where r is the radial coordinate in a cylindrical reference frame aligned with the string, and $T^{\mu\nu}$

is the energy-momentum tensor computed from the microscopic field configuration, in the particular case of a straight string. For an ordinary relativistic string we have $T^{tt} + T^{zz} = 0$; hence we recover the equation of state for a structureless Goto–Nambu string, where U (the energy per unit length) and T are constant and equal to each other. For more general vortex-forming field theories, the corresponding cosmic string model will be characterized by variable tension and energy per unit length which, in a generic state, will be related by an inequality of the form $T \leq U$. *See* duality in elastic string models, energy per unit length (cosmic string), equation of state (cosmic string), Goto–Nambu string.

tension head (ψ)   There is tension on pore water in the unsaturated zone because water is held to the mineral grains by surface-tension forces. It is conventional to measure pressure (p) relative to atmospheric pressure, thus in the unsaturated zone $p < 0$ and $\psi > 0$. Negative pressure is often called suction or tension, and ψ is called *tension head* when $p < 0$. Tension head is also known as capillary-pressure head and moisture potential. Tension in the unsaturated zone generally varies from about -0.35 m at field capacity to $-3{,}100$ m, where water is adsorbed by mineral grains directly from the air. The units are length (height) in meters.

tensor   A generalization of the concepts of vector and of one-forms as operators on functions and vectors.

A rank-2 contravariant *tensor* T is a linear sum of operators of the form

$$T = \mathbf{A} \otimes \mathbf{B}(\cdot, \cdot) \; ;$$

where $\mathbf{A}, \mathbf{B}$ are vectors, and the notation $\otimes$ indicates that the vector $\mathbf{A}$ acts on the function that is the first argument producing a number, and the vector $\mathbf{B}$ acts on the function that is the second argument producing a number, and the results of these operations is multiplied (ordinary multiplication). For instance

$$T(g, h) = \mathbf{A}(g)\mathbf{B}(h) \; ;$$

where $\mathbf{A}$ acts on g, and $\mathbf{B}$ acts on h. Since $\mathbf{A}, \mathbf{B}$ can be thought of as fields defined at least in a region of space, and the arguments f, g are

scalar functions of position, the tensor, and the result of this operation, are functions of position.

If we have a set of basis contravariant vectors, then the general rank 2 contravariant vector can be expressed:

$$T = T^{ij}\mathbf{e_i} \otimes \mathbf{e_j} \; .$$

Higher rank contravariant tensors are formed by repeatedly multiplying vectors using $\otimes$.

A rank-2 covariant tensor is constructed similarly to act on a pair of vectors; the result is the ordinary product of the results for each factor. A general covariant rank-2 tensor can be written as

$$g = g_{ij}\sigma^i \otimes \sigma^j \; ,$$

where σ^i constitutes a set of basis one-forms.

Higher rank covariant tensors, or tensors of mixed covariant/contravariant rank can be created by repeated use of $\otimes$. *See* summation convention.

terminal velocity   For an object falling through a normal fluid (e.g., air or water) the retarding force increases with the velocity of the object. Hence, there is a speed at which the retarding force equals the weight of the body less any buoyant forces. At this speed the body will not accelerate to fall faster; this is the *terminal velocity.*

terminator   The line on a planet or other solar system object, between sunlit and night sides of the planet, the sunrise/sunset line. Extended to bodies in other solar systems.

Terrestrial Coordinate Time (TCG)   Terrestrial Dynamical Time (TDT) has been deemed by the International Astronomical Union (IAU) to require supplementation with another standard because the progress of TDT depends, in a sense, on the gravitational potential on the geoid. The IAU, therefore, established in 1991 a time standard representing what an SI clock would measure in a coordinate system, such that the barycenter of the Earth was stationary in this nearly inertial system, as was the clock, but the clock was so far removed from the Earth (but not the sun and planets) that it suffered no effect of the Earth's gravity or rotation. That time is $TCG =$

$TT + L_G \cdot (JD - 2443144.5) \cdot 86400$ sec, where $L_G = 6.96929 \cdot 10^{-10}$ by definition as of 1992, JD stands for the Julian Date in TDT, and TT stands for Terrestrial Time, which is essentially the same as TDT. Presumably, the "constant", L_G, is subject to revision when and if the potential on the geoid is redetermined. For details, see *IERS Technical Note 13, The IERS Standards* (Ed.: D.D. McCarthy) (U.S. Naval Observatory, Washington, 1992).

Terrestrial Dynamical Time (TDT)   In 1977, *Dynamical Time* was introduced in two forms, TDT and Barycentric Dynamical Time (TDB). The difference in these two consists of periodic terms due to general relativity and does not exceed about 1.7 milliseconds in any year. TDT is used for the determination of the orbits of objects orbiting the Earth. It advances at the same rate as International Atomic Time (TAI), being equal, for all practical purposes, to TAI + 32.184 s. Both TAI and TDT contain relativistic effects of the Earth's motion around the sun, which reflect the fact that any reference frame centered on Earth's center is not inertial. Any of TDT, TAI, and TCG are thus subject to further special relativistic corrections in their relationship to any barycentric time standard, such as TCB. For details, see *IERS Technical Note 21, The IERS Conventions* (Ed.: D.D. McCarthy) (U.S. Naval Observatory, Washington, 1996). *See* dynamical time.

terrestrial heat flow   Heat flux from the interior of the Earth. It usually means the conductive heat flux measured at the surface of the Earth, and it is usually called heat flow or heat flow density. The present global average is about 0.06 W/m^2.

terrestrial planets   Those planets that are similar to the Earth in their characteristics. There are four terrestrial planets in our solar system: Mercury, Venus, Earth, and Mars. Characteristics of the *terrestrial planets* include small size ($< 13,000$ km in diameter), high density (> 3000 kg/m^3, indicating rocky compositions), close to the sun (< 1.5 astronomical units), few or no moons, and no rings. These characteristics differ considerably from those of the Jovian planets, which dominate the outer solar system. Although Pluto is similar to the terrestrial planets with its small size, one moon, and lack of rings, its large distance from the sun and its lower density (about 2000 kg/m^3, indicating an icy composition) place it in a category of its own, where it shares characteristics with many of the large icy bodies in the outer solar system. The terrestrial planets are sometimes also called the inner planets, since they are located in the inner part of the solar system.

Terrestrial Time (TT)   In 1991, the International Astronomical Union (IAU) defined an "ideal form of Terrestrial Dynamical Time", and designated *Terrestrial Time*. This definition appears to be primarily intended to extend the scope of TDT off the geoid, by imposing certain restrictions on relativistic coordinate systems. It is used in dynamical theories of solar system motions, and it could take into consideration the need to bring dynamical measures of time (from the motions of spacecraft orbiting the Earth) into accord with atomic time. For practical purposes, it is identical with Terrestrial Dynamical Time. For details, see *IERS Technical Note 13, The IERS Standards* (Ed.: D.D. McCarthy) (U.S. Naval Observatory, Washington, 1992).

tesla   A unit of magnetic flux equal to one weber per square meter.

Tethys   Moon of Saturn, also designated SIII. Discovered by Cassini in 1684. Its surface shows a crater, Odysseus, with a radius of 200 km. This is some 40% of the radius of *Tethys* and is probably at the limit of what the moon could sustain without breakup. The crater is now flattened and conforms to Tethys' spherical shape. There is also a large valley, Ithaca Chasma, which is 2000 km long and extends 3/4 of the way around Tethys. It is some 100 km wide and 3 to 5 km deep. Tethys' orbit has an eccentricity of 0, an inclination of 1.86°, a precession of 72.25° yr^{-1}, and a semimajor axis of 2.95×10^5 km. Its radius is 530 km, its mass 7.55×10^{20} kg, and its density 1.21 g cm^{-3}. It has a geometric albedo of 0.9, and orbits Saturn once every 1.888 Earth days.

Tethys Ocean   In geophysics, the continents of Laurasia and Gondwanaland were separated by the *Tethys Ocean* approximately 200 million years ago. This ocean has closed, and the Mediterranean Sea is a remnant. *See* continental drift.

texture   *See* cosmic texture.

Thalassa   Moon of Neptune also designated NVII. Discovered by Voyager 2 in 1989, it is a small, irregularly shaped body approximately 40 km in radius. Its orbit has an eccentricity of 0.00016, an inclination of 0.21°, a precession of $551°$ yr^{-1}, and a semimajor axis of 5.01×10^4 km. Its mass has not been measured. It orbits Neptune once every 0.3115 Earth days.

Tharsis Province   A broad Martian topographic rise centered on the equator at longitude 115° W. It stands as much as 10 km above the reference datum, measures ≈ 8000 km across, and it occupies $\approx 25\%$ of the surface area of Mars. It is the most pronounced region of central vent volcanism on Mars. It has four large shield volcanoes (Olympus Mons, Ascraeus Mons, Pavonis Mons, and Arsia Mons, the summits of which are concordant, standing 27 km above the reference datum). Olympus Mons stands 25 km above the plains, and the other three stand 17 km above the plains. All are considered to have formed by successive eruptions of low-viscosity lavas. The volcanoes account for half the planet's volcanic production from the late Hesperian until the Amazonian.

The province is asymmetrical being twice as steep at the northwestern limit as it is on the southeastern limit. At the NW, it forms a continuous slope (only broken by remnants of old cratered terrain, e.g., Tempe Terra) with the sparsely cratered northern lowland terrain, and on the SE, it forms a continuous slope grading into the high-standing, highly cratered southern hemisphere terrain. It demonstrates a complex and extended tectonic record, being at the center of a vast radial fracture system that affects half the planet, and a substantial free-air gravity anomaly. Additionally, the huge equatorial canyons of Noctis Labyrinthus start at the center of the bulge and extend down the eastern flank to form Valles Marineris.

Thebe   Moon of Jupiter, also designated JXV. Discovered by S. Synnott in 1979, its orbit has an eccentricity of 0.015, an inclination of 0.8°, and a semimajor axis of 2.22×10^5 km. Its size is 55×45 km, its mass 7.60×10^{17} kg, and its density 1.6 g cm^{-3}. It has a geometric albedo of 0.05 and orbits Jupiter once every 0.674 Earth days.

thermal anisotropy   *See* kinetic temperature, plasma stress tensor.

thermal bar   Due to cabbeling, mixing of two water masses with identical density but different temperature and salinity generates denser water. The *thermal bar* relates to the special case of mixing two water bodies with temperatures below and above 4°C. The mixed water has close to maximum density and subsequently sinks as thermal plumes. The term "bar" refers to the phenomenon that the original waters cannot cross the sinking plumes.

thermal boundary layer   A portion on the edge of a body with a high thermal gradient. In convective systems (such as the Earth's mantle), this is commonly associated with the edges of each convecting layer. The oceanic lithosphere is one example of a *thermal boundary layer* and illustrates clearly the importance of this boundary layer for the convective process: the layer thickness is very small at mid-ocean ridges where new boundary layer material is erupted from beneath, but the thickness increases moving away from the ridge as heat is removed from the plate by diffusion. This production of a cool boundary layer causes negative buoyancy, which allows the oceanic slab to plunge down into the interior of the convective region, transporting a heat deficit inwards (and by implication, causing a net transport of heat towards the surface). It is quite likely that there is another thermal boundary layer at the base of the mantle, where the seismically distinct D" layer is to be found and where perhaps a reverse process is taking place except that boundary layer material appear to detach from the core-mantle boundary in the form of blobs rather than slabs.

thermal bremsstrahlung   *See* bremsstrahlung [thermal].

thermal conductivity   The property of a material characterizing the ease with which heat is conducted through the material at steady state. *Thermal conductivity* λ appears in Fourier's law of heat conduction: $q = -\lambda \cdot \nabla T$, where q is the heat flux vector, and ∇T is the temperature gradient. For an anisotropic material, λ is a tensor.

thermal diffusive sublayer   The layer along a contiguous boundary, within which momentum (shear stress) is transferred by molecular processes; i.e., $\tau = \nu \partial u / \partial z$.

thermal diffusivity   A property that determines the rate of the diffusion of heat in a transient state, defined as $\alpha = \lambda / \rho c$, where λ is the thermal conductivity, ρ is density, and c is the specific heat. For an anisotropic material, α is a tensor.

thermal Doppler line broadening   Broadening of a spectral line due to the random motion of gas particles in a Maxwellian distribution, which thus depends on temperature. The average speed of a molecule or atom of gas is given by

$$v_0 = \sqrt{\frac{2kT}{\mu}}$$

where k is Boltzmann's constant, T is the kinetic temperature, and μ is the atomic weight in molecules. Then the number of gas particles with a velocity in the range of v from $v_0 + \Delta v$ to $v_0 - \Delta v$ is given by

$$\frac{dN}{N} = \frac{1}{v_0 \sqrt{\pi}} e^{-(v/v_0)^2} dv \;.$$

This expression can be combined with the Doppler shift and an expression for the Doppler half width as the distance from the line center where the intensity of the line falls to 1/e of the maximum:

$$\Delta \lambda_D = \lambda_0 \frac{v_0}{c} \;.$$

Then the normalized Doppler profile is

$$\phi_\lambda = \frac{1}{\Delta \lambda_D \sqrt{\pi}} e^{-\Delta \lambda^2 / \Delta \lambda_D^2} d\Delta \lambda \;.$$

In general, the thermal Doppler broadening for heavy elements is less than for light elements at the same temperature, since heavier elements have lower mean velocities at any given temperature. For example, at a temperature of $T = 6000$ K, the $n = 3$ to $n = 2$ electronic transition of neutral hydrogen (Balmer α at 6563 Å) will have a thermal Doppler line width of $\Delta \lambda / \lambda \approx v/c \approx 1/4$ Å.

thermal expansivity   Coefficient α expressing the relative change of density due to the change in potential temperature at constant salinity and pressure; i.e., $\alpha = -\rho^{-1} (\partial \rho / \partial \Theta)_{S,p}$ [K^{-1}]. α is strongly temperature-dependent and is always positive: $\alpha > 0$ for ocean water; in low-saline water $\alpha < 0$ below temperature of maximum density. See actual values in UNESCO [1983], Algorithms for computation of fundamental properties of seawater, *Unesco Tech. Pap. Mar. Sci.*, 44, 53 pp.

thermal wind relationship   The relation between density gradient (on an isobaric surface) and wind shear in the hydrostatic and geostrophic flow. This relationship takes the form

$$f \partial v / \partial z = -g \rho^{-1} (\partial \rho / \partial x)_p,$$
$$f \partial u / \partial z = g \rho^{-1} (\partial \rho / \partial y)_p$$

where u, v are eastward and northward wind components, respectively, g is the gravity, ρ is density, f is Coriolis parameter, and x, y, z are eastward, northward, and upward coordinates, respectively.

In the atmosphere when temperature decreases toward the poles, winds become more westerly with height. For an ideal atmosphere, the relation is given by

$$f \partial v / \partial z = g T^{-1} (\partial T / \partial x)_p,$$
$$f \partial u / \partial z = -g T^{-1} (\partial T / \partial y)_p$$

where T is temperature.

thermobaric effect   *See* thermobaricity.

thermobaricity   The compressibility of water depends on potential temperature as well as salinity. Laterally displacing water without performing work against gravity (following neutral tracks), and will lead to displacements relative to isopycnal surfaces. As a result, meso-scale eddies lead to a diapycnal flux. In contrast to cabbeling, where mixing is involved, the thermobaric effect arises from displacement only and does not require mixing.

thermocline   The region of large temperature gradient in oceans or lakes. Often there is a region of large temperature gradient near the surface of the ocean that appears only in summer and autumn; this is called seasonal *thermocline*. In low and middle latitudes, there is an ocean thermocline present all the time at depths between 200 and 1000 m, called the main or permanent thermocline. The e-folding thermocline depth scale in the ocean is about 1 km; in lakes it is much less and depends on the clarity of the water.

thermohaline circulation   A circulation that is driven by the buoyancy force. *See also* wind-driven circulation.

thermosphere   *See* ionosphere.

Theta aurora   Rare form of the aurora extending across the polar cap from night- to dayside. Viewed from a high flying satellite, this arc combined with the auroral oval closely resembles the Greek letter Θ (Theta). Observations of *Theta aurora* are limited to time periods when the interplanetary magnetic field has a northward component. *See* polar cap arc.

thick-target   A plasma in which a non-thermal population of energetic electrons is thermalized while generating radiation. The thermalization may be due to Coulomb collisions of the electrons with ambient particles or collective interactions with each other.

thin-target   A plasma that has no appreciable effect on an injected spectrum of non-thermal particles passing through it. A *thin-target* scenario would be applicable to electrons injected outwards through the corona.

30 Doradus   A star formation region in the Large Magellanic Cloud. It is located at RA = 5.6^h and dec = $-69.1°$, and the main cluster subtends approximately 7 arcmin. This is a re-

gion of very active and current star formation, often referred to as a "starburst" and is the closest and most visible example of such a region. It contains a large collection of very early O-type and Wolf–Rayet stars. The core of the cluster, R 136 (HD 38268), was once thought to be a single supermassive star but has now been resolved into a very dense cluster of young stars.

Thompson circulation theorem   *See* Kelvin circulation theorem.

Thomson scattering   Scattering of electromagnetic radiation by a charged particle that moves nonrelativistically in the process. For unpolarized incident radiation:

$$d\sigma/d\Omega = \left(e^2/mc^2\right)^2 (1/2)(1 + \cos\theta) ,$$

where θ is the scattering angle, and e and m are the charge and mass of the scatterer.

The total Thomson cross-section is

$$\sigma_T = 8\pi/3 \left(e^2/mc^2\right)^2$$
$$= 0.665 \times 10^{-24} cm^2$$

for electrons.

t'Hooft–Polyakov monopole (1974)   A particular exactly describable magnetic-like monopole involving a vector Higgs field ϕ^a $a = 1, 2, 3$ (connected to a phase transition from a higher temperature configuration), in which each of the field components ϕ^a is equal in value to the corresponding spatial coordinate x^a. *See* cosmic topological defect, inflation, monopole, monopole excess problem, winding number.

Thorpe displacement   The distance a water parcel must be moved vertically so that it is in stable equilibrium with the surrounding water. Turbulence can generate local overturns of water parcels that lead to inversions of the density profile. If a water parcel at depth z_1 must be moved to depth z_2 to generate a monotonically increasing density profile, the displacement $d_1 = z_2 - z_1$ is called the *Thorpe displacement*. Thorpe displacements are useful as an aid for defining the vertical extent of oceanic mixing events and overturns. However, since ocean turbulence is not two-dimensional, the displacement d_1 is not necessarily the distance the water parcel has traveled vertically (*see* Thorpe scale).

Thorpe displacements are computed from measured vertical space series of density. A sorting algorithm is applied to the sampled density profile, with ordering beginning at the shallowest depth, which re-orders the samples in ascending order. If a sample at position n is moved to position m, the associated entry in the Thorpe displacement series is then d_n.

Thorpe scale Measure of the mean eddy size in turbulent oceanic flows. The *Thorpe scale L_T* is defined as the root-mean-square of the Thorpe displacement, d,

$$L_T = \left(\overline{d^2}\right)^{1/2}$$

where the over-bar signifies an appropriate spatial average that depends on the vertical extent of the turbulent process. If the mean vertical density gradient is much larger than the mean horizontal gradient, the Thorpe scale is proportional to the mean eddy size of the turbulence. *See also* Ellison scale. Since experimental evidence indicates that the Thorpe scale is nearly identical to the Ozmidov scale L_O, the Thorpe scale is often considered as the maximum size of the vertically overturning eddies. *See* Thorpe displacement.

thrust fault A dip-slip fault upon which the deformation is compressional. *See* reverse fault.

Tibetan Plateau Located to the southwest of China. Its area is about 200×10^4 km^2, and average sea level elevation is about 4 km. Since the *Tibetan Plateau* is at the middle level of the troposphere (about 600 hPa level), it strongly affects the atmospheric circulations by its orographic thermal and dynamic effects. In the summer, almost every meteorological observation element over the Tibetan Plateau has the most strong diurnal variations in the world; and cumulus convective activities are very frequent and active due to the large orographic heating effects. In the summer, the low levels of the Tibetan Plateau are a thermal low pressure region, and the high level at 200 hPa is the south Asian high, which is maintained by the high temperature and high vertical moisture transport from the lower levels. Such vertical transport is mainly carried out by the strong and frequent cumulus convective activities over the plateau. In contrast, in winter, at low levels they are cold high pressure systems, and at high levels they are cold low pressure systems. The opposite plateau pressure systems in winter and summer cause the special plateau monsoon. To the atmosphere, the plateau is a heat source during both winter and summer; and the atmosphere over the plateau to the atmosphere around it is a heat source in summer and a cold source in winter. The thermal effects of the plateau in the winter enhance the Hadley circulation over the plateau; in the summer, they create the monsoon meridional and zonal circulation systems, which all come from the ascending air over the plateau and flow to the east Pacific, northern Africa, and southern hemisphere. The seasonal change of the cold and heat source is also an important factor to cause the seasonal change (jump) of the east Asian circulations.

tidal bore A translating wave found in coastal areas that resembles a hydraulic jump. Found at locations with very large ($O(10\,m)$) tide range.

tidal currents A current that is driven by pressure gradients within the wave that results in tidal fluctuations in coastal areas. Often bidirectional (ebb and flood) in estuaries or rotary in open water.

tidal delta A deposit of sediments transported by a combination of waves and tidal currents at a tidal inlet. Many inlets have both ebb and flood deltas.

tidal energy dissipation Conversion of work done to a celestial body by tidal forces into heat due to the anelastic tidal response of the body. For example, the total rotational energy of the Earth–moon system decreases as a result of *tidal energy dissipation,* while the angular momentum is conserved. The rate of lunar tidal energy dissipation is about 3×10^{12} W.

tidal forces Differences in the gravitational force on opposite ends of an extended body caused by the different distances of those ends from the source of the gravitational field. The tidal acceleration is a relative acceleration (locally) proportional to the gradient of the gravitational acceleration, equivalent to the second derivative of the potential, $\phi_{,ij}$, and is also proportional to the separation ϵ^i between the points considered. Here the subscript "," denotes partial derivative.

Hence

$$(\delta a_{\text{tidal}})_j = -\epsilon^i \phi_{,ij} .$$

A similar expression, involving the Riemannian tensor, is found in general relativity (so $\phi_{,ij}$ is called the Newtonian (analog of the) Reimannian tensor).

Because they are proportional to the gradient of the gravitational acceleration, *tidal forces* are proportional to r^{-3}, and so rise sharply at small distances but become negligible at large distances. Typically, tidal forces deform planets and moons. On the Earth, both the solid body of the Earth and, more noticeably, the surface of water in the oceans reacts to these relative accelerations, in the form of tides, because of interactions with both the sun and the moon.

In fact, because it is closer, the moon generates higher tides on the Earth than does the sun. Far from the ocean shores, the height of the lunar tidal wave is 65 cm, and the height of the solar tidal wave is 35 cm. These waves travel around the Earth because the Earth rotates. The tides become much higher at the shores because of local topography (when the tidal wave goes into a funnel-shaped bay, its width is decreased, and so the height must increase) or because of resonances. The latter happens in the Bay of Fundy in Canada, where tidal waves reach the height of 16 m.

Because tidal forces deform a solid body (a planet or moon), tides dissipate energy and act to decrease rotation. As a result, the rotation is decelerated until it becomes synchronous with the orbital motion — which means that afterwards the planet or moon faces its companion body always with the same side. This has happened to our moon in its orbit around the Earth.

tidal formation of solar system   A theory attributing the formation of the solar system to a close tidal encounter with another star, which drew material out of the sun to condense into planets. This theory is now out of favor because it suggests solar systems are rare, since such encounters are rare, while recent observations provide evidence for planetary systems around a number of local stars, and even around neutron stars.

tidal friction   As a result of the anelastic response of the Earth, the peak of its tidal bulge lags the maximum tidal force by about 12 min. Because the Earth rotates faster than the moon orbits the Earth, the tidal bulge leads the Earth-moon axis by about 3°. This lag angle causes a torque acting on the Earth to slow down its rotation, resulting in a length-of-day increase of over 2 msec per century. This effect is called *tidal friction*. The same torque acting on the moon increases its orbiting speed and hence the Earth–moon distance. There is also tidal friction between the Earth and the sun, about 1/5 of the strength of the lunar tidal friction.

tidal heating   The process by which a body's interior heat is generated by the gravitational tidal forces of external bodies. Tidal deformation leads to friction, which in turn creates internal heat. For the Earth's moon, tidal heating occurred for only a short period, until its orbit became circularized and synchronous (when the rotation period exactly equals the orbital period). However, if the moon is in a resonance with another moon, as in the case of several moons of Jupiter, the rotation can fail to become locked, and the *tidal heating* stage can last much longer. This is the situation with Jupiter's moons of Io and Europa. The combination of Io's proximity to massive Jupiter and its orbital resonance with Europa causes tremendous tidal heating and gives rise to the active volcanism seen on Io's surface. Europa is slightly further from Jupiter than Io, so it feels less of Jupiter's tidal forces, but it also is in a resonance (with Ganymede), and this situation causes some tidal heating on Europa, which may allow for a liquid water ocean to exist under its icy crust.

tidal inlet   An opening between the sea and a sheltered estuary or river.

tidal period   The time between two points of equal phase in a tidal curve. Commonly measured between two high or two low tides.

tidal prism   The volume of water that is exchanged between a tidal estuary and the sea in one tidal period. Equivalent to the average tide range within the estuary times the area of the estuary. Since the range will vary in space and time, the volume is not simply a horizontal slice, but a complicated function.

tidal radius   The radius within which all the luminous matter of a cluster or a galaxy is contained. The *tidal radius* can be measured for globular clusters and for galaxies belonging to clusters, which are found to have well-definite outer limits (in contrast with brightness profiles of isolated elliptical galaxies, described by de Vaucouleurs or Hubble's–Reynolds law). The name arises from the understanding that in the case of a cluster galaxy, repeated encounters with nearby galaxies can lead to tidal stripping of the outer stars, which are loosely gravitationally bound, and to the evaporation of the outer envelope, leaving only stars which are inside the tidal radius.

tidal stripping   The escape of gas and stars gravitationally bound to a system, such as a galaxy or a globular cluster, due to tidal forces exerted by an object external to the system. For example, in a cluster of galaxies, *tidal stripping* may remove loosely bound stars from the galaxy outer envelope; in a close encounter between galaxies, stars and gas can be transferred from one galaxy to the other.

tidal tail   A highly elongated feature produced by tidal forces exerted on a spiral galaxy by a companion galaxy. A most notable example of *tidal tails* is observed in the "Antennae" pair of galaxies (NGC 4038 and NGC 4039), where the tidal tail extends for a projected linear size of ≈ 100 kpc, much larger than the size of the galaxies themselves. Computational models show that very extended tails, like the ones in the Antennae, are produced by a prograde encounter between galaxies, i.e., an encounter between a spiral galaxy and an approaching companion galaxy which moves in the same sense of the spiral rotation.

tide   The response of the solid or fluid components of a planet or other astronomical body, under the influence of tidal forces.

tide range   The vertical distance between high tide and low tide at a point. Will vary in time because of temporal variation in the tide signal.

tilt angle   In solar magnetohydrodynamics, on the source surface, a neutral line separates the two hemispheres of the sun with opposing magnetic polarity. The maximum excursion of this neutral line with respect to the heliographic equator is called *tilt angle*. Since this neutral line is carried outwards as the heliospheric current sheet, the tilt angle also defines the waviness of the current sheet and, therefore, is an important parameter in the modulation of the galactic cosmic radiation.

In terrestrial magnetospheric research, it is the angle between the z-axis in GSM coordinates and the dipole axis of the Earth. The GSM x-y plane can be viewed as providing an approximate north–south symmetry plane of the magnetosphere (*see* equatorial surface). A tilt angle $\psi = 0^0$, therefore, signifies a magnetosphere with the dipole axis perpendicular to the equator, and the larger ψ, the more the axis departs from the perpendicular. Sunward inclination of the dipole gives $\psi > 0$, tailward inclinations $\psi < 0$.

In the Earth's magnetosphere, ψ can vary within $\pm 35°$, while for Uranus and Neptune all values are possible. For Earth the tilt shifts the location of the polar cusps (both near the magnetopause and in the ionosphere) and causes warping of the plasma sheet. *See* modulation of galactic cosmic rays, source surface.

time dilatation   The slowing-down of clocks in a system moving with respect to a given observer. Suppose a system S (for example, an interstellar spacecraft) moves with respect to observer O with the velocity v along the straight line OS.

Suppose an observer in the system S measures the time-interval T between two events, say two ticks (1 sec apart) of his clock, which is at rest in his frame. Then the observer O will find that the time-interval T' between the same two events on O's own clock is $T' = T/\sqrt{1 - v^2/c^2}$, where c is the velocity of light. Hence, $T' > T$, and the observer O will decide that the clock in S runs slow. This formula follows from the experimentally verified fact that the velocity of light is the same in every inertial reference frame. The time dilatation is a relative effect: the observer in S will also observe that O's clock goes more slowly. This is one of the most famous predictions of the Special Relativity Theory and is very directly or accurately verified. Many elementary particles observed in laboratories and cosmic rays are unstable; they decay into other particles. For instance, the mean decay-time for μ-mesons is $2.3 \cdot 10^{-6}$ sec. Such μ-mesons, formed at the top of the Earth's atmosphere by cosmic ray collisions, would move close to the velocity of light $c = 3 \cdot 10^{10}$ cm/s. Thus, in the absence of time dilatation they would be able to fly, on average, the distance of only 690 m before decaying, and only a minute fraction of the initial number would survive the journey to the surface of the Earth. In fact, μ-mesons created in high layers of Earth's atmosphere reach detectors on the surface of the Earth in copious quantities. They are seen to live longer because they move with a large velocity with respect to us. *See* special relativity.

time dilation   *See* time dilitation.

time-distance helioseismology   The study of the solar interior using the direct measurement of travel times and distances of individual acoustic waves. Carried out using temporal cross-correlations of the intensity fluctuations on the solar surface.

time in semiclassical gravity   Current approaches to quantization of gravity do not produce an obvious time variable. However, one can introduce time if the semiclassical approximation is assumed to hold for (at least some of) the gravitational degrees of freedom, i.e., if these degrees of freedom behave classically, while the rest of the system is quantized. The introduced time is associated with the change of such observables; gravity is treated as a clock. In a more refined model, one could imagine some macroscopic apparatus which couples to the specified gravitational degrees of freedom and gives the value of the time. In this sense, the rotation of the Earth or any other geodesic motion in a curved space-time is a good and well-known example. *See* Wheeler–DeWitt equation.

time-like infinity $(i^{\pm})$. The distant future and distant past limits of timelike curves. *See* conformal infinity.

time-like vector   An element t of a linear space with a Lorentzian metric g of signature $(-, +, +, +)$, such that $g(t, t) = g_{ab}t^a t^b < 0$. In the theory of general relativity, a time-like four-vector represents the velocity of propagation of a particle with rest-mass. *See* metric, signature.

time zone   One of 24 zones approximately $15°$ (1 h) of longitude in width and centered in multiples of $15°$ from Greenwich, England, throughout which the standard time is constant and one hour earlier than the zone immediately to the east, except at the International Date Line at longitude $180°$, when 24 h is subtracted from the date in moving westward.

Titan   Moon of Saturn, also designated SVI. It was discovered by Huygens in 1655. Its orbit has an eccentricity of 0.029, an inclination of $0.33°$, a semimajor axis of 1.22×10^6 km, and a precession of $0.521°$ yr^{-1}. The radius of its solid body is 2575 km, but its thick atmosphere extends more than 100 km above this surface. Its mass is 1.35×10^{23} kg, and its density is 1.89 g cm^{-3}. Its geometric albedo in the visible is about 0.2, and it orbits Saturn once every 15.95 Earth days. Titan is the only solar system satellite with a significant atmosphere.

Titania   Moon of Uranus, also designated UIII. Discovered by Herschel in 1787, its surface is a mixture of craters and interconnected valleys. *Titania* shows signs of resurfacing and may have once undergone melting. Possibly the resultant volume change upon cooling caused the observed cracks and valleys. Its orbit has an eccentricity of 0.0022, an inclination of $0.14°$, a precession of $2.0°$ yr^{-1}, and a semimajor axis of 4.36×10^5 km. Its radius is 789 km, its mass 3.49×10^{21} kg, and its density 1.66 g cm^{-3}. It has a geometric albedo of 0.27 and orbits Uranus once every 8.707 Earth days.

Tolman model   Also called the Tolman–Bondi model and the Lemaître–Tolman model. An inhomogeneous cosmological model containing pressureless fluid (dust). The most completely researched of the inhomogeneous models of the universe. It results from Einstein's equations if it is assumed that the spacetime is spherically symmetric, the matter in it is dust, and that at any given moment different spherical shells of matter have different radii. (If the radii are the same, then a complementary model results whose spaces of constant time have the geometry of a deformed 3-dimensional cylinder; in the Lemaître–Tolman model the spaces are curved deformations of the Euclidean space or of a 3-dimensional sphere.) The model was first derived in 1933 by G. Lemaître, but today it is better known under the name of the Tolman, or the Tolman–Bondi model. The Oppenheimer–Snyder model is a specialization of such models to homogeneous dust sources. *See* Oppenheimer–Snyder model, comoving frame.

tombolo   A sand spit in the lee of an island that forms a bridge between the island and the mainland. Referred to as a salient if the connection is not complete.

Tomimatsu–Sato metrics (1973)   An infinite series of metrics describing the exterior gravitational field of stationary spinning sources. The δth *Tomimatsu–Sato metric* has the form

$$ds^2 = \frac{B}{\delta^2 p^{2\delta - 2}(a-b)^{\delta^2 - 1}} \left(\frac{dy^2}{b} - \frac{dx^2}{a} \right)$$
$$+ g_{ik}\, dx^i\, dx^k$$

where (the other two coordinates are x^3, x^4)

$$g_{33} = \frac{bD}{\delta^2 B}$$
$$g_{34} = 2q\frac{bC}{B}$$
$$g_{44} = \frac{A}{B}$$

and

$$a = x^{2-1}, \quad b = 1 - y^2, \quad p^2 + q^2 = 1.$$

The polynomials A, B, C, and D are all constructed from the order-δ Hankel matrix elements

$$M_{ik} = f(i + k - 1)$$

where $f(k) = p^2 a^k + q^2 b^k$. The first, $\delta = 1$, member is the Kerr metric. The $\delta = 2, 3$, and 4 members were found by A. Tomimatsu and H. Sato in 1973. The full theory was established by S. Hori (1978) and M. Yamazaki (1982). The singularities of the δth Tomimatsu–Sato spacetime are located at δ concentric rings in the equatorial plane.

Toomre's stability parameter Q (1964)   A numerical parameter describing the stability (or lack of stability) of a system of self-gravitating stars with a Maxwellian velocity distribution.

By consideration of the dispersion relation for small perturbations in an infinite, uniformly rotating gaseous disk of zero thickness and constant surface density Σ, at angular velocity Ω one finds the stability criterion

$$Q_g = v \cdot \frac{2\Omega}{\pi G \Sigma} > 1$$

where G is Newton's constant, and v is the speed of sound in the gas.

A generalization of the above statement to a differentially rotating gaseous disk in the tight winding approximation leads to a local stability condition at radius r: $Q_g(r) \equiv v(r)\kappa(r)/\pi G \Sigma(r) > 1$, where $\kappa(r)$ is the local epicyclic frequency at radius r. The single unstable wavelength in the $Q_g(r) = 1$ gaseous disk is $\lambda_*(r) = 2\pi^2 G\Sigma(r)/\kappa^2(r)$. [These gaseous and stellar disks are shown to be stable to all local non-axisymmetric perturbations (Julian and Toomre, 1966).] A similar local stability condition $Q_s(r) \equiv \sigma_R(r)\kappa(r)/3.36G$ $\Sigma(r) > 1$ and the single unstable wavelength in the $Q_s(r) = 1$ stellar disk $\lambda_*(r) = (2\pi \times 3.36/\sqrt{0.948})G\Sigma(r)/\kappa^2(r)$ can be derived from a dispersion relation for axisymmetric perturbations in a differentially rotating stellar disk with the local radial velocity dispersion $\sigma_R(r)$.

TOPEX/POSEIDON   A cooperative project between the U.S. and France to develop and operate a satellite dedicated to observing the Earth's oceans. Since its launch in 1992, this satellite has used an advanced radar altimeter to measure sea surface height over 90% of the world's ice-free oceans. By determining the orbit of the spacecraft to within a few centimeters, the project has produced global maps of ocean topography significantly more accurate than any previous satellite altimeter. A second goal of the project was the production of accurate global tide maps. Consequently, the *TOPEX/POSEIDON* orbit was designed to avoid the aliasing of the solar tides into undesirable frequencies, and this repeat orbit covers the globe every 10 days. *See also* satellite altimetry.

Top hat detector   A toroidal-geometry focusing detector for low energy ions and electrons (e.g., 100 to 30,000 eV) widely used in magnetospheric research. It gives high counting rates and uses fast electronic circuits to sort particles by their time of flight through the instrument.

topocentric system   This spherical coordinate system (range, altitude, azimuth) serves to specify a direction in space relative to the local horizontal, vertical, and North, as well as the distance from the location of the observer. *See* altitude, azimuth, zenith, zenith angle, nadir.

topographic wave   An orographic wave; a wave in moving air driven by irregularities in the surface topography of the Earth.

topography   The elevation of the Earth's surface above sea level. Topography is created by active mountain building processes and is destroyed by erosion.

topological defect   *See* cosmic topological defect.

topology of space   The set of properties of a space that can be established without using the notion of distance. Two spaces are said to be topologically equivalent if one of them can be obtained from the other by stretching, squeezing, and any other deformation that does not involve disruption or gluing points together. For example, the outer surface of an open bottle is topologically equivalent to the top surface of a coin. However, the surface of a bicycle tube is not topologically equivalent to the coin-surface. The topology of the space we live in is usually imagined to be that of the Euclidean space, i.e., infinite in every direction, with no identification of points. However, general relativity and Einstein's equations allow more complicated topologies. The simplest (e.g., Robertson–Walker) solutions to Einstein's equations for cosmology admit 3-spaces that are Euclidean, or 3-hyperboloidal, or 3-spherical. More complicated spaces are admitted; for example, cosmological space could be that of a three-torus, a three-dimensional analog of the bicycle tube surface, which has the property that it can be circled around in three different directions after covering only a finite distance. The attractive feature of such a space (called a "small Universe") would be that, given a sufficiently long but finite time, all of matter existing in the universe would come into view of every observer. Whether our real universe is small is a question that is in principle decidable by (very difficult) observations.

tornado   An intense cyclonic wind (windspeeds up to 450 km/h) in contact with the ground and locally associated with a thunderstorm. Most frequently occurring in summer and the early autumn in the midwestern U.S. and in Australia. Typically characterized by a funnel-shaped condensation cloud extending toward the ground, though simply a debris cloud beneath a thunderstorm is evidence of a tornado.

torque   The product of a force **F** times the lever arm from a fulcrum, measured perpendicular to the direction of the force.

$$\text{torque } \boldsymbol{\tau} = \mathbf{r} \times \mathbf{F} ,$$

where $\times$ denotes the vector cross product. $\boldsymbol{\tau}$ is a pseudovector, with direction given by the right-hand rule: right-hand fingers in direction of $\mathbf{r}$, curl fingers toward $\mathbf{F}$, thumb then points in direction of $\boldsymbol{\tau}$. Its magnitude is

$$|\boldsymbol{\tau}| = |\mathbf{r}_\perp| |\mathbf{F}| ,$$

where $|\mathbf{r}|_\perp$ is the perpendicular distance from the origin, or which is equivalent to

$$|\boldsymbol{\tau}| = |\mathbf{r}||\mathbf{F}| \sin\theta ,$$

where θ is the angle $\mathbf{r}$ makes with $\mathbf{F}$.

torsion　A tensor quantity expressing the fact that covariant derivatives do not commute when applied to a scalar function:

$$\nabla_\alpha \nabla_\beta f - \nabla_\beta \nabla_\alpha f = T^\sigma_{\alpha\beta} \nabla_\sigma f .$$

Here ∇_α is the covariant derivative along basis vector α, and $T^\sigma_{\alpha\beta}$ are the components of the torsion tensor. In the formalism of general relativity, the torsion vanishes. *See* covariant derivative.

torsional oscillation　A type of motion thought to occur in the Earth's core, where the flow is akin to the solid body rotation of cylinders coaxial with the Earth's rotation axis about that axis. The flow would be of this form if the density in the core were constant; however, in the real Earth, rather than the azimuthal component of velocity u_ϕ, it should be the azimuthal component of momentum ρu_ϕ (where ρ is the density) that is constant on these cylinders. Different cylinders rotate at different speeds, stretching any magnetic field that links cylinders, which generates a torque resisting the motion through the tension in the magnetic field. This torque on the cylinders causes an acceleration, which leads to oscillations. In the real Earth, such motions would be superimposed upon a much more slowly evolving "basic state", likely a flow in a magnetostrophic balance. There is observational evidence of oscillatory flows at the surface of the core with the spatial form of *torsional oscillations* with periods of around 75 years, although these oscillations seem to be fairly heavily damped, perhaps by resistive coupling at the core-mantle boundary. *See* core flow.

total electron content (TEC)　The total number of free electrons in a unit area column from the ground to a height well above the level of peak ionization. The units are electrons m^{-2} and 1 TEC unit = 10^{16} electrons m^{-2}. Typically, TEC varies from 1 to 200 TEC units. The TEC is usually measured by observing the transmissions along a slant path to a satellite. The slant TEC is then converted to a vertical TEC by using a geometric correction and assuming that the ionosphere is a thin shell about the Earth. The TEC can be used to provide estimates of Faraday rotation and time delay on transionospheric propagation paths.

total radiance mean cosine　The average cosine of the polar angle of all photons; it equals the ratio of the net plane irradiance to the total scalar irradiance.

towards polarity　The condition in interplanetary space when the local magnetic field lines, when followed to the sun, point toward it. *See* interplanetary magnetic sector, away polarity.

tracers　Identifiable substances or sets of ocean properties that do not affect the density of sea water. *Tracers* have no impact on water movement but can be used to infer currents. In addition to the classical tracers (oxygen and nutrients), oceanographers use tracers introduced or enriched by human activity, such as carbon, cesium, chlorofluorocarbons, plutonium, strontium, and tritium.

traction　A method of transporting material by pushing, rolling, or sliding that material across a surface. The forces causing *traction* to occur are usually the result of eolian and/or fluvial processes. Traction is the primary mode of transport for particles larger than about 2 mm in diameter.

transcurrent (transform) plate boundary
The boundary, which is a strike-slip fault or a

shear zone, between two lithospheric plates that both move parallel with the boundary but in opposite directions.

transform boundary   Represents the boundary where two plates slide past each other. This boundary is characterized by strike-slip faulting and can be a left-lateral or a right-lateral fault, depending on the movement of the plates involved. A *transform boundary* is characterized by shallow earthquakes but no volcanic activity. An example of a transform boundary is the San Andreas Fault which runs through California.

transform fault   When two surface plates slide past each other, the fault on which this occurs is a *transform fault*. A transform fault is also a strike-slip fault. An accretional plate margin is made up of an orthogonal pattern of spreading centers (ocean ridges) and transform faults. The San Andreas fault is an example of a major transform fault. Transform faults are usually very active seismically.

transform push   Normal force that one lithospheric plate exerts on another plate across a transform plate boundary.

transient shock   Shock propagating through a medium, either as a blast wave shock or as a driven shock in front of an obstacle moving faster than the local signal speed of the medium. Transient shocks are also called traveling shocks; examples are the super-sonic bang in front of an aircraft, or the shock piling up in front of a fast coronal mass ejection or a fast magnetic cloud.

transitional depth   A water depth between shallow water and deep water. The term depth, as used here, is a relative term and denotes actual depth divided by wavelength.

transition region   In solar physics, the thin (some hundred kilometers) layer in the solar atmosphere at a height of a few thousand kilometers above the photosphere, where the temperature increases quickly from about 6000 K to more than 10^5 while the density drops by about one order of magnitude, and the collision times increase by more than four orders of magnitude. Thus, below the transition region, the solar plasma basically behaves like a gas with the bulk motion of the plasma carrying around the frozen-in magnetic field, while above the transition region collisions are rare, and the magnetic field determines the particle motion. It is observed mainly in EUV emission lines between 10^4 and 10^6 K.

transition region and coronal explorer (TRACE)   A NASA funded spacecraft launched in April 1998. It explores the magnetic field in the solar atmosphere by studying the 3-dimensional field structure, its temporal evolution in response to photospheric flows, the time-dependent coronal fine structure, and the coronal and transition region thermal topology. *TRACE* has a spatial resolution of 1 arcsec (700 km).

translatory wave   A wave that exhibits a net motion, as opposed to a standing wave.

transmission coefficient   In essentially 1-dimensional wave motion, a (complex) dimensionless parameter giving the ratio of the transmitted wave (amplitude and phase) at some barrier, to the incident wave. In some contexts, the term is used to mean the absolute value (modulus) of this quantity. For instance, in water wave descriptions, the meaning is the ratio of a transmitted wave height to the corresponding incident wave height. Used to describe the effectiveness of a coastal structure, such as a breakwater for reduction of wave height.

transmissivity (T)   A measure of the ability of a saturated aquifer of thickness b and hydraulic conductivity K to transmit water: $T = Kb$, with units of square meters per day [L^2/T].

transonic string model   *See* elastic string model.

transparency   The degree of lack of absorption of a wavelength of interest as light traverses the atmosphere from the source to an Earth-based detector, or as light traverses any semi-transparent medium.

transpiration   The process whereby plants give off water vapor through their leaves back to the atmosphere, via absorption of soil water by plant roots and translocation through the vascular system to stomata in the leaves, where evaporation takes place.

transverse Doppler shift   A shift arising because of the Doppler effect when a photon is emitted from a source moving perpendicularly to the line of sight. The transverse shift is not present in the classical Doppler effect (where a frequency shift is possible only if there is a nonzero component of the velocity along the direction of emission of the photon). It is a purely special relativistic effect that arises because of time dilatation, i.e., because the observer in a different frame measures a different time interval than an observer co-moving with the source.

transverse wave   Wave propagation in which the associated local forces or motions are perpendicular to the direction of the wave. In materials supporting shear (such as elastic metals or rock), mechanical waves (called shear waves) can propagate with transverse modes. In electromagnetism, propagating fields (e.g., light, X-rays, radio waves) are transverse in that they propagate magnetic and electric fields that are perpendicular to the direction of propagation. Gravitational radiation in general relativity induces relative motions between test particles that are perpendicular to the direction of propagation. Hence, gravitational radiation as described by general relativity is a *transverse wave*.

trapped surface   A C^2 space-like two-surface $\mathcal{T}$, such that the two families of null geodesics orthogonal to $\mathcal{T}$ are converging at $\mathcal{T}$. A closed *trapped surface* is a compact trapped surface without boundary. *See* apparent horizon, future/past event horizon.

traveling ionospheric disturbance (TID)
Traveling ionospheric disturbances, or TIDs, are readily observed on ionograms and in a variety of observations from other systems depending on reflections from the ionosphere. The TID is due to an atmospheric gravity wave traveling in the neutral atmosphere. The wave moves the neutral atmosphere, which in turn moves the ionization. TIDs can be relatively large-scale structures in the F region, with horizontal wavelengths of the order of 100 to 1000 km traveling with speeds between 50 m s^{-1} and 1000 m s^{-1}, and have periods of minutes to more than 1 h. Some, often the faster large scale TIDs, are associated with magnetic storms and originate in the auroral zone. These may travel great distances. Others appear to be more localized, possibly originating in the troposphere or lower E region. TIDs appear as a rippled electron density surface to a radio wave and can cause focusing and defocusing if the TID wavelength is greater than the beamwidth. Thus, TIDs can give rise to fading in radio systems and misleading angle of arrival measurements. *See* F region.

travel time curve   A plot of the arrival time of a seismic phase as a function of distance of the receiver from the source. A single graph may contain many such curves representing the arrival of different seismic phases. The distance between the seismic station and the source is commonly quoted in terms of the angle subtended by the two from the center of the Earth, Δ. Not all phases will be observed at all Δ, because in some places within the Earth wave velocity decreases with depth, causing rays to refract around a shadow zone, and some waves may be entirely blocked. The Earth's outer core causes both of these effects, blocking the passage of S waves, because the fluid outer core is unable to support shear waves on seismic timescales, and causing P waves to refract around a shadow zone because the P wave velocity is much reduced. If the wave speed within a depth range strongly increases with depth (as in the transition zone), then the resulting refraction patterns may cause a triplication, i.e., a phase may be observed at a particular Δ at three different times corresponding to the arrival of waves at three different takeoff angles that are refracted to arrive at the same site.

trench   Generally refers to *trenches* on the sea floor. These are deep valleys also known as subduction zones. The oceanic lithosphere bends and descends into the interior of the Earth at ocean trenches.

Also, the bathymetric low at a subduction zone, related to the flexure of the subducting plate.

trench excavation   A method of investigation for active faults to deduce when large earthquakes occurred on the fault. Digging a trench several meters in depth so as to cross the active fault discloses disturbance of strata outcroppings on the trench wall and their age of formation. From such investigations, active periods of large earthquakes over the past several thousand years, recurrence time, and magnitude of large earthquakes can be estimated. Such information is very useful to forecast a forthcoming large earthquake on the active fault. Recently, Japanese archeological excavations have discovered evidence for large earthquakes, such as liquefaction of a ground and traces of ground cracks.

triggered star formation   The process where star formation in one area can induce star formation in a nearby region. This process can occur by various mechanisms (e.g., stellar winds, supernovae) in a star cluster which can compress the nearby interstellar medium to densities sufficient for new stars to condense.

triple-alpha process   The set of nuclear reactions whereby three helium atoms (alpha particles) are converted to one carbon atom with the liberation of about 10^{18} ergs per gram of energy. The *triple-alpha process* is a major energy source in stars that have exhausted hydrogen in their cores (*see* horizontal branch, clump stars, asymptotic giant branch), and the main source of carbon in the universe. It occurs at a temperature near 10^8K and only at relatively high density, because the three helium atoms must approach each other very closely, since the expected intermediate nucleus, ^8B, is unstable to dissociation back into two alpha particles. Another reaction, $^{12}C(\alpha, \gamma)^{16}O$, occurs at very slightly higher temperature, so that helium fusion normally results in a mix of carbon and oxygen (more oxygen at higher temperatures). Some of the nuclear reaction cross-sections involved are still not adequately measured in the laboratory or calculated, so that prediction of the C/O ratio in the products remains somewhat uncertain. *See* asymptotic giant branch star, clump star, horizontal branch star.

triple junction   In geophysics, a point on the Earth where three plate boundaries meet.

triple point   In physical chemistry, the temperature and pressure under which gas, liquid, and solid phases of a substance can coexist in equilibrium. For water, at a standard pressure of one atmosphere, this occurs at a temperature of 273.16 K (0.01°C). The (saturated) water vapor pressure is 6.11 millibar under these conditions.

In meteorology, a junction of three distinct air masses, leading to unstable conditions.

tripton   Inanimate particulate matter in natural waters.

tritium ^{3}H; the isotope of hydrogen whose nucleus consists of one proton and two neutrons. Radioactive to beta decay to ^{3}He with a half-life of 12.26 years.

Tritium Unit   A unit used to scale the amount of tritium present in a particular sample of ocean water. It is defined by convention as 10^{18} times the atom ratio of tritium to normal hydrogen. *See also* tracers.

Triton   Moon of Neptune, also designated NI. It was discovered by Lassell in 1846. Its orbit has an eccentricity of 0.00002, an inclination of 157.4°, a precession of 0.5232° yr^{-1}, and a semimajor axis of 3.55×10^5 km. Its radius is 1350 km, its mass is 2.14×10^{22} kg, and its density is 2.1 g cm^{-3}. Its high albedo and large heliocentric distance gives it a very low surface temperature (34.5 K) and almost no atmosphere. It orbits Neptune, retrograde, once every 5.877 Earth days. The peculiar orbit (retrograde and highly inclined) makes it likely that *Triton* is a captured body, probably from the Kuiper belt.

trochoidal wave theory   A finite-amplitude wave theory attributed to Gerstner (1802). The water surface profile is a trochoid, which represents a good match to many observed waves, but the resulting predicted velocities do not match observations well.

Trojans   A class of asteroids that orbit at the L4 and L5 Lagrangian points. These two points, 60° before and after the position of a planet in its orbit around the sun, are stable points in a three body system (sun – planet – asteriod). For the case of the planet Jupiter, a number of asteriods have been found in this region and are all named after heroes of the Trojan wars.

Trojans asteroids   Asteroids located near Jupiter's Lagrange points (60° ahead and behind Jupiter in its orbit).

tropical cyclone   A large low pressure system that originates over the tropical oceans; including tropical depressions, tropical storms, and hurricanes (cyclones, typhoons), with winds sometimes up to 300 km/h.

tropical instability waves   Oceanic disturbances of meridional scale of 500 km, zonal wavelength of 1000 km and a period of 30 days that travel westward along 2-3°N in the equatorial Atlantic and Pacific Oceans. Instabilities of velocity shears among equatorial currents are their cause. They start to grow in June and decay toward the end of a year, in correlation with the intensity of equatorial currents. They are well visible from satellite infrared images as they cause meanders of a strong sea surface temperature (SST) front between the equator and 5°N. Often called the Legeckis waves after the discoverer of the waves' signature in SST. These wave-induced SST variations excite atmospheric waves of significant amplitudes in the marine boundary layer.

tropical month   *See* month.

tropical storm   A storm originating in the tropics; particularly severe *tropical storms* become hurricanes (sustained wind speeds > 120 km/h).

tropical year   *See* year.

Tropic of Cancer   The parallel of latitude 23°27′N. The sun lies in the zenith (overhead at noon) at this latitude, at the summer solstice, around June 22. North of this latitude, the sun is always south of directly overhead.

Tropic of Capricorn   The parallel of latitude 23°27′S. The sun lies in the zenith (overhead at noon) at this latitude, at the winter solstice, around December 22. South of this latitude, the sun is always north of directly overhead.

tropopause   The boundary layer between the troposphere and the stratosphere. In the troposphere, air temperature generally decreases with height, and in the lower parts of the stratosphere, air temperature remains constant or increases (0.1 to 0.2°C/100 m). The altitude of the *tropopause* varies with the variations of sea — surface temperature, season, latitude, and weather systems, such as the passage of cyclones and anti-cyclones. It has its highest height (about 17 to 18 km) over the equator and lowest (about 6 to 8 km, average 10 to 12 km) over the poles. There are two different types of tropopause. One is the pole-region tropopause with height lower than 150 hPa; another is the tropical-subtropical tropopause with height higher than 150 hPa. Because of this distinct height difference, when these two different types of air mass systems approach, the tropopause will appear as a broken phenomenon. Under such a broken tropopause, there is often a high level jet which is related to the front.

troposphere   The layer of the atmosphere closest to the planet's surface. The *troposphere* is characterized by a decrease in temperature with altitude, caused by a decrease in the amount of warming due to the Greenhouse Effect with altitude. Hadley Cell circulation and the Greenhouse Effect dominate the temperature profile and weather patterns within the troposphere. On Earth, the troposphere extends to about 25 km above the surface. The exact height varies with latitude and season. The boundary between the troposphere and the stratosphere is called the tropopause. The troposphere is also often called the lower atmosphere.

trough   A long, relatively wide *trough* running on sea bottoms, whose maximum depth of the water is less than 6000 m. This definition is applied irrespective of the origin of troughs. There are some troughs whose structure and origin are basically the same as those of trenches. Such an example is the Nankai trough situated on the Pacific side of the Japanese islands. For such a trough, as is the case for a trench, seismic activity and crustal movements associated with subduction of an oceanic plate and formation of accretionary prisms on the continental slope are marked.

true anomaly   In Newtonian dynamics, the angle between an orbiting body and periapse.

true celestial equator   The *true celestial equator* or celestial equator of date is the great circle on the celestial sphere perpendicular to the true celestial pole. q.v. Its intersections with the ecliptic define the vernal equinox of date and the autumnal equinox of date.

true celestial pole   The *true north celestial pole* is the direction of Earth's instantaneous rotation pole. It differs from the mean north celestial pole due to the short timescale (days to decades) variations called nutation. The nutation itself, of order $10''$ arc, is normally described by a theory, such as the IAU 1980 nutation theory, which suffices to predict the pole position within a few hundreths of an arc second, and smaller observed corrections called the celestial pole offset. See http://hpiers.obspm.fr/webiers/general/Earthor/precnut/PNUT.html.

true north   Geographic north defined by the rotational pole of the Earth, as opposed to magnetic north defined by the geomagnetic north pole.

true polar wander   The Earth's hot spot system appears to define a nearly rigid deep mantle frame. The motion of the hot spot system with respect to the rotational axis of the Earth is called *true polar wander*, as opposed to the apparent polar wander as a result of continental drift. *See* continental drift, hot spot, polar wander.

tsunami   Water wave in the ocean or large lakes caused by underwater earthquakes, landslides, or volcanic activity. In the open ocean, *tsunamis* can have wave lengths of hundreds of kilometers, traveling at speeds of several hundreds of kilometers per hour. They are often called tidal waves, although they have no relation with tides.

tsunami earthquake   Generally means an earthquake accompanying tsunami (tidal wave), or an earthquake that causes a large tsunami for its magnitude. Typical examples are the 1896 Sanriku-oki, Japan earthquake and the 1946 Aleutian earthquake. Existence of *tsunami earthquakes* indicates that we cannot predict tsunami height at a coast precisely only from the magnitude of earthquakes, which makes tsunami difficult to forecast. Extremely slow faulting and generation of submarine landslide have been considered as mechanisms of tsunami earthquakes.

TT   *See* Terrestrial Time.

T Tauri object   A class of variable stars whose brightness varies irregularly, with broad and very intense emission lines. T Tauris, also known as class II protostars, are very young pre-main sequence stars that gain mass through an accretion disk as evinced by an excess of infrared emission, but whose surface is optically visible. Also characterized by extensive and violent ejections of mass and magnetic activity, T Tauri stars are named after the first of their class and have masses of roughly 0.2 to 5 $M_\odot$.

T Tauri star   The last stage of stellar evolution before the main sequence. *T Tauri stars* are characterized by emission lines, rapid variability, and X-ray emission, associated with gas being both accreted and lost in bipolar outflows and with relatively strong magnetic fields and rapid rotation (*see* stellar activity). The name is frequently, but not always, restricted to stars of roughly 0.5 to 2.5 solar masses, more massive ones being called Herbig Ze/Be stars, and the most massive being seen only as radio sources called compact H II regions.

Tully–Fisher law   An empirical law which relates the width of the 21-cm neutral hydrogen spectral emission line to the luminosity of a spiral galaxy, proposed by R.B. Tully and J.R. Fisher (1977): the total galaxy luminosity is proportional to the fourth power of the width of the 21-cm line. The relationship is best (i.e., data points show less scatter) if the luminosity is measured in the infrared. The infrared luminosity depends little on Hubble type and is correlated to the number of old disk and intermediate disk population stars, which make up the largest fraction of the mass in the disk of a spiral galaxy. The physical basis of the *Tully–Fisher law* resides in equating two observable parameters that are independent estimates of the mass of a galaxy. Using the Tully–Fisher law, the intrinsic luminosity of a galaxy can be obtained from the measurement of the broadening of the Hɪ 21-cm line, which does not depend on distance. The Tully–Fisher relationship can be used, therefore, as a distance indicator. *See* Faber–Jackson law.

turbidity   A measure of the amount of suspended debris within a body of water. Different units are used for description of *turbidity,* including Jackson Turbidity Units (JTU), Silica Standard Turbidity Units (TU), and Nephelometric Turbidity Units (NTU).

turbulence   Condition of fluid flow characterized by irregular and aperiodic distribution of velocity and vorticity components. Random motions cause a distribution of scalar properties at rates much higher than molecular diffusion processes. The length scales (eddy sizes) associated with active, three-dimensional turbulence range between approximately 0.6η and $11L_N$, where η is the Kolmogorov microscale describing the length scale where velocity fluctuations are stopped by viscosity, and L_N is the buoyancy scale, which represents the largest vertical size of turbulent eddies allowed by the ambient stratification.

turbulence cascade   This term implies that turbulent eddies decay successively to smaller scales starting from the energy-containing scale (where the energy input occurs) to the smallest scales, given by the Kolmogorov Scale $L_K = (\nu^3/\varepsilon)^{1/4}$, where the turbulent kinetic energy is converted into heat by molecular viscosity ν. If the turbulent fluid is not stratified or if the Ozmidov-scale is substantially larger than the Kolmogorov Scale, an inertial subrange may develop. The source of the non-linear term of the momentum.

turbulent cascade   In a turbulent shear flow, the largest eddies have roughly the same size as the width of the shear flow itself.   In the ocean or the atmosphere this is often the height of the boundary layer. The largest eddies extract kinetic energy directly from the mean flow field. Smaller eddies are advected in the velocity field of the large eddies, extracting kinetic energy from the larger eddies. Thus, through a series of increasingly smaller eddies, kinetic energy is transferred ("cascaded") from larger to smaller scales by a mechanism that arises from the non-linear terms in the equations motion of turbulent flows (vortex stretching). The smaller eddies do not directly interact with the mean flow field or the larger eddies. Kinetic energy is cascaded from larger to smaller scales until the turbulent motions are stopped by viscous effects. The size of the dissipating eddies is given by the Kolmogorov microscale.

turbulent flow   Occurs when fluid particles follow irregular paths in either space or time because inertial forces are much greater than viscous forces. *Turbulent flow* is empirically defined as flow with a high Reynolds number.

turbulent flux   The total flux of a property $< a_i u_i >$ can be interpreted as the sum of the flux due to the mean flow $< a_i >< u_i >$ plus the *turbulent flux* $< a_i' u_i' >$ due to the covariance of the turbulent fluctuations of a_i' and u_i'.

turbulent kinetic energy   The energy e [J kg^{-1}] contained in the turbulent velocity fluctuations per unit mass of fluid, defined by $e = 0.5 < u'^2 + v'^2 + w'^2 >$, where u', v', w' are the three velocity fluctuations components (*see* Reynolds decomposition for definitions).

twilight   The interval of time preceding sunrise and following sunset during which the sky is partially illuminated. "Civil twilight" is the interval when the zenith distance (angle), referred to the center of the Earth, of the center of the sun's disk is between $90°50'$ and $96°$. "Nautical twilight" is the interval between $96°$ and $102°$. "Astronomical twilight" is the interval between $102°$ and $108°$.

twin paradox   Because of the relativity of simultaneity, clocks in relative motion that have a relative velocity v run at different rates. An object that is carried away at speed v (from an identical object that remains at rest) to a different location and is returned to its original location accumulates less total time interval than does its nonmoving "twin". In principle, if the motions are at a substantial fraction of the speed of light and the distances are substantial (light years), one could demonstrate this effect with human twins. For instance, if one newborn twin travels 5 light years at a speed of $0.99c(c =$ speed of light) and then returns at the same speed, the stay-at-home twin would see the return at age 10, while the elapsed time for (the age of) the traveling twin would be

$$\text{Age} = t\sqrt{1 - \frac{v^2}{c^2}}$$
$$= 10 \text{ yr.} \times 0.14$$
$$= 1.4 \text{ yr}.$$

The paradox is with everyday experience, where these effects are so small as to be unnoticeable. The effect has been verified in atomic clocks flown around the Earth, in the operation of atomic accelerators, and in the correct operation of the global positioning satellite system. *See* Lorentz transformation, special relativity.

twistor   An element of the four-dimensional spinor representation space of the group SU(2,2). Following work in 1962–1963 with R.P. Kerr, the notion of a *twistor* was put forward by R. Penrose as a possible mathematical tool for formulating the quantum theory of gravitation. In the theory of relativity, a twistor represents the state of a massless particle.

two-flow equations   The two coupled differential equations for irradiances obtained by integrating the radiative transfer equation over the hemispheres of downward and upward directions.

2-form   A rank-2 antisymmetric covariant tensor. A basis for 2-forms is the antisymmetric tensor product of pairs of 1-form basis forms. For instance, if $\{dx^\beta\}$ is a basis for 1-forms, then $\{dx^\beta \wedge dx^\alpha\}$ is a basis for 2-forms. Here $\wedge$ is the antisymmetric tensor product, i.e.,

$$dx^\beta \wedge dx^\gamma = \frac{1}{2}\left(dx^\beta \otimes dx^\gamma - dx^\gamma \otimes dx^\beta\right)$$

with $\otimes$ the tensor product. *See* tensor.

two-ribbon flare   A large flare that has developed as a pair of bright strands (ribbons) on both sides of the main inversion ("neutral") line of the magnetic field of the active region, usually most prominent in Hα. *Two-ribbon flares* result in a substantial reconfiguration of the coronal magnetic field.

two-stream approximation   The two-flow approximation. *See* two-flow equations.

two-stream instability   A broad term referring to any of a number of resonant and non-resonant instabilities driven by the streaming of charged particles relative to the main body of a collisionless plasma.

type I radio burst   The metric type I burst is continuous radio emission from the sun, basically the normal solar radio noise. It can be enhanced during the late phase of a flare. The radio noise is unstructured, no frequency drift can be identified.

type II radio burst   In the metric type II burst, a relatively slow frequency drift can be observed, indicating a propagation of the electron source at a speed of about 1000 km/s. It is interpreted as evidence for a shock wave propagating through the solar corona, but it should be noted that the shock itself is not generating the radio emission but rather the electrons accelerated at the shock. As these electrons stream away from the shock, they generate small type III like structures, giving the burst the appearance of a herringbone in the frequency time diagram with the type II burst as the backbone and the type III structures as fish-bones. The type II burst is often split into two parallel frequency bands, interpreted as forward and reverse shocks.

Type II radio bursts can also be observed with frequencies in the kilometer range (kilometric type II radio burst). The drift characteristic is similar to that of the metric counterpart, the frequency indicates that these bursts are excited in interplanetary space. Consequently, the kilometric type II burst is interpreted as the signature of a shock propagating through interplanetary space. It should be noted, however, that only the fastest (speeds exceeding 1000 km/s) and strongest shocks excite kilometric type II bursts; thus, most interplanetary shocks do not excite these bursts.

type III radio burst   Short time scale decreasing frequency solar radio features occurring in association with a type II radio burst.

type IV radio burst   Metric type IV emission is a continuum, directly following the metric type II burst. It is generated by gyro-synchrotron emission of electrons with energies of about 100 keV. It consists of two components: a non-drifting part generated by electrons captured in closed magnetic loops, and a propagating type IV burst generated by electrons moving higher in the corona.

type V radio burst   Metric type V emission is a continuum, directly following the metric type III burst. It is stationary, showing no frequency drift. Most likely, the metric type V burst is radiation of the plasma itself.

typhoon   A tropical cyclone occurring in the (western) Pacific Ocean, equivalent to a hurricane in the Atlantic Ocean.

U

ultrasound "Sound" waves of frequency above 20,000 Hz (hence inaudible to humans).

ultraviolet Referring to that invisible part of the electromagnetic spectrum with radiation of wavelength slightly shorter than violet-colored visible light.

Ultraviolet Coronagraph Spectrometer (UVCS) An ultraviolet spectrometer/coronagraph on board the SOHO spacecraft. The purpose of the *UVCS* instrument is to locate and characterize the coronal source regions of the solar wind, to identify and understand the dominant physical processes that accelerate the solar wind, to understand how the coronal plasma is heated in solar wind acceleration regions, and to increase the knowledge of coronal phenomena that control the physical properties of the solar wind as determined by *in situ* measurements.

Ulysses Mission The first spacecraft to explore interplanetary space at high solar latitudes, specifically the high latitude heliosphere away from the plane of the ecliptic where expanding gases from the solar corona dominate the properties of interstellar space. The spacecraft was launched on October 6, 1990 by the shuttle Discovery. Ulysses flew by Jupiter in February 1992, where a gravity-assisted maneuver placed the spacecraft in a solar polar orbit, allowing it to fly over the south pole of the sun in 1994 and over the north pole in 1995. High latitude observations obtained then were during the quiet (minimum) portion of the 11-year solar cycle. The spacecraft's orbital period is 6 years and, if kept in operation, high latitude observations will also be obtained during the active (maximum) portion of the solar cycle.

The primary results of the mission have been to discover the properties of the solar corona, the solar wind, the heliospheric magnetic field, solar energetic particles, galactic cosmic rays, solar radio bursts, and plasma waves. Other investigations pertained to the study of cosmic dust, gamma ray bursts, and studies of the Jovian magnetosphere acquired during the Jupiter fly-by.

Ulysses is a joint endeavor of the European Space Agency (ESA) and the National Aeronautics and Space Administration (NASA).

umbra The central dark part of a sunspot. During an eclipse, the part of the shadow from which no part of the sun is visible.

Umbriel Moon of Uranus, also designated UII. It was discovered by Lassell in 1851. Its orbit has an eccentricity of 0.005, an inclination of 0.36°, a precession of 3.6° yr^{-1}, and a semimajor axis of 2.66×10^5 km. Its radius is 585 km, its mass 1.27×10^{21} kg, and its density 1.44 g cm^{-3}. It has a geometric albedo of 0.18 and orbits Uranus once every 4.144 Earth days. *Umbriel* is the darkest of the major Uranian satellites.

unconformities Discontinuities in the sedimentary geological record. In general, younger sediments overlie older sediments in sedimentary basins, and the fossils found in the sediments define the geological record. At an *unconformity*, the overlying sediments are significantly younger than the underlying sediments, and unconformities are used to define the geologic epochs; the boundary between the Tertiary and the Cretaceous and the boundary between the Jurassic and Triassic are unconformities. The origin of unconformities is not fully understood, but some are attributed to low stands of sea level. Some major unconformities are associated with extinctions of species, and these in turn are associated with catastrophic meteorite impacts.

undertow The mean offshore current within the surf zone. Breaking waves drive water shoreward in the upper part of the water column, leading to a net offshore flow in the lower part of the water column.

unification of active galactic nuclei Theories attempting to explain the variegated phenomenology of active galactic nuclei on the basis of a few key parameters. In unification schemes, the central engine is considered to be basically the same for all active galactic nuclei (AGN), i.e., a supermassive black hole surrounded by a hot accretion disk, with orientation, black hole mass, accretion rate, and spin of the black hole (or, alternatively, the morphology of the host galaxy) accounting for all differences observed between AGN types. In the unification scheme for radio quiet AGN, different orientation of the accretion disk with respect to the line of sight could give rise to a Seyfert-1 galaxy (if the disk is seen at intermediate inclination), or to a Seyfert-2 galaxy because of obscuration: If the disk is seen at large inclination, the line emitting regions should be obscured by a molecular torus. Unification of radio-loud AGN relies on the effect of relativistic beaming to explain the different appearance of radio-loud AGN and to establish a link between the beamed and unbeamed (also called the parent) population. For instance, a low luminosity radio-loud AGN is classified as BL Lac in the optical and as a compact core in the radio if the disk is seen face-on. On the basis of the radio morphology, the same object would be classified as a low luminosity Fanaroff–Riley type I radio source if the line of sight is oriented at a large angle from the radio jet. The validity of unification models and the identification of the parent population for some AGN classes is a subject of current debate.

unified soil classification   A classification of sediment by particle size (diameter) commonly used by engineers. Based on the Casagrande Classification of soils. Ranges from cobbles (>76 mm) down through clays (<0.0039 mm).

unipolar magnetic region   Large-scale regions of the solar surface which extend over hundreds of thousand of kilometers in both latitude and longitude. They contain flux elements of predominantly one polarity and are long-lived with lifetimes of 1 year or more. Coronal holes can be found above some large unipolar regions.

unit weight   Weight per unit volume.

universal gravitation, Newton law of   Every particle in the universe attracts every other particle with a force directed along the line between them, proportional to the product of the masses of the particles and inversely proportional to the square of the distance between the particles.

Universal Time (UT or UT1)   Starting in 1972, Greenwich Mean Solar Time was split into Coordinated Universal Time (UTC), used for civil timekeeping, and *Universal Time (UT1)*, a measure of Earth rotation, which is 12^h at Greenwich mean solar noon, i.e., UT1 is Greenwich Mean Time (GMT, also called Zulu Time). *Universal Time (UT)* could refer to any of UT0, UT1, or UTC, but in the absence of other specifiers it means UT1. UT0 is latitude and longitude dependent and is of mostly historical interest; for details see the *Astronomical Almanac*. UT1 is derived, nowadays, by VLBI observations of quasars. Several national agencies contribute their data to the International Earth Rotation Service (IERS), which integrates the data into published values of the difference UT1 – UTC. Both predictive and definitive data are routinely available. The difference, UT1 – UTC, tends to slowly decrease over the passage of months, due to the cumulative slowing of Earth rotation, but before this difference reaches −0.9 s due to oceanic and atmospheric tidal friction, the IERS inserts a leap second, which brings UT1 – UTC positive again, starting a new cycle of decrease.

universe   The totality of all physical reality. In astronomy it is often used to refer to the scale larger than galaxies, which contains galactic-scale structures. Also used to refer to a particular (inevitably simplified) model of the physical cosmos. Dynamical 4-dimensional solutions to Einstein's equations, which are in principle descriptions of possible gravitational fields, are often called *universes*.

unstable equilibrium   In mechanics, a configuration in which the system experiences no net force (equilibrium) but slight displacements from this state cause forces moving the system away from the equilibrium state. The paradigm is an inverted pendulum at equilibrium in the vertical position. *See* neutral equilibrium, stable equilibrium.

upper mantle   The region of rock in the Earth's interior reaching roughly 5700 to 6330 km in radius.

upper neutral atmospheric regions   These regions are classified on the basis of the distribution of temperatures and constituents with height. The lowest region closest to the ground is called troposphere, marked by a temperature decrease with height. At the tropopause, the temperature ceases to decrease and starts increasing in the stratosphere to the stratopause. The stratosphere includes the ozone layer, which shields the Earth from dangerous extreme ultraviolet rays. Above the stratopause the temperature again starts decreasing in the mesosphere until the mesopause is reached. Above the mesopause, the temperature starts increasing in the thermosphere until the thermopause is reached, above which it stays almost constant with height.

In terms of species, the region around 30 km where ozone is abundant is called ozonosphere. The region below 110 km is called homosphere, where species are well mixed by turbulence. The region above 110 km is called the heterosphere where various species are not well mixed as a result of prevalance of diffusive separation and many photochemical reactions. The region above about 500 km is known as the exosphere or geocorona, where atomic hydrogen predominates. In this region the atoms only occasionally collide and move in ballistic orbits and may escape. The critical level above which the region can be designated as the exosphere is defined where the mean free path is in the order of the scale height of the atomic hydrogen.

Uranus   The seventh planet from the sun. Named after the Greek elder god. It has a mass $M = 8.683 \times 10^{28}$ g and an equatorial radius of $R = 25,559$ km, giving it a mean density of 1.29 g cm^{-3} and a surface gravity of 0.91 that of Earth. Its rotational period is 17.24 h, and its rotation axis has an obliquity of 97.9°. The planet's oblateness is 0.023. *Uranus'* orbit around the sun is characterized by a mean distance of 19.2 AU, an eccentricity of $e = 0.047$, and an orbital inclination of $i = 0.772°$. Its synodic period is 370 days, and its sidereal period is 84.01 years. It has an average albedo of 0.51, and a temperature at 1 bar of around 70 K. Unlike the other giant planets, it does not have a measurable internal heat source. The bulk of Uranus' composition is probably a mixture of ice and rock, although about 20% of the mass is hydrogen and helium. The atmosphere also contains significant amounts (2%) of methane, which is the cause of its blue color. Uranus has 15 known satellites, although a number of additional candidates have been reported. In addition, it has a system of 9 narrow rings. The larger named moons include Cordelia, Ophelia, Bianca, Cressida, Desdemona, Juliet, Portia, Rosalind, Belinda, Puck, Miranda, Ariel, Umbriel, Titania, Oberon, Caliban, and Sycorax.

Urca process   The high densities in the cores of massive stars can force energetic electrons to capture onto nuclei, emitting an electron neutrino ($(Z, A) + e^- \rightarrow (Z - 1, A) + \nu_e$). This new nucleus then beta decays ($(Z - 1, A) \rightarrow (Z, A) + e^- + \bar{\nu}_e$). The net result of this cycle is the emission of two neutrinos and the loss of energy. This cycle is known as the *Urca process,* named after the Casino d'Urca in Brazil where the net result of gambling has a similar effect on money as the Urca process has on energy. Cooling from the Urca process is the cause of the extremely short burning timescales of post main-sequence evolution of massive stars and triggers the collapse of the iron core which leads to supernovae explosions.

Urey ratio   The heat flow to the surface of the Earth from its interior has two contributions, (1) the heat generated within the Earth by the decay of the radioactive isotopes of uranium, thorium, and potassium, and (2) the secular cooling of the Earth. The *Urey ratio,* named after the eminent geochemist Harold Urey, is the fraction of the total heat flow attributed to radioactive decay. The present estimate for the value of the Urey ratio is 70 ± 10%.

UT   *See* Universal Time.

V

V471 Tauri stars Binary systems in which the more massive star has ended its life as a white dwarf, the companion is still on the main sequence, and no material is being transferred between the stars. Such systems are not very conspicuous, but a dozen or so have been identified. Further evolution of the main sequence star will transform them into cataclysmic variables.

vacuum In field theory, and as applied to condensed matter physics and to cosmology, the lowest state of energy in which a system can be. In many cases the vacuum depends on external parameters, e.g., the temperature. Whether the vacuum is a particular configuration, or whether a number of different configurations give the same lowest energy has a large influence on the physics associated with the field.

vacuum fluctuation The process, as a consequence of the uncertainty principle, $\Delta E \Delta t \gtrsim h$, in which pairs of particles and antiparticles spontaneously appear and disappear in space and time. From the uncertainty principle a fluctuation ΔE can last at most $\Delta t = h/\Delta E$, where ΔE is the energy or mass $\times c^2$ involved in the fluctuation.

vacuum manifold In theoretical physics applied to condensed matter, and to cosmology, the result of spontaneous symmetry breaking. At high temperatures the system has a symmetry described by an invariance group, G. At lower temperatures, the lowest energy state (the vacuum) may be less symmetric (described by a subgroup H resulting in a structure for the vacuum which is not necessarily trivial). The vacuum is topologically equivalent (isomorphic) to the quotient space $\mathcal{M} \sim G/H$. The *vacuum manifold* $\mathcal{M}$ and its topological features (like connectedness) will unambiguously determine the kind of topological defect that will be produced during the phase transition. Such residual defects (e.g., cosmic strings or monopoles) may have had a significant effect on the early evolution of the universe. *See* cosmic topological defect, spontaneous symmetry breaking.

vacuum polarization If vacuum fluctuations occur in the neighborhood of an electrically charged particle, the members of the created pairs with opposite charge to that particle will be attracted towards it, and members with identical charge will be repelled from it. This migration is called *vacuum polarization*. *See* vacuum fluctuation.

Väisälä frequency *See* buoyancy frequency.

Valles Marineris A vast system of interconnected canyons that exists to the east of the Tharsis Province, just south of the equator. It extends from Noctis Labyrinthus, at 5°S 100°W, eastwards for 4000 km until it merges with the chaotic terrain. Along the entire length are multiple, parallel canyons, chains of craters, and graben. The canyons are better integrated towards Noctis Labyrinthus and more segmented towards the chaotic terrain. In most places individual canyons are over 3 km deep, and 100 km wide, although in the central section three parallel canyons merge to form a depression over 7 km deep and 600 km wide. Therefore, the canyons significantly larger than the Grand Canyon, Arizona, U.S. on Earth, which is 450 km long, with a maximum depth of 2 km and a maximum width of 30 km.

The canyon walls are steep and gullied and in many places have collapsed in gigantic landslides. In contrast, the canyon floors are generally flat and possess outcrops of layered terrain. The landslides, the layered terrain, and the debris, all imply a fluidizing mechanism in their formation, but besides these features there is evidence only for slow erosion in the canyons, independent of fluvial activity. It is generally agreed that the formation of *Valles Marineris* is "in some way" related to the evolution of the Tharsis Province, because canyon location correlates with the Tharsis circumferential extensional stress field. The favored mechanism that is considered to be responsible for, or to have contributed to, the formation of Valles Marineris and Noctis Labyrinthus is incipient or aborted tectonic rifting (by analogy with Earth), with collapse along the crest of the bulge due to subsurface withdrawal of material.

valley networks   The small dendritic channels found on Mars. These channels are found on the flanks of volcanos and in the ancient terrain of the planet. Although similar in some respects to terrestrial channels formed by rainfall, the detailed morphologies of these channels suggest they were formed by sapping, whereby groundwater is removed and the overlying surface collapses along the path of the underground river, producing the channel shape. Those found on the flanks of volcanos may have been produced by hydrothermal activity. *Valley networks* are typically about 1 km in width and tens to hundreds of kilometers in length.

Van Allen Belts   Named after their discoverer, James Van Allen. First detected in 1958 by Explorer 1; James Van Allen correctly interpreted the results. Two distinct toroidal belts; the inner one, located between about 1.1 to 3.3 Earth radii, near the equatorial plane, contains primarily protons with energies exceeding 10 MeV. Flux maximum is at about 2 Earth radii. This population varies with 11-year solar cycle and is subject to occasional perturbations due to geomagnetic storms. The protons in this region are produced from cosmic rays striking the atmosphere. The outer belt (between 3 to 9 Earth radii, with a maximum around 4 Earth radii) contains mainly electrons with energies up to 10 MeV arising from solar wind electron injection and acceleration in geomagnetic storms. The population of this belt thus shows day/night variation and is sensitive to solar activity. The charged particles contained in the *Van Allen belts* remain trapped along field lines of the Earth. The particles drift and spiral around the Earth's magnetic field lines. As the particles approach the converging field lines near the poles, they are reflected back towards the opposite pole. "Horns" of especially the outer belt dip sharply in toward the polar caps.

VAN method   A method of earthquake prediction proposed by three Greek researchers, Varotsos, Alexopouls, and Nomicos in the mid 1980s. According to *VAN method,* earthquakes could be predicted based on the observed facts that abnormal signals of Earth current with a duration time of several minutes to hours (SES) and continuing more than several days appear weeks before generations of disastrous earthquakes. VAN claim that the magnitude of earthquakes, epicentral distance, and occurrence time of earthquakes are respectively predicted empirically from maximum amplitude of SES, station distribution where SES appears, and kinds of SES. VAN method was tested throughout Greece. Some researchers reported that prediction of disastrous earthquakes succeeded with substantially high probability, but others are critical owing to lack of objectivity of the method.

vapor   A gas whose temperature is less than its critical temperature. In such a situation an increase of pressure at constant temperature will cause the gas to condense (liquify or solidify); an example is as water vapor in air at any temperature below 374°C. (The corresponding pressure is 221 bar.) More colloquially, a dispersion of molecules of a substance that is a liquid or a solid in its normal state at the given temperature, through a substrate gas.

vapor concentration   The *vapor concentration* or absolute humidity is the mass of vapor per unit of volume of moist air.

vapor pressure   The partial pressure of the water vapor in moist air. It is a measurement of water vapor content in air. The pressure of moist air is equal to the partial pressure of the dry air in the moist air plus the vapor pressure, according to Dalton's law (the additivity of partial pressures). *Vapor pressure e* is directly proportional to vapor density ρ_w and temperature T as

$$e = \frac{R}{m_w} \rho_w T$$

where R is the gas constant, $R = 8.314 \, J/mol \cdot K$, and m_w is the water vapor molecular weight. For a given temperature, a certain volume of moist air can only contain a limited content of water vapor. If vapor content reaches this limit, the air is said to be saturated, and its vapor pressure is the saturated vapor pressure or the maximum vapor pressure. As temperature increases, saturated vapor pressure increases.

variable star   A star that changes its flux over time, which can be hours or years. Variable stars are classified as either extrinsic or intrinsic, depending on the cause for the variability being due to the environment or interior processes of the star, respectively. Three major groups of variable stars are: eclipsing, cataclysmic, and pulsating. An eclipsing variable is really two stars with different temperature types that, due to fortuitous line of sight with Earth, orbit in front of and in back of each other, thus changing the light the system puts out for observers on Earth. Pulsating variables alternately expand and contract their atmosphere and the pulsations are fueled by a driver under the photosphere that operates on the opacity changes of partially ionized H and He. The cataclysmic variables, which include novae, dwarf novae, recurrent novae, and flare stars, can change their brightness by 10 magnitudes or more. The novae cataclysmic variables are binary stars whose increase in brightness arises because of the hydrogen-rich material of the companion star igniting (fusion) on the surface of the hot white dwarf. Other minor types of variables include the R CrB stars that are believed to puff out great clouds of dark carbon soot periodically, causing their light to dim.

variable stars [cataclysmic variables]   *See* cataclysmic variables.

variable stars [geometric variables]   Binary systems whose stars actually present constant luminosity, but are periodically eclipsing each other when seen from Earth, thus exhibiting apparent variability.

variable stars [peculiar variables]   Stars that present an inhomogenous surface luminosity and, therefore, are observed as variables due to their rotation.

variable stars [pulsating variables]   Stars that present variable luminosity due to instabilities in their internal structure.

variational principle   A way of connecting an integral definition of some physical orbit, with the differential equations describing evolution along that orbit. Varying a path or varying the values of field parameters around an extremizing configuration in an action (which is defined via an integral) leads to the condition that the first-order variation vanishes (since we vary around an extremum). The vanishing of this first-order variation is typically a differential equation, the equation defining the field configuration which produces the extremum. In point mechanics, one may compute the action:

$$I = \int_{x_i, t_i C}^{x_f, t_f} L\left(\mathbf{x}, \frac{d\mathbf{x}}{dt}, \frac{d^2\mathbf{x}}{dt^2} \cdots \right) dt$$

(where $\mathbf{x} = x^i, i = 1 \cdots N$, the dimension of the space). By demanding that the curve $\mathcal{C}$ extremize the quantity I, compared to other curves that pass through the given endpoints at the given times, one is led to differential conditions on the integrand

$$\frac{\partial L}{\partial x^i} - \frac{d}{dt}\frac{\partial L}{\partial \dot{x}^i} + \frac{d^2}{dt^2}\frac{\partial^2 L}{\partial \ddot{x}}$$
$$- \cdots = 0, i = 1 \cdots N.$$

Here the partial derivatives relate to the explicit appearance of $\dot{x}^i = \frac{dx^i}{dt}, \ddot{x}^i = \frac{d^2x^i}{dt^2}, \cdots$. Newton's laws for the motion of point in conservative field follow from the simple Lagrangian

$$L = T - V,$$

where $T = T_{ij}(x^l)\dot{x}^i\dot{x}^j$ is quadratic in the $\dot{x}^i$, and V is a function of the x^i only. Then, one obtains

$$-\frac{\partial V}{\partial x^i} - \frac{d}{dt}\left(T_{il}\frac{dx^l}{dt}\right) = 0.$$

For the simplest case of $T_{il} = \frac{1}{2}m\delta_{il}$, one finds: $m\ddot{x}^i = -\frac{\partial V}{\partial x^i}$.

The quantity $I = \int L dt$ is called the action.

For classical and quantum field theories, one defines a Lagrangian density. The Lagrangian is then defined as an integral of this density. For instance, the action for a free (real) scalar field satisfying a massless wave equation is

$$I = \int \mathcal{L}\sqrt{|g|}d^4x,$$

where

$$\mathcal{L} = \left(\nabla_\alpha\phi\nabla_\beta\phi g^{\alpha\beta}\right)$$

with $g^{\alpha\beta}$ the components of the inverse metric tensor ($\alpha = 0, 1, 2, 3$, corresponding to t, x, y, z), $\eta^{\alpha\beta} = \text{diagonal}(-1, 1, 1, 1)$ in the usual rectangular coordinates (thus $\sqrt{|g|} = 1$ in these coordinates).

In this case the variations are of the form $\phi(t, x, y, z) \rightarrow \phi(t, x, y, z) + \delta\phi(t, x, y, z)$, and an additional requirement is that the integration be considered over a finite 4-dimensional domain, and the variation vanish on the spatial boundary of the domain (the field values are held fixed there), as well as at the time end values. The equation of motion is then

$$\nabla_\alpha \nabla^\alpha \phi \equiv \Box\phi =$$
$$\left(-\frac{\partial^2}{\partial t^2} + \frac{\partial^2}{\partial x^2} + \frac{\partial^2}{\partial y^2} + \frac{\partial^2}{\partial z^2}\right)\phi = 0 .$$

See action, Lagrangian.

Vasyliunas theorem   *See* Grad–Vasyliunas theorem.

vector   An abstraction of the concept of displacement from one position to the other. In Euclidean 3-space, for instance, one can introduce coordinates x, y, z, and the displacement vector has components $(\delta x, \delta y, \delta z)$. Formally, the *vector* is abstracted from the concept of tangent vector to a curve. A curve, parameterized by a parameter t, has a tangent, which can be thought of as a derivative along the curve: d/dt acting on a function f can be thought of in elementary terms as

$$\left(t^i \partial/\partial x^i\right)(f) \stackrel{\text{def}}{=} df/dt$$

where the left side is written in a locally defined coordinate system $\{x^i\}$. The $\{\partial/\partial x^i\}$ can be thought of as a set of basis vectors in this coordinate system. The geometrical quantity is the vector itself, independent of which coordinate system is used to express its components, however. Vectors in general can be viewed as directional derivative operators. For our tangent vector example, one writes

$$\mathbf{t} = t^i \frac{\partial}{\partial x^i} = t^{i'} \frac{\partial}{\partial x^{i'}}$$

where $\{x^{i'}\}$ is a new coordinate system. Using the chain rule:

$$\frac{\partial}{\partial x^{i'}} = \left(\frac{\partial x^j}{\partial x^{i'}}\right)\frac{\partial}{\partial x^j}$$

produces the rule to obtain the components of $\mathbf{t}$ in one coordinate basis, given the components in another basis:

$$t^{i'}\left(\frac{\partial x^j}{\partial x^{i'}}\right) = t^j .$$

A vector is sometimes called a contravariant vector, to distinguish it from a one-form, which is sometimes called a covariant vector. *See* curve, one-form, tensor.

vector cross product   For two 3-space vectors $\mathbf{A}$, $\mathbf{B}$, of lengths $|A|$, $|B|$, making an angle θ to one another; the vector of length $|A||B|\sin\theta$, pointing in the direction perpendicular to both $\mathbf{A}$ and $\mathbf{B}$, with direction given by the right-hand rule.

vector field   A smooth collection of vectors in some region of space such that the integral curves they define have no intersections in that region.

vector magnetograph   Magnetograph which uses Zeeman-sensitive lines in the solar atmosphere (mostly in the photosphere) to measure all three components of the magnetic field on the surface of the sun.

velocity   The (limit of the) vector difference in position at two instants, divided by the time between the instants. Here the limit is taken as the time interval goes to zero, so the definition defines *velocity* $\mathbf{v}$ as the vector time derivative of the position vector $\mathbf{x}$. Units: cm/sec, m/sec.

velocity curve   A plot of the velocity of stars observed in a spiral galaxy, measured along a major axis of the light distribution. Near the center the velocity is zero. On one side of the center the velocity is away from the observer, on the other side it is toward. The observed speed typically initially grows linearly with distance from the center. There follows a long plateau of constant velocity, typically $\sim$ 200 to 300 km/sec.

A drop with increasing distance, as expected for Newtonian motion around a central condensed mass, is almost never seen.

velocity dispersion   A parameter describing the range of velocities around a mean velocity value in a system of stars or galaxies. For example, the radial *velocity dispersion* can be measured for the stars in an elliptical galaxy and for the galaxies in a cluster. In the first case, the radial velocity dispersion is derived from the width of spectral lines, whose broadening is due to the motion of a large number of unresolved stars in the galaxy; in the second case the radial velocity dispersion is derived from the redshift of each individual galaxy. From the measurement of the velocity dispersion, the mass of the aggregation can be estimated.

velocity distribution   In kinetic theory, the density f of particles in velocity space. The *velocity distribution* (or velocity distribution function) is a function of the velocity $\mathbf{v}$ and spatial coordinates $\mathbf{x}$ and the time t and is commonly normalized so that $f(\mathbf{v}, \mathbf{x}, t)d^3v d^3x$ is the number of particles contained in the volume element $d^3v d^3x$. In a plasma each charge species (electrons and the various ions) is characterized by its own velocity distribution. The velocity distribution must satisfy an appropriate kinetic equation, such as the Boltzmann equation or Vlasov equation.

velocity strengthening   Under certain conditions, the coefficient of static friction of the interface between two rocks increases with increasing speed of relative slip. This phenomenon of the sliding surface becoming stronger with slip rate is called *velocity strengthening*. If the coefficient of friction decreases with increasing slip rate, the phenomenon is called velocity weakening. Velocity weakening is an unstable process that results in fast slips such as in earthquake faulting.

velocity weakening   *See* velocity strengthening.

ventifacts   Rocks that display a flat surface due to eolian erosion. When the environment has one or two prevailing wind directions, rocks will be sandblasted with wind-blown material until the face of the rock becomes smooth. *Ventifacts* often have two smooth faces (sometimes more, if the rock has been moved from its original orientation) separated by sharp edges.

ventilated thermocline   A layered model of the upper ocean, in which interfaces between some layers intersect the surface. This model better describes the three-dimensional structure of the thermocline than unventilated quasi-geostrophic theories of the upper ocean, which describe the vertical structure of the thermocline, but ignore the considerable horizontal structure present in the ocean. A *ventilated thermocline* model results in more realistic flow domains. A constantly refreshed, ventilated region contains fluid layers all in motion. On the periphery of the ventilated region are shadow zones which bound regions of constant potential vorticity or stagnant regions. The final possible ventilated zone is a region of Ekman upwelling. *See also* thermocline, Ekman pumping.

Venturi tube   A device consisting of a tube of smaller diameter in the middle than at the ends. When a fluid flows through such a tube, the pressure in the central portion is reduced according to Bernouli's principle. Measurement of the pressure drop gives a measure of the fluid flux through the tube. A *Venturi tube* may also be used to entrain other fluids into the main stream: atmospheric pressure forces the secondary fluid into an opening at the low pressure point in the Venturi tube. Applications in practice include internal combustion carburetors (gasoline vapor into air) and natural gas cook stoves (air into natural gas).

Venus   The second planet from the sun. Named after the Roman god of love, Venus. It has a mass $M = 4.869 \times 10^{27}$ g and a radius of $R = 6052$ km, giving it a mean density of 5.24 g cm^{-3} and a surface gravity of 0.91 that of Earth. Its rotational period of 243 days is longer than its orbital period around the sun, 224.7 days. The rotation axis has an obliquity of 177.3°, so its rotation is retrograde. The slow rotation means that the planet's oblateness is very nearly 0. Venus' orbit around the sun is characterized by a mean distance of 0.723 AU, an eccentricity of $e = 0.0068$, and an orbital inclination of $i = 3.395°$. Its synodic period is 584 days. The average albedo of 0.75 refers to the cloud tops. These clouds, predominantly sulfuric acid drops, are embedded in an atmosphere primarily composed of CO_2. A strong greenhouse effect maintains an average surface temperature of close to 750 K. Venus' interior structure is probably similar to that of the Earth, with a core composed of a mixture of Fe, Ni, and S, surrounded by a silicate mantle. Radar maps of the surface show it to be uncratered and relatively young ($< 10^9$ years at the oldest) and covered with lava flows that are still active in some spots, but with no evidence of plate tectonics. *Venus* has been the subject of many missions including Mariner 2, Pioneer Venus, Venera 7, Venera 9, and Magellan. Because its moment of inertia is not known, the relative sizes of the core and mantle can only be estimated from the mean density of the planet. Venus has no satellites.

Venusian Tessera   A distinct type of highland region on the surface of Venus belonging to the Plateau Highlands class (as opposed to Volcanic Rises). Plateau Highlands are topographically rugged, characterized by steep margins, and lack volcanic landforms (lava flows are rare; small volcanic constructs do not occur regularly and there are no shield volcanoes). Tessera form complex ridged terrain within the highlands. They may have been created by shortening the lithosphere so that the surface fractures and buckles, and the crust thickens, similar to the mountain-building process on Earth. In the absence of plate tectonics on Venus, however, a tessera-like landscape is formed.

vernal equinox   The epoch at the end of the Northern hemisphere winter on which the sun is located at the intersection of the celestial equator and the ecliptic; on this day the night and day are of equal length throughout the Earth. The date of the *vernal equinox* is at the end of the Southern hemisphere summer. The *vernal equinox* also refers to the direction to the sun at this instant: $0^h RA$, $0°$ declination. *See* autumnal equinox.

Very Long Baseline Interferometry (VLBI)
A technique for measurement of small scale (milliarc second) features in radio sources by carrying out interferometry between widely separated radio telescope sites, up to thousands of kilometers apart.

Vesta   The 3rd largest asteroid known; the fourth asteroid to be discovered, in 1807. It is located in the Asteroid Belt between Mars and Jupiter. *Vesta* has a mean diameter of 576 km and mass of 2.76×10^{20} kg and rotates every 5.34 hours. As it rotates, Vesta's brightness changes which suggests that the surface is not uniform either in terrain or composition. Images of Vesta from the Hubble Space Telescope show dark and light patches indicating a very diverse terrain, consisting of exposed mantle, lava flows, and a large impact basin. Spectra suggest that the compound pyroxene is present, which is typically found in lava flows. The composition of Vesta is sufficiently different from that of other known asteroids, so that Vesta has its own class. Orbit: semimajor axis 2.3615 AU, eccentricity 0.09, inclination to the ecliptic $7°.13405$, period 3.629 years.

virga   Rain (or perhaps snow) falling from a cloud but evaporating before it reaches the ground.

Virgo   A gravitational wave observatory under construction near the Italian town Cascina. The device is a laser interferometer, with two arms at right angles. The length of each arm of the laser interferometer device is 3 km. Operation is expected to start in 2001. *See* LIGO.

Virgocentric flow   The motion of the galaxies in the local group toward the Virgo Cluster. The gravitational force exerted by the mass of the Virgo attracts all surrounding galaxies, including the galaxies of the local group. The velocity of approach toward Virgo lies in the range $v_r \approx 100$ to 400 km s^{-1}. A correction for *Virgocentric flow* should be applied to the radial velocity measured for nearby galaxies, especially if the recessional velocity is used as an indicator of distance according to Hubble's Law.

Virgo constellation   **(1.)** A zodiac constellation most visible in spring, covering the area of sky approximately ranging from 12 to 15 h in right ascension and from $\approx -15°$ to $\approx +15°$

in declination. Virgo (Latin for virgin) can be identified by looking for a "Y"- shaped configuration formed by four stars between Libra and Leo. The brightest star of Virgo (α Virginis, Spica, apparent magnitude = 1.0) is located at the lower tip of the "Y". The sky region of the *Virgo constellation* includes the Virgo cluster of galaxies. *See also* Virgo Supercluster.

(**2.**) Sixth sign of the zodiac; constellation on the ecliptic around RA 13^h; the second largest constellation in the sky. Brightest star , Spica (α−Virginis) at RA $13^h25^m11^s.5$, dec $-11°9^m41^s$, spectral class B, V_B 0.98. The sun passes through this constellation from late September through October. The Virgo Supercluster of galaxies overlaps the northern border of Virgo into the neighboring constellation of Coma Berenices.

Virgo Supercluster (**1.**) A supercluster roughly centered on the Virgo Cluster of galaxies, which includes the galaxy and which is accordingly known as the local supercluster. The *Virgo Supercluster,* first introduced by G. de Vaucouleurs, has a clumpy structure which includes several groups and clusters of galaxies, and whose somewhat flattened distribution shows a preferential plane defining a "supergalactic equator" and a "supergalactic" system of coordinates. The galaxy is located in the outskirts of the Local Supercluster, at approximately 15 to 20 Mpc from the Virgo cluster. *See also* supercluster.

(**2.**) The *Virgo Supercluster* contains about 2000 member galaxies. The giant elliptical (E1) galaxy M87 (NGC4486), M_V 8.6 at RA $12^h30^m.8$, dec $+12°24'$, $18Mpc$ distant, is at the physical center of the Virgo cluster and of the Virgo (or Coma–Virgo) Supercluster. The supercluster's enormous mass modifies the local Hubble flow, thus causing an effective matter flow towards itself (the so-called Virgo-centric flow). Our local group has apparently acquired a Virgocentric velocity (a modification of our Hubble flow) of up to 400 km/sec toward the Virgo cluster.

The actual recession velocity of the Virgo Supercluster is of order 1100 km/sec. However, the Virgo supercluster is very tightly bound, with high peculiar velocities, to over 1500 km/sec with respect to the cluster's center of mass.

Hence, some of the supercluster's members show blue shifts (moving toward us); the largest blue shift is exhibited by IC3258, which is approaching us at 517 km/sec; while others show large red shifts, up to 2535 km/sec (NGC4388).

virtual geomagnetic pole If the Earth's magnetic field was a pure geocentric dipole, then knowledge of the direction of the magnetic field at any position on the surface (i.e., declination D, the angle that the horizontal component of the field makes with a line pointing north, and inclination I, the angle the field makes with the horizontal) can be used to calculate the position of the magnetic north pole:

$$\theta_m = \cot^{-1}\left(\frac{1}{2}\tan I\right)$$

$$\theta_p = \cos^{-1}\left(\cos\theta\cos\theta_m + \sin\theta\sin\theta_m\cos D\right)$$

$$\alpha = \sin^{-1}\left(\frac{\sin\theta_m\sin D}{\sin\theta_p}\right)$$

$$\phi_p - \phi = \alpha \qquad \cos\theta_m > \cos\theta\cos\theta_p$$

$$\pi + \phi - \phi_p = \alpha \qquad \cos\theta_m < \cos\theta\cos\theta_p$$

where (ϕ, θ) are the coordinates (longitude, colatitude) of observation, and (ϕ_p, θ_p) are the coordinates of the magnetic north pole. It is possible using various techniques to estimate from measurements the original inclination and declination of the magnetic field from a rock at the time of its formation. By assuming that the field at the time was a geocentric dipole, one may infer the orientation of the magnetic field at that time. By further assuming that the dipole was oriented parallel to the Earth's rotation axis, one may infer the latitude of the rock at the time of its formation (θ_m) as well as the horizontal orientation. There are various techniques for correcting for the effects of folding, and it can be argued that it is likely that the Earth's field, when averaged over time, is sufficiently close to an axial geocentric dipole that hence using many samples, an accurate estimate of paleolatitude and orientation can be made.

viscoelastic material A material that deforms as an elastic solid at certain time scales but as a viscous fluid at other time scales. The application of a constant stress causes an immediate deformation that disappears if the stress is quickly removed but increases for a time and becomes permanent if the stress is maintained. The deformation is generally a combination of the elastic deformation and viscous flow. The simplest *viscoelastic materials* are Maxwell and Kelvin materials in which both the elastic and viscous deformation obey linear constitutive laws. The general expression for the relation between deviatoric stress (σ'_{ij}) and deviatoric strain (ε'_{ij}) for a linearly viscoelastic rheology is

$$P\sigma'_{ij} = Q\varepsilon'_{ij}$$

where

$$P = \frac{\partial^m}{\partial t^m} + a_{m-1}\frac{\partial^{m-1}}{\partial t^{m-1}} + \cdots + a_1\frac{\partial}{\partial t} + a_0$$

$$Q = b_n\frac{\partial^n}{\partial t^n} + b_{n-1}\frac{\partial^{n-1}}{\partial t^{n-1}} + \cdots + b_1\frac{\partial}{\partial t} + b_0$$

and $a_0, a_1, \ldots a_{m-1}$, and $b_0, b_1, \ldots b_n$ are constants. Nonlinear viscoelastic rheology involves nonlinear elastic or viscous deformation.

viscosity   The kinematic viscosity ν [m^{2}s^{-1}], expressing the molecular viscosity (μ; [kgm^{-1} s^{-1}],) of a fluid, is defined by $\nu = \mu/\rho$ (typically $1.0 \cdot 10^{-6}$ to $1.5 \cdot 10^{-6}$ m^2 s^{-1} in water). The turbulent viscosity ν_t of a fluid is a property of the eddy-like motions in the fluid and usually not dependent on the kinematic viscosity ν which is solely a physical property of the fluid. Turbulence implies $\nu_t \gg \nu$.

viscosity (η)   The proportionality factor between shear rate and shear stress, defined through the equation $f = \eta A(dv/dx)$, where F is the tangential force required to move a planar surface of area A relative to a parallel surface. The result depends on dv/dx, the rate of increase of velocity with distance from the fixed plate. Sometimes called dynamic or absolute viscosity. The term kinematic viscosity (symbol ν) is defined as η divided by the mass density. η is also denoted by μ, the molecular viscosity.

viscous-like force   *See* quasi-viscous force.

viscous shear stress   The stress resulting from the intermolecular forces (proportional to viscosity) between water parcels of different velocities (shear). In a Newtonian fluids, such as water, the *viscous shear stress,* in general 9 tensor terms τ_{ij} (= force per unit area in i-direction acting on the surface that is normal to the j-direction) is linearly dependent on the shear. In turbulent fluids, the viscous shear stress is much smaller than the turbulent Reynolds stresses and is usually neglected for large scale processes. Viscous shear stress becomes the dominant force at small scales, such as microstructure (*see* Kolmogorov scale) and in viscous sublayer.

viscous sublayer   The layer along a contiguous boundary, within which momentum (shear stress) is transferred by intermolecular forces; i.e., $\tau = \nu\rho\partial u/\partial z$.

visible light   Light that is detectable by the normal human eye.

visible wavelengths   Approximately 400 to 700 nm.

visual binary system   An orbiting pair of stars far enough apart from each other and close enough to us that two points of light can be resolved in the sky (generally only with a telescope). This requires a separation of nearly 1 arcsec. For instance, a pair of stars like the sun with an orbit period of 20 years would have to be less than about 10 pc from us to be seen as a visual binary. Most orbit periods of visual binaries are long as a result, and analysis requires data to be collected over centuries or more. Once the distance to the system is known, the masses of the two stars can be found directly from the observed orbits of the stars using Kepler's third law of motion, $(M_1 + M_2)P^2 = a^3 M_1/M_2 = a_2/a_1$, where P is the orbit period in years, a_1 and a_2 are the semi-major axes of the two orbits around their center of mass in astronomical units, $a = a_1 + a_2$, and M_1 and M_2 are the larger and smaller masses in solar masses. Only about 100 *visual binary* systems have been studied well enough to yield masses with 15% accuracy, and most of them are stars of roughly solar mass.

VLA   Very Large Array, a "Y"-shaped array of 27 dish antennas, each of 25 m aperture, mounted on railway tracks that can be moved to form an interferometer up to 36 km across.

Located near Socorro, NM at 34°04′43″.497N, 107°37′03″.819W. The array has a maximum resolution of 0″.04 and can be used over the frequency range 300 to 50,000 MHz (90 to 0.7 cm).

Vlasov equation The kinetic equation governing the velocity distributions of electrons and ions in an environment where Coulomb collisions may be neglected. For non-relativistic particles of charge q and mass m in the presence of an electric field $\mathbf{E}(\mathbf{x}, t)$ and a magnetic field $\mathbf{B}(\mathbf{x}, t)$, the *Vlasov equation* is (cgs units)

$$\left(\frac{\partial}{\partial t} + \mathbf{v} \cdot \frac{\partial}{\partial \mathbf{x}} + \frac{q}{m} \left[\mathbf{E} + \frac{\mathbf{v}}{c} \times \mathbf{B} \right] \cdot \frac{\partial}{\partial \mathbf{v}} \right) f = 0 \,,$$

where $f(\mathbf{v}, \mathbf{x}, t)$ is the velocity distribution. Here $\mathbf{v}$ and $\mathbf{x}$ are, respectively, the velocity and spatial coordinates, and t is the time.

Moment equations for macroscopic quantities are obtained by multiplying the Vlasov equation by 1, $\mathbf{v}$, etc. and integrating over velocity space. For example, the zero-order moment equation is

$$\frac{\partial n}{\partial t} + \nabla \cdot (n\mathbf{V}) = 0 \,,$$

where the zeroth moment of f

$$n = \int f d^3 v$$

is the particle number density, and the first moment (divided by n)

$$\mathbf{V} = \frac{1}{n} \int \mathbf{v} f d^3 v$$

is the mean vector velocity. Another useful moment is the stress (or pressure) tensor

$$\mathcal{P} = m \int (\mathbf{v} - \mathbf{V})(\mathbf{v} - \mathbf{V}) f d^3 v \,;$$

note that the stress tensor need not be isotropic in a collisionless plasma.

The electric charge and current densities in the plasma are

$$\sum_\alpha q_\alpha \int f_\alpha d^3 v$$

and

$$\sum_\alpha q_\alpha \int \mathbf{v} f_\alpha d^3 v \,,$$

respectively, where the summation is taken over all charge species α.

Vlasov–Maxwell equations The fundamental equations governing the behavior of a plasma in which Coulomb collisions are negligible. A solution of the Vlasov–Maxwell system is comprised of velocity distributions for all charge-species and the associated electric charge and current densities, together with electric and magnetic fields, all consistent with the Maxwell equations governing electrodynamics. *See* Vlasov equation.

voids Large regions of the universe in which the matter-density is distinctly lower than the average. *Voids* are surrounded by relatively thin layers into which nearly all galaxies are crowded. Typical diameters of voids are between $20h^{-1}$Mpc and $60h^{-1}$Mpc (h is the dimensionless Hubble parameter, i.e., H_0/100/km/sec/Mpc), and the mass-density within a void is believed to be about 10% of the cosmic average matter-density. Hence, the large-scale distribution of matter in the universe is somewhat similar to a foam, with the voids being analogs of the air-bubbles and the layers with galaxies being analogs of the soap films.

Voids were observationally discovered in 1978–1979 and hailed at that time as a surprising discovery. However, papers were already published in 1934 that predicted that the Friedmann–Lemaître cosmological models were unstable against the formation of local minima in matter-density. Had the faith in the cosmological principle not prevailed later, voids would have been, in fact, expected.

Voigt body Also called Kelvin body. A kind of material with both properties of elastic body and viscous fluid. A *Voigt body* is represented by the sum of two terms for which stresses are proportional to elastic strain and viscous strain rate. Most elastic bodies containing many pores filled with viscous fluid are examples of Voigt bodies.

Voigt profile An expression that describes the shape of an absorption line considering the effects of natural and thermal Doppler broadening mechanisms. The total absorption coefficient, α, is then

$$\alpha = \alpha(\text{natural}) * \alpha(\text{thermal})$$

where $*$ indicates convolution. Convolving the natural and Doppler absorption coefficients, we have the normalized *Voigt profile,*

$$V(\Delta\nu, \Delta\nu_D, \gamma) =$$
$$\int_{-\infty}^{+\infty} \frac{\gamma/(4\pi^2)}{[(\Delta\nu - \Delta\nu)^2 - (\gamma/4\pi)^2]} \frac{1}{\Delta\nu_D\sqrt{\pi}}$$
$$e^{-(\Delta\nu_1/\Delta\nu_D)^2} d(\Delta\nu_1) ,$$

where $\Delta\nu_D$ is the Doppler width at half maximum, and γ is the radiation damping constant (*see* natural line broadening).

The absorption coefficient becomes

$$\alpha = \frac{\pi e^2}{m_e c} f V(\Delta\nu, \Delta\nu_D) .$$

This expression can be calculated numerically or simplified slightly by expressing it as a Hjerting function. In any case, the profile of a line is then formed by integrating α over all frequencies. This simplified form is used in calculating model atmospheres for hot stars.

volatile   Easily evaporated, or in geology, elements or compounds with low molecular weight and low melting points that are easily driven from formations by heating. *See* refractory.

volcanic front   A front line of areas of volcano distribution nearest to a trench at a subduction zone. The closer to the *volcanic front,* the greater the density of volcano distribution and the number of eruptions. Although most volcanic fronts are almost parallel to a trench, there are some exceptions such as the northern part of the central American arc. In many cases, there exists a deep seismic plane at around a depth of 110 km just below a volcanic front. In general, the amount of alkali in volcanic rocks tends to increase as we move away from a volcanic front toward the back-arc side.

volcanic line   A well-defined line of volcanos generally is associated with subduction zones. One example is the line of volcanos extending from Mount Shasta in California to Mount Baker in Washington. These volcanos systematically lie close to 125 km above the subducted oceanic plate.

volcanism   Planetary interiors are often quite hot due to a number of heat producing processes, including radioactive decay, heat from accretion, and differentiation. This heat can cause the interior rocks to melt, producing a magma, which can make its way to the surface and erupt in a volcanic event. The types of volcanic features resulting from this eruption depend on the rate of the eruption and the viscosity (i.e., the stickiness) of the lava. Low viscosity, fluid lavas will produce flat volcanic flows (i.e., Columbia River Basalts, Washington). Slightly stickier lavas will produce low-sloped shield volcanoes (i.e., Hawaii). Magma with substantial amounts of gas incorporated in it will produce cinder cones (i.e., Sunset Crater, Arizona). Explosive eruptions are associated with stratovolcanoes (composite volcanoes) which consist of alternate episodes of quiet lava flows and explosive ash eruptions (i.e., Mt. St. Helens, Washington). The most explosive eruptions are associated with ignimbrites, which produce large ash flows (i.e., Long Valley Caldera, California). Magma composed of molten rock is the most common type of volcanic material on Earth, but elsewhere in the solar system magmas composed of sulfur and various types of volatiles (such as water, ammonia, etc.) are seen.

volcanos   *Volcanos* are mountains, generally conical in shape, created by magmas erupted from the interior of the Earth. Most of the famous volcanos are associated with subduction. Examples include Mount Fuji in Japan, Mount St. Helens in the U.S., Penetubo in the Philippines, and Etna in Italy. This class of volcanos often has explosive eruptions. Volcanos are also associated with ocean ridges, but almost all of these volcanos are in the deep oceans and cannot be observed. An exception is the group of volcanos in Iceland. A third class of volcanos occur within plate interiors and are known as hotspots. These volcanos are generally associated with mantle plumes, and their eruptions are usually quiescent; an example is Hawaii.

volume scattering function (VSF) The ratio of the scattered intensity [W sr^{-1}] to the incident irradiance [W m^{-2}] per unit volume [m^3], given in [m^{-1} sr^{-1}]; the integral of the *volume scattering function* over all directions and all final wavelengths is the scattering coefficient; the VSF can be written as the product of the phase function [sr^{-1}] and the scattering coefficient [m^{-1}].

volumetric water content (or simply water content) (θ) The ratio of liquid water volume to the total soil or rock volume: $\theta = V_w/V_s$, where V_w is the volume of water in the sample, and V_s is the total volume of soil or rock in the sample. The theoretical range of the *volumetric water content* is from 0 (completely dry) to saturation (porosity or ϕ), but the range in natural systems is generally much less than this.

von Mises criterion The condition by which a polycrystal can deform coherently to large strains by dislocation slip. *von Mises criterion* points out that five independent slip systems are needed in order to deform a polycrystalline material by crystallographic slip. This is because arbitrary shape change must be possible to satisfy the displacement compatibility at grain boundaries.

vortex defect of the vacuum *See* cosmic string.

vortex street The series of vortices shed systematically downstream from a body in a rapid fluid flow (sufficiently high Reynolds number). For a cylindrically symmetrical body, a pair of symmetrically placed opposite vortices (with a sign which can be deduced from the streamflow) form behind the body on either side of the downstream direction. Small disturbances then lead to a shedding of one of the two vortices, and subsequently alternate shedding of vortices of opposite sign. The trail of vortices behind the body is the *vortex street.*

vorticity A vector measure, $\boldsymbol{\omega}$, of local rotation in a fluid flow:

$$\boldsymbol{\omega} = \nabla \times \mathbf{v}.$$

vorton Equilibrium configurations of conducting cosmic string loops, where the string tension, tending to make the loop collapse, is balanced by the centrifugal repulsion due to rotation. Symmetry reasons (Lorentz invariance along the string world sheet) forbid rotating configurations for Goto–Nambu strings and, therefore, vortons can only arise in theories where cosmic strings have a microscopic internal structure, like currents flowing along the core.

The existence of vorton equilibrium configurations would prevent the loops from decaying away in the form of radiation. They would then evolve like ordinary (eventually charged) non-relativistic matter very early on in the history of the universe, eventually disrupting some of the predictions of standard cosmology.

This possible incompatibility is known as the vorton excess problem. Many mechanisms aiming at diluting vorton overdensity are currently being studied. If successful, vortons could stop being a problem and would become one more interesting candidate for the non-baryonic dark matter that combined observations and theoretical models tend to require. *See* CHUMP, conducting string, cosmic string, Goto–Nambu string.

Wadati–Benioff zones (Benioff zones)
Zones of deep earthquakes in the subducted oceanic plates. These earthquakes were first identified by Japanese seismologist K. Wadati in 1928. H. Benioff recognized the inclined planar shape of the zones of such deep seismicity.

wake   The region downstream from a body moving through a fluid. Vorticity generated as the flow separates from the body is concentrated in the *wake*. Visible as surface waves in the case of boats moving through water.

wake (cosmic string)   As a consequence of the conical structure of the space around a cosmic string, particle trajectories will be deflected from their original directions when passing close to the string. The deviation angle is $4\pi GU$ (half of the deficit angle on each side of the string) and is due to a pure kinematic effect. Here $U =$ mass per unit length of the string.

Consider now a moving string traversing a given distribution of collisionless matter, like stardust, initially at rest. From the viewpoint of the string it is the dust that moves and dust particle geodesics on opposite sides of the string trajectory will tend to converge. Dust velocities perpendicular to the string motion will be of order $4\pi GU v_s \gamma_s$, where v_s is the velocity of the string, and γ_s is the corresponding Lorentz factor $(1 - v_s^2/c^2)^{-1/2}$. This produces a region of relatively large matter density behind the string that after subsequent gravitational clumping will lead to the formation of a sheet-like distribution of matter known as a string wake. Wakes are one of the most characteristic signatures in the generation of large scale structure with fast moving cosmic strings. *See* Abelian string, cosmic string, cosmic topological defect, deficit angle (cosmic string), wiggle (cosmic string).

Walker circulation   A longitude-vertical circulation cell of the tropical atmosphere, with rising motion and high convective activity over the equatorial Indian and western Pacific Oceans. The downdraft is located over the eastern equatorial Pacific, a direct result of cold sea surface underneath. During an El Niño event when the eastern equatorial Pacific Ocean warms, the rising branch of the *Walker circulation* shifts eastward to the central Pacific, causing drought over the maritime continent and Australia.

wall terracing   The walls of large impact craters are often formed at a steep angle (greater than the angle of repose for the material) and are subject to collapse under the influence of gravity. This collapse of the wall material creates a terraced appearance to the interior walls of the crater. The collapse of the walls causes the final crater to be larger than the original crater and decreases the height of the crater rim. The onset crater diameter for *wall terracing* provides important information about the strength of the material in which the crater formed.

warm front   Along a temperature discontinuous surface, warm air advances to replace cold air being pushed forwards so the warm air slides up and over the cold air to produce steady rain. Slopes of warm fronts are typically about 1/100 to 1/300, and the ascent of air is gradual. Ahead of warm fronts for several hundreds of kilometers there will appear cirrus; then as the warm fronts approach, cirrostratus, lower and thickening altostratus, and heavy nimbostratus will appear one by one. If the warm air mass is unstable, shower or thunderstorm activity may be present. Passage of *warm fronts* is marked by a rise of temperature, decreasing of relative humidity, clearing of precipitation, cloud or fair, and veering wind.

warping (of the neutral sheet or plasma sheet)
The deformation of the equatorial surface ("neutral sheet") in the plasma sheet regions of the magnetosphere, caused by the tilt of the Earth's dipole axis. The midnight section of that surface is displaced northward or southward, in a way that matches the displacement of the midnight section of the equatorial plane of the tilted Earth dipole. However, since the two lobes have equal magnetic fluxes and are expected to have equal magnetic intensity (and hence, equal magnetic pressure), the sections of the neutral sheet near

the magnetopause are displaced in the direction opposite to that of the midnight section.

wash load   Sediment that is carried in suspension in a water column through a section of channel or river, which does not originate in that section of channel. *Compare with* suspended load, bed load.

water H_2O, colorless, odorless, tasteless liquid; freezing point (to ice) 0°C, boiling point (to steam) 100°C (at standard atmospheric pressure). Pure water has a maximum density approximately $1000 \, kg/m^3$ at 4°C and frozen water (ice) is much less dense than water, with a density of $920 \, kg/m^3$. The density of liquid water at 0°C is $999.85 \, kg/m^3$ so ice floats in ice-cold water. Under normal conditions the latent heat of fusion of water is 3.33×10^5 J/kg, and the heat of vaporization is $2.26 \times 10^6$6 J/kg. These very large values act to make the oceans substantial reservoirs of heat and buffers of temperature, and ocean currents can distribute this heat, thereby stabilizing the temperature of the Earth.

water mass   In physical oceanography, a body of water with a common formation history. *Water masses* are usually identified through physical relationships on a temperature-salinity (T-S) diagram. In addition, some assumption about the degree of spatial and temporal variability during a water mass's formation is almost always needed as well.

water mega-masers   Astrophysical masers of a spectral line at 22.235 GHz due to a rotational transition of the water vapor molecule. *Water mega-masers* are more luminous than the most luminous galactic masers by a factor of 100. Approximately 20 water mega-masers are known as of early 1998; they have been observed in several active galaxies. Interferometric observations have shown that mega-masers are formed within 1 pc from the active nucleus of a galaxy, indicating that the source of pumping radiation needed to sustain the population inversion is provided by the active nucleus itself.

waterspout   A tornado that occurs over the ocean.

water table   A surface separating the saturated and unsaturated zones of the subsurface, defined as a surface at which the fluid pressure is atmospheric (or zero gage pressure). The *water table* is generally measured by installing wells or piezometers into the saturated zone and measuring the water level in those wells.

watt (W)   The SI unit of power, equal to J/s.

wave   A disturbance propagating through a continuous medium. It gives rise to a periodic motion. A *wave* transports energy and information but not matter. Depending on the oscillatory motion, waves can be longitudinal (such as sound waves) or transversal (such as surface waves or electromagnetic waves). A wave is characterized by its amplitude (or wave height in surface waves), the wavelength or wave number, the wave period or (angular) frequency, and the phase or group speed.

wave-affected surface layer (WASL)   The top layer of the surface boundary layer which is significantly affected by the action of the waves. The extent is typically $(2k)^{-1} = \lambda/4\pi$ where k is the wavenumber and λ is the wavelength at the peak of the surface wave spectrum.

wave crest   The high point on a periodic wave; a point where phase angle is an integer multiple of 2π.

wavefront   A 2-surface on which the wave has the same phase (e.g., a maximum of field strength) at a particular time. A wave propagating at a great distance from its source through a vacuum or a uniform medium has very nearly plane *wavefronts*.

wave-function of the universe   Any solution of the Wheeler–DeWitt equation in the superspace (or minisuperspace) of 3-metrics of the whole space-time.

One usually considers the minisuperspace case corresponding to homogeneous and isotropic cosmological models (*see* Robertson–Walker metric) as a manageable approximation of the full quantum theory of gravity, valid in an early stage of our universe when typical particle energies had already fallen below the Planck mass but quantum effects were not yet negligible. Such effects include fluctuations around classical trajectories of the scale factor a of the Robertson–Walker metric and tunneling of a between disconnected Lorentzian regions by passing through a classically forbidden Euclidean stage. What the boundary conditions for the wave-function could be is still an open controversy among several schools of thought. *See* minisuperspace, Robertson–Walker cosmological models, superspace, Wheeler–DeWitt equation.

wavelength   The distance between successive maxima of a wave field at any particular instant, or more precisely of one Fourier component of such a field.

wave number   The *wave number* k gives the number of waves per unit length. It is related to the wavelength λ as $k = 2\pi/\lambda$. *See* wave vector.

wave-particle interactions   An important component, generally described by quasi-linear theory, in the evolution of particles and waves in a plasma.

wave rose   A polar histogram that shows the directional distribution of waves for a site. Generally also scaled to indicate wave height as well.

wave trough   The low point in a periodic wave; a point where phase angle is $(2n + 1)\pi$, $n = 0, 1, 2, \ldots$.

wave vector   The *wave vector* $\vec{k}$ gives the direction of propagation of the wave; its length gives the number of waves per unit length, the wave number. The wave vector is related to the phase speed $\vec{v}_{ph}$ by $\vec{v}_{ph} = \omega\vec{k}/k^2$.

weak anthropic principle   Minimal statement that since we observe the universe, parameters must be such as to support carbon-based life. *See* anthropic principle.

weak interaction   Any nuclear interaction involving neutrinos or antineutrinos. Weak rates are typically much slower than strong interaction rates, at temperatures and energies relevant to astrophysics. Hence in any reaction chain, the rate will be determined by the *weak interaction* rate.

weakly asymptotically simple space-time
A space-time $(\mathcal{M}, g)$ is weakly asymptotically simple if there exists an asymptotically simple space-time, such that in it a neighborhood of $\mathcal{I}$ (null infinity) is an isometric with an open set of $\mathcal{M}$. *See* asymptotically simple space-time.

weather   At a given time or over a short period, measured in days, the state of the atmosphere, defined by measurement of every meteorological element, such as air temperature, pressure, humidity, wind speed and direction, clouds, fog, precipitation, visibility, etc.; the synthetical status of the distributions of the meteorological elements and their accompanying phenomena; the state of the sky which affects the everyday life of humans, such as cloudy, fair, cold, warm, dry, humid, etc. In aviation meteorological observation, *weather* means precipitation phenomena such as rain, snow, etc.

weather forecast   The prediction of future weather based on the meteorological theory and the basis of observations made from satellite data, radar data, and observation networks on both land and sea, which contains the observed reports from the ground surface to high level atmosphere. With the development of computational technology, numerical *weather forecast* is the most important weather forecast method. It is a complete objective forecast method and is based on atmospheric dynamical and thermal-dynamical physics laws implemented on powerful supercomputers to simulate the future atmospheric for the prediction results.

weather modification   The intentional change in weather conditions by human action, such as artificial precipitation, artificial hail, artificial cloud dispersal, artificial fog clearing, artificial hurricane weakening, artificial thunder – lightning suppression, artificial frost prevention, etc. Inadvertent change in weather conditions by human action, such as urban heat island effects, is excluded from the definition of *weather modification*. Currently, weather modification methods are mainly based on the microphysical instability properties of cloud and fog. For example, there are abundant unfrozen water droplets in cold clouds whose temperature is lower than 0°C, so seeding glacigenic catalysts can cause the unfrozen water droplets to freeze to ice crystals and speed the forming processes of waterdrops. The released condensational latent heat in such processes can change the thermal and dynamical structure of clouds, to enhance precipitation, to reduce hail, to disperse clouds, to clear fog, or to reduce wind speed of hurricanes. Seeding appropriate sized salt powder can enhance the forming process of rain drops in warm clouds, and thus can enhance precipitation or disperse cloud and fog. Currently, the techniques of artificial cloud dispersal and fog clearing are mature and used widely on airports, while other weather modification methods often lead to unexpected results.

Weber–Davis Model   The magnetohydrodynamic theory of angular momentum transport a steady, axially symmetric, magnetized, rotating astrophysical wind.

The theory was originally developed as a tool for analysis of the transfer of angular momentum in the solar wind but has found much broader application in theoretical investigations of stellar winds and star formation. In the most general form of the theory, the flow velocity and magnetic field are separated into poloidal and toroidal parts, $\mathbf{B} = \mathbf{B}_p + B_t \mathbf{e}_\phi$ and $\mathbf{V} = \mathbf{V}_p + V_t \mathbf{e}_\phi$, where $\mathbf{e}_\phi$ is the unit vector in the azimuthal direction; these quantities are related through conservation laws corresponding to conservation of mass and magnetic flux,

$$\mathbf{B}_p = \lambda \rho \mathbf{V}_p$$

conservation of total (mechanical plus magnetic) angular momentum

$$\varpi \left(V_t - \lambda B_t / 4\pi \right) = \mathcal{J}(\lambda) ,$$

and Faraday's law for a frozen-in magnetic field

$$\frac{1}{\varpi} \left(B_t / \lambda \rho - V_t \right) = \mathcal{Q}(\lambda) .$$

These invariants hold for each streamline, which may be labeled by λ. ϖ is distance from the rotation axis, ρ is the mass density, and $\mathcal{J}$ and $\mathcal{Q}$

are functions of λ to be determined from boundary conditions.

If the flow describes expansion of a wind, the velocity on each streamline passes through an "Alfvén point" $\varpi_A(\lambda)$ where the (poloidal) flow and Alfvén speeds are equal, $\mathbf{V}_p = \mathbf{B}_p / \sqrt{4\pi\rho}$. It can be shown that the constant of the second conservation equation is $\mathcal{J}(\lambda) = \varpi_A^2 \Omega$, where Ω is the rotation rate of the central object, and that the angular momentum per unit mass of gas approaches $\mathcal{J}(\lambda)$ infinitely far from the source of the wind. Thus, the magnetic field enhances angular momentum transport by providing an effective lever arm of length ϖ_A.

weight   The force of gravity; the force with which an object is attracted to the center of a planet;

$$\mathbf{W} = m\mathbf{g} ,$$

where g is the magnitude of the local acceleration of gravity $g = GM/r^2$, where M is the mass, and r is the distance to the center of the planet; G is Newton's gravitational constant

$$g \sim 980 \text{ cm/sec}^2 = 9.8 \text{ m/sec}^2 .$$

near the surface of the Earth.

weir   A structure placed on the bottom of a channel, across which water flows, to control depth or to facilitate measurement of flowrate. A variety of designs are available, including broad-crested and short-crested.

well-graded   A term used to describe the distribution of particle sizes in a soil sample. A *well-graded* soil has a wide range of particle sizes.

wentworth size classification   A classification of sediments based on size (diameter). Ranges from boulders (>256 mm) down to colloids (<0.0024 mm).

Wesselink method   *See* Baade–Wesselink method.

westerlies   The global wind system blowing west to east in both hemispheres of the Earth, between approximately 35° and 65° (north or south) from the equator.

western boundary currents Ocean currents flowing on the western boundary at speeds much higher than those in the rest of an ocean basin. This western intensification of ocean currents is a combined effect of rotation and surface curvature of the Earth, or the so-called beta-effect. The Gulf Stream and Kuroshio Current are the surface *western boundary currents* of the North Atlantic and North Pacific, respectively. In the Atlantic, deep western boundary currents are also observed, transporting deep water formed in the Nordic and Labrador Seas.

western boundary intensification The intensification of current toward the western boundary of the ocean due to the variation of the Coriolis parameter with latitude (the β-effect).

west Greenland current A branch of the east Greenland current that branches northwestward along the southwest coast of Greenland.

westward drift The movement of features of the geomagnetic field to the west. This was first noticed by Halley in the 17th century, who hypothesized that the Earth's magnetic field emanates from magnetized layers in the Earth's interior that are in rotation with respect to the surface. It is now understood that the magnetic field is generated in a molten iron outer core, and that changes in the magnetic field observed at the Earth's surface are related either to flows at the top of the core, or diffusion of the magnetic field. If the former is responsible for the *westward drift*, then it would be related to a general westward flow of the core's surface. The magnitude of the drift is around 0.2 to 0.4° per year, although it appears to vary with time and latitude. It is better determined in some places than others: for example, the field in the Pacific region is relatively plain, so it is difficult to tell whether or not it is drifting, and some models of core flow show flows to the east there. Part of the time variation of the westward drift may be related to torsional oscillations. *See* core flow.

wet-bulb temperature The temperature to which a parcel of air is cooled by evaporating water into it gradually, adiabatically, and at constant pressure until it is saturated. It is measured directly by a thermometer whose bulb is covered by a moist cloth over which air is drawn.

wetness (or degree of saturation) (S) The proportion of pore space that contains water:

$$S = \frac{V_w}{V_a + V_w} = \frac{\theta}{\phi}$$

where V_w is the volume of water in the sample, and V_a is the volume of air in the sample. When a soil or rock sample is saturated with water the volume of air goes to zero, the volumetric water content equals porosity, and *wetness* equals one.

wetted perimeter A linear measure of the length of a river or canal cross-section that is wetted by flowing fluid.

Weyl space-times (1917) The static and axially symmetric vacuum metrics of the form

$$ds^2 = -e^{2U}dt^2 + e^{-2U} \left[e^{2\gamma}(d\rho^2 + dz^2) + \rho^2 d\phi^2 \right]$$

where ρ and z are generalized cylindrical coordinates. The function $U = U(\rho, z)$ satisfies the Laplace equation $\Delta U = 0$ as follows from the vacuum Einstein equations. The linearity of the Laplace equation allows a superposition of solutions. The function $\gamma = \gamma(\rho, z)$ is determined by the line integral in the (ρ, z) plane

$$\gamma = \int \rho \left[\left(\left(\frac{\partial U}{\partial \rho} \right)^2 - \left(\frac{\partial U}{\partial z} \right)^2 \right) d\rho + 2 \frac{\partial U}{\partial z} \frac{\partial U}{\partial \rho} dz \right].$$

The solution $U = m/(\rho^2 + z^2)^{1/2}$ yields the axisymmetric Curzon metric (rather than the spherical Schwarzschild solution).

Weyl tensor A tensor, components of which are the following linear combination of the components of the curvature tensor:

$$C_{\alpha\beta\rho\sigma} = R_{\alpha\beta\rho\sigma} - \frac{2}{n-2} \left(R_{\alpha[\rho} g_{\sigma]\beta} - R_{\beta[\rho} g_{\sigma]\alpha} \right) + \frac{2R}{(n-1)(n-2)} g_{\alpha[\rho} g_{\sigma]\beta}.$$

The *Weyl tensor* shares all the permutation symmetries with the Riemann tensor $R_{\alpha\beta\rho\sigma}$. Moreover, Weyl tensor possesses some additional features. It is traceless $C^{\alpha}{}_{\beta\alpha\sigma} = 0$ and does not transform under a local conformal transformation

$$g_{\mu\nu} = \bar{g}_{\mu\nu} \cdot e^{2\sigma(x)}, \qquad C^{\alpha}{}_{\beta\rho\sigma} = \bar{C}^{\alpha}{}_{\beta\rho\sigma} .$$

Hence, it is also called conformal tensor. The minimal dimension of a manifold in which the Weyl-tensor exists is four. (The conformal structure of a 3-manifold is carried by the three-index Cotton tensor.) In the theory of general relativity, the stress-energy tensor of a given matter distribution determines the traces of the curvature locally by Einstein's gravitational equations, and the Weyl tensor is the carrier of the intrinsic degrees of freedom of the gravitational field. *See* curvature tensor, Riemann tensor.

Wheeler–DeWitt equation   The quantized form of the Hamiltonian constraint obtained by replacing the classical canonical variables γ_{ij} and their momenta π^{ij} with operators acting on a Hilbert space of states. The simplest way to employ quantization is to map

$$\gamma_{ij} \quad \rightarrow \quad \hat{\gamma}_{ij} = \gamma_{ij} \times$$
$$\pi^{ij} \quad \rightarrow \quad \hat{\pi}^{ij} = -i\,\hbar\,\frac{\partial}{\partial\gamma_{ij}} ,$$

in analogy with what is done with coordinates and momenta of a particle in ordinary quantum mechanics.

This procedure is, however, plagued by several formal problems, e.g., the operator ordering ambiguity: both the super-Hamiltonian and the super-momentum densities (which should be quantized as well) contain products of elements of the metric tensor times momenta. But no unique way of writing the quantized version of such products is given *a priori,* since any choice would lead to the same classical expressions. This reflects the possibility of defining canonical operators differing from the ones shown above and thus different quantum theories for the same 3+1 splitting of space-time.

Whatever form of quantization is chosen, the *Wheeler–DeWitt equation* can be formally written as

$$\hat{\mathcal{H}}\,\Psi = 0 ,$$

where $\Psi = \Psi[\gamma, \phi]$ is often referred to as the wave function of the universe and is a functional of the metric tensor γ and the matter fields ϕ on the hypersurface Σ_t. This equation is in the form of a time-independent Schrödinger equation which selects out the zero energy eigenstates as physical states. Indeed, one of the most intriguing puzzles of this approach to quantum gravity is the lack of a time variable. One possible way of solving this puzzle is to resort to the semiclassical approximation (*see* time in semiclassical gravity).

Physical states Ψ, which satisfy both the Wheeler–DeWitt equation and the quantized super-momentum constraints, are functionals of the 3-metric of space (*see* Superspace). In order to build a Hilbert space out of them, one must also face the problem of defining a conserved scalar product (the constraints are functional differential equations). Since solutions of the constraints are not available in general and because of all the formal problems mentioned above, one usually imposes some symmetry in order to reduce the number of degrees of freedom of the metric before quantizing. In some cases this is sufficient to reduce the metric tensor and the matter fields to a form that contains only a finite number N of functions of time only (*see* minisuperspace). Therefore, the quantized secondary constraints become ordinary differential equations for Ψ as a function of those N variables.

It is by now generally accepted that the Wheeler–DeWitt approach to quantizing gravity cannot lead to the ultimate quantum theory of the gravitational interaction and better candidates are to be found among the theories of extended objects (strings and n-dimensional branes). *See* Hamiltonian and momentum constraints in general relativity, semiclassical gravity, wave-function of the universe.

whimper   A space-time singularity not accompanied by infinities in physical quantities like density and pressure (such infinities are essential in a Big Bang singularity), though some curvature components may be infinite.

whistlers   Pulses of electromagnetic wave energy with frequencies in 0.3 to 35 kHz (VLF) range. When converted to audio signals, they are heard as "whistles" with steadily falling pitch that lasts for a few tenths of a second (short whistlers) to several seconds (long whistlers). *Whistlers* are generated by lightning sources or atmospherics. The electromagnetic energy produced in the lightning sources travels along the Earth's magnetic lines in the ionosphere almost without any attenuation to the other hemisphere. A part of this energy in the other hemisphere is reflected back along the same magnetic lines producing whistlers in a radio receiver in the hemisphere of origin, and another part enters the ionosphere and propagates in the Earth – ionosphere waveguide mode to the hemisphere of origin producing short whistlers. Since the losses in the ionosphere are small, many echoes may occur in what is known as an "echo" train before they are lost in the background noise. The changing pitch indicates that the path time "t" along the magnetic line of force is related to the frequency "f" by the dispersion relation

$$D = t\sqrt{f}$$

where D is the dispersion and may range from about 10 to 200 $\sqrt{\sec}$. Since t will depend on the line of force along which the energy travels and the electron density which decreases with height, D is found to increase with increase in electron density. The whistler properties can be explained by the application of the magnetoionic theory to low frequency wave propagation in the ionosphere. Whistlers are used as a research tool to study the electron density and associated properties in the Earth's magnetosphere.

whistler waves   A low-frequency plasma wave which propagates parallel to the magnetic field at a frequency less than the electron-cyclotron frequency. *Whistler waves* are circularly polarized, rotating in the same sense as the electrons in the plasma. Whistlers are so-named because of their characteristic audio frequency tone.

white dwarf   A star supported entirely by electron degeneracy pressure, formed by contraction when internal fusion energy sources are exhausted.

white hole   The time-reverse of a black hole, i.e., a region of the spacetime that can only eject

matter and radiation but can accrete none of it. A *white hole* would be a naked singularity, with a past event horizon, which does not shield it from our view. Light from a white hole would be blue-shifted. A white hole can be imagined as a Big Bang taking place in a limited volume of space, while the remaining part of the space already exists. White holes are natural in inhomogeneous models of the universe — they are places where the Big Bang occurs later and was still going on while it had already been completed elsewhere. Since literally everything can come out of these holes, thus strongly violating causality, a cosmic censorship principle has been invoked to avoid their existence in our universe. There is currently no evidence that they exist. *See* black hole, future/past event horizon, naked singularity.

white light   Solar radiation integrated over the visible portion of the spectrum (4000 to 7000 Å) to produce a broad-band (*white light*) signal. Best used to observe the photosphere and also in eclipse or coronagraph observations of the solar corona.

white light flare   A major flare in which compact regions become visible in white light. Such flares are usually strong X-ray, radio, and particle emitters.

Widmanstätten pattern   A distinctive pattern of crystal faces found in nickel-rich iron meteorites. When an iron-nickel mixture is heated to near melting and slowly cooled, two types of crystals are formed: a nickel-poor phase (kamacite) and a nickel-rich phase (taenite). The pattern formed by the growth of these two crystal phases can be seen when the meteorite is polished and etched with acid.

Wien's displacement law   (Wien law) The wavelength λ_m at which the maximum radiation intensity (ergs/sec/cm^2/steradian/wavelength interval) of a black body radiator occurs is inversely proportional to the absolute temperature T, so:

$$T\lambda_m = C ,$$

where C is a constant, equal to 0.28978 cm-K.

Wien distribution law A relation between the monochromatic emittance F (erg/sec/wavelength/cm^2/steradian) of an ideal black body and that body's temperature T:

$$F \propto \lambda^{-5} f(T\lambda) .$$

Wien's law *See* Wien's displacement law.

wiggle (cosmic string) After cosmological phase transitions in which cosmic strings are produced, the strings will trace regions where the Higgs field departs from the low temperature vacuum manifold of the theory. These regions have random spatial locations and, therefore, the ensuing string has no reason to be straight. The string will, in general, possess an irregular shape and small scale structures in the form of *wiggles* will accumulate all along its length. Further interactions will make the number of these small irregularities increase. Universal expansion does not eliminate these wiggles; hence, wiggly strings are the most natural outcome during network evolution.

This structure can be analytically approximated by an effective equation of state describing how a distant observer would perceive the string. The effective energy per unit length $\tilde{U}$ (larger than the Goto–Nambu energy U, as there is more string matter in a given segment due to the wiggles) and the effective tension $\tilde{T}$ (smaller than the tension of a Goto–Nambu string) will satisfy the relation $\tilde{U}\tilde{T} = U^2$.

Furthermore, the space around the wiggly string is no longer locally flat, and thus the string will behave like a standard gravitational attractor. Now particle geodesics in the vicinity of a wiggly string will be deflected by two mechanisms: the conical deficit angle deviation (from far away the string looks pretty straight and featureless) and the standard gravitational attraction. The relative velocity between two test particles on different sides of the string will be

$$\delta v = 8\pi G \tilde{U} v_s \gamma_s + 4\pi G$$
$$\left(\tilde{U} - \tilde{T}\right) / (v_s \gamma_s)$$

where v_s is the velocity of the string, γ_s is the corresponding Lorentz factor $(1 - v_s^2/c^2)^{-1/2}$, and G is Newton's constant. The first of these

effects dominates for fast moving strings and would lead to the formation of wakes. The second one is relevant for slow strings and could be at the root of the generation of filamentary distributions of astrophysical large scale structures. *See* cosmic phase transition, cosmic string, cosmic topological defect, deficit angle (cosmic string), wake (cosmic string).

Wilson cycle The Atlantic ocean is presently "opening", growing wider due to the creation of new sea flow at the Mid-Atlantic Ridge. It is expected that in the future, subduction zones will form on the boundaries of the Atlantic, and the ocean will close resulting in a continental collision between the Americas and Europe and Africa. This opening and closing of the Atlantic has happened twice in the past and is known as the *Wilson cycle,* in honor of J. Tuzo Wilson who first proposed this behavior. The last closing of the Atlantic Ocean created the Appalachian Mountains and occurred about 200 million years ago.

wind The general term for moving air, typically driven by naturally arising pressure gradients depending on altitude differences, solar heating, surface temperature, and the Coriolis force due to the Earth's rotation.

wind chill factor The perceived sensation of temperature, T_{WC}, in the presence of a wind. Based on studies of the rate of water freezing under different conditions.

The following equations can be used to determine the *wind chill factor* T_{WC}:
Wind speed V given in mph:

$$T_{WC} = T_s - ((T_s - T)(.474266 + \left(.303\sqrt{V}\right) - .02 * V))$$

Wind speed in knots:

$$T_{WC} = T_s - ((T_s - T)(.474266 + \left(.325518\sqrt{V}\right) - .0233 * V))$$

Wind speed in m/s:

$$T_{WC} = T_s - ((T_s - T)(.474266 + \left(.4538\sqrt{V}\right) - .045384 * V))$$

Wind speed in km/h:

$$T_{WC} = T_s - ((T_s - T)(.474266$$
$$+ \left(.2392\sqrt{V}\right) - .01261 * V)) \ .$$

In every case T_s is the skin temperature, $T_S = 91.4°F$ or $33°C$; the effective wind chill factor gives a result in the same units.

Note that these formulae are inapplicable below very low wind speeds (equivalently 4 mph, 1.78816 m/s, 6.437 km/sec, 3.476 knots). Below this speed the wind chill temperature is taken to be the same as the actual temperature (although the formula will actually give temperatures above the measured temperature for winds below this lower limit).

Also notice that above about 40 mph of windspeed, there is little additional wind chill effect.

wind-driven circulation   The circulation that is driven by the wind stress forcing in the ocean or lake.

wind evaporation-sea surface temperature feedback   An ocean-atmospheric interaction mechanism that causes equatorial asymmetries in the climate system. The most notable examples of its effect are the northward departure of the Pacific and Atlantic intertropical convergence zone and the north-south seesaw oscillation in the tropical Atlantic climate. Suppose that there are a pair of sea surface temperature (SST) anomalies, positive to the north and negative to the south of the equator. The lower atmosphere responds with anomalous southerly winds blowing across the equator. The induced Coriolis force generates westerly (easterly) wind anomalies to the north (south) of the equator. Superimposed on easterly trade winds in the basic state, these zonal wind anomalies weaken (enhance) the trade winds and surface evaporation, raising (lowering) SST to the north (south) of the equator and amplifying the initial SST anomalies.

winding number   In solid state physics and in cosmology, topological defects may appear when, during a phase transition, the symmetry of the full Hamiltonian is not respected by the vacuum of the theory. Then, for example, in the case of a Higgs field self-potential with the form

of a paraboloid at high temperatures, but which turns into a potential with "Mexican hat" shape at low temperatures, the initial $U(1)$ (change of phase) symmetry is not present after the potential changes shape; in this latter case, the complex field, attempting to minimize its energy, will choose one particular location at the bottom of the potential, in so doing breaking the symmetry present before. Further, the particular potential location chosen may be a function of position.

Now, it may well happen that when we follow a closed path in physical space, the Higgs field also performs a complete circle around the minima of its potential. It is in this case that its phase will develop a non-trivial winding of 2π, signaling the presence of the vortex line. The field solution far away from the string will then take the form $\phi \propto |\phi|e^{in\theta}$ with $n = 1$ and θ the azimuthal angle. More generally, the field may also wrap around the bottom of its potential n times during just one closed path in physical space. Thus n, the *winding number,* can exceed unity. In the cosmic string example described above, for example, n always takes integer values and is directly related to the magnetic flux quantization of vortex lines.

Other topological defects produced during the breakdown of continuous symmetries will also carry a winding number; examples are magnetic monopoles and cosmic textures. In the texture case, in particular, the defect configuration can be characterized by a non-integer (fractional, for example) winding number. *See* cosmic texture, cosmic topological defect, cosmic string, homotopy group, monopole, Nielsen–Olesen vortex.

wind rose   A polar histogram that shows the directional distribution of winds for a site. Generally also scaled to indicate wind speeds as well.

wind scale   A system similar to the Beaufort scale, used in the U.S. to describe the speed of winds. *See* Beaufort wind scale.

wind shear   The spatial variation of wind velocity (speed and/or direction) in a predetermined direction. Generally can be divided as the horizontal wind shear and vertical wind

Wind Condition	Speed (MPH)	Scale Number
Light	Less than 1	0
	1–3	1
	4–7	2
Gentle	8–12	3
Moderate	13–18	4
Fresh	19–24	5
Strong	25–31	6
	32–38	7
Gale	39–46	8
	47–54	9
Whole gale	55–63	10
	64–75	11
Hurricane	Above 75	12

shear. Also it can be divided as the cyclonic wind shear and anti-cyclonic wind shear.

winter anomaly   In ionospheric physics, the F region peak electron densities in the winter hemisphere at middle latitudes are usually enhanced above the summer peak electron densities. Often, the equinoctial electron densities are enhanced above both solstices. However, it is still more common to refer to this as a *winter anomaly*. The phenomenon is most evident during daytime solar maximum. The exact cause of the anomaly has been attributed to a number of sources, more recently to changes in atmospheric chemistry associated with upper atmosphere winds. *See* F region.

winter solstice   The point that lies on the ecliptic midway between the vernal and autumnal equinoxes and at which the sun, in its apparent annual motion, is at its greatest angular distance south of the celestial equator. On the day of the *winter solstice,* which occurs on about December 21, the length of night time darkness is at its maximum in the northern hemisphere. After the winter solstice, the length of daylight time in the northern hemisphere will increase until the summer solstice. Because of complex interactions with heat reservoirs in the atmosphere, soil, and oceans, the northern hemisphere temperature continues to decrease for a period of time after the winter solstice. In the southern hemisphere, the winter solstice (about December 21) is the beginning of summer, the day of the year with the longest period of sunlight.

Witten conducting string   Extensions of the simplest grand unified models with the formation of cosmic strings in general involve extra degrees of freedom which are coupled to the vortex-forming Higgs field. These may be, for example, other microscopic fields that could be responsible for the generation of currents, bound to the core of the vortices, and that may play a fundamental role in the dynamics of the defects.

The Witten-type bosonic superconductivity model is one in which the fundamental Lagrangian is invariant under the action of a $U(1) \times U(1)$ symmetry group. The first $U(1)$ is spontaneously broken through the usual Higgs mechanism in which the Higgs field acquires a nonvanishing vacuum expectation value. Hence, at an energy scale $\sim m$ we are left with a network of ordinary cosmic strings with tension and energy per unit length $T \sim U \sim m^2$, as dictated by the Kibble mechanism. The Higgs field is coupled not only with its associated gauge vector field but also with a second charged scalar boson, the current carrier field, which in turn obeys a quartic potential. A second phase transition breaks the second $U(1)$ gauge (or global, in the case of neutral currents) group and, at an energy scale $\sim m_*$ (in general $m_* \lesssim m$), the generation of a current-carrying condensate in the vortex makes the tension no longer constant, but dependent on the magnitude of the current, with the general feature that $T \leq m^2 \leq U$, breaking, therefore, the degeneracy of the Goto–Nambu strings.

The fact that the absolute value of the current carrier field is nonvanishing in the string results in that either electromagnetism (in the case that the associated gauge vector is the electromagnetic potential) or the global $U(1)$ is spontaneously broken in the core, with the resulting Goldstone bosons carrying charge up and down the string. Once generated, these currents are not affected by any resistance mechanism in the string core (unlike what would be the case for ordinary conducting wires) and hence are persistent.

Thus these strings are endowed with some superconducting properties, although the fact of being defined inside a core no bigger than the Higgs Compton wavelength (of order the electromagnetic penetration depth) complicates somewhat the interpretation (e.g., the Meissner effect) in usual condensed matter terms. *See* Abelian Higgs model, conducting string, Nielsen–Olesen vortex, spontaneous symmetry breaking, winding number.

wolf number   An historic term for the sunspot number.

work   In Newtonian mechanics, the product of an applied force and the displacement in the direction of the force.

$$\text{work } W = \mathbf{F} \cdot \mathbf{d} = |F||d| \sin \theta \ ,$$

where θ is the angle between the force and the displacement.

Work has units of Joules: 1 Joule = 1 Newton·meter, or of ergs: 1 erg = 1 dyne·cm.

worldsheet charge   *See* electric regime (string).

worldsheet geometry   *See* fundamental tensors of a worldsheet.

wormhole   A connection between two locations in space-time. This involves a curvature of space-time which acts as a bridge between two distant points that may lie in the same region or in different regions and which are otherwise inaccessible to each other. There is currently no evidence of their existence. *See* curved space-time, Einstein–Roser bridge.

wrinkle ridges   Features that occur in the northern plains of Mars, probably of volcanic origin. They are classically described as consisting of a rise, a broad arch, and an associated narrow sinuous ridge. They may be segmented to form an "en echelon" arrangement. Detailed analysis shows that almost all *wrinkle ridges* mark a regional elevation offset and so are asymmetrical. Dimensions are ≈ 10 km wide, ≈ 100 m high, with a regional elevation offset range between ≈ 10 to 1000 m where detectable.

W Ursa Majoris (W UMa)   *See* contact binary.

WWSSN (World-Wide Standardized Seismograph Network)   In 1960, three-component Press–Ewing type seismographs and short-period Benioff type seismographs with identical properties were deployed at about 120 stations throughout the world by U.S. Coast and Geodetic Survey (USCGS), mainly aiming at detection of underground nuclear tests. This world-wide seismograph network is called *WWSSN*. Seismic waveforms recorded by the seismograph network have contributed substantially to research on the source process of large earthquakes and structure of the Earth's interior using surface waves. In the 1970s, digital recording type seismograph networks began to be deployed. To date, there are IDA, SRO, DWWSSN, RSTN, and GEOSCOPE as seismograph networks deployed. Recently, principal observation stations of WWSSN have been moved to IRIS (Incorporated Research Institutions for Seismology).

X

xenoliths Rocks carried to the surface of the Earth in volcanic eruptions. These rocks originate both from the deep crust and from the mantle. They are a primary source of information on the composition of these regions. Diamonds are carried from the interior of the Earth in Kimberlitic eruptions and are one class of *xenoliths*.

xenon Rare noble gas, Xe, atomic number 54, naturally occurring atomic weight 131.29. Melting point 161.4 K, boiling point 165.1 K. Naturally occurring isotopes include those with atomic weights 124, 126, 128, 129 (the most abundant), 130, 131, 132, 134, 136. *Xenon* is present in the atmosphere at about 50 ppb and in the Martian atmosphere at 80 ppb.

X-ray The part of the electromagnetic spectrum whose radiation has energies in the approximate range from 1 to 150 keV. Typically used to observe the solar corona and, in particular, solar flares.

X-ray binary A binary star in which one component is a neutron star or black hole and the other is a main sequence star or red giant. Material is transferred from the latter to the former through Roche lobe overflow or a stellar wind, sometimes via an accretion disk. The accreting gas gets very hot, liberating gravitational potential energy of up to 10^{20} ergs per gram (about 10% of mc^2). In the case of a neutron star accretor, there can be variability associated with the neutron star's rotation period and with explosive nuclear reactions on the surface. Variability in the black hole systems is associated with changes in accretion rate and disk structure. Either kind of accretor can occur with either high mass (stellar wind) or low mass (Roche lobe overflow) donor. Most of the 10 or so known black hole systems have black hole masses of 5 to 10 solar masses. Most of the neutron star masses are 1.5 to 2.0 solar masses, permitting a clean separation of the classes. Luminosities can reach 10^5 solar luminosities.

X-ray bright point X-ray bright points are small, compact, short-lived solar X-ray brightenings that are most easily seen in coronal holes.

X-ray burst A temporary enhancement of the X-ray emission of the sun. The time-intensity profile of soft *X-ray bursts* is similar to that of the Hα profile of an associated flare.

X-ray burster An irregular X-ray source, believed to occur when matter from an evolved large companion star accretes onto the surface of a neutron star in a binary system. Hydrogen accumulates on the surface, and when the density is high enough, explosively burns (at very high temperature) on the surface. This is similar to a nova in mechanism, but the energy associated with the burster is much higher because the gravitational field of the neutron star is so much stronger than that of a white dwarf.

X-ray flare classification Rank of a flare based on its X-ray energy output. Flares are classified by the order of magnitude of the peak burst intensity (I) measured at the Earth in the 1 to 8 Å band as follows:

Class	(Wm^{-2})
B	$I < 10^{-6}$
C	$10^{-6} \leq I \leq 10^{-5}$
M	$10^{-5} \leq I \leq 10^{-4}$
X	$I \geq 10^{-4}$

X-ray source In astrophysics, any detectable emitter of X-radiation (for instance the sun). In terms of frequency of occurrence and intrinsic brightness, X-ray binaries and the X-ray background are the major detectable examples. It is believed that all strong *X-ray sources* involve accretion onto a compact object, as onto a neutron star or onto a black hole, from a normal star in a X-ray binary. The background X-ray flux is believed to be made up principally from sources such as active galactic nuclei and/or quasars, which are powered by accretion from a disk onto a supermassive black hole.

X-ray stars   Stars whose X-ray emission is strong enough to be detected from Earth. This is rather rare and generally associated with youth or rapid rotation (*see* stellar activity, coronae) or presence of a companion (*see* RS Canum Venaticorum stars).   The first to be discovered was Capella (with an energetic corona). Brighter X-ray sources are associated with mass transfer to white dwarfs, neutron stars, or black holes from companions and are called X-ray binaries.

Y

yardangs   Streamlined elongated hills oriented parallel to the wind and created by erosion of adjacent friable material. The hills are aerodynamically shaped by the wind and resemble inverted boat hulls. The highest and widest part of the ridge is about 1/3 of the distance from the upwind to the downwind side. The downwind side usually has a gently tapering bedrock surface or may have a sand "tail". *Yardangs* range in size from about 1 to 200 m in height and a few meters to a few kilometers in length. They typically form in linear clusters and have been identified on both Earth and Mars.

year   Unit of time equal to one revolution of the Earth about the sun. "A tropical year" is a measure of mean solar time, specifically the time between successive passages of the vernal equinox by the mean sun (365.242119 days of mean solar time). A "sidereal year" is a measure of sidereal time, namely the period between successive passages through the hour circle of a fixed star (365.25636556 days of mean solar time). An "anomalistic year" is the mean interval between passages of the Earth through perihelion (365.25964134 days of mean solar time). Calendar years approximate tropical years and consist of 365 days plus 1 day added every leap year. A "Besselian year" is defined as the period between successive passages by the mean sun of the right ascension 18h 40m and is, hence, independent of an observer's location. A "Julian year" consists of 365.25 days of mean solar time. Some types of years refer to longer periods, e.g., the "platonic year" is the period of the precession of the equinoxes (25725 tropical years relative to the fixed ecliptic).

yellow line   A coronal line observed at 5694 Å resulting from a forbidden transition in highly ionized calcium atoms (Ca XV). Important for the study of coronal structures at temperatures of order 3.5 MK.

Yerkes classification scheme of galaxies   A classification scheme of galaxies, conceived by W.W. Morgan and based on the central concentration of light, which closely correlates to the stellar population of a galaxy. The concentration class of a galaxy is indicated with the same letter used for stellar spectral types, but written in lower case (i.e., k for ellipticals and a for Sc spirals whose spectrum is dominated by K giants and A stars, respectively). A second parameter describes the "form family" of galaxies: S indicates a spiral; B a barred spiral; E an elliptical; I an irregular; R a galaxy having rotational symmetry without prominent spiral arms (an S0 galaxy according to the Hubble scheme); and D indicates an elliptical galaxy with an extended envelope. A third symbol (a number from 0 to 7) describes the apparent flattening of a galaxy. For example, M31 is classified as kS5. Albeit no longer widely in use, the Yerkes scheme is still used to denote supergiant elliptical galaxies often found at the center of clusters and groups of galaxies (cD galaxies).

Yohkoh ("sunbeam")   A Japanese satellite launched in August 1991 designed to study the dynamics, energetics, and morphology of solar flares. *Yohkoh* is comprised of four scientific instruments: the Soft X-ray Telescope (SXT) for imaging in the 3 to 60 Å wavelength range, the Bragg Crystal Spectrometer (BCS) for spectroscopic observations in the 1 to 5 Å range, the Hard X-ray Telescope (HXT) for imaging at energies above ~ 14 keV, and the Wide Band Spectrometer (WBS) for spectroscopic observations over a wide range of energies from soft X-rays to γ-rays.

Yoshida jet   *See* jet.

Young's modulus   The ratio $Y = T/(\Delta l/l)$ for a material under compression or tension. Here T is the stress $[N/m^2]$, and $\Delta l/l$ is the strain, the fractional length extension or compression. Hence, Young's modulus has dimensions of stress, $[N/m^2]$.

Young's modulus   In an elastic isotropic material, stress σ is proportional to strain ε: $\sigma = Y\varepsilon$. The quantity Y is a property of the material and is called *Young's modulus*.

Yukawa coupling Particle physics theories predict that fermions will acquire their masses through the Higgs mechanism, by means of a so-called Yukawa term in the Lagrangian, where fermions couple with a Higgs field ϕ. Specifically, a fermion ψ is said to have a mass m when the Lagrangian density that describes its dynamics contains the term

$$\mathcal{L}_m = m\bar{\psi}\psi .$$

In some cases, however, it is not possible to have such a term, and the corresponding fermions would then appear massless (the symmetry rules obeyed by the total Lagrangian are responsible for this). Instead, a coupling such as

$$\mathcal{L}_m = f\bar{\psi}\phi\psi$$

(the *Yukawa coupling*) with f the coupling constant, may always be set. When the Higgs field gets its nonzero vacuum expectation value $\langle|\phi|\rangle$ in a phase transition, the above coupling reduces to a simple mass term, and thus the fermion acquires a mass proportional to $\langle|\phi|\rangle$.

This same mechanism also allows the possibility of fermionic currents along topological defects in the form of zero modes. *See* cosmic phase transition, fermionic zero mode, Higgs mechanism.

Z

Zeeman effect   The splitting of spectral lines into groups of closely spaced lines in the presence of a strong magnetic field. The *Zeeman effect* occurs in the spectra of sunspots and stars. This effect demonstrates the existence of magnetic fields in celestial bodies, and since the degree of the splitting depends on the magnitude of the magnetic fields, it allows measurement of the field to be made.

Zeldovich process   A process of energy extraction from a rotating black hole via incident electromagnetic or gravitational radiation. When radiation hits a black hole, part of it is absorbed and part is carried away. Normally, the absorbed part has positive energy and the amplitude of the transmitted wave is reduced. However, in the case of cylindrical waves, the absorbed wave may carry negative energy, in which case the transmitted wave is amplified. This is the variant with electromagnetic radiation of the Penrose process. *See* Penrose process.

zenith   The point directly overhead an observer, as defined by the local horizontal. Technically, there are three *zenith* directions: geocentric zenith, the direction directly away from the Earth's center; geodetic zenith, the direction of the local upwards normal from the Earth ellipsoid; and astronomical zenith, the direction of the upwards normal from the geoid. Normally, when used without another qualifier, zenith refers to either the geodetic or astronomical zenith, which are within a few arc seconds of each other. *See also* altitude, nadir.

zenith angle   The angle from the vertical, or zenith, to a given direction. The *zenith angle* is the complement of the altitude.

zero mode   *See* fermionic zero mode.

zero up-crossing   The point where the water surface in a water wave profile crosses the zero line and is trending upward. Used in the definition of wavelength.

zodiac   The apparent path of the sun through the background stars over the course of the year; the intersection of the plane of the ecliptic with the celestial sphere.

zodiacal light   A band of diffuse light seen along the ecliptic near the sun immediately after sunset or before sunrise. It is created by sunlight reflecting off the interplanetary dust particles, which are concentrated along the ecliptic plane. To see the *zodiacal light,* you must have very dark skies; under the best conditions it rivals the Milky Way in terms of brightness. The material within the zodiacal light is slowly spiraling inward toward the sun, due to the Poynting–Robertson Effect; hence, it must be continuously replaced by asteroid collisions and the debris constituting comet tails. The zodiacal light is the name given to the band of light seen close to the sun while gegenschein is the term applied to the same band of light located 180° from the sun.

zonal   In the direction parallel to the equator, i.e., east-west. *See also* meridional.

zooplankton   Animal forms of plankton.

Z-string   *See* embedded defect.

Zulu Time   *See* Universal Time.

Zwicky compact galaxies   Galaxies originally defined by F. Zwicky as distinguishable from stars on the Palomar 1.2 m Schmidt telescope plate and with angular diameter between 2 and 5 sec of arc. "Compact" is, in astronomy, a loose synonym of "not fully resolved". The notion of compactness of a galaxy therefore depends on the resolving power of the observational equipment, although compact galaxies have, in general, high surface brightness and sharp borders. F. Zwicky circulated seven lists of compact galaxies including about 200 objects in the mid-1960s. Most of them are of blue color and show emission lines in their spectra.

Zwicky's compact galaxies have turned out to be a rather heterogeneous class, which included star-forming dwarf galaxies, as well as several active galaxies. For example, the object I Zw 1 (the first object of Zwicky's first list) is a nearby quasar.